ESSENTIAL ALGEBRA
WITH PROBLEM SOLVING

ESSENTIAL ALGEBRA
WITH PROBLEM SOLVING

MARVIN L. BITTINGER
Indiana University—Purdue University at Indianapolis

MERVIN L. KEEDY
Purdue University

ADDISON-WESLEY PUBLISHING COMPANY

Reading, Massachusetts ● Menlo Park, California
New York ● Don Mills, Ontario ● Wokingham, England
Amsterdam ● Bonn ● Sydney ● Singapore ● Tokyo
Madrid ● Bogotá ● Santiago ● San Juan

Sponsoring Editor • *Elizabeth Burr*
Production Supervisor • *Herbert Nolan*
Editorial and Production Service • *Quadrata, Inc.*
Text Designer • *Vanessa Piñeiro, Piñeiro Design Associates*
Illustrators • *VAP Group Limited, Belanger Associates*
Art Consultant • *Joseph Vetere*
Manufacturing Supervisor • *Ann DeLacey*
Cover Designer • *Marshall Henrichs*

Photo Credits

1, Michael Hayman/Stock, Boston **30,** Belanger Associates **64,** Belanger Associates, with permission of Sozio, Revere, MA **69,** Belanger Associates **75,** Owen Franklin/Stock, Boston **92,** Lionel J-M Delevingne/Stock, Boston **100,** AP/Wide World Photos **102,** Belanger Associates **104,** Boston Museum of Science **117,** Peter Manzel/Stock, Boston **120,** Phyllis Graber Jensen/Stock, Boston **122,** Barbara Alper/Stock, Boston **147,** Belanger Associates **170,** Bohdan Hrynewych/Stock, Boston **171,** Joseph A. Kovacs/Stock, Boston **187,** Mark Antman/Stock, Boston **215,** Geoffrey Clements/Collection of the Whitney Museum of American Art, NY **233,** AP/Wide World Photos **235,** W. B. Finch/Stock, Boston **236,** Belanger Associates **243,** AP/Wide World Photos **245,** Belanger Associates **247,** Belanger Associates **254,** Tim Carlson/Stock, Boston **267,** Ellis Herwig/Stock, Boston **268,** Belanger Associates **269,** AP/Wide World Photos **273,** Belanger Associates **277,** Ira Kirschenbaum/Stock, Boston **278,** Belanger Associates **280,** NASA **283,** AP/Wide World Photos **287,** Bohdan Hrynewych/Stock, Boston **312,** Ellis Herwig/Stock, Boston **313,** Dean Abramson/Stock, Boston **315,** Robert V. Eckert/Stock, Boston **321,** Rick Smolan/Stock, Boston **331,** Phyllis Craber Jensen/Stock, Boston **335,** Belanger Associates **371,** Belanger Associates **374,** Bohdan Hrynewych/Stock, Boston **375,** AP/Wide World Photos **392,** Frank Siteman/Stock, Boston **399,** Jeff Albertson/Stock, Boston **421,** Belanger Associates **422,** H. Morgan/Stock, Boston

Library of Congress Cataloging-in-Publication Data

Bittinger, Marvin L.
   Essential algebra with problem solving.

   Includes index.
   Adaptation of: A problem solving approach to introductory algebra/Mervin L. Keedy, Marvin L. Bittinger. 2nd ed. c1986.
   1. Algebra. 2. Algebra—Problems, exercises, etc. I. Keedy, Mervin Laverne. II. Keedy, Mervin Laverne. Problem solving approach to introductory algebra. III. Title.
QA152.2.B58 1987   512.9   87-982
ISBN 0-201-14293-7 (pbk.)

ABCDEFGHIJ-MU-8987

PREFACE

This text is a special edition prepared for the Department of Mathematics of the European Division of the University of Maryland. Many faculty in the department made a significant contribution to its creation, and we wish to thank the following people for their hard work and dedication to the development of this text: Dr. Samuel B. Thompson, Elaine L. Stelter, Mohsen Emami, Lindsey G. Court, David L. Powell, and Ronald Souverein.

MATHEMATICAL CONTENT AND LEVEL

The mathematical content of this text is the core of intermediate algebra. It is intended for use by students who have a firm background in arithmetic and some basics of algebra, although many such topics are reviewed in the early chapters. This text is written to provide sufficient background for students to begin the text *Algebra and Trigonometry: A Functions Approach* by the same authors.

WORKTEXT FEATURES

This book is a worktext, which is distinguished by its thorough and integrated treatment of problem solving. Following are some additional features of the text.

- *Objectives*. The objectives for each section are stated in the margin and are easy to identify by their symbol **i**. That same symbol is placed in the text and also in the exercises beside the corresponding material, making the text easy to read and use for self-study and review.

- *Margin exercises*. While working through the material the student is encouraged to do exercises located in the margins. Answers to these exercises are at the back of the book. The margin exercises are designed to get the student *actively involved* in the development of each section of material, and can also serve as a self-study guide. They are very much like the examples in the text as well as the homework exercises, enabling the student to practice as he or she is learning.

- *Exercise sets.* The exercises are carefully indexed to the objectives and identified by the symbols ▮ i ▮. This coding occurs in the basic exercises at the beginning of each exercise set, enabling students to refer to appropriate sections of the text when review is needed.

- *Features of the answer section.* The answer section of the text has much to offer the student. The answers to all the margin exercises are included, and most of the answers to the exercises in the exercise sets also appear. More specifically, every answer appears except those to every third even-numbered exercise. For example, answers to exercises numbered 2, 8, 14, 20, 26, and so on, are not in the text.

    As the student checks answers he or she will note that certain odd-numbered answers are underlined. The underline indicates that those exercises have worked-out solutions in the latter part of the answer section. These can be especially helpful to a student who must be away from the classroom. Worked-out solutions are given for every third odd-numbered exercise beginning with the first. Thus, exercises 1, 7, 13, 19, 25, and so on, have worked-out solutions.

READABILITY AND TEACHABILITY

The style and format of the book are designed to make the text easy to read and comprehend. Examples to illustrate each skill and concept have been chosen carefully to ensure that they are appropriate and to the point. The exercises in the exercise sets are very much like the examples. Exercises are paired and graded.

The exercises that follow the symbolism

are generally different from the examples in the text, and require more than a simple understanding of the material at hand, thus offering a challenge to students. There are also exercises that are designed to be done with a calculator. Those exercises are marked ▦.

PROBLEM SOLVING

The primary distinguishing feature of this book is its treatment of and emphasis on problem solving. The approach to problem solving is different here, not only because of the increased space devoted to it, but also because of the way students are guided in formulating problems and translating the situation of a problem to mathematical language. Problem-solving techniques are introduced in Chapter 2 and applied throughout the text, with problem-solving practice included in most exercise sets.

SKILL MAINTENANCE

Skill maintenance is featured throughout the text.

- The problem-solving practice in the exercise sets covers problem solving from earlier parts of the book, constantly reinforcing the student's thinking skills.

- Each chapter ends with a Summary and Review section designed to give students maximum assistance when they review the material. The chapter objectives are stated in boldface type, followed by exercises testing students' comprehension of them. Answers are given at the back of the book together with section references so the student can restudy material as appropriate.

- A practice test follows the Summary and Review; practice-test answers are in the Instructor's Manual, but are not in the text.

SUPPLEMENTS

The following supplementary materials are available.

- *Instructor's Manual.* Contains one alternate form of each chapter test and final exam and some black-line grid masters for graphing.
- *Instructional Software for Algebra.* A highly interactive computer software supplement that covers most of the topics treated in the text. It consists of guided exercises and quizzes, and can be used either to supplement the text or for individualized instruction.
- *Computerized Testing* (AWTEST). Contains test questions for all summary objectives in the text. The computer can generate an infinite variety of test questions with just a few keystrokes by inserting random numbers into model problems for each objective. The problems are constrained so that only reasonable variations are generated.
- *Videotape Cassettes.* Keyed to objectives in the text. They can supplement the regular lectures or can be used in a number of teaching situations, such as learning labs, or fill-in lectures when instructors are absent. John Jobe of Oklahoma State University speaks to students and works out examples with lucid explanations on more than 35 thirty-minute tapes. There are also two thirty-minute tapes on problem solving, which cover the entire scope of the material in the text.
- *Placement Test.* Statistically validated, this test can be used to place students properly in either this book or other books by the same authors, such as *Algebra and Trigonometry: A Functions Approach*, 4th ed. (1986). More specific placement in this book can be attained using the final exam in the *Instructor's Manual.*

ACKNOWLEDGMENTS

The authors wish to thank many people without whose committed efforts our work could not have been completed successfully. Gloria Schnippel and Karen Anderson and those faculty mentioned earlier at the University of Maryland, European Division, did a tremendous job checking the manuscript. Judy Beecher, Judy Penna, and Julie Stephenson were superb in their work on the supplements.

Many instructors provided reviewing information. We thank them for their many suggestions for improvement. They are Ruth Afflack (California State University at Long Beach), Don Albers (Menlo College), Jean Berdon (Cañada College), Randy Davidson (University of New Orleans), Arthur Dull (Diablo Valley College), Robert Limburg (St. Louis Community College at Florissant Valley), John Samoylo (Delaware County Community College), Dick Spangler (Tacoma Community College), Louise Dyson (Clark College), Eleanor Kendrick (San Jose City College), Roger Judd (Chemeketa Community College), and Kathleen Thestia (Phoenix College).

M. L. B.
M. L. K.

CONTENTS

3

POLYNOMIALS   117

4

POLYNOMIALS AND FACTORING   147

5

GRAPHS, LINEAR EQUATIONS, AND SLOPE   187

6

SYSTEMS OF EQUATIONS AND PROBLEM SOLVING   215

11

QUADRATIC EQUATIONS AND FUNCTIONS   371

1

BASIC CONCEPTS
OF ALGEBRA

A football team makes many gains and losses of yards during a game. Integers can be used to solve the problem of finding the total gain or loss.

This book covers those concepts of algebra essential for a course in college algebra and trigonometry. Basic to a study of algebra is the study of arithmetic. For the first part of this chapter we will both review arithmetic and begin to study the kinds of numbers used in algebra. In the second part of the chapter we get into algebra by studying the kinds of symbols used in algebra as well as important number properties used to manipulate these symbols.

1.1

THE NUMBER LINE

In this section we become acquainted with the set of real numbers and its various subsets.

The Set of Integers

The numbers used in algebra are the so-called *real numbers*. There is a real number for every point on a number line.

$$-2.5 \quad -\tfrac{1}{2} \quad \tfrac{1}{2} \quad \sqrt{2} \quad \pi$$
$$-5\ -4\ -3\ -2\ -1\ \ 0\ \ 1\ \ 2\ \ 3\ \ 4\ \ 5$$

The set of real numbers is made up of several subsets. We will consider these and build up to a description of the real numbers.

> The set of *natural numbers* is 1, 2, 3, and so on.

When 0 is included in the set of natural numbers, we obtain the set of whole numbers.

> The set of *whole numbers* is 0, 1, 2, 3, and so on.

When fractions, such as $\frac{1}{2}$, $\frac{2}{3}$, and $\frac{9}{5}$, are included in the set of whole numbers, we obtain the set of numbers of arithmetic.

> The *numbers of arithmetic* consist of the whole numbers and the fractions, such as $\frac{1}{2}$, $\frac{2}{3}$, and $\frac{9}{5}$. All the numbers of arithmetic can be named by fractional notation a/b, where b is not 0.

We obtain negative numbers by going back to the set of whole numbers. We create a new set, called the *integers*, by starting with the whole numbers, 0, 1, 2, 3, and so on. For each natural number 1, 2, 3, and so on, we invent a new number:

For the number 1 there will be a new number named -1 (negative 1).

For the number 2 there will be a new number named -2 (negative 2), and so on.

The *integers* consist of the whole numbers and these new numbers. We picture them on a number line as follows.

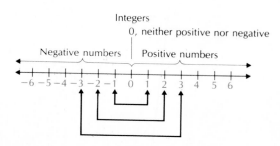

Integers
0, neither positive nor negative

Negative numbers | Positive numbers

$$-6\ -5\ -4\ -3\ -2\ -1\ \ 0\ \ 1\ \ 2\ \ 3\ \ 4\ \ 5\ \ 6$$

We call the newly invented numbers *negative* integers. The natural numbers are called *positive* integers. Zero is neither positive nor negative.

i   Integers and the Real World

Integers can be associated with many real-world situations and used in problem solving.

EXAMPLE 1   Tell which integer corresponds to this situation: The temperature is 3 degrees below zero.

3° below zero is −3°

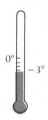

EXAMPLE 2   Tell which integer corresponds to this situation: Getting set 21 points in a card game.

Getting set 21 points in a card game gives you −21 points.

EXAMPLE 3   Tell which integer corresponds to this situation: Death Valley is 280 ft below sea level.

The integer −280 corresponds to the situation. The elevation is −280 ft.

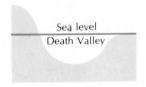

EXAMPLE 4   Tell which integers correspond to this situation: A salesperson made $78 on Monday, but lost $57 on Tuesday.

The integers 78 and −57 correspond to the situation. The integer 78 corresponds to the profit on Monday and −57 corresponds to the loss on Tuesday.

DO EXERCISES 1–3 AT THE RIGHT.

ii   Order on the Number Line

Numbers are named in order on the number line, with larger numbers named further to the right. For any two numbers on the line, the one to the left is less than the one to the right.

Tell which integer corresponds to each situation.

1. A student owes $12 to the bookstore. A friend has $235 in a savings account.

A $ −12.00
B $ +235.00

2. The halfback gained 8 yd on first down. The quarterback was sacked for a 5-yd loss on second down.

A. +8 yd
B. −5 yd.

3. At 10 sec before liftoff ignition occurs. At 63 sec after liftoff the first-stage engine stops.

A. +10 sec.
B. −63 sec.

Use either < or > to write a true sentence.

4. 14 $>$ 7

5. 11 $>$ −2

6. −15 $<$ −5

7. −3 $>$ −19

Translate to mathematical language.

8. Zero is greater than negative four.

O > −4

9. A temperature of −45° is lower than a temperature of −13°.

−45 < −13

10. A gain of 14 yd is better than a loss of 5 yd.

14 > 5

We use the symbol < to mean "is less than." The sentence −6 < 8 means "−6 is less than 8." The symbol > means "is greater than." The sentence −3 > −7 means "−3 is greater than −7."

EXAMPLES Use either < or > to write true sentences.

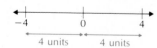

5. 2 $<$ 9 Since 9 is to the right of 2, 9 is greater than 2 or 2 is less than 9. The answer is 2 < 9.

6. −7 $<$ 3 Since −7 is to the left of 3, we have −7 < 3.

7. 6 $>$ −12 Since 6 is to the right of −12, then 6 > −12.

8. −18 $<$ −5 Since −18 is to the left of −5, we have −18 < −5.

DO EXERCISES 4–7.

In the following examples we see some ways in which integers are used in translating to mathematical language.

EXAMPLE 9 Translate to mathematical language: Negative five is less than negative three.

The translation is −5 < −3.

EXAMPLE 10 Translate to mathematical language: A temperature of −20° is higher than a temperature of −36°.

The translation is −20 > −36.

EXAMPLE 11 Translate to mathematical language: A debt of $150 is worse than a debt of $100.

The translation is −150 < −100.

Note that all positive integers are greater than zero and all negative integers are less than zero.

DO EXERCISES 8–10.

iii Absolute Value

From the number line we see that numbers like 4 and −4 are the same distance from zero. We call the distance from zero the *absolute value* of the number.

The *absolute value* of a number is its distance from 0 on a number line. We use the symbol |n| to represent the absolute value of a number n.

To find absolute value:

1. If a number is negative, make it positive.
2. If a number is positive or zero, leave it alone.

EXAMPLES Find the absolute value.

12. $|-7|$ The distance of -7 from 0 is 7, so $|-7|$ is 7.

13. $|12|$ The distance of 12 from 0 is 12, so $|12|$ is 12.

14. $|0|$ The distance of 0 from 0 is 0, so $|0|$ is 0.

DO EXERCISES 11–14.

iv The Rational Numbers

We created the set of integers by inventing a negative number for each natural number. To create a bigger number system, called the set of *rational numbers,* we invent negative numbers that correspond with the nonzero numbers of arithmetic.

   For $\frac{2}{3}$ there will be a new number named $-\frac{2}{3}$ (negative $\frac{2}{3}$).

   For $\frac{13}{5}$ there will be a new number named $-\frac{13}{5}$ (negative $\frac{13}{5}$).

   For 1.5 there will be a new number named -1.5 (negative 1.5).

   For 73 there will be a new number named -73 (negative 73).

   And so on.

   Note that this new set of numbers, the rational numbers, contains the whole numbers, the integers, and the numbers of arithmetic.

DO EXERCISE 15.

v Graphing Rational Numbers

We picture the rational numbers on a number line as follows. There is a point on the line for every rational number.

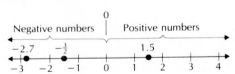

To *graph* a number means to find and mark its point on the line. Some numbers are graphed in the figure above.

EXAMPLE 15 Graph: $\frac{5}{2}$.

   The number $\frac{5}{2}$ can be named $2\frac{1}{2}$ or 2.5. Its graph is halfway between 2 and 3.

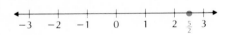

EXAMPLE 16 Graph: -3.2.

   The graph of -3.2 is $\frac{2}{10}$ of the way from -3 to -4.

EXAMPLE 17 Graph: $\frac{13}{8}$.

   The number $\frac{13}{8}$ can be named $1\frac{5}{8}$ or 1.625. The graph is about $\frac{6}{10}$ of the way from 1 to 2.

Find the absolute value.

11. $|17|$

N = 17

12. $|-8|$

N = 8

13. $|-14|$

N = 14

14. $|0|$

N = 0

15. List five examples of a rational number.

5 = −5

2/3 = −2/3

1/5 = −1/5

10 = −10

200 = −200

Graph the rational number.

16. $\frac{12}{5}$

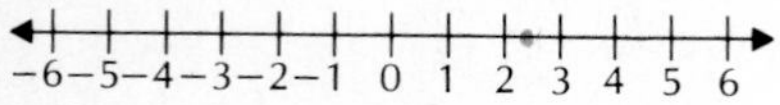

17. −4.8

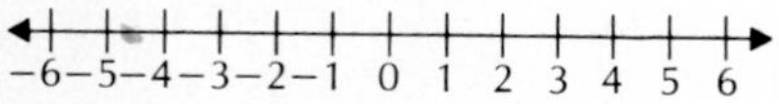

18. $-\frac{18}{5}$

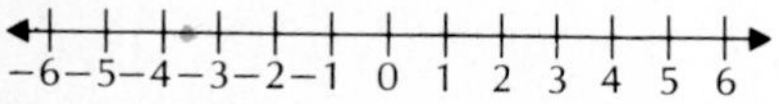

19. 5.3

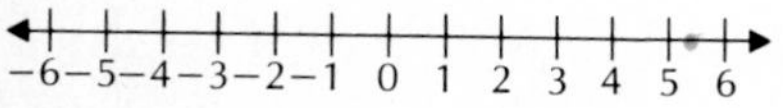

Find decimal notation.

20. $\frac{7}{8}$

21. $\frac{7}{11}$

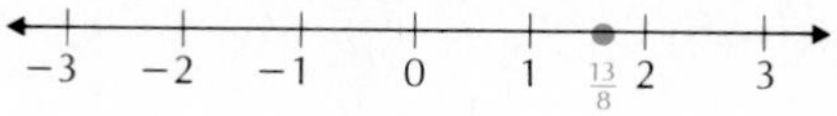

DO EXERCISES 16–19.

Table 1, located on the inside front cover, contains many common fractional–decimal equivalents. It would be most helpful to you to memorize this table.

vi Notation for Rational Numbers

The rational numbers *can be named* as quotients of integers.

Fractional Notation. *Fractional notation* consists of symbolism such as the following:

$$\frac{2}{3},\quad \frac{12}{-7},\quad \frac{-17}{15},\quad -\frac{9}{7}.$$

Decimal Notation. The rational numbers can also be named using *decimal notation.*

EXAMPLE 18 Find decimal notation for $\frac{5}{8}$.

Since $\frac{5}{8}$ means $5 \div 8$, we divide:

$$
\begin{array}{r}
0.625\\
8\,\overline{)5.000}\\
\underline{4\,8}\\
20\\
\underline{16}\\
40\\
\underline{40}
\end{array}
$$

Decimal notation for $\frac{5}{8}$ is thus 0.625.

Decimal notation for some numbers repeats.

EXAMPLE 19 Find decimal notation for $\frac{6}{11}$.

We divide:

$$
\begin{array}{r}
0.5454\ldots\\
11\,\overline{)6.0000}\\
\underline{5\,5}\\
50\\
\underline{44}\\
60\\
\underline{55}\\
50\\
\underline{44}\\
6
\end{array}
$$

Repeating decimal notation can be abbreviated by writing a bar over the repeating part, in this case, $0.\overline{54}$.

DO EXERCISES 20 AND 21.

EXAMPLE 20 Find fractional notation for 3.6. You need not simplify.

$$3.6 = \frac{36}{10}$$

The last decimal place is *tenths* so the denominator is 10.

EXAMPLE 21 Find fractional notation for 0.75. You need not simplify.

$$0.75 = \frac{75}{100}$$

The denominator is 100 because the last decimal place is *hundredths*.

DO EXERCISES 22 AND 23.

vii Order of the Rational Numbers

The relations < (is less than) and > (is greater than) are the same for rational numbers as they are for integers. Recall that numbers on the line increase from left to right.

EXAMPLES Use either < or > to write a true sentence.

22. 1.38   1.83   The answer is $1.38 < 1.83$.

23. -3.45 1.32   The answer is $-3.45 < 1.32$ because -3.45 is to the left of 1.32.

24. -3.33 -4.44 The answer is $-3.33 > -4.44$.

25. $\frac{5}{8}$ $\frac{7}{11}$ We convert to decimal notation: $\frac{5}{8} = 0.625, \frac{7}{11} = 0.6363, \ldots$ Thus $\frac{5}{8} < \frac{7}{11}$.

DO EXERCISES 24–27.

viii Absolute Value

The absolute value of a rational number is its distance from zero, just as for integers.

EXAMPLES Find the absolute value.

26. $\left|\frac{3}{2}\right| = \frac{3}{2}$

27. $\left|-2.73\right| = 2.73$

DO EXERCISES 28–30.

ix Real Numbers

The number line has a point for every rational number. Is there a rational number for every point of the line? The answer is *no*. There are some points of the line for which there is no rational number. These points correspond to what are called *irrational numbers*.

We can create a number system in which all points have numbers. That new number system is called the system of *real numbers*. The *real numbers* consist of the *rational numbers* and the *irrational numbers*.

The following figure shows the relationship between various kinds of numbers.

Find fractional notation. You need not simplify.

22. 5.9 $\frac{59}{10}$

23. 0.125 $\frac{125}{1000}$

Write a true sentence using < or >.

24. 4.62 $>$ 4.26

25. 3.11 $>$ -9.56

26. $\frac{6}{7}$ $\frac{13}{15}$

$.857 < .866$

$\frac{6}{7} < \frac{13}{15}$

27. $-\frac{3}{5}$ $-\frac{5}{9}$

$-.6 < -.55$

$-\frac{3}{5} < -\frac{5}{9}$

Find the absolute value.

28. $|4.1|$ 4.1

29. $\left|-\frac{8}{3}\right|$ $\frac{8}{3}$

30. $|-3.5|$ 3.5

Graph each real number on a number line.

31. $\sqrt{2}$ = 1.4142

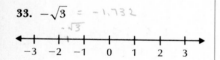

32. $\pi \approx 3.1416$

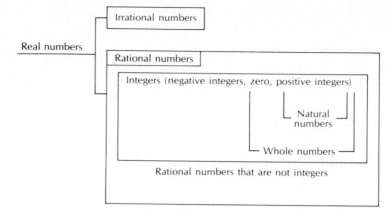

33. $-\sqrt{3}$ = -1.732

An example of an irrational number is $\sqrt{3}$. This is the positive number which when multiplied by itself gives 3. There is no rational number that we can multiply by itself to get 3. But the following are rational approximations:

1.7 is an approximation of $\sqrt{3}$ because $(1.7)^2 = 2.89$.

1.73 is a better approximation of $\sqrt{3}$ because $(1.73)^2 = 2.9929$.

1.732 is an even better approximation of $\sqrt{3}$ because

$$(1.732)^2 = 2.999824.$$

We can find rational approximations for square roots using a calculator. Decimal notation for rational numbers either terminates or repeats. Decimal notation for irrational numbers neither terminates nor repeats. Some examples of irrational numbers are π, $\sqrt{3}$, $-\sqrt{8}$, $\sqrt{11}$, and $0.121221222122221\ldots$. Whenever we take the square root of a number that is not a perfect square we will get an irrational number. We will study square roots and irrational numbers in greater detail in Chapter 10.

EXAMPLE 28 Graph the real number $\sqrt{3}$ on a number line.

We use a calculator or Table 2 and approximate $\sqrt{3} \approx 1.732$ ("$\approx$" means "approximately equal to"). Then we locate this number on a number line.

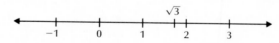

DO EXERCISES 31–33.

EXERCISE SET 1.1

At the back of the book you will find certain answers as well as worked-out solutions to certain exercises. All answers to the Margin Exercises and the Chapter Summary and Reviews appear there. For the Exercise Sets:

1. All the answers to the odd-numbered exercises appear.

2. There are worked-out solutions to every third odd-numbered exercise beginning with the first. These are found in the next part of the back matter. You will know that an exercise has a worked-out solution when its number is underlined in the answer section.

3. The even-numbered exercise answer following one that has a worked-out solution is *not* included in the text. Thus,

   1, 7, 13, 19, 25, 31, and so on, have worked-out solutions, and

   2, 8, 14, 20, 26, 32, and so on, do not have answers in the text.

i Tell which integers correspond to each situation.

1. In a game a player won 5 points. In the next game a player lost 12 points.

2. The temperature on Wednesday was 18° above zero. On Thursday it was 2° below zero.

3. A family owes $17. The same family has $12 in its bank account.

4. A business earned $1200 one week and lost $560 the next.

5. The Dead Sea, between Jordan and Israel, is 1286 ft below sea level, whereas Mt. Everest is 29,028 ft above sea level.

6. In bowling, team A is 34 pins behind team B after the first game. Describe this situation in two ways.

7. A student deposited $750 in a savings account. Two weeks later the student withdrew $125.

8. During a certain time period, the United States had a deficit of $3 million in foreign trade.

9. During a video game a player intercepted a missile worth 20 points, lost a starship worth 150 points, and captured a base worth 300 points.

10. 3 seconds before liftoff of a rocket occurs. 128 seconds after the liftoff of a rocket.

ii Write a true sentence using $<$ or $>$.

11. 5   0

12. 9   0

13. -9 5

14. 8 -8

15. -6 6

16. 0 -7

17. -8 -5

18. -4 -3

19. -5 -11

20. -3 -4

21. -6 -5

22. -10 -14

Translate to mathematical language.

23. Negative five is greater than negative eight.

24. Negative seven is less than negative one.

25. In pinochle a score of 120 is better than a score of -20.

26. A deposit of $20 in a savings account is better than a withdrawal of $25.

27. In trade, a deficit of $500,000 is worse than an excess of $1,000,000.

28. In bowling, it is better to be 60 pins ahead than to be 20 pins ahead.

29. On a test, it is better to have 2 points taken off than 10 points taken off.

iii Find the absolute value.

30. $|-3|$

31. $|-7|$

32. $|10|$

33. $|11|$

34. $|0|$

35. $|-4|$

36. $|-24|$

37. $|325|$

38. $|x|$ when $x = 5$

39. $|b|$ when $b = -3$

40. $|-224|$

iv

41. List ten examples of rational numbers.

42. List ten examples of rational numbers that are not integers.

v , **ix** Graph each number on a number line.

43. $\frac{10}{3}$

44. $-\frac{17}{5}$

45. -4.3

46. 6.45

47. $\sqrt{5}$

48. $-\sqrt{8}$

49. $-\sqrt{2}$

50. $-\pi$

51. $-\frac{1}{2}$

52. $-\frac{3}{4}$

vi Find decimal notation.

53. $\frac{3}{8}$

54. $\frac{1}{8}$

55. $\frac{5}{3}$

56. $\frac{5}{6}$

57. $\frac{7}{6}$

58. $\frac{5}{12}$

59. $\frac{2}{3}$

60. $\frac{1}{4}$

61. $\frac{1}{2}$

62. $\frac{5}{8}$

63. $\frac{1}{10}$

64. $\frac{7}{20}$

Find fractional notation. You need not simplify.

65. 2.7

66. 13.91

67. 0.145

68. 0.0213

69. 0.34

70. 0.02

71. 1.4

72. 234.7

vii Write a true sentence using $<$ or $>$.

73. 2.14   1.24

74. -3.3 -2.2

75. -14.5 0.011

76. 17.2 -1.67

77. -12.88 -6.45

78. -14.34 -17.88

79. $\frac{5}{12}$ $\frac{11}{25}$

80. $-\frac{14}{17}$ $-\frac{27}{35}$

viii Find the absolute value.

81. $\left|-\frac{2}{3}\right|$

82. $\left|-\frac{10}{7}\right|$

83. $\left|\frac{0}{4}\right|$

84. $|14.8|$

85. $|3|$

86. $|-8|$

87. $\left|-\frac{7}{8}\right|$

88. $\left|\frac{0}{13}\right|$

List each set of integers from least to greatest.

89. 13, −12, 5, −17

90. −23, 4, 0, −17

91. Explain why in golf or dieting it is better to be −3, rather than +3, with respect to par or a certain weight.

Write a true sentence using $<$, $>$, or $=$.

92. $|-3|$ 5

93. 2 $|-4|$

94. 0 $|0|$

95. $|-5|$ $|-2|$

96. $|4|$ $|-7|$

97. $|-8|$ $|8|$

List in order from least to greatest.

98. 7, −5, $|-6|$, 4, $|3|$, −100, 0, 1, $\frac{14}{4} = 3.5$

99. What number corresponds to point A in the figure?

100. What number corresponds to point B in the figure?

101. We know that 0.3333... is $\frac{1}{3}$ and 0.6666... is $\frac{2}{3}$. What rational number is named by each of the following?

a) 0.9999...

b) 0.1111...

OBJECTIVES

After finishing Section 1.2, you should be able to:

i Factor numbers, find prime factorizations, find common multiples, and find the LCM of two or more numbers.

ii Write an expression equivalent to a given one, simplify fractional notation, and multiply, add, subtract, and divide using fractional notation.

iii Convert between percent notation and decimal notation, and between percent notation and fractional notation.

iv Find estimates of sums, products, differences, and quotients.

Factor each number.

1. 9 1×9 3×3

2. 16 1×16 8×2
   4×4

Write several factorizations for each number.

3. 18 1×18, 2×9 3×6

4. 20 1×20 4×5
   2×10

1.2

USING FRACTIONAL, DECIMAL, AND PERCENT NOTATION

In algebra the notion of factoring is quite important. You will eventually learn to factor algebraic expressions. For now, we review factoring of numbers in order to review addition and subtraction using fractional notation.

i Finding Least Common Multiples

Factoring

In this section we will be considering the set of *natural numbers*:

$$1, 2, 3, 4, 5, \text{ and so on.}$$

To *factor* a number means to name it as a product.

EXAMPLE 1 Factor the number 8.

The number 8 can be named as a product in several ways:

$$2 \cdot 4, \quad 1 \cdot 8, \quad 2 \cdot 2 \cdot 2.$$

DO EXERCISES 1 AND 2.

A symbol that names a number as a product is called a *factorization* of the number.

EXAMPLE 2 Write several factorizations of the number 12.

$$1 \cdot 12, \quad 2 \cdot 6, \quad 3 \cdot 4, \quad 2 \cdot 2 \cdot 3$$

DO EXERCISES 3 AND 4.

Prime Numbers and Prime Factorizations

Some numbers have exactly two factors, themselves and 1. Such numbers are called *prime.*

> A *prime* number is a natural number that has exactly two different factors.

EXAMPLE 3 Which of these numbers are prime?   7, 4, 11, 16, 1

   7 is prime. It has exactly two different factors, 7 and 1.

   4 is not prime. It has three different factors, 1, 2, and 4.

   11 is prime. It has exactly two different factors, 11 and 1.

   16 is not prime. It has factors 1, 2, 4, 8, and 16.

   1 is not prime. It has only itself as a factor.

The following is a table of the prime numbers from 2 to 101. There are more extensive tables, but these prime numbers will be the most helpful to you in this book.

> **A TABLE OF PRIMES**
>
> 2, 3, 5, 7, 11, 13, 17, 19, 23, 29, 31, 37, 41,
> 43, 47, 53, 59, 61, 67, 71, 73, 79, 83, 89, 97, 101

DO EXERCISE 5.

Prime Factorizations

If a natural number, other than 1, is not prime, we call it *composite.* Every composite number can be factored into prime numbers. Such a factorization is called a *prime factorization.*

EXAMPLE 4 Find the prime factorization of 36.

   We begin by factoring 36 any way we can. One way is like this:

$$36 = 4 \cdot 9.$$

The factors 4 and 9 are not prime, so we factor them:

$$36 = 4 \cdot 9 = \underbrace{2 \cdot 2}_{4} \cdot \underbrace{3 \cdot 3}_{9}$$

The factors in the last factorization are all prime, so we have the *prime factorization* of 36.

EXAMPLE 5 Find the prime factorization of 60.

   This time, we use our list of primes from the table.

Divide 60 by 2: $2\overline{)60}^{\,30}$

$$60 = 2 \cdot 30$$

---

5. Which of these numbers is or are prime?

   8, 6, 13, 14, 1

Find the prime factorization of each number.

6. 48

$48 = 2 \cdot 24$

$2 \cdot 3 \cdot 8$

$48 = 2 \cdot 3 \cdot 2 \cdot 2 \cdot 2$

7. 50

$50 = 25 \cdot 2$

$50 = 5 \cdot 5 \cdot 2$

8. 770

$770 = 11 \cdot 70$

$770 = 11 \cdot 7 \cdot 10$

$770 = 11 \cdot 7 \cdot 5 \cdot 2$

9. Find several of the common multiples of 3 and 5 by making lists of multiples.

$3 \cdot 6 \cdot 9 \cdot 12 \cdot 15 \cdot 18 \cdot 21 \cdot 24 \cdot$

$27 \cdot \boxed{30} \cdot 33 \; 36 \; 39 \cdot 42 \cdot \boxed{45}$

$5 \cdot 10 \cdot 15 \cdot 20 \cdot 25 \; \boxed{30} \; 35 \cdot 40$

$\boxed{45} \; 50$

Divide 30 by 2: $2\,\overline{)\,30\,}$ with quotient 15

$$60 = 2 \cdot \underbrace{2 \cdot 15}_{30}$$

We cannot divide 15 by 2 without remainder. So we try dividing by the next prime,

3:   $3\,\overline{)\,15\,}$ with quotient 5

$$60 = 2 \cdot 2 \cdot \underbrace{3 \cdot 5}_{15} \qquad \text{This is the prime factorization.}$$

DO EXERCISES 6–8.

Multiples

The *multiples* of a number all have that number as a factor. For example, the multiples of 2 are

$$2, 4, 6, 8, 10, 12, 14, 16, \ldots .$$

We could name them in such a way as to show 2 as a factor, as follows:

$$2 \cdot 1, 2 \cdot 2, 2 \cdot 3, 2 \cdot 4, 2 \cdot 5, 2 \cdot 6, \ldots .$$

The multiples of 3 all have 3 as a factor. They are

$$3, 6, 9, 12, 15, 18, \ldots .$$

We could also name them as follows:

$$3 \cdot 1, 3 \cdot 2, 3 \cdot 3, 3 \cdot 4, 3 \cdot 5, \ldots .$$

Common Multiples.   Two or more numbers always have a great many multiples in common. From lists of multiples, we can find common multiples.

EXAMPLE 6   Find some of the multiples that 2 and 3 have in common.

   We make lists of their multiples and circle the multiples that appear in both lists.

$$2, 4, \boxed{6}, 8, 10, \boxed{12}, 14, 16, \boxed{18}, 20, 22, \boxed{24}, 26, 28, \boxed{30}, 32, 34, \boxed{36}, \ldots$$
$$3, \boxed{6}, 9, \boxed{12}, 15, \boxed{18}, 21, \boxed{24}, 27, \boxed{30}, 33, \boxed{36}, \ldots$$

The common multiples of 2 and 3 are

$$6, 12, 18, 24, 30, 36, \ldots .$$

DO EXERCISE 9.

Least Common Multiples

The numbers 2 and 3 have many common multiples. The *least*, or smallest, of those common multiples is 6. We abbreviate *least common multiple* as LCM.

   Here is a method of finding LCMs of two or more numbers.

To find the LCM of two or more numbers:

a) **Determine whether the largest of the numbers is a multiple of the others. If it is, then it is the LCM.**

b) **If not, make a list of multiples of the largest number, until you get one that is a multiple of the other numbers. That number is the LCM.**

EXAMPLE 7 Find the LCM of 12 and 15.

a) 15 is not a multiple of 12.

b) We list multiples of 15, the larger number.

$15 \cdot 2 = 30$, but 30 is not a multiple of 12.

$15 \cdot 3 = 45$, but 45 is not a multiple of 12.

$15 \cdot 4 = 60$, and 60 is a multiple of 12, because $60 = 5 \cdot 12$.

The LCM of 12 and 15 is 60.

DO EXERCISE 10.

EXAMPLE 8 Find the LCM of 4 and 20.

a) 20 is a multiple of 4, because $20 = 4 \cdot 5$.

The LCM of 4 and 20 is 20.

EXAMPLE 9 Find the LCM of 6, 10, and 12.

a) The largest number, 12, is not a multiple of both 6 and 10.

b) We list multiples of 12.

$12 \cdot 2 = 24$, and 24 is a multiple of 6 but not a multiple of 10.

$12 \cdot 3 = 36$, and 36 is a multiple of 6 but not a multiple of 10.

$12 \cdot 4 = 48$, and 48 is a multiple of 6 but not a multiple of 10.

$12 \cdot 5 = 60$, and 60 is a multiple of both 6 and 10.

The LCM of 6, 10, and 12 is 60.

DO EXERCISES 11–14.

Finding LCMs by Factoring

The method we have just used works for natural numbers, but it does not work in algebra. We now learn another method that will work in arithmetic *and also in algebra.* To show how the other method works, we make lists, but we write factorizations. Let's look for the LCM of 9 and 15.

The multiples of 9 are

$3 \cdot 3 \cdot 2, 3 \cdot 3 \cdot 3, 3 \cdot 3 \cdot 4, 3 \cdot 3 \cdot 5, 3 \cdot 3 \cdot 6, \ldots .$

The multiples of 15 are

$3 \cdot 5 \cdot 2, 3 \cdot 5 \cdot 3, 3 \cdot 5 \cdot 4, 3 \cdot 5 \cdot 5, \ldots .$

The LCM of 9 and 15 is $3 \cdot 3 \cdot 5$. The LCM must have all of the factors of 9, and it must have all of the factors of 15.

10. Find the LCM of 8 and 10.

$8 \cdot 2 = 16$ NO
$8 \cdot 3 = 24$ NO
$8 \cdot 4 = 32$ NO
$8 \cdot 5 = 40$ yes $10 \cdot 4 = 40$

LCM of 8, 10 = 40

Find the LCM.

11. 5 and 24

$24 \cdot 2 = 48$ NO
$24 \cdot 3 = 72$ NO
$24 \cdot 4 = 96$ NO
$24 \cdot 5 = 120$ yes

LCM of 5, 24 = 120

12. 8 and 48

$8 \cdot 6 = 48$ $48 \cdot 1 = 48$

48 is LCM

13. 12 and 16

$16 \cdot 2 = 32$
$16 \cdot 3 = 48$

LCM = 48

$16 \cdot 3$, $12 \cdot 4$

14. 20, 40, and 80

LCM = 80

$20 \cdot 4 = 80$
$40 \cdot 2 = 80$
$80 \cdot 1 = 80$

Find the LCM by factoring.

15. 8 and 10

$8 = 2 \cdot 2 \cdot 2$

$10 = 2 \cdot 5$

$LCM = 2 \cdot 2 \cdot 2 \cdot 5, \text{ or } 40$

16. 18 and 40

$18 = 2 \cdot 3 \cdot 3$

$40 = 2 \cdot 2 \cdot 5 \cdot 2$

$LCM = 2 \cdot 2 \cdot 2 \cdot 3 \cdot 3 \cdot 5 = 360$

17. Find the LCM of 24, 35, and 45.

$24 = 2 \cdot 2 \cdot 3 \cdot 2$

$35 = 5 \cdot 7$

$45 = 5 \cdot 3 \cdot 3$

$LCM = 2 \cdot 2 \cdot 2 \cdot 3 \cdot 3 \cdot 5 \cdot 7 = 2520$

Find the LCM.

18. 3, 18

$18 = 2 \cdot 3 \cdot 3$

$LCM = 2 \cdot 3 \cdot 3 = 18$

19. 12, 24

$12 = 2 \cdot 3 \cdot 2$

$24 = 2 \cdot 2 \cdot 3 \cdot 2$

$LCM = 2 \cdot 2 \cdot 2 \cdot 3 = 24$

Find the LCM.

20. 4, 9

$4 = 2 \cdot 2$

$9 = 3 \cdot 3$

$LCM = 2 \cdot 2 \cdot 3 \cdot 3 = 36$

21. 5, 6, 7

$5 = 5$

$6 = 2 \cdot 3$

$7 = 7$

$LCM = 2 \cdot 3 \cdot 5 \cdot 7 = 210$

> **To find the LCM of several numbers:**
> 1. **Write the prime factorization of each number.**
> 2. **Write each factor the greatest number of times it appears in any one factorization.**

EXAMPLE 10   Find the LCM of 24 and 36.

a)   We find the prime factorizations.

$$24 = 2 \cdot 2 \cdot 2 \cdot 3$$
$$36 = 2 \cdot 2 \cdot 3 \cdot 3$$

b)   We write 2 as a factor 3 times (the greatest number of times it occurs).

c)   We write 3 as a factor 2 times (the greatest number of times it occurs).

The LCM is $2 \cdot 2 \cdot 2 \cdot 3 \cdot 3$, or 72.

DO EXERCISES 15 AND 16.

EXAMPLE 11   Find the LCM of 27, 90, and 84.

a)   We factor.

$$27 = 3 \cdot 3 \cdot 3$$
$$90 = 2 \cdot 3 \cdot 3 \cdot 5$$
$$84 = 2 \cdot 2 \cdot 3 \cdot 7$$

b)   We write 2 as a factor 2 times, 3 as a factor 3 times, 5 one time, and 7 one time.

The LCM is $2 \cdot 2 \cdot 3 \cdot 3 \cdot 3 \cdot 5 \cdot 7$, or 3780.

DO EXERCISE 17.

EXAMPLE 12   Find the LCM of 7 and 21.

Since 7 is prime, it has no prime factorization. We still need it as a factor.

$$7 = \quad 7$$
$$21 = 3 \cdot 7$$

The LCM is $3 \cdot 7$, or 21.

If one number is a factor of another, the LCM is the larger of the numbers.

DO EXERCISES 18 AND 19.

EXAMPLE 13   Find the LCM of 8 and 9.

$$8 = 2 \cdot 2 \cdot 2$$
$$9 = 3 \cdot 3$$

The LCM is $2 \cdot 2 \cdot 2 \cdot 3 \cdot 3$, or 72.

If two or more numbers have no common prime factor, the LCM is the product of the numbers.

DO EXERCISES 20 AND 21.

ii Fractional Notation

Properties of 0 and 1

Some simple but powerful properties of numbers are the properties of 0 and 1. We will see some ways to use both of these properties in algebra and arithmetic.

THE IDENTITY PROPERTY OF 0

For any real number a,

$$a + 0 = a.$$

(Adding 0 to any number gives that same number.)

THE PROPERTY OF 1

For any real number a,

$$1 \cdot a = a.$$

(Multiplying a number by 1 gives that same number.)

Here are some of the many ways to name the number 1:

$$\frac{5}{5}, \qquad \frac{3}{3}, \qquad \frac{26}{26}.$$

For any real number a, except 0,

$$\frac{a}{a} = 1.$$

We can use the property of 1 to write equivalent expressions.

EXAMPLE 14 Write an expression equivalent to $\frac{2}{3}$ by multiplying by 1. Use $\frac{5}{5}$ for 1.

Recall from arithmetic that to multiply with fractional notation we multiply numerators and denominators.

$$\frac{2}{3} = \frac{2}{3} \cdot 1 \qquad \text{Using the property of 1}$$
$$= \frac{2}{3} \cdot \frac{5}{5} \qquad \text{Using } \frac{5}{5} \text{ for 1}$$
$$= \frac{10}{15} \qquad \text{Multiplying numerators and denominators}$$

EXAMPLE 15 Write an expression equivalent to $\frac{5}{2}$ by multiplying by $\frac{8}{8}$.

$$\frac{5}{2} = \frac{5}{2} \cdot 1 \qquad \text{Using the property of 1}$$
$$= \frac{5}{2} \cdot \frac{8}{8} \qquad \text{Using } \frac{8}{8} \text{ for 1}$$
$$= \frac{40}{16}$$

DO EXERCISES 22 AND 23.

Simplifying Expressions

We know that $\frac{1}{2}, \frac{2}{4}, \frac{4}{8}$, etc., all name the same number. Any number of arithmetic can be named in many ways. The *simplest fractional notation* is the notation that has the smallest numerator and denominator. We call the process of finding the simplest

22. Write an expression equivalent to $\frac{8}{5}$. Use $\frac{12}{12}$ for 1.

$$\frac{8}{5} \cdot \frac{12}{12} = \frac{96}{60} \text{ or } \frac{48}{30} \text{ or } \frac{24}{15}$$

23. Write an expression equivalent to $\frac{4}{23}$. Use $\frac{7}{7}$ for 1.

$$4/23 \cdot 7/7 = \frac{28}{161}$$

Simplify.

24. $\dfrac{18}{27} = \dfrac{2}{3}$

$\dfrac{18}{27} = \dfrac{3 \cdot 2 \cdot 3}{3 \cdot 3 \cdot 3}$

25. $\dfrac{48}{18} = \dfrac{16}{6}$ or $\dfrac{8}{3}$

$\dfrac{48}{18} = \dfrac{2 \cdot 3 \cdot 2 \cdot 2 \cdot 2}{2 \cdot 3 \cdot 3} = \dfrac{8}{3}$

26. $\dfrac{56}{49}$

$\dfrac{56}{49} = \dfrac{2 \cdot 2 \cdot 2 \cdot 7}{7 \cdot 7}$

$\dfrac{2 \cdot 2 \cdot 2}{7} = \dfrac{8}{7}$

Simplify.

27. $\dfrac{27}{54}$

$\dfrac{27}{54} = \dfrac{3 \cdot 3 \cdot 3 \cdot 1}{3 \cdot 3 \cdot 3 \cdot 2} = \dfrac{1}{2}$

28. $\dfrac{48}{12} = 4$

$\dfrac{48}{12} = \dfrac{2 \cdot 2 \cdot 2 \cdot 2 \cdot 3}{2 \cdot 2 \cdot 3 \cdot 1 \cdot 1}$

---

fractional notation *simplifying*. We begin by factoring the numerator and the denominator. Then we factor the fractional expression and use the property of 1.

EXAMPLE 16 Simplify: $\frac{10}{15}$.

$$\frac{10}{15} = \frac{2 \cdot 5}{3 \cdot 5} \qquad \text{Factoring the numerator and the denominator}$$

$$= \frac{2}{3} \cdot \frac{5}{5} \qquad \text{Factoring the fractional expression}$$

$$= \frac{2}{3} \qquad \text{Using the property of 1 (removing a factor of 1)}$$

EXAMPLE 17 Simplify: $\frac{36}{24}$.

$$\frac{36}{24} = \frac{6 \cdot 6}{4 \cdot 6} \qquad \text{Factoring the numerator and the denominator}$$

$$= \frac{3 \cdot 2 \cdot 6}{2 \cdot 2 \cdot 6} \qquad \text{Further factoring}$$

$$= \frac{3}{2} \cdot \frac{2 \cdot 6}{2 \cdot 6} \qquad \text{Factoring the fractional expression}$$

$$= \frac{3}{2} \qquad \text{Using the property of 1}$$

DO EXERCISES 24–26.

The number of factors in the numerator and the denominator may not always be the same. If not, we can always insert the number 1 as a factor. The property of 1 allows us to do that.

EXAMPLE 18 Simplify: $\frac{18}{72}$.

$$\frac{18}{72} = \frac{2 \cdot 9}{4 \cdot 2 \cdot 9} = \frac{1 \cdot 2 \cdot 9}{4 \cdot 2 \cdot 9} \qquad \text{Using the property of 1 to insert the factor 1}$$

$$= \frac{1}{4} \cdot \frac{2 \cdot 9}{2 \cdot 9} \qquad \text{Factoring the fractional expression}$$

$$= \frac{1}{4}$$

EXAMPLE 19 Simplify: $\frac{72}{9}$.

$$\frac{72}{9} = \frac{8 \cdot 9}{1 \cdot 9} \qquad \text{Factoring and inserting a factor of 1 in the denominator}$$

$$= \frac{8}{1} \cdot \frac{9}{9} = \frac{8}{1}$$

$$= 8$$

DO EXERCISES 27 AND 28.

Multiplication, Addition, Subtraction, and Division

After we have performed an operation of multiplication, addition, subtraction, or division, the answer may not be simplified. In such cases, we should simplify. We continue as we did in Examples 16–19.

EXAMPLE 20 Multiply and simplify: $\frac{5}{6} \cdot \frac{9}{25}$.

$$\frac{5}{6} \cdot \frac{9}{25} = \frac{5 \cdot 9}{6 \cdot 25} \qquad \text{Multiplying numerators and denominators}$$

$$= \frac{1 \cdot 5 \cdot 3 \cdot 3}{2 \cdot 3 \cdot 5 \cdot 5} \qquad \text{Factoring the numerator and the denominator}$$

$$= \frac{3 \cdot 5 \cdot 1 \cdot 3}{3 \cdot 5 \cdot 2 \cdot 5} = \frac{3 \cdot 5}{3 \cdot 5} \cdot \frac{1 \cdot 3}{2 \cdot 5} \qquad \text{Factoring the fractional expression}$$

$$= \frac{3}{10} \qquad \text{"Removing" a factor of 1}$$

DO EXERCISES 29 AND 30.

When denominators are the same, we can add by adding the numerators and keeping the same denominator. That is,

$$\frac{a}{c} + \frac{b}{c} = \frac{a + b}{c}.$$

When denominators are different, we use the property of 1 and multiply to find a common denominator. The smallest such denominator is called the *least common denominator*. That number is the least common multiple of the original denominators.

EXAMPLE 21 Add and simplify: $\frac{3}{8} + \frac{5}{12}$.

The least common multiple of the denominators is 24. We multiply by 1 to obtain the least common multiple for each denominator:

$$\frac{3}{8} + \frac{5}{12} = \frac{3}{8} \cdot \frac{3}{3} + \frac{5}{12} \cdot \frac{2}{2} \qquad \text{Multiplying by 1. Since } 3 \cdot 8 = 24, \text{ we multiply the first number by } \frac{3}{3}. \text{ Since } 2 \cdot 12 = 24, \text{ we multiply the second number by } \frac{2}{2}.$$
$$= \frac{9}{24} + \frac{10}{24}$$
$$= \frac{19}{24}.$$

DO EXERCISES 31 AND 32.

EXAMPLE 22 Subtract and simplify: $\frac{9}{8} - \frac{4}{5}$.

$$\frac{9}{8} - \frac{4}{5} = \frac{9}{8} \cdot \frac{5}{5} - \frac{4}{5} \cdot \frac{8}{8} \qquad \text{The LCM of 8 and 5 is 40.}$$
$$= \frac{45}{40} - \frac{32}{40}$$
$$= \frac{13}{40}$$

In arithmetic you usually write $1\frac{1}{8}$ rather than $\frac{9}{8}$. In algebra you will find that the so-called *improper* symbols such as $\frac{9}{8}$ are more useful and are quite *proper*.

EXAMPLE 23 Subtract and simplify: $\frac{7}{10} - \frac{1}{5}$.

$$\frac{7}{10} - \frac{1}{5} = \frac{7}{10} - \frac{1}{5} \cdot \frac{2}{2} \qquad \text{The LCM of 10 and 5 is 10.}$$
$$= \frac{7}{10} - \frac{2}{10}$$
$$= \frac{5}{10} = \frac{1}{2} \cdot \frac{5}{5} = \frac{1}{2}$$

DO EXERCISES 33 AND 34.

Multiply and simplify.

29. $\frac{6}{5} \cdot \frac{25}{12}$

$\frac{2 \cdot 3 \cdot 5 \cdot 5}{5 \cdot 3 \cdot 2 \cdot 2} = \frac{5}{2}$

30. $\frac{3}{8} \cdot \frac{5}{3} \cdot \frac{7}{2}$

$\frac{3 \cdot 5 \cdot 7 \cdot 1 \cdot 1}{2 \cdot 2 \cdot 2 \cdot 3 \cdot 2} \qquad \frac{35}{16}$

Add and simplify.

31. $\frac{5}{6} + \frac{7}{10}$

$\frac{5}{6} \cdot \frac{5}{5} + \frac{7}{10} \cdot \frac{3}{3} =$

$\frac{25}{30} + \frac{21}{30} = \frac{46}{30}$

$\frac{46}{30} = \boxed{\frac{23}{15}}$

32. $\frac{1}{4} + \frac{1}{3}$

$\frac{1}{4} \cdot \frac{3}{3} + \frac{1}{3} \cdot \frac{4}{4} = \frac{3}{12} + \frac{4}{12} = \boxed{\frac{7}{12}}$

Subtract and simplify.

33. $\frac{4}{5} - \frac{4}{6}$

$\frac{4}{5} \cdot \frac{6}{6} - \frac{4}{6} \cdot \frac{5}{5} = \frac{24}{30} - \frac{20}{30} = \frac{7}{30}$

$\boxed{\frac{2}{15}}$

34. $\frac{5}{12} - \frac{2}{9}$

$\frac{5}{12} \cdot \frac{3}{3} - \frac{2}{9} \cdot \frac{4}{4} = \frac{15}{36} - \frac{8}{36}$

$\boxed{\frac{7}{36}}$

Divide by multiplying by 1.

35. $\dfrac{\frac{3}{5}}{\frac{4}{7}}$ *(handwritten: $\frac{7}{4} \times \frac{7}{4}$ $\frac{3}{5} \times \frac{7}{4} = \frac{21}{20}$)*

36. $\dfrac{\frac{5}{4}}{\frac{3}{2}}$ *(handwritten: $\times \frac{2}{3}$ $\frac{2}{3}$ $\frac{5}{4} \times \frac{2}{3} = \frac{10}{12}$)*

37. $\dfrac{\frac{9}{7}}{\frac{4}{5}}$ *(handwritten: $\times \frac{5}{4}$ $\times \frac{5}{4}$ $\frac{9}{7} \times \frac{5}{4} = \frac{45}{28}$)*

Divide by multiplying by the reciprocal of the divisor.

38. $\dfrac{4}{3} \div \dfrac{7}{2}$ *(handwritten: $\frac{4}{3} \cdot \frac{2}{7} = \frac{8}{21}$)*

39. $\dfrac{3}{5} \div \dfrac{7}{4}$ *(handwritten: $\frac{3}{5} \cdot \frac{4}{7} = \frac{12}{35}$)*

40. $\dfrac{\frac{2}{9}}{\frac{5}{7}}$ *(handwritten: $\frac{2}{9} \cdot \frac{7}{5} = \frac{14}{45}$)*

Reciprocals

Two numbers whose product is 1 are called *reciprocals* of each other. All the numbers of arithmetic except zero have reciprocals.

EXAMPLES

24. The reciprocal of $\frac{2}{3}$ is $\frac{3}{2}$ because $\frac{2}{3} \cdot \frac{3}{2} = \frac{6}{6} = 1$.

25. The reciprocal of 9 is $\frac{1}{9}$ because $9 \cdot \frac{1}{9} = \frac{9}{9} = 1$.

26. The reciprocal of $\frac{1}{4}$ is 4 because $\frac{1}{4} \cdot 4 = 1$.

Reciprocals and Division

The number 1 and reciprocals can be used to explain division of numbers of arithmetic. To divide, we can multiply by 1, choosing carefully the symbol for 1.

EXAMPLE 27   Divide $\frac{2}{3}$ by $\frac{7}{5}$.

$$\dfrac{\frac{2}{3}}{\frac{7}{5}} = \dfrac{\frac{2}{3}}{\frac{7}{5}} \times \dfrac{\frac{5}{7}}{\frac{5}{7}} \qquad \text{Multiplying by } \dfrac{\frac{5}{7}}{\frac{5}{7}}. \text{ We use } \tfrac{5}{7} \text{ because it is the reciprocal of } \tfrac{7}{5}.$$

$$= \dfrac{\frac{2}{3} \times \frac{5}{7}}{\frac{7}{5} \times \frac{5}{7}} \qquad \text{Multiplying numerators and denominators}$$

$$= \dfrac{\frac{10}{21}}{1} = \dfrac{10}{21} \qquad \text{Simplifying}$$

After multiplying we got 1 for a denominator. That was because we used the reciprocal of the divisor, $\frac{7}{5}$, for both the numerator and the denominator of the symbol for 1.

DO EXERCISES 35–37.

When multiplying by 1 to divide, we get a denominator of 1. What do we get in the numerator? In Example 27, we got $\frac{2}{3} \times \frac{5}{7}$. This is the product of $\frac{2}{3}$, the dividend, and $\frac{5}{7}$, the reciprocal of the divisor.

> **To divide, multiply by the reciprocal of the divisor:**
> $$\dfrac{a}{b} \div \dfrac{c}{d} = \dfrac{a}{b} \cdot \dfrac{d}{c}.$$

EXAMPLE 28   Divide by multiplying by the reciprocal of the divisor: $\frac{1}{2} \div \frac{3}{5}$.

$$\tfrac{1}{2} \div \tfrac{3}{5} = \tfrac{1}{2} \cdot \tfrac{5}{3} = \tfrac{5}{6} \qquad \tfrac{5}{3} \text{ is the reciprocal of } \tfrac{3}{5}$$

After dividing, simplification is often possible and should be done.

EXAMPLE 29   Divide: $\frac{2}{3} \div \frac{4}{9}$.

$$\tfrac{2}{3} \div \tfrac{4}{9} = \tfrac{2}{3} \cdot \tfrac{9}{4} = \dfrac{2 \cdot 3 \cdot 3}{3 \cdot 2 \cdot 2} = \dfrac{2 \cdot 3}{2 \cdot 3} \cdot \tfrac{3}{2} = \tfrac{3}{2}$$

Simplifying

DO EXERCISES 38–40.

Complex Fractional Expressions

The following is a *complex fractional expression*:

$$\frac{\frac{2}{3} + \frac{3}{4}}{\frac{7}{8} - \frac{5}{6}}.$$

To simplify such an expression we first add or subtract, if necessary, to get a single fractional expression in both the numerator and the denominator. Then we divide by multiplying by the reciprocal of the denominator.

EXAMPLE 30   Simplify: $\frac{\frac{2}{3} + \frac{3}{4}}{\frac{7}{8} - \frac{5}{6}}$.

$$\frac{\frac{2}{3} + \frac{3}{4}}{\frac{7}{8} - \frac{5}{6}} = \frac{\frac{2}{3} \cdot \frac{4}{4} + \frac{3}{4} \cdot \frac{3}{3}}{\frac{7}{8} \cdot \frac{3}{3} - \frac{5}{6} \cdot \frac{4}{4}} \left.\begin{array}{l}\end{array}\right\}$$

⟶ The LCM is 12. We multiply by 1 to obtain the LCM in the numerator.
⟶ The LCM is 24. We multiply by 1 to obtain the LCM in the denominator.

$$= \frac{\frac{8}{12} + \frac{9}{12}}{\frac{21}{24} - \frac{20}{24}}$$ ⟶ Adding in the numerator
⟶ Subtracting in the denominator

$$= \frac{\frac{17}{12}}{\frac{1}{24}}$$

$$= \frac{17}{12} \div \frac{1}{24}$$

$$= \frac{17}{12} \cdot \frac{24}{1} = \frac{17 \cdot 2 \cdot 2 \cdot 2 \cdot 3}{2 \cdot 2 \cdot 3 \cdot 1} = \frac{2 \cdot 2 \cdot 3}{2 \cdot 2 \cdot 3} \cdot \frac{17 \cdot 2}{1} = 34$$

⟶ Simplifying

DO EXERCISES 41 AND 42.

▦ Percent Notation

Converting to Decimal Notation

On the average, a family spends 26% of its income for food. What does this mean? It means that out of every $100 earned, $26 is spent for food. Thus 26% is a ratio of 26 to 100.

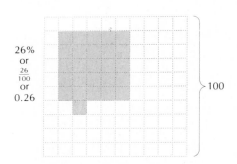

26%
or
$\frac{26}{100}$
or
0.26

⟩100

The percent symbol % means "per hundred." We can regard the percent symbol as part of a name for a number. For example,

26%   is defined to mean 26×0.01 or $26 \times \frac{1}{100}$.

Simplify.

41. $\dfrac{\frac{5}{6} - \frac{2}{3}}{\frac{4}{5} + \frac{7}{3}}$

$\frac{15}{18} - \frac{12}{18} = \frac{3}{18}$

$\frac{12}{15} + \frac{35}{15} = \frac{47}{15}$

$\frac{3}{18} \cdot \frac{15}{47}$ or $\frac{1}{6} \cdot \frac{15}{47} = \frac{15}{1692}$

$\boxed{\frac{5}{564}}$

42. $\dfrac{\frac{5}{8} + \frac{7}{10}}{\frac{13}{16}}$

$\frac{25}{40} + \frac{28}{40}$ $\frac{53}{40}$

$\frac{53}{40} \cdot \frac{16}{13} = \frac{848}{520}$

$\frac{424}{260}$ or $\boxed{\frac{106}{65}}$

Find decimal notation.

43. 46.2%

.462

44. 100%

1.0

Find fractional notation.

45. 67%

$\frac{67}{100}$

46. 45.6%

$\frac{456}{1000}$

47. $\frac{1}{4}$%   .25

$\frac{25}{100}$

Find percent notation.

48. 6.77

677%

49. 0.9944

99.44%

In general,

$$n\% \quad \text{means} \quad n \times 0.01, \quad \text{or} \quad n \times \tfrac{1}{100}.$$

EXAMPLE 31   Find decimal notation for 78.5%.

$$78.5\% = 78.5 \times 0.01 \qquad \text{Replacing \% by} \times 0.01$$
$$= 0.785$$

DO EXERCISES 43 AND 44.

Converting to Fractional Notation

EXAMPLE 32   Find fractional notation for 88%.

$$88\% = 88 \times \tfrac{1}{100} \qquad \text{Replacing \% by} \times \tfrac{1}{100}$$
$$= \tfrac{88}{100} \qquad \text{You need not simplify.}$$

EXAMPLE 33   Find fractional notation for 34.7%.

$$34.7\% = 34.7 \times \tfrac{1}{100} \qquad \text{Replacing \% by} \times \tfrac{1}{100}$$
$$= \tfrac{34.7}{100}$$
$$= \tfrac{34.7}{100} \cdot \tfrac{10}{10} \qquad \text{Multiplying by 1 to get a whole number in the numerator}$$
$$= \tfrac{347}{1000}$$

There is a table of decimal and percent equivalents at the back of the book. If you do not already know these facts, you should memorize them.

DO EXERCISES 45–47.

Converting from Decimal Notation to Percent Notation

By applying the definition of % in reverse, we can convert from decimal notation to percent notation. We multiply by 1, naming it 100×0.01.

EXAMPLE 34   Find percent notation for 0.93.

$$0.93 = 0.93 \times 1$$
$$= 0.93 \times (100 \times 0.01) \qquad \text{Replacing 1 by } 100 \times 0.01$$
$$= (0.93 \times 100) \times 0.01 \qquad \text{Using associativity}$$
$$= 93 \times 0.01$$
$$= 93\% \qquad \text{Replacing} \times 0.01 \text{ by \%}$$

EXAMPLE 35   Find percent notation for 0.002.

$$0.002 = 0.002 \times (100 \times 0.01)$$
$$= (0.002 \times 100) \times 0.01$$
$$= 0.2 \times 0.01$$
$$= 0.2\% \qquad \text{Replacing} \times 0.01 \text{ by \%}$$

DO EXERCISES 48 AND 49.

Converting from Fractional Notation to Percent Notation

We can also convert from fractional notation to percent notation. Again we multiply by 1, but this time we use $100 \times \frac{1}{100}$.

EXAMPLE 36 Find percent notation for $\frac{5}{8}$.

$$\frac{5}{8} = \frac{5}{8} \times (100 \times \frac{1}{100})$$
$$= (\frac{5}{8} \times 100) \times \frac{1}{100}$$
$$= \frac{500}{8} \times \frac{1}{100}$$
$$= \frac{500}{8}\%, \quad \text{or} \quad 62.5\%$$

The result of this example says that the ratio of 5 to 8 is the same as the ratio of 62.5 to 100.

DO EXERCISES 50–52.

Table 1, located on the inside front cover, contains many common fractional–decimal–percent equivalents. It would be most helpful to memorize that table.

iv Estimating

Estimating is a useful tool in mathematics. One way it can be used is to determine if an answer you have made by hand or by a calculator is correct. In actual practice an estimated answer is often sufficient for our uses in problem solving. For example, to the question "How many stars are there in the universe?" an estimated answer would be quite sufficient.

Keep in mind that there are no absolute ways of carrying out estimating, and there are no absolute answers when you are asked to estimate. Sometimes we round, and sometimes we make an estimate by analyzing the problem.

EXAMPLE 37 Estimate: $347 + 279 + 543$.
Method 1. We can round each number to the nearest hundred and then add:

$$347 + 279 + 543 \approx 300 + 300 + 500 = 1100.$$

We can also do this estimate by rounding to the nearest ten:

$$347 + 279 + 543 \approx 350 + 280 + 540 = 1170.$$

Method 2. Note that 347 is about 350 and 543 is about 550, so when these numbers are added we get 900. Then 279 is about 300, so we add 300 to 900 and get 1200 as an estimate.

Note that we got three answers to the estimation and all are reasonable.

DO EXERCISES 53 AND 54.

EXAMPLE 38 Estimate: $\frac{4035}{1400}$.
Method 1. We can round to the nearest thousand and divide:

$$\frac{4035}{1400} \approx \frac{4000}{1000} = 4.$$

Method 2. We note that 3 times 1400 is 4200, and 4200 is close to 4035. Thus,

$$\frac{4035}{1400} \approx \frac{4200}{1400} = 3.$$

Find percent notation.

50. $\frac{1}{4}$ $\quad \frac{100}{4}$ or 25%

51. $\frac{3}{8}$ $\quad \frac{300}{8} = 37.5\%$

52. $\frac{2}{3}$ $\quad \frac{200}{3} = 66.67\%$

Estimate. Keep in mind that answers may vary.

53. $162 + 584 + 273$
$160 + 580 + 270 \approx 1010$

54. $1.24 + 23.5 + 0.07 + 0.009$
$1.25 + 23. + .05 + .010$
≈ 24.31

Estimate. Keep in mind that answers may vary.

55. $\dfrac{5923}{1479}$ **56.** $\dfrac{0.65}{0.06} \times \dfrac{100}{100}$

$\dfrac{6000}{1500} \approx 4$ $\dfrac{65}{6} \approx \dfrac{70}{5}$

$\dfrac{70}{5} \approx 14$

EXAMPLE 39   Estimate: $\dfrac{0.34}{0.019}$.

We first multiply by 1, using $\frac{1000}{1000}$, to clear of decimals.

$$\frac{0.34}{0.019} = \frac{340}{19} \approx \frac{340}{20} = 17$$

DO EXERCISES 55 AND 56.

EXERCISE SET 1.2

i Write at least one factorization of each number.

1. 21   **2.** 45   **3.** 49   **4.** 28   **5.** 76   **6.** 56   **7.** 93   **8.** 144

Find the prime factorization of each number.

9. 14   **10.** 33   **11.** 9   **12.** 49   **13.** 18   **14.** 40   **15.** 90   **16.** 210

Find the prime factorization of each number. Then find the LCM.

17. 12, 18   **18.** 18, 30   **19.** 45, 72   **20.** 30, 36
21. 30, 50   **22.** 24, 36   **23.** 30, 40   **24.** 13, 23
25. 18, 24   **26.** 12, 28   **27.** 35, 45   **28.** 60, 70
29. 2, 3, 5   **30.** 3, 5, 7   **31.** 24, 36, 12   **32.** 8, 16, 22
33. 5, 12, 15   **34.** 12, 18, 40   **35.** 6, 12, 18   **36.** 18, 24, 30

ii Simplify.

37. $\frac{18}{45}$ **38.** $\frac{16}{56}$ **39.** $\frac{49}{14}$ **40.** $\frac{72}{27}$ **41.** $\frac{6}{42}$ **42.** $\frac{13}{104}$ **43.** $\frac{56}{7}$ **44.** $\frac{132}{11}$

Compute and simplify. If exercises are given in fractional notation, give fractional notation for the answer. If exercises are given in mixed numerals, give the answer in mixed numerals.

45. $\frac{12}{5} \cdot \frac{9}{8}$ **46.** $\frac{16}{15} \cdot \frac{5}{4}$ **47.** $\frac{10}{9} \cdot \frac{7}{5}$ **48.** $\frac{25}{12} \cdot \frac{4}{3}$ **49.** $\frac{9}{5} \div \frac{4}{5}$ **50.** $\frac{5}{12} \div \frac{25}{36}$
51. $120 \div \frac{5}{6}$ **52.** $360 \div \frac{8}{7}$ **53.** $\frac{3}{10} + \frac{1}{100}$ **54.** $\frac{9}{10} + \frac{3}{100}$ **55.** $\frac{5}{12} + \frac{4}{15}$ **56.** $\frac{3}{16} + \frac{1}{12}$
57. $\frac{7}{15} - \frac{3}{25}$ **58.** $\frac{18}{25} - \frac{4}{35}$ **59.** $\frac{99}{100} - \frac{9}{10}$ **60.** $\frac{78}{100} - \frac{11}{20}$ **61.** $8\frac{3}{4} + 5\frac{5}{6}$ **62.** $4\frac{3}{8} + 6\frac{5}{12}$
63. $3\frac{2}{5} + 8\frac{7}{10}$ **64.** $5\frac{1}{2} + 3\frac{7}{10}$ **65.** $34\frac{1}{3} - 12\frac{5}{8}$ **66.** $23\frac{5}{16} - 16\frac{3}{4}$ **67.** $21 - 8\frac{3}{4}$ **68.** $42 - 3\frac{7}{8}$
69. $8 \times 2\frac{5}{6}$ **70.** $5 \times 3\frac{3}{4}$ **71.** $3\frac{2}{5} \cdot 2\frac{7}{8}$ **72.** $4\frac{1}{5} \cdot 5\frac{1}{4}$ **73.** $4\frac{7}{10} \cdot 5\frac{3}{10}$ **74.** $20\frac{1}{2} \cdot 10\frac{1}{5}$
75. $20 \div 3\frac{1}{5}$ **76.** $8\frac{2}{5} \div 7$ **77.** $4\frac{3}{4} \div 1\frac{1}{3}$ **78.** $5\frac{4}{5} \div 2\frac{1}{2}$ **79.** $1\frac{7}{8} \div 1\frac{2}{3}$ **80.** $4\frac{3}{8} \div 2\frac{5}{6}$
81. $\frac{1}{4} \cdot \frac{1}{2}$ **82.** $\frac{11}{2} \cdot \frac{8}{5}$ **83.** $\frac{17}{2} \cdot \frac{3}{4}$ **84.** $\frac{11}{12} \cdot \frac{12}{11}$ **85.** $\frac{1}{2} + \frac{1}{2}$ **86.** $\frac{1}{2} + \frac{1}{4}$
87. $\frac{4}{9} + \frac{13}{18}$ **88.** $\frac{4}{5} + \frac{8}{15}$ **89.** $\frac{3}{10} + \frac{8}{15}$ **90.** $\frac{9}{8} + \frac{7}{12}$ **91.** $\frac{5}{4} - \frac{3}{4}$ **92.** $\frac{12}{5} - \frac{2}{5}$
93. $\frac{13}{18} - \frac{4}{9}$ **94.** $\frac{13}{15} - \frac{8}{45}$ **95.** $\frac{11}{12} - \frac{2}{5}$ **96.** $\frac{15}{16} - \frac{2}{3}$ **97.** $\frac{7}{5} \div \frac{3}{5}$ **98.** $\frac{7}{5} \div \frac{3}{4}$
99. $\frac{8}{9} \div \frac{4}{15}$ **100.** $\frac{3}{4} \div \frac{3}{7}$ **101.** $\frac{1}{4} \div \frac{1}{2}$ **102.** $\frac{1}{10} \div \frac{1}{5}$ **103.** $\dfrac{\frac{13}{12}}{\frac{39}{5}}$ **104.** $\dfrac{\frac{17}{6}}{\frac{3}{8}}$
105. $100 \div \frac{1}{5}$ **106.** $78 \div \frac{1}{6}$ **107.** $\frac{3}{4} \div 10$ **108.** $\frac{5}{6} \div 15$

Simplify.

109. $\dfrac{1 - \frac{3}{4}}{1 + \frac{9}{16}}$ **110.** $\dfrac{3 + \frac{1}{2}}{9 - \frac{1}{4}}$ **111.** $\dfrac{\frac{1}{2} + \frac{2}{3}}{\frac{5}{6} - \frac{3}{8}}$ **112.** $\dfrac{\frac{5}{8} + \frac{3}{4}}{\frac{2}{3} - \frac{5}{9}}$
113. $\dfrac{1 + \frac{1}{5}}{1 - \frac{3}{5}}$ **114.** $\dfrac{\frac{5}{27} + 5}{\frac{1}{3} + 1}$ **115.** $\dfrac{5 - \frac{1}{3}}{3 + \frac{1}{4}}$ **116.** $\dfrac{\frac{3}{5} + 5}{\frac{7}{4} + 7}$

iii Find decimal notation.

117. 76%   **118.** 54%   **119.** 54.7%   **120.** 96.2%
121. 100%   **122.** 1%   **123.** 0.61%   **124.** 125%

Find fractional notation.

125. 20% **126.** 80% **127.** 78.6% **128.** 13.5%

129. $12\frac{1}{2}\%$ **130.** 120% **131.** 0.042% **132.** 0.68%

Find percent notation.

133. 4.54 **134.** 1 **135.** 0.998 **136.** 0.73

137. 2 **138.** 0.0057 **139.** 0.072 **140.** 1.34

Find percent notation.

141. $\frac{1}{8}$ **142.** $\frac{1}{3}$ **143.** $\frac{17}{25}$ **144.** $\frac{11}{20}$ **145.** $\frac{17}{100}$ **146.** $\frac{119}{100}$

147. $\frac{7}{10}$ **148.** $\frac{8}{10}$ **149.** $\frac{3}{5}$ **150.** $\frac{17}{50}$ **151.** $\frac{2}{3}$ **152.** $\frac{1}{8}$

153. $\frac{7}{4}$ **154.** $\frac{7}{8}$ **155.** $\frac{3}{4}$ **156.** $\frac{99.4}{100}$

iv Estimate.

157. $0.39 \div 0.02$ **158.** $0.26 \div 0.004$ **159.** $0.049 \div 0.006$ **160.** $\dfrac{0.31 \times 6.2}{18.3 \times 4.51}$

161. $\dfrac{13}{8} \times \dfrac{5}{3} \times \dfrac{6}{11}$ **162.** $3.47 + 27.9 + 6.09$ **163.** $2359 - 1402$ **164.** $1.462 - 0.5936$

165. $613 \cdot 11$ **166.** $306 \cdot 19$ **167.** $\dfrac{49.2}{2.1}$ **168.** $\dfrac{1.882}{9.302}$

169. Consider 8 and 12. Determine whether each of the following is the LCM of 8 and 12. Tell why or why not.

a) $2 \cdot 2 \cdot 3 \cdot 3$ **b)** $2 \cdot 2 \cdot 3$ **c)** $2 \cdot 3 \cdot 3$ **d)** $2 \cdot 2 \cdot 2 \cdot 3$

Use your calculator and the list-of-multiples method to find the LCM of each pair of numbers.

170. 288, 324 **171.** 2700, 7800

172. A cigar company uses two sizes of boxes, 6 in. and 8 in. long. These are packed in bigger cartons to be shipped. What is the shortest length carton that will accommodate boxes of either size without any room left over? (Each carton can contain only boxes of one size; no mixing is allowed.)

Planet Orbits and LCMs. The earth, Jupiter, Saturn, and Uranus all revolve around the sun. The earth takes 1 year, Jupiter takes 12 years, Saturn takes 30 years, and Uranus takes 84 years. On a certain night you look at all the planets, and you wonder how many years it will be before they have the same position again. To find out, you find the LCM of 12, 30, and 84. It will be that number of years.

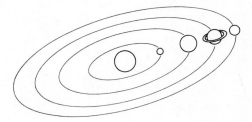

173. How often will Jupiter and Saturn appear in the same direction?

174. How often will Saturn and Uranus appear in the same direction?

175. How often will Jupiter, Saturn, and Uranus appear in the same direction?

Compute and simplify.

176. $\dfrac{4\frac{3}{14} + 2\frac{1}{3}}{5\frac{1}{7} - 4\frac{2}{3}}$ **177.** $\dfrac{10\frac{2}{3} - 8\frac{1}{5}}{24\frac{5}{8} - 6\frac{5}{7}}$ **178.** $\frac{3}{4} \cdot (\frac{4}{5} + \frac{3}{2})$ **179.** $\frac{3}{4} + (\frac{4}{5} \cdot \frac{3}{2})$

180. In the following table, the top number has been factored in such a way that the sum of the factors is the bottom number. For example, in the first column 56 has been factored as $7 \cdot 8$, and $7 + 8 = 15$, the bottom number. Find the missing numbers in the table.

Product	56	63	36	72	140	96	48	168	110	90	432	63
Factor	7	7	18	2	14	12	6	21	11	9	24	3
Factor	8	9	2	36	10	8	8	8	10	10	18	21
Sum	15	16	20	38	24	20	14	29	21	19	42	24

OBJECTIVES

After finishing Section 1.3, you should be able to:

i Add real numbers, find the additive inverse of a real number, change the sign of a number, and solve problems involving addition of real numbers.

ii Subtract real numbers, solve problems involving subtraction of real numbers, and simplify combinations of additions and subtractions.

1.3

ADDITION AND SUBTRACTION OF REAL NUMBERS

In this section we consider addition and subtraction of real numbers.

i Addition of Real Numbers

We first consider addition of real numbers. To gain an understanding, we begin by adding using a number line. Then we consider rules for addition.

Addition on a Number Line

Addition of numbers can be illustrated on a number line. To do the addition $a + b$, we start at a, and then move according to b.

a) If b is positive, we move to the right.

b) If b is negative, we move to the left.

c) If b is 0, we stay at a.

EXAMPLE 1 Add: $3 + (-5)$. Parentheses are used to indicate that this is the sum of 3 and -5. Otherwise, $3 + -5$ would create confusion of signs.

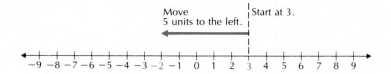

$$3 + (-5) = -2$$

EXAMPLE 2 Add: $-4 + (-3)$.

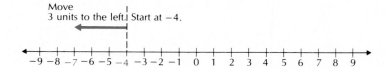

$$-4 + (-3) = -7$$

EXAMPLE 3 Add: $-4 + 9$.

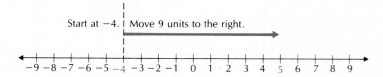

$$-4 + 9 = 5$$

EXAMPLE 4 Add: $-5.2 + 0$.

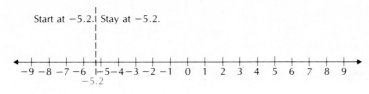

Start at −5.2. Stay at −5.2.

−5.2

$-5.2 + 0 = -5.2$

DO EXERCISES 1–6.

Adding Without a Number Line

You may have noticed some patterns in the preceding examples. These lead us to rules for adding without using a number line.

RULES FOR ADDITION

1. *Positive numbers:* Add the same as numbers of arithmetic.
2. *Negative numbers:* Add absolute values. Make the answer negative.
3. *A positive and a negative number:* Subtract absolute values. Then:
   a) If the positive number has the greater absolute value, make the answer positive.
   b) If the negative number has the greater absolute value, make the answer negative.
   c) If the numbers have the same absolute value, make the answer 0.
4. *One number is zero:* The sum is the other number.

Rule 4 is the *Identity Property of 0*, which we considered in Section 1.2.

EXAMPLES Add without using a number line.

5. $-12 + (-7)$ Two negatives. Think: Add the absolute values, getting 19. Make the answer *negative*, -19.

6. $-1.4 + 8.5$ The absolute values are 1.4 and 8.5. The difference is 7.1. The positive addend has the larger absolute value, so the answer is *positive*, 7.1.

7. $-36 + 21$ The absolute values are 36 and 21. The difference is 15. The negative addend has the larger absolute value, so the answer is *negative*, -15.

8. $1.5 + (-1.5)$ The sum is 0.

9. $-\frac{7}{8} + 0$ The sum is $-\frac{7}{8}$.

EXAMPLES Add.

10. $-9.2 + 3.1 = -6.1$
11. $-\frac{3}{2} + \frac{9}{2} = \frac{6}{2} = 3$
12. $-\frac{2}{3} + \frac{5}{8} = -\frac{16}{24} + \frac{15}{24} = -\frac{1}{24}$

DO EXERCISES 7–20.

Suppose we wish to add several numbers, some positive and some negative, as follows. How can we proceed?

$$13 + (-4) + 17 + (-5) + (-2)$$

Add using a number line.

1. $0 + (-8) = -8$

2. $1 + (-4) = -3$

3. $-3 + (-5) = -8$

4. $-3 + 7 = +4$

5. $-5.4 + 5.4 = 0$

6. $-\frac{5}{2} + \frac{1}{2} = -\frac{4}{2}$

Add without using a number line.

7. $-5 + (-6) = -11$

8. $-9 + (-3) = -12$

9. $-4 + 6 = +2$

10. $-7 + 3 = -4$

11. $5 + (-7) = -2$

12. $-20 + (-14) = -34$

13. $-11 + (-11) = -22$

14. $10 + (-7) = +3$

15. $-0.17 + 0.7 = -.10$

16. $-6.4 + 8.7 = 2.3$

17. $-4.5 + (-3.2) = -7.7$

18. $-8.6 + 2.4 = -6.2$

19. $\frac{5}{9} + \left(-\frac{7}{9}\right) = -\frac{2}{9}$

20. $-\frac{1}{5} + \left(-\frac{3}{4}\right) = $

$-\frac{4}{20} + -\frac{15}{20} = -\frac{19}{20}$

Add.

21. $(-15) + (-37) + 25 + 42 + (-59) + (-14)$ $= -58$

22. $42 + (-81) + (-28) + 24 + 18 + (-31) = -56$

23. $-2.5 + (-10) + 6 + (-7.5) = -14$

24. -35
   17
   14
   -27
   31
   -12
   -12

Find the additive inverse.

25. $-4 = +4$

26. $8.7 = -8.7$

27. $-7.74 = +7.74$

28. $-\dfrac{8}{9} = +8/9$

29. $0 = 0$

30. $12 = -12$

Find $-x$ and $-(-x)$ when x is:

31. $14 = -14, +14$

32. $1 = -1, +1$

33. $-19 = +19, -19$

34. $-1.6 = +1.6, -1.6$

35. $\dfrac{2}{3} = -2/3, +2/3$

36. $-\dfrac{9}{8} = 9/8, -9/8$

We can change grouping and order as we please. For instance, we can group the positive numbers together and the negative numbers together and add them separately. Then we add the two results.

EXAMPLE 13   Add: $15 + (-2) + 7 + 14 + (-5) + (-12)$.

a) $15 + 7 + 14 = 36$ Adding the positive numbers

b) $-2 + (-5) + (-12) = -19$ Adding the negative numbers

c) $36 + (-19) = 17$ Adding the results

DO EXERCISES 21–24.

Additive Inverses

When two numbers such as 6 and -6 are added, the result is 0. Such numbers are called *additive inverses* of each other. Every real number has an additive inverse.

> Two numbers whose sum is 0 are called *additive inverses* (or simply, *inverses*) of each other.

EXAMPLES   Find the additive inverse of each number.

14. 34   The inverse of 34 is -34 because $34 + (-34) = 0$.

15. -8 The inverse of -8 is 8 because $-8 + 8 = 0$.

16. 0   The inverse of 0 is 0 because $0 + 0 = 0$.

17. $-\dfrac{7}{8}$ The inverse of $-\dfrac{7}{8}$ is $\dfrac{7}{8}$ because $-\dfrac{7}{8} + \dfrac{7}{8} = 0$.

DO EXERCISES 25–30.

To name the additive inverse we use the symbol $-$, as follows.

> The additive inverse of a number a can be named $-a$ (read "the inverse of a").

A symbol such as -8 is usually read "negative 8." It could be read "the inverse of 8," because the inverse of 8 is negative 8. Thus a symbol like -8 can be read in two ways. A symbol like $-x$, however, should be read "the inverse of x" and *not* "negative x," because we do not know whether it represents a positive number, a negative number, or 0.

Note that if we take a number, say, 8, and find its inverse, -8, and then find the inverse of the result, we will have the original number, 8, again.

> The inverse of the inverse of a number is the number itself. That is, for any number a,
> $$-(-a) = a.$$

EXAMPLE 18   Find $-x$ and $-(-x)$ when $x = 16$.

a) If $x = 16$, then $-x = -16$ The inverse of 16 is -16.

b) If $x = 16$, then $-(-x) = -(-16) = 16$.   The inverse of the inverse of 16 is 16.

EXAMPLE 19   Find $-x$ and $-(-x)$ when $x = -3$.

a) If $x = -3$, then $-x = -(-3) = 3$.

b) If $x = -3$, then $-(-x) = -(-(-3)) = -3$.

DO EXERCISES 31–36.

We can use the symbolism $-a$ for the additive inverse of a to restate the definition of additive inverse.

> For any real number a,
> $$a + (-a) = (-a) + a = 0.$$
> Thus, a and $-a$ are additive inverses.

Signs of Numbers. A negative number is sometimes said to have a "negative sign." A positive number is said to have a "positive sign." When we replace a number by its additive inverse, we can say that we have "changed its sign."

EXAMPLES Change the sign. (Find the additive inverse.)

20. -3 $-(-3) = 3$
21. -10 $-(-10) = 10$
22. 0 $-0 = 0$
23. 14 $-(14) = -14$ The parentheses here indicate that we are finding the additive inverse of the positive number, 14. The answer is negative 14.

DO EXERCISES 37–40.

Problem Solving

Let us now see how we can use addition of real numbers to solve problems.

EXAMPLE 24 A business made a profit of $18 on Monday. There was a loss of $7 on Tuesday. On Wednesday there was a loss of $5, and on Thursday there was a profit of $11. Find the total profit or loss.

We can represent a loss with a negative number and a profit with a positive number. Thus we have the following losses and profits:

$$18, \quad -7, \quad -5, \quad 11.$$

To translate to mathematical language we write the following expression, which represents the total profit or loss:

$$18 + (-7) + (-5) + 11.$$

We then add the numbers:

$$18 + (-7) + (-5) + 11 = 29 + (-12) = 17.$$

The total profit was $17.

DO EXERCISE 41.

ii Subtraction of Real Numbers

We now consider subtraction of real numbers. Subtraction is defined as follows.

> The difference $a - b$ is the number which when added to b gives a.

EXAMPLE 25 Subtract: $5 - 8$.

Think: $5 - 8$ is the number which when added to 8 gives 5. What number can we add to 8 to get 5? The number has to be negative. The number is -3; $5 - 8 = -3$.

DO EXERCISES 42–44.

Change the sign. (Find the additive inverse.)

37. -4 = 4

38. -13.4 = 13.4

39. 0 = 0

40. $\frac{1}{4}$ = $-1/4$

41. A lab technician had 500 mL (milliliters) of acid in a beaker. After 16.5 mL had been poured out and 27.3 were then added, how much was in the beaker?

$500 + (-16.5) + 27.3 =$
510.8 mL

Subtract.

42. $-6 - 4$ *Think:* $-6 - 4$ is the number which when added to 4 gives -6. What number can be added to 4 to get -6?

-10

43. $-7 - (-10)$ *Think:* $-7 - (-10)$ is the number which when added to -10 gives -7. What number can be added to -10 to get -7?

$-7(+)10 = 3$

44. $-7 - (-2)$ *Think:* $-7 - (-2)$ is the number which when added to -2 gives -7. What number can be added to -2 to get -7?

$-7(+)2 = -5$

Complete each addition and compare with the subtraction.

45. $4 + (-6) = \underline{-2}$;
$4 - 6 = -2$

46. $-3 + (-8) = \underline{-11}$;
$-3 - 8 = -11$

47. $-5 + 9 = \underline{4}$;
$-5 - (-9) = 4$

48. $-5 + 3 = \underline{-2}$;
$-5 - (-3) = -2$

Read each of the following. Then subtract by adding the inverse of the subtrahend.

49. $4 - 9$

$4 + -9 \qquad = -5$

50. $6 - (-4)$

$= 10$

51. $-4 - 17$

$= -21$

52. $-3 - (-12)$

$= +9$

53. $-7.3 - (-2.1)$

$= -5.2$

54. $\frac{1}{5} - \frac{1}{2}$ $\frac{2}{10} - \frac{5}{10} = -\frac{3}{10}$

Subtract.

55. $-3 - 6$ $= -9$

56. $\frac{7}{10} - \frac{9}{10}$ $= -\frac{2}{10}$

57. $-8.8 - (-1.3) = -7.5$

58. $-\frac{2}{3} - \left(-\frac{5}{8}\right) =$

$-\frac{16}{24}$

$+\frac{15}{24} = \frac{-1}{24}$

The definition above is *not* the most efficient way to do subtraction. From that definition a faster way can be developed. Look for a pattern in the following examples. The subtractions on the left were done in either Example 25 above or in Margin Exercises 42–44.

Subtractions	*Adding an Inverse*
$5 - 8 = -3$	$5 + (-8) = -3$
$-6 - 4 = -10$	$-6 + (-4) = -10$
$-7 - (-10) = 3$	$-7 + 10 = 3$
$-7 - (-2) = -5$	$-7 + 2 = -5$

DO EXERCISES 45–48.

Perhaps you have noticed in the preceding examples and exercises that we can subtract by adding an additive inverse. This can always be done.

> For any real numbers a and b,
>
> $$a - b = a + (-b).$$
>
> **(To subtract, we can add the additive inverse of the subtrahend.)**

The preceding is the method one normally uses for quick subtraction of real numbers.

EXAMPLES Subtract. Check by addition.

26. $2 - 6 = 2 + (-6) = -4$ Changing the sign of 6 and changing the subtraction to addition

  Check: $6 + (-4) = 2$ $2 - 6$, or -4, is that number which when added to 6 gives 2.

27. $-4 - (-9) = -4 + 9 = 5$ Changing the sign of -9 and changing the subtraction to addition

  Check: $-9 + 5 = -4$ $-4 - (-9)$, or 5, is that number which when added to -9 gives -4.

28. $-4.2 - (-3.6) = -4.2 + 3.6 = -0.6$

  Check: $-3.6 + (-0.6) = -4.2$

29. $-\frac{1}{2} - \left(-\frac{3}{4}\right) = -\frac{1}{2} + \frac{3}{4} = \frac{1}{4}$

  Check: $\left(-\frac{3}{4}\right) + \frac{1}{4} = -\frac{1}{2}$

DO EXERCISES 49–54.

EXAMPLES Subtract.

30. $3 - 5 = 3 + (-5) = -2$

31. $\frac{1}{8} - \frac{7}{8} = \frac{1}{8} + \left(-\frac{7}{8}\right) = -\frac{6}{8}$, or $-\frac{3}{4}$

32. $-4.6 - (-9.8) = -4.6 + 9.8 = 5.2$

33. $-\frac{3}{4} - \frac{7}{5} = -\frac{15}{20} + \left(-\frac{28}{20}\right) = -\frac{43}{20}$

DO EXERCISES 55–58.

When several additions and subtractions occur together, we can make them all additions.

EXAMPLES Simplify.

34. $8 - (-4) - 2 - (-4) + 2 = 8 + 4 + (-2) + 4 + 2 = 16$

35. $8.2 - (-6.1) + 2.3 - (-4) = 8.2 + 6.1 + 2.3 + 4 = 20.6$

DO EXERCISES 59–61.

Problem Solving

Let us now see how we can use subtraction of real numbers to solve problems.

EXAMPLE 36 The lowest point in Asia is the Dead Sea, which is 400 meters (m) below sea level. The lowest point in the United States is Death Valley, which is 86 meters below sea level. How much higher is Death Valley than the Dead Sea?

In this case it is helpful to draw a picture of the situation.

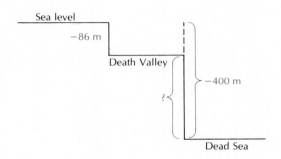

We see that -86 is the higher altitude at Death Valley and -400 is the lower altitude at the Dead Sea. To find how much higher Death Valley is we subtract. This gives the subtraction

$$-86 - (-400).$$

We carry out the subtraction:

$$-86 - (-400) = -86 + 400 = 314.$$

Death Valley is 314 m higher than the Dead Sea.

DO EXERCISE 62.

Simplify.

59. $-6 - (-2) - (-4) - 12 + 3$

$-6 + 2 + 4 - 12 + 3$
$-6 + 6 - 12 + 3$
$-12 + 3 = -9$

60. $3 - (-7.1) + 6.3 - (-5.2)$

$3 + 7.1 + 6.3 + 5.2 =$
21.6

61. $-8 - (-3.4) + 2.8 - (-13)$

$-8 + 3.4 + 2.8 + 13 =$
3.4
2.8
11.2 13.0
$\overline{19.2}$
-8.0
$.11\,2$

62. Your total assets are $734. You borrow $950.78 for a plane ticket. What are your total assets now? NONE

-216.78
in debt.

EXERCISE SET 1.3

i Add using a number line.

1. $-9 + 2$ **2.** $2 + (-5)$ **3.** $-10 + 6$ **4.** $8 + (-3)$

5. $-8 + 8$ **6.** $6 + (-6)$ **7.** $-3 + (-5)$ **8.** $-4 + (-6)$

Add without using a number line.

9. $-7 + 0$ **10.** $-13 + 0$ **11.** $0 + (-27)$ **12.** $0 + (-35)$

13. $17 + (-17)$ **14.** $-15 + 15$ **15.** $-17 + (-25)$ **16.** $-24 + (-17)$

17. $18 + (-18)$ **18.** $-13 + 13$ **19.** $-18 + 18$ **20.** $11 + (-11)$

21. $8 + (-5)$ **22.** $-7 + 8$ **23.** $-4 + (-5)$ **24.** $10 + (-12)$

25. $13 + (-6)$ **26.** $-3 + 14$ **27.** $11 + (-9)$ **28.** $-14 + (-19)$

29. $-20 + (-6)$ **30.** $19 + (-19)$ **31.** $-15 + (-7)$ **32.** $23 + (-5)$

33. $40 + (-8)$ **34.** $-23 + (-9)$ **35.** $-25 + 25$ **36.** $40 + (-40)$

37. $63 + (-18)$ **38.** $85 + (-65)$ **39.** $-6.5 + 4.7$ **40.** $-3.6 + 1.9$

41. $-2.8 + (-5.3)$ **42.** $-7.9 + (-6.5)$ **43.** $-\frac{3}{5} + \frac{2}{5}$ **44.** $-\frac{4}{3} + \frac{2}{3}$

45. $-\frac{3}{7} + (-\frac{5}{7})$ **46.** $-\frac{4}{9} + (-\frac{6}{9})$ **47.** $-\frac{5}{8} + \frac{1}{4}$ **48.** $-\frac{5}{6} + \frac{2}{3}$

49. $-\frac{3}{7} + (-\frac{2}{5})$ **50.** $-\frac{5}{8} + (-\frac{1}{3})$ **51.** $-\frac{3}{5} + (-\frac{2}{15})$ **52.** $-\frac{5}{9} + (-\frac{1}{18})$

53. $-5.7 + (-7.2) + 6.6$ **54.** $-10.3 + (-7.5) + 3.1$

55. $-8.5 + 7.9 + (-3.7)$ **56.** $-9.6 + 8.4 + (-11.8)$

57. $-\frac{7}{16} + \frac{7}{8}$ **58.** $-\frac{3}{28} + \frac{5}{42}$

59. $75 + (-14) + (-17) + (-5)$ **60.** $28 + (-44) + 17 + 31 + (-94)$

61. $-44 + (-\frac{3}{8}) + 95 + (-\frac{5}{8})$ **62.** $24 + 3.1 + (-44) + (-8.2) + 63$

63. $98 + (-54) + 113 + (-998) + 44 + (-612) + (-18) + 334$

64. $-455 + (-123) + 1026 + (-919) + 213 + 111 + (-874)$

Find the additive inverse of each number.

65. 24 **66.** -64 **67.** -9 **68.** $\frac{7}{2}$ **69.** -26.9 **70.** 48.2

Find $-x$ when x is:

71. 9 **72.** -26 **73.** $-\frac{14}{3}$ **74.** $\frac{1}{328}$ **75.** 0.101 **76.** 0

Find $-(-x)$ when x is:

77. -65 **78.** 29 **79.** $\frac{5}{3}$ **80.** -9.1

Change the sign. (Find the additive inverse.)

81. -1 **82.** -7 **83.** 7 **84.** 10

85. -14 **86.** -22.4 **87.** 0 **88.** $-\frac{7}{8}$

Problem solving

89. In a football game, the quarterback attempted passes with the following results.

First try	13-yd gain
Second try	incomplete
Third try	12-yd loss (tackled behind the line)
Fourth try	21-yd gain
Fifth try	14-yd loss

Find the total gain (or loss).

90. The following table shows the profits and losses of a company over a five-year period. Find the profit or loss after this period of time.

Year	Profit or loss
1983	+32,056
1984	−2,925
1985	+81,429
1986	−19,365
1987	−13,875

91. The barometric pressure at Omaha was 1012 millibars (mb). The pressure dropped 6 mb; then it rose 3 mb. After that it dropped 14 mb and then rose 4 mb. What was the pressure then?

92. One day the value of a share of IBM stock was $253\frac{1}{4}$. That day it rose in value $\$\frac{5}{8}$. The next day it lost $\$\frac{3}{8}$. What was the value of the stock at the end of the two days?

ii Subtract by adding the inverse of the subtrahend.

93. $3 - 7$ **94.** $4 - 9$ **95.** $0 - 7$ **96.** $0 - 10$

97. $-8 - (-2)$ **98.** $-6 - (-8)$ **99.** $-10 - (-10)$ **100.** $-8 - (-8)$

101. $7 - 7$ **102.** $9 - 9$ **103.** $7 - (-7)$ **104.** $4 - (-4)$

105. $8 - (-3)$ **106.** $-7 - 4$ **107.** $-6 - 8$ **108.** $6 - (-10)$

109. $-4 - (-9)$ **110.** $-14 - 2$ **111.** $2 - 9$ **112.** $2 - 8$

113. $-6 - (-5)$ **114.** $-4 - (-3)$ **115.** $8 - (-10)$ **116.** $5 - (-6)$

117. $0 - 5$ **118.** $0 - 6$ **119.** $-5 - (-2)$ **120.** $-3 - (-1)$

121. $-7 - 14$ **122.** $-9 - 16$ **123.** $0 - (-5)$ **124.** $0 - (-1)$

125. $-8 - 0$ **126.** $-9 - 0$ **127.** $7 - (-5)$ **128.** $8 - (-3)$

129. $2 - 25$ **130.** $18 - 63$ **131.** $-42 - 26$ **132.** $-18 - 63$

133. $-71 - 2$ **134.** $-49 - 3$ **135.** $24 - (-92)$ **136.** $48 - (-73)$

137. $-50 - (-50)$ **138.** $-70 - (-70)$ **139.** $\frac{3}{8} - \frac{5}{8}$ **140.** $\frac{3}{9} - \frac{9}{9}$

141. $\frac{3}{4} - \frac{2}{3}$ **142.** $\frac{5}{8} - \frac{3}{4}$ **143.** $-\frac{3}{4} - \frac{2}{3}$ **144.** $-\frac{5}{8} - \frac{3}{4}$

145. $-\frac{5}{8} - (-\frac{3}{4})$ **146.** $-\frac{3}{4} - (-\frac{2}{3})$ **147.** $6.1 - (-13.8)$ **148.** $1.5 - (-3.5)$

149. $-3.2 - 5.8$ **150.** $-2.7 - 5.9$ **151.** $0.99 - 1$ **152.** $0.87 - 1$

153. $-79 - 114$ **154.** $-197 - 216$ **155.** $0 - (-500)$ **156.** $500 - (-1000)$

157. $-2.8 - 0$ **158.** $6.04 - 1.1$ **159.** $7 - 10.53$ **160.** $8 - (-9.3)$

161. $\frac{1}{6} - \frac{2}{3}$ **162.** $-\frac{3}{8} - (-\frac{1}{2})$ **163.** $-\frac{4}{7} - (-\frac{10}{7})$ **164.** $\frac{12}{5} - \frac{12}{5}$

165. $-\frac{7}{10} - \frac{10}{15}$ **166.** $-\frac{4}{18} - (-\frac{2}{9})$ **167.** $\frac{1}{13} - \frac{1}{12}$ **168.** $-\frac{1}{7} - (-\frac{1}{6})$

Simplify.

169. $18 - (-15) - 3 - (-5) + 2$ **170.** $22 - (-18) + 7 + (-42) - 27$ **171.** $-31 + (-28) - (-14) - 17$

172. $-43 - (-19) - (-21) + 25$ **173.** $-34 - 28 + (-33) - 44$ **174.** $39 + (-88) - 29 - (-83)$

175. $-93 - (-84) - 41 - (-56)$ **176.** $84 + (-99) + 44 - (-18) - 43$

Problem solving

177. Your total assets are $619.46. You borrow $950 for the purchase of a stereo system. What are your total assets now?

178. You owe a friend $420. The friend decides to cancel $156 of the debt. How much do you owe now?

179. You are in debt $215.50. How much money will you need to make your total assets $450?

180. On a winter night the temperature dropped from $-5°C$ to $-12°C$. How many degrees did it drop?

181. The lowest point in Africa is Lake Assal, which is 156 m below sea level. The lowest point in South America is the Valdes Peninsula, which is 40 m below sea level. How much lower is Lake Assal than the Valdes Peninsula?

182. The deepest point in the Pacific Ocean is the Marianas Trench with a depth of 10,415 m. The deepest point in the Atlantic Ocean is the Puerto Rico Trench with a depth of 8,648 m. How much higher is the Puerto Rico Trench than the Marianas Trench?

1.4

MULTIPLICATION AND DIVISION OF REAL NUMBERS

We now consider multiplication and division of real numbers.

i Multiplication of Real Numbers

Multiplication of real numbers is very much like multiplication of whole numbers. The only difference is that we must determine whether the answer is positive or negative.

OBJECTIVES

After finishing Section 1.4, you should be able to:

i Multiply real numbers.

ii Divide integers, find the reciprocal of a real number, and divide real numbers by multiplying by a reciprocal.

1. Complete, as in the example.

$$4 \cdot 10 = 40$$
$$3 \cdot 10 = 30$$
$$2 \cdot 10 = 20$$
$$1 \cdot 10 = 10$$
$$0 \cdot 10 = 0$$
$$-1 \cdot 10 = -10$$
$$-2 \cdot 10 = -20$$
$$-3 \cdot 10 = -30$$
$$-4 \cdot 10 = -40$$

Multiply.

2. $-3 \cdot 6 = -18$

3. $20 \cdot (-5) = -100$

4. $4.5(-20) = -15.5$

5. $\left(-\frac{2}{3}\right)\left(\frac{9}{4}\right) = -\frac{18}{12}$

6. Complete, as in the example.

$$3 \cdot (-10) = -30$$
$$2 \cdot (-10) = -20$$
$$1 \cdot (-10) = -10$$
$$0 \cdot (-10) = 0$$
$$-1 \cdot (-10) = 10$$
$$-2 \cdot (-10) = 20$$
$$-3 \cdot (-10) = 30$$

To see how to multiply a positive number and a negative number, consider the pattern in the following.

This number decreases by 1 each time.   This number decreases by 5 each time.

$$4 \cdot 5 = 20$$
$$3 \cdot 5 = 15$$
$$2 \cdot 5 = 10$$
$$1 \cdot 5 = 5$$
$$0 \cdot 5 = 0$$
$$-1 \cdot 5 = -5$$
$$-2 \cdot 5 = -10$$
$$-3 \cdot 5 = -15$$

DO EXERCISE 1.

According to this pattern, it looks as though the product of a negative number and a positive number is negative. That is the case, and we have the first part of the rule for multiplying real numbers.

> **To multiply a positive number and a negative number, multiply their absolute values. The answer is negative.**

EXAMPLES   Multiply.

1. $8(-5) = -40$

2. $-\frac{1}{3} \cdot \frac{5}{7} = -\frac{5}{21}$

3. $(-7.2)5 = -36$

DO EXERCISES 2–5.

How do we multiply two negative numbers? We again look for a pattern.

This number decreases by 1 each time.   This number decreases by 5 each time.

$$4 \cdot (-5) = -20$$
$$3 \cdot (-5) = -15$$
$$2 \cdot (-5) = -10$$
$$1 \cdot (-5) = -5$$
$$0 \cdot (-5) = 0$$
$$-1 \cdot (-5) = 5$$
$$-2 \cdot (-5) = 10$$
$$-3 \cdot (-5) = 15$$

DO EXERCISE 6.

According to the pattern it looks as though the product of two negative numbers should be positive. That is actually so, and we have the second part of the rule for multiplying real numbers.

> **To multiply two negative numbers, multiply their absolute values. The answer is positive.**

We already know how to multiply two positive numbers, of course. The only case we have not considered is multiplying by 0. The product of any real number and 0 is 0.

THE MULTIPLICATIVE PROPERTY OF ZERO

For any real number n,

$$n \cdot 0 = 0.$$

(The product of 0 and any real number is 0.)

The symbolism $-3(-4)$ means $-3 \cdot (-4)$, that is, -3 times -4. We agree not to write the multiplication sign.

EXAMPLES Multiply.

4. $-3(-4) = 12$

5. $-1.6(-2) = 3.2$

6. $-19 \cdot 0 = 0$

7. $\left(-\frac{5}{6}\right)\left(-\frac{1}{9}\right) = \frac{5}{54}$

DO EXERCISES 7–10.

We can choose order and grouping as we please when multiplying.

EXAMPLES Multiply.

8. $-8 \cdot 2(-3) = -16(-3)$ Multiplying the first two numbers
             $= 48$ Multiplying the results

9. $-8 \cdot 2(-3) = 24 \cdot 2$ Multiplying the negatives
             $= 48$

10. $-3(-2)(-5)(4) = 6(-5)(4)$ Multiplying the first two numbers
                 $= (-30)4$
                 $= -120$

11. $\left(-\frac{1}{2}\right)(8)\left(-\frac{2}{3}\right)(-6) = (-4)4$
                        $= -16$

12. $-5 \cdot (-2) \cdot (-3) \cdot (-6) = 10 \cdot 18$ Multiplying the first two numbers and
                    $= 180$ the last two numbers

13. $(-3)(-5)(-2)(-3)(-6) = (-30)(18) = -540$

DO EXERCISES 11–16.

We can see the following pattern in the results of Examples 12 and 13.

The product of an even number of negative numbers is positive.
The product of an odd number of negative numbers is negative.

---

ii Division of Real Numbers

Division is defined in terms of multiplication. We will see that this results in rules for division very much like those for multiplication.

Multiply.

7. $(-5)(-4) = 20$

8. $(-8)(-9) = 72$

9. $(-4.2)(-3)$ 12.6

10. $\left(-\frac{3}{8}\right)\left(-\frac{1}{7}\right) = \frac{3}{56}$

Multiply.

11. $(4)(-7)(5) = -140$

12. $(-5)(-6)(-3) = -90$

13. $(-4)(5)(-3)(2) = 120$

14. $-3 \times (-4.1) \times (-2.5) = -30.75$

15. $-2 \cdot (-5) \cdot (-4) \cdot (-3) = +120$

16. $-\frac{1}{2} \cdot \left(-\frac{4}{3}\right) \cdot \left(-\frac{5}{2}\right) = -\frac{20}{12}$

Divide, if possible. Check your answer.

17. $-24 \div 8$ = −3

18. $15 \div (-3)$ = −5

19. $\dfrac{-15}{-3}$ = 5

20. $\dfrac{32}{-4}$ = −8

21. $\dfrac{-30}{5}$ = −6

22. $\dfrac{-105}{0}$ = 0

23. $\dfrac{0}{-4}$ = 0

no even division

24. $\dfrac{-24}{5}$ −4.8

Division of Integers

Division is defined as follows.

> The quotient $\dfrac{a}{b}$ (or $a \div b$) is the number, if there is one, which when multiplied by b gives a.

Let us use the definition to divide integers.

EXAMPLES Divide, if possible. Check your answer.

14. $14 \div (-7)$ We look for a number which when multiplied by -7 gives 14. That number is -2. Thus $14 \div (-7) = -2$. *Check:* $(-2)(-7) = 14$.

15. $\dfrac{-32}{-4}$ We look for a number which when multiplied by -4 gives -32. That number is 8. Thus $\dfrac{-32}{-4} = 8$. *Check:* $8(-4) = -32$.

16. $\dfrac{-10}{7}$ We look for a number which when multiplied by 7 gives -10. That number is $-\dfrac{10}{7}$. *Check:* $-\dfrac{10}{7} \cdot 7 = -10$.

17. $-17 \div 0$ We look for a number which when multiplied by 0 gives -17. There is no such number because the product of 0 and *any* number is 0.

> When we divide a positive number by a negative or a negative number by a positive, the answer is negative. When we divide two negative numbers, the answer is positive.

DO EXERCISES 17–24.

Example 17 shows why we cannot divide -17 by 0. We can use the same argument to show why we cannot divide any nonzero number b by 0. For $b \div 0$ we look for a number which when multiplied by 0 gives b. There is no such nonzero number because the product of 0 and any number is 0.

On the other hand, if we divide 0 by 0, we look for a number r, such that $r \cdot 0 = 0$. But, $0 \cdot r = 0$ for any number r. Thus it appears that $0 \div 0$ could be any number we choose. This would be very confusing, getting any answer we want when we divide 0 by 0. Thus we agree to exclude division by zero.

> We never divide any number by 0.

Reciprocals

When two numbers such as $\frac{1}{2}$ and 2 are multiplied, the result is 1. Such numbers are called *reciprocals* of each other. Every nonzero real number has a reciprocal, also called a *multiplicative inverse*.

> Two numbers whose product is 1 are called *reciprocals* of each other.

EXAMPLES Find the reciprocal of each number.

18. $\frac{7}{8}$ The reciprocal of $\frac{7}{8}$ is $\frac{8}{7}$ because $\frac{7}{8} \cdot \frac{8}{7} = 1$.

19. -5 The reciprocal of -5 is $-\frac{1}{5}$ because $-5(-\frac{1}{5}) = 1$.

20. 3.9   The reciprocal of 3.9 is $\frac{1}{3.9}$ because $3.9(\frac{1}{3.9}) = 1$.

21. $-\frac{1}{2}$ The reciprocal of $-\frac{1}{2}$ is -2 because $(-\frac{1}{2})(-2) = 1$.

22. $-\frac{2}{3}$ The reciprocal of $-\frac{2}{3}$ is $-\frac{3}{2}$ because $(-\frac{2}{3})(-\frac{3}{2}) = 1$.

23. $\frac{1}{3/4}$ The reciprocal of $\frac{1}{3/4}$ is $\frac{3}{4}$ because $(\frac{1}{3/4})(\frac{3}{4}) = 1$.

> The reciprocal of a nonzero number a can be named $\dfrac{1}{a}$.
>
> The reciprocal of a nonzero number $\dfrac{a}{b}$ can be named $\dfrac{b}{a}$.

The reciprocal of a positive number is also a positive number, because their product must be a positive number, 1. The reciprocal of a negative number is also a negative number, because their product must be the positive number 1.

DO EXERCISES 25–29.

It is important *not* to confuse *additive inverse* with *reciprocal*. Keep in mind that the additive inverse of a number is what we add to it to get 0, whereas a reciprocal is what we multiply the number by to get 1. Compare the following.

Number	Additive inverse	Reciprocal
$-\frac{3}{8}$	$\frac{3}{8}$	$-\frac{8}{3}$
19	-19	$\frac{1}{19}$
$\frac{18}{7}$	$-\frac{18}{7}$	$\frac{7}{18}$
-7.9	7.9	$-\frac{1}{7.9}$ or $-\frac{10}{79}$
0	0	Does not exist

DO EXERCISE 30.

Division of Real Numbers

We know we can subtract by adding an inverse. Similarly, we can divide by multiplying by a reciprocal.

> For any real number a and any real number b,
> $$\frac{a}{b} = a \cdot \frac{1}{b}.$$
> (To divide, we can multiply by a reciprocal.)

EXAMPLES Rewrite each division as multiplication.

24. $-4 \div 3$ $-4 \div 3$ is the same as $-4 \cdot \dfrac{1}{3}$

25. $\dfrac{6}{-7}$ $\dfrac{6}{-7} = 6\left(-\dfrac{1}{7}\right)$

26. $\dfrac{3}{5} \div \left(-\dfrac{9}{7}\right)$ $\dfrac{3}{5} \div \left(-\dfrac{9}{7}\right) = \dfrac{3}{5}\left(-\dfrac{7}{9}\right)$

DO EXERCISES 31–34.

Find the reciprocal.

25. $\dfrac{2}{3} = \dfrac{3}{2}$

26. $-\dfrac{5}{4} = -\dfrac{4}{5}$

27. $-3 = -\dfrac{1}{3}$

28. $-\dfrac{1}{5} = -\dfrac{5}{1}$

29. $-\dfrac{7}{8} = -\dfrac{8}{7}$

30. Complete the following table.

Number	Additive inverse	Reciprocal
$\frac{2}{3}$	$-\frac{2}{3}$	$\frac{3}{2}$
$-\frac{5}{4}$	$\frac{5}{4}$	$-\frac{4}{5}$
0	0	0
1	-1	1
-4.5	4.5	$-\frac{1}{4.5}$

Rewrite each division as multiplication.

31. $-6 \div \dfrac{1}{5}$ $-6 \cdot \dfrac{5}{1}$

32. $\dfrac{-5}{7}$ $-5 \cdot \dfrac{1}{7}$

33. $\dfrac{13}{\frac{-2}{3}}$ $13\left(-\dfrac{3}{2}\right)$

34. $-\dfrac{4}{7} \div \left(-\dfrac{3}{5}\right) =$ $-\dfrac{4}{7}\left(-\dfrac{5}{3}\right)$

Divide.

35. $-\dfrac{3}{5} \div \left(-\dfrac{12}{11}\right) = -\dfrac{3}{5} \cdot -\dfrac{11}{12} = $

$\dfrac{-11}{20}$

36. $-\dfrac{8}{5} \div \dfrac{2}{3} = -\dfrac{8}{5} \cdot \dfrac{3}{2} = \dfrac{12}{5}$

37. $-64.8 \div 4 = -64.8 \cdot \dfrac{1}{4} = $

-16.2

38. $-78.6 \div (-3) = -78.6 \div -\dfrac{1}{3} = $

$= 26.2$

When actually doing division calculations, we sometimes multiply by a reciprocal and we sometimes divide directly. With fractional notation, it is usually better to multiply by a reciprocal. With decimal notation, it is usually better to divide directly.

EXAMPLES Divide.

27. $\dfrac{2}{3} \div \left(-\dfrac{5}{4}\right) = \dfrac{2}{3} \cdot \left(-\dfrac{4}{5}\right) = -\dfrac{8}{15}$

28. $-\dfrac{5}{6} \div \left(-\dfrac{3}{4}\right) = -\dfrac{5}{6} \cdot \left(-\dfrac{4}{3}\right) = \dfrac{20}{18} = \dfrac{10 \cdot 2}{9 \cdot 2} = \dfrac{10}{9} \cdot \dfrac{2}{2} = \dfrac{10}{9}$

Be careful not to change the sign when taking a reciprocal!

29. $-\dfrac{3}{4} \div \dfrac{3}{10} = -\dfrac{3}{4} \cdot \left(\dfrac{10}{3}\right) = -\dfrac{30}{12} = -\dfrac{5}{2} \cdot \dfrac{6}{6} = -\dfrac{5}{2}$

30. $-27.9 \div (-3) = \dfrac{-27.9}{-3} = 9.3$ Do the long division $3\,)\overline{27.9}$.

31. $-6.3 \div 2.1 = -3$ Do the long division $2.1\,)\overline{6.3}$. Make the answer negative.

DO EXERCISES 35–38.

EXERCISE SET 1.4

i Multiply.

1. $-8 \cdot 2$
2. $-2 \cdot 5$
3. $-7 \cdot 6$
4. $-9 \cdot 2$
5. $8 \cdot (-3)$

6. $9 \cdot (-5)$
7. $-9 \cdot 8$
8. $-10 \cdot 3$
9. $-8 \cdot (-2)$
10. $-2 \cdot (-5)$

11. $-7 \cdot (-6)$
12. $-9 \cdot (-2)$
13. $15 \cdot (-8)$
14. $-12 \cdot (-10)$
15. $-14 \cdot 17$

16. $-13 \cdot (-15)$
17. $-25 \cdot (-48)$
18. $39 \cdot (-43)$
19. $-3.5 \cdot (-28)$
20. $97 \cdot (-2.1)$

21. $9 \cdot (-8)$
22. $7 \cdot (-9)$
23. $4 \cdot (-3.1)$
24. $3 \cdot (-2.2)$
25. $-6 \cdot (-4)$

26. $-5 \cdot (-6)$
27. $-7 \cdot (-3.1)$
28. $-4 \cdot (-3.2)$
29. $\frac{2}{3} \cdot (-\frac{3}{5})$
30. $\frac{5}{7} \cdot (-\frac{2}{3})$

31. $-\frac{3}{8} \cdot (-\frac{2}{9})$
32. $-\frac{5}{8} \cdot (-\frac{2}{5})$
33. -6.3×2.7
34. -4.1×9.5
35. $-\frac{5}{9} \cdot \frac{3}{4}$

36. $-\frac{8}{3} \cdot \frac{9}{4}$
37. $7 \cdot (-4) \cdot (-3) \cdot 5$
38. $9 \cdot (-2) \cdot (-6) \cdot 7$

39. $-\frac{2}{3} \cdot \frac{1}{2} \cdot (-\frac{6}{7})$
40. $-\frac{1}{8} \cdot (-\frac{1}{4}) \cdot (-\frac{3}{5})$
41. $-3 \cdot (-4) \cdot (-5)$

42. $-2 \cdot (-5) \cdot (-7)$
43. $-2 \cdot (-5) \cdot (-3) \cdot (-5)$
44. $-3 \cdot (-5) \cdot (-2) \cdot (-1)$

45. $\frac{1}{5}(-\frac{2}{9})$
46. $-\frac{3}{5}(-\frac{2}{7})$
47. $-7 \cdot (-21) \cdot 13$

48. $-14 \cdot (34) \cdot 12$
49. $-4 \cdot (-1.8) \cdot 7$
50. $-8 \cdot (-1.3) \cdot (-5)$

51. $-\frac{1}{9}(-\frac{2}{3})(\frac{5}{7})$
52. $-\frac{7}{2}(-\frac{5}{7})(-\frac{2}{5})$
53. $4 \cdot (-4) \cdot (-5) \cdot (-12)$

54. $-2 \cdot (-3) \cdot (-4) \cdot (-5)$
55. $0.07 \cdot (-7) \cdot 6 \cdot (-6)$
56. $80 \cdot (-0.8) \cdot (-90) \cdot (-0.09)$

57. $(-\frac{5}{6})(\frac{1}{8})(-\frac{3}{7})(-\frac{1}{7})$
58. $(\frac{4}{5})(-\frac{2}{3})(-\frac{15}{7})(\frac{1}{2})$
59. $(-14) \cdot (-27) \cdot 0$

60. $7 \cdot (-6) \cdot 5 \cdot (-4) \cdot 3 \cdot (-2) \cdot 1 \cdot 0$
61. $(-8)(-9)(-10)$
62. $(-7)(-8)(-9)(-10)$

63. $(-6)(-7)(-8)(-9)(-10)$
64. $(-5)(-6)(-7)(-8)(-9)(-10)$

ii Divide, if possible. Check each answer.

65. $36 \div (-6)$
66. $\dfrac{28}{-7}$
67. $\dfrac{26}{-2}$
68. $26 \div (-13)$
69. $\dfrac{-16}{8}$

70. $-22 \div (-2)$
71. $\dfrac{-48}{-12}$
72. $-63 \div (-9)$
73. $\dfrac{-72}{9}$
74. $\dfrac{-50}{25}$

75. $-100 \div (-50)$
76. $\dfrac{-200}{8}$
77. $-108 \div 9$
78. $\dfrac{-64}{-7}$
79. $\dfrac{200}{-25}$

80. $-300 \div (-13)$ **81.** $\dfrac{75}{0}$ **82.** $\dfrac{0}{-5}$ **83.** $\dfrac{88}{-9}$ **84.** $\dfrac{-145}{-5}$

Find the reciprocal.

85. $\dfrac{15}{7}$ **86.** $\dfrac{3}{8}$ **87.** $-\dfrac{47}{13}$ **88.** $-\dfrac{31}{12}$ **89.** 13

90. -10 **91.** 4.3 **92.** -8.5 **93.** $-\dfrac{1}{7.1}$ **94.** $\dfrac{1}{-4.9}$

Rewrite each division as multiplication.

95. $3 \div 19$ **96.** $4 \div (-9)$ **97.** $\dfrac{6}{-13}$ **98.** $-\dfrac{12}{41}$ **99.** $\dfrac{13.9}{-1.5}$ **100.** $-\dfrac{47.3}{21.4}$

Divide.

101. $\dfrac{3}{4} \div \left(-\dfrac{2}{3}\right)$ **102.** $\dfrac{7}{8} \div \left(-\dfrac{1}{2}\right)$ **103.** $-\dfrac{5}{4} \div \left(-\dfrac{3}{4}\right)$ **104.** $-\dfrac{5}{9} \div \left(-\dfrac{5}{6}\right)$ **105.** $-\dfrac{2}{7} \div \left(-\dfrac{4}{9}\right)$

106. $-\dfrac{3}{5} \div \left(-\dfrac{5}{8}\right)$ **107.** $-\dfrac{3}{8} \div \left(-\dfrac{8}{3}\right)$ **108.** $-\dfrac{5}{8} \div \left(-\dfrac{6}{5}\right)$ **109.** $-6.6 \div 3.3$ **110.** $-44.1 \div (-6.3)$

111. $\dfrac{-11}{-13}$ **112.** $\dfrac{-1.9}{20}$ **113.** $\dfrac{48.6}{-3}$ **114.** $\dfrac{-17.8}{3.2}$

Compute. Parentheses tell us what to do first. Always calculate within parentheses first. If there are no parentheses, we multiply or divide, first, working from left to right, and then add or subtract, working from left to right.

115. $7(-9) + 3$
[*Hint:* Multiply 7 and -9. Then add 3.]

116. $-(-3 + 2)$
[*Hint:* Add -3 and 2. Then take the additive inverse.]

117. $\left(2\frac{1}{3}\right)\left(-5\frac{3}{4}\right)$

118. $\left(1\frac{2}{5}\right)\left(-\frac{4}{7}\right)\left(3\frac{1}{8}\right)$

119. $\frac{3}{4} \cdot \frac{2}{9} + \frac{1}{3}$ **120.** $12\left(\frac{5}{6} - \frac{1}{3}\right)$ **121.** $\frac{2}{5}\left(\frac{8}{9} + 4\frac{2}{3}\right)$ **122.** $\frac{1}{3} - 4$

123. $|3 - 7| - 4$ **124.** $-5|2| + 3|4|$ **125.** $-|39 - (-12) - 60|$ **126.** $\dfrac{(-9)(-8) + (-3)}{25}$

127. $\dfrac{-3(-9) + 7}{-4}$ **128.** $\left(-5\frac{3}{7}\right) \div 4\frac{2}{5}$ **129.** $\frac{10}{7} \div (-0.25)$ **130.** $\left(-2\frac{2}{3}\right) \div \left(-\frac{40}{15}\right)$

131. $-6[(-5) + (-7)]$ **132.** $7[(-16) + 9]$ **133.** $-3[(-8) + (-6)]\left(-\frac{1}{7}\right)$ **134.** $8[17 - (-3)]\left(-\frac{1}{4}\right)$

135. $-\pi + \pi$ **136.** $\sqrt{2} - \sqrt{2}$ **137.** $\sqrt{3} + \sqrt{3}$ **138.** $\dfrac{\sqrt{2}}{\sqrt{2}}$

139. $3\pi + \pi$ **140.** $3\pi - \pi$

1.5

INTRODUCTION TO ALGEBRA AND EXPRESSIONS

This section will introduce you to algebra. We will study evaluating expressions and translating to expressions of the type used in algebra.

i Algebraic Expressions

In arithmetic you have worked with expressions such as

$$23 + 54, \quad 4 \times 5, \quad 16 - 7, \quad \text{and} \quad \frac{5}{8}.$$

In algebra we use certain letters, or *variables*, for numbers and work with *algebraic expressions* such as

$$23 + x, \quad 4 \times t, \quad 16 - y, \quad \text{and} \quad \frac{a}{8}.$$

OBJECTIVES

After finishing Section 1.5, you should be able to:

i Evaluate an algebraic expression by substitution.

ii Translate phrases to algebraic expressions.

iii Tell the meaning of exponential notation such as 3^5 and n^4 and of expressions with exponents of 0 and 1, and write exponential notation for a product such as *yyy*.

iv Evaluate expressions containing exponential notation.

Translate this problem to an equation. Use the chart at the right.

1. How many more monthly flights are there on the Dallas–Houston route than on the New York–Boston route?

$2,128 + x = 2,866$

$x = 2,866 - 2,128$

$x = 738$

How do these expressions arise? Most often they arise in problem solving. For example, consider the following chart, which you might see in a magazine.

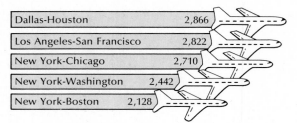

Taking flight in traffic

Here's how many flights are made monthly on the busiest air routes:

Dallas-Houston	2,866
Los Angeles-San Francisco	2,822
New York-Chicago	2,710
New York-Washington	2,442
New York-Boston	2,128

Suppose we want to know how many more flights there are on the Los Angeles–San Francisco route than on the New York–Washington route.

In algebra we translate this problem to an equation. It might be done as follows:

$$\underbrace{\text{(Flights NY–Wash)}}_{2442} \;\; \underbrace{\text{plus}}_{+} \;\; \underbrace{\text{(How many?)}}_{x} \;\; \underbrace{\text{is}}_{=} \;\; \underbrace{\text{(Flights LA–SF)}}_{2822}$$

Note that we have an algebraic expression on the left. To find the number x, we can subtract 2442 on both sides of the equation:

$$x = 2822 - 2442.$$

Then we carry out the subtraction and obtain the answer, 380.

In arithmetic, you probably would do this subtraction right away without considering an equation. In algebra, you will find most problems extremely difficult to solve without first translating to an equation.

DO EXERCISE 1.

An *algebraic expression* consists of variables, numerals, and operation signs. When we replace a variable by a number, we say that we are *substituting* for the variable. When we calculate the results, we get a number. This process is called *evaluating the expression.*

EXAMPLE 1   Evaluate $x + y$ for $x = 23$ and $y = 54$.

We substitute 23 for x and 54 for y and carry out the addition:

$$x + y = 23 + 54 = 77.$$

The number 77 is called the *value* of the expression.

EXAMPLE 2   Evaluate a/b for $a = 56$ and $b = 7$.

We substitute 56 for a and 7 for b and carry out the division:

$$\frac{a}{b} = \frac{56}{7} = 8.$$

Go back and consider the equation related to the problem regarding flights of airplanes in certain airports:

$$2442 + x = 2822.$$

The expression $2442 + x$ takes on different values depending on what real number we substitute for x. What we look for is that value which will result in a true equa-

tion. That number, 380, is called the *solution* of the equation. It is the abstraction to unknowns such as *x* that makes algebra a higher level of operation than arithmetic.

DO EXERCISES 2–4.

In arithmetic, we often use × for a multiplication. We also use a dot.

$$4 \cdot 9 \quad \text{means} \quad 4 \times 9.$$

In algebra, when two letters or a number and a letter are written together, that also means that they are to be multiplied. Numbers to be multiplied are called *factors*. We usually write a factor that is a number before any factor named by a letter. For example,

$$3y \quad \text{means} \quad 3 \cdot y \quad \text{or} \quad 3 \times y$$

and

$$ab \quad \text{means} \quad a \cdot b \quad \text{or} \quad a \times b.$$

Now we evaluate expressions involving products.

EXAMPLE 3 Evaluate $3y$ for $y = 14$.

We substitute 14 for *y* and carry out the multiplication:

$$3y = 3 \cdot 14 = 42.$$

EXAMPLE 4 The area of a rectangle of length *l* and width *w* is *lw*. Find the area when *l* is 24.5 in. and *w* is 16 in.

We substitute 24.5 for *l* and 16 for *w* and carry out the multiplication:

$$lw = 24.5 \times 16 = 392 \text{ in}^2.$$
("in²" means "square inches")

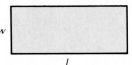

DO EXERCISES 5 AND 6.

EXAMPLE 5 Evaluate $12m/n$ for $m = 8$ and $n = 16$.

We substitute 8 for *m* and 16 for *n* and carry out the calculation:

$$\frac{12m}{n} = \frac{12 \cdot 8}{16} = \frac{96}{16} = 6.$$

DO EXERCISE 7.

ii Translating to Algebraic Expressions

In algebra we translate problems to equations. The parts of equations are translations of phrases to algebraic expressions. To help you become familiar with algebra and to make such translations easier when we do them later, we now practice translating.

EXAMPLE 6 Translate to an algebraic expression: seven less than some number.

Although we can use any variable we wish, such as *x*, *y*, *m*, or *n*, we let *t* represent the number. If we knew the number to be 23, then the translation would be $23 - 7$. If we knew the number to be 345, then the translation would be $345 - 7$. Since we are using a variable for the number, the translation is

$$t - 7.$$

DO EXERCISE 8.

---

2. Evaluate $a + b$ for $a = 38$ and $b = 26$.

$$38 + 26 = 64$$

3. Evaluate $x - y$ for $x = 57$ and $y = 29$.

$$57 - 29 = 28$$

4. Evaluate a/b for $a = 200$ and $b = 8$.

$$200/8 \text{ or } 25$$

5. Evaluate $4t$ for $t = 15$.

$$4(15) = 60$$

6. Find the area of a rectangle when *l* is 24 ft and *w* is 8 ft.

$$24 \times 8 = 192$$

7. Evaluate $10p/q$ when $p = 40$ and $q = 25$.

$$10(40)(25) = 10,000$$

Translate to an algebraic expression.

8. Twelve less than some number

$$x - 12$$

Translate to an algebraic expression.

9. Twelve more than some number

$x + 12$

Translate to an algebraic expression.

10. Four less than some number

$x - 4$

11. Half of some number

$1/2 x$

12. Six more than eight times some number

$8x + 6$

13. The difference of two numbers

$a - b$ or $b - a$

14. Fifty-nine percent of some number

$59\% x$

15. Two hundred less than the product of two numbers

$ab - 200$

16. The sum of two numbers

$a + b$

17. Write exponential notation for $10 \cdot 10 \cdot 10 \cdot 10$.

10^4

18. What is the meaning of 5^4?

$5 \cdot 5 \cdot 5 \cdot 5$

19. What is the meaning of x^5?

$x \cdot x \cdot x \cdot x \cdot x$

20. a) What is the meaning of $(-5)^2$? 25

b) What is the meaning of -5^2? -25

EXAMPLE 7 Translate to an algebraic expression: fifteen more than some number.

This time we let y represent the number. If we knew the number to be 47, then the translation would be $47 + 15$, or $15 + 47$. If we knew the number to be 1.688, then the translation would be $1.688 + 15$, or $15 + 1.688$. Since we are using a variable, the translation is

$$y + 15, \quad \text{or} \quad 15 + y.$$

DO EXERCISE 9.

EXAMPLE 8 Translate each of the following to an algebraic expression.

Phrase	Algebraic expression
Eight more than some number	$m + 8$, or $8 + m$
Eight less than some number	$a - 8$
The difference of two numbers	$a - b$, or $b - a$
One-fourth of a number	$1/4x$, or $x/4$
Three more than five times a number	$3 + 5t$, or $5t + 3$
Seven less than the product of two numbers	$pq - 7$
Ninety-four percent of some number	$94\%z$, or $0.94z$

DO EXERCISES 10–16.

iii Exponential Notation

Algebraic expressions can be written using exponential notation. In this section we learn how to write exponential notation and to evaluate algebraic expressions involving exponential notation.

Shorthand notation for $10 \cdot 10 \cdot 10$ is called *exponential notation*.

$$\text{For } \underbrace{10 \cdot 10 \cdot 10}_{3 \text{ factors}} \text{ we write } 10^3.$$

This is read "ten cubed" or "ten to the third power." We call the number 3 an *exponent* and we say that 10 is the *base*. For $10 \cdot 10$ we write 10^2, read "ten squared," or "ten to the second power."

EXAMPLE 9 Write exponential notation for $10 \cdot 10 \cdot 10 \cdot 10 \cdot 10$.

$$10 \cdot 10 \cdot 10 \cdot 10 \cdot 10 = 10^5$$

An exponent of 2 or greater tells us how many times the base is used as a factor.

EXAMPLE 10 What is the meaning of 3^5? n^4? $(2n)^3$? $(50x)^2$? $(-2)^2$? -2^2?

3^5 means $3 \cdot 3 \cdot 3 \cdot 3 \cdot 3$, or 243;

n^4 means $n \cdot n \cdot n \cdot n$;

$(2n)^3$ means $2n \cdot 2n \cdot 2n$;

$(50x)^2$ means $50x \cdot 50x$;

$(-2)^2$ means $(-2)(-2)$, or 4;

-2^2 means $-(2^2)$, or $-(2)(2)$, or -4

DO EXERCISES 17–20.

1 and 0 as Exponents

Look for a pattern.

$$10 \cdot 10 \cdot 10 \cdot 10 = 10^4$$
$$10 \cdot 10 \cdot 10 = 10^3$$
$$10 \cdot 10 = 10^2$$
$$10 = 10^?$$
$$1 = 10^?$$

We are dividing by 10 each time. The exponents decrease by 1 each time. To continue the pattern we would say that

$$10 = 10^1 \quad \text{and} \quad 1 = 10^0.$$

We read exponential notation as follows:

b^n is read the *nth power of b*, or simply *b to the nth*, or *b to the n.*

We may also read b^2 as *b-squared.*

b^3 may also be read *b-cubed.*

We now summarize our definition of exponential notation.

> b^0 means 1, for any number b, except 0.
>
> b^1 means b, for any number b.
>
> If n is a whole number greater than 1,
>
> $$n \text{ factors}$$
> $$b^n \text{ means } \overbrace{b \cdot b \cdot b \cdot b \cdots b}.$$

DO EXERCISES 21 AND 22.

EXAMPLE 11 Write exponential notation for each of the following.

$7 \cdot 7 \cdot 7 \cdot 7$ $7 \cdot 7 \cdot 7 \cdot 7$ means 7^4;

$2y \cdot 2y \cdot 2y$ $2y \cdot 2y \cdot 2y$ means $(2y)^3$;

$n \cdot n \cdot n \cdot n \cdot n \cdot n$ $n \cdot n \cdot n \cdot n \cdot n \cdot n$ means n^6;

$10x \cdot 10x$ $10x \cdot 10x$ means $(10x)^2$.

iv Evaluating Expressions

We now consider evaluating expressions containing exponential notation.

EXAMPLE 12 Evaluate x^4 for $x = 2$.

$$x^4 = 2^4 \qquad \text{Substituting}$$
$$= 2 \cdot 2 \cdot 2 \cdot 2$$
$$= 16$$

EXAMPLE 13 The area of a circle is given by $A = \pi r^2$, where r is the radius. Find the area of a circle with a radius of 10 cm. Use 3.14 for π.

$$A = \pi r^2 \approx 3.14 \times (10 \text{ cm})^2$$
$$= 3.14 \times 100 \text{ cm}^2$$
$$= 314 \text{ cm}^2$$

Here "cm^2" means "square centimeters," and "$\approx$" means "approximately equal."

21. What is 4^1?   4

22. What is m^0?   1

$A = \pi R^2$

23. Find the area of a circle when $r = 52$ cm. Use 3.14 for π.

$A \approx 3.14 \times (52)^2 = 3.14 \times 2704$

$A = 8490.56$

24. Evaluate $200 - n^4$ for $n = 3$.

$200 - n^4 = 200 - (3)^4$

$200 - 81 = 119$

EXAMPLE 14 Evaluate $m^3 + 5$ for $m = 4$.

We agree to evaluate $m^3 + 5$ by evaluating m^3 first and then adding 5:

$$m^3 + 5 = 4^3 + 5 \qquad \text{Substituting}$$
$$= (4 \cdot 4 \cdot 4) + 5$$
$$= 64 + 5$$
$$= 69$$

DO EXERCISES 23 AND 24.

EXERCISE SET 1.5

i Substitute to find values of the expressions.

1. Chris is 4 years younger than her brother Lowell. Suppose the variable x stands for Lowell's age. Then $x - 4$ stands for Chris's age. How old is Chris when Lowell is 14? 29? 52?

2. Employee A took five times as long to do a job as employee B. Suppose t stands for the time it takes B to do the job. Then $5t$ stands for the time it takes A. How long did it take A if B took 30 sec? 90 sec? 2 min?

3. The area of a triangle with base b and height h is $\frac{1}{2}bh$. Find the area when $b = 16$ yd and $h = 9$ yd.

4. The area of a parallelogram with base b and height h is bh. Find the area of a parallelogram with a height of 15.4 cm (centimeters) and a base of 6.5 cm.

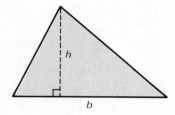

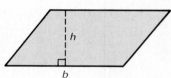

Evaluate.

5. $7y$ for $y = 8$

6. $6x$ for $x = 9$

7. $\dfrac{x}{y}$ for $x = 9$ and $y = 3$

8. $\dfrac{m}{n}$ for $m = 14$ and $n = 2$

9. $\dfrac{m - n}{8}$ for $m = 20$ and $n = 4$

10. $\dfrac{p + q}{5}$ for $p = 10$ and $q = 20$

11. $\dfrac{5z}{y}$ for $z = 8$ and $y = 2$

12. $\dfrac{18m}{n}$ for $m = 4$ and $n = 18$

ii Translate to an algebraic expression.

13. 6 more than m

14. 8 more than t

15. 9 less than c

16. 4 less than d

17. 6 greater than q

18. 11 greater than z

19. b more than a

20. c more than d

21. x less than y

22. c less than b

23. 98% of x

24. Forty-five percent of m

25. The sum of r and s

26. The sum of d and f

27. Twice x

28. Four times p

29. 5 times t

30. 9 times d

31. The difference of 3 and b

32. The difference of p and q

33. Six more than some number

34. One more than some number

35. Four less than some number

36. Forty-three less than some number

37. A number x plus three times y

38. A number a plus 2 plus b

iii What is the meaning of each expression?

39. 2^4

40. 5^3

41. $(1.4)^5$

42. $(2.5)^0$

43. n^1

44. m^6

45. $(7p)^2$

46. $(11c)^3$

47. $(19k)^4$

48. $(104d)^5$

49. $(10pq)^0$

50. $(24ct)^3$

Write exponential notation for each of the following.

51. $10 \times 10 \times 10 \times 10 \times 10 \times 10$

52. $6 \times 6 \times 6 \times 6$

53. $x \cdot x \cdot x \cdot x \cdot x \cdot x \cdot x$

54. $y \cdot y \cdot y$

55. $3y \cdot 3y \cdot 3y \cdot 3y$

56. $5m \cdot 5m \cdot 5m \cdot 5m \cdot 5m$

iv Evaluate each expression.

57. m^3 for $m = 3$

58. x^6 for $x = 2$

59. x^2 for $x = -3$

60. $-x^2$ for $x = -3$

61. x^4 for $x = 4$

62. y^{15} for $y = 1$

63. n^0 for $n = 5$

64. $y^2 - 7$ for $y = 10$

65. $z^5 + 5$ for $z = 2$

66. Find the area of a circle when $r = 34$ ft. Use 3.14 for π.

67. The area of a square with sides of length s is given by $A = s^2$. Find the area of a square with sides of length 24 m (meters).

Write an algebraic expression.

68. A number x plus three times y

69. A number y plus two times x

70. A number a plus 2 plus b

71. A number that is 3 less than twice x

72. Your age in 5 years, if you are a years old now

73. Your age two years ago, if you are b years old now

74. A number x increased by itself

75. A number that is 98% of x

76. The perimeter of a square with side s

77. Evaluate $\dfrac{x + y}{4}$ when $y = 8$ and x is twice y.

78. Evaluate $\dfrac{x - y}{7}$ when $y = 35$ and x is twice y.

79. Evaluate $\dfrac{y - x}{3}$ when $x = 9$ and y is three times x.

80. Evaluate $\dfrac{256y}{32x}$ for $y = 1$ and $x = 4$.

81. Evaluate $\dfrac{y + x}{2} + \dfrac{3 \cdot y}{x}$ for $x = 2$ and $y = 4$.

Answer each question with an algebraic expression.

82. If $w + 3$ is a whole number, what is the next whole number after it?

83. If $d + 2$ is an odd whole number, what is the preceding odd number?

84. The difference between two numbers is 3. One number is t. What are two possible values for the other number?

85. Two numbers are $v + 2$ and $v - 2$. Write an expression for their sum.

86. Translate to an algebraic expression: the perimeter of a rectangle of length l and width w.

87. You invest n dollars at 10% interest. Write an expression for the number of dollars in the bank a year from now.

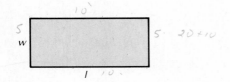

Write each of the following with a single exponent. For example,

$$\frac{3^5}{3^3} = \frac{3 \cdot 3 \cdot 3 \cdot 3 \cdot 3}{3 \cdot 3 \cdot 3} = 3 \cdot 3 = 3^2.$$

88. $\dfrac{10^5}{10^3}$

89. $\dfrac{10^7}{10^2}$

90. $\dfrac{5^4}{5^2}$

91. $\dfrac{2^6}{8^2}$

92. Evaluate $x^3 y^2 + zx$ for $x = 2$, $y = 1$, and $z = 3$.

93. Evaluate $c^2 a^3 + ba$ for $a = 3$, $b = 1$, and $c = 2$.

94. Evaluate $x^2 + 2xy + y^2$ for $x = 7$ and $y = 8$.

$(3n)^3$ means $3n \cdot 3n \cdot 3n$; $3n^3$ means $3 \cdot n \cdot n \cdot n$. Evaluate $(3n)^3$ and $3 \cdot n^3$ for each of the following.

95. When $n = 2$ **96.** When $n = 3$ **97.** When $n = -4$ **98.** When $n = -5$

Evaluate $(5p)^2$ and $5 \cdot p^2$ for each of the following.

99. When $p = -4$ **100.** When $p = -7$ **101.** When $p = 11$ **102.** When $p = 26$

Write exponential notation for each of the following.

103. $x \cdot x \cdot x \cdot y \cdot y \cdot y$ **104.** $3a \cdot 3a \cdot 3a \cdot 2b \cdot 2b$

105. Find $x^{149}y$ for $x = 13$ and $y = 0$.   **106.** Find $x^{410}y^2$ for $x = 1$ and $y = 3$.

107. 10^{127} is one followed by how many zeros?   **108.** Find $(x^2)^2$ if $x = 3$.

109. 🖩 What powers of 6 are too large for your calculator readout?

Which of the following pairs of expressions are equivalent?

110. $\dfrac{5n}{n}$, $5n^0$ $(n \neq 0)$ **111.** n^3, $\dfrac{n \cdot n \cdot n \cdot n}{n}$ $(n \neq 0)$ **112.** $\dfrac{4^5}{2}$, 2^5

113. 2^2, $2 \cdot 2 \cdot 2^0$ **114.** xy^2, $x \cdot x \cdot y \cdot y$ **115.** $4x$, $x \cdot x \cdot x \cdot x$

OBJECTIVES

After finishing Section 1.6, you should be able to:

i Using a commutative law, write an expression equivalent to a given one.

ii Calculate expressions such as $15 - 2 \times 5 + 3$ and $3 \times 4^2 - 29$, and evaluate expressions such as $(4x)^3 - 2$ for $x = 2$.

iii Using an associative law, write an expression equivalent to a given one.

iv Using the commutative and associative laws, write an expression equivalent to a given one.

v Identify the terms of an expression, and use the distributive laws to multiply two expressions where one expression has several terms.

vi Factor when terms have a common factor.

vii Collect like terms.

1.6

ALGEBRAIC EXPRESSIONS AND PROPERTIES OF REAL NUMBERS

When you translate a problem situation to mathematical language, you may have an algebraic expression. Then what do you do? One thing you may wish to do in the course of solving a problem is to find a simpler expression that has the same values as the one to which you translated. Algebraic expressions can be manipulated in other ways to help give answers to problems. To handle expressions we study the properties of the real numbers.

i **The Commutative Laws**

Let us examine the expressions $x + y$ and $y + x$.

EXAMPLE 1 Evaluate $x + y$ and $y + x$ for $y = 7$ and $x = 4$.

We substitute 7 for y and 4 for x in both expressions:

$$x + y = 4 + 7 = 11; \qquad y + x = 7 + 4 = 11.$$

EXAMPLE 2 Evaluate $x + y$ and $y + x$ for $y = 12$ and $x = 23$.

We substitute 12 for y and 23 for x in both expressions:

$$x + y = 23 + 12 = 35; \qquad y + x = 12 + 23 = 35.$$

Note that the expressions

$$x + y \quad \text{and} \quad y + x$$

have the same values, no matter what the variables stand for. We know that because when we add two numbers, it does not matter which comes first. Similarly, xy and yx have the same values, no matter what the variables stand for.

These are examples of general patterns or laws that hold for the real numbers.

THE COMMUTATIVE LAWS

Addition. **For any numbers** *a* **and** *b*,

$$a + b = b + a.$$

(We can change the order when adding without affecting the answer.)

Multiplication. **For any numbers** *a* **and** *b*,

$$ab = ba.$$

(We can change the order when multiplying without affecting the answer.)

Expressions that have the same value for all replacements are called *equivalent expressions.*

Using a commutative law, we know that $x + 2$ and $2 + x$ are equivalent. Similarly, $3 \cdot x$ and $x \cdot 3$ are equivalent.

EXAMPLE 3 Use a commutative law to write an expression equivalent to $y + 5$, xy, and $7 + ab$.

An expression equivalent to $y + 5$ is $5 + y$ by the commutative law of addition.

An expression equivalent to xy is yx by the commutative law of multiplication.

An expression equivalent to $7 + ab$ is $ab + 7$ by the commutative law of addition.

Another expression equivalent to $7 + ab$ is $7 + ba$ by the commutative law of multiplication.

An expression equivalent to $7 + ab$ is $ba + 7$ by both commutative laws.

DO EXERCISES 1–3.

ii Parentheses

What does $5 \times 2 + 4$ mean? If we multiply 5 by 2 and add 4, we get 14. If we add 2 and 4 and multiply by 5, we get 30. To tell which operation to do first, we use parentheses. For example,

$$(3 \times 5) + 6 \quad \text{means} \quad 15 + 6, \text{ or } 21$$

and

$$3 \times (5 + 6) \quad \text{means} \quad 3 \times 11, \text{ or } 33.$$

Parentheses tell us what to do first. If there are no parentheses, we multiply or divide first working from left to right, and then add or subtract working from left to right.

EXAMPLE 4 Calculate: $15 - 2 \times 5 + 3$.

$$15 - 2 \times 5 + 3$$
$$15 - \quad 10 \quad + 3 \qquad \text{Multiplying}$$
$$\quad\quad 5 \quad\quad + 3 \qquad \text{Subtracting and adding, from left to right}$$
$$\quad\quad\quad 8$$

DO EXERCISES 4–6.

Write an expression equivalent to each of the following. Use a commutative law.

1. $x + 9$

2. pq

3. $xy + t$

Calculate.

4. $5 + 2 \times 3$

5. $6 \times 2 + 3 \times 5$

6. $7 - 3 \times 2 + 5$

Calculate.

7. $(3 \times 5)^2$

8. 3×5^2

9. 4×2^2

10. $(4 + 2)^2$

11. $(5 - 1)^2$

12. $5^2 - 1^2$

Evaluate each expression.

13. $(4y)^2 - 5$ for $y = 3$

14. $6(x + 12)$ for $x = 8$

15. $\dfrac{t + 6}{5t^3}$ for $t = 2$

16. $(x - 4)^3$ for $x = 6$

17. $(4 + y) \cdot (x - 3)$ for $y = 3$ and $x = 12$

Calculate and compare.

18. $4 + (5 + 11)$ and $(4 + 5) + 11$

19. $6 \cdot (3 \cdot 4)$ and $(6 \cdot 3) \cdot 4$

Always calculate within parentheses first. When there are exponents and no parentheses, simplify powers before multiplying or dividing.

EXAMPLE 5 Calculate: $(3 \times 4)^2$.

$$(3 \times 4)^2 = 12^2 \qquad \text{Working with parentheses first}$$
$$= 144$$

EXAMPLE 6 Calculate: $3 \times 4^2 - 29$.

$$3 \times 4^2 - 29 = 3 \times 16 - 29 \qquad \text{There are no parentheses, so find } 4^2 \text{ first}$$
$$= 48 - 29 \qquad \text{Multiplying second}$$
$$= 19 \qquad \text{Subtracting}$$

DO EXERCISES 7–12.

EXAMPLE 7 Evaluate $(3x)^3 - 2$ for $x = 2$.

$$(3x)^3 - 2 = (3 \cdot 2)^3 - 2 \qquad \text{Substituting}$$
$$= 6^3 - 2 \qquad \text{Multiplying within parentheses first}$$
$$= 216 - 2$$
$$= 214$$

EXAMPLE 8 Calculate $(2 + x) \cdot (y - 1)$ for $x = 3$ and $y = 5$.

$$(2 + x) \cdot (y - 1) = (2 + 3) \cdot (5 - 1)$$
$$= 5 \cdot 4 \qquad \text{Working within parentheses first}$$
$$= 20$$

DO EXERCISES 13–17.

iii Using the Associative Laws

EXAMPLE 9 Calculate and compare: $3 + (8 + 5)$ and $(3 + 8) + 5$.

$$3 + (8 + 5) = 3 + 13 \qquad \text{Calculating within parentheses first}$$
$$= 16$$

$$(3 + 8) + 5 = 11 + 5 \qquad \text{Calculating within parentheses first}$$
$$= 16$$

The two expressions are equivalent. Moving the parentheses did not affect the expression.

EXAMPLE 10 Calculate and compare: $3 \cdot (4 \cdot 2)$ and $(3 \cdot 4) \cdot 2$.

$$3 \cdot (4 \cdot 2) = 3 \cdot 8 \qquad\qquad (3 \cdot 4) \cdot 2 = 12 \cdot 2$$
$$= 24 \qquad\qquad\qquad\qquad = 24$$

DO EXERCISES 18 AND 19.

When only addition is involved, parentheses can be placed any way we please without affecting the answer. When only multiplication is involved, parentheses can be placed any way we please without affecting the answer.

THE ASSOCIATIVE LAWS

Addition. **For any numbers *a*, *b*, and *c*,**

$$a + (b + c) = (a + b) + c.$$

(Numbers can be grouped in any manner for addition.)

Multiplication. **For any numbers *a*, *b*, and *c*,**

$$a \cdot (b \cdot c) = (a \cdot b) \cdot c.$$

(Numbers can be grouped in any manner for multiplication.)

EXAMPLE 11 Use an associative law to write an expression equivalent to $y + (z + 3)$.

An equivalent expression is

$$(y + z) + 3$$

by the associative law of addition.

EXAMPLE 12 Use an associative law to write an expression equivalent to $8 \cdot (x \cdot y)$.

An equivalent expression is

$$(8 \cdot x) \cdot y$$

by the associative law of multiplication.

When only additions or only multiplications are involved, parentheses may be placed any way we please, so we often omit them. For example,

$$x + (y + 7) \quad \text{means} \quad x + y + 7, \quad \text{and} \quad l(wh) \quad \text{means} \quad lwh.$$

DO EXERCISES 20 AND 21.

iv Using the Laws Together

If addition or multiplication is the only operation in an expression, then the associative and commutative laws allow us to group and change order as we please. For instance, in a calculation like $(5 + 2) + (3 + 5) + 8$, addition is the only operation. So we can change grouping and order to make easy combinations: $5 + 5 + 2 + 8 + 3 = 10 + 10 + 3 = 23$.

EXAMPLE 13 Use the commutative and associative laws to write at least three expressions equivalent to $(x + 5) + y$.

a) $(x + 5) + y = x + (5 + y)$ Using the associative law first and then using the
$= x + (y + 5)$ commutative law

b) $(x + 5) + y = y + (x + 5)$ Using the commutative law and then the
$= y + (5 + x)$ commutative law again

c) $(x + 5) + y = 5 + (x + y)$ Using the commutative law first and then the
associative law

EXAMPLE 14 Use the commutative and associative laws to write at least three expressions equivalent to $(3 \cdot x) \cdot y$.

a) $(3 \cdot x) \cdot y = 3 \cdot (x \cdot y)$ Using the associative law first and then the
$= 3 \cdot (y \cdot x)$ commutative law

b) $(3 \cdot x) \cdot y = y \cdot (x \cdot 3)$ Using the commutative law twice

c) $(3 \cdot x) \cdot y = x \cdot (y \cdot 3)$ Using the commutative law, then the associative,
and then the commutative law again

DO EXERCISES 22 AND 23.

Use an associative law to write an equivalent expression.

20. $a + (b + 2)$

21. $3 \cdot (v \cdot w)$

Use the commutative and associative laws to write at least three equivalent expressions.

22. $4 \cdot (t \cdot u)$

23. $r + (2 + s)$

Calculate and compare.

24. $4 \cdot (2 + 5)$ and
$(4 \cdot 2) + (4 \cdot 5)$

25. $(7 \cdot 8) + (7 \cdot 4)$ and
$7 \cdot (8 + 4)$

26. $6 \cdot (3 + 2 + 4)$ and
$(6 \cdot 3) + (6 \cdot 2) + (6 \cdot 4)$

What are the terms of the expression?

27. $5x - 4y + 3$

28. $-4y - 2x + 3z$

v **The Distributive Laws**

The distributive laws are the basis of many procedures in both arithmetic and algebra. They are probably the most important laws that we use to manipulate algebraic expressions.

If we wish to multiply a sum of several numbers by a factor, we can either add and then multiply, or multiply and then add.

EXAMPLE 15 Compute two ways: $5 \cdot (4 + 8)$.

a) $5 \cdot (\underbrace{4 + 8})$ Adding within parentheses first, and then multiplying

$\quad 5 \quad \cdot \quad 12$

$\quad\quad 60$

b) $(\underline{5 \cdot 4}) + (\underline{5 \cdot 8})$ Distributing the multiplication to terms within parentheses first and then adding

$\quad\quad 20 \quad + \quad 40$

$\quad\quad\quad\quad 60$

DO EXERCISES 24–26.

> **THE DISTRIBUTIVE LAW OF MULTIPLICATION OVER ADDITION**
> **For any real numbers a, b, and c.**
>
> $$a(b + c) = ab + ac.$$

There is another distributive law that relates multiplication and subtraction. This law says that to multiply by a difference we can either subtract and then multiply or multiply and then subtract.

> **THE DISTRIBUTIVE LAW OF MULTIPLICATION OVER SUBTRACTION**
> **For any real numbers a, b, and c,**
>
> $$a(b - c) = ab - ac.$$

What do we mean by the *terms* of an expression? When they are all separated by plus signs, it is easy to tell. If there are subtraction signs, we can find an equivalent expression that uses addition signs.

EXAMPLE 16 What are the terms of $3x - 4y + 2z$?

$$3x - 4y + 2z = 3x + (-4y) + 2z \qquad \text{Separating parts with } + \text{ signs}$$

The terms are $3x$, $-4y$, and $2z$.

DO EXERCISES 27 AND 28.

The distributive laws are a basis for a procedure called *multiplying*. In an expression such as $8(a + 2b - 7)$, we multiply each term inside the parentheses by 8:

$$8 \cdot (a + 2b - 7) = 8 \cdot a + 8 \cdot 2b - 8 \cdot 7 = 8a + 16b - 56.$$

EXAMPLES   Multiply.

17. $9(x - 5) = 9x - 9(5)$ Using the distributive law of multiplication
          $= 9x - 45$ over subtraction

18. $\frac{4}{3}(s - t + w) = \frac{4}{3}s - \frac{4}{3}t + \frac{4}{3}w$ Using both distributive laws

19. $-4(x - 2y + 3z) = -4 \cdot x - (-4)(2y) + (-4)(3z)$
          $= -4x - (-8y) + (-12z)$
          $= -4x + 8y - 12z$

DO EXERCISES 29–33.

vi Factoring

Factoring is the reverse of multiplying. To factor, we can use the distributive laws in reverse:

$$ab + ac = a(b + c),$$

$$ab - ac = a(b - c).$$

Look at Example 17. To *factor* $9x - 45$, we find an equivalent expression that is a product, $9(x - 5)$. When all the terms of an expression have a factor in common, we can "factor it out" using the distributive laws. Note the following.

$9x$ has the factors $9, x, 3, -3, 1, -1, -9$;

-45 has the factors $45, -1, 9, -5, -9, 5, 3, -3, 15, -15, -45, 1$.

We usually remove the largest common factor. In the case of $9x - 45$, that factor is 9.

Remember that an expression is factored when we find an equivalent expression that is a product.

EXAMPLES   Factor.

20. $5x - 10 = 5 \cdot x - 5 \cdot 2$
          $= 5(x - 2)$

21. $ax - ay + az = a(x - y + z)$

22. $9x + 27y - 9 = 9 \cdot x + 9 \cdot 3y - 9 \cdot 1$
          $= 9(x + 3y - 1)$

Note that $3(3x + 9y - 3)$ is also equivalent to $9x + 27y - 9$, but it is *not* the desired form. Factor out the largest common factor.

EXAMPLES   Factor.

23. $5x - 5y = 5(x - y)$

24. $-3x + 6y - 9z = -3(x - 2y + 3z)$

25. $18z - 12x - 24 = 6(3z - 2x - 4)$

26. $\frac{1}{2}x + \frac{3}{2}y - \frac{1}{2} = \frac{1}{2}(x + 3y - 1)$

DO EXERCISES 34–39.

vii Collecting Like Terms

The process of *collecting like terms* is also based on the distributive laws. In particular, the distributive law of multiplication over subtraction allows us to collect like terms without rewriting subtractions as additions. We can apply the distributive law "on the right" because of the commutative law of multiplication.

Multiply.

29. $3(x - 5)$

30. $5(x - y + 4)$

31. $-2(x - 3)$

32. $b(x - 2y + 4z)$

33. $\frac{5}{6}(x - y - z)$

Factor.

34. $4x - 8$

35. $3x - 6y - 15$

36. $-2y + 8z - 2$

37. $12z - 16x - 4$

38. $bx - by + bz$

39. $\frac{1}{4}m + \frac{3}{4}n - \frac{5}{4}$

Collect like terms.

40. $6x - 3x$

41. $7y - y$

42. $m - 0.41m$

43. $5x + 4y - 2x - y$

44. $3x - 7x - 11 + 8y - 4 - 13y$

EXAMPLES   Collect like terms.

27. $4x - 2x = (4 - 2)x = 2x$ Factoring out the x

28. $2x + 3y - 5x - 2y = 2x - 5x + 3y - 2y$
$$= (2 - 5)x + (3 - 2)y = -3x + y$$

29. $3x - x = (3 - 1)x = 2x$

30. $x - 0.24x = 1 \cdot x - 0.24x = (1 - 0.24)x = 0.76x$

31. $x - 6x = 1 \cdot x - 6 \cdot x = (1 - 6)x = -5x$

32. $4x - 7y + 9x - 5 + 3y - 8 = 13x - 4y - 13$

DO EXERCISES 40–44.

EXERCISE SET 1.6

i   Write an expression equivalent to each of the following. Use a commutative law.

1. $y + 8$ **2.** $x + 3$ **3.** mn **4.** ab

5. $9 + xy$ **6.** $11 + ab$ **7.** $ab + c$ **8.** $rs + t$

ii   Calculate.

9. $7 + 2 \times 6$ **10.** $11 + 4 \times 4$ **11.** $8 \times 7 + 6 \times 5$ **12.** $10 \times 5 + 1 \times 1$

13. $19 - 5 \times 3 + 3$ **14.** $14 - 2 \times 6 + 7$ **15.** $9 \div 3 + 16 \div 8$ **16.** $32 - 8 \div 4 - 2$

17. $7 + 10 - 10 \div 2$ **18.** $(5 \cdot 4)^2$ **19.** $(6 \cdot 3)^2$ **20.** $3 \cdot 2^3$

21. $4 \cdot 5^2$ **22.** $(8 + 2)^2$ **23.** $(5 + 3)^3$ **24.** $7 + 2^2$

25. $6 + 4^2$ **26.** $(5 - 2)^2$ **27.** $(3 - 2)^2$ **28.** $10 - 3^2$

29. $12 - 2^3$ **30.** $20 + 4^3 \div 8$ **31.** $2 \times 10^3 - 500$ **32.** $7 \times 3^4 + 18$

33. $80 - 6^2 \div 9$

Evaluate each expression.

34. $3 \cdot (a + 10)$ for $a = 12$

35. $b \cdot (7 + b)$ for $b = 5$

36. $(t + 3)^3$ for $t = 4$

37. $(12 - w)^3$ for $w = 7$

38. $(x + 5) \cdot (12 - x)$ for $x = 7$

39. $(y - 4) \cdot (y + 6)$ for $y = 10$

40. $(5y)^3 - 75$ for $y = 2$

41. $(7x)^2 + 59$ for $x = 3$

42. $\dfrac{y + 3}{2y}$ for $y = 5$

43. $\dfrac{(4x) + 2}{2x}$ for $x = 5$

44. $\dfrac{w^2 + 4}{5w}$ for $w = 4$

45. $\dfrac{b^2 + b}{2b}$ for $b = 5$

46. $(x - 4) \cdot (8 + y)$ for $x = 12$ and $y = 2$

47. $(y + 6) \cdot (9 - x)$ for $x = 7$ and $y = 10$

iii   Use the associative laws to write an expression equivalent to each of the following.

48. $(a + b) + 3$ **49.** $(5 + x) + y$ **50.** $3 \cdot (a \cdot b)$ **51.** $(6 \cdot x) \cdot y$

iv   Use the commutative and associative laws to write three equivalent expressions.

52. $(a + b) + 2$ **53.** $(3 + x) + y$ **54.** $5 + (v + w)$ **55.** $6 + (x + y)$ **56.** $(x \cdot y) \cdot 3$

57. $(a \cdot b) \cdot 5$ **58.** $7 \cdot (a \cdot b)$ **59.** $5 \cdot (x \cdot y)$ **60.** $2 \cdot c \cdot d$

v   Give the terms of each expression.

61. $4x + 3z$ **62.** $8x - 1.4y$ **63.** $7x + 8y - 9z$

64. $8a + 10b - 18c$ **65.** $12x - 13.2y + \frac{5}{8}z - 4.5$ **66.** $-7.8a - 3.4y - 8.7z - 12.4$

Multiply.

67. $7(4 - 3)$
68. $15(8 - 6)$
69. $-3(3 - 7)$
70. $1.2(5 - 2.1)$
71. $4.1(6.3 - 9.4)$
72. $-\frac{8}{9}(\frac{2}{3} - \frac{5}{3})$
73. $7(x - 2)$
74. $5(x - 8)$
75. $-7(y - 2)$
76. $-9(y - 7)$
77. $-9(-5x - 6y + 8)$
78. $-7(-2x - 5y + 9)$
79. $-4(x - 3y - 2z)$
80. $8(2x - 5y - 8z)$
81. $3.1(-1.2x + 3.2y - 1.1)$
82. $-2.1(-4.2x - 4.3y - 2.2)$

vi Factor.

83. $8x - 24$
84. $10x - 50$
85. $32 - 4y$
86. $24 - 6m$
87. $8x + 10y - 22$
88. $9a + 6b - 15$
89. $ax - 7a$
90. $by - 9b$
91. $ax - ay - az$
92. $cx + cy - cz$

vii Collect like terms.

93. $11x - 3x$
94. $9t - 17t$
95. $6n - n$
96. $y - 17y$
97. $9x + 2y - 5x$
98. $8y - 3z + 4y$
99. $11x + 2y - 4x - y$
100. $13a + 9b - 2a - 4b$
101. $2.7x + 2.3y - 1.9x - 1.8y$
102. $6.7a + 4.3b - 4.1a - 2.9b$
103. $\frac{1}{5}x + \frac{4}{5}y + \frac{2}{5}x - \frac{1}{5}y$
104. $\frac{7}{8}x + \frac{5}{8}y + \frac{1}{8}x - \frac{3}{8}y$
105. $4a + 5a$
106. $9x + 3x$
107. $8b - 11b$
108. $9c - 12c$
109. $14y + y$
110. $13x + x$
111. $12a - a$
112. $15x - x$
113. $t - 9t$
114. $x - 6x$
115. $5x - 3x + 8x$
116. $3x - 11x + 2x$
117. $5x - 8y + 3x$
118. $9a - 10b + 4a$
119. $7c + 8d - 5c + 2d$
120. $12a + 3b - 5a + 6b$
121. $4x - 7 + 18x + 25$
122. $13p + 5 - 4p + 7$
123. $13x + 14y - 11x - 47y$
124. $17a + 17b - 12a - 38b$

Tell which of the following expressions are equivalent. Also, tell why.

125. $3t + 5$ and $3 \cdot 5 + t$
126. $4x$ and $x + 4$
127. $5m + 6$ and $6 + 5m$
128. $(x + y) + z$ and $z + (x + y)$
129. $bxy + bx$ and $yxb + bx$
130. $ab + bc$ and $ac + db$
131. $a + c + e + g$ and $ae + cg$
132. $abc \cdot de$ and $a \cdot b \cdot c \cdot ed$

Find a value of the variable that shows that the two expressions are *not* equivalent.

133. $3x^2$; $(3x)^2$
134. $(a + 2)^3$; $a^3 + 2^3$
135. $\frac{x + 2}{2}$; x
136. $\frac{y^6}{y^3}$; y^2

Write an algebraic expression for each of the following.

137. A number squared plus 7

138. A number plus the square of 7

139. The square of a sum of 7 and some number

140. A number squared plus 7 squared

141. The numerator is 3 more than some number and the denominator is the square of the numerator.

142. Two numbers are multiplied. One of them is 5 more than the other.

143. Carole is twice as old as Victor was a year ago. Victor's age is now x. Write an expression for Carole's age.

144. Is there an associative law for subtraction of whole numbers? If not, give a counterexample.

Factor.

145. $2\pi r + \pi r s$

146. $\frac{1}{2}ah + \frac{1}{2}bh$

Write an algebraic expression. Simplify each expression, if possible.

147. A principal of P dollars was invested in a savings account at 8% simple interest. How much was in the account after 1 year?

148. The population of a town is P. After a 6% increase, what was the new population?

149. Eight times the difference of x and y

150. Nine times the difference of y and z, increased by $3z$

151. Three times the sum of a and b, decreased by $7a$

152. The total cost if you buy x cassette tapes at $2.95 on Monday and y cassettes at the same price on Wednesday

153. The total intake of a store when branch A sells x microcomputers at $2500 and branch B sells y microcomputers at the same price

154. An investor has 5420 shares of a stock bought at $41\frac{1}{8}$. The stock is now worth $37\frac{3}{4}$. Show two ways of determining how much has been lost. Solve.

Simplify.

155. $\dfrac{128}{192}$

156. $\dfrac{pqrs}{qrst}$

157. $\dfrac{33sba}{2(11a)}$

158. $\dfrac{4 \cdot 9 \cdot 16}{2 \cdot 8 \cdot 15}$

159. $\dfrac{36 \cdot (2rh)}{8 \cdot (9hg)}$

160. $\dfrac{3 \cdot (4xy) \cdot (5)}{2 \cdot (3x) \cdot (4y)}$

161. Is there a commutative law for division of whole numbers? If not, give a counterexample to show that there is no such law.

Write an expression equivalent to each of the following. Use the indicated name for 1.

162. $\dfrac{y}{10} \left(\text{Use } \dfrac{z}{z} \text{ for } 1. \right)$

163. $\dfrac{s}{20} \left(\text{Use } \dfrac{t}{t} \text{ for } 1. \right)$

164. $\dfrac{m}{3n} \left(\text{Use } \dfrac{p}{p} \text{ for } 1. \right)$

OBJECTIVES

After finishing Section 1.7, you should be able to:

i Rename an additive inverse without parentheses, where an expression has several terms.

ii Simplify expressions by removing parentheses and collecting like terms.

iii Simplify expressions with parentheses inside parentheses.

iv Simplify expressions using the rules for order of operations, and convert from algebraic symbolism to BASIC notation and from BASIC notation to algebraic symbolism.

Rename each additive inverse without parentheses.

1. $-(x + 2)$

2. $-(5x + 2y + 8)$

1.7

SIMPLIFICATION AND ORDER OF OPERATIONS

We expand our ability to manipulate expressions in this section. We first consider additive inverses of sums and differences. Then we simplify expressions involving parentheses.

i Inverses of Sums

What happens when we multiply a real number by -1?

EXAMPLES Multiply.

1. $-1 \cdot 7$ $-1 \cdot 7 = -7$

2. $-1(-5)$ $-1 \; -5 \; = 5$

3. $-1 \cdot 0$ $-1 \cdot 0 = 0$

From these examples it appears that when we multiply a number by -1, we get the additive inverse of that number.

THE PROPERTY OF -1

For any real number a,

$$-1 \cdot a = -a.$$

(Negative one times a is the additive inverse of a.)

The property of -1 enables us to find certain expressions equivalent to additive inverses of sums.

EXAMPLES Rename each additive inverse without parentheses.

4. $\begin{aligned} -(3 + x) &= -1(3 + x) &&\text{Using the property of } -1 \\ &= -1 \cdot 3 + (-1)x &&\text{Using a distributive law} \\ &= -3 + (-x) &&\text{Using the property of } -1 \\ &= -3 - x \end{aligned}$

5. $\begin{aligned} -(3x + 2y + 4) &= -1(3x + 2y + 4) &&\text{Using the property of } -1 \\ &= -1(3x) + (-1)(2y) + (-1)4 &&\text{Using a distributive law} \\ &= -3x - 2y - 4 &&\text{Using the property of } -1 \end{aligned}$

DO EXERCISES 1 AND 2.

Examples 4 and 5 illustrate an important property of real numbers.

THE INVERSE OF A SUM

For any real numbers a and b,

$$-(a + b) = -a + (-b).$$

(The inverse of a sum is the sum of the inverses.)

If we want to remove parentheses in an expression like

$$-(x - 5),$$

we can do so by replacing each term in the parentheses by its additive inverse ("change the sign of every term"). Doing so for $-(x - 5)$ we obtain $-x + 5$ as an equivalent expression. The same is true for a sum or difference of more than two terms.

EXAMPLES Rename each additive inverse without parentheses.

6. $-(5 - y) = -5 + y$ Changing the sign of each term

7. $-(2a - 7b - 6) = -2a + 7b + 6$

DO EXERCISES 3–6.

⬛ii⬛ Removing Parentheses and Simplifying

When a sum is added as in $5x + (2x + 3)$ we can simply remove, or drop, the parentheses and collect like terms. On the other hand, when a sum is subtracted, as in $3x - (4x + 2)$, we can subtract by adding an inverse, as usual. We then remove parentheses by changing the sign of each term inside the parentheses and collect like terms.

EXAMPLE 8 Remove parentheses and simplify.

$$3x - (4x + 2) = 3x + (-(4x + 2))$$ Adding the inverse of $(4x + 2)$
$$= 3x + (-4x - 2)$$ Changing the sign of each term inside the parentheses
$$= 3x - 4x - 2$$
$$= -x - 2$$ Collecting like terms

EXAMPLES Remove parentheses and simplify.

9. $5y - (3y + 4) = 5y - 3y - 4$ Removing parentheses by changing the sign of every term
$$= 2y - 4$$

10. $3y - 2 - (2y - 4) = 3y - 2 - 2y + 4$
$$= y + 2$$

DO EXERCISES 7–9.

Next, consider subtracting an expression consisting of several terms preceded by a number other than 1 or -1.

EXAMPLES Remove parentheses and simplify.

11. $x - 3(x + y) = x + (-3(x + y))$ Adding the inverse of $3(x + y)$
$$= x + (-3x - 3y)$$ Multiplying $x + y$ by -3
$$= x - 3x - 3y$$ Removing parentheses
$$= -2x - 3y$$ Collecting like terms

Rename each additive inverse without parentheses. Try to do this in one step.

3. $-(6 - t)$

4. $-(x - y)$

5. $-(-4a + 3t - 10)$

6. $-(18 - m - 2n + 4z)$

Remove parentheses and simplify.

7. $5x - (3x + 9)$

8. $8y - 2 - (2y - 4)$

9. $5x - 2y - (2y - 3x - 4)$

Remove parentheses and simplify.

10. $y - 9(x + y)$

11. $5a - 3(7a - 6)$

Simplify.

12. $[24 \div (-2)] \div (-2)$

13. $[9 - (6 + 4)]$

14. $3(4 + 2) - \{7 - [4 - (6 + 5)]\}$

15. $[3(x + 2) + 2x]$
$\quad -[4(y + 2) - 3(y - 2)]$

---

12. $3y - 2(4y - 5) = 3y + (-2(4y - 5))$ Adding the inverse of $2(4y - 5)$
$\qquad\qquad\qquad = 3y + (-8y + 10)$
$\qquad\qquad\qquad = 3y - 8y + 10$
$\qquad\qquad\qquad = -5y + 10$

DO EXERCISES 10 AND 11.

iii Parentheses Within Parentheses

Sometimes parentheses occur within parentheses. When that happens we can use parentheses of different shapes, such as [ ], called "brackets," or { }, called "braces."

> When parentheses occur within parentheses, do the computations in the innermost ones first. Then work from the inside out.

EXAMPLES Simplify.

13. $[3 - (7 + 3)] = [3 - 10]$ Working with the innermost parentheses first;
computing $7 + 3$
$\qquad\qquad\qquad = -7$

14. $\{8 - [9 - (12 + 5)]\} = \{8 - [9 - 17]\}$ Computing $12 + 5$
$\qquad\qquad\qquad\qquad = \{8 - [-8]\}$ Computing $9 - 17$
$\qquad\qquad\qquad\qquad = 16$

15. $[(-4) \div (-\frac{1}{4})] \div \frac{1}{4} = [(-4) \cdot (-4)] \div \frac{1}{4}$ Working with the innermost parentheses first
$\qquad\qquad\qquad\qquad = 16 \div \frac{1}{4}$ Computing $(-4) \div (-\frac{1}{4})$
$\qquad\qquad\qquad\qquad = 16 \cdot 4$
$\qquad\qquad\qquad\qquad = 64$

16. $4(2 + 3) - \{7 - [4 - (8 + 5)]\}$
$\quad = 4 \cdot 5 - \{7 - [4 - 13]\}$ Working with the innermost parentheses first
$\quad = 20 - \{7 - [-9]\}$ Computing $4 \cdot 5$ and $4 - 13$
$\quad = 20 - 16$ Computing $7 - [-9]$
$\quad = 4$

17. $[5(x + 2) - 3x] - [3(y + 2) - 7(y - 3)]$
$\quad = [5x + 10 - 3x] - [3y + 6 - 7y + 21]$ Working with the innermost parentheses first
$\quad = [2x + 10] - [-4y + 27]$ Collecting like terms within parentheses
$\quad = 2x + 10 + 4y - 27$ Removing parentheses
$\quad = 2x + 4y - 17$ Collecting like terms

DO EXERCISES 12–15.

iv Computer Language and Order of Operations

When several operations are to be done in a calculation or a problem, in what order should they be done? We have agreements about such calculations, and we also use parentheses. In handwritten algebra, little, if any, difficulty arises in knowing in which order operations are to be performed. Consider, for example,

$$8 + 5^3.$$

It is rather natural to first find the power 5^3, and then add it to 8:

$$8 + 5^3 = 8 + 125 = 133.$$

With computers, however, there may be some problem in knowing in which order operations are to be performed because the symbolism is different. The order in which a computer performs operations is the same as it is in ordinary algebra, but because of the different symbolism it is important to give some thought to order of operations when using a computer language. The computer symbols we will use to illustrate are from the computer language known as BASIC.

BASIC notation	Algebra notation
A + B	$a + b$
A − B	$a - b$
A * B	ab, or $a \cdot b$
A/B	$\dfrac{a}{b}$, or $a \div b$
A^B	a^b
( )	( )

We now state the rules for order of operations.

RULES FOR ORDER OF OPERATIONS

1. **Do all calculations within parentheses before operations outside.**
2. **Evaluate all exponential expressions.**
3. **Do all multiplications and divisions in order from left to right.**
4. **Do all additions and subtractions in order from left to right.**

If you are computing by hand, it can be helpful to think of each division as a multiplication by a reciprocal. Then, by the commutative and associative laws, you know that you can do the results in any order you prefer. Nevertheless, we will do these operations in order from left to right as a computer would do them. A similar statement holds with respect to the order of addition and subtraction.

Let us apply these rules to expressions we see in arithmetic or algebra and then with expressions in BASIC language.

EXAMPLE 18 Simplify: $-34 \cdot 56 - 17$.

There are no parentheses or powers so we start with the third step.

$-34 \cdot 56 - 17 = -1904 - 17$ Carrying out all multiplications and divisions in order from left to right

$= -1921$ Carrying out all additions and subtractions in order from left to right

EXAMPLE 19 Simplify: $2^4 + 51 \cdot 4 - (37 + 23 \cdot 2)$.

$2^4 + 51 \cdot 4 - (37 + 23 \cdot 2) = 2^4 + 51 \cdot 4 - (37 + 46)$ Carry out all operations inside parentheses first. Multiply 23 by 2.

$= 2^4 + 51 \cdot 4 - 83$ Complete the addition inside parentheses.

$= 16 + 51 \cdot 4 - 83$ Evaluate exponential expressions.

$= 16 + 204 - 83$ Do all multiplications.

$= 220 - 83$ Do all additions and subtractions in order from left to right.

$= 137$

DO EXERCISES 16 AND 17.

Simplify.

16. $23 - 42 \cdot 30$

17. $52 \cdot 5 + 5^3 - (4^2 - 48 \div 4)$

Simplify.

18. $5 * 6 - (2 + 3)\verb|^|3 - 4 * 10$

Write BASIC notation.

19. $\left(\dfrac{1}{4 \cdot 5}\right)^2$

20. $a^2 + b^2 - 2ab$

21. $\dfrac{x}{y} - \dfrac{t}{s}$

Write algebraic symbolism.

22. $2/(A + 3)$

23. $A\verb|^|2 - 2 * A * B + B\verb|^|2$

Now let us apply the rules to BASIC language.

EXAMPLE 20   Simplify: $3 + 4 * 7 - (7 + 5)\verb|^|2$.

$3 + 4 * 7 - (7 + 5)\verb|^|2 = 3 + 4 * 7 - 12\verb|^|2$ Carry out all operations inside parentheses first.

$\qquad\qquad = 3 + 4 * 7 - 144$ Evaluate exponential expressions.

$\qquad\qquad = 3 + 28 - 144$ Do the multiplication.

$\qquad\qquad = 31 - 144$ Do all additions and subtractions in order from left to right.

$\qquad\qquad = -113$

DO EXERCISE 18.

(*Optional*) To program a computer to perform a task, it is important to be able to write BASIC notation for algebraic symbolism.

EXAMPLES   Write BASIC notation for each of the following.

21. $\left(\dfrac{54 \cdot 37}{2}\right)^3$ $(54 * 37/2)\verb|^|3$

22. $b^2 - 4ac$ $B\verb|^|2 - 4 * A * C$

23. $c + d - \dfrac{a^2}{b}$ $C + D - A\verb|^|2/B$

DO EXERCISES 19–21.

It is also important to be able to convert from BASIC notation to algebraic symbolism.

EXAMPLES   Write algebraic symbolism for each of the following.

24. $(5 * 6)\verb|^|2 - (7 * 8)\verb|^|3$ $(5 \cdot 6)^2 - (7 \cdot 8)^3$

25. $(A\verb|^|2 - B\verb|^|2)/(A - B)$ $\dfrac{a^2 \cdot b^2}{a - b}$

DO EXERCISES 22 AND 23.

EXERCISE SET 1.7

i   Rename each additive inverse without parentheses.

1. $-(2x + 7)$ **2.** $-(3x + 5)$ **3.** $-(5x - 8)$ **4.** $-(6x - 7)$

5. $-(4a - 3b + 7c)$ **6.** $-(5x - 2y - 3z)$ **7.** $-(6x - 8y + 5)$ **8.** $-(8x + 3y + 9)$

9. $-(3x - 5y - 6)$ **10.** $-(6a - 4b - 7)$ **11.** $-(-8x - 6y - 43)$ **12.** $-(-2a + 9b - 5c)$

ii Remove parentheses and simplify.

13. $9x - (4x + 3)$

14. $7y - (2y + 9)$

15. $2a - (5a - 9)$

16. $11n - (3n - 7)$

17. $2x + 7x - (4x + 6)$

18. $3a + 2a - (4a + 7)$

19. $2x - 4y - 3(7x - 2y)$

20. $3a - 7b - 1(4a - 3b)$

21. $15x - y - 5(3x - 2y + 5z)$

22. $4a - b - 4(5a - 7b + 8c)$

iii Simplify.

23. $[9 - 2(5 - 4)]$

24. $[6 - 5(8 - 4)]$

25. $8[7 - 6(4 - 2)]$

26. $10[7 - 4(7 - 5)]$

27. $[4(9 - 6) + 11] - [14 - (6 + 4)]$

28. $[7(8 - 4) + 16] - [15 - (7 + 3)]$

29. $[10(x + 3) - 4] + [2(x - 1) + 6]$

30. $[9(x + 5) - 7] + [4(x - 12) + 9]$

31. $[7(x + 5) - 19] - [4(x - 6) + 10]$

32. $[6(x + 4) - 12] - [5(x - 8) + 11]$

33. $3\{[7(x - 2) + 4] - [2(2x - 5) + 6]\}$

34. $4\{[8(x - 3) + 9] - [4(3x - 7) + 2]\}$

35. $[(-24) \div (-3)] \div (-\frac{1}{2})$

36. $[32 \div (-2)] \div (-2)$

37. $16 \cdot (-24) + 50$

38. $10 \cdot 20 - 15 \cdot 24$

39. $2^4 + 2^3 - 10$

40. $40 - 3^2 - 2^3$

41. $5^3 + 26 \cdot 71 - (16 + 25 \cdot 3)$

42. $4^3 + 10 \cdot 20 + 8^2 - 23$

43. 🖩 $3000 \cdot (1 + 0.16)^3$

44. 🖩 $2000 \cdot (3 + 1.14)^2$

iv Simplify.

45. $4 * 5 - 2 * 6 + 4$

46. $8 * (7 - 3)/4$

47. $4 * (6 + 8)/(4 + 3)$

48. $4\hat{}3/8$

49. $(2 * (5 - 3))\hat{}2$

50. $5\hat{}3 - 7\hat{}2$

Write BASIC notation for each of the following.

51. $a^2 + 2ab + b^2$

52. $a^3 - b^3$

53. $\dfrac{2(3 - b)}{c}$

54. $\dfrac{a + b}{c - d}$

55. $\dfrac{a}{b} - \dfrac{c}{d}$

56. $\left(\dfrac{a^2}{b^3}\right)^3$

Write algebraic symbolism for each of the following.

57. $2 * A + 7$

58. $A + 1/B$

59. $3 * A\hat{}2 - 5$

60. $(A + 4)/(2 * A)$

61. $(A + B)\hat{}2$

62. $A\hat{}2 - 3 * A + 2$

○ ────────────────────────────────

Find an equivalent expression by enclosing the last three terms in parentheses preceded by a minus sign.

63. $6y + 2x - 3a + c$

64. $x - y - a - b$

65. $6m + 3n - 5m + 4b$

66. If $-(a + b)$ is $-a + (-b)$ what should be the sum of $(a + b)$ and $-a + (-b)$? Show that your answer is correct.

Simplify.

67. $z - \{2z - [3z - (4z - 5z) - 6z] - 7z\} - 8z$

68. $\{x - [f - (f - x)] + [x - f]\} - 3x$

69. $x - \{x - 1 - [x - 2 - (x - 3 - \{x - 4 - [x - 5 - (x - 6)]\})]\}$

If $n > 0$, $m > 0$, and $n \neq m$, which of the following are true?

70. $-n + m = n - m$

71. $-n + m = -(n + m)$

72. $-n - m = -(n + m)$

73. $-n - m = -(n - m)$

74. $n(-n - m) = -n^2 + nm$

75. $-m(n - m) = -(mn + m^2)$

76. $-m(-n + m) = m(n - m)$

77. $-n(-n - m) = n(n + m)$

78. Determine whether it is true that, for any real number x, $(-x)^2 = x^2$. Explain why or why not.

79. Determine whether it is true that, for any real numbers a and b, $ab = (-a)(-b)$. Explain why or why not.

80. Determine whether it is true that, for any real numbers a and b, $-(ab) = (-a)b = a(-b)$. Explain why or why not.

Simplify.

81. $4 \cdot 4 \cdot 4 \div 8 \cdot 8 \cdot 8 \div 2 \cdot 2 \cdot 2 \cdot 2 \cdot 2$

82. $6 \cdot 6 \cdot 6 \cdot 6 \div 3 \cdot 3 \cdot 3 \cdot 3 \div 3 \cdot 3 \cdot 3$

OBJECTIVES

After finishing Section 1.8, you should be able to:

i Rename a number with or without negative exponents.

ii Multiply with exponential notation by adding the exponents.

iii Divide with exponential notation by subtracting the exponents in the appropriate manner.

iv Raise a power to a power by multiplying the exponents.

v Solve problems involving compound interest.

Explain the meaning of each of the following without using negative exponents.

1. 4^{-3}

2. 5^{-2}

3. 2^{-4}

Rename using negative exponents.

4. $\dfrac{1}{3^2}$

5. $\dfrac{1}{5^4}$

6. $\dfrac{1}{7^3}$

Rename using positive exponents.

7. 5^{-3}

8. 7^{-5}

9. 10^{-4}

1.8

INTEGERS AS EXPONENTS: APPLICATIONS

Algebraic expressions like the following are called *monomials*:

$$5x^3, \quad 7y^4, \quad \tfrac{1}{4}t^2, \quad x^1, \quad x^0, \quad 8, \quad 34, \quad 0.$$

Each expression is a number or a number times a variable to some power. More formally, a monomial is an expression of the type ax^n, where n is a whole number.

We will learn to multiply and divide monomials and to raise a power to a power.

i Negative Exponents

Negative numbers can be used as exponents. Look for a pattern in the following.

$$\begin{aligned}
10^3 &= 10 \cdot 10 \cdot 10 & 8^3 &= 8 \cdot 8 \cdot 8 \\
10^2 &= 10 \cdot 10 & 8^2 &= 8 \cdot 8 \\
10^1 &= 10 & 8^1 &= 8 \\
10^0 &= 1 & 8^0 &= 1 \\
10^{-1} &= ? & 8^{-1} &= ? \\
10^{-2} &= ? & 8^{-2} &= ?
\end{aligned}$$

In the first case we divided by 10 each time. In the second we divided by 8. Continuing the pattern, we have

$$10^{-1} = \frac{1}{10} = \frac{1}{10^1} \quad \text{and} \quad 8^{-1} = \frac{1}{8} = \frac{1}{8^1};$$

$$10^{-2} = \frac{1}{10 \cdot 10} = \frac{1}{10^2} \quad \text{and} \quad 8^{-2} = \frac{1}{8 \cdot 8} = \frac{1}{8^2}.$$

We make the following agreement. It is a definition.

If n is any positive integer,

$$b^{-n} \text{ is given the meaning } \frac{1}{b^n}.$$

In other words, b^n and b^{-n} are reciprocals.

EXAMPLE 1 Explain the meaning of 3^{-4} without using negative exponents.

$$3^{-4} \quad \text{means} \quad \frac{1}{3^4}, \quad \text{or} \quad \frac{1}{3 \cdot 3 \cdot 3 \cdot 3}, \quad \text{or} \quad \frac{1}{81}. \qquad 3^{-4} \text{ is } not \text{ a negative number.}$$

EXAMPLE 2 Rename $\dfrac{1}{5^2}$ using a negative exponent.

$$\frac{1}{5^2} = 5^{-2}$$

EXAMPLE 3 Rename 4^{-3} using a positive exponent.

$$4^{-3} = \frac{1}{4^3}$$

DO EXERCISES 1–9.

ii Multiplying Using Exponents

Consider an expression with exponents, such as $a^3 \cdot a^2$. To simplify it, recall the definition of exponents:

$$a^3 \cdot a^2 \quad \text{means} \quad (a \cdot a \cdot a)(a \cdot a)$$

and $(a \cdot a \cdot a)(a \cdot a) = a^5$. The exponent in a^5 is the sum of those in $a^3 \cdot a^2$. Suppose one exponent is positive and one is negative:

$$a^5 \cdot a^{-2} \quad \text{means} \quad (a \cdot a \cdot a \cdot a \cdot a) \cdot \left(\frac{1}{a \cdot a}\right),$$

which simplifies as follows:

$$a \cdot a \cdot a \cdot \left(\frac{a \cdot a}{1}\right)\left(\frac{1}{a \cdot a}\right) = a \cdot a \cdot a \cdot \frac{a \cdot a}{a \cdot a} = a \cdot a \cdot a = a^3.$$

If we add the exponents, we again get the correct result. Next suppose that both exponents are negative:

$$a^{-3} \cdot a^{-2} \quad \text{means} \quad \frac{1}{a \cdot a \cdot a} \cdot \frac{1}{a \cdot a}.$$

This is equal to

$$\frac{1}{a \cdot a \cdot a \cdot a \cdot a}, \quad \text{or} \quad \frac{1}{a^5}, \quad \text{or} \quad a^{-5}.$$

Again, adding the exponents gives the correct result. The same is true if one or both exponents are zero.

> In multiplication with exponential notation, we can add exponents if the bases are the same:
>
> $$a^m \cdot a^n = a^{m+n}.$$

EXAMPLES Multiply and simplify.

4. $8^4 \cdot 8^3 = 8^{4+3}$ Adding exponents
     $= 8^7$

5. $7^{-3} \cdot 7^6 = 7^{-3+6}$
     $= 7^3$

6. $x \cdot x^8 = x^{1+8}$
     $= x^9$

7. $x^4 \cdot x^{-3} = x^{4+(-3)}$
     $= x^1$
     $= x$

DO EXERCISES 10–15.

iii Dividing Using Exponents

Consider dividing with exponential notation.

$$\frac{5^4}{5^2} \quad \text{means} \quad \frac{5 \cdot 5 \cdot 5 \cdot 5}{5 \cdot 5},$$

which is $\quad 5 \cdot 5 \cdot \dfrac{5 \cdot 5}{5 \cdot 5} \quad$ and we have $\quad 5 \cdot 5$, or 5^2.

Multiply and simplify.

10. $3^5 \cdot 3^3$

11. $5^{-2} \cdot 5^4$

12. $6^{-3} \cdot 6^{-4}$

13. $x \cdot x^{-5}$

14. $y^2 \cdot y^{-4}$

15. $x^{-2} \cdot x^{-6}$

Divide and simplify.

16. $\dfrac{4^5}{4^2}$

17. $\dfrac{7^{-2}}{7^3}$

18. $\dfrac{a^2}{a^{-5}}$

19. $\dfrac{b^{-2}}{b^{-3}}$

20. $\dfrac{x}{x^{-3}}$

21. $\dfrac{x^8}{x}$

> **In division, we subtract exponents if the bases are the same:**
>
> $$\frac{a^m}{a^n} = a^{m-n}.$$

This is true whether the exponents are positive, negative, or zero.

EXAMPLES   Divide and simplify.

8. $\dfrac{5^4}{5^{-2}} = 5^{4-(-2)}$ Subtracting exponents

$\qquad = 5^6$

9. $\dfrac{x}{x^7} = x^{1-7}$

$\qquad = x^{-6}$, or $\dfrac{1}{x^6}$

10. $\dfrac{b^{-4}}{b^{-5}} = b^{-4-(-5)}$

$\qquad = b^1$

$\qquad = b$

> **In exercises such as Examples 8–10 above, it may help to think as follows: After writing the base, write the top exponent. Then write a subtraction sign. Then write the bottom exponent. Then do the subtraction. For example,**
>
> $$\frac{x^{-3}}{x^{-5}} = x^{-3-(-5)}$$
>
> | Writing the base and the top exponent | Writing a subtraction sign | Writing the bottom exponent |

DO EXERCISES 16–21.

iv   **Raising a Power to a Power**

Consider raising a power to a power.

EXAMPLE 11

$$(3^2)^4 \text{ means } 3^2 \cdot 3^2 \cdot 3^2 \cdot 3^2,$$
$$\text{or } 3 \cdot 3 \cdot 3 \cdot 3 \cdot 3 \cdot 3 \cdot 3 \cdot 3, \text{ or } 3^8.$$

We could have multiplied the exponents in $(3^2)^4$. Suppose the exponents are not positive.

EXAMPLE 12

$$(5^{-2})^3 \text{ means } \frac{1}{5^2} \cdot \frac{1}{5^2} \cdot \frac{1}{5^2},$$
$$\text{or } \frac{1}{5 \cdot 5} \cdot \frac{1}{5 \cdot 5} \cdot \frac{1}{5 \cdot 5}, \text{ or } \frac{1}{5^6}, \text{ or } 5^{-6}.$$

Again, we could have multiplied the exponents. This works for any integer exponents.

To raise a power to a power we can multiply the exponents. For any exponents *m* and *n*,

$$(a^m)^n = a^{mn}.$$

EXAMPLES   Simplify.

13. $(3^5)^4 = 3^{5 \cdot 4}$ Multiply exponents.

$\qquad = 3^{20}$

14. $(y^{-5})^7 = y^{-5 \cdot 7}$

$\qquad = y^{-35}$, or $\dfrac{1}{y^{35}}$

15. $(x^4)^{-2} = x^{4(-2)}$

$\qquad = x^{-8}$, or $\dfrac{1}{x^8}$

16. $(a^{-4})^{-6} = a^{(-4)(-6)}$

$\qquad = a^{24}$

DO EXERCISES 22–25.

When several factors are in parentheses, raise each to the given power:

$$(a^m b^n)^t = (a^m)^t (b^n)^t = a^{mt} b^{nt}.$$

⎡*Caution!* Be sure to raise *every* factor in
⎣parentheses to the power.

EXAMPLES   Simplify.

17. $(5x^2 y^{-2})^3 = 5^3 (x^2)^3 (y^{-2})^3 = 125 x^6 y^{-6}$, or $125 \dfrac{x^6}{y^6}$

18. $(3x^3 y^{-5} z^2)^4 = 3^4 (x^3)^4 (y^{-5})^4 (z^2)^4 = 81 x^{12} y^{-20} z^8$, or $81 \dfrac{x^{12} z^8}{y^{20}}$

The following is a summary of the laws of exponents considered in this section.

$$a^m a^n = a^{m+n},$$

$$\frac{a^m}{a^n} = a^{m-n},$$

$$(a^m)^n = a^{mn},$$

$$(a^m b^n)^t = a^{mt} b^{nt}$$

DO EXERCISES 26–28.

v   **Application: Interest Compounded Annually**

Suppose we invest P dollars at an interest rate of 8%, compounded annually. The amount to which this grows at the end of one year is given by

$$P + 8\% P = P + 0.08P \qquad \text{By definition of percent}$$

$$= (1 + 0.08)P \qquad \text{Factoring}$$

$$= 1.08P. \qquad \text{Simplifying}$$

Going into the second year, the new principal is $(1.08)P$ dollars since interest has been added to the account. By the end of the second year, the following amount

Simplify.

22. $(3^4)^5$

23. $(x^{-3})^4$

24. $(y^{-5})^{-3}$

25. $(x^{-4})^8$

Simplify.

26. $(2x^5 y^{-3})^4$

27. $(5x^5 y^{-6} z^{-3})^2$

28. $(3y^{-2} x^{-5} z^8)^3$

29. Suppose $2000 is invested at 16%, compounded annually. How much is in the account at the end of 3 years?

will be in the account:

$$(1.08)[\ (1.08)P\ ], \quad \text{or} \quad (1.08)^2P. \qquad \text{New principal}$$

Going into the third year, the principal will be $(1.08)^2P$ dollars. At the end of the third year, the following amount will be in the account:

$$(1.08)[\ (1.08)^2P\ ], \quad \text{or} \quad (1.08)^3P. \qquad \text{New principal}$$

Note the pattern: At the end of years 1, 2, and 3 the amount is

$$(1.08)P, \qquad (1.08)^2P, \qquad (1.08)^3P, \quad \text{and so on.}$$

> **If principal P is invested at interest rate r, compounded annually, in t years it will grow to the amount A given by**
>
> $$A = P(1 + r)^t.$$

Compare the formula $A = P(1 + r)^t$ with the formula for simple interest, $A = P(1 + rt)$.

EXAMPLE 19 Suppose $1000 is invested at 8%, compounded annually. How much is in the account at the end of 3 years?

Substituting 1000 for P, 0.08 for r, and 3 for t, we get

$$A = P(1 + r)^t$$
$$= 1000(1 + 0.08)^3 = 1000(1.08)^3 = 1000(1.259712) \approx \$1259.71.$$

DO EXERCISE 29.

EXERCISE SET 1.8

i Explain the meaning of each without using negative exponents.

1. 3^{-2} **2.** 2^{-3} **3.** 10^{-4} **4.** 5^{-6}

Rename using negative exponents.

5. $\dfrac{1}{4^3}$ **6.** $\dfrac{1}{5^2}$ **7.** $\dfrac{1}{x^3}$ **8.** $\dfrac{1}{y^2}$ **9.** $\dfrac{1}{a^4}$ **10.** $\dfrac{1}{t^5}$ **11.** $\dfrac{1}{p^n}$ **12.** $\dfrac{1}{m^n}$

Rename using positive exponents.

13. 7^{-3} **14.** 5^{-2} **15.** a^{-3} **16.** x^{-2} **17.** y^{-4} **18.** t^{-7} **19.** z^{-n} **20.** h^{-m}

ii Multiply and simplify.

21. $2^4 \cdot 2^3$ **22.** $3^5 \cdot 3^2$ **23.** $3^{-5} \cdot 3^8$ **24.** $5^{-8} \cdot 5^9$ **25.** $x^{-2} \cdot x$ **26.** $x \cdot x$

27. $x^4 \cdot x^3$ **28.** $x^9 \cdot x^4$ **29.** $x^{-7} \cdot x^{-6}$ **30.** $y^{-5} \cdot y^{-8}$ **31.** $t^8 \cdot t^{-8}$ **32.** $m^{10} \cdot m^{-10}$

iii Divide and simplify.

33. $\dfrac{7^5}{7^2}$ **34.** $\dfrac{4^7}{4^3}$ **35.** $\dfrac{x}{x^{-1}}$ **36.** $\dfrac{x^6}{x}$ **37.** $\dfrac{x^7}{x^{-2}}$ **38.** $\dfrac{t^8}{t^{-3}}$

39. $\dfrac{z^{-6}}{z^{-2}}$ **40.** $\dfrac{y^{-7}}{y^{-3}}$ **41.** $\dfrac{x^{-5}}{x^{-8}}$ **42.** $\dfrac{y^{-4}}{y^{-9}}$ **43.** $\dfrac{m^{-9}}{m^{-9}}$ **44.** $\dfrac{x^{-8}}{x^{-8}}$

iv Simplify.

45. $(2^3)^2$ **46.** $(3^4)^3$ **47.** $(5^2)^{-3}$ **48.** $(9^3)^{-4}$

49. $(x^{-3})^{-4}$ **50.** $(a^{-5})^{-6}$ **51.** $(x^4y^5)^{-3}$ **52.** $(t^5x^3)^{-4}$

53. $(x^{-6}y^{-2})^{-4}$ **54.** $(x^{-2}y^{-7})^{-5}$ **55.** $(3x^3y^{-8}z^{-3})^2$ **56.** $(2a^2y^{-4}z^{-5})^3$

v Solve.

57. Suppose $2000 is invested at 12%, compounded annually. How much is in the account at the end of 2 years?

58. Suppose $2000 is invested at 15%, compounded annually. How much is in the account at the end of 3 years?

59. ▦ Suppose $10,400 is invested at 16.5%, compounded annually. How much is in the account at the end of 5 years?

60. ▦ Suppose $20,800 is invested at 20.5%, compounded annually. How much is in the account at the end of 6 years?

61. ▦ Suppose $10,400 is invested at $8\frac{3}{4}$%, compounded annually. How much is in the account at the end of 5 years?

62. ▦ Suppose $20,800 is invested at $9\frac{1}{2}$% compounded annually. How much is in the account at the end of 6 years?

63. Write $4^3 \cdot 8 \cdot 16$ as a power of 2.

64. Write $2^8 \cdot 16^3 \cdot 64$ as a power of 4.

65. Determine whether $(5y)^0$ and $5y^0$ are equivalent expressions.

66. Simplify: $(y^{2x})(y^{3x})$.

Simplify.

67. $\dfrac{(5^{12})^2}{5^{25}}$

68. $\dfrac{a^{20+20}}{(a^{20})^2}$

69. $\dfrac{(3^5)^4}{3^5 \cdot 3^4}$

70. $\dfrac{(7^5)^{14}}{(7^{14})^5}$

71. $\dfrac{a^{22}}{(a^2)^{11}}$

72. $\dfrac{49^{18}}{7^{35}}$ [*Hint:* Study Exercise 71.]

73. Simplify: $\dfrac{(\frac{1}{2})^4}{(\frac{1}{2})^5}$.

74. Simplify: $\dfrac{(0.4)^5}{((0.4)^3)^2}$.

75. How might you define 3^{-2} so that your definition is consistent with our rules for operating with exponents?

76. Determine whether $(a + b)^n = a^n + b^n$ is true for all numbers a, b, and n. (*Hint:* Substitute and evaluate.)

77. Determine whether $(a^b)^c = a^{bc}$ for all numbers a, b, and c. (*Hint:* Try different values. Remember to work inside parentheses first.)

78. Solve for x: $\dfrac{w^{50}}{w^x} = w^x$.

79. Solve for a: $\dfrac{(9x)^{12}}{(9x)^{14}} = \dfrac{1}{ax^2}$.

Insert $>$, $<$, or $=$ between each pair of numbers to make a true sentence.

80. $3^5 \quad 3^4$

81. $4^2 \quad 4^3$

82. $4^3 \quad 5^3$

83. $4^3 \quad 3^4$

Determine whether each of the following is true for any pair of whole numbers m and n and any positive numbers x and y.

84. $x^m \cdot y^n = (xy)^{mn}$

85. $x^m \cdot y^m = (xy)^m$

86. $x^m \cdot y^m = (xy)^{2m}$

87. $x^m \cdot x^n = x^{mn}$

88. $\left(\dfrac{x}{y}\right)^n = \dfrac{x^n}{y^n}$

89. $(x - y)^m = x^m - y^m$

Simplify.

90. $\dfrac{(0.2)^3}{5.21 - 4.41}$

91. $\dfrac{(0.2)(0.3)^3}{(0.03)(0.5)^2}$

92. $\dfrac{-24x^6y^7}{18x^{-3}y^9}$

93. $\dfrac{14a^4b^{-3}}{-8a^8b^{-5}}$

94. $\dfrac{-18x^{-2}y^3}{-12x^{-5}y^5}$

95. $\dfrac{11^{b+2}}{11^{3b-3}}$

96. $\dfrac{9a^{x-2}}{3a^{2x+2}}$

97. $\dfrac{-12x^{a+1}}{4x^{2-a}}$

98. $\dfrac{45x^{2a+4}y^{b+1}}{-9x^{a+3}y^{2+b}}$

99. $\dfrac{-28x^{b+5}y^{4+c}}{7x^{b-5}y^{c-4}}$

100. $\left(\dfrac{-4x^4y^{-2}}{5x^{-1}y^4}\right)^{-4}$

101. $(7^a)^{2b}$

102. $(3^{a+2})^a$

103. $(12^{3-a})^{2b}$

104. $(x^{a-1})^{3b}$

105. $(5x^{a-1}y^{b+1})^{2c}$

106. $(4x^{3a}y^{2b})^{5c}$

107. $\dfrac{4x^{2a+3}y^{2b-1}}{2x^{a+1}y^{b+1}}$

108. $\dfrac{25x^{a+b}y^{b-a}}{-5x^{a-b}y^{b+a}}$

Simplify.

109. $(-2)^0 - (-2)^3 - (-2)^{-1} + (-2)^4 - (-2)^{-2}$

110. $2(6^1 \cdot 6^{-1} - 6^{-1} \cdot 6^0)$

111. $\dfrac{(-8)^{-2} \cdot (8 - 8^0)}{2^{-6}}$

112. $\left[\dfrac{1}{(-3)^{-2}} - (-3)^1\right] \cdot [(-3)^2 + (-3)^{-2}]$

113. $\dfrac{(2^{-2})^a \cdot (2^b)^{-a}}{(2^{-2})^{-b}(2^b)^{-2a}}$

114. $\{[(8^{-a})^{-2}]^b\}^{-c} \cdot [(8^0)^a]^c$

115. $\left[\dfrac{(-3x^{-2a}y^{5b})^{-2}}{(2x^{4a}y^{-8b})^{-3}}\right]^2$

116. $\left[\left(\dfrac{a^{-2c}}{b^{7c}}\right)^{-3}\left(\dfrac{a^{4c}}{b^{-3c}}\right)^2\right]^{-a}$

1. Convert 460,000,000,000 to scientific notation.

2. The distance from the earth to the sun is about 93,000,000 miles. Write scientific notation for this number.

1.9

SCIENTIFIC NOTATION

There are many kinds of symbolism, or *notation*, for numbers. You are already familiar with fractional notation, decimal notation, and percent notation. In this section, we study another notation, one that is especially useful for work with very large numbers or very small numbers. It is thus important for problem solving. It is also useful in estimating.

The following are examples of scientific notation:

6.4×10^{13} means   64000000000000;

4.6×10^{-6} means   0.0000046;

3.75×10^{-2} means   0.0375.

> *Scientific notation* for a number consists of exponential notation for a power of 10, and, if needed, decimal notation for a number a between 1 and 10 and a multiplication sign. The following are both scientific notation:
>
> $$a \times 10^b; \quad 10^b.$$

i   Conversions

We can convert to scientific notation by multiplying by 1, choosing a name like $10^b \cdot 10^{-b}$ for the number 1.

EXAMPLE 1   Light travels about 9,460,000,000,000 km in one year. Write scientific notation for the number.

We want to move the decimal point 12 places, between the 9 and the 4, so we choose $10^{-12} \times 10^{12}$ as a name for 1. Then we multiply.

$$9,460,000,000,000 \times 10^{-12} \times 10^{12} \quad \text{Multiplying by 1}$$

$$= 9.46 \times 10^{12} \quad \text{The } 10^{-12} \text{ moved the decimal point 12 places to the left and we have scientific notation.}$$

DO EXERCISES 1 AND 2.

EXAMPLE 2   Write scientific notation for 0.0000000000156.

We want to move the decimal point 11 places. We choose $10^{11} \times 10^{-11}$ as a name for 1, and then multiply.

$0.0000000000156 \times 10^{11} \times 10^{-11}$ Multiplying by 1

$= 1.56 \times 10^{-11}$ The 10^{11} moved the decimal point 11 places to the right and we have scientific notation.

You should try to make conversions to scientific notation mentally as much as possible.

DO EXERCISES 3 AND 4.

EXAMPLES   Convert mentally to decimal notation.

3. $7.893 \times 10^5 = 789,300$ Moving the decimal point 5 places to the right

4. $4.7 \times 10^{-8} = 0.000000047$ Moving the decimal point 8 places to the left

DO EXERCISES 5 AND 6.

ii Multiplying and Dividing

Multiplying and dividing with scientific notation is easy because we can use the properties of exponents.

EXAMPLE 5   Multiply and write scientific notation for the answer.

$$(3.1 \times 10^5)(4.5 \times 10^{-3})$$

We apply the commutative and associative laws to get

$$(3.1 \times 4.5)(10^5 \times 10^{-3}) = 13.95 \times 10^2.$$

To find scientific notation for the result, we convert 13.95 to scientific notation and then simplify:

$$13.95 \times 10^2 = (1.395 \times 10^1) \times 10^2 = 1.395 \times 10^3.$$

DO EXERCISES 7 AND 8.

EXAMPLE 6   Divide and write scientific notation for the answer:

$$\frac{6.4 \times 10^{-7}}{8.0 \times 10^6}.$$

We solve as follows:

This shows two divisions.

$$\frac{6.4 \times 10^{-7}}{8.0 \times 10^6} = \frac{6.4}{8.0} \times \frac{10^{-7}}{10^6} \quad \text{Factoring}$$

$$= 0.8 \times 10^{-13} \quad \text{Doing the divisions separately}$$

$$= (8.0 \times 10^{-1}) \times 10^{-13} \quad \text{Converting 0.8 to scientific notation}$$

$$= 8.0 \times 10^{-14}.$$

DO EXERCISES 9 AND 10.

In the preceding example, you may be tempted to quit when you have obtained 0.8×10^{-13}. That would not be an incorrect answer, except that it is not yet scientific notation, because 0.8 is not a number between 1 and 10.

3. Convert 0.00000001235 to scientific notation.

4. The mass of a hydrogen atom is about 0.00000000000000000000000017 gram. Write scientific notation for this number.

Convert to decimal notation.

5. 7.893×10^{11}

6. 5.67×10^{-5}

Multiply and write scientific notation for the answer.

7. $(9.1 \times 10^{-17})(8.2 \times 10^3)$

8. $(1.12 \times 10^{-8})(5 \times 10^{-7})$

Divide and write scientific notation for the answer.

9. $\dfrac{4.2 \times 10^5}{2.1 \times 10^2}$

10. $\dfrac{1.1 \times 10^{-4}}{2.0 \times 10^{-7}}$

11. Estimate. Answers will vary depending on how you round.

$$\frac{830{,}000{,}000 \times 0.0000012}{3{,}100{,}000}$$

Estimate the order of magnitude of each computation.

12. $38{,}514 \times 2412$

13. $\dfrac{0.0034 \times 673}{1240 \times 317 \times 14.7}$

14. The mass of an electron, in grams, is in the order of magnitude -28. What does this tell you, in terms of scientific notation, about the mass of an electron?

iii **Estimating and Approximating**

Scientific notation is helpful in estimating or approximating.

EXAMPLE 7 Estimate: $\dfrac{780{,}000{,}000 \times 0.00071}{0.000005}$.

We first convert to scientific notation:

$$\frac{(7.8 \times 10^8)(7.1 \times 10^{-4})}{5 \times 10^{-6}}.$$

Next, we group as follows:

$$\frac{7.8 \times 7.1}{5} \times \frac{(10^8)(10^{-4})}{10^{-6}}.$$

We round the numbers that are between 1 and 10 and calculate:

$$\frac{8 \times 7}{5} = \frac{56}{5} \approx \frac{55}{5} = 11.$$

This answer will vary, depending on how you round. Next, we consider the powers of 10:

$$\frac{10^8 \times 10^{-4}}{10^{-6}} = \frac{10^4}{10^{-6}} = 10^{10}.$$

We now have an approximation of 11×10^{10}, or 1.1×10^{11}.

DO EXERCISE 11.

With some practice, you will be able to do this kind of estimating mentally. It is important that you acquire this skill to use in problem solving.

It is helpful to remember the following:

$10^3 = 1000$ (thousand);

$10^6 = 1{,}000{,}000$ (million);

$10^9 = 1{,}000{,}000{,}000$ (billion);

$10^{12} = 1{,}000{,}000{,}000{,}000$ (trillion).

Often the exponent in scientific notation is the only estimate desired. Such an answer is referred to as the *order of magnitude*. For example, the star Alpha Centauri is about 2.4×10^{13} miles from earth. One might just say that "the distance of Alpha Centauri from earth is in the order of magnitude 13." You get the idea of such a large distance from such a statement.

EXAMPLE 8 Estimate the order of magnitude of the product $3{,}012{,}000 \times 41{,}210$.

We estimate as before:

$$\begin{aligned}
3{,}012{,}000 \times 41{,}210 &\approx (3 \times 10^6) \times (4 \times 10^4) \\
&= 12 \times 10^{10} \\
&= (1.2 \times 10^1) \times 10^{10} \\
&= 1.2 \times 10^{11}.
\end{aligned}$$

The order of magnitude is 11.

DO EXERCISES 12–14.

SOME EXAMPLES USING SCIENTIFIC NOTATION

The distance from the planet Pluto to the sun is 3.664×10^9 mi.

The mass of an electron is 9.11×10^{-28} g.

The population of the United States was 2.04×10^8.

The wavelength of a certain red light is 6.6×10^{-5} cm.

The gross national product (GNP) one year was $\$9.32 \times 10^{11}$.

An electron has a charge of 4.8×10^{-10} electrostatic units.

15. A sheet of plastic has a thickness of 0.0003 mm. The sheet, which is 1.34 m wide, is rolled into a roll. The length of the plastic on the roll is 103.4 m. Use scientific notation to find the volume of the plastic sheet.

iv Scientific Notation in Problems

The following example shows how scientific notation can be useful in problem solving.

EXAMPLE 9   A certain kind of wire will be used to construct 350 km of transmission line. The wire has a diameter of 1.2 cm. What is the volume of wire needed for one wire 350 km long?

Drawing a picture, we see that we have a cylinder (a very *long* one). Its length is 350 km and the base has a diameter of 1.2 cm.

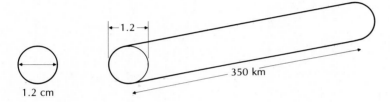

At this stage, you may need to look up a formula for the volume of a cylinder. It is

$$V = \pi r^2 h,$$

where r is the radius of the base and h is the height (in this case, the length of the wire.) We will use the volume formula, but it will help to make the units the same. Let's put everything into meters:

Length: 350 km = 350,000 m,   or 3.5×10^5 m;

Diameter: 1.2 cm = 0.012 m,   or 1.2×10^{-2} m.

The radius, which we will need in the formula, is half the diameter:

Radius: 0.6×10^{-2} m,   or 6×10^{-3} m.

We now substitute into the formula:

$$V = \pi(6 \times 10^{-3})^2 \cdot 3.5 \times 10^5.$$

We do the calculation. We use 3.14 for π:

$$
\begin{aligned}
V &= 3.14 \times (6 \times 10^{-3})^2 \times 3.5 \times 10^5 \\
&= 3.14 \times 6^2 \times 10^{-6} \times 3.5 \times 10^5 \\
&= (3.14 \times 6^2 \times 3.5) \times (10^{-6} \times 10^5) \\
&= 395.64 \times 10^{-1} \\
&\approx 3.96 \times 10^1, \quad \text{or} \quad 39.6. \qquad \approx \text{means "is approximately"}
\end{aligned}
$$

The answer is that the volume of wire is about 39.6 m^3 (cubic meters).

DO EXERCISE 15.

16. Consider the following problem.

There are 300,000 words in the English language. The average person knows 10,000 of them. What part of the total number of words does the average person know?

Suppose someone asserts that the answer is about 3×10^1. Is this a correct answer? Explain.

Scientific notation can be useful in checking the possibility of a wrong answer.

EXAMPLE 10 Consider the following problem.

The distance from the earth to the sun is about 1.5×10^{11} m. The distance from the earth to the moon is about 3.8×10^8 m. How many times farther is it to the sun than to the moon?

Suppose someone asserted that the correct answer is about 2.5×10^{-3}. Is this a correct answer? Explain.

This is actually the reciprocal of the correct answer and is a number that is less than 1. We know this is incorrect because we know that the earth is farther away from the sun than it is from the moon.

DO EXERCISE 16.

EXERCISE SET 1.9

i Convert to scientific notation.

1. 47,000,000,000

2. 2,600,000,000,000

3. 863,000,000,000,000,000

4. 957,000,000,000,000,000

5. 0.000000016

6. 0.000000263

7. 0.00000000007

8. 0.00000000009

9. 407,000,000,000

10. 3,090,000,000,000

Convert to decimal notation.

11. 4×10^{-4}

12. 5×10^{-5}

13. 6.73×10^8

14. 9.24×10^7

15. 8.923×10^{-10}

16. 7.034×10^{-2}

17. 9.03×10^{10}

18. 1.01×10^{12}

ii Multiply and write scientific notation for the answer.

19. $(2.3 \times 10^6)(4.2 \times 10^{-11})$

20. $(6.5 \times 10^3)(5.2 \times 10^{-8})$

21. $(2.34 \times 10^{-8})(5.7 \times 10^{-4})$

22. $(3.26 \times 10^{-6})(8.2 \times 10^{-6})$

23. $(3.2 \times 10^6)(2.6 \times 10^4)$

24. $(3.11 \times 10^3)(1.01 \times 10^{13})$

25. $(3.01 \times 10^{-5})(6.5 \times 10^7)$

26. $(4.08 \times 10^{-10})(7.7 \times 10^5)$

Divide and write scientific notation for the answer.

27. $\dfrac{8.5 \times 10^8}{3.4 \times 10^5}$

28. $\dfrac{5.1 \times 10^6}{3.4 \times 10^3}$

29. $\dfrac{4.0 \times 10^{-6}}{8.0 \times 10^{-3}}$

30. $\dfrac{7.5 \times 10^{-9}}{2.5 \times 10^{-4}}$

31. $\dfrac{12.6 \times 10^8}{4.2 \times 10^{-3}}$

32. $\dfrac{3.2 \times 10^{-7}}{8.0 \times 10^8}$

33. $\dfrac{2.42 \times 10^5}{1.21 \times 10^{-5}}$

34. $\dfrac{9.36 \times 10^{-11}}{3.12 \times 10^{11}}$

iii Estimate.

35. $\dfrac{(6.1 \times 10^4)(7.2 \times 10^{-6})}{9.8 \times 10^{-4}}$

36. $\dfrac{(8.05 \times 10^{-11})(5.9 \times 10^7)}{3.1 \times 10^{14}}$

37. $\dfrac{780,000,000 \times 0.00071}{0.000005}$

38. $\dfrac{830,000,000 \times 0.12}{3,100,000}$

39. $\dfrac{43,000,000 \times 0.095}{63,000}$

40. $\dfrac{0.0073 \times 0.84}{0.000006}$

41. $\dfrac{40,000 \times 0.29}{0.057 \times 160,000}$

42. $\dfrac{80,000 \times 0.64}{0.032 \times 450,000}$

Estimate the order of magnitude of each computation.

43. $3,014,000 \times 0.0000072$

44. 0.000213×0.502

45. $\dfrac{47,812}{6.45}$

46. $\dfrac{5917}{0.076}$

47. $\dfrac{0.0125 \times 39,412}{375 \times 4112 \times 0.142}$

48. $\dfrac{4814 \times 0.0031 \times 7.2}{861 \times 17.6}$

iv *Problem solving*

Solve. Give answers in scientific notation.

49. A certain thin plastic sheet is used in many applications of building and landscaping. The sheet is packaged in rolls 1 m wide and 30 m long. The thickness of the sheet is 0.8 mm. Find the volume of plastic in a roll.

50. The distance that light travels in 100 yr is approximately 5.88×10^{14} mi. How far does light travel in 13 weeks?

51. The average distance of the earth from the sun is about 9.3×10^7 mi. About how far does the earth travel in a yearly orbit about the sun?

52. A *mil* is one thousandth of a dollar. The taxation rate in a certain school district is 5 mils for every dollar of assessed valuation. The assessed valuation for the district is 13.4 million dollars. How much tax revenue will be raised?

53. There are 2864 members of the Professional Bowlers Association. There are 234 million people in the United States. What part of the population are members of the Professional Bowlers Association? Write scientific notation for the answer.

54. Americans drink 3 million gallons of orange juice in one day. How much orange juice is consumed by Americans in one year? Write scientific notation for the answer.

55. The average discharge of water at the mouth of the Amazon River is 4,200,000 cubic feet per second. How much water is discharged in one hour?

56. There are 300,000 words in the English language. The exceptional person knows 20,000. What part of the total number of words does the exceptional person know?

Consider the following problem. Determine whether the asserted approximate answer seems correct. Explain.

Americans eat 6.5 million gallons of popcorn each day. How much popcorn do they eat in one year?

57. Asserted approximate answer: 2.8×10^7

58. Asserted approximate answer: 1.8×10^4

59. Asserted approximate answer: 2.8×10^8

60. Asserted approximate answer: 2.8×10^9

61. A *light year* is the distance traveled by light in one year. One light year is 5.88×10^{12} mi. The star Alpha Centauri is 2.4×10^{13} mi from earth. About how many light years is Alpha Centauri from earth?

62. The star Beta Centauri is 130 light years from earth. How many miles is Beta Centauri from the earth?

63. Compare $8 \cdot 10^{-90}$ and $9 \cdot 10^{-91}$. Which is the larger value? How much larger? Write scientific notation for the difference.

64. Find 8^{-x} when $8^{3x} = 27$.

65. 🔢 Evaluate: $(4096)^{0.05}(4096)^{0.2}$.

66. What is the unit's digit in $(513)^{127}$?

67. Given that $a = 4 \cdot 3^b$ and $c = 2 - 3^{-b}$, find c in terms of a.

Simplify.

68. $[7y(7-8)^{-2} - 8y(8-7)^{-2}]^{-(-2)^2}$

69. $\dfrac{3^{q+3} - 3^2(3^q)}{3(3^{q+4})}$

70. $[3^{-(3s+1)} - 3^{-(3s-1)} + 3^{-3s}]^{-2^2}$

SUMMARY AND REVIEW: CHAPTER 1

The following contains a summary of what you should be able to do after completing this chapter. The review exercises are for practice. Answers are at the back of the book. If you miss an exercise, restudy the section indicated alongside the answer.

You should be able to:

Evaluate simple algebraic expressions.

Evaluate.

1. $3a$ for $a = 5$

2. $\dfrac{x}{y}$ for $x = 12$ and $y = 2$

3. $\dfrac{2 \cdot p}{q}$ for $p = 20$ and $q = 8$

4. $\dfrac{x - y}{3}$ for $x = 17$ and $y = 5$

5. $n^3 + 1$ for $n = 2$

6. x^0 for $x = 6$

7. $(x + 1)^2$ for $x = 4$

8. $(y - 1)(y + 3)$ for $y = 5$

Calculate with and without parentheses.

Calculate and compare answers to 9–11.

9. $120 - 6^2 \div 4 + 8$

10. $(120 - 6^2) \div 4 + 8$

11. $(120 - 6^2) \div (4 + 8)$

Translate phrases to algebraic expressions.

Translate to an algebraic expression.

12. 8 less than z

13. Three times x

14. Nineteen percent of some number

15. A number x is 1 more than a smaller number. Write an expression for the smaller number.

Tell the meaning of exponential notation such as 4^6 and write expressions such as $y \cdot y \cdot y \cdot y$ in exponential notation.

16. What is the meaning of the expression $(2m)^3$?

17. Write $6z \cdot 6z$ in exponential notation.

Write equivalent expressions using the commutative laws, the associative laws, and the property of 1.

Use a commutative law to write an expression equivalent to each of the following.

18. $4 + y$

19. ab

20. $pq + 2$

Use an associative law to write an expression equivalent to each of the following.

21. $(3 + x) + 1$

22. $m \cdot (4 \cdot n)$

Use the commutative and associative laws to write three expressions equivalent to each of the following.

23. $(1 + m) + n$

24. $4 \cdot (x \cdot y)$

Find the prime factorization of a number and find the least common multiple of two or more numbers using both the list-of-multiples method and the prime-factorization method.

Find the prime factorization of each number.

25. 92

26. 1400

Find the LCM.

27. 12, 32

28. 5, 18, 45

Simplify fractional notation and multiply, add, subtract, and divide with fractional notation.

Simplify.

29. $\dfrac{20}{48}$

30. $\dfrac{10}{18}$

Compute and simplify.

31. $\dfrac{4}{9} + \dfrac{5}{12}$

32. $\dfrac{3}{4} \div 3$

33. $\dfrac{2}{3} - \dfrac{1}{15}$

34. $\dfrac{9}{10} \cdot \dfrac{16}{5}$

Convert between fractional, decimal, and percent notation.

35. Find decimal notation: 4.7%.

36. Find fractional notation: 60%.

37. Find percent notation and fractional notation: 0.886.

Find decimal and percent notation.

38. $\dfrac{5}{8}$

39. $\dfrac{29}{25}$

Tell which integers correspond to real-world situations.

40. Tell which integers correspond to this situation: Mike has a debt of $45 and Joe has $72 in his savings account.

Find the absolute value of any real number.

Find the absolute value.

41. $|-38|$

42. $|q|$ when $q = 7$

43. $\left|\dfrac{5}{2}\right|$

Graph real numbers on a number line and tell which of two numbers is greater writing a true sentence using < or >.

Graph each number on a number line.

44. -2.5

45. $\dfrac{8}{9}$

Write a true sentence using < or >.

46. $-3 \quad 10$

47. $-1 \quad -6$

48. $0.126 \quad -12.6$

49. $-\dfrac{2}{3} \quad -\dfrac{1}{10}$

Find the additive inverse and the reciprocal of a real number.

Find the additive inverse of each.

50. 3.8

51. $-\dfrac{3}{4}$

52. Find $-x$ when x is -34.

53. Find $-(-x)$ when x is 5.

Find the reciprocal.

54. $\dfrac{3}{8}$

55. -7

56. $-\dfrac{9}{8}$

Add, subtract, multiply, and divide real numbers.

Compute and simplify.

57. $4 + (-7)$

58. $-\dfrac{2}{3} + \dfrac{1}{12}$

59. $6 + (-9) + (-8) + 7$

60. $-3.8 + 5.1 + (-12) + (-4.3) + 10$

61. $-3 - (-7)$

62. $-\dfrac{9}{10} - \dfrac{1}{2}$

63. $-3.8 - 4.1$

64. $-9 \cdot (-6)$

65. $-2.7(3.4)$

66. $\dfrac{2}{3} \cdot \left(-\dfrac{3}{7}\right)$

67. $3 \cdot (-7) \cdot (-2) \cdot (-5)$

68. $35 \div (-5)$

69. $-5.1 \div 1.7$

70. $-\dfrac{3}{5} \div \left(-\dfrac{4}{5}\right)$

Solve problems using real numbers.

71. On first, second, and third downs a football team had these gains and losses: 5-yd gain, 12-yd loss, and 15-yd gain, respectively. Find the total gain (or loss).

72. Your total assets are $170. You borrow $300. What are your total assets now?

Use the distributive laws to multiply and factor algebraic expressions and to simplify expressions by collecting like terms.

Multiply.

73. $5(3x - 7)$

74. $-2(4x - 5)$

75. $10(0.4x + 1.5)$

76. $-8(3 - 6x)$

Factor.

77. $2x - 14$

78. $6x - 6$

79. $5x + 10$

80. $12 - 3x$

Collect like terms.

81. $11a + 2b - 4a - 5b$

82. $7x - 3y - 9x + 8y$

83. $6x + 3y - x - 4y$

84. $-3a + 9b + 2a - b$

Simplify expressions by removing parentheses and collecting like terms.

Remove parentheses and simplify.

85. $13 - 4 + 8 - (-2)$

86. $20 - 17 - 12 + 13 - (-4)$

87. $-5 - 3x + 8 - (-9x)$

88. $4y - 19 - (-7y) + 3$

89. $2a - (5a - 9)$

90. $3(b + 7) - 5b$

91. $3[11 - 3(4 - 1)]$

92. $2[6(y - 4) + 7]$

93. $[8(x + 4) - 10] - [3(x - 2) + 4]$

94. $5\{[6(x - 1) + 7] - [3(3x - 4) + 8]\}$

Rename a number with or without negative exponents, and use exponents in multiplying, dividing, and raising a power to a power.

95. Rename $\dfrac{1}{y^4}$ using a negative exponent.

96. Rename 5^{-3} using a positive exponent.

Simplify.

97. $x^{-6} \cdot x^4$

98. $\dfrac{t^{-2}}{t^{-11}}$

99. $7^{-5} \cdot 7^{-5}$

100. $\dfrac{4^{-7}}{4^8}$

101. $(8^3)^3$

102. $(3a^{-6})^4$

103. $(x^{-2}yz^7)^{-5}$

104. Suppose \$4000 is invested at 9%, compounded annually. How much is in the account at the end of 3 years?

Use scientific notation to multiply and divide large and small numbers.

Multiply.

105. $(4.5 \times 10^3)(2.4 \times 10^{-6})$

106. $(2.18 \times 10^8)(1.21 \times 10^{-10})$

Divide.

107. $\dfrac{2.4 \times 10^{-5}}{4.8 \times 10^6}$

108. $\dfrac{2.8 \times 10^{-3}}{7.0 \times 10^{-5}}$

Estimate.

109. $\dfrac{(2.03 \times 10^{-9})(5.5 \times 10^4)}{2.5 \times 10^{12}}$

110. $\dfrac{21{,}000{,}000 \times 0.027}{44{,}000}$

Solve problems using scientific notation.

Solve.

111. The average distance from the planet Jupiter to the sun is 4.8×10^8 mi. About how far does Jupiter travel in one orbit around the sun?

112. Simplify: $-\left|\frac{7}{8} - \left(-\frac{1}{2}\right) - \frac{3}{4}\right|$.

113. If $0.090909\ldots = \frac{1}{11}$ and $0.181818\ldots = \frac{2}{11}$, what rational number is named by each of the following?

a) $0.272727\ldots$ **b)** $0.909090\ldots$

114. Evaluate $a^{50} - 20a^{25}b^4 + 100b^8$ when $a = 1$ and $b = 2$.

115. Simplify: $(|2.7 - 3| + 3^2 - |-3|) \div (-3)$.

Simplify.

116. $\dfrac{8b^{2x-1}}{-2b^{x+4}}$

117. $(5x^{2a}y^{b-1})^{4c}$

TEST: CHAPTER 1

Evaluate.

1. $\dfrac{3x}{y}$ for $x = 10$ and $y = 5$

2. $x^2 - 5$ for $x = -8$

3. $(y + 3)(y - 4)$ for $y = 6$

4. x^1 for $x = 8$

5. Calculate: $200 - 2^3 + 5 + 10$.

6. Write an algebraic expression for 8 less than n.

Use a commutative law to write an expression equivalent to each of the following.

7. pq

8. $z + 3$

Use an associative law to write an expression equivalent to each of the following.

9. $x \cdot (4 \cdot y)$

10. $a + (b + 1)$

11. Write an expression equivalent to $\frac{3}{7}$ using $\frac{7}{7}$ as a name for 1.

12. Find the prime factorization of 300.

13. Find the LCM: 15, 24, 60.

Simplify.

14. $\dfrac{16}{24}$

15. $\dfrac{9}{15}$

Compute and simplify.

16. $\dfrac{10}{27} \div \dfrac{8}{3}$

17. $\dfrac{9}{10} - \dfrac{5}{8}$

Write a true sentence using $<$ or $>$.

18. $-4 \quad 0$

19. $-3 \quad -8$

20. $-0.78 \quad -0.87$

21. $-\dfrac{1}{8} \quad \dfrac{1}{2}$

Find the absolute value.

22. $|-7|$

23. $\left|\dfrac{9}{4}\right|$

Find the additive inverse of each.

24. $\dfrac{2}{3}$

25. -1.4

26. Find $-x$ when x is -8.

Find the reciprocal.

27. -2

28. $\dfrac{4}{7}$

Compute and simplify.

29. $3.1 + (-4.7)$

30. $-8 + 4 + (-7) + 3$

31. $-\dfrac{1}{5} + \dfrac{3}{8}$

32. $2 - (-8)$

33. $3.2 - 5.7$

34. $\dfrac{1}{8} - \left(-\dfrac{3}{4}\right)$

35. $4 \cdot (-12)$

36. $-\dfrac{1}{2} \cdot \left(-\dfrac{3}{8}\right)$

37. $-45 \div 5$

38. $-\dfrac{3}{5} \div \left(-\dfrac{4}{5}\right)$

39. $4.864 \div (-0.5)$

40. $6 + 7 - 4 - (-3)$

41. Wendy had \$43 in her savings account. She withdrew \$25. Then she made a deposit of \$30. How much was in her savings account?

42. Find decimal notation for $\frac{17}{22}$.

43. Find percent notation for $\frac{37}{40}$.

44. Find decimal notation for $66\frac{2}{3}\%$.

45. Find fractional notation for 73.1.

Multiply.

46. $3(6 - x)$

47. $-5(y - 1)$

Factor.

48. $12 - 22x$

49. $7x + 21 + 14y$

Simplify.

50. $5x - (3x - 7)$

51. $4(2a - 3b) + a - 7$

52. $-5[2(x - 3) + 4] + 6(x - 5)$

53. $4\{3[5(y - 3) + 9] + 2(y + 8)\}$

54. $x^{-6} \cdot x^{12}$

55. $(2x^{-2}y^5)^4$

56. $\dfrac{x^{-14}}{x^{-7}}$

57. $(9 + 1)^2 - 23 \cdot 5$

Multiply.

58. $(2.9 \times 10^8)(6.1 \times 10^{-4})$

59. $(9.05 \times 10^{-3})(2.22 \times 10^{-5})$

60. $\dfrac{3.6 \times 10^7}{7.2 \times 10^{-4}}$

61. $\dfrac{1.8 \times 10^{-4}}{4.8 \times 10^{-7}}$

62. Estimate: $\dfrac{43{,}000{,}000 \times 0.064}{27{,}000}$.

63. The average distance from the planet Venus to the sun is 6.7×10^7 mi. About how far does Venus travel in one orbit around the sun?

Simplify.

64. $\dfrac{-27a^{x-1}}{3a^{x-2}}$

65. $\dfrac{(-2)^7}{(-4)^2}$

66. $\dfrac{9xy}{15yz}$

67. $\dfrac{13{,}860}{42{,}000}$

68. $4 \cdot 4 \cdot 4 \div 4 \cdot 4 \div 2 \cdot 2$

69. Evaluate $\dfrac{5y^2 - x}{4}$ when $x = 20$ and $y = -4$.

2

SOLVING EQUATIONS AND PROBLEMS

The price of an automobile was decreased to a sale price of $13,559. This was a 9% reduction. What was the original price of the automobile? An equation can be used to solve such a problem.

In this chapter we formulate several equation-solving principles. Then we apply these equation-solving skills throughout the remainder of the chapter. We first apply them to the manipulation of formulas. Formulas arise in many kinds of applications. The skill of solving a formula for a certain letter is particularly important. In the last three sections of the chapter we use our new equation-solving ability to solve problems. First, we consider a five-step problem-solving process and apply it to some fairly easy problems. Then in the next section we apply it to the solving of percent problems. In the last section we consider more complicated problems.

OBJECTIVES

After finishing Section 2.1, you should be able to:

i Solve equations using the addition principle.

ii Solve equations involving absolute value like $|x| + a = b$.

iii Solve equations using the multiplication principle.

iv Solve equations involving absolute value like $|ax| = b$.

Translate each problem to an equation.

1. 536.25 is 8.25 times what number?

2. In a recent year Mike Schmidt, third baseman for the Philadelphia Phillies, made $40,000 more than Jim Rice, outfielder for the Boston Red Sox. Schmidt made $2,130,000. How much did Rice make?

Determine whether the equation is true, false, or neither.

3. $3 \cdot 5 + 2 = 13$

4. $4 \cdot 2 - 3 = 5$

5. $y + 5 = 6$

6. $34t = 76$

2.1

THE ADDITION AND MULTIPLICATION PRINCIPLES

You have had practice in translating to algebraic expressions. In this section, you will extend that skill. You will learn to translate a problem situation to a mathematical sentence called an *equation*. The skills you have already learned will be most useful, and when you learn this extension of those skills, you will be in a stronger position to attack problem solving.

First, we define what we mean by an equation. An *equation* is a number sentence with an equals sign, $=$, for its verb. Here are some examples:

$$3 + 2 = 5, \qquad 7 - 3 = 4, \qquad x + 6 = 13, \qquad 3x - 2 = 7 - x.$$

It often happens that an equation will have some algebraic expression with variables on one or both sides.

Translating to Equations

We now look at some examples of problem situations and how they can be translated to an equation.

EXAMPLE 1 Translate the following problem to an equation.

What number plus 478 is 1019?

We will use a variable to represent "what number." In this example, the translation comes almost directly from the English sentence.

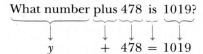

$$y \qquad + \quad 478 = 1019$$

Note that "is" translates to "$=$" and "plus" translates to "$+$".

Sometimes it helps to reword a problem before translating.

EXAMPLE 2 When 54 is multiplied by a number, the result is 7896. Find the number.

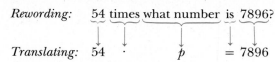

Rement: 54 times what number is 7896?

Translating: 54 $\cdot$ p $= 7896$

Note that "times" translates to "$\cdot$".

DO EXERCISES 1 AND 2.

i Solutions of Equations

An equation says that the symbols on either side of the equals sign, $=$, stand for the same number. Equations may be true. They may be false. They may be neither true nor false.

EXAMPLES Determine whether these equations are true, false, or neither.

3. $3 + 2 = 5$ The equation is *true*.

4. $7 - 2 = 4$ The equation is *false*.

5. $x + 6 = 13$ The equation is *neither* true nor false, because we do not know what number x represents.

DO EXERCISES 3–6.

Any replacement for the variable that makes an equation true is called a *solution* of the equation. To solve an equation means to find *all* of its solutions.

One way to solve equations is by trial.

EXAMPLE 6   Solve $x + 6 = 13$ by trial.

If we replace x by 2 we get $2 + 6 = 13$, a false equation.
If we replace x by 8 we get $8 + 6 = 13$, a false equation.
If we replace x by 7 we get $7 + 6 = 13$, a true equation.

No other number makes the equation true, so the only solution is the number 7.

DO EXERCISES 7–11.

An equation $a = b$ says that a and b stand for the same number. Suppose this is true, and we add a number c to the number a. We get the same answer if we add c to b, because a and b are the same number.

THE ADDITION PRINCIPLE

If an equation $a = b$ is true, then

$$a + c = b + c$$

is true for any number c.

The idea in using this principle is to obtain an equation for which you can *see* what the solution is.

EXAMPLE 7   Solve: $x + 5 = -7$.

$$x + 5 = -7$$
$$x + 5 + (-5) = -7 + (-5) \quad \text{Using the addition principle;}$$
$$\text{adding } -5 \text{ on both sides}$$
$$x + 0 = -7 + (-5) \quad \text{Simplifying}$$
$$x = -12$$

We can see that the solution of $x = -12$ is the number -12. To check the answer we substitute -12 in the original equation.

Check:
$$\begin{array}{c|c} x + 5 = -7 \\ \hline -12 + 5 & -7 \\ -7 & \end{array}$$

The solution of the original equation is -12.

In Example 7, to get x alone, we added the inverse of 5. This "got rid of" the 5 on the left. When using the addition principle, we sometimes say we "add the same number on both sides of an equation." We started with $x + 5 = -7$, and using the addition principle we derived a simple equation $x = -12$ for which it was easy to "*see*" the solution. Equations with the same solutions, such as $x + 5 = -7$ and $x = -12$, are called *equivalent equations*.

DO EXERCISE 12.

---

7. Find three replacements that make the equation $x + 5 = 12$ false.

8. Find the replacement that makes the equation $x + 5 = 12$ true.

Solve by trial.

9. $x + 4 = 10$

10. $3x = 12$

11. $3y + 1 = 16$

Solve using the addition principle.

12. $x + 7 = 2$

Solve using the addition principle.

13. $n - 4.5 = 8.7$

14. $x + \dfrac{1}{2} = -\dfrac{3}{2}$

15. $7.6 = -3.2 + y$

The addition principle also allows us to subtract on both sides of an equation. This is because subtracting is the same as adding an inverse. In Example 7 we could have subtracted 5 on both sides of the equation.

Now we "undo" a subtraction using the addition principle.

EXAMPLE 8 Solve: $y - 8.4 = -6.5$.

$$y - 8.4 = -6.5$$
$$y - 8.4 + 8.4 = -6.5 + 8.4 \qquad \text{Adding 8.4 to get rid of } -8.4 \text{ on the left}$$
$$y = 1.9$$

Check:

$$\begin{array}{c|c} y - 8.4 = -6.5 \\ \hline 1.9 - 8.4 & -6.5 \\ -6.5 & \end{array}$$

The solution is 1.9.

EXAMPLE 9 Solve: $x - \frac{2}{3} = 2\frac{1}{2}$.

$$x - \tfrac{2}{3} = 2\tfrac{1}{2}$$
$$x - \tfrac{2}{3} + \tfrac{2}{3} = 2\tfrac{1}{2} + \tfrac{2}{3} \qquad \text{Adding } \tfrac{2}{3}$$
$$x = 2\tfrac{1}{2} + \tfrac{2}{3} \qquad \text{Simplifying}$$
$$x = \tfrac{5}{2} + \tfrac{2}{3} \qquad \text{Converting to fractional notation to carry out the addition}$$
$$x = \tfrac{5}{2} \cdot \tfrac{3}{3} + \tfrac{2}{3} \cdot \tfrac{2}{2} \qquad \text{Multiplying by 1 to obtain the least common denominator}$$
$$x = \tfrac{15}{6} + \tfrac{4}{6}$$
$$x = \tfrac{19}{6}$$

The check is left to the student. The solution is $\frac{19}{6}$.

Note that we can solve the equation $2\frac{1}{2} = x - \frac{2}{3}$ in the same manner as Example 9, except that at the end we would have $\frac{19}{6} = x$. The solution is still $\frac{19}{6}$. In general, if an equation $a = b$ is true, then it is reversible. That is, $b = a$ is also true.

DO EXERCISES 13–15.

ii Solving Equations with Absolute Value

There are equations that have more than one solution. Examples are equations with absolute value.

EXAMPLE 10 Solve: $|x| = 4$.

There are two numbers whose absolute value is 4. They are -4 and 4. We can see this on a number line. We are determining those numbers whose distance from 0 is 4.

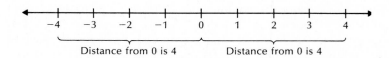

The solutions are -4 and 4.

THE ABSOLUTE VALUE PRINCIPLE

For any nonnegative real number a, if

$$|x| = a, \quad \text{then} \quad x = -a \text{ or } x = a.$$

The solutions are $-a$ and a.

Solve.

16. $|x| = 3$

When $a = 0$, there is only one solution, 0. Otherwise, there are two solutions.

EXAMPLE 11 Solve: $|x| + 5 = 9$.

We first use the addition principle to get $|x|$ by itself. Then we use the absolute value principle:

$$|x| + 5 = 9$$
$$|x| + 5 - 5 = 9 - 5 \qquad \text{Adding } -5$$
$$|x| = 4$$
$$x = -4 \quad \text{or} \quad x = 4. \qquad \text{Using the absolute value principle}$$

The solutions are -4 and 4.

17. $|x| + 10 = 37$

DO EXERCISES 16–18.

▓ The Multiplication Principle

Another principle for solving equations, similar to the addition principle, uses multiplication.

Suppose $a = b$ is true, and we multiply a by some number c. We get the same answer if we multiply b by c, because a and b are the same number.

THE MULTIPLICATION PRINCIPLE

If an equation $a = b$ is true, then

$$a \cdot c = b \cdot c$$

is true for any number c.

18. $|x| - \dfrac{1}{4} = 1$

When using the multiplication principle, we sometimes say that we "multiply on both sides by the same number."

EXAMPLE 12 Solve: $3x = 9$.

$$3x = 9$$
$$\tfrac{1}{3} \cdot 3x = \tfrac{1}{3} \cdot 9 \qquad \text{Using the multiplication principle; multiplying by } \tfrac{1}{3} \text{ on both sides}$$
$$1 \cdot x = 3 \qquad \text{Simplifying}$$
$$x = 3$$

It is easy to see that the solution of $x = 3$ is 3.

Check:

$$\begin{array}{c|c} 3x = 9 \\ \hline 3 \cdot 3 & 9 \\ 9 & \end{array}$$

The solution of the original equation is 3.

Solve using the multiplication principle.

19. $5x = 25$

20. $4x = -7$

21. $-3y = -42$

Solve using the multiplication principle.

22. $\dfrac{5}{6} = -\dfrac{2}{3}x$

Solve.

23. $-x = 6$

In Example 12, to get x alone, we multiplied by the reciprocal of 3. When we multiplied we got $1 \cdot x$, which simplified to x. This enabled us to "get rid of" the 3 on the left. The multiplication principle also allows us to divide on both sides of an equation by a nonzero number. This is because division by a number c is the same as multiplying by a reciprocal.

DO EXERCISES 19–21.

EXAMPLE 13   Solve: $\frac{3}{8} = -\frac{5}{4}x$.

$$\frac{3}{8} = -\frac{5}{4}x$$

$$-\frac{4}{5} \cdot \frac{3}{8} = -\frac{4}{5} \cdot \left(-\frac{5}{4}x\right) \qquad \text{Multiplying by } -\frac{4}{5} \text{ to get rid of } -\frac{5}{4} \text{ on the right (this is the same as dividing by } -\frac{5}{4})$$

$$-\frac{3}{10} = x \qquad \text{Simplifying}$$

Check:

$$\begin{array}{c|c} \frac{3}{8} = -\frac{5}{4}x \\ \hline \frac{3}{8} & -\frac{5}{4}\left(-\frac{3}{10}\right) \\ & \frac{3}{8} \end{array}$$

The solution is $-\frac{3}{10}$.

DO EXERCISE 22.

EXAMPLE 14   Solve: $-x = 9$.

$$-x = 9$$

$$-1 \cdot x = 9 \qquad \text{Using the property of } -1$$

$$-1 \cdot (-1 \cdot x) = -1 \cdot 9 \qquad \text{Multiplying on both sides by } -1, \text{ the reciprocal of itself, or dividing by } -1$$

$$1 \cdot x = -9$$

$$x = -9$$

Check:

$$\begin{array}{c|c} -x = 9 \\ \hline -(-9) & 9 \\ 9 & \end{array}$$

The solution is -9.

DO EXERCISE 23.

Now we "undo" a division using the multiplication principle.

EXAMPLE 15   Solve: $\dfrac{-y}{9} = 14$.

$$\frac{-y}{9} = 14$$

$$9 \cdot \frac{-y}{9} = 9 \cdot 14 \qquad \text{Think of } \frac{-y}{9} \text{ as } \frac{1}{9} \cdot -y; \text{ multiply by 9 on both sides.}$$

$$-y = 126$$

$$y = -126$$

Check: $\dfrac{-y}{9} = 14$

$$\dfrac{-(-126)}{9} \,\Big|\, 14$$

$$\dfrac{126}{9}$$

$$14$$

The solution is -126.

DO EXERCISES 24 AND 25.

EXAMPLE 16 Solve: $1.16y = 9744$.

$$1.16y = 9744$$

$$\tfrac{1}{1.16} \cdot (1.16y) = \tfrac{1}{1.16} \cdot (9744) \qquad \text{Multiplying by } \tfrac{1}{1.16} \text{ (or dividing by 1.16)}$$

$$y = \dfrac{9744}{1.16}$$

$$y = 8400$$

Check: $1.16y = 9744$

$$\overline{1.16(8400) \,\Big|\, 9744}$$

$$9744 \,\Big|$$

The solution is 8400.

DO EXERCISE 26.

Note that equations are reversible. That is, if $a = b$ is true, then $b = a$ is true. Thus, when we solve $15 = 3x$, we can reverse it and solve $3x = 15$ if we wish.

iv Other Equations with Absolute Value

EXAMPLE 17 Solve: $|3x| = 6$.

To solve this equation, we first use the absolute value principle. Then we solve the equations separately.

$$|3x| = 6$$

$$3x = -6 \quad \text{or} \quad 3x = 6 \qquad \text{Using the absolute value principle}$$

$$x = -2 \quad \text{or} \quad x = 2 \qquad \text{Solving each equation; multiplying by } \tfrac{1}{3}$$

The solutions are -2 and 2.

DO EXERCISES 27 AND 28.

Solve.

24. $\dfrac{-x}{5} = 10$

25. $-14 = -\dfrac{y}{2}$

Solve.

26. $-2.1y = 6.3$

Solve.

27. $|4x| = 12$

28. $|25x| = 16$

EXERCISE SET 2.1

i Solve using the addition principle. Don't forget to check!

1. $x + 2 = 6$

2. $x + 5 = 8$

3. $x + 15 = -5$

4. $y + 9 = 43$

5. $x + 6 = -8$

6. $t + 9 = -12$

7. $x + 16 = -2$

8. $y + 25 = -6$

9. $x - 9 = 6$

10. $x - 8 = 5$

11. $x - 7 = -21$

12. $x - 3 = -14$

13. $5 + t = 7$

14. $8 + y = 12$

15. $-7 + y = 13$

16. $-9 + z = 15$

17. $-3 + t = -9$

18. $-6 + y = -21$

19. $r + \tfrac{1}{3} = \tfrac{8}{3}$

20. $t + \tfrac{3}{8} = \tfrac{5}{8}$

21. $m + \tfrac{5}{6} = -\tfrac{11}{12}$

22. $x + \tfrac{2}{3} = -\tfrac{5}{6}$

23. $x - \tfrac{5}{6} = \tfrac{7}{8}$

24. $y - \tfrac{3}{4} = \tfrac{5}{6}$

25. $-\frac{1}{5} + z = -\frac{1}{4}$ **26.** $-\frac{1}{8} + y = -\frac{3}{4}$ **27.** $x + 2.3 = 7.4$ **28.** $y + 4.6 = 9.3$

29. $x - 4.8 = 7.6$ **30.** $y - 8.3 = 9.5$ **31.** $-9.7 = -4.7 + y$ **32.** $-7.8 = 2.8 + x$

33. $5\frac{1}{6} + x = 7$ **34.** $5\frac{1}{4} = 4\frac{2}{3} + x$ **35.** $q + \frac{1}{3} = -\frac{1}{7}$ **36.** $47\frac{1}{8} = -76 + z$

iii Solve using the multiplication principle. Don't forget to check!

37. $6x = 36$ **38.** $3x = 39$ **39.** $5x = 45$ **40.** $9x = 72$

41. $84 = 7x$ **42.** $56 = 8x$ **43.** $-x = 40$ **44.** $100 = -x$

45. $-x = -1$ **46.** $-68 = -r$ **47.** $7x = -49$ **48.** $9x = -36$

49. $-12x = 72$ **50.** $-15x = 105$ **51.** $-21x = -126$ **52.** $-13x = -104$

53. $\frac{t}{7} = -9$ **54.** $\frac{y}{-8} = 11$ **55.** $\frac{3}{4}x = 27$ **56.** $\frac{4}{5}x = 16$

57. $\frac{-t}{3} = 7$ **58.** $\frac{-x}{6} = 9$ **59.** $-\frac{m}{3} = \frac{1}{5}$ **60.** $\frac{1}{9} = -\frac{z}{7}$

61. $-\frac{3}{5}r = -\frac{9}{10}$ **62.** $-\frac{2}{5}y = -\frac{4}{15}$ **63.** $-\frac{3}{2}r = -\frac{27}{4}$ **64.** $\frac{5}{7}x = -\frac{10}{14}$

65. $6.3x = 44.1$ **66.** $2.7y = 54$ **67.** $-3.1y = 21.7$ **68.** $-3.3y = 6.6$

69. $38.7m = 309.6$ **70.** $29.4x = 235.2$ **71.** $-\frac{2}{3}y = -10.6$ **72.** $-\frac{9}{7}y = 12.06$

ii , **iv** Solve.

73. $|x| = 5$ **74.** $|x| = 4.5$ **75.** $|x| + 7 = 18$ **76.** $|x| - 2 = 6.3$

77. $678 = 289 + |t|$ **78.** $-567 = -1000 + |x|$ **79.** $|5x| = 40$ **80.** $|2y| = 18$

81. $|3x| - 4 = 17$ **82.** $|6x| + 8 = 62$

---

83. 🖩 Solve: $-356.788 = -699.034 + t$.

Solve.

84. $8 - 25 = 8 + x - 21$ **85.** $16 + x - 22 = -16$ **86.** $x + x = x$

87. $x + 3 = 3 + x$ **88.** $x + 4 = 5 + x$

Solve for x.

89. $x + 7 = b + 10$ **90.** $1 - c = a + x$ **91.** $x - 4 = a$ **92.** $|x| = -5$

93. Solve: $-\frac{3}{2} + x = -\frac{5}{17} - \frac{3}{2}$.   **94.** If $x - 4720 = 1634$, find $x + 4720$.

95. Explain why it is not necessary to prove a subtraction principle: If $a = b$ is true, then

$$a - c = b - c \text{ is true,}$$

for any number c.

96. Solve: $|x| + 17 = 4$.

97. 🖩 Solve: $-0.2344m = 2028.732$.

Solve.

98. $0 \cdot x = 0$ **99.** $0 \cdot x = 9$

Solve for x.

100. $ax = 5a$ **101.** $3x = \frac{b}{a}$ **102.** $cx = a^2 + 1$

103. $\frac{a}{b}x = 4$ **104.** $|x| + 37 = 4$ **105.** $2|x| = -12$

106. Determine whether you can square both sides of an equation and get an equivalent equation.

107. A student makes a calculation and gets an answer of 22.5. On the last step the student multiplies by 0.3 when a division by 0.3 should have been done. What should the correct answer be?

Solve.

108. $-\frac{4}{5} + \frac{7}{10} = x - \frac{3}{4}$ **109.** $-\frac{2}{3} - \frac{4}{15} = x + \frac{4}{5}$

2.2

USING THE PRINCIPLES TOGETHER

In this section we consider equation solving where we may need to use both the addition and multiplication principles. We also consider equations where collecting like terms is useful.

i Applying Both Principles

Let's consider an equation where we apply both principles. We usually apply the addition principle first. Then we apply the multiplication principle.

EXAMPLE 1 Solve: $3x + 4 = 13$.

$$3x + 4 = 13$$
$$3x + 4 + (-4) = 13 + (-4) \quad \text{Using the addition principle;}$$
$$\text{adding } -4 \text{ on both sides}$$
$$3x = 9 \quad \text{Simplifying}$$
$$\tfrac{1}{3} \cdot 3x = \tfrac{1}{3} \cdot 9 \quad \text{Using the multiplication principle;}$$
$$\text{multiplying by } \tfrac{1}{3} \text{ on both sides}$$
$$x = 3 \quad \text{Simplifying}$$

Check:
$$\begin{array}{c|c} 3x + 4 = 13 \\ \hline 3 \cdot 3 + 4 & 13 \\ 9 + 4 & \\ 13 & \end{array}$$

The solution is 3.

DO EXERCISE 1.

EXAMPLE 2 Solve: $-5x - 6 = 16$.

$$-5x - 6 = 16$$
$$-5x - 6 + 6 = 16 + 6 \quad \text{Adding 6 on both sides}$$
$$-5x = 22$$
$$-\tfrac{1}{5} \cdot (-5x) = -\tfrac{1}{5} \cdot 22 \quad \text{Multiplying by } -\tfrac{1}{5} \text{ on both sides}$$
$$x = -\tfrac{22}{5} \quad \text{or} \quad -4\tfrac{2}{5}$$

Check:
$$\begin{array}{c|c} -5x - 6 = 16 \\ \hline -5(-\tfrac{22}{5}) - 6 & 16 \\ 22 - 6 & \\ 16 & \end{array}$$

The solution is $-\tfrac{22}{5}$.

DO EXERCISES 2 AND 3.

Solve.

1. $9x + 6 = 51$

Solve.

2. $-8x - 4 = 28$

3. $-\dfrac{1}{2}x + 3 = 1$

Solve.

4. $-18 - x = -57$

Solve.

5. $-4 - 8x = 8$

6. $41.68 = 4.7 - 8.6y$

Solve.

7. $4x + 3x = -21$

8. $x - 0.09x = 728$

EXAMPLE 3 Solve: $45 - x = 13$.

$$45 - x = 13$$
$$-45 + 45 - x = -45 + 13 \qquad \text{Adding } -45 \text{ on both sides}$$
$$-x = -32$$
$$-1 \cdot x = -32 \qquad \text{Using the property of } -1: -x = -1 \cdot x$$
$$x = \frac{-32}{-1} \qquad \text{Dividing on both sides by } -1. \text{ (You could have multiplied on both sides by } -1 \text{ instead. That would change the sign on both sides.)}$$
$$x = 32$$

Check:

$$\begin{array}{c|c} 45 - x = 13 \\ \hline 45 - 32 & 13 \\ 13 & \end{array}$$

The solution is 32.

DO EXERCISE 4.

EXAMPLE 4 Solve: $16.3 - 7.2y = -8.18$.

$$16.3 - 7.2y = -8.18$$
$$-7.2y = -16.3 + (-8.18) \qquad \text{Adding } -16.3 \text{ on both sides}$$
$$-7.2y = -24.48$$
$$y = \frac{-24.48}{-7.2} \qquad \text{Dividing by } -7.2 \text{ on both sides}$$
$$y = 3.4$$

Check:

$$\begin{array}{c|c} 16.3 - 7.2y = -8.18 \\ \hline 16.3 - 7.2(3.4) & -8.18 \\ 16.3 - 24.48 & \\ -8.18 & \end{array}$$

The solution is 3.4.

DO EXERCISES 5 AND 6.

ⅱ Collecting Like Terms

If there are like terms on one side of the equation, we collect them before using the principles.

EXAMPLE 5 Solve: $3x + 4x = -14$.

$$3x + 4x = -14$$
$$7x = -14 \qquad \text{Collecting like terms}$$
$$\tfrac{1}{7} \cdot 7x = \tfrac{1}{7} \cdot (-14)$$
$$x = -2$$

The number -2 checks, so the solution is -2.

DO EXERCISES 7 AND 8.

If there are like terms on opposite sides of the equation, we get them on the same side by using the addition principle. Then we collect them.

EXAMPLE 6 Solve: $2x - 2 = -3x + 3$.

$$2x - 2 = -3x + 3$$

$$2x - 2 + 2 = -3x + 3 + 2 \qquad \text{Adding 2}$$

$$2x = -3x + 5 \qquad \text{Simplifying}$$

$$2x + 3x = -3x + 3x + 5 \qquad \text{Adding } 3x$$

$$5x = 5 \qquad \text{Collecting like terms and simplifying}$$

$$\tfrac{1}{5} \cdot 5x = \tfrac{1}{5} \cdot 5 \qquad \text{Multiplying by } \tfrac{1}{5}$$

$$x = 1 \qquad \text{Simplifying}$$

Check:

$$
\begin{array}{c|c}
\multicolumn{2}{c}{2x - 2 = -3x + 3} \\
\hline
2 \cdot 1 - 2 & -3 \cdot 1 + 3 \\
2 - 2 & -3 + 3 \\
0 & 0
\end{array}
$$

The solution is 1.

In Example 6, we used the addition principle to get all terms with a variable on one side and all numbers on the other side. Then we collected like terms and proceeded as before. If there are like terms on one side at the outset, they should be collected first.

DO EXERCISE 9.

EXAMPLE 7 Solve: $6x + 5 - 7x = 10 - 4x + 3$.

$$6x + 5 - 7x = 10 - 4x + 3$$

$$-x + 5 = 13 - 4x \qquad \text{Collecting like terms}$$

$$-x + 4x = 13 - 5 \qquad \text{Adding } 4x \text{ and subtracting 5 to get all terms with variables on one side and all other terms on the other}$$

$$3x = 8 \qquad \text{Collecting like terms}$$

$$\tfrac{1}{3} \cdot 3x = \tfrac{1}{3} \cdot 8 \qquad \text{Multiplying by } \tfrac{1}{3}$$

$$x = \tfrac{8}{3} \qquad \text{Simplifying}$$

The number $\tfrac{8}{3}$ checks, so it is the solution.

DO EXERCISES 10–12.

iii Using the Multiplication Principle First

We have stated that we generally use the addition principle first. In some situations it is to our advantage to use the multiplication principle first. If we multiply by 4 on both sides of $\tfrac{1}{2}x = \tfrac{3}{4}$, we get $2x = 3$, which has no fractions. We have "cleared the fractions." If we multiply by 10 on both sides of $2.3x = 5$, we get $23x = 50$, which has no decimal points. We have "cleared the decimals." The equations are then easier to solve. It is your choice whether to clear the fractions or decimals, but doing so eases computations.

In what follows we use the multiplication principle first to "clear," or "get rid of," fractions or decimals. For fractions, the number we multiply by is either the product of all the denominators or the least common multiple of all the denominators.

Solve.

9. $7y + 5 = 2y + 10$

Solve.

10. $5 - 2y = 3y - 5$

11. $7x - 17 + 2x = 2 - 8x + 15$

12. $3x - 15 = 5x + 2 - 4x$

Solve.

13. $\dfrac{7}{8}x - \dfrac{1}{4} + \dfrac{1}{2}x = \dfrac{3}{4} + x$

EXAMPLE 8 Solve: $\frac{2}{3}x - \frac{1}{6} + \frac{1}{2}x = \frac{7}{6} + 2x$.

The number 6 is the least common multiple of all the denominators. We multiply on both sides by 6.

$$6(\tfrac{2}{3}x - \tfrac{1}{6} + \tfrac{1}{2}x) = 6(\tfrac{7}{6} + 2x) \qquad \text{Multiplying by 6 on both sides}$$

$$6 \cdot \tfrac{2}{3}x - 6 \cdot \tfrac{1}{6} + 6 \cdot \tfrac{1}{2}x = 6 \cdot \tfrac{7}{6} + 6 \cdot 2x \qquad \begin{array}{l}\text{Using the distributive laws (\textit{Caution}! Be} \\ \text{sure to multiply all the terms by 6.)}\end{array}$$

$$4x - 1 + 3x = 7 + 12x \qquad \begin{array}{l}\text{Simplifying. Note that the fractions} \\ \text{are cleared.}\end{array}$$

$$7x - 1 = 7 + 12x \qquad \text{Collecting like terms}$$

$$7x - 12x = 7 + 1 \qquad \begin{array}{l}\text{Subtracting } 12x \text{ and adding 1 to get all} \\ \text{the terms with variables on one side} \\ \text{and all the other terms on the other}\end{array}$$

$$-5x = 8 \qquad \text{Collecting like terms}$$

$$-\tfrac{1}{5} \cdot (-5x) = -\tfrac{1}{5} \cdot 8 \qquad \text{Multiplying by } -\tfrac{1}{5} \text{ or dividing by } -5$$

$$x = -\tfrac{8}{5}$$

The number $-\frac{8}{5}$ checks and is the solution.

DO EXERCISE 13.

Here is a procedure for solving the equations in this section.

---

1. **Multiply on both sides to clear the equation of fractions or decimals. (This is optional, but it can ease computations.)**
2. **Collect like terms on each side, if necessary.**
3. **Get all terms with variables on one side and all the other terms on the other side.**
4. **Collect like terms again, if necessary.**
5. **Multiply or divide to solve for the variable.**

Solve.

14. $41.68 = 4.7 - 8.6y$

We illustrate this by repeating Example 4, but we clear the equation of decimals first.

EXAMPLE 9 Solve: $16.3 - 7.2y = -8.18$.

The greatest number of decimal places in any one number is two. Multiplying by 100, which has two 0s, will clear the decimals.

$$100(16.3 - 7.2y) = 100(-8.18) \qquad \text{Multiplying by 100 on both sides}$$

$$100(16.3) - 100(7.2y) = 100(-8.18) \qquad \text{Using a distributive law}$$

$$1630 - 720y = -818 \qquad \text{Simplifying}$$

$$-720y = -818 - 1630 \qquad \text{Subtracting 1630 on both sides}$$

$$-720y = -2448 \qquad \text{Collecting like terms}$$

$$y = \tfrac{-2448}{-720} = 3.4 \qquad \text{Dividing by } -720 \text{ on both sides}$$

The number 3.4 checks and is the solution.

DO EXERCISE 14.

EXAMPLE 10 Solve: $|3x + 2| = 17$.

We first use the absolute value principle. Then we solve each equation individually using the addition and multiplication principles.

$$|3x + 2| = 17$$

$3x + 2 = -17$ or	$3x + 2 = 17$	Using the absolute value principle
$3x = -19$ or	$3x = 15$	Adding -2
$x = \frac{-19}{3}$ or	$x = \frac{15}{3}$	Multiplying by $\frac{1}{3}$
$x = -\frac{19}{3}$ or	$x = 5$	

The solutions are $-\frac{19}{3}$ and 5.

DO EXERCISES 15 AND 16.

Solve.

15. $|5x + 4| = 11$

16. $|3x - 2| = 20$

EXERCISE SET 2.2

i Solve and check.

1. $5x + 6 = 31$ **2.** $3x + 6 = 30$ **3.** $8x + 4 = 68$ **4.** $7z + 9 = 72$

5. $4x - 6 = 34$ **6.** $6x - 3 = 15$ **7.** $3x - 9 = 33$ **8.** $5x - 7 = 48$

9. $7x + 2 = -54$ **10.** $5x + 4 = -41$ **11.** $6y + 3 = -45$ **12.** $9t + 8 = -91$

13. $-4x + 7 = 35$ **14.** $-5x - 7 = 108$ **15.** $-7x - 24 = -129$ **16.** $-6z - 18 = -132$

17. $-4x + 71 = -1$ **18.** $-8y + 83 = -85$

ii Solve and check.

19. $5x + 7x = 72$ **20.** $4x + 5x = 45$ **21.** $8x + 7x = 60$ **22.** $3x + 9x = 96$

23. $4x + 3x = 42$ **24.** $6x + 19x = 100$ **25.** $4y - 2y = 10$ **26.** $8y - 5y = 48$

27. $-6y - 3y = 27$ **28.** $-4y - 8y = 48$ **29.** $-7y - 8y = -15$ **30.** $-10y - 3y = -39$

31. $10.2y - 7.3y = -58$ **32.** $6.8y - 2.4y = -88$ **33.** $x + \frac{1}{3}x = 8$ **34.** $x + \frac{1}{4}x = 10$

35. $8y - 35 = 3y$ **36.** $4x - 6 = 6x$ **37.** $4x - 7 = 3x$ **38.** $9x - 6 = 3x$

39. $8x - 1 = 23 - 4x$ **40.** $5y - 2 = 28 - y$ **41.** $2x - 1 = 4 + x$ **42.** $5x - 2 = 6 + x$

43. $6x + 3 = 2x + 11$ **44.** $5y + 3 = 2y + 15$ **45.** $5 - 2x = 3x - 7x + 25$

46. $10 - 3x = 2x - 8x + 40$ **47.** $4 + 3x - 6 = 3x + 2 - x$ **48.** $5 + 4x - 7 = 4x + 3 - x$

49. $4y - 4 + y = 6y + 20 - 4y$ **50.** $5y - 7 + y = 7y + 21 - 5y$

iii Solve and check. Clear fractions or decimals first.

51. $\frac{7}{2}x + \frac{1}{2}x = 3x + \frac{3}{2} + \frac{5}{2}x$ **52.** $\frac{7}{8}x - \frac{1}{4} + \frac{3}{4}x = \frac{1}{16} + x$ **53.** $\frac{2}{3} + \frac{1}{4}t = \frac{1}{3}$

54. $-\frac{3}{2} + x = -\frac{5}{6} - \frac{4}{3}$ **55.** $\frac{2}{3} + 3y = 5y - \frac{2}{15}$ **56.** $\frac{1}{2} + 4m = 3m - \frac{5}{2}$

57. $\frac{5}{3} + \frac{2}{3}x = \frac{25}{12} + \frac{5}{4}x + \frac{3}{4}$ **58.** $1 - \frac{2}{3}y = \frac{9}{5} - \frac{y}{5} + \frac{3}{5}$ **59.** $2.1x + 45.2 = 3.2 - 8.4x$

60. $0.96y - 0.79 = 0.21y + 0.46$ **61.** $1.03 - 0.62x = 0.71 - 0.22x$ **62.** $1.7t + 8 - 1.62t = 0.4t - 0.32 + 8$

63. $0.42 - 0.03y = 3.33 - y$ **64.** $0.7n - 15 + n = 2n - 8 - 0.4n$ **65.** $\frac{2}{7}x + \frac{1}{2}x = \frac{3}{4}x + 1$

66. $\frac{5}{16}y + \frac{3}{8}y = 2 + \frac{1}{4}y$ **67.** $\frac{4}{5}x - \frac{3}{4}x = \frac{3}{10}x - 1$ **68.** $\frac{8}{5}y - \frac{2}{3}y = 23 - \frac{1}{15}y$

69. $|4x - 3| = 25$ **70.** $|7x + 2| = 23$ **71.** $|4 - 5x| = 16$

72. $100 = |2 + 8x|$

73. 🔢 Solve: $0.008 + 9.62x - 42.8 = 0.944x + 0.0083 - x$.

Solve the first equation for x. Then substitute this number into the second equation and solve it for y.

74. $9x - 5 = 22$, **75.** $9x + 2 = -1$, **76.** $0.2x + 0.12 = 0.146$,

$4x + 2y = 2$ $4x - y = \frac{11}{3}$ $0.17x + 0.03y = 0.01238$

Solve for y.

77. $\dfrac{y - 2}{3} = \dfrac{2 - y}{5}$ **78.** $\dfrac{y}{a} - 3y = 1$ **79.** $0 = y - (-14) - (-3y)$

80. $\dfrac{5 + 2y}{3} = \dfrac{25}{12} + \dfrac{5y + 3}{4}$

81. $0.05y - 1.82 = 0.708y - 0.504$

82. Solve the equation $4x - 8 = 32$ by first using the addition principle. Then solve it by first using the multiplication principle.

Solve.

83. $3x = 4x$

84. $-2y + 5y = 6y$

85. $\dfrac{4 - 3x}{7} = \dfrac{2 + 5x}{49} - \dfrac{x}{14}$

86. $\dfrac{2x - 5}{6} + \dfrac{4 - 7x}{8} = \dfrac{10 + 6x}{3}$

OBJECTIVE

After finishing Section 2.3, you should be able to:

| i | Solve simple equations containing parentheses.

Solve.

1. $2(2y + 3) = 14$

2. $5(3x - 2) = 35$

Solve.

3. $3(7 + 2x) = 30 + 7(x - 1)$

4. $4(3 + 5x) - 4 = 3 + 2(x - 2)$

2.3

EQUATIONS CONTAINING PARENTHESES

| i | Here we consider certain kinds of equations that contain parentheses. To solve such equations we use the distributive laws to first remove the parentheses. Then we proceed as before.

EXAMPLE 1   Solve: $4x = 2(12 - 2x)$.

$$4x = 2(12 - 2x)$$
$$4x = 24 - 4x \qquad \text{Using a distributive law to multiply and remove parentheses}$$
$$4x + 4x = 24 \qquad \text{Adding } 4x \text{ to get all } x\text{-terms on one side}$$
$$8x = 24 \qquad \text{Collecting like terms}$$
$$x = 3 \qquad \text{Multiplying by } \tfrac{1}{8}$$

Check:

$$
\begin{array}{c|c}
\multicolumn{2}{c}{4x = 2(12 - 2x)} \\
\hline
4 \cdot 3 & 2(12 - 2 \cdot 3) \\
12 & 2(12 - 6) \\
 & 2 \cdot 6 \\
 & 12
\end{array}
$$

The solution is 3.

DO EXERCISES 1 AND 2.

EXAMPLE 2   Solve: $3(x - 2) - 1 = 2 - 5(x + 5)$.

$$3(x - 2) - 1 = 2 - 5(x + 5)$$
$$3x - 6 - 1 = 2 - 5x - 25 \qquad \text{Using the distributive laws to multiply and remove parentheses}$$
$$3x - 7 = -5x - 23 \qquad \text{Simplifying}$$
$$3x + 5x = -23 + 7 \qquad \text{Adding } 5x \text{ and also 7, to get all } x\text{-terms on one side and all other terms on the other side}$$
$$8x = -16 \qquad \text{Simplifying}$$
$$x = -2 \qquad \text{Multiplying by } \tfrac{1}{8}$$

Check:

$$
\begin{array}{c|c}
\multicolumn{2}{c}{3(x - 2) - 1 = 2 - 5(x + 5)} \\
\hline
3(-2 - 2) - 1 & 2 - 5(-2 + 5) \\
3 \cdot (-4) - 1 & 2 - 5(3) \\
-12 - 1 & 2 - 15 \\
-13 & -13
\end{array}
$$

The solution is -2.

DO EXERCISES 3 AND 4.

EXERCISE SET 2.3

i Solve the following equations. Check.

1. $3(2y - 3) = 27$

2. $4(2y - 3) = 28$

3. $40 = 5(3x + 2)$

4. $9 = 3(5x - 2)$

5. $2(3 + 4m) - 9 = 45$

6. $3(5 + 3m) - 8 = 88$

7. $5r - (2r + 8) = 16$

8. $6b - (3b + 8) = 16$

9. $3g - 3 = 3(7 - g)$

10. $3d - 10 = 5(d - 4)$

11. $6 - 2(3x - 1) = 2$

12. $10 - 3(2x - 1) = 1$

13. $5(d + 4) = 7(d - 2)$

14. $3(t - 2) = 9(t + 2)$

15. $3(x - 2) = 5(x + 2)$

16. $5(y + 4) = 3(y - 2)$

17. $8(2t + 1) = 4(7t + 7)$

18. $7(5x - 2) = 6(6x - 1)$

19. $3(r - 6) + 2 = 4(r + 2) - 21$

20. $5(t + 3) + 9 = 3(t - 2) + 6$

21. $19 - (2x + 3) = 2(x + 3) + x$

22. $13 - (2c + 2) = 2(c + 2) + 3c$

23. $\frac{1}{4}(8y + 4) - 17 = -\frac{1}{2}(4y - 8)$

24. $\frac{1}{3}(6x + 24) - 20 = -\frac{1}{4}(12x - 72)$

25. $2[4 - 2(3 - x)] - 1 = 4[2(4x - 3) + 7] - 25$

26. $5[3(7 - t) - 4(8 + 2t)] - 20 = -6[2(6 + 3t) - 4]$

27. $\frac{2}{3}(2x - 1) = 10$

28. $\frac{4}{5}(3x + 4) = 20$

29. $\frac{3}{4}(3 + 2x) + 1 = 13$

30. $\frac{7}{8}(5 - 4x) - 17 = 38$

31. $\frac{3}{4}(3x - \frac{1}{2}) - \frac{2}{3} = \frac{1}{3}$

32. $\frac{2}{3}(\frac{7}{8} - 4x) - \frac{5}{8} = \frac{3}{8}$

33. $0.7(3x + 6) = 1.1 - (x + 2)$

34. $0.9(2x + 8) = 20 - (x + 5)$

35. $a + (a - 3) = (a + 2) - (a + 1)$

36. $0.8 - 4(b - 1) = 0.2 + 3(4 - b)$

Solve for x.

37. $475(54x + 7856) + 9762 = 402(83x + 975)$

38. $-2[3(x - 2) + 4] = 4(1 - x) + 8$

39. $x(x - 4) = 3x(x + 1) - 2(x^2 + x - 5)$

40. $3(x + 4) = 3(4 + x)$

41. $4(x - a) = 16$

42. ▦ $30,000 + 20,000x = 55,000(1 + 12,500x)$

An *identity* is an equation that is true for all sensible replacements. Determine which of the following are identities.

43. $2(x - 3) + 5 = 3(x - 2) + 5$

44. $3(x - 4) = 3x - 4$

45. $5(x + 3) = 5x + 15$

46. $\dfrac{6y + 4}{2} = 6y + 2$

2.4

FORMULAS

i A formula is a kind of "recipe" for doing a certain kind of calculation. In this section we learn how to calculate with formulas and how to solve a formula for a certain letter. For example, the formula

$$L = 4D$$

is used to determine the safest position of a ladder, where L is the distance along the ladder to where it rests against a wall or other object and D is the distance that the bottom of the ladder is out from the wall.

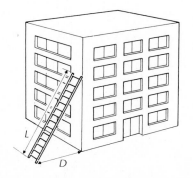

OBJECTIVE

After finishing Section 2.4, you should be able to:

i Substitute values into formulas, and solve a formula for a specified letter.

1. Suppose we know that $D = 7$ ft. Find L.

Suppose we know that $D = 3$ ft. Then we can find L by substituting 3 for D in the formula:

$$L = 4(3) = 12 \text{ ft.}$$

Thus, if a ladder is to be 3 ft out from a wall, the safest position of the ladder occurs when length L is 12 ft.

DO EXERCISE 1.

Suppose we know the length L and want to know how far out from the wall, that is, D, we should set the ladder. For example, say we have a 40-ft ladder that is to touch a wall at the top of the ladder. How far out from the wall should the bottom of the ladder be placed to be safest? To find out, we substitute 40 for L and solve for D:

$$40 = 4D$$
$$\tfrac{1}{4} \cdot 40 = \tfrac{1}{4} \cdot 4D \qquad \text{Multiplying by } \tfrac{1}{4}$$
$$10 = D.$$

If we had to do this many times, it might be faster to find a formula with D by itself on one side. Then computations could be done directly. Consider the equation

$$L = 4D.$$

We multiply on both sides by $\tfrac{1}{4}$ and get a formula for D in terms of L:

$$L = 4D \qquad \text{We want to get this letter alone.}$$
$$\tfrac{1}{4} \cdot L = \tfrac{1}{4}(4D) \qquad \text{Multiplying by } \tfrac{1}{4}$$
$$\tfrac{1}{4}L = D.$$

Now to find D when $L = 40$, we substitute 40 for L in the formula $D = \tfrac{1}{4}L$:

$$D = \tfrac{1}{4}(40) = 10.$$

EXAMPLE 1 A formula for the circumference C of a circle of radius r is

$$C = 2\pi r.$$

a) Find the circumference when the radius is 5 ft.

b) Solve the formula for the radius r.

c) Use the formula in part (b) to find the radius when the circumference is 300 cm.

a) To find the circumference when the radius is 5 ft, we substitute 5 for r in the formula $C = 2\pi r$:

$$C = 2\pi r$$
$$= 2\pi(5)$$
$$= 10\pi.$$

To get an approximation for the answer we could use 3.14 for π. Then

$$C = 10\pi \approx 10(3.14) = 31.4 \text{ ft.}$$

b)
$$C = 2\pi r \qquad \text{We want this letter alone.}$$
$$\frac{1}{2\pi} \cdot C = \frac{1}{2\pi} \cdot 2\pi r \qquad \text{Multiplying by } \frac{1}{2\pi}$$
$$\frac{C}{2\pi} = \frac{2\pi}{2\pi} \cdot r = r$$

We can write the answer as

$$r = \frac{C}{2\pi}.$$

c) To find the radius when the circumference is 300 cm, we substitute 300 for C:

$$r = \frac{C}{2\pi} = \frac{300}{2\pi} = \frac{150}{\pi}.$$

DO EXERCISES 2 AND 3.

Remember, formulas are equations. We can use the same principles in solving a formula that we use for any other equations. To see how the principles apply to formulas, compare the following.

A. Solve.

$$5x + 2 = 12$$
$$5x = 12 - 2$$
$$5x = 10$$
$$x = \frac{10}{5}$$
$$x = 2$$

B. Solve.

$$5x + 2 = 12$$
$$5x = 12 - 2$$
$$x = \frac{12 - 2}{5}$$

C. Solve for x.

$$ax + b = c$$
$$ax = c - b$$
$$x = \frac{c - b}{a}$$

In (A) we solved as we did before. In (B) we did not carry out the calculations. In (C) we could not carry out the calculations because we had unknown numbers.

With the formulas in this section we can use a procedure like that described in Section 2.2.

To solve a formula for a given letter, identify the letter and:

1. **Multiply on both sides to clear fractions or decimals, if that is needed.**
2. **Collect like terms on each side, if necessary.**
3. **Get all terms with the letter to be solved for on one side of the equation and all other terms on the other side.**
4. **Collect like terms again, if necessary.**
5. **Solve for the letter in question.**

EXAMPLE 2 Solve for a: $A = \dfrac{a + b + c}{3}$.

This is a formula for the average A of three numbers a, b, and c.

$$A = \frac{a + b + c}{3} \qquad \text{We want this letter alone.}$$

$$3A = a + b + c \qquad \text{Multiplying by 3 to clear fractions}$$

$$3A - b - c = a$$

DO EXERCISE 4.

EXAMPLE 3 Solve for C: $Q = \dfrac{100M}{C}$.

This is a formula used in psychology for finding the intelligence quotient Q, where M is mental age and C is chronological, or actual, age.

---

2. A formula for the circumference C of a circle of diameter D is

$$C = \pi D.$$

   a) Find the circumference when the diameter is 20 m.

   b) Solve the formula for the diameter D.

   c) Use the formula in (b) to find the diameter when the circumference is 400 m.

3. Solve for I: $E = IR$.

4. Solve for c: $A = \dfrac{a + b + c + d}{4}$.

5. Solve for I: $A = \dfrac{9R}{I}$.

(This is a formula for computing the earned run average A of a pitcher who has given up R earned runs in I innings of pitching.)

We solve as follows:

$$Q = \frac{100M}{C} \qquad \text{We want this letter alone.}$$

$$CQ = 100M \qquad \text{Multiplying by } C \text{ to clear fractions}$$

$$C = \frac{100M}{Q}. \qquad \text{Multiplying by } \frac{1}{Q}$$

DO EXERCISE 5.

EXERCISE SET 2.4

i

1. The area A of a rectangle of length l and width w is given by

$$A = lw.$$

w

l

a) Find the area when the length is 17 ft and the width is 4 ft.
b) Solve the formula for w.
c) Using the formula found in part (b), find the width w when the area is 48 cm^2 (square centimeters) and the length is 6 cm.

3. The area A of a triangle with a base of length b and a height of length h is given by

$$A = \tfrac{1}{2}bh.$$

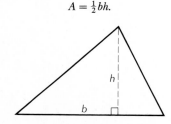

2. The simple interest I on an investment of principal P at interest rate r for time t, in years, is given by

$$I = Prt.$$

a) Find the interest I when principal P is \$2000, interest rate r is 12%, and the time is 3 years.
b) Solve the formula for t.
c) Using the formula found in part (b), find the time t that it takes for a principal of \$5000, at simple interest rate 11%, to earn \$2200 on interest.

a) Find the area of a triangle with a base of length $5\tfrac{1}{2}$ m (meters) and a height of $6\tfrac{1}{4}$ m.
b) Solve the formula for h.
c) Using the formula found in part (b), find the height h of a triangle that has an area of 504 yd^2 and a base of length 12 yd.

4. The perimeter P of a rectangle of length l and width w is given by

$$P = 2l + 2w.$$

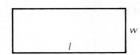

a) Find the perimeter of a rectangle of length 44 cm and width 25 cm.
b) Solve the formula for l.
c) Using the formula found in part (b), find the length of a rectangle whose perimeter is 250 mi and whose width is 14 mi.

Solve.

5. $A = bh$, for b. (This is the formula for the area of a parallelogram of base b and height h.)

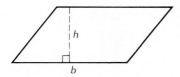

6. $A = bh$, for h.

7. $d = rt$, for r. (This is a formula for distance in terms of speed r and time t.)

8. $d = rt$, for t.

9. $I = Prt$, for P.

10. $I = Prt$, for r.

11. $F = ma$, for a. (This is a physics formula for force F in terms of mass m and acceleration a.)

12. $F = ma$, for m.

13. $P = 2l + 2w$, for w.

14. $A = \frac{1}{2}bh$, for b.

15. $A = \pi r^2$, for r^2. (This is the formula for the area A of a circle of radius r.)

16. $A = \pi r^2$, for π.

17. $E = mc^2$, for m. (This is a relativity formula from physics.)

18. $E = mc^2$, for c^2.

19. $A = \dfrac{a + b + c}{3}$, for b.

20. $A = \dfrac{a + b + c}{3}$, for c.

21. $v = \dfrac{3k}{t}$, for t.

22. $P = \dfrac{ab}{c}$, for c.

23. $A = \frac{1}{2}ah + \frac{1}{2}bh$, for b.

24. $A = \frac{1}{2}ah + \frac{1}{2}bh$, for a.

25. The formula

$$H = \dfrac{D^2 N}{2.5}$$

is used to find the horsepower H of an N-cylinder engine. Solve for D^2.

26. Solve for N:

$$H = \dfrac{D^2 N}{2.5}.$$

27. The area of a sector of a circle is given by

$$A = \dfrac{\pi r^2 S}{360},$$

where r is the radius and S is the angle measure of the sector. Solve for S.

28. Solve for r^2:

$$A = \dfrac{\pi r^2 S}{360}.$$

29. The formula

$$R = -0.0075t + 3.85$$

can be used to estimate the world record in the 1500-m run t years after 1930. Solve for t.

30. The formula

$$F = \tfrac{9}{5}C + 32$$

can be used to convert from Celsius, or Centigrade, temperature C to Fahrenheit temperature F. Solve for C.

31. In $P = 2a + 2b$, P doubles. Do a and b both double?

32. In $A = lw$, l and w both double. What happens to A?

33. In $A = \frac{1}{2}bh$, b increases by 4 units, and h does not change. What happens to A?

34. In $T = 1.2a + 1.09b$, does an increase in a or an increase in b have more effect on T?

Solve.

35. $ax + b = c$, for b

36. $ax + b = c$, for a

37. $ax + b = 0$, for x

38. $A = \frac{1}{R}$, for R

39. $\frac{s}{t} = \frac{t}{v}$, for s

40. $\frac{a}{b} = \frac{c}{d}$, for $\frac{a}{c}$

41. $g = 40n + 20k$, for k

42. $r = 2h - \frac{1}{4}f$, for f

43. $y = a - ab$, for a

44. $x = a + b - 2ab$, for a

45. $d = \frac{1}{e + f}$, for f

46. $x = \frac{\left(\frac{y}{z}\right)}{\left(\frac{z}{t}\right)}$, for y

47. $m = ax^2 + bx + c$, for b

48. If $a^2 = b^2$, does $a = b$?

The sum A of the measures of the interior angles of a polygon of n sides is given by the formula

$$A = 180°(n - 2).$$

49. A polygon has 6 sides. What is the sum of the interior angle measures?

50. The sum of the measures of the angles of a polygon is 1440°. How many sides does it have?

51. Solve $A = 180°(n - 2)$ for n.

Density. The density D of a substance is defined to be its mass M divided by its volume V:

$$D = \frac{M}{V}.$$

Suppose you had a sponge, a softball, and a shotput, all of about the same volume. The sponge would be less dense than the softball, and the softball would be less dense than the shotput.

52. Solve the density formula for the mass M.

53. Solve the density formula for the volume V.

54. The density of iron is 7.5 grams per cubic centimeter (cm^3). Find the volume of a piece of iron that has a mass of 908 g.

55. The density of iron is 7.5 g/cm^3. Find the density of a piece of iron that has a volume of 363 cm^3.

56. The mass of 200 cm^3 of alcohol is 158 g. Find the density of alcohol.

57. The mass of 400 cm^3 of gold is 7720 g. Find the density of gold.

58. The density of copper is 8.93 g/cm^3. A cylindrical copper wire with a diameter of 1 cm has a mass of 428 g. Find its length.

OBJECTIVE

After finishing Section 2.5, you should be able to:

i Use the five-step problem-solving process to solve simple problems with algebra.

2.5

INTRODUCTION TO PROBLEM SOLVING

There are many kinds of problems. Generally speaking, when you have a problem to solve, there is some kind of question. To solve the problem, in some fashion or other, you find an answer to the question. Here are some examples.

EXAMPLES OF PROBLEMS

1. **How do I get a business degree at this university?**
2. **What is the best route to drive to Pittsburgh?**
3. **How can I make my apartment burglar proof?**
4. **If I get a 16% raise, will I be able to afford a motorcycle?**
5. **How many dimples are on a golf ball?**
6. **Is a golf ball with more dimples better than one with fewer dimples?**
7. **I have 80 ft of fencing. How can I fence a rectangular garden with the greater area?**
8. **Is there a number which when multiplied by itself gives 10?**

Although the preceding problems are quite different, there are also some similarities. We will look at some of the similarities and see if we can come up with the beginnings of a plan, or *process*, for attacking problems. As you will see later, not all these examples can be solved using algebra. Nonetheless, problem solving with algebra has its similarities with problem solving without algebra.

The First Step in Problem Solving: To Familiarize

To solve any problem, the first step is to become familiar with the situation. You should make sure that you know what information is available and that you know what the question or unknown is. Sometimes you must find information for yourself. The following are some things that you might do to help familiarize yourself with a problem situation. You might not need to use them all for a specific problem.

THE FIRST STEP IN PROBLEM SOLVING WITH ALGEBRA

Familiarize yourself with a problem situation.

1. **If a problem is given in words, read it carefully.**
2. **List the information given and the question to be answered.**
3. **Find further information.**
4. **Make a table of the information given and the information you have collected.**
5. **Make a drawing and label it with known information. Also indicate unknown information.**
6. **Guess or estimate the answer.**

Once you have become familiar with a problem situation, you are well on the way toward solving the problem. In this section, you will get some practice in this all-important first step. Later you will learn the complete problem-solving process.

EXAMPLE 1   How might you familiarize yourself with the situation of Problem 1: "How do I get a business degree at this university?"

Obviously further information is needed. You will have to find it. You might:

a) Talk to an upperclassman who is a business major.

b) Get a university catalog and study it.

c) Talk to a counselor in the business department.

When the information is known, it would be a good idea to list it in a table.

DO EXERCISE 1.

How might you familiarize yourself with this problem situation?

1. How much do Americans spend in one year on lottery tickets?

How might you familiarize yourself with the situation of the following problem?

2. Is there a number which when multiplied by itself gives 10?

The problem of Example 1 is a bit unusual because once you have familiarized yourself with the situation, the problem is already solved.

EXAMPLE 2 How might you familiarize yourself with the situation of Problem 7, "If I have 80 feet of fencing, how can I fence a rectangular garden with the greatest area?"

The problem is given in words. You should by all means read it carefully, perhaps several times. Probably the next thing to do is to list the information.

Fencing—80 ft available

Garden—to be rectangular

Area—to be the greatest possible

A drawing is highly important in this case. You should make a sketch and mark the information on it.

Rectangular garden

How can you find the area? You probably already know that the area of a rectangle is found by multiplying the length by the width. If you ever need a formula that you do not know, you should look it up. (In doing so, you are finding further information.) It would be a *great* idea, at this point, to add this information to your drawing.

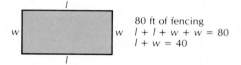

From your drawing you see that $l + w$ will have to be 40 ft. At this point you might guess what the length and width might be to give the greatest area. You could try several guesses and compare them. You might complete a table as follows.

① Pick values of l and w whose sum is 40.
② Multiply to find area and compare.

l	w	Area
15	25	375
18	22	396
13.4	26.6	356.44
20	20	400

Go back and read the problem again. Now do you feel more familiar with the problem?

DO EXERCISE 2.

Remember that problems can be very, very different. There are no strict rules for solving problems; we can only give you guidelines. The first and most important one is to thoroughly familiarize yourself with the situation. You can even familiarize yourself with a problem in many different ways.

The Second Step in Problem Solving: To Translate

After familiarizing yourself with a problem situation, you are ready for the next step in problem solving. In many problems, a mathematical expression corresponds to the situation of the problem. The second step in the problem-solving process is to *translate* the situation to mathematical language of some kind. In Example 2 above, for example, we wrote an algebraic expression. We translated the conditions of the problem to mathematical language.

THE SECOND STEP IN PROBLEM SOLVING WITH ALGEBRA

Translate the situation of the problem to mathematical language. In some cases translation can be done by writing an algebraic expression.

When you wrote algebraic expressions in Section 1.5, you were practicing the skill of translating. In the next section we will expand on that skill when we translate problems to equations.

The Third Step in Problem Solving: To Carry Out

The first step in problem solving is to *familiarize* yourself with the situation. The second step is to *translate* to mathematical language. The third step is to *carry out* some mathematical manipulation. If you have translated to an equation, that step will consist of solving the equation.

THE THIRD STEP IN PROBLEM SOLVING WITH ALGEBRA

Carry out some mathematical manipulation. If you have translated to an equation, this means to solve the equation.

Solving equations is an important part of problem solving. Equations can sometimes be complex and not too easy to solve. In this chapter we have seen how to solve many kinds of equations.

The Five-Step Process for Problem Solving

We have discussed three steps in problem solving with algebra, the most recent being to carry out some mathematical manipulation. Often that means to solve an equation. There are two more steps in the problem-solving process. The next one is to *check* your possible answer by going to the conditions of the original problem. That way you will know whether you have an answer to the problem itself. The last step is to *state* clearly the answer to the problem. We list the five steps below. You should learn them well, and remember to apply them as you work problems in algebra.

THE FIVE STEPS IN PROBLEM SOLVING IN ALGEBRA

1. *Familiarize* yourself with the problem situation.
2. *Translate* to mathematical language.
3. *Carry out* some mathematical manipulation.
4. *Check* your possible answer in the original problem.
5. *State* the answer clearly.

Before we begin to solve problems some comments regarding the *check* are in order. Suppose you have solved an equation. Is the solution necessarily a solution to the original problem? *Not* always! Your goal is to find solutions to the original problem, and you must go back and check in the original problem to be *sure* you have solved it. You will often discover situations in algebra where a solution to an equation may not be a solution to the original problem. For example, suppose you are trying to find the width of a sidewalk, and you find a negative solution to the equation. That number cannot be a solution to the problem because length is a nonnegative number.

There are partial checks that you can do, so long as you realize they are not certain. Retracing the computations in the solution to the equation is a partial check. Another partial check is to consider the reasonableness of an answer. For example, suppose you want to find how much farther it is from the earth to the sun than from the earth to the moon. The answer has to be a number larger than 1 because it is farther from the earth to the sun than from the earth to the moon. If you solved an equation and got an answer less than 1, your check of its being a reasonable answer would have told you that that solution to the equation cannot be a solution to the problem. There are many problems in algebra where we can solve an equation, but a solution to the equation may not be a solution to the problem.

i Problem Solving

Thus far in this text we have reviewed some arithmetic skills, introduced some algebraic tools, and presented a five-step process for problem solving. Now we solve some problems using these skills. The problems in this section are simple ones. You may be able to solve some of them without using equations, but it is better if you do not. You are learning how algebra works. As we proceed through the text, you will continue to learn methods of algebra that will allow you to solve harder problems.

EXAMPLE 3   What number plus 478.6 is 1019.2?

1. **Familiarize** yourself with the situation. We identify all the pertinent information. There are two numbers in the problem, 478.6 and 1019.2. We want to find what number added to 478.6 gives 1019.2. We can try some possibilities:

$$478.6 + 500 = 978.6, \qquad 478.6 + 600 = 1078.6, \qquad 478.6 + 580.2 = 1058.8.$$

You might find the answer this way, but let us continue with the process.

2. **Translate** the problem to mathematical language. We translate as follows:

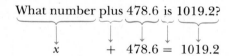

$$x + 478.6 = 1019.2$$

3. **Carry out** some mathematical manipulation. The translation gives us the equation

$$x + 478.6 = 1019.2.$$

To carry out some mathematical manipulation, we solve the equation:

$$x = 1019.2 - 478.6 \qquad \text{Subtracting 478.6}$$
$$x = 540.6.$$

4. **Check** the answer in the original problem. To do this we add 540.6 to 478.6:

$$540.6 + 478.6 = 1019.2.$$

We see that 540.6 checks in the original problem.

5. **State** the answer clearly. The answer is 540.6.

DO EXERCISE 3.

EXAMPLE 4 Three-fourths of what number is thirty-five?

1. **Familiarize** yourself with the situation. We identify all the pertinent information. We sketch and compare the "three-fourths" and "thirty-five." We try to draw a picture of the situation.

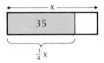

We have let x represent the unknown number.

2. **Translate** the problem to mathematical language.

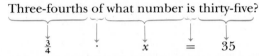

3. **Carry out** some mathematical manipulation. The translation gives us the equation

$$\tfrac{3}{4}x = 35.$$

We solve the equation:

$$x = \frac{35}{\frac{3}{4}} \qquad \text{Dividing by } \tfrac{3}{4}$$

$$x = 35 \cdot \tfrac{4}{3} = \tfrac{140}{3}.$$

4. **Check** the answer in the original problem. To check, we find out if $\tfrac{3}{4}$ of this number is 35.

$$\frac{3}{4} \cdot \frac{140}{3} = \frac{3 \cdot 140}{4 \cdot 3}$$

$$= \frac{3 \cdot 35 \cdot 4}{4 \cdot 3}$$

$$= 35.$$

5. **State** the answer clearly. The number $\tfrac{140}{3}$ is the answer.

Recall that in translating, *is* translates to =. The word *of* translates to ·, and the unknown number translates to a variable.

DO EXERCISES 4 AND 5.

EXAMPLE 5 Hank Aaron hit 755 home runs in his career. Babe Ruth hit 714 in his career. How many more home runs did Aaron hit than Ruth?

Solve.

3. What number plus 397 is 1821?

4. Sixteen times what number is 496?

5. Two thirds of what number is 27?

6. The population of Canada is 24,882,000. The population of the United States is 234,249,000. How many more people live in the United States than in Canada?

Aaron hit 755 home runs   Ruth hit 714 home runs

1. | Familiarize | The pertinent information is that Aaron hit 755 home runs and Ruth 714. We can draw a picture to show this. What we are trying to find out is how many more home runs Aaron hit than Ruth. We are asking what we should add to 714 to get 755.

714	y
755	

2. | Translate | Sometimes it helps to reword a problem before translating.

Rewording: Ruth's home runs plus how many is Aaron's home runs

Translating: 714   + y =   755

3. | Carry out | We solve the equation as follows:

$$y = 755 - 714 \qquad \text{Subtracting 714}$$
$$y = 41.$$

4. | Check | We check by adding 41 to 714: $714 + 41 = 755$.

5. | State | Aaron hit 41 more home runs than Ruth.

DO EXERCISE 6.

EXAMPLE 6   Which of the positions of these ladders appears to be safest?

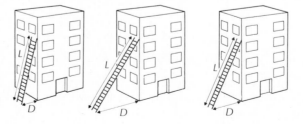

Actually it has been determined that the safest position of a ladder is when it satisfies the condition

$$L = 4D,$$

where L is the distance along the ladder to where it rests against a wall, or other object, and D is the distance of the bottom of the ladder out from the wall.

Suppose the distance L is 30 ft. How far should the bottom of the ladder be from the wall to be safest?

1. **Familiarize** ▷ We can familiarize ourselves with the problem situation by drawing a picture.

We can also make some computations for various values of D. For $D = 5$ ft, we get $L = 4(5) = 20$ ft. For $D = 10$ ft, $L = 4(10) = 40$ ft. Thus it seems reasonable that the answer to our problem is between 5 ft and 10 ft.

2. **Translate** ▷ We know that $L = 4D$. We substitute 30 for L and obtain the equation

$$30 = 4D.$$

3. **Carry out** ▷ We solve the equation:

$$30 = 4D$$

$$\frac{30}{4} = \frac{4D}{4} \qquad \text{Dividing by 4 on both sides}$$

$$7.5 = D.$$

4. **Check** ▷ We check by substituting 7.5 for D in the equation $L = 4D$:

$$L = 4(7.5) = 30.$$

5. **State** ▷ If the distance L is 30 ft, the distance D should be 7.5 ft.

DO EXERCISE 7.

7. In 1985, the value of a 1957 Corvette, restored and in good condition, was about $19,935.90. This was 5.1 times its value in 1957. What was the original cost of the 1957 Corvette?

EXERCISE SET 2.5

i Solve these problems. Even though you might find the answer quickly some other way, practice using the five-step problem-solving process.

1. What number added to 60 gives 112?

2. What number added to 45.3 gives 53.1?

3. The result of adding 29 to a number is 171. Find the number.

4. The result of adding 123 to a number is 987. Find the number.

5. Seven times what number is 2233?

6. Four times what number is 8944?

7. When 42 is multiplied by a number the result is 2352. Find the number.

8. When 48 is multiplied by a number the result is 624. Find the number.

9. Two-thirds of what number is forty-eight?

10. One-eighth of what number is fifty-six?

11. A student missed a perfect quiz paper by 5 problems. There were 8 problems on the quiz. How many did the student get right?

12. A football player caught 3 passes for a total of 55 yards. The first two were for 23 and 8 yards. Find the third.

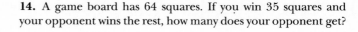

13. The New York Yankees won 37 more games than the Minnesota Twins. The Yankees won 101. How many did the Twins win?

14. A game board has 64 squares. If you win 35 squares and your opponent wins the rest, how many does your opponent get?

15. Cash register A contains $48 less than cash register B. If B has $115, how much does A have?

16. There are 352,198 people in a city. 187,804 are at least 28 years old. How many have not reached age 28?

17. A dozen bagels cost $3.12. How much is each bagel?

18. A movie theater took in $438.75 from 117 customers. All the tickets were the same price. What was the price of a ticket?

19. A consultant charges $80 an hour. How many hours did the consultant work to make $53,400?

20. The area of Lake Superior is about four times the area of Lake Ontario. The area of Lake Superior is 78,114 km². What is the area of Lake Ontario?

21. It takes a 60-watt bulb about 16.6 hours to use one kilowatt-hour of electricity. That is about 2.5 times as long as it takes a 150-watt bulb to use one kilowatt-hour. How long does it take a 150-watt bulb to use one kilowatt-hour?

22. The area of Alaska is about 483 times the area of Rhode Island. The area of Alaska is 1,519,202 km². What is the area of Rhode Island?

23. The boiling point of ethyl alcohol is 78.3°C. That is 13.5°C higher than the boiling point of methyl alcohol. What is the boiling point of methyl alcohol?

24. The height of the Eiffel Tower is 295 m. It is about 203 m higher than the Statue of Liberty. What is the height of the Statue of Liberty?

25. The distance from the earth to the sun is about 150,000,000 km. That is about 391 times the distance from the earth to the moon. What is the distance from the earth to the moon?

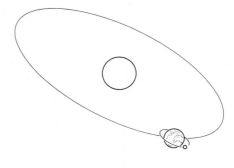

Solve.

26. In baseball, "batting average" times "at bats" equals hits. Reggie has 125 at bats and 36 hits. Find his batting average.

27. The equation for converting Celsius temperature to Fahrenheit is $F = 1.8C + 32$. Find F if the temperature is 15°C.

28. In three-way light bulbs, the highest wattage is a sum of the two lower wattages. If the lowest is 30 watts and the highest is 150 watts, what is the middle wattage?

29. A roll of film cost $3.14 and development cost $6.13. What was the cost for each of the 36 prints?

30. Franklin Laundry dryers cost a dime for 7 minutes. How many dimes will it take to dry your clothes in 45 minutes?

31. One inch = 2.54 cm. A meter is 100 cm. Find the number of inches in a meter.

32. If sound travels at 1087 feet per second, how long does it take the sound an airplane makes to travel to your ear when it is 10,000 feet overhead?

These problems are impossible to solve because some piece of information is missing. Tell what you would need to know to solve the problem.

33. A person makes three times the salary of ten years ago. What was the salary ten years ago?

34. Records were on sale for 75¢ off the marked price. After buying four records, a person has $8.72 left. How much was there to begin with?

35. The absolute value of a number is −19. Find the number.

2.6

PROBLEM SOLVING INVOLVING PERCENTS

i Let's solve some problems involving percents. Our five-step problem-solving process will again be helpful.

EXAMPLE 1 What percent of 45 is 15?

1. | Familiarize | This type of problem is stated so explicitly that we can go right to the translation.

2. | Translate | We translate as follows.

$$\text{What percent of 45 is 15?}$$
$$x \qquad \% \qquad \cdot \ 45 = 15$$

3. | Carry out | We solve the equation:

$$x\% \cdot 45 = 15$$
$$(x \times 0.01) \times 45 = 15$$
$$x(0.45) = 15$$
$$x = \frac{15}{0.45} \qquad \text{Dividing by 0.45}$$
$$x = \frac{15}{0.45} \times \frac{100}{100} = \frac{1500}{45}$$
$$x = 33\tfrac{1}{3}.$$

4. | Check | We check by finding $33\tfrac{1}{3}\%$ of 45:

$$33\tfrac{1}{3}\% \cdot 45 = \tfrac{1}{3} \cdot 45 = 15. \qquad \text{See Table 1.}$$

5. | State | The answer is $33\tfrac{1}{3}\%$.

DO EXERCISES 1–4.

EXAMPLE 2 3 is 16 percent of what?

1. | Familiarize | This problem is stated so explicitly that we can go right to the translation.

2. | Translate |

$$\text{3 is 16 percent of what?}$$
$$3 = 16 \quad \% \quad \cdot \quad y$$

OBJECTIVE

After finishing Section 2.6, you should be able to:

i Solve problems involving percents.

1. What percent of 50 is 16?

2. 15 is what percent of 60?

3. What is 23% of 48?

4. 25% of 40 is what?

5. 45 is 20 percent of what?

3. **Carry out** We solve the equation:

$$3 = 16\% \cdot y$$
$$3 = 16 \times 0.01 \times y$$
$$3 = 0.16y$$
$$0.16y = 3$$
$$y = \frac{3}{0.16} \qquad \text{Dividing by 0.16}$$
$$y = 18.75.$$

4. **Check** We check by finding 16% of 18.75:

$$16\% \times 18.75 = 0.16 \times 18.75 = 3.$$

5. **State** The answer is 18.75.

DO EXERCISES 5 AND 6.

Perhaps you have noticed in Examples 1 and 2 that to handle percents in such problems, you can convert to decimal notation and then go ahead.

6. 120 percent of what is 60?

EXAMPLE 3 Blood is 90% water. The average adult has 5 quarts of blood. How much water is in the average adult's blood?

7. The area of Arizona is 19% of the area of Alaska. The area of Alaska is 586,400 sq mi. What is the area of Arizona?

1. **Familiarize** The translation step is quite easy, and we can go to it without familiarization.

2. **Translate** Sometimes it is helpful to reword the problem before translating.

Rewording: 90% of 5 is what?

Translating: 90% · 5 = x

3. **Carry out** We solve the equation:

$$90\% \cdot 5 = x$$
$$90 \times 0.01 \times 5 = x$$
$$0.90 \times 5 = x \qquad \text{Converting 90\% to decimal notation}$$
$$4.5 = x.$$

4. **Check** The check is actually the computation we use to solve the equation:

$$90\% \cdot 5 = 0.90 \times 5 = 4.5.$$

5. **State** The answer is that there are 4.5 quarts of water in a human being who has 5 quarts of blood.

DO EXERCISE 7.

EXAMPLE 4   An investment is made at 8% simple interest for 1 year. It grows to $783. How much was originally invested (the principal)?

1. |Familiarize⟩   Suppose that $100 was invested. Recalling the formula for simple interest, $I = Prt$, we know that the interest for 1 year on $100 at 8% simple interest is given by $I = \$100 \cdot 8\% \cdot 1 = \$8$. Then, at the end of the year, the *amount* in the account is found by adding principal and interest:

$$\text{(Principal)} + \text{(Interest)} = \text{Amount}$$
$$\$100 \quad + \quad \$8 \quad = \quad \$108.$$

In this problem we are working backwards. We are trying to find the principal.

2. |Translate⟩   We reword the problem and then translate:

Rewording: $\underbrace{\text{(Principal)}} + \underbrace{\text{(Interest)}} = \underbrace{\text{Amount}}$

Translating: $x \quad + \quad 8\%x \quad = \quad 783$ Interest is 8% of the principal

3. |Carry out⟩   We solve the equation:

$$x + 8\%x = 783$$
$$x + 0.08x = 783 \qquad \text{Converting}$$
$$1.08x = 783 \qquad \text{Collecting like terms}$$
$$x = \tfrac{783}{1.08} \qquad \text{Dividing by 1.08}$$
$$x = 725.$$

4. |Check⟩   We check by taking 8% of $725 and adding it to $725:

$$8\% \times \$725 = 0.08 \times 725 = \$58.$$

Then $\$725 + \$58 = \$783$, so $725 checks.

5. |State⟩   The original investment was $725.

DO EXERCISE 8.

EXAMPLE 5   The price of an automobile was decreased to a sale price of $13,559. This was a 9% reduction. What was the former price?

1. |Familiarize⟩   Suppose the former price was $16,000. A 9% reduction would be found by taking 9% of $16,000, that is,

$$9\% \text{ of } \$16,000 = 0.09(\$16,000) = \$1440.$$

Then the sale price is found by subtracting the amount of reduction:

$$\text{(Former price)} - \text{(Reduction)} = \text{(Sale price)}$$
$$\$16,000 \quad - \quad \$1440 \quad = \quad \$14,560$$

Our guess of $16,000 was too high; we are getting familiar with the problem.

8. An investment is made at 7% simple interest for one year. It grows to $8988. How much was originally invested (the principal)?

9. On sale, the price of a compact disc player was reduced 20% to $180. What was the former price?

2. [Translate] We reword and then translate.

$$\underbrace{\text{(Former price)}}_{} - \text{(Reduction)} = \underbrace{\text{Sale price}}_{} \qquad \text{Rewording}$$

$$x \qquad - \qquad 9\%x \qquad = \$13,559 \qquad \text{Translating}$$

3. [Carry out] We solve the equation:

$$x - 9\%x = 13,559$$
$$x - 0.09x = 13,559 \qquad \text{Converting to decimal notation}$$
$$0.91x = 13,559 \qquad \text{Collecting like terms}$$
$$x = \tfrac{13,559}{0.91} \qquad \text{Dividing by 0.91}$$
$$x = \$14,900.$$

4. [Check] To check we find 9% of $14,900 and subtract:

$$9\% \times \$14,900 = 0.09 \times \$14,900 = \$1341$$
$$\$14,900 - \$1341 = \$13,559.$$

Since we get the sale price, $13,559, the $14,900 checks.

5. [State] The former price was $14,900.

This problem is easy with algebra. Without algebra it is not. A common error in a problem like this is to take 9% of the sale price and subtract.

DO EXERCISE 9.

EXERCISE SET 2.6

i Solve.

1. What percent of 68 is 17?

2. What percent of 75 is 36?

3. What percent of 125 is 30?

4. What percent of 300 is 57?

5. 45 is 30% of what number?

6. 20.4 is 24% of what number?

7. 0.3 is 12% of what number?

8. 7 is 175% of what number?

9. What number is 65% of 840?

10. What number is 1% of a million?

11. What percent of 80 is 100?

12. What percent of 10 is 205?

13. What is 2% of 40?

14. What is 40% of 2?

15. 2 is what percent of 40?

16. 40 is 2% of what number?

17. On a test of 88 items, a student got 76 correct. What percent were correct?

18. A baseball player had 13 hits in 25 times at bat. What percent were hits?

19. A family spent $208 one month for food. This was 26% of its income. What was their monthly income?

20. The sales tax rate in New York City was 8%. How much would be charged on a purchase of $428.86? How much will the total cost of the purchase be?

21. Water volume increases 9% when it freezes. If 400 cubic centimeters of water is frozen, how much will its volume increase? What will be the volume of the ice?

22. An investment is made at 9% simple interest for 1 year. It grows to $8502. How much was originally invested?

23. An investment is made at 8% simple interest for 1 year. It grows to $7776. How much was originally invested?

24. Due to inflation the price of an item rose 8%, which was 12¢. What was the old price? the new price?

25. After a 40% reduction, a shirt is on sale at $9.60. What was the marked price (that is, the price before reduction)?

26. After a 34% reduction, a blouse is on sale at $9.24. What was the marked price?

27. Money is invested in a savings account at 12% simple interest. After 1 year there is $4928 in the account. How much was originally invested?

28. Money is borrowed at 10% simple interest. After 1 year $7194 pays off the loan. How much was originally borrowed?

29. The population of the world in 1980 was 4.4 billion. This was a 23% increase over the population in 1970. What was the population in 1970, to the nearest tenth of a billion?

30. The population of the United States in 1980 was 224 million. This was a 48% increase over the population in 1950. What was the population in 1950, to the nearest million?

31. A bottle factory has 59 breaks on the assembly line during a business day. The expected breakage rate is 1.3%. About how many bottles were produced?

32. Which is higher, if either?

a) x is increased by 25%; then that amount is decreased 25%.

b) x is decreased by 25%; then that amount is increased 25%.

Explain.

Simplify.

33. $12\% + 14\%$

34. $84\% - 16\%$

35. $1 - 10\%$

36. $81\% - 10\%$

37. $12 \times 100\%$

38. $42\% - (1 - 58\%)$

39. $3(1 + 15\%)$

40. $7(1\% + 13\%)$

41. $\dfrac{100\%}{40}$

42. Twenty-seven people make a certain amount of money at a sale. What percentage does each receive if they share the profit equally?

43. A meal came to $16.41 without tax. Calculate 6% sales tax and then calculate a 15% tip based on the sum of the meal price and the tax. What is the total paid?

44. Rollie's Records charges $7.99 for an album. Warped Records charges $9.95 but you have a coupon for $2 off. 7% sales tax is charged on the *regular* prices. How much does the record cost at each store?

45. The weather report is "a 60% chance of showers during the day, 30% tonight, and 5% tomorrow morning." What are the chances it won't rain during the day? tonight? tomorrow morning?

46. If x is 160% of y, y is what % of x?

47. The new price of a car is 25% higher than the old price of $8800. The old price is what percent lower than the new price?

48. A distributing company gives successive discounts to dealers and computes prices as follows:

List price (price printed in the catalog) less successive discounts of 10%, 20%, and 10%. To find the actual price, take 10% off. Then take 20% off what is left and then 10% off that amount.

A list price is $140. What is the actual price?

49. One number is 25% of another. The larger number is 12 more than the smaller. What are the numbers?

50. In a basketball league, the Falcons won 15 out of their first 20 games. How many more games will they have to play where they win only half the time in order to win 60% of the total games?

51. In one city, a sales tax of 9% was added to the price of gasoline as registered on the pump. Suppose a driver asked for $10 worth of regular. The attendant filled the tank until the pump read $9.10 and charged the driver $10. Something was wrong. Use algebra to correct the error.

2.7

MORE ON PROBLEM SOLVING

i In this section we continue to practice our problem-solving skills. We will consider problems that are a bit more complicated, and we will learn some additional tips to improve your problem-solving skills. The following tip may seem somewhat simple, but is quite helpful.

PROBLEM-SOLVING TIP

To be good at problem solving, do lots and lots of problems!

---

1. An 8-ft board is cut into two pieces. One piece is 2 ft longer than the other. How long are the pieces?

Do as many as you can in this book, and if time permits, do problems in other books.

Let us review the process for solving problems in algebra.

> **THE FIVE STEPS IN PROBLEM SOLVING IN ALGEBRA**
>
> 1. *Familiarize* yourself with the problem situation.
> 2. *Translate* to mathematical language.
> 3. *Carry out* some mathematical manipulation.
> 4. *Check* your possible answer in the original problem.
> 5. *State* the answer clearly.

EXAMPLE 1 A 6-ft board is cut into two pieces, one twice as long as the other. How long are the pieces?

1. Familiarize We first draw a picture. Note that we have let x represent the length of one piece, and $2x$, the length of the other.

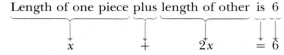

2. Translate From the figure, we can see that the lengths of the two pieces add up to 6 ft. That gives us our translation.

$$\underbrace{\text{Length of one piece}}_{x} \underbrace{\text{plus}}_{+} \underbrace{\text{length of other}}_{2x} \underbrace{\text{is}}_{=} \underbrace{6}_{6}$$

3. Carry out We solve the equation:

$$x + 2x = 6$$
$$3x = 6 \qquad \text{Collecting like terms}$$
$$x = 2. \qquad \text{Multiplying by } \tfrac{1}{3} \text{ on both sides}$$

4. Check Do we have an answer to the *problem*? If one piece is 2 ft long, then the other, to be twice as long, must be 4 ft long. The lengths of the pieces add up to 6 ft. This checks.

5. State One piece is 2 ft long and the other is 4 ft long.

DO EXERCISE 1.

EXAMPLE 2 Five plus three more than a number is nineteen. What is the number?

1. Familiarize This problem is stated explicitly enough that we can go right to the translation.

2. Translate

$$\underbrace{\text{Five}}_{5} \underbrace{\text{plus}}_{+} \underbrace{\text{three more than a number}}_{(x + 3)} \underbrace{\text{is}}_{=} \underbrace{\text{nineteen}}_{19}$$

3. **Carry out** We solve the equation:

$$5 + (x + 3) = 19$$
$$x + 8 = 19 \quad \text{Collecting like terms}$$
$$x = 11. \quad \text{Adding } -8$$

4. **Check** Three more than 11 is 14. Adding 5 to 14, we get 19. This checks.

5. **State** The number is 11.

DO EXERCISE 2.

EXAMPLE 3 The sum of two consecutive integers is 29. What are the integers?

1. **Familiarize** *Consecutive* integers are next to each other, such as 3 and 4 or -6 and -5. The larger is 1 plus the smaller. Thus, if x represents the smaller number, then $x + 1$ represents the larger number. Another way such numbers could be represented is to let y represent the larger number, and $y - 1$, the smaller.

To get more familiar with the problem we can make a table. How do we get the entries in the table? First, we just guess a value for x. Then we find $x + 1$. Finally, we add the two numbers and see what happens. You might actually solve the problem this way, even though we want you to practice using algebra.

x	$x + 1$	Sum of x and $x + 1$
3	4	7
-6	-5	-11
19	20	39
13	14	27
-1	0	-1

2. **Translate** We reword the problem and translate as follows.

$$\underbrace{\text{First integer}}_{x} + \underbrace{\text{Second integer}}_{(x + 1)} = \underbrace{29}_{29} \quad \text{Rewording}$$

$$x + (x + 1) = 29 \quad \text{Translating}$$

We have let x represent the smaller integer. Then $x + 1$ represents the larger. Note that it is a good idea to write down what your letters represent in a problem.

3. **Carry out** We solve the problem:

$$x + (x + 1) = 29$$
$$2x + 1 = 29 \quad \text{Collecting like terms}$$
$$2x = 28 \quad \text{Adding } -1$$
$$x = 14. \quad \text{Multiplying by } \frac{1}{2}$$

Now if x is 14, then $x + 1$ is 15.

4. **Check** Our possible answers are 14 and 15. These are consecutive integers. Their sum is 29, so the answers check in the *original problem*.

5. **State** The consecutive integers are 14 and 15.

DO EXERCISE 3.

2. When 5 is subtracted from 3 times a certain number, the result is 10. What is the number?

3. The sum of two consecutive even integers is 38. (Consecutive even integers are next to each other, such as 4 and 6 or -20 and -18. The larger is 2 more than the smaller.) What are the integers?

4. Acme also rents compact cars at a daily rate of $34.95 plus 27¢ per mile. What mileage will allow the businessperson to stay within a daily budget of $100?

EXAMPLE 4 Acme Rent-A-Car rents an intermediate-size car (such as a Chevrolet, Ford, or Plymouth) at a daily rate of $44.95 plus 29¢ per mile. A businessperson is not to exceed a daily budget of $100. What mileage will allow the businessperson to stay within budget?

ACME
Rent-a-Car
$44.95
Plus 29¢ Per Mile

1. **Familiarize** Suppose the businessperson drives 75 miles. Then the cost is

Daily charge plus mileage charge

or

($44.95) plus (Cost per mile) times (Number of miles driven)
$44.95   + $0.29   ·   (75)

which is $44.95 + $21.75, or $66.70. This familiarizes us with the way in which a calculation is made. Note that we converted 29¢ to $0.29 so that $44.95 and 29¢ are in the same units. Otherwise, we would get an incorrect answer after solving the equation.

2. **Translate** We reword the problem and translate as follows.

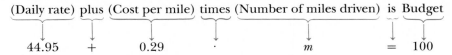

(Daily rate) plus (Cost per mile) times (Number of miles driven) is Budget

$$44.95 \quad + \quad 0.29 \quad \cdot \quad m \quad = \quad 100$$

We have let m represent the number of miles driven.

3. **Carry out** We solve the equation:

$$44.95 + 0.29m = 100$$
$$100(44.95 + 0.29m) = 100(100) \qquad \text{Multiplying by 100 on both sides to clear the decimals}$$
$$100(44.95) + 100(0.29m) = 10{,}000 \qquad \text{Using a distributive law}$$
$$4495 + 29m = 10{,}000$$
$$29m = 5505 \qquad \text{Adding } -4495$$
$$m = \frac{5505}{29} \qquad \text{Dividing by 29}$$
$$m \approx 189.8. \qquad \text{Rounding to the nearest tenth}$$

4. **Check** We check in the original problem. We multiply 189.8 by $0.29, obtaining $55.042. Then we add $55.042 to $44.95 and get $99.992, which is just under $100, the budget. At least the businessperson now knows to stay under this mileage.

5. **State** The businessperson should stay under 189.8 miles in order not to exceed the budget.

DO EXERCISE 4.

EXAMPLE 5 The perimeter of a rectangle is 150 cm (centimeters). The length is 15 cm greater than the width. Find the dimensions.

1. **Familiarize** We first draw a picture. We have let x represent the width, and $x + 15$, the length.

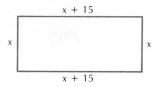

The perimeter of a polygon is the distance around it. (You may need to look up this word in a geometry book.) We can find the perimeter by adding the lengths of the sides.

2. **Translate** The definition of perimeter leads us to a rewording of the problem and a translation.

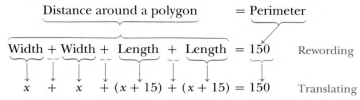

3. **Carry out** We solve the equation:

$$x + x + (x + 15) + (x + 15) = 150$$
$$4x + 30 = 150$$
$$4x = 120$$
$$x = 30.$$

Possible dimensions are $x = 30$ and $x + 15 = 45$.

4. **Check** If the width is 30 and the length is $30 + 15$, or 45, then the perimeter is $30 + 30 + 45 + 45$, or 150. This checks.

5. **State** The width is 30 cm and the length is 45 cm.

DO EXERCISE 5.

EXAMPLE 6 The second angle of a triangle is twice as large as the first. The measure of the third angle is $20°$ greater than that of the first angle. How large are the angles?

1. **Familiarize** We draw a picture. We use x for the measure of the first angle. The second is twice as large, so its measure will be $2x$. The third angle is $20°$ greater than the first angle, so its measure will be $x + 20$.

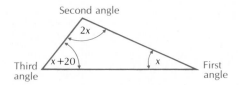

2. **Translate** To translate we need to recall a geometric fact (you might, as part of step (1), look it up in a geometry book or in the list of formulas at the back of this book). The measures of any triangle add up to $180°$.

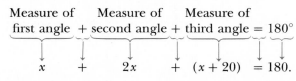

5. The length of a rectangle is twice the width. The perimeter is 60 m. Find the dimensions.

As part of the familiarization, use algebraic expressions to label this rectangle.

6. The second angle of a triangle is three times as large as the first. The third angle measures 30° more than the first angle. Find the measures of the angles.

3. [Carry out] We solve:

$$x + 2x + (x + 20) = 180$$
$$4x + 20 = 180$$
$$4x = 160$$
$$x = 40.$$

Possible answers for the angle measures are as follows:

First angle: $\quad x = 40°,$
Second angle: $\quad 2x = 80°,$
Third angle: $\quad x + 20 = 60°.$

4. [Check] Consider 40°, 80°, and 60°. The second is twice the first, and the third is 20° greater than the first. The sum is 180°. These numbers check.

5. [State] The measures are 40°, 80°, and 60°.

DO EXERCISE 6.

The examples in this section illustrate another tip when solving problems.

PROBLEM-SOLVING TIP

When translating to an equation, or some other mathematical language, consider the dimensions of the variables and constants in the equation. The variables that represent length should all involve the same unit, those that represent money should all be in dollars or all in cents, and so on.

Look back over the examples in this section. Note that in:

Example 1, the units were all in *feet*;

Examples 2 and 3, only numbers were involved and no units were needed;

Example 4, all units of money were expressed in *dollars* (or all units could have been expressed in *cents*);

Example 5, all units were expressed in *centimeters*;

Example 6, all units were expressed in *degrees*.

EXERCISE SET 2.7

1. When 18 is subtracted from six times a certain number, the result is 96. What is the number?

2. When 28 is subtracted from five times a certain number, the result is 232. What is the number?

3. If you double a number and then add 16, you get $\frac{2}{5}$ of the original number. What is the original number?

4. If you double a number and then add 85, you get $\frac{3}{4}$ of the original number. What is the original number?

5. If you add two-fifths of a number to the number itself, you get 56. What is the number?

6. If you add one-third of a number to the number itself, you get 48. What is the number?

7. A 180-m rope is cut into three pieces. The second piece is twice as long as the first. The third piece is three times as long as the second. How long is each piece of rope?

8. A 480-m wire is cut into three pieces. The second piece is three times as long as the first. The third piece is four times as long as the second. How long is each piece?

9. Consecutive odd integers are next to each other, such as 5 and 7. The larger is 2 plus the smaller. The sum of two consecutive odd integers is 76. What are the integers?

10. The sum of two consecutive odd integers is 84. What are the integers?

11. Consecutive even integers are next to each other, like 6 and 8. The larger is 2 plus the smaller. The sum of two consecutive even integers is 114. What are the integers?

12. The sum of two consecutive even integers is 106. What are the integers?

13. The sum of three consecutive integers is 108. What are the integers?

14. The sum of three consecutive integers is 126. What are the integers?

15. The sum of three consecutive odd integers is 189. What are the integers?

16. The sum of three consecutive odd integers is 255. What are the integers?

17. The perimeter of a rectangle is 310 m. The length is 25 m greater than the width. Find the width and length of the rectangle.

18. The perimeter of a rectangle is 304 cm. The length is 40 cm greater than the width. Find the width and length of the rectangle.

19. The perimeter of a rectangle is 152 m. The width is 22 m less than the length. Find the width and the length.

20. The perimeter of a rectangle is 280 m. The width is 26 m less than the length. Find the width and the length.

21. The second angle of a triangle is four times as large as the first. The third angle is 45° less than the sum of the other two angles. Find the measure of the first angle.

22. The second angle of a triangle is three times as large as the first. The third angle is 25° less than the sum of the other two angles. Find the measure of the first angle.

23. In Churchill, Manitoba, the average daily low temperature in January is $-31°C$. This is 50° less than the average daily low temperature in Key West, Florida. What is the average daily low temperature in Key West in January?

24. After depositing paychecks of $232.58 and $486.79, a family had a balance of $1279.88 in its checking account. How much was in the account before the deposits?

25. Apollo 10 reached a speed of 24,790 miles per hour. That was 37 times the speed of the first supersonic flight in 1947. What was the speed of the first supersonic flight?

26. Roger Staubach completed 1685 passes in his pro football career. This is about 57% of the number he attempted. How many did he attempt?

27. Badger Rent-A-Car rents an intermediate-size car at a daily rate of $34.95 plus 10¢ per mile. A businessperson is not to exceed a daily car rental budget of $80. What mileage will allow the businessperson to stay within budget?

28. Badger also rents compact cars at $43.95 plus 10¢ per mile. What mileage will allow the businessperson to stay within the budget of $90?

29. The second angle of a triangle is three times as large as the first. The measure of the third angle is 40° greater than that of the first angle. How large are the angles?

30. One angle of a triangle is 32 times as large as another. The measure of the third angle is 10° greater than that of the smallest angle. How large are the angles?

31. The equation

$$R = -0.028t + 20.8$$

can be used to predict the world record in the 200-meter dash. R stands for the record in seconds, and t stands for the number of years since 1920. In what year will the record be 19.0 seconds?

32. The equation

$$F = \tfrac{1}{4}N + 40$$

can be used to determine temperatures given how many times a cricket chirps per minute, where F represents temperature in degrees and N is the number of chirps per minute. Determine the chirps per minute necessary in order for the temperature to be 80°.

33. A 12-ft piece of rope is to be cut into two pieces, one piece 4 ft longer than the other. How should the rope be cut?

34. A piece of wire 10 m long is to be cut into two pieces, one of them $\frac{2}{3}$ as long as the other. How should the wire be cut?

35. One angle of a triangle is three times as great as a second angle. The third angle measures 12° less than twice the second angle. Find the measures of the angles.

36. One angle of a triangle is four times as great as a second angle. The third angle measures 5° more than twice the second angle. Find the measures of the angles.

37. Find three consecutive odd integers such that the sum of the first, two times the second, and three times the third is 70.

38. Find two consecutive even integers such that two times the first plus three times the second is 76.

39. A piece of wire 100 cm long is to be cut into two pieces and those pieces are each to be bent to make a square. The length of a side of one square is to be 2 cm greater than the length of a side of the other. How should the wire be cut?

40. A student's scores on four tests are 85, 91, 75, and 83. What must the score be on the fifth test so that the average will be 84?

41. Three numbers are such that the second is six less than three times the first and the third is two more than two-thirds of the second. The sum of the three numbers is 172. Find the largest number.

42. An appliance store is having a sale on 13 TV sets. They are displayed in order of increasing price from left to right. The price of each set differs by $20 from either set next to it. For the price of the set at the extreme right, a customer can buy both the second and seventh sets. What is the price of the least expensive set?

43. A student's scores on five tests are 93, 89, 72, 80, and 96. What must the score be on the next test so that the average will be 88?

44. The changes in population of a city for three consecutive years are, respectively, 20% increase, 30% increase, and 20% decrease. What is the percent of total change for those three years?

45. Abraham Lincoln's 1863 Gettysburg Address refers to the year 1776 as "Four *score* and seven years ago." Write an equation and find what a *score* is.

46. Bowling at Chan's Bowling Lanes cost Steve and Trina $9.00. This included shoe rental at 75¢ a pair for each of them. Steve bowled 3 games and Trina bowled 2 games. How much was each game?

47. If the daily rental for a car is $18.90 plus a certain price per mile and a person must drive 190 miles and stay within a $55.00 budget, what is the highest price per mile the person can afford?

48. A student scored 78 on a test that had 4 seven-point fill-ins and 24 three-point multiple choice questions. The student had one fill-in wrong. How many multiple choice questions did the student get right?

49. The width of a rectangle is $\frac{3}{4}$ the length. The perimeter of the rectangle becomes 50 cm when the length and width are each increased by 2 cm. Find the length and width.

50. Apples are collected in a basket for six people. One third, one fourth, one eighth, and one fifth are given to four people, respectively. The fifth person gets ten apples with one apple remaining for the sixth person. Find the original number of apples in the basket.

51. The buyer of a piano priced at $2000 is given the choice of paying cash at the time of purchase or $2150 at the end of one year. What rate of interest is the buyer being charged if payment is made at the end of one year?

52. A student has an average score of 82 on three tests. The student's average score on the first two tests is 85. What was the score on the third test?

53. A storekeeper goes to the bank to get $10 worth of change. The storekeeper requests twice as many quarters as half dollars, twice as many dimes as quarters, three times as many nickels as dimes, and no pennies or dollars. How many of each coin did the storekeeper get?

54. 🖩 The area of this triangle is 2.9047 in². Find *x*.

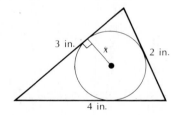

55. The perimeter of a rectangle is 640 ft. The length is 6 in. more than the width. Find the area of the rectangle.

The following problem will illustrate the importance of checking.

56. The sum of two consecutive even integers is 279. Find the integers.

SUMMARY AND REVIEW: CHAPTER 2

The following contains a summary of what you should be able to do after completing this chapter. The review exercises are for practice. Answers are at the back of the book. If you miss an exercise, restudy the section indicated alongside the answer.

You should be able to:

Solve equations using the addition principle, the multiplication principle, the absolute value principle, the addition and multiplication principles together, and the distributive laws to first remove parentheses.

Solve.

1. $x + 5 = -17$

2. $-8x = -56$

3. $-\dfrac{x}{4} = 48$

4. $n - 7 = -6$

5. $15x = -35$

6. $x - 11 = 14$

7. $-\frac{2}{3} + x = -\frac{1}{6}$

8. $\frac{4}{5}y = -\frac{3}{16}$

9. $y - 0.9 = 9.09$

10. $5 - x = 13$

11. $5t + 9 = 3t - 1$

12. $7x - 6 = 25x$

13. $\frac{1}{4}x - \frac{5}{8} = \frac{3}{8}$

14. $14y = 23y - 17 - 10$

15. $0.22y - 0.6 = 0.12y + 3 - 0.8y$

16. $\frac{1}{4}x - \frac{1}{8}x = 3 - \frac{1}{16}x$

17. $4(x + 3) = 36$

18. $3(5x - 7) = -66$

19. $8(x - 2) = 5(x + 4)$

20. $-5x + 3(x + 8) = 16$

21. $|x| + 11 = 23$

22. $|7x - 4| = 45$

Solve a formula for a certain letter.

Solve.

23. $C = \pi d$, for d.

24. $V = \frac{1}{3}Bh$, for B.

25. $A = \dfrac{a + b}{2}$, for a.

Solve problems using the five-step problem-solving process.

Solve.

26. A color TV sold for $629 in May. This was $38 more than the January cost. Find the January cost.

27. Selma gets a $4 commission for each appliance that she sells. One week she got $108 in commissions. How many appliances did she sell?

28. State the five-step problem-solving process.

29. What number added to 35 is 102?

30. Three-fifths of what number is 30?

31. An artist charges $75 for each drawing. How many drawings did the artist complete to earn $2400?

32. 25 is 10% of what number?

33. 40% of 75 is what number?

34. A government employee received an 8% raise. The new salary is $15,336. What was the original salary?

35. An 8-m board is cut into two pieces. One piece is 2 m longer than the other. How long are the pieces?

36. If 14 is added to three times a certain number, the result is 41. Find the number.

37. The sum of two consecutive odd integers is 116. Find the integers.

38. The perimeter of a rectangle is 56 cm. The width is 6 cm less than the length. Find the width and the length.

39. After a 30% reduction, an item is on sale for $154. What was the marked price (the price before reducing)?

40. A businessperson's salary is $30,000. That is a 15% increase over the previous year's salary. What was the previous salary (to the nearest dollar)?

41. The measure of the second angle of a triangle is 50° more than that of the first. The measure of the third angle is 10° less than twice the first. Find the measures of the angles.

42. The total length of the Nile and Amazon Rivers is 13,108 km. If the Amazon were 234 km longer, it would be as long as the Nile. Find the length of each river.

43. One cashier works at a rate of 3 minutes per customer and a second express cashier works at a rate of 2 customers per minute. How many customers are served in an hour?

44. Consumer experts advise us never to pay the sticker price for a car. A rule of thumb is to pay the sticker price minus 20% of the sticker price, plus $200. A car is purchased for $11,520 using the rule. What was the sticker price?

45. Solve using this information.

a) "Everyone will get 0.4% of the profits," says Big Louie. How many people must there be for all the profits to be shared equally?

b) It turned out that Louie was taking 10% of the profits. Everyone else received 0.4% of the 90% that remained. Including Louie, how many people were there?

46. There must be 6 parts per billion (ppb) of chlorine in a swimming pool. What percent of total volume is this?

47. Which is better, a discount of 40% or successive discounts of 20% and 20%?

48. Which is better, successive discounts of 10%, 10%, and 20% or of 20%, 10%, and 10%?

49. The sum of two consecutive odd integers is 467. Find the integers.

50. Solve for a: $y = 2a - ab + 3$.

TEST: CHAPTER 2

Solve.

1. $x + 7 = 15$

2. $t - 9 = 17$

3. $3x = -18$

4. $-\frac{4}{7}x = -28$

5. $3t + 7 = 2t - 5$

6. $\frac{1}{2}x - \frac{3}{5} = \frac{2}{5}$

7. $8 - y = 16$

8. $-\frac{2}{5} + x = -\frac{3}{4}$

9. $3(x + 2) = 27$

10. $-3x + 6(x + 4) = 9$

11. $0.4p + 0.2 = 4.2p - 7.8 - 0.6p$

12. $|6 - 5x| = 51$

Problem solving

13. State the five-step problem-solving process.

14. What number added to 43 is 60?

15. Four times what number is 56?

16. 16 is 25% of what number?

17. 62% of 125 is what number?

18. Tom's salary is seven-eighths of Jeff's. Tom's salary is $14,700. What is Jeff's salary?

19. On a test of 100 points, student A missed 18 more than student B. If B's score is 83, what is A's score?

20. An investment is made at 12% simple interest for 1 year. It grows to $28,000. How much was originally invested?

21. The perimeter of a rectangle is 36 cm. The length is 4 cm greater than the width. Find the width and length.

22. If you triple a number and then subtract 14, you get $\frac{2}{3}$ of the original number. What is the original number?

23. The sum of three consecutive odd integers is 249. Find the integers.

24. Money is invested in a savings account at 12% simple interest. After one year there is $840 in the account. How much was originally invested?

Solve the formulas for the given letter.

25. Solve $A = 2\pi rh$, for r.

26. Solve $w = \dfrac{P - 2l}{2}$, for l.

27. Solve $c = \dfrac{1}{a - d}$, for d.

28. Solve: $3|w| - 8 = -37$.

29. A movie theater had a certain number of tickets to give away. Five people got the tickets. The first got $\frac{1}{3}$ of the tickets, the second got $\frac{1}{4}$ of the tickets, and the third got $\frac{1}{5}$ of the tickets. The fourth person got eight tickets, and there were five tickets left for the fifth person. Find the total number of tickets given away.

30. Solve using this information.

a) Your boss says, "I'll raise your $1000 salary 50% this month but lower that salary 50% next month." Would you receive more money by taking the offer or by keeping your old salary for two months?

b) Would you receive more money by continuing to alternate 50% increases and decreases indefinitely or by keeping your old salary? (*Hint:* Calculate each for six months.)

3

POLYNOMIALS

Polynomial equations can be used to solve problems concerning interest on investments.

An algebraic expression like $3x^2 - 7x + 5$ is called a *polynomial*. One of the most important parts of introductory algebra is the study of polynomials. In this chapter we learn to add, subtract, and multiply polynomials.

Of particular importance in this chapter is the study of fast ways to find special products of polynomials, which will be helpful not only in this book but also in more advanced mathematics.

THE ASVAB TEST: SAMPLE QUESTIONS

The ASVAB, Armed Services Vocational Aptitude Battery, is a series of tests designed as a measure of your academic ability in relation to admittance to military service. Different branches of the service require different degrees of proficiency. There are three mathematical parts of the ASVAB: (1) *numerical operations* measures your ability to do 50 simple arithmetic operations in three minutes; (2) *arithmetic reasoning* measures your ability to do problem solving using skills of arithmetic; the problems require more careful thinking and may involve more than one step; and (3) *mathematics knowledge* measures your knowledge of the high school subjects of algebra and geometry as well as arithmetic skills.

Periodically throughout this text will be samples of the types of mathematical questions appearing on the ASVAB exam. When an arithmetic question of type (1) or (2) appears, it is assumed that you have had enough mathematics either in this book or in a prior course in order to do the mathematics. If you miss such a question, you may need to consult an arithmetic book. For most of the questions of type (3), you will have had the prerequisite mathematics in this book or in the course that precedes it. Some questions, however, may require review from a geometry or college algebra book. The sample questions that occur here are not all-inclusive. Further review of the mathematical parts of ASVAB may still be necessary, but your work here will certainly enhance your chances of being successful on that test.

ARITHMETIC REASONING

1. The cost of a Navy F-14 Tomcat jet is $24 million. Eventually each jet has to be renovated at 2% of its original cost. How much is that cost?

  A $48 million
  B $4.8 million
  C $12 million
  D $480,000

2. How many days are there in 172,800 seconds?

  A 48
  B 4
  C 2880
  D 2

3. It takes 2.4 gal of paint to cover a side of a house that has an area of 280 ft². How much paint will it take for the whole house if the total area is 1400 ft²?

  A 12 gal
  B 9.6 gal
  C 5 gal
  D 3.6 gal

4. Jack got 32 mi/gal while driving on temporary tour of duty to Heidelberg. Due to a mechanical problem he got only 23 mi/gal on the return trip. What was the percent decrease?

  A 11.43%
  B 9%
  C 28.13%
  D 27%

OBJECTIVES

After finishing Section 3.1, you should be able to:

  i Evaluate a polynomial for a given value of a variable, and use evaluation of polynomials in problem solving.

  ii Identify the terms of a polynomial.

  iii Identify like terms of a polynomial.

  iv Collect like terms of a polynomial.

3.1

POLYNOMIALS

We have already learned how to evaluate certain kinds of algebraic expressions. We have also learned to collect like terms for certain kinds of algebraic expressions. Now we learn to evaluate polynomials and to collect like terms for polynomials.

Algebraic expressions like the following are *polynomials*.

$$\tfrac{3}{4}y^5, \quad -2, \quad 5y + 3, \quad 3x^2 + 2x - 5, \quad -7a^3 + \tfrac{1}{2}a, \quad 6x, \quad 37p^4, \quad x, \quad 0.$$

A *polynomial* is a monomial or a combination of sums and/or differences of monomials.

The following algebraic expressions are *not* polynomials:

$$\frac{x+3}{x-4}, \qquad 5x^3 - 2x^2 + \frac{1}{x}, \quad \text{and} \quad \frac{1}{x^3 - 2}.$$

Polynomials are used often in algebra, especially in equation solving. We will learn how to add, subtract, and multiply polynomials.

DO EXERCISE 1.

■ i Evaluating Polynomials and Polynomials in Problem Solving

When we replace the variable in a polynomial by a number, the polynomial then represents a number. Finding that number is called *evaluating the polynomial.*

EXAMPLES Evaluate each polynomial for $x = 2$.

1. $3x + 5$: $\quad 3 \cdot 2 + 5 = 6 + 5$
$$= 11$$

2. $2x^2 - 7x + 3$: $\quad 2 \cdot 2^2 - 7 \cdot 2 + 3 = 2 \cdot 4 - 14 + 3$
$$= 8 - 14 + 3$$
$$= -3$$

DO EXERCISES 2–5.

Polynomials arise in many real-world situations and are used in problem solving. The following examples are two such applications. Although the examples are problem solving in nature, they involve only the evaluation of a polynomial. For that reason we do not apply the entire problem-solving strategy, although you can if you wish.

EXAMPLE 3 The volume of a cube with side of length x is given by the polynomial
$$x^3.$$

Find the volume of a cube with side of length 5 cm.

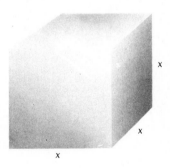

We evaluate the polynomial for $x = 5$:
$$x^3 = 5^3 = 125.$$

The volume is 125 cm^3 (cubic centimeters).

EXAMPLE 4 In a sports league of n teams in which each team plays every other team twice, the total number of games to be played is given by the polynomial
$$n^2 - n.$$

1. Write three polynomials.

Evaluate each polynomial for $x = 3$.

2. $-4x - 7$

3. $-5x^3 + 7x + 10$

Evaluate each polynomial for $y = -4$.

4. $5y + 9$

5. $2y^2 + 5y - 4$

6. In the situation of Example 4, what is the total number of games to be played in a league of 15 teams?

7. The area of a circle of radius r is given by the polynomial

$$\pi r^2.$$

Find the area of a circle of radius 10 mi.

Find an equivalent polynomial using only additions.

8. $-9x^3 - 4x^5$

9. $-2y^3 + 3y^7 - 7y$

Identify the terms of the polynomial.

10. $3x^2 + 6x + \frac{1}{2}$

11. $-4y^5 + 7y^2 - 3y - 2$

A slow-pitch softball league has 10 teams. What is the total number of games to be played?

We evaluate the polynomial for $n = 10$:

$$n^2 - n = 10^2 - 10 = 100 - 10 = 90.$$

The league plays 90 games.

DO EXERCISES 6 AND 7.

ii Identifying Terms

Subtractions can be rewritten as additions. We showed this in Section 2.4. For any polynomial we can find an equivalent polynomial using only additions.

EXAMPLES Find an equivalent polynomial using only additions.

5. $-5x^2 - x = -5x^2 + (-x)$

6. $4x^5 - 2x^6 - 4x = 4x^5 + (-2x^6) + (-4x)$

DO EXERCISES 8 AND 9.

When a polynomial has only additions, the parts being added are called *terms*.

EXAMPLE 7 Identify the terms of the polynomial

$$4x^3 + 3x + 12 + 8x^3 + 5x.$$

Terms: $4x^3$, $3x$, 12, $8x^3$, $5x$

If there are subtractions you can think of them as additions without rewriting.

EXAMPLE 8 Identify the terms of the polynomial

$$3t^4 - 5t^6 - 4t + 2.$$

Terms: $3t^4$, $-5t^6$, $-4t$, 2

Note in an expression like $3x + 5y$ that the terms are the things added: $3x$ and $5y$. Note also that each term is a product.

DO EXERCISES 10 AND 11.

iii Like Terms

Terms that have the same variable and the same exponent are called *like terms*, or *similar terms*.

EXAMPLES Identify the like terms in each polynomial.

9. $4x^3 + 5x - 4x^2 + 2x^3 + x^2$

Like terms: $4x^3$ and $2x^3$ Same exponent and variable
Like terms: $-4x^2$ and x^2 Same exponent and variable

10. $6 - 3a^2 + 8 - a - 5a$

Like terms: 6 and 8 No variable at all
Like terms: $-a$ and $-5a$

DO EXERCISES 12 AND 13.

iv Collecting Like Terms

We can often simplify polynomials by *collecting like terms*, or *combining similar terms*. To do this we use the distributive laws.

EXAMPLES Collect like terms.

11. $2x^3 - 6x^3 = (2 - 6)x^3$ Using a distributive law
            $= -4x^3$

12. $5x^2 + 7 + 4x^4 + 2x^2 - 11 - 2x^4 = (5 + 2)x^2 + (4 - 2)x^4 + (7 - 11)$
                                   $= 7x^2 + 2x^4 - 4$

Note that using the distributive laws in this manner allows us to collect like terms by adding or subtracting the coefficients.

DO EXERCISES 14–16.

In collecting like terms we may get zero.

EXAMPLES Collect like terms.

13. $5x^3 - 5x^3 = (5 - 5)x^3 = 0x^3 = 0$

14. $3x^4 + 2x^2 - 3x^4 = (3 - 3)x^4 + 2x^2 = 0x^4 + 2x^2 = 2x^2$

DO EXERCISES 17–19.

Multiplying a term of a polynomial by 1 does not change the polynomial, but it may make it easier to factor.

EXAMPLES Collect like terms.

15. $5x^2 + x^2 = 5x^2 + 1x^2$ Replacing x^2 by $1x^2$
           $= (5 + 1)x^2$ Using a distributive law
           $= 6x^2$

16. $5x^4 - 6x^3 - x^4 = 5x^4 - 6x^3 - 1x^4$ $x^4 = 1x^4$
                    $= (5 - 1)x^4 - 6x^3$
                    $= 4x^4 - 6x^3$

17. $\frac{2}{3}x^4 - x^3 - \frac{1}{6}x^4 + \frac{2}{5}x^3 - \frac{3}{10}x^3 = \frac{3}{6}x^4 - \frac{9}{10}x^3$
                                     $= \frac{1}{2}x^4 - \frac{9}{10}x^3$

DO EXERCISES 20–23.

Identify the like terms in the polynomial.

12. $4x^3 - x^3 + 2$

13. $4t^4 - 9t^3 - 7t^4 + 10t^3$

Collect like terms.

14. $3x^2 + 5x^2$

15. $4t^3 - 2t^3 + 2 + 5$

16. $\frac{1}{2}y^5 - \frac{3}{4}y^5 + 4y^2 - 2y^2$

Collect like terms.

17. $57 - 4y^3 - 57$

18. $7x^3 - 8x^5 + 8x^5$

19. $-2m^4 + 16 + 2m^4 + 9 - 3m^5$

Collect like terms.

20. $7y - y$

21. $5x^3 - x^3 + 4$

22. $\frac{3}{4}p^3 + 4p^2 - p^3 + 7$

23. $8t^2 - t^2 + t^3 - 1 - 4t^2 + 10$

EXERCISE SET 3.1

i Evaluate each polynomial for $x = 4$.

1. $-5x + 2$ **2.** $-3x + 1$ **3.** $2x^2 - 5x + 7$ **4.** $3x^2 + x + 7$ **5.** $x^3 - 5x^2 + x$ **6.** $7 - x + 3x^2$

The daily number of accidents (average number of accidents per day) involving drivers of age a is approximated by the polynomial $0.4a^2 - 40a + 1039$.

7. Evaluate the polynomial for $a = 18$ to find the number of daily accidents involving a 18-year-old driver.

8. Evaluate the polynomial for $a = 20$ to find the number of daily accidents involving a 20-year-old driver.

Evaluate each polynomial for $x = -1$.

9. $3x + 5$ **10.** $6 - 2x$ **11.** $x^2 - 2x + 1$

12. $5x - 6 + x^2$ **13.** $-3x^3 + 7x^2 - 3x - 2$ **14.** $-2x^3 - 5x^2 + 4x + 3$

The perimeter of a square of side x is given by the polynomial

$$4x.$$

15. Find the perimeter of a square of side 17 ft.

16. Find the perimeter of a square of side 28.5 m.

The distance s, in feet, traveled by a body falling freely from rest in t seconds is approximated by the polynomial

$$16t^2.$$

17. A stone is dropped from a cliff and takes 8 sec to hit the ground. How high is the cliff?

18. A brick is dropped from a building and takes 3 sec to hit the ground. How high is the building?

The amount of water, in gallons, in a tub after it has drained is given by the polynomial

$$400 - 200t + 25t^2,$$

where t is the time, in minutes, that the water has drained, and where t is greater than or equal to 0 and less than or equal to 4.

19. How much water was in the tub before it began draining? **20.** How much water was in the tub after 1 min? 2 min? 3 min?

ii Identify the terms of each polynomial.

21. $2 - 3x + x^2$ **22.** $2x^2 + 3x - 4$

iii Identify the like terms in each polynomial.

23. $5x^3 + 6x^2 - 3x^2$ **24.** $3x^2 + 4x^3 - 2x^2$ **25.** $2x^4 + 5x - 7x - 3x^4$ **26.** $-3t + t^3 - 2t - 5t^3$

iv Collect like terms.

27. $2x - 5x$

28. $2x^2 + 8x^2$

29. $x - 9x$

30. $x - 5x$

31. $5x^3 + 6x^3 + 4$

32. $6x^4 - 2x^4 + 5$

33. $5x^3 + 6x - 4x^3 - 7x$

34. $3a^4 - 2a + 2a + a^4$

35. $6b^5 + 3b^2 - 2b^5 - 3b^2$

36. $2x^2 - 6x + 3x + 4x^2$

37. $\frac{1}{4}x^5 - 5 + \frac{1}{2}x^5 - 2x - 37$

38. $\frac{1}{3}x^3 + 2x - \frac{1}{6}x^3 + 4 - 16$

39. $6x^2 + 2x^4 - 2x^2 - x^4 - 4x^2$

40. $8x^2 + 2x^3 - 3x^3 - 4x^2 - 4x^2$

41. $\frac{1}{4}x^3 - x^2 - \frac{1}{6}x^2 + \frac{3}{8}x^3 + \frac{5}{16}x^3$

42. $\frac{1}{5}x^4 + \frac{1}{5} - 2x^2 + \frac{1}{10} - \frac{3}{15}x^4 + 2x^2 - \frac{3}{10}$

Combine like terms.

43. $3x^2 + 2x - 2 + 3x^0$

44. $\frac{9}{2}x^8 + \frac{1}{9}x^2 + \frac{1}{2}x^9 + \frac{9}{2}x^1 + \frac{9}{2}x^9 + \frac{8}{9}x^2 + \frac{1}{2}x - \frac{1}{2}x^8$

45. $(3x^2)^3 + 4x^2 \cdot 4x^4 - x^4(2x)^2 + ((2x)^2)^3 - 100x^2(x^2)^2$

Solve.

46. 🖩 Evaluate $s^2 - 50s + 675$ and $-s^2 + 50s - 675$ for $s = 18$, $s = 25$, and $s = 32$.

47. 🖩 The daily number of accidents involving drivers of age a is approximated by the polynomial $0.4a^2 - 40a + 1039$. For what age is the number of daily accidents smallest?

3.2

MORE ON POLYNOMIALS

We now consider the basic concepts regarding polynomials. Such concepts are important not only in future work in this chapter but also in other areas of mathematics.

i Descending Order

Note in the following polynomial that the exponents decrease. We say that the polynomial is arranged in *descending order*:

$$8x^4 - 2x^3 + 5x^2 - x + 3.$$

The term with the largest exponent is first. The term with the next largest exponent is second, and so on. The associative and commutative laws allow us to arrange the terms of a polynomial in descending order.

EXAMPLES Arrange each polynomial in descending order.

1. $6x^5 + 4x^7 + x^2 + 2x^3 = 4x^7 + 6x^5 + 2x^3 + x^2$

2. $\frac{2}{3} + 4x^5 - 8x^2 + 5x - 3x^3 = 4x^5 - 3x^3 - 8x^2 + 5x + \frac{2}{3}$

We usually arrange polynomials in descending order. The opposite order is called *ascending*.

DO EXERCISES 1–3.

ii Collecting Like Terms and Descending Order

EXAMPLE 3 Collect like terms and then arrange in descending order.

$$2x^2 - 4x^3 + 3 - x^2 - 2x^3 = x^2 - 6x^3 + 3 \quad \text{Collecting like terms}$$
$$= -6x^3 + x^2 + 3 \quad \text{Arranging in descending order}$$

DO EXERCISES 4 AND 5.

OBJECTIVES

After finishing Section 3.2, you should be able to:

i Arrange a polynomial in descending order.

ii Collect like terms of a polynomial and arrange in descending order.

iii Identify the degrees of terms of polynomials and degrees of polynomials.

iv Identify the coefficients of the terms of a polynomial.

v Identify the missing terms of a polynomial.

vi Tell whether a polynomial is a monomial, binomial, or trinomial, or none of these.

Arrange the polynomial in descending order.

1. $x + 3x^5 + 4x^3 + 5x^2 + 6x^7 - 2x^4$

2. $4x^2 - 3 + 7x^5 + 2x^3 - 5x^4$

3. $-14 + 7t^2 - 10t^5 + 14t^7$

Collect like terms and then arrange in descending order.

4. $3x^2 - 2x + 3 - 5x^2 - 1 - x$

5. $-x + \frac{1}{2} + 14x^4 - 7x - 1 - 4x^4$

Identify the degree of each term and the degree of the polynomial.

6. $-6x^4 + 8x^2 - 2x + 9$

iii Degrees

The degree of the term $5x^3$ is 3. The *degree* of a term is its exponent.

EXAMPLE 4   Identify the degree of each term of $8x^4 + 3x + 7$.

The degree of $8x^4$ is 4.

The degree of $3x$ is 1.   Recall that $x = x^1$.

The degree of 7 is 0.   Think of 7 as $7x^0$. Recall that $x^0 = 1$.

The *degree of a polynomial* is its largest exponent, unless it is the polynomial 0. The polynomial 0 is a special case. Mathematicians agree that it has *no* degree either as a term or as a polynomial.

EXAMPLE 5   Identify the degree of $5x^3 - 6x^4 + 7$.

$$5x^3 - 6x^4 + 7 \qquad \text{The largest exponent is 4.}$$

The degree of the polynomial is 4.

DO EXERCISE 6.

Identify the coefficient of each term.

7. $5x^9 + 6x^3 + x^2 - x + 4$

iv Coefficients

The coefficient of the term $5x^3$ is 5. In the following polynomial the color numbers are the *coefficients*:

$$3x^5 - 2x^3 + 5x + 4.$$

EXAMPLE 6   Identify the coefficient of each term in the polynomial

$$3x^4 - 4x^3 + 7x^2 + x - 8.$$

The coefficient of the first term is 3.

The coefficient of the second term is -4.

The coefficient of the third term is 7.

The coefficient of the fourth term is 1.

The coefficient of the fifth term is -8.

DO EXERCISE 7.

Identify the missing terms in the polynomial.

8. $2x^3 + 4x^2 - 2$

9. $-3x^4$

10. $x^3 + 1$

11. $x^4 - x^2 + 3x + 0.25$

v Missing Terms

If a coefficient is 0, we usually do not write the term. We say that we have a *missing term*.

EXAMPLE 7   In

$$8x^5 - 2x^3 + 5x^2 + 7x + 8,$$

there is no term with x^4. We say that the x^4-term (or the *fourth-degree term*) is missing.

We could write missing terms with zero coefficients or leave space. For example, we could write the polynomial $3x^2 + 9$ as

$$3x^2 + 0x + 9 \quad \text{or} \quad 3x^2 + \qquad 9,$$

but ordinarily we do not.

DO EXERCISES 8–11.

vi Monomials, Binomials, and Trinomials

Polynomials with just one term are called *monomials*. Polynomials with just two terms are called *binomials*. Those with just three terms are called *trinomials*.

EXAMPLE 8

Monomials	Binomials	Trinomials
$4x^2$	$2x + 4$	$3x^3 + 4x + 7$
9	$3x^5 + 6x$	$6x^7 - 7x^2 + 4$
$-23x^{19}$	$-9x^7 - 6$	$4x^2 - 6x - \frac{1}{2}$

DO EXERCISES 12–15.

Tell whether the polynomial is a monomial, binomial, trinomial, or none of these.

12. $5x^4$

13. $4x^3 - 3x^2 + 4x + 2$

14. $3x^2 + x$

15. $3x^2 + 2x - 4$

EXERCISE SET 3.2

i Arrange the polynomial in descending order.

1. $x^5 + x + 6x^3 + 1 + 2x^2$

2. $3 + 2x^2 - 5x^6 - 2x^3 + 3x$

3. $5x^3 + 15x^9 + x - x^2 + 7x^8$

4. $9x - 5 + 6x^3 - 5x^4 + x^5$

5. $8y^3 - 7y^2 + 9y^6 - 5y^8 + y^7$

6. $p^8 - 4 + p + p^2 - 7p^4$

ii Collect like terms and then arrange in descending order.

7. $3x^4 - 5x^6 - 2x^4 + 6x^6$

8. $-1 + 5x^3 - 3 - 7x^3 + x^4 + 5$

9. $-2x + 4x^3 - 7x + 9x^3 + 8$

10. $-6x^2 + x - 5x + 7x^2 + 1$

11. $3x + 3x + 3x - x^2 - 4x^2$

12. $-2x - 2x - 2x + x^3 - 5x^3$

13. $-x + \frac{3}{4} + 15x^4 - x - \frac{1}{2} - 3x^4$

14. $2x - \frac{5}{6} + 4x^3 + x + \frac{1}{3} - 2x$

iii Identify the degree of each term of the polynomial and the degree of the polynomial.

15. $2x - 4$

16. $6 - 3x$

17. $3x^2 - 5x + 2$

18. $5x^3 - 2x^2 + 3$

19. $-7x^3 + 6x^2 + 3x + 7$

20. $5x^4 + x^2 - x + 2$

21. $x^2 - 3x + x^6 - 9x^4$

22. $8x - 3x^2 + 9 - 8x^3$

iv Identify the coefficient of each term of the polynomial.

23. $-3x + 6$

24. $2x - 4$

25. $5x^2 + 3x + 3$

26. $3x^2 - 5x + 2$

27. $-7x^3 + 6x^2 + 3x + 7$

28. $5x^4 + x^2 - x + 2$

29. $-5x^4 + 6x^3 - 3x^2 + 8x - 2$

30. $7x^3 - 4x^2 - 4x + 5$

v Identify the missing terms in the polynomial.

31. $x^3 - 27$

32. $x^5 + x$

33. $x^4 - x$

34. $5x^4 - 7x + 2$

35. $2x^3 - 5x^2 + x - 3$

36. $-6x^3$

vi Tell whether the polynomial is a monomial, binomial, trinomial, or none of these.

37. $x^2 - 10x + 25$

38. $-6x^4$

39. $x^3 - 7x^2 + 2x - 4$

40. $x^2 - 9$

41. $4x^2 - 25$

42. $2x^4 - 7x^3 + x^2 + x - 6$

43. $40x$

44. $4x^2 + 12x + 9$

Problem-solving practice

45. A family spent $2011 to drive a car one year, during which the car was driven 7400 miles. The family spent $972 for insurance and $114 for a license registration fee. The only other cost was for gasoline. How much did gasoline cost per mile?

46. Three tired campers stopped for the night. All they had to eat was a bag of apples. During the night one awoke and ate one third of the apples. Later, a second camper awoke and ate one third of the apples that remained. Much later, the third camper awoke and ate one third of those apples yet remaining after the other two had eaten. When they got up the next morning, 8 apples were left. How many did they have to begin with?

47. Construct a polynomial in x (meaning that x is the variable) of degree 5 with four terms and coefficients that are integers.

48. Construct a trinomial in y of degree 4 with coefficients that are rational numbers.

49. What is the degree of $(5m^5)^2$?

50. Construct three like terms of degree 4.

51. A polynomial in x has degree 3. The coefficient of x^2 is 3 less than the coefficient of x^3. The coefficient of x is 3 times the coefficient of x^2. The remaining coefficient is 2 more than the coefficient of x^3. The sum of the coefficients is -4. Find the polynomial.

OBJECTIVES

After finishing Section 3.3, you should be able to:

i Add polynomials.

ii Solve problems using addition of polynomials.

Add.

1. $3x^2 + 2x - 2$ and $-2x^2 + 5x + 5$

2. $-4y^5 + 3y^3 + 4$ and $7y^4 + 2y^2$

3. $31x^4 + x^2 + 2x - 1$ and $-7x^4 + 5x^3 - 2x + 2$

4. $17x^3 - x^2 + 3x + 4$ and $-15x^3 + x^2 - 3x - \frac{2}{3}$

Add mentally. Try to just write the answer.

5. $(4x^2 - 5x + 3) + (-2x^2 + 2x - 4)$

6. $(3t^3 - 4t^2 - 5t + 3) + (5t^3 + 2t^2 - 3t - \frac{1}{2})$

3.3

ADDITION OF POLYNOMIALS

We now consider addition of polynomials. This addition is based on collecting like terms. We then use addition of polynomials in problem solving.

i Addition

To add two polynomials we could think of writing a plus sign between them and then collecting like terms. Depending on the situation, you may see polynomials written in descending order, ascending order, or neither. Generally, if an exercise is written in one kind of order, we write the answer in that same order.

EXAMPLE 1 Add: $-3x^3 + 2x - 4$ and $4x^3 + 3x^2 + 2$.

$$(-3x^3 + 2x - 4) \quad (4x^3 + 3x^2 + 2)$$
$$= (-3 + 4)x^3 + 3x^2 + 2x + (-4 + 2) \qquad \text{Collecting like terms}$$
$$\qquad\qquad\qquad\qquad\qquad\qquad\qquad (\textit{No} \text{ signs are changed.})$$
$$= x^3 + 3x^2 + 2x - 2$$

EXAMPLE 2 Add: $\frac{2}{3}x^4 + 3x^2 - 2x + \frac{1}{2}$ and $-\frac{1}{3}x^4 + 5x^3 - 3x^2 + 3x - \frac{1}{2}$.

$$(\tfrac{2}{3}x^4 + 3x^2 - 2x + \tfrac{1}{2}) + (-\tfrac{1}{3}x^4 + 5x^3 - 3x^2 + 3x - \tfrac{1}{2})$$
$$= (\tfrac{2}{3} - \tfrac{1}{3})x^4 + 5x^3 + (3 - 3)x^2 + (-2 + 3)x + (\tfrac{1}{2} - \tfrac{1}{2}) \qquad \text{Collecting like terms}$$
$$= \tfrac{1}{3}x^4 + 5x^3 + x$$

DO EXERCISES 1–4.

We can add polynomials as we do because they represent numbers. After some practice you will be able to add mentally.

EXAMPLE 3 Add: $3x^2 - 2x + 2$ and $5x^3 - 2x^2 + 3x - 4$.

$$(3x^2 - 2x + 2) + (5x^3 - 2x^2 + 3x - 4)$$
$$= 5x^3 + (3 - 2)x^2 + (-2 + 3)x + (2 - 4) \qquad \text{You might do this step mentally.}$$
$$= 5x^3 + x^2 + x - 2 \qquad \text{Then you would write only this.}$$

DO EXERCISES 5 AND 6.

We can also add polynomials by writing like terms in columns.

EXAMPLE 4 Add: $9x^5 - 2x^3 + 6x^2 + 3$ and $5x^4 - 7x^2 + 6$ and $3x^6 - 5x^5 + x^2 + 5$.

Arrange the polynomials with like terms in columns.

$$9x^5 \qquad -2x^3 + 6x^2 + 3$$
$$5x^4 \qquad -7x^2 + 6 \qquad \text{We leave spaces for missing terms.}$$
$$\underline{3x^6 - 5x^5 \qquad\qquad + \ x^2 + 5}$$
$$3x^6 + 4x^5 + 5x^4 - 2x^3 \qquad\quad + 14$$

We write the answer as $3x^6 + 4x^5 + 5x^4 - 2x^3 + 14$ without the missing space.

DO EXERCISES 7 AND 8.

ii Problem Solving

The first two steps in our problem-solving process are [Familiarize] and [Translate].
In Example 5 we consider only these steps. You will see how a problem can be translated to a polynomial.

EXAMPLE 5 Find a polynomial for the sum of the areas of these rectangles.

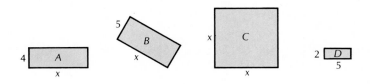

1. [Familiarize] Recall that the area of a rectangle is the product of the length and the width. (You may need to look up such a formula.)

2. [Translate] We translate the problem to mathematical language. The sum of the areas is a sum of products. We find these products and then collect like terms.

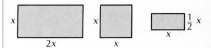

$$\text{area of } A \text{ plus } \text{area of } B \text{ plus } \text{area of } C \text{ plus } \text{area of } D$$

$$4x \qquad + \qquad 5x \qquad + \qquad x \cdot x \qquad + \qquad 2 \cdot 5$$
$$= 4x + 5x + x^2 + 10$$
$$= x^2 + 9x + 10$$

A polynomial for the sum of the areas is $x^2 + 9x + 10$.

DO EXERCISES 9 AND 10.

Add.

7.
$$-2x^3 + 5x^2 - 2x + 4$$
$$x^4 \qquad + 6x^2 + 7x - 10$$
$$-9x^4 + 6x^3 + \ x^2 \qquad - 2$$

8. $-3x^3 + 5x + 2$ and
$x^3 + x^2 + 5$ and $x^3 - 2x - 4$

9. Find a polynomial for the sum of the areas of these rectangles.

10. Find the sum of the areas in Exercise 9 when $x = 8$.

EXERCISE SET 3.3

i Add.

1. $3x + 2$ and $-4x + 3$

2. $5x^2 + 6x + 1$ and $-7x + 2$

3. $-6x + 2$ and $x^2 + x - 3$

4. $6x^4 + 3x^3 - 1$ and $4x^2 - 3x + 3$

5. $3x^5 + 6x^2 - 1$ and $7x^2 + 6x - 2$

6. $7x^3 + 3x^2 + 6x$ and $-3x^2 - 6$

7. $-4x^4 + 6x^2 - 3x - 5$ and $6x^3 + 5x + 9$

8. $5x^3 + 6x^2 - 3x + 1$ and $5x^4 - 6x^3 + 2x - 5$

9. $(1 + 4x + 6x^2 + 7x^3) + (5 - 4x + 6x^2 - 7x^3)$

10. $(3x^4 - 6x - 5x^2 + 5) + (6x^2 - 4x^3 - 1 + 7x)$

11. $5x^4 - 6x^3 - 7x^2 + x - 1$ and $4x^3 - 6x + 1$

12. $8x^5 - 6x^3 + 6x + 5$ and $-4x^4 + 3x^3 - 7x$

13. $9x^8 - 7x^4 + 2x^2 + 5$ and $8x^7 + 4x^4 - 2x$

14. $4x^5 - 6x^3 - 9x + 1$ and $6x^3 + 9x^2 + 9x$

15. $\frac{1}{4}x^4 + \frac{2}{3}x^3 + \frac{5}{8}x^2 + 7$ and $-\frac{3}{4}x^4 + \frac{3}{8}x^2 - 7$

16. $(\frac{1}{3}x^9 + \frac{1}{5}x^5 - \frac{1}{2}x^2 + 7) + (-\frac{1}{5}x^9 + \frac{1}{4}x^4 - \frac{3}{5}x^5 + \frac{3}{4}x^2 + \frac{1}{2})$

17. $0.02x^5 - 0.2x^3 + x + 0.08$ and $-0.01x^5 + x^4 - 0.8x - 0.02$

18. $(0.03x^6 + 0.05x^3 + 0.22x + 0.05) + (\frac{7}{100}x^6 - \frac{3}{100}x^3 + 0.5)$

19.
$$-3x^4 + 6x^2 + 2x - 1$$
$$-3x^2 + 2x + 1$$

20.
$$-4x^3 + 8x^2 + 3x - 2$$
$$-4x^2 + 3x + 2$$

21.
$$3x^5 \qquad - 6x^3 \qquad + 3x$$
$$-3x^4 + 3x^3 + x^2$$

22.
$$4x^5 \qquad - 5x^3 \qquad + 2x$$
$$-4x^4 + 2x^3 + 2x^2$$

23.
$$-3x^2 + x$$
$$5x^3 - 6x^2 \qquad + 1$$
$$3x - 8$$

24.
$$-4x^2 + 2x$$
$$3x^3 - 5x^2 \qquad + 3$$
$$5x - 5$$

25.
$$-\tfrac{1}{2}x^4 - \tfrac{3}{4}x^3 \qquad + 6x$$
$$\tfrac{1}{2}x^3 + x^2 + \tfrac{1}{4}x$$
$$\tfrac{3}{4}x^4 \qquad + \tfrac{1}{2}x^2 + \tfrac{1}{2}x + \tfrac{1}{4}$$

26.
$$-\tfrac{1}{4}x^4 - \tfrac{1}{2}x^3 \qquad + 2x$$
$$\tfrac{3}{4}x^3 - x^2 + \tfrac{1}{2}x$$
$$\tfrac{1}{2}x^4 \qquad + \tfrac{1}{2}x^2 + \tfrac{1}{2}x + \tfrac{1}{2}$$

27.
$$-4x^2$$
$$4x^4 - 3x^3 + 6x^2 + 5x$$
$$6x^3 - 8x^2 \qquad + 1$$
$$-5x^4$$
$$6x^2 - 3x$$

28.
$$3x^2$$
$$5x^4 - 2x^3 + 4x^2 + 5x$$
$$5x^3 - 5x^2 \qquad + 2$$
$$-7x^4$$
$$3x^2 - 2x$$

29.
$$3x^4 - 6x^2 + 7x$$
$$3x^2 - 3x + 1$$
$$-2x^4 + 7x^2 + 3x$$
$$5x - 2$$

30.
$$5x^4 - 8x^2 + 4x$$
$$5x^2 - 2x + 3$$
$$-3x^4 + 3x^2 + 5x$$
$$3x - 5$$

31.
$$3x^5 - 6x^4 + 3x^3 \qquad - 1$$
$$6x^4 - 4x^3 + 6x^2$$
$$3x^5 \qquad + 2x^3$$
$$-6x^4 \qquad - 7x^2$$
$$-5x^5 \qquad + 3x^3 \qquad + 2$$

32.
$$4x^5 - 3x^4 + 2x^3 \qquad - 2$$
$$6x^4 + 5x^3 + 3x^2$$
$$5x^5 \qquad + 4x^3$$
$$-6x^4 \qquad - 5x^2$$
$$-3x^5 \qquad + 2x^3 \qquad + 5$$

33.
$$-x^3 + 6x^2 + 3x + 5$$
$$x^4 \qquad - 3x^2 \qquad + 2$$
$$-5x + 3$$
$$6x^4 \qquad + 4x^2 \qquad - 1$$
$$-x^3 \qquad + 6x$$

34.
$$-2x^3 + 3x^2 + 5x + 3$$
$$x^4 \qquad - 5x^2 \qquad + 1$$
$$-7x + 4$$
$$4x^4 \qquad + 6x^2 \qquad - 2$$
$$-x^3 \qquad + 5x$$

35.
$$1 + 5x - 6x^2 + 6x^3 - 3x^4$$
$$-5x \qquad - 3x^3 \qquad + 5x^5$$
$$-8 \qquad + 7x^2 \qquad + 4x^4$$
$$1 + 3x \qquad - 2x^5$$

36.
$$2 + 3x - 7x^2 + 4x^3 - 5x^4$$
$$-6x \qquad - 7x^3 \qquad + 3x^5$$
$$-3 + 5x \qquad + 5x^3 + 3x^4$$
$$4 \qquad + 10x^2 \qquad - 5x^5$$

37.
$$0.15x^4 + 0.10x^3 - 0.9x^2$$
$$-0.01x^3 + 0.01x^2 + x$$
$$1.25x^4 \qquad + 0.11x^2 \qquad + 0.01$$
$$0.27x^3 \qquad + 0.99$$
$$-0.35x^4 \qquad + 15x^2 \qquad - 0.03$$

38.
$$0.05x^4 + 0.12x^3 - 0.5x^2$$
$$-0.02x^3 + 0.02x^2 + 2x$$
$$1.5x^4 \qquad + 0.01x^2 \qquad + 0.15$$
$$0.25x^3 \qquad + 0.85$$
$$-0.25x^4 \qquad + 10x^2 \qquad - 0.04$$

ii *Problem solving*

39. Solve.

a) Find a polynomial for the sum of the areas of these rectangles.

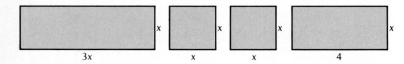

b) Find the sum of the areas when $x = 3$ and $x = 8$.

40. Solve.

a) Find a polynomial for the sum of the areas of these circles.

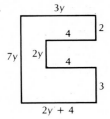

b) Find the sum of the areas when $r = 5$ and $r = 11.3$.

Find a polynomial for the perimeter of each figure.

41.

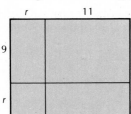

42.

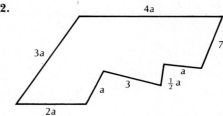

For each figure find two algebraic expressions for the area.

43.

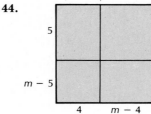

44.

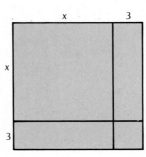

45. Find $(x + 3)^2$ using the four areas of the square shown here.

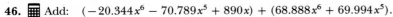

46. ▦ Add: $(-20.344x^6 - 70.789x^5 + 890x) + (68.888x^6 + 69.994x^5)$.

47. Addition of real numbers is commutative. That is, $a + b = b + a$, where a and b are any real numbers.

a) Show that addition of binomials such as $(ax + b)$ and $(cx + d)$ is commutative.

b) Show that addition of trinomials such as $(ax^2 + bx + c)$ and $(dx^2 + ex + f)$ is commutative.

48. The sum of a number and 2 is multiplied by the number and then 3 is subtracted from the result. Find a polynomial for the final result.

49. Three brothers have ages that are consecutive multiples of five. The sum of their ages two years ago was 69. Find their ages now.

50. Find four consecutive multiples of four when the sum of the first two is the fourth.

OBJECTIVES

After finishing Section 3.4, you should be able to:

i Find two equivalent polynomials for the additive inverse of a polynomial, and simplify an expression such as $-(3x^2 + x - 2)$ by replacing each term by its additive inverse.

ii Subtract polynomials.

iii Solve problems using subtraction and/or addition of polynomials.

Find two equivalent expressions for the additive inverse of the polynomial.

1. $12x^4 - 3x^2 + 4x$

2. $-4x^4 + 3x^2 - 4x$

3. $-13x^6 + 2x^4 - 3x^2 + x - \frac{5}{13}$

4. $-7y^3 + 2y^2 - y + 3$

Simplify.

5. $-(4x^3 - 6x + 3)$

6. $-(5x^4 + 3x^2 + 7x - 5)$

7. $-(14x^{10} - \frac{1}{2}x^5 + 5x^3 - x^2 + 3x)$

3.4

SUBTRACTION OF POLYNOMIALS

We now consider subtraction of polynomials. To do so we first consider the concept of the additive inverse of a polynomial.

i Additive Inverses of Polynomials

We know that two numbers are additive inverses of each other if their sum is zero. The same definition is true of polynomials.

> Two polynomials are *additive inverses* of each other if their sum is zero.

To find a way to determine an additive inverse, look for a pattern in the following examples:

a) $2x + (-2x) = 0$;

b) $-6x^2 + 6x^2 = 0$;

c) $(5t^3 - 2) + (-5t^3 + 2) = 0$;

d) $(7x^3 - 6x^2 - x + 4) + (-7x^3 + 6x^2 + x - 4) = 0$.

Since $(5t^3 - 2) + (-5t^3 + 2) = 0$, we know that the additive inverse of $(5t^3 - 2)$ is $(-5t^3 + 2)$. To say the same thing with purely algebraic symbolism, consider

The additive inverse of $(5t^3 - 2)$ is $-5t^3 + 2$.

$$-(5t^3 - 2) = -5t^3 + 2.$$

> We can find an equivalent polynomial for the additive inverse of a polynomial by replacing each term by its additive inverse (changing the sign of every term).

EXAMPLE 1 Find two equivalent expressions for the additive inverse of $4x^5 - 7x^3 - 8x + \frac{5}{6}$.

a) $-(4x^5 - 7x^3 - 8x + \frac{5}{6})$

b) $-4x^5 + 7x^3 + 8x - \frac{5}{6}$ Changing the sign of every term

DO EXERCISES 1–4.

EXAMPLE 2 Simplify: $-(-7x^4 - \frac{5}{9}x^3 + 8x^2 - x + 67)$.

$$-(-7x^4 - \frac{5}{9}x^3 + 8x^2 - x + 67) = 7x^4 + \frac{5}{9}x^3 - 8x^2 + x - 67$$

DO EXERCISES 5–7.

ii Subtraction of Polynomials

Recall that we can subtract a rational number by adding its additive inverse: $a - b = a + (-b)$. This allows us to find an equivalent expression for the difference of two polynomials.

131

EXAMPLE 3 Subtract: $(9x^5 + x^3 - 2x^2 + 4) - (2x^5 + x^4 - 4x^3 - 3x^2)$.

$(9x^5 + x^3 - 2x^2 + 4) - (2x^5 + x^4 - 4x^3 - 3x^2)$

$\quad = (9x^5 + x^3 - 2x^2 + 4) + [\ (2x^5 + x^4 - 4x^3 - 3x^2)]$ Adding an inverse

$\quad = (9x^5 + x^3 - 2x^2 + 4) + (-2x^5 - x^4 + 4x^3 + 3x^2)$ Finding the inverse by changing the sign of *every* term

$\quad = 7x^5 - x^4 + 5x^3 + x^2 + 4$ Collecting like terms

DO EXERCISES 8 AND 9.

After some practice you will be able to subtract mentally.

EXAMPLE 4 Subtract: $(9x^5 + x^3 - 2x) - (-2x^5 + 5x^3 + 6)$.

$(9x^5 + x^3 - 2x) - (-2x^5 + 5x^3 + 6)$

$\quad = (9x^5 + 2x^5) + (x^3 - 5x^3) - 2x - 6$ Subtract the like terms mentally.

$\quad = 11x^5 - 4x^3 - 2x - 6$ Write only this.

DO EXERCISES 10 AND 11.

We can use columns to subtract. We replace coefficients by their inverses, as shown in Example 3. You can also do it mentally.

EXAMPLE 5 Subtract: $(5x^2 - 3x + 6) - (9x^2 - 5x - 3)$.

a) $5x^2 - 3x + 6$ Writing similar terms in columns
$\quad 9x^2 - 5x - 3$

b) $\quad 5x^2 - 3x + 6$
$\quad -9x^2 \pm 5x \pm 3$ Changing signs

c) $\quad 5x^2 - 3x + 6$
$\quad -9x^2 + 5x + 3$ Adding
$\quad \overline{-4x^2 + 2x + 9}$

If you can do so without error, you should skip step (b). Just write the answer.

EXAMPLE 6 Subtract: $(x^3 + x^2 + 2x - 12) - (2x^3 + x^2 - 3x)$.

$\quad\quad x^3 + x^2 + 2x - 12$
$\quad\quad 2x^3 + x^2 - 3x$
$\quad\overline{-x^3 \quad\quad + 5x - 12}$

DO EXERCISES 12 AND 13.

iii Problem Solving

EXAMPLE 7 A 4-ft by 4-ft sandbox is placed on a square lawn x ft on a side. Find a polynomial for the remaining area.

Subtract.

8. $(7x^3 + 2x + 4) - (5x^3 - 4)$

9. $(-3y^2 + 5y - 4)$
$\quad - (-4y^2 + 11y - 2)$

Subtract mentally. Try to write just the answer.

10. $(-6x^4 + 3x^2 + 6)$
$\quad - (2x^4 + 5x^3 - 5x^2 + 7)$

11. $(\frac{3}{2}x^3 - \frac{1}{2}x^2 + 0.3)$
$\quad - (\frac{1}{2}x^3 + \frac{1}{2}x^2 + \frac{4}{3}x + 1.2)$

12. Subtract the second polynomial from the first. Use columns.

$\quad 4t^3 + 2t^2 - 2t - 3,$
$\quad 2t^3 - 3t^2 + 2$

13. Subtract.

$\quad\quad 2x^3 + x^2 - 6x + 2$
$\quad x^5 + 4x^3 - 2x^2 - 4x$

14. Find a polynomial for the area shown in color.

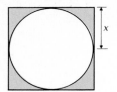

1. [Familiarize] We draw a picture of the situation as follows.

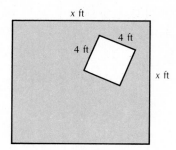

2. [Translate] We reword the problem and translate as follows.

$$\underbrace{(\text{Area of lawn})} - \underbrace{(\text{Area of sandbox})} = \text{Area left over}$$

$$x \cdot x \quad - \quad 4 \cdot 4 \quad = \text{Area left over}$$

3. [Carry out] We carry out the manipulations by multiplying the numbers:

$$x^2 - 16 = \text{Area left over.}$$

4. [Check] We could check the problem by assigning some value to x, say, 10, and carrying out the computation of the area two ways:

a) Area of lawn $= 10 \cdot 10 = 100$;

b) Area of sandbox $= 4 \cdot 4 = 16$;

c) Area left over $= 100 - 16 = 84$.

This is the same as substituting 10 for x in $x^2 - 16$:

$$10^2 - 16 = 100 - 16 = 84.$$

5. [State] The area left over is $x^2 - 16$.

DO EXERCISE 14.

EXERCISE SET 3.4

---

i Find two equivalent expressions for the additive inverse of each polynomial.

1. $-5x$

2. $x^2 - 3x$

3. $-x^2 + 10x - 2$

4. $-4x^3 - x^2 - x$

5. $12x^4 - 3x^3 + 3$

6. $4x^3 - 6x^2 - 8x + 1$

Simplify.

7. $-(3x - 7)$

8. $-(-2x + 4)$

9. $-(4x^2 - 3x + 2)$

10. $-(-6a^3 + 2a^2 - 9a + 1)$

11. $-(-4x^4 - 6x^2 + \frac{3}{4}x - 8)$

12. $-(-5x^4 + 4x^3 - x^2 + 0.9)$

ii Subtract.

13. $(5x^2 + 6) - (3x^2 - 8)$

14. $(7x^3 - 2x^2 + 6) - (7x^2 + 2x - 4)$

15. $(6x^5 - 3x^4 + x + 1) - (8x^5 + 3x^4 - 1)$

16. $(\frac{1}{2}x^2 - \frac{3}{2}x + 2) - (\frac{3}{2}x^2 + \frac{1}{2}x - 2)$

17. $(6x^2 + 2x) - (-3x^2 - 7x + 8)$

18. $7x^3 - (-3x^2 - 2x + 1)$

19. $(\frac{5}{8}x^3 - \frac{1}{4}x - \frac{1}{3}) - (-\frac{1}{8}x^3 + \frac{1}{4}x - \frac{1}{3})$

20. $(\frac{1}{5}x^3 + 2x^2 - 0.1) - (-\frac{2}{5}x^3 + 2x^2 + 0.01)$

21. $(0.08x^3 - 0.02x^2 + 0.01x) - (0.02x^3 + 0.03x^2 - 1)$

22. $(0.8x^4 + 0.2x - 1) - (\frac{7}{10}x^4 + \frac{1}{5}x - 0.1)$

23. $x^2 + 5x + 6$
$x^2 + 2x$ ___

24. $x^3 \quad\quad + 1$
$x^3 + x^2$ ___

25. $x^4 \quad\quad - 3x^2 + x + 1$
$x^4 - 4x^3$ ___

26. $3x^2 - 6x + 1$
$6x^2 + 8x - 3$

27. $\quad 5x^4 + 6x^3 - 9x^2$
$-6x^4 - 6x^3 \quad\quad + 8x + 9$

28. $5x^4 \quad\quad + 6x^2 - 3x + 6$
$\quad\quad 6x^3 + 7x^2 - 8x - 9$

29. $\quad\quad 3x^4 + 6x^2 + 8x - 1$
$4x^5 - 6x^4 \quad\quad - 8x - 7$

30. $6x^5 \quad\quad + 3x^2 - 7x + 2$
$10x^5 + 6x^3 - 5x^2 - 2x + 4$

31. $x^5 \quad\quad\quad\quad - 1$
$x^5 - x^4 + x^3 - x^2 + x - 1$

32. $x^5 + x^4 - x^3 + x^2 - x + 2$
$x^5 - x^4 + x^3 - x^2 - x + 2$

---

iii *Problem solving*

Find a polynomial for the color area of each figure.

33.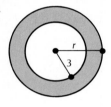

34.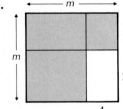

35.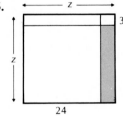

36.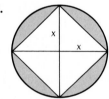

37.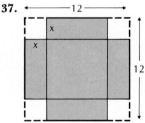

38. Find $(y - 2)^2$ using the four parts of this square.

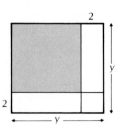

Simplify.

39. $(y + 4) + (y - 5) - (y + 8)$

40. $(7y^2 - 5y + 6) - (3y^2 + 8y - 12) + (8y^2 - 10y + 3)$

41. $(4a^2 - 3a) + (7a^2 - 9a - 13) - (6a - 9)$

42. $(3x^2 - 4x + 6) - (-2x^2 + 4) + (-5x - 3)$

43. $(-8y^2 - 4) - (3y + 6) - (2y^2 - y)$

44. $(5x^3 - 4x^2 + 6) - (2x^3 + x^2 - x) + (x^3 - x)$

45. $(-y^4 - 7y^3 + y^2) + (-2y^4 + 5y - 2) - (-6y^3 + y^2)$

46. $(-4 + x^2 + 2x^3) - (-6 - x + 3x^3) - (-x^2 - 5x^3)$

47. 🖩 Subtract:
$(345.099x^3 - 6.178x) - (-224.508x^3 + 8.99x).$

48. Does replacing each occurrence of the variable x in $5x^3 - 3x^2 + 2x$ with its additive inverse result in the additive inverse of the polynomial?

OBJECTIVES

After finishing Section 3.5, you should be able to:

i Multiply two monomials.

ii Multiply a monomial and a binomial.

iii Multiply two binomials.

iv Multiply any polynomials.

Multiply.

1. $3x$ and -5

2. $-x$ and x

3. $-x$ and $-x$

4. $-x^2$ and x^3

5. $3x^5$ and $4x^2$

6. $4y^5$ and $-2y^6$

7. $-7y^4$ and $-y$

8. $7x^5$ and 0

Multiply.

9. $4x$ and $2x + 4$

10. $3t^2$ and $-5t + 2$

Multiply.

11. $x + 8$ and $x + 5$

12. $(x + 5)(x - 4)$

3.5

MULTIPLICATION OF POLYNOMIALS

We now multiply polynomials using techniques based, for the most part, on the distributive laws. As we proceed in this chapter we will develop special ways to find certain products.

i Multiplying Monomials

To find an equivalent expression for the product of two monomials, we multiply the coefficients and then use properties of exponents. Recall that a product such as $(3x) \cdot (4x)$ can be written as $(3x)(4x)$.

EXAMPLES Multiply.

1. $(3x)(4x) = (3 \cdot 4)(x \cdot x)$ Multiplying the coefficients
$= 12x^2$ Simplifying

2. $(3x)(-x) = (3x)(-1x)$
$= (3)(-1)(x \cdot x) = -3x^2$

3. $(-7x^5)(4x^3) = (-7 \cdot 4)(x^5 \cdot x^3)$
$= -28x^{5+3}$
$= -28x^8$ Adding exponents and simplifying

After some practice you can do this mentally. Multiply the coefficients and add the exponents. Write only the answer.

DO EXERCISES 1–8.

ii Multiplying a Monomial and a Binomial

To find an equivalent expression for the product of a monomial, such as $2x$, by a binomial, such as $5x + 3$, we use a distributive law.

EXAMPLE 4 Multiply: $2x$ and $5x + 3$.
$$(2x)(5x + 3) = (2x)(5x) + (2x)(3)$$ Using a distributive law
$$= 10x^2 + 6x$$ Multiplying the monomials

DO EXERCISES 9 AND 10.

iii Multiplying Two Binomials

To find an equivalent expression for the product of two binomials, we use the distributive laws more than once. In Example 5 we use a distributive law three times.

EXAMPLE 5 Multiply: $x + 5$ and $x + 4$.
$$(x + 5)(x + 4) = (x + 5)x + (x + 5)4$$ Using a distributive law
$$= x \cdot x + 5 \cdot x + x \cdot 4 + 5 \cdot 4$$ Using a distributive law on each part
$$= x^2 + 5x + 4x + 20$$ Multiplying the monomials
$$= x^2 + 9x + 20$$ Collecting like terms

DO EXERCISES 11 AND 12.

EXAMPLE 6 Multiply: $4x + 3$ and $x - 2$.

$$(4x + 3)(x - 2) = (4x + 3)x + (4x + 3)(-2) \qquad \text{Using a distributive law}$$
$$= (4x) \cdot x + (3) \cdot x + (4x)(-2) + (3)(-2) \qquad \text{Using a distributive law on each part}$$
$$= 4x^2 + 3x - 8x - 6 \qquad \text{Multiplying the monomials}$$
$$= 4x^2 - 5x - 6 \qquad \text{Collecting like terms}$$

DO EXERCISES 13 AND 14.

iv Multiplying Any Polynomials

Let us consider the product of a binomial and a trinomial. We again use a distributive law three times.

EXAMPLE 7 Multiply: $(x^2 + 2x - 3)(x^2 + 4)$.

$$(x^2 + 2x - 3)(x^2 + 4) = (x^2 + 2x - 3)x^2 + (x^2 + 2x - 3)4$$
$$= (x^2)(x^2) + 2x(x^2) - 3(x^2) + (x^2) \cdot 4 + (2x) \cdot 4 + (-3) \cdot 4$$
$$= x^4 + 2x^3 - 3x^2 + 4x^2 + 8x - 12$$
$$= x^4 + 2x^3 + x^2 + 8x - 12$$

DO EXERCISES 15 AND 16.

Perhaps you have discovered the following in the preceding examples.

> **To multiply two polynomials, multiply each term of one by every term of the other. Then add the results.**

We can use columns for long multiplications. We multiply each term at the top by every term at the bottom. We locate like terms in columns, and then we add the results.

EXAMPLE 8 Multiply: $(4x^2 - 2x + 3)(x + 2)$.

$$
\begin{array}{r}
4x^2 - 2x + 3 \\
x + 2 \\
\hline
4x^3 - 2x^2 + 3x \qquad \text{Multiplying the top row by } x \\
8x^2 - 4x + 6 \qquad \text{Multiplying the top row by 2} \\
\hline
4x^3 + 6x^2 - x + 6 \qquad \text{Adding}
\end{array}
$$

EXAMPLE 9 Multiply: $(5x^3 - 3x + 4)(-2x^2 - 3)$.

$$
\begin{array}{r}
5x^3 - 3x + 4 \\
-2x^2 - 3 \\
\hline
-10x^5 + 6x^3 - 8x^2 \qquad \text{Multiplying by } -2x^2 \\
-15x^3 + 9x - 12 \qquad \text{Multiplying by } -3 \\
\hline
-10x^5 - 9x^3 - 8x^2 + 9x - 12 \qquad \text{Adding}
\end{array}
$$

When we multiplied $-2x^2$ by $-3x$, the power dropped from x^5 to x^3 so we left a space for the missing x^4 term.

In addition, we leave spaces for "missing terms."

DO EXERCISES 17 AND 18.

Multiply.

13. $5x + 3$ and $x - 4$

14. $(2x - 3)(3x - 5)$

Multiply.

15. $(x^2 + 3x - 4)(x^2 + 5)$

16. $(2y^3 - 2y + 5)(3y^2 - 7)$

Multiply.

17. $3x^2 - 2x + 4$
   $\quad\ x + 5$

18. $-5x^2 + 4x + 2$
   $\quad -4x^2 - 8$

Multiply.

19. $3x^2 - 2x - 5$
$2x^2 + x - 2$

EXAMPLE 10 Multiply: $(2x^2 + 3x - 4)(2x^2 - x + 3)$.

$$
\begin{array}{r}
2x^2 + 3x \ - 4 \\
2x^2 - \ x \ + 3 \\
\hline
4x^4 + 6x^3 - 8x^2 \\
- 2x^3 - 3x^2 + \ 4x \\
6x^2 + \ 9x - 12 \\
\hline
4x^4 + 4x^3 - 5x^2 + 13x - 12
\end{array}
$$

Multiplying by $2x^2$

Multiplying by $-x$

Multiplying by 3

Adding

DO EXERCISE 19.

EXERCISE SET 3.5

i Multiply.

1. $6x^2$ and 7

2. $5x^2$ and -2

3. $-x^3$ and $-x$

4. $-x^4$ and x^2

5. $-x^5$ and x^3

6. $-x^6$ and $-x^2$

7. $3x^4$ and $2x^2$

8. $5x^3$ and $4x^5$

9. $7t^5$ and $4t^3$

10. $10a^2$ and $3a^2$

11. $-0.1x^6$ and $0.2x^4$

12. $0.3x^3$ and $-0.4x^6$

13. $-\frac{1}{5}x^3$ and $-\frac{1}{3}x$

14. $-\frac{1}{4}x^4$ and $\frac{1}{5}x^8$

15. $-4x^2$ and 0

ii Multiply.

16. $3x$ and $-x + 5$

17. $2x$ and $4x - 6$

18. $4x^2$ and $3x + 6$

19. $5x^2$ and $-2x + 1$

20. $-6x^2$ and $x^2 + x$

21. $-4x^2$ and $x^2 - x$

22. $3y^2$ and $6y^4 + 8y^3$

23. $4y^4$ and $y^3 - 6y^2$

iv Multiply.

24. $3x^4$ and $14x^{50} + 20x^{11} + 6x^{57} + 60x^{15}$

25. $5x^6$ and $4x^{32} - 10x^{19} + 5x^8$

26. $-4a^7$ and $20a^{19} + 6a^{15} - 5a^{12} + 14a$

27. $-6y^8$ and $11y^{100} - 7y^{50} + 11y^{41} - 60y^4 + 9$

iii Multiply.

28. $(x + 6)(x + 3)$

29. $(x + 5)(x + 2)$

30. $(x + 5)(x - 2)$

31. $(x + 6)(x - 2)$

32. $(x - 4)(x - 3)$

33. $(x - 7)(x - 3)$

34. $(x + 3)(x - 3)$

35. $(x + 6)(x - 6)$

36. $(5 - x)(5 - 2x)$

37. $(3 + x)(6 + 2x)$

38. $(2x + 5)(2x + 5)$

39. $(3x - 4)(3x - 4)$

40. $(3y - 4)(3y + 4)$

41. $(2y + 1)(2y - 1)$

42. $(x - \frac{5}{2})(x + \frac{2}{5})$

43. $(x + \frac{4}{3})(x + \frac{3}{2})$

iv Multiply.

44. $(x^2 + x + 1)(x - 1)$

45. $(x^2 - x + 2)(x + 2)$

46. $(2x^2 + 6x + 1)(2x + 1)$

47. $(4x^2 - 2x - 1)(3x - 1)$

48. $(3y^2 - 6y + 2)(y^2 - 3)$

49. $(y^2 + 6y + 1)(3y^2 - 3)$

50. $(x^3 + x^2 - x)(x^3 + x^2)$

51. $(x^3 - x^2 + x)(x^3 - x^2)$

52. $(-5x^3 - 7x^2 + 1)(2x^2 - x)$

53. $(-4x^3 + 5x^2 - 2)(5x^2 + 1)$

54. $(1 + x + x^2)(-1 - x + x^2)$

55. $(1 - x + x^2)(1 - x + x^2)$

56. $(2x^2 + 3x - 4)(2x^2 + x - 2)$

57. $(2x^2 - x - 3)(2x^2 - 5x - 2)$

58. $(2t^2 - t - 4)(3t^2 + 2t - 1)$

59. $(3a^2 - 5a + 2)(2a^2 - 3a + 4)$

60. $(2x^2 + x - 2)(-2x^2 + 4x - 5)$

61. $(3x^2 - 8x + 1)(-2x^2 - 4x + 2)$

62. $(x - x^3 + x^5)(x^2 - 1 + x^4)$

63. $(x - x^3 + x^5)(3x^2 + 3x^6 + 3x^4)$

64. $(x^3 + x^2 + x + 1)(x - 1)$

65. $(x^3 - x^2 + x - 2)(x - 2)$

66. $(x^3 + x^2 - x - 3)(x - 3)$

67. $(x^3 - x^2 - x + 4)(x + 4)$

Multiply.

68. $(a + b)^2$

69. $(a - b)^2$

70. $(2x + 3)^2$

71. $(5y + 6)^2$

Problem Solving

Find an expression for each shaded area.

72.

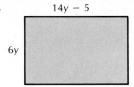

73.

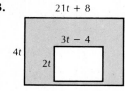
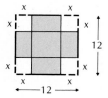

74. A box with a square bottom is to be made from a 12-inch-square piece of cardboard. Squares with side x are cut out of the corners and the sides are folded up. Find polynomials for the volume and the outside surface area of the box.

75. An open wooden box is a cube with side x cm. The wood from which the box is made is 1 cm thick. Find a polynomial for the interior volume of the cube.

76. The height of a triangle is 4 ft longer than its base. The area of the triangle is 30 ft². Find its height and base.

77. A rectangular garden is twice as long as it is wide. It is surrounded by a sidewalk 4 ft wide. The area of the garden and sidewalk together is 256 ft² more than the area of the garden alone. Find the dimensions of the garden.

Compute.

78. a) $(x + 3)(x + 6) + (x + 3)(x + 6)$
b) $(x + 4)(x + 5) - (x + 4)(x + 5)$

80. a) $(x + 5)(x - 3) + (x + 5)(x - 3)$
b) $(x + 9)(x - 4) - (x + 9)(x - 4)$

79. a) $(x - 2)(x - 7) + (x - 2)(x - 7)$
b) $(x - 6)(x - 2) - (x - 6)(x - 2)$

81. a) $(x + 7)(x - 8) + (x - 7)(x + 8)$
b) $(x + 2)(x - 5) - (x - 2)(x + 5)$

3.6

SPECIAL PRODUCTS OF POLYNOMIALS

We now consider special products of polynomials. These are products that we encounter so often that it is helpful to have methods of computing them that are faster than multiplying each term of one by each term of the other.

i   Product of a Monomial and Any Polynomial

There is a quick way to multiply a monomial and any polynomial. We use the distributive law mentally, multiplying every term inside by the monomial. Just write the answer.

EXAMPLE 1 Multiply: $5x(2x^2 - 3x + 4)$.

$$5x(2x^2 - 3x + 4) = 10x^3 - 15x^2 + 20x$$

DO EXERCISES 1 AND 2.

OBJECTIVES

After finishing Section 3.6, you should be able to:

i Multiply a monomial and a polynomial mentally.

ii Multiply two binomials mentally.

Multiply. Just write the answer.

1. $4x(2x^2 - 3x + 4)$

2. $2y^3(5y^3 + 4y^2 - 5y)$

Multiply mentally. Just write the answer.

3. $(x + 3)(x + 4)$

4. $(x + 3)(x - 5)$

5. $(2t + 1)(t + 4)$

6. $(2x^2 - 3)(x - 2)$

7. $(6x^2 + 5)(2x^3 + 1)$

8. $(y^3 + 7)(y^3 - 7)$

9. $(2x^4 + x^2)(-x^3 + x)$

Multiply.

10. $(x + \frac{4}{5})(x - \frac{4}{5})$

11. $(x^3 - 0.5)(x^2 + 0.5)$

12. $(2 + 3x^2)(4 - 5x^2)$

13. $(6y^3 - 3y^2)(5y^2 + 2y)$

ii Products of Two Binomials

To multiply two binomials, we multiply each term of one by every term of the other, as shown here:

$$(A + B)(C + D) = AC + AD + BC + BD$$

1. Multiply First terms: AC
2. Multiply Outside terms: AD
3. Multiply Inside terms: BC
4. Multiply Last terms: BD

       FOIL   This will help you remember the rule.

EXAMPLE 2 Multiply: $(x + 8)(x^2 + 5)$.

$$\begin{array}{cccc} \text{F} & \text{O} & \text{I} & \text{L} \end{array}$$
$$(x + 8)(x^2 + 5) = x^3 + 5x + 8x^2 + 40$$

Often we can collect like terms after we multiply.

EXAMPLES Multiply.

3. $(x + 6)(x - 6) = x^2 - 6x + 6x - 36$ Using FOIL
$$= x^2 - 36 \qquad \text{Collecting like terms}$$

4. $(y + 3)(y - 2) = y^2 - 2y + 3y - 6$
$$= y^2 + y - 6$$

5. $(x^3 + 5)(x^3 - 5) = x^6 - 5x^3 + 5x^3 - 25$
$$= x^6 - 25$$

6. $(4t^3 + 5)(3t^2 - 2) = 12t^5 - 8t^3 + 15t^2 - 10$

DO EXERCISES 3–9.

EXAMPLES Multiply.

7. $(x - \frac{2}{3})(x + \frac{2}{3}) = x^2 + \frac{2}{3}x - \frac{2}{3}x - \frac{4}{9}$
$$= x^2 - \frac{4}{9}$$

8. $(x^2 - 0.3)(x^2 - 0.3) = x^4 - 0.3x^2 - 0.3x^2 + 0.09$
$$= x^4 - 0.6x^2 + 0.09$$

9. $(3 - 4x)(7 - 5x^3) = 21 - 15x^3 - 28x + 20x^4$
$$= 21 - 28x - 15x^3 + 20x^4$$

(*Note:* If the original polynomials are in ascending order, it is natural to write the product in ascending order, but this is not a "must.")

10. $(5x^4 + 2x^3)(3x^2 - 7x) = 15x^6 - 35x^5 + 6x^5 - 14x^4$
$$= 15x^6 - 29x^5 - 14x^4$$

DO EXERCISES 10–13.

EXERCISE SET 3.6

i Multiply. Write only the answer.

1. $4x(x + 1)$

2. $3x(x + 2)$

3. $-3x(x - 1)$

4. $-5x(-x - 1)$

5. $x^2(x^3 + 1)$

6. $-2x^3(x^2 - 1)$

7. $3x(2x^2 - 6x + 1)$

8. $-4x(2x^3 - 6x^2 - 5x + 1)$

ii Multiply. Write only the answer.

9. $(x + 1)(x^2 + 3)$ **10.** $(x^2 - 3)(x - 1)$ **11.** $(x^3 + 2)(x + 1)$ **12.** $(x^4 + 2)(x + 12)$

13. $(y + 2)(y - 3)$ **14.** $(a + 2)(a + 2)$ **15.** $(3x + 2)(3x + 3)$ **16.** $(4x + 1)(2x + 2)$

17. $(5x - 6)(x + 2)$ **18.** $(x - 8)(x + 8)$ **19.** $(3t - 1)(3t + 1)$ **20.** $(2m + 3)(2m + 3)$

21. $(4x - 2)(x - 1)$ **22.** $(2x - 1)(3x + 1)$ **23.** $(p - \frac{1}{4})(p + \frac{1}{4})$ **24.** $(q + \frac{3}{4})(q + \frac{3}{4})$

25. $(x - 0.1)(x + 0.1)$ **26.** $(3x^2 + 1)(x + 1)$ **27.** $(2x^2 + 6)(x + 1)$ **28.** $(2x^2 + 3)(2x - 1)$

29. $(-2x + 1)(x + 6)$ **30.** $(3x + 4)(2x - 4)$ **31.** $(a + 7)(a + 7)$ **32.** $(2y + 5)(2y + 5)$

33. $(1 + 2x)(1 - 3x)$ **34.** $(-3x - 2)(x + 1)$ **35.** $(x^2 + 3)(x^3 - 1)$ **36.** $(x^4 - 3)(2x + 1)$

37. $(x^2 - 2)(x - 1)$ **38.** $(x^3 + 2)(x - 3)$ **39.** $(3x^2 - 2)(x^4 - 2)$ **40.** $(x^{10} + 3)(x^{10} - 3)$

41. $(3x^5 + 2)(2x^2 + 6)$ **42.** $(1 - 2x)(1 + 3x^2)$ **43.** $(8x^3 + 1)(x^3 + 8)$ **44.** $(4 - 2x)(5 - 2x^2)$

45. $(4x^2 + 3)(x - 3)$ **46.** $(7x - 2)(2x - 7)$ **47.** $(4y^4 + y^2)(y^2 + y)$ **48.** $(5y^6 + 3y^3)(2y^6 + 2y^3)$

Multiply.

49. $4y(y + 5)(2y + 8)$

50. $8x(2x - 3)(5x + 9)$

51. $[(x + 1) - x^2][(x - 2) + 2x^2]$

52. $[(2x - 1)(2x + 1)](4x^2 + 1)$

Solve.

53. $(x + 2)(x - 5) = (x + 1)(x - 3)$

54. $(2x + 5)(x - 4) = (x + 5)(2x - 4)$

55. $(x + 1)(x + 2) = (x + 3)(x + 4)$

56. $(x + 3)(x + 1) + (2x - 3)(x - 2) = (3x + 4)(x - 1)$

Problem solving

The height of a box is one more than its length l, and the length is one more than its width w. Find a polynomial for the volume in terms of each of the following.

57. The width w.   **58.** The length l.   **59.** The height h.

Find two expressions for each color region.

60. 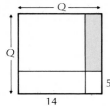   **61.** 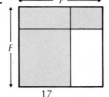   **62.**

63. A cab company charges 70¢ for the first $\frac{1}{7}$ mile and 10¢ each additional $\frac{1}{7}$ mile per trip. A person takes x number of 4-mile trips every month for 11 months, and 6 trips in August before going on vacation. How much did a year of taxi service cost?

3.7

MORE SPECIAL PRODUCTS

Now we find special products that are sums and differences and those that are squares of binomials.

i Multiplying Sums and Differences of Two Expressions

A product of the sum and difference of two expressions, such as $(x + 2)(x - 2)$, occurs quite often. To find a faster way to compute such products, look for a pattern in the following:

a) $(x + 2)(x - 2) = x^2 - 2x + 2x - 4$
$= x^2 - 4;$

OBJECTIVES

After finishing Section 3.7, you should be able to:

i Multiply the sum and difference of two expressions mentally.

ii Square a binomial mentally.

iii Find special products such as those above and those in Section 3.6, mentally, when they are mixed together.

Multiply.

1. $(x + 5)(x - 5)$

2. $(2y - 3)(2y + 3)$

Multiply.

3. $(t + 2)(t - 2)$

4. $(x - 7)(x + 7)$

5. $(4 - 5t)(4 + 5t)$

6. $(2x^3 - 1)(2x^3 + 1)$

Multiply.

7. $(x + 8)(x + 8)$

8. $(x - 5)(x - 5)$

b) $(3x - 5)(3x + 5) = 9x^2 + 15x - 15x - 25$
$$= 9x^2 - 25.$$

DO EXERCISES 1 AND 2.

Perhaps you discovered the following.

> The product of the sum and difference of two expressions is the square of the first expression minus the square of the second:
> $$(A + B)(A - B) = A^2 - B^2.$$

It is helpful to memorize this rule in both words and symbols.

EXAMPLES   Multiply. (Carry out the rule, and say the words as you go.)

$$(A + B)(A - B) = A^2 - B^2$$

1. $(x + 4)(x - 4) = x^2 - 4^2$ "The square of the first expression x^2 minus the square of the second 4^2."

$$= x^2 - 16$$ Simplifying

2. $(5 + 2w)(5 - 2w) = 5^2 - (2w)^2$
$$= 25 - 4w^2$$

3. $(3x^2 - 7)(3x^2 + 7) = (3x^2)^2 - 7^2$
$$= 9x^4 - 49$$

4. $(-4x - 10)(-4x + 10) = (-4x)^2 - 10^2$
$$= 16x^2 - 100$$

DO EXERCISES 3–6.

ii Squaring Binomials

In this special product we multiply a binomial by itself. This is also called "squaring a binomial." Look for a pattern below:

a) $(x + 3)^2 = (x + 3)(x + 3)$
$$= x^2 + 3x + 3x + 9 = x^2 + 6x + 9;$$

b) $(5 + 3x)^2 = (5 + 3x)(5 + 3x)$
$$= 25 + 15x + 15x + 9x^2 = 25 + 30x + 9x^2;$$

c) $(x - 3)^2 = (x - 3)(x - 3)$
$$= x^2 - 3x - 3x + 9 = x^2 - 6x + 9;$$

d) $(3x - 5)^2 = (3x - 5)(3x - 5)$
$$= 9x^2 - 15x - 15x + 25 = 9x^2 - 30x + 25.$$

DO EXERCISES 7 AND 8.

Perhaps you discovered a quick way to square a binomial.

> The square of a sum or difference of two expressions is the square of the first expression plus or minus twice the product of the two expressions, plus the square of the last:
> $$(A + B)^2 = A^2 + 2AB + B^2;$$
> $$(A - B)^2 = A^2 - 2AB + B^2.$$

EXAMPLES Multiply. (Carry out the rule, and say the words as you go.)

$$(A + B)^2 = A^2 + 2 \quad A \quad B + B^2$$

5. $(x + 3)^2 = x^2 + 2 \cdot x \cdot 3 + 3^2$
$$= x^2 + 6x + 9$$

6. $(t - 5)^2 = t^2 - 2 \cdot t \cdot 5 + 5^2$
$$= t^2 - 10t + 25$$

7. $(2x + 7)^2 = (2x)^2 + 2 \cdot 2x \cdot 7 + 7^2 = 4x^2 + 28x + 49$

8. $(5x - 3x^2)^2 = (5x)^2 - 2 \cdot 5x \cdot 3x^2 + (3x^2)^2$
$$= 25x^2 - 30x^3 + 9x^4$$

Note carefully in these examples that the square of a sum is *not* the sum of squares:

$$(A + B)^2 \neq A^2 + B^2.$$

To see this, note that

$$(20 + 5)^2 = 25^2 = 625,$$

but

$$20^2 + 5^2 = 400 + 25 = 425 \neq 625.$$

DO EXERCISES 9–14.

iii Multiplications of Various Types

Now that we have considered how to quickly multiply certain kinds of polynomials, let us try several kinds mixed together so we can learn to sort them out. When you multiply, first see what kind of multiplication you have. Then use the best method. The formulas and methods you have used so far are as follows.

> 1. $(A + B)(A + B) = (A + B)^2 = A^2 + 2AB + B^2$
>
> 2. $(A - B)(A - B) = (A - B)^2 = A^2 - 2AB + B^2$
>
> 3. $(A - B)(A + B) = A^2 - B^2$
>
> 4. **FOIL**
>
> 5. **The product of a monomial and any polynomial is found by multiplying each term of the polynomial by the monomial.**

Note that FOIL will work for any of the first three rules, but it is faster to learn to use them as they are given.

EXAMPLE 9 Multiply: $(x + 3)(x - 3)$.

$$(x + 3)(x - 3) = x^2 - 9 \qquad \text{Using method 3 (the product of the sum and difference of two expressions)}$$

EXAMPLE 10 Multiply: $(t + 7)(t - 5)$.

$$(t + 7)(t - 5) = t^2 + 2t - 35 \qquad \text{Using method 4 (the product of two binomials, but neither the square of a binomial nor the product of the sum and difference of two expressions)}$$

EXAMPLE 11 Multiply: $(x + 7)(x + 7)$.

$$(x + 7)(x + 7) = x^2 + 14x + 49 \qquad \text{Using method 1 (the square of a binomial sum)}$$

Multiply.

9. $(x + 2)^2$

10. $(a - 4)^2$

11. $(2x + 5)^2$

12. $(4x^2 - 3x)^2$

13. $(7 + y)(7 + y)$

14. $(3x^2 - 5)(3x^2 - 5)$

Multiply.

15. $(x + 5)(x + 6)$

16. $(t - 4)(t + 4)$

17. $4x^2(-2x^3 + 5x^2 + 10)$

18. $(9x^2 + 1)^2$

19. $(2a - 5)(2a + 8)$

20. $(5x + \frac{1}{2})^2$

21. $(2x - \frac{1}{2})^2$

EXAMPLE 12 Multiply: $2x^3(9x^2 + x - 7)$.

$$2x^3(9x^2 + x - 7) = 18x^5 + 2x^4 - 14x^3$$

Using method 5 (the product of a monomial and a trinomial; multiply each term of the trinomial by the monomial)

EXAMPLE 13 Multiply: $(3x^2 - 7x)^2$.

$$(3x^2 - 7x)^2 = 9x^4 - 42x^3 + 49x^2$$

Using method 2 (the square of a binomial difference)

EXAMPLE 14 Multiply: $(3x + \frac{1}{4})^2$.

$$(3x + \tfrac{1}{4})^2 = 9x^2 + 2(3x)(\tfrac{1}{4}) + \tfrac{1}{16}$$

Using method 1 (the square of a binomial sum. To get the middle term, we multiply $3x$ by $\frac{1}{4}$ and double.)

$$= 9x^2 + \tfrac{3}{2}x + \tfrac{1}{16}$$

EXAMPLE 15 Multiply: $(4x - \frac{3}{4})^2$.

$$(4x - \tfrac{3}{4})^2 = 16x^2 - 2(4x)(\tfrac{3}{4}) + \tfrac{9}{16}$$
$$= 16x^2 - 6x + \tfrac{9}{16}$$

DO EXERCISES 15–21.

EXERCISE SET 3.7

i Multiply mentally.

1. $(x + 4)(x - 4)$
2. $(x + 1)(x - 1)$
3. $(2x + 1)(2x - 1)$
4. $(x^2 + 1)(x^2 - 1)$
5. $(5m - 2)(5m + 2)$
6. $(3x^4 + 2)(3x^4 - 2)$
7. $(2x^2 + 3)(2x^2 - 3)$
8. $(6x^5 - 5)(6x^5 + 5)$
9. $(3x^4 - 4)(3x^4 + 4)$
10. $(t^2 - 0.2)(t^2 + 0.2)$
11. $(x^6 - x^2)(x^6 + x^2)$
12. $(2x^3 - 0.3)(2x^3 + 0.3)$
13. $(x^4 + 3x)(x^4 - 3x)$
14. $(\frac{3}{4} + 2x^3)(\frac{3}{4} - 2x^3)$
15. $(x^{12} - 3)(x^{12} + 3)$
16. $(12 - 3x^2)(12 + 3x^2)$
17. $(2y^8 + 3)(2y^8 - 3)$
18. $(m - \frac{2}{3})(m + \frac{2}{3})$

ii Multiply mentally.

19. $(x + 2)^2$
20. $(2x - 1)^2$
21. $(3x^2 + 1)^2$
22. $(3x + \frac{3}{4})^2$
23. $(a - \frac{1}{2})^2$
24. $(2a - \frac{1}{5})^2$
25. $(3 + x)^2$
26. $(x^3 - 1)^2$
27. $(x^2 + 1)^2$
28. $(8x - x^2)^2$
29. $(2 - 3x^4)^2$
30. $(6x^3 - 2)^2$
31. $(5 + 6t^2)^2$
32. $(3p^2 - p)^2$

iii Multiply mentally.

33. $(3 - 2x^3)^2$
34. $(x - 4x^3)^2$
35. $4x(x^2 + 6x - 3)$
36. $8x(-x^5 + 6x^2 + 9)$
37. $(2x^2 - \frac{1}{2})(2x^2 - \frac{1}{2})$
38. $(-x^2 + 1)^2$
39. $(-1 + 3p)(1 + 3p)$
40. $(-3q + 2)(3q + 2)$
41. $3t^2(5t^3 - t^2 + t)$
42. $-6x^2(x^3 + 8x - 9)$
43. $(6x^4 + 4)^2$
44. $(8a + 5)^2$
45. $(3x + 2)(4x^2 + 5)$
46. $(2x^2 - 7)(3x^2 + 9)$
47. $(8 - 6x^4)^2$
48. $(\frac{1}{5}x^2 + 9)(\frac{3}{5}x^2 - 7)$

49. ▦ Multiply: $(67.58x + 3.225)^2$.

Calculate as the difference of squares.

50. 18×22 (*Hint:* $(20 - 2)(20 + 2)$.)
51. 93×107

Multiply. (Do not collect like terms before multiplying.)

52. $[(2x - 1)(2x + 1)](4x^2 + 1)$
53. $[(a + 5) + 1][(a + 5) - 1]$
54. $[3a - (2a - 3)][3a + (2a - 3)]$
55. $[(x + 3) + 2]^2$

Solve.

56. $x^2 = (x + 10)^2$

57. $x^2 = (x - 12)^2$

58. $(x + 4)^2 = (x + 8)(x - 8)$

59. $(4x - 1)^2 - (3x + 2)^2 = (7x + 4)(x - 1)$

Problem solving

60. Consider the following rectangle.

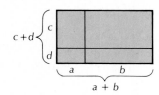

a) Find a polynomial for each of the areas of the four smaller rectangles.

b) Find a polynomial for the sum of the areas in part (a).

c) Find a polynomial for the area of the entire rectangle. Compare your result with the answer to part (b).

61. A polynomial for the color region in this rectangle is $(a + b)(a - b)$.

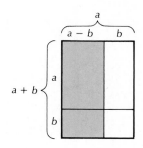

a) Find a polynomial for the area of the entire rectangle.

b) Find a polynomial for the sum of the two small uncolored rectangles.

c) Find a polynomial for the area in part (a) minus the area in part (b).

d) Find a polynomial for the area of the color region and compare this with the polynomial found in part (c).

62. Find three consecutive integers, the sum of whose squares is 65 more than 3 times the square of the smallest integer.

63. Find $(10x + 5)^2$. Use your result to show how to mentally square any two-digit number ending in 5.

SUMMARY AND REVIEW: CHAPTER 3

The following contains a summary of what you should be able to do after completing this chapter. The review exercises are for practice. Answers are at the back of the book. If you miss an exercise, restudy the section indicated alongside the answer.

You should be able to:

Evaluate a polynomial for a given value of a variable.

Evaluate for $x = -1$.

1. $7x - 10$

2. $x^2 - 3x + 6$

3. $x^3 - 3x^2 + 2x - 20$

Evaluate for $x = 2$.

4. $x^2 - 4$

5. $(x + 2)(x - 2)$

6. $x^3 - 3x^2 + 2x - 20$

Evaluate for $x = -5$.

7. $x^2 + 25$

8. $(x + 5)(x - 5)$

9. $(x + 5)^2$

10. $x^2 - 10x + 25$

11. $x^2 + 10x + 25$

12. $x^2 - 25$

For a polynomial, identify the terms, the missing terms, the coefficients of each term, the degree of each term, and the degree of the polynomial, and classify the polynomial as a monomial, binomial, trinomial, or none of these.

Identify the terms of the polynomial.

13. $3x^2 + 6x + \frac{1}{2}$

14. $-4y^5 + 7y^2 - 3y - 2$

15. Identify the missing terms in $x^3 + x$.

Identify the coefficient of each term of the polynomial.

16. $6x^2 + 17$

17. $4x^3 + 6x^2 - 5x + \frac{5}{3}$

Identify the degree of each term and the degree of the polynomial.

18. $x^3 + 4x - 6$

19. $3 - 2x^4 + 3x^9 + x^6 - \frac{3}{4}x^3$

Tell whether the polynomial is a monomial, binomial, trinomial, or none of these.

20. $4x^3 - 1$

21. $4 - 9t^3 - 7t^4 + 10t^2$

22. $7y^2$

Collect like terms of a polynomial and arrange the terms in descending order.

Collect like terms.

23. $5x - x^2 + 4x$

24. $\frac{3}{4}x^3 + 4x^2 - x^3 + 7$

25. $-2x^4 + 16 + 2x^4 + 9 - 3x^5$

Collect like terms and then arrange in descending order.

26. $3x^2 - 2x + 3 - 5x^2 - 1 - x$

27. $-x + \frac{1}{2} + 14x^4 - 7x^2 - 1 - 4x^4$

Add and subtract polynomials and find polynomials for certain perimeters and areas of figures.

Add.

28. $(3x^4 - x^3 + x - 4) + (x^5 + 7x^3 - 3x^2 - 5) + (-5x^4 + 6x^2 - x)$

29. $(3x^5 - 4x^4 + x^3 - 3) + (3x^4 - 5x^3 + 3x^2) + (4x^5 + 4x^3) + (-5x^5 - 5x^2) + (-5x^4 + 2x^3 + 5)$

30. $-\frac{3}{4}x^4 + \frac{1}{2}x^3 \qquad\qquad + \frac{7}{8}$
$\qquad\qquad -\frac{1}{4}x^3 - x^2 - \frac{7}{4}x$
$\qquad \frac{3}{2}x^4 \qquad\qquad + \frac{2}{3}x^2 \qquad\quad - \frac{1}{2}$

Subtract.

31. $(5x^2 - 4x + 1) - (3x^2 + 7)$

32. $(3x^5 - 4x^4 + 2x^2 + 3) - (2x^5 - 4x^4 + 3x^3 + 4x^2 - 5)$

33. $2x^5 \qquad - x^3 \qquad + x + 3$
$\quad 3x^5 - x^4 + 4x^3 + 2x^2 - x + 3$

34. The length of a rectangle is 4 m greater than its width.

a) Find a polynomial for the perimeter.

b) Find a polynomial for the area.

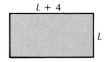

$L + 4$

L

Multiply two monomials, a monomial and a binomial, two binomials, a binomial and a trinomial, and any polynomials. Find special products such as the sum and difference of two expressions and the square of binomials.

Multiply.

35. $3x(-4x^2)$

36. $(7x + 1)^2$

37. $(x + \frac{2}{3})(x + \frac{1}{2})$

38. $(1.5x - 6.5)(0.2x + 1.3)$

39. $(4x^2 - 5x + 1)(3x - 2)$

40. $(x - 9)^2$

41. $5x^4(3x^3 - 8x^2 + 10x + 2)$

42. $(x + 4)(x - 7)$

43. $(x - 0.3)(x - 0.75)$

44. $(x^4 - 2x + 3)(x^3 + x - 1)$

45. $(3y^2 - 2y)^2$

46. $(2t^2 + 3)(t^2 - 7)$

47. $(x^3 - 2x + 3)(4x^2 - 5x)$

48. $(3x^2 + 4)(3x^2 - 4)$

49. $(2 - x)(2 + x)$

50. $(13x - 3)(x - 13)$

51. If a and b are positive, how many terms are there in each of the following?

a) $(x - a)(x - b) + (x - a)(x - b)$

b) $(x + a)(x - b) + (x - a)(x + b)$

c) $(x + a)(x - b) - (x + a)(x - b)$

52. Collect like terms:

$-3x^5 \cdot 3x^3 - x^6(2x)^2 + (3x^4)^2 + (2x^2)^4 - 40x^2(x^3)^2$.

53. A polynomial has degree 4. The x^2-term is missing. The coefficient of x^4 is 2 times the coefficient of x^3. The coefficient of x is 3 less than the coefficient of x^4. The remaining coefficient is 7 less than the coefficient of x. The sum of the coefficients is 15. Find the polynomial.

54. Multiply: $[(x - 4) - x^3][(x + 4) + 4x^3]$.

55. Solve: $(x - 7)(x + 10) = (x - 4)(x - 6)$.

TEST: CHAPTER 3

1. Evaluate the polynomial:

$$x^2 + 5x - 1 \quad \text{for} \quad x = -2.$$

2. Identify the coefficient of each term of the polynomial:

$$\tfrac{1}{3}x^5 - x + 7.$$

3. Identify the degree of each term and the degree of the polynomial:

$$2x^3 - 4 + 5x + 3x^6.$$

4. Tell whether the polynomial is a monomial, binomial, trinomial, or none of these:

$$7 - x.$$

Collect like terms.

5. $4a^2 - 6 + a^2$

6. $y^2 - 3y - y + \tfrac{3}{4}y^2$

7. Collect like terms and then arrange in descending order:

$$3 - x^2 + 2x^3 + 5x^2 - 6x - 2x + x^5.$$

Add.

8. $(3x^5 + 5x^3 - 5x^2 - 3) + (x^5 + x^4 - 3x^3 - 3x^2 + 2x - 4)$

9. $(x^4 + \tfrac{2}{3}x + 5) + (4x^4 + 5x^2 + \tfrac{1}{3}x)$

Subtract.

10. $(2x^4 + x^3 - 8x^2 - 6x - 3) - (6x^4 - 8x^2 + 2x)$

11. $(x^3 - 0.4x^2 - 12) - (x^5 + 0.3x^3 + 0.4x^2 + 9)$

Multiply.

12. $-3x^2(4x^2 - 3x - 5)$

13. $(x - \tfrac{1}{3})^2$

14. $(3x + 10)(3x - 10)$

15. $(3b + 5)(b - 3)$

16. $(x^6 - 4)(x^8 + 4)$

17. $(8 - y)(6 + 5y)$

18. $(2x + 1)(3x^2 - 5x - 3)$

19. $(5t + 2)^2$

20. The height of a box is one less than its length, and the length is 2 more than its width. Find the volume in terms of the length.

21. Solve:

$$x^2 + (x - 7)(x + 4) = 2(x - 6)^2.$$

THE ASVAB TEST: SAMPLE QUESTIONS

CHAPTER 3

NUMERICAL OPERATIONS

Time: 3 minutes; 50 questions

Allow yourself 3 minutes. Work as carefully as you can and try to get as many correct as possible. See the discussion of the ASVAB test at the beginning of Chapter 3.

1. $5 + 4 =$
A 11
B 9
C 1
D 20

2. $6 - 3 =$
A 3
B 4
C 9
D 2

3. $2 \times 8 =$
A 10
B 12
C 16
D 28

4. $9 + 6 =$
A 3
B 15
C 14
D 13

5. $9 \div 3 =$
A 6
B 12
C 3
D 4

6. $8 - 8 =$
A 16
B 8
C 1
D 0

7. $24 \div 4 =$
A 28
B 6
C 20
D 8

8. $5 \times 7 =$
A 30
B 35
C 37
D 40

9. $6 + 1 =$
A 5
B 6
C 7
D 8

10. $0 \times 4 =$
A 2
B 1
C 0
D 4

11. $8 - 2 =$
A 4
B 6
C 10
D 16

12. $7 + 8 =$
A 13
B 16
C 15
D 17

13. $18 \div 2 =$
A 9
B 16
C 10
D 20

14. $2 \times 4 =$
A 4
B 12
C 8
D 6

15. $6 \times 1 =$
A 0
B 1
C 7
D 6

16. $7 + 5 =$
A 1
B 12
C 11
D 13

17. $9 - 3 =$
A 3
B 7
C 6
D 12

18. $9 + 9 =$
A 18
B 1
C 16
D 0

19. $5 + 0 =$
A 0
B 1
C 4
D 5

20. $9 \times 3 =$
A 27
B 26
C 24
D 21

21. $7 \div 7 =$
A 0
B 1
C 7
D 14

22. $4 + 9 =$
A 14
B 12
C 13
D 11

23. $10 \times 4 =$
A 6
B 14
C 40
D 20

24. $6 - 1 =$
A 4
B 5
C 6
D 7

25. $6 \times 3 =$
A 12
B 9
C 3
D 18

26. $0 \div 5 =$
A 0
B 1
C 5
D 10

27. $8 + 3 =$
A 5
B 11
C 13
D 24

28. $6 + 2 =$
A 4
B 12
C 3
D 8

29. $70 \div 7 =$
A 10
B 77
C 63
D 11

30. $10 - 6 =$
A 16
B 6
C 3
D 4

31. $8 \times 8 =$
A 1
B 8
C 64
D 16

32. $63 \div 7 =$
A 70
B 8
C 56
D 9

33. $5 + 3 =$
A 8
B 2
C 9
D 7

34. $3 + 2 =$
A 1
B 2
C 5
D 6

35. $4 \times 7 =$
A 11
B 18
C 21
D 28

36. $4 - 3 =$
A 0
B 1
C 7
D 2

37. $5 \times 2 =$
A 15
B 7
C 10
D 3

38. $3 \div 1 =$
A 4
B 1
C 2
D 3

39. $9 + 2 =$
A 11
B 7
C 10
D 12

40. $9 \times 5 =$
A 4
B 14
C 40
D 45

41. $7 - 5 =$
A 1
B 2
C 3
D 4

42. $7 \times 6 =$
A 43
B 13
C 42
D 44

43. $80 \div 10 =$
A 8
B 9
C 88
D 90

44. $8 \times 4 =$
A 28
B 4
C 32
D 12

45. $9 + 8 =$
A 1
B 19
C 17
D 16

46. $7 - 0 =$
A 0
B 1
C 6
D 7

47. $5 \times 4 =$
A 20
B 9
C 11
D 22

48. $25 \div 5 =$
A 4
B 5
C 20
D 30

49. $7 + 1 =$
A 5
B 6
C 7
D 8

50. $8 \times 6 =$
A 86
B 24
C 48
D 14

4

POLYNOMIALS AND FACTORING

How can we find the area of a beverage can? A polynomial can be evaluated to solve such a problem.

Factoring is the reverse of multiplication. To *factor* a polynomial, or other algebraic expression, is to find an equivalent expression that is a product. In this chapter we study factoring polynomials. To learn to factor quickly, we study quick methods for multiplication.

At the end of the chapter we get the payoff for learning to factor. We have certain new equations containing second-degree polynomials that we can now solve. This then allows us to solve problems that we could not have solved before.

OBJECTIVES

After finishing Section 4.1, you should be able to:

i Factor monomials.

ii Factor polynomials when the terms have a common factor.

iii Factor certain expressions with four terms by grouping.

1. a) Multiply: $(3x)(4x)$

 b) Factor: $12x^2$.

2. a) Multiply: $(2x)(8x^2)$.

 b) Factor: $16x^3$.

Find three factorizations of each monomial.

3. $8x^4$

4. $21x^2$

5. $6x^5$

6. a) Multiply: $3(x + 2)$.

 b) Factor: $3x + 6$.

7. a) Multiply: $2x(x^2 + 5x + 4)$.

 b) Factor: $2x^3 + 10x^2 + 8x$.

4.1

FACTORING POLYNOMIALS

In this section we learn to factor polynomials with one term. We also learn how to factor out a common factor.

i Factoring Monomials

To factor a monomial we find two monomials whose product is equivalent to that monomial. Compare.

	Multiplying	*Factoring*
a)	$(4x)(5x) = 20x^2$	$20x^2 = (4x)(5x)$
b)	$(2x)(10x) = 20x^2$	$20x^2 = (2x)(10x)$
c)	$(-4x)(-5x) = 20x^2$	$20x^2 = (-4x)(-5x)$
d)	$(x)(20x) = 20x^2$	$20x^2 = (x)(20x)$

The monomial $20x^2$ thus has many factorizations. There are still other ways to factor $20x^2$.

DO EXERCISES 1 AND 2.

To factor a monomial, factor the coefficient first. Then shift some of the letters to one factor and some to the other.

EXAMPLE 1 Find three factorization of $15x^3$.

a) $15x^3 = (3 \cdot 5)x^3$
$= (3x)(5x^2)$

b) $15x^3 = (3 \cdot 5)x^3$
$= (3x^2)(5x)$

c) $15x^3 = (-1) \cdot (-15)x^3$
$= (-x)(-15x^2)$

DO EXERCISES 3–5.

ii Factoring When Terms Have a Common Factor

To multiply a monomial and a polynomial with more than one term, we multiply each term by the monomial. To factor, we do the reverse. Compare.

Multiply	*Factor*
$3x(x^2 + 2x - 4)$	$3x^3 + 6x^2 - 12x$
$= 3x \cdot x^2 + 3x \cdot 2x + 3x(-4)$	$= 3x \cdot x^2 + 3x \cdot 2x + 3x \cdot (-4)$
$= 3x^3 + 6x^2 - 12x$	$= 3x(x^2 + 2x - 4)$

Consider the following:

$$3x^3 + 6x^2 - 12x = 3 \cdot x \cdot x \cdot x + 6 \cdot x \cdot x - 4 \cdot 3x.$$

The parts, or terms, of the expression have been factored but the expression itself has not been factored. This is not a correct factorization.

DO EXERCISES 6 AND 7.

We are finding a factor common to all the terms. There may not always be one other than 1. When there is, we generally use the factor with the largest possible coefficient and the largest exponent. In this way we "factor completely."

EXAMPLE 2 Factor: $3x^2 + 6$.

$$3x^2 + 6 = 3 \cdot x^2 + 3 \cdot 2$$
$$= 3(x^2 + 2) \qquad \text{Factoring out the common factor, 3}$$

EXAMPLE 3 Factor: $16x^3 + 20x^2$.

$$16x^3 + 20x^2 = (4x^2)(4x) + (4x^2)(5)$$
$$= 4x^2(4x + 5) \qquad \text{Factoring out } 4x^2$$

EXAMPLE 4 Factor: $15x^5 - 12x^4 + 27x^3 - 3x^2$.

$$15x^5 - 12x^4 + 27x^3 - 3x^2 = (3x^2)(5x^3) - (3x^2)(4x^2) + (3x^2)(9x) - (3x^2)(1)$$
$$= 3x^2(5x^3 - 4x^2 + 9x - 1) \qquad \text{Factoring out } 3x^2$$

If you can spot the common factor without factoring each term, you should write just the answer.

EXAMPLE 5 Factor: $\frac{4}{5}x^2 + \frac{1}{5}x + \frac{2}{5}$.

$$\frac{4}{5}x^2 + \frac{1}{5}x + \frac{2}{5} = \frac{1}{5}(4x^2 + x + 2)$$

DO EXERCISES 8–12.

iii Factoring by Grouping

The method that we are about to consider is called *factoring by grouping*. Certain polynomials with four terms can be factored using this method. Consider

$$x^2 + 3x + 4x + 12.$$

There is no factor common to all the terms other than 1. But we can factor $x^2 + 3x$ and $4x + 12$:

$$x^2 + 3x = x(x + 3); \qquad \text{Factoring } x^2 + 3x$$
$$4x + 12 = 4(x + 3). \qquad \text{Factoring } 4x + 12$$

Then

$$x^2 + 3x + 4x + 12 = x(x + 3) + 4(x + 3).$$

Note the common *binomial* factor $x + 3$. We can use the distributive law again like this:

$$x(x + 3) + 4(x + 3) = (x + 4)(x + 3). \qquad \text{Factoring out the common factor, } x + 3$$

EXAMPLES Factor. For purposes of learning this method, do not collect like terms.

6. $x^2 + 7x + 2x + 14 = (x^2 + 7x) + (2x + 14)$ Separating into two binomials
$$= x(x + 7) + 2(x + 7) \qquad \text{Factoring each binomial}$$
$$= (x + 2)(x + 7) \qquad \text{Factoring out the common factor, } x + 7$$

7. $5x^2 - 10x + 2x - 4 = (5x^2 - 10x) + (2x - 4)$ Separating into two binomials
$$= 5x(x - 2) + 2(x - 2) \qquad \text{Factoring each binomial}$$
$$= (5x + 2)(x - 2) \qquad \text{Factoring out the common factor, } x - 2$$

Factor.

8. $x^2 + 3x$

9. $3x^6 - 5x^3 + 2x^2$

10. $9x^4 - 15x^3 + 3x^2$

11. $\frac{3}{4}x^3 + \frac{5}{4}x^2 + \frac{7}{4}x + \frac{1}{4}$

12. $35x^7 - 49x^6 + 14x^5 - 63x^3$

Factor.

13. $x^2 + 5x + 2x + 10$

14. $y^2 - 4y + 3y - 12$

15. $4x^3 - 6x^2 - 6x + 9$

16. $8t^3 + 2t^2 + 12t + 3$

17. $3m^5 - 15m^3 + 2m^2 - 10$

8. $x^2 + 3x - x - 3 = (x^2 + 3x) + (-x - 3)$
$$= x(x + 3) - 1(x + 3) \qquad \text{Factoring} - 1 \text{ out of the second binomial}$$
$$= (x - 1)(x + 3)$$

9. $2x^2 - 12x - 3x + 18 = (2x^2 - 12x) + (-3x + 18)$
$$= 2x(x - 6) - 3(x - 6)$$
$$= (2x - 3)(x - 6)$$

10. $12x^5 + 20x^2 - 21x^3 - 35 = (12x^5 + 20x^2) + (-21x^3 - 35)$
$$= 4x^2(3x^3 + 5) - 7(3x^3 + 5)$$
$$= (4x^2 - 7)(3x^3 + 5)$$

This method is called *factoring by grouping*. Not all expressions with four terms can be factored by this method.

DO EXERCISES 13–17.

EXERCISE SET 4.1

i Find three factorizations for each monomial.

1. $6x^3$ **2.** $9x^4$ **3.** $-9x^5$ **4.** $-12x^6$ **5.** $24x^4$ **6.** $15x^5$

ii Factor.

7. $x^2 - 4x$ **8.** $x^2 + 8x$ **9.** $2x^2 + 6x$ **10.** $3x^2 - 3x$

11. $x^3 + 6x^2$ **12.** $4x^4 + x^2$ **13.** $8x^4 - 24x^2$ **14.** $5x^5 + 10x^3$

15. $2x^2 + 2x - 8$ **16.** $6x^2 + 3x - 15$ **17.** $17x^5 + 34x^3 + 51x$ **18.** $16x^6 - 32x^5 - 48x$

19. $6x^4 - 10x^3 + 3x^2$ **20.** $5x^5 + 10x^2 - 8x$ **21.** $x^5 + x^4 + x^3 - x^2$ **22.** $x^9 - x^7 + x^4 + x^3$

23. $2x^7 - 2x^6 - 64x^5 + 4x^3$ **24.** $10x^3 + 25x^2 + 15x - 20$ **25.** $1.6x^4 - 2.4x^3 + 3.2x^2 + 6.4x$

26. $2.5x^6 - 0.5x^4 + 5x^3 + 10x^2$ **27.** $\frac{5}{3}x^6 + \frac{4}{3}x^5 + \frac{1}{3}x^4 + \frac{1}{3}x^3$ **28.** $\frac{5}{7}x^7 + \frac{3}{7}x^5 - \frac{6}{7}x^3 - \frac{1}{7}x$

iii Factor by grouping.

29. $y^2 + 4y + y + 4$ **30.** $x^2 + 5x + 2x + 10$ **31.** $x^2 - 4x - x + 4$ **32.** $a^2 + 5a - 2a - 10$

33. $6x^2 + 4x + 9x + 6$ **34.** $3x^2 - 2x + 3x - 2$ **35.** $3x^2 - 4x - 12x + 16$ **36.** $24 - 18y - 20y + 15y^2$

37. $35x^2 - 40x + 21x - 24$ **38.** $8x^2 - 6x - 28x + 21$ **39.** $4x^2 + 6x - 6x - 9$ **40.** $2x^4 - 6x^2 - 5x^2 + 15$

41. $2x^4 + 6x^2 + 5x^2 + 15$ **42.** $4x^4 - 6x^2 - 6x^2 + 9$ **43.** $2x^3 + 6x^2 + x + 3$ **44.** $3x^3 + 2x^2 + 3x + 2$

45. $8x^3 - 12x^2 + 6x - 9$ **46.** $10x^3 - 25x^2 + 4x - 10$ **47.** $12x^3 - 16x^2 + 3x - 4$

48. $18x^3 - 21x^2 + 30x - 35$ **49.** $x^3 + 8x^2 - 3x - 24$ **50.** $2x^3 + 12x^2 - 5x - 30$

○ _____

Two polynomials are *relatively prime* if they have no common factors other than constants. Tell which pairs are relatively prime.

51. $5x, x^2$ **52.** $3x, ax - 3$ **53.** $x + x^2, 3x^3$

54. $y - 6, y$ **55.** $7a, a$ **56.** $2p^2 + 2, 2p$

57. $t^2 - 4t, t^2 - 4$ **58.** $4x^5 + 8x^3 - 6x, 8x^3 + 12x^2 + 24x - 16$

59. $b^3 + 3b^2, b^3 - a^2$ **60.** $ax^2 + a^2x, ax - 2a$

Factor.

61. $4x^5 + 6x^3 + 6x^2 + 9$ **62.** $4x^5 + 6x^4 + 6x^3 + 9x^2$ **63.** $x^6 + x^4 + x^2 + 1$ **64.** $x^{13} + x^7 + x^6 + 1$

65. Subtract $(x^2 + 1)^2$ from $x^2(x + 1)^2$ and factor the result.

4.2

DIFFERENCES OF SQUARES

The following polynomials are differences of squares:

$$x^2 - 9, \quad 4t^2 - 49.$$

In this section we learn how to factor differences of squares.

i Recognizing Differences of Squares

A difference of squares is an expression like the following:

$$A^2 - B^2.$$

How can we recognize such expressions? Look at $A^2 - B^2$. In order for a binomial to be a difference of squares:

a) There must be two expressions, both squares, such as

$$4x^2, \quad 9, \quad 25t^4, \quad 1, \quad x^6.$$

b) There must be a minus sign between them.

EXAMPLE 1 Is $9x^2 - 64$ a difference of squares?

a) The first expression is a square: $9x^2 = (3x)^2$.
The second expression is a square: $64 = (8)^2$.

b) There is a minus sign between them.

So we have a difference of squares.

DO EXERCISE 1.

ii Factoring Differences of Squares

To factor a difference of squares we can use the following equation:

$$A^2 - B^2 = (A - B)(A + B).$$

Where does this equation come from? If you multiply out the two expressions on the right, you do get the expression on the left. So the two expressions are equivalent. In effect, we are going to use the equation $(A - B)(A + B) = A^2 - B^2$ in reverse so we can factor.

To factor a difference of squares $A^2 - B^2$, we find A and B, the square roots of the expressions A^2 and B^2. We write a plus sign one time and a minus sign one time.

EXAMPLE 2 Factor: $x^2 - 4$.

$$x^2 - 4 = x^2 - 2^2 = (x - 2)(x + 2)$$
$$A^2 - B^2 = (A - B)(A + B)$$

OBJECTIVES

After finishing Section 4.2, you should be able to:

i Recognize differences of squares.
ii Factor differences of squares.
iii Factor completely polynomials that are differences of squares.

1. Which of the following are differences of squares?
   a) $x^2 - 25$
   b) $t^2 - 24$
   c) $y^2 + 36$
   d) $4x^2 - 15$
   e) $16x^4 - 49$
   f) $9w^6 - 1$
   g) $-49 + 25t^2$

Factor.

2. $x^2 - 9$

3. $t^2 - 64$

4. $32y^2 - 8y^6$

5. $64x^4 - 25x^6$

6. $5 - 20x^6$
(*Hint:* $1 = 1^2$, $x^6 = (x^3)^2$.)

Factor.

7. $81x^4 - 1$

8. $49t^4 - 25t^{10}$

EXAMPLE 3 Factor: $18x^2 - 50x^6$.

Always look first for a factor common to all terms. This time there is one:

$$18x^2 - 50x^6 = 2x^2(9 - 25x^4) = 2x^2[3^2 - (5x^2)^2]$$
$$= 2x^2(3 - 5x^2)(3 + 5x^2)$$

EXAMPLE 4 Factor: $49x^4 - 9x^6$.

$$49x^4 - 9x^6 = x^4(49 - 9x^2) = x^4(7 - 3x)(7 + 3x)$$

Note carefully in these examples that a difference of squares is *not* the square of the difference; that is,

$$A^2 - B^2 \neq (A - B)^2.$$

For example,

$$(45 - 5)^2 = 40^2 = 1600,$$

but

$$45^2 - 5^2 = 2025 - 25 = 2000.$$

DO EXERCISES 2–6.

iii Factoring Completely

If a factor with more than one term can still be factored, you should do so. When no factor can be factored further, you have *factored completely*. Always factor completely even when the directions do not say so.

EXAMPLE 5 Factor: $1 - 16x^{12}$.

$$1 - 16x^{12} = (1 + 4x^6)(1 - 4x^6) \qquad \text{Factoring a difference of squares}$$
$$= (1 + 4x^6)(1 - 2x^3)(1 + 2x^3) \qquad \text{Factoring further (The factor } 1 - 4x^6 \text{ is a difference of squares.)}$$

FACTORING HINTS

1. **Always look first for a common factor.**
2. **Always factor completely.**
3. **Check by multiplying.**
4. **Never try to factor a sum of squares, $A^2 + B^2$.**

DO EXERCISES 7 AND 8.

EXERCISE SET 4.2

---

i Which of the following are differences of squares?

1. $x^2 - 4$ **2.** $x^2 - 36$ **3.** $x^2 + 36$ **4.** $x^2 + 4$ **5.** $x^2 - 35$

6. $x^2 - 50$ **7.** $16x^2 - 25$ **8.** $36x^2 - 1$ **9.** $49x^2 - 2$

ii Factor. Remember to look first for a common factor.

10. $x^2 - 4$ **11.** $x^2 - 36$ **12.** $x^2 - 9$ **13.** $x^2 - 1$ **14.** $16a^2 - 9$

15. $25x^2 - 4$ **16.** $4x^2 - 25$ **17.** $9a^2 - 16$ **18.** $8x^2 - 98$ **19.** $24x^2 - 54$

20. $36x - 49x^3$ **21.** $16x - 81x^3$ **22.** $16x^2 - 25x^4$ **23.** $x^{16} - 9x^2$ **24.** $49a^4 - 81$

25. $25a^4 - 9$ **26.** $a^{12} - 4a^2$ **27.** $121a^8 - 100$ **28.** $81y^6 - 25$ **29.** $100y^6 - 49$

iii Factor.

30. $x^4 - 1$ **31.** $x^4 - 16$ **32.** $4x^4 - 64$ **33.** $5x^4 - 80$ **34.** $1 - y^8$

35. $x^8 - 1$ **36.** $x^{12} - 16$ **37.** $x^8 - 81$ **38.** $\frac{1}{16} - y^2$ **39.** $\frac{1}{25} - x^2$

40. $25 - \frac{1}{49}x^2$ **41.** $4 - \frac{1}{9}y^2$ **42.** $16 - t^4$ **43.** $1 - a^4$

Problem-solving practice

44. In a recent year, 29,090 people were arrested for counterfeiting. This was down 1.2% from the year before. How many were arrested the year before?

45. The first angle of a triangle is four times as large as the second. The measure of the third angle is 30° greater than that of the second. How large are the angles?

○ ————————————————————————

Factor.

46. $4x^4 - 4x^2$ **47.** $3x^5 - 12x^3$ **48.** $3x^2 - \frac{1}{3}$ **49.** $18x^3 - \frac{8}{25}x$

50. $x^2 - 2.25$ **51.** $x^3 - \dfrac{x}{1.69}$ **52.** $3.24x^2 - 0.81$ **53.** $0.64x^2 - 1.21$

54. $1.28x^2 - 2$ **55.** $(x + 3)^2 - 9$ **56.** $(y - 5)^2 - 36$ **57.** $(3a + 4)^2 - 49$

58. $(2y - 7)^2 - 1$ **59.** $y^8 - 256$ **60.** $x^{16} - 1$ **61.** $x^2 - \left(\dfrac{1}{x}\right)^2$

A polynomial is called *irreducible* if it cannot be factored except for removing a common constant factor. If the coefficient of the leading term is 1, the irreducible polynomial is called *prime*. Which of these polynomials are irreducible? Which are prime?

62. $3x^3 + 9x$ **63.** $4x^2 + 2y$ **64.** $4x^2 + 16y^2$ **65.** $x^2 + y$ **66.** $16x^3 - 9x$ **67.** $-25y^2 - 49$

4.3

TRINOMIAL SQUARES

Recall that a trinomial is a polynomial with just three terms. Some trinomials are squares of binomials. For example, the trinomial $x^2 + 10x + 25$ is the square of $(x + 5)$. To see this we can calculate $(x + 5)^2$. It is $x^2 + 2 \cdot 5 \cdot x + 5^2$, or $x^2 + 10x + 25$.

A trinomial that is the square of a binomial is called a *trinomial square*.

i Recognizing Trinomial Squares

We use the equations for squaring a binomial in reverse to factor trinomial squares:

$$A^2 + 2AB + B^2 = (A + B)^2;$$
$$A^2 - 2AB + B^2 = (A - B)^2.$$

How can we recognize when an expression to be factored is a trinomial square? Look at $A^2 + 2AB + B^2$ and $A^2 - 2AB + B^2$. In order for an expression to be a trinomial square:

a) Two of the terms, A^2 and B^2, must be squares, such as

$$4, \quad x^2, \quad 25x^4, \quad 16t^2.$$

OBJECTIVES

After finishing Section 4.3, you should be able to:

i Recognize trinomial squares.

ii Factor trinomial squares.

1. Which of the following are trinomial squares?

a) $x^2 + 8x + 16$

b) $25 + x^2 - 10x$

c) $t^2 - 12t + 4$

d) $25 + 20y + 4y^2$

e) $5x^2 + 16 - 14x$

f) $16x^2 + 40x + 25$

g) $m^2 + 6m - 9$

h) $25x^2 - 9 - 30x$

Factor.

2. $x^2 + 2x + 1$

3. $1 - 2x + x^2$

4. $4 + t^2 + 4t$

5. $25x^2 - 70x + 49$

6. $49 - 56y + 16y^2$

7. $48m^2 + 75 + 120m$

b) There must be no minus sign before A^2 or B^2.

c) If we multiply A and B (the square roots of these expressions) and double the result, we get the remaining term $2 \cdot A \cdot B$, or its additive inverse, $-2 \cdot A \cdot B$.

EXAMPLE 1 Determine whether $x^2 + 6x + 9$ is a trinomial square.

a) We know that x^2 and 9 are squares.

b) There is no minus sign before x^2 or 9.

c) If we multiply the square roots, x and 3, and double the product, we get the remaining term: $2 \cdot 3 \cdot x = 6x$.

Thus $x^2 + 6x + 9$ is the square of a binomial.

EXAMPLE 2 Determine whether $x^2 + 6x + 11$ is a trinomial square.

The answer is *no*, because only one term is a square.

EXAMPLE 3 Determine whether $16x^2 + 49 - 56x$ is a trinomial square.

a) We know that $16x^2$ and 49 are squares.

b) There is no minus sign before $16x^2$ or 49.

c) If we multiply the square roots, $4x$ and 7, and double the product, we get the additive inverse of the remaining term: $2 \cdot 4x \cdot 7 = 56x$; and $56x$ is the additive inverse of $-56x$.

Thus $16x^2 + 49 - 56x$ is a trinomial square.

DO EXERCISE 1.

ii Factoring Trinomial Squares

To factor trinomial squares we use the following equations:

$$A^2 + 2AB + B^2 = (A + B)^2;$$
$$A^2 - 2AB + B^2 = (A - B)^2.$$

We use the square roots of the squared terms and the sign of the remaining term.

EXAMPLE 4 Factor: $x^2 + 6x + 9$.

$$x^2 + 6x + 9 = x^2 + 2 \cdot x \cdot 3 + 3^2 = (x + 3)^2$$

The sign of the middle term is positive.

$$A^2 + 2AB + B^2 = (A + B)^2$$

EXAMPLE 5 Factor: $x^2 + 49 - 14x$.

$$x^2 + 49 - 14x = x^2 - 14x + 49 \qquad \text{Changing order}$$
$$= x^2 - 2 \cdot x \cdot 7 + 7^2$$
$$= (x - 7)^2 \qquad \text{The sign of the middle term is negative.}$$

EXAMPLE 6 Factor: $16x^2 - 40x + 25$.

$$16x^2 - 40x + 25 = (4x)^2 - 2 \cdot 4x \cdot 5 + 5^2 = (4x - 5)^2$$

$$A^2 - 2AB + B^2 = (A - B)^2$$

DO EXERCISES 2–7.

EXERCISE SET 4.3

i Determine whether each of the following is a trinomial square.

1. $x^2 - 14x + 49$ **2.** $x^2 - 16x + 64$ **3.** $x^2 + 16x - 64$ **4.** $x^2 - 14x - 49$ **5.** $x^2 - 3x + 9$

6. $x^2 + 2x + 4$ **7.** $8x^2 + 40x + 25$ **8.** $9x^2 - 36x + 24$ **9.** $36x^2 - 24x + 16$

ii Factor. Remember to look first for a common factor.

10. $x^2 - 14x + 49$ **11.** $x^2 - 16x + 64$ **12.** $x^2 + 16x + 64$ **13.** $x^2 + 14x + 49$

14. $x^2 - 2x + 1$ **15.** $x^2 + 2x + 1$ **16.** $4 + 4x + x^2$ **17.** $4 + x^2 - 4x$

18. $y^2 - 6y + 9$ **19.** $y^2 + 6y + 9$ **20.** $2x^2 - 4x + 2$ **21.** $2x^2 - 40x + 200$

22. $x^3 - 18x^2 + 81x$ **23.** $x^3 + 24x^2 + 144x$ **24.** $20x^2 + 100x + 125$ **25.** $12x^2 + 36x + 27$

26. $49 - 42x + 9x^2$ **27.** $64 - 112x + 49x^2$ **28.** $5y^4 + 10y^2 + 5$ **29.** $a^4 + 14a^2 + 49$

30. $y^6 + 26y^3 + 169$ **31.** $y^6 - 16y^3 + 64$ **32.** $16x^{10} - 8x^5 + 1$ **33.** $9x^{10} + 12x^5 + 4$

34. $1 + 4x^4 + 4x^2$ **35.** $1 - 2a^3 + a^6$ **36.** $\frac{1}{81}x^6 + \frac{8}{27}x^3 + \frac{16}{9}$ **37.** $\frac{1}{9}a^2 + \frac{1}{3}a + \frac{1}{4}$

Problem-solving practice

38. About 5 L of oxygen can be dissolved in 100 L of water at 0°C. This is 1.6 times the amount that can be dissolved in the same volume of water at 20°C. How much oxygen can be dissolved at 20°C?

39. The perimeter of a rectangle is 540 m. The width is 19 m less than the length. Find the width and the length.

Factor, if possible.

40. $49x^2 - 216$ **41.** $27x^3 - 13x$ **42.** $x^2 + 22x + 121$ **43.** $4x^2 + 9$

44. $x^2 - 5x + 25$ **45.** $18x^3 + 12x^2 + 2x$ **46.** $63x - 28$ **47.** $162x^2 - 82$

48. $x^4 - 9$ **49.** $8.1x^2 - 6.4$ **50.** $x^8 - 2^8$ **51.** $3^4 - x^4$

Factor.

52. $(y + 3)^2 + 2(y + 3) + 1$ **53.** $(a + 4)^2 - 2(a + 4) + 1$ **54.** $4(a + 5)^2 + 20(a + 5) + 25$

55. $49(x + 1)^2 - 42(x + 1) + 9$ **56.** $(x + 7)^2 - 4x - 24$ **57.** $(a + 4)^2 - 6a - 15$

58. Is $(y + 2)^2(y - 2)^2$ a factorization of $y^4 - 8y^2 + 16$? Prove your answer.

59. Is $(x + 3)^2(x - 3)^2$ a factorization of $x^4 + 18x^2 + 81$? Prove your answer.

Factor.

60. $9x^{18} + 48x^9 + 64$

61. $x^{2n} + 10x^n + 25$

Factor the trinomial square, and then the difference of two squares.

62. $a^2 + 2a + 1 - 9$

63. $y^2 + 6y + 9 - x^2 - 8x - 16$

Find c so that the polynomial will be the square of a binomial.

64. $cy^2 + 6y + 1$

65. $cy^2 - 24y + 9$

66. Show that the difference of the squares of two consecutive integers is the sum of the integers. (*Hint:* Use x for the smaller number.)

67. Find the value of a if $x^2 + a^2x + a^2$ factors into $(x + a)^2$.

4.4

FACTORING TRINOMIALS OF THE TYPE $x^2 + bx + c$

i Some trinomials are not trinomial squares, as in the following examples:

$$x^2 + 5x + 6 \quad \text{and} \quad x^2 + 3x - 10.$$

To try to factor such trinomials, we use a trial-and-error procedure.

OBJECTIVE

After finishing Section 4.4, you should be able to:

i Factor trinomials of the type $x^2 + bx + c$ by examining the last coefficient c.

Factor.

1. $x^2 + 7x + 12$

Constant Term Positive

Recall the FOIL method of multiplying two binomials:

$$(x + 2)(x + 5) \overset{\text{F} \quad \text{O} \quad \text{I} \quad \text{L}}{=} x^2 + \underbrace{5x + 2x} + 10$$

$$= x^2 + \quad 7x \quad + 10.$$

The product is a trinomial. In the example, the term of highest degree, called the leading term, has a coefficient of 1. The constant term is positive. To factor $x^2 + 7x + 10$, we think of FOIL in reverse. We multiplied x times x to get the first term of the trinomial. So the first term of each binomial factor is x:

$$(x + \underline{\quad})(x + \underline{\quad}).$$

To get the middle term and the last term of the trinomial we look for two numbers whose product is 10 and whose sum is 7. Those numbers are 2 and 5. Thus the factorization is

$$(x + 2)(x + 5).$$

EXAMPLE 1 Factor: $x^2 + 5x + 6$.

Think of FOIL in reverse. The first term of each factor is x:

$$(x + \underline{\quad})(x + \underline{\quad}).$$

Then look for two numbers whose product is 6 and whose sum is 5. Since both 5 and 6 are positive, we need only consider positive factors.

2. $x^2 + 13x + 36$

Pairs of factors	Sums of factors
1, 6	7
2, 3	5

The numbers we want are 2 and 3. The factorization is $(x + 2)(x + 3)$. We can check by multiplying to see whether we get the original trinomial.

DO EXERCISES 1 AND 2.

Consider this multiplication:

$$(x - 3)(x - 4) \overset{\text{F} \quad \text{O} \quad \text{I} \quad \text{L}}{=} x^2 - \underbrace{4x - 3x} + 12$$

$$= x^2 - \quad 7x \quad + 12.$$

When the constant term of a trinomial is positive, we look for two numbers with the same sign. The sign is that of the middle term:

$$(x^2 - 7x + 12) = (x - 3)(x - 4).$$

EXAMPLE 2 Factor: $x^2 - 8x + 12$.

Since the constant term is positive and the coefficient of the middle term is negative, we look for a factorization of 12 in which both factors are negative. Their sum must be -8.

Pairs of factors	Sums of factors
$-1, -12$	-13
$-2, -6$	-8
$-3, -4$	-7

The numbers we want are -2 and -6. The factorization is $(x - 2)(x - 6)$.

DO EXERCISES 3 AND 4.

Constant Term Negative

Sometimes when we use FOIL, the product has a negative constant term. Consider these multiplications:

a) $(x - 5)(x + 2) = x^2 + \underset{\text{O}}{2x} - \underset{\text{I}}{5x} - 10$

$$= x^2 - 3x - 10;$$

b) $(x + 5)(x - 2) = x^2 - \underset{\text{O}}{2x} + \underset{\text{I}}{5x} - 10$

$$= x^2 + 3x - 10.$$

When the constant term is negative, the middle term may be positive or negative. In these cases, we still look for two factors whose product is -10. One of them must be positive and the other negative. Their sum must still be the coefficient of the middle term.

EXAMPLE 3 Factor: $x^2 - 8x - 20$.

Since the constant term is negative, we look for a factorization of -20 in which one factor is positive and one factor is negative. Their sum must be -8.

Pairs of factors	Sums of factors
$-1, 20$	19
$1, -20$	-19
$-2, 10$	8
$2, -10$	-8
$-5, 4$	-1
$5, -4$	1

The numbers we want are 2 and -10. The factorization is $(x + 2)(x - 10)$.

Factor.

3. $x^2 - 8x + 15$

4. $t^2 - 9t + 20$

Factor.

5. $x^2 + 4x - 12$

6. $y^2 - 4y - 12$

7. $t^2 + 5t - 14$

8. $x^2 - 30 - x$

EXAMPLE 4 Factor: $x^2 + 5x - 24$.

We look for a factorization of -24 in which one factor is positive and the other is negative. Their sum must be 5.

Pairs of factors	Sums of factors
1, −24	−23
−1, 24	23
2, −12	−10
−2, 12	10
−8, 3	−5
8, −3	5
4, −6	−2
−4, 6	2

The numbers we want are 8 and -3. The factorization is $(x + 8)(x - 3)$.

EXAMPLE 5 Factor: $x^2 - x - 110$.

Since the constant term is negative, we look for a factorization of -110 in which one factor is positive and one factor is negative. Their sum must be -1. The numbers we want are 10 and -11. The factorization is

$$(x + 10)(x - 11).$$

DO EXERCISES 5–8.

Some trinomials cannot be factored. An example is

$$x^2 - x + 5.$$

EXERCISE SET 4.4

i Factor.

1. $x^2 + 8x + 15$

2. $x^2 + 5x + 6$

3. $x^2 + 7x + 12$

4. $x^2 + 9x + 8$

5. $x^2 - 6x + 9$

6. $y^2 + 11y + 28$

7. $x^2 + 9x + 14$

8. $a^2 + 11a + 30$

9. $b^2 + 5b + 4$

10. $x^2 - \frac{2}{5}x + \frac{1}{25}$

11. $x^2 + \frac{2}{3}x + \frac{1}{9}$

12. $z^2 - 8z + 7$

13. $d^2 - 7d + 10$

14. $x^2 - 8x + 15$

15. $y^2 - 11y + 10$

16. $x^2 - 2x - 15$

17. $x^2 + x - 42$

18. $x^2 + 2x - 15$

19. $x^2 - 7x - 18$

20. $y^2 - 3y - 28$

21. $x^2 - 6x - 16$

22. $x^2 - x - 42$

23. $y^2 - 4y - 45$

24. $x^2 - 7x - 60$

25. $x^2 - 2x - 99$

26. $x^2 - 72 + 6x$

27. $c - 56 + c^2$

28. $b^2 + 5b - 24$

29. $a^2 + 2a - 35$

30. $2 - x - x^2$

Factor.

31. $x^2 + 20x + 100$

32. $x^2 + 20x + 99$

33. $x^2 - 21x - 100$

34. $x^2 - 20x + 96$

35. $x^2 - 21x - 72$

36. $4x^2 + 40x + 100$

37. $x^2 - 25x + 144$

38. $y^2 - 21y + 108$

39. $a^2 + a - 132$

40. $a^2 + 9a - 90$

41. $120 - 23x + x^2$

42. $96 + 22d + d^2$

43. $108 - 3x - x^2$

44. $112 + 9y - y^2$

45. Find all integers m for which $y^2 + my + 50$ can be factored.

46. Find all integers b for which $a^2 + ba - 50$ can be factored.

Factor.

47. $x^2 - \frac{1}{2}x - \frac{3}{16}$

48. $x^2 - \frac{1}{4}x - \frac{1}{8}$

49. $x^2 + \frac{30}{7}x - \frac{25}{7}$

50. $\frac{1}{3}x^3 + \frac{1}{3}x^2 - 2x$

Find a polynomial in factored form for each shaded region. (Leave answers in terms of π.)

51.

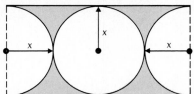

52.

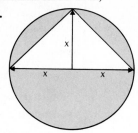

4.5

FACTORING TRINOMIALS OF THE TYPE $ax^2 + bx + c, a \neq 1$

i In Section 4.4 we learned to factor trinomials of the type $x^2 + bx + c$. In this section we consider two methods of factoring trinomials where the leading, or x^2, coefficient is not 1. You may choose the one that works best for you, or you may use the one that your instructor chooses for you. Both methods involve trial and error, but the first requires trial and error in only one step.

Method 1

We know how to factor the trinomial $x^2 + 5x + 6$. We look for factors of the constant term, 6, whose sum is the coefficient, 5, of the middle term:

$$x^2 + 5x + 6$$

① Factor: $6 = 2 \cdot 3$.

② Sum of factors: $2 + 3 = 5$.

What happens when the coefficient of the first, or x^2, term is not 1? Consider the trinomial $6x^2 + 23x + 20$. The method we use is similar to what we used for the preceding trinomial, but we need two more steps. We first multiply the leading coefficient 6 and the constant 20, and get 120. Then we look for a factorization of 120 in which the sum of the factors is the coefficient, 23, of the middle term. Next we write the middle term as a sum using these factors.

$$6x^2 + 23x + 20$$

① Multiply 6 and 20: $6 \cdot 20 = 120$.

② Factor 120: $120 = 8 \cdot 15$ and $8 + 15 = 23$.

③ Write the middle term as a sum: $23x = 8x + 15x$.

④ Then factor by grouping.

We factor by grouping as follows:

$$
\begin{aligned}
6x^2 + 23x + 20 &= 6x^2 + 8x + 15x + 20 \\
&= 2x(3x + 4) + 5(3x + 4) \qquad \text{Factoring by grouping} \\
&\qquad\qquad\qquad\qquad\qquad\qquad\quad \text{(see Section 4.1)} \\
&= (2x + 5)(3x + 4).
\end{aligned}
$$

It does not matter which way we write the middle term as a sum. We still get the same factorization, although the factors may be in a different order. Note the following:

$$
\begin{aligned}
6x^2 + 23x + 20 &= 6x^2 + 15x + 8x + 20 \\
&= 3x(2x + 5) + 4(2x + 5) \\
&= (3x + 4)(2x + 5).
\end{aligned}
$$

OBJECTIVE

After finishing Section 4.5, you should be able to:

i Factor trinomials of the type $ax^2 + bx + c, a \neq 1$.

Factor.

1. $2x^2 - x - 15$

The method we used to factor trinomials of the type $x^2 + bx + c$ was based on FOIL. So too is the method we have just introduced. Before we show why, we state the method more formally and consider other examples.

To factor $ax^2 + bx + c$:

a) First look for a common factor.

b) Multiply the leading coefficient a and the constant c.

c) Try to factor the product ac so that the sum of the factors is b. That is, find p and q such that $ac = pq$ and $p + q = b$.

d) Write the middle term, bx, as a sum: $bx = px + qx$. (We can think of this as "splitting" the middle term.)

e) Then factor by grouping. Regroup $ax^2 + px$ and $qx + c$.

EXAMPLE 1   Factor: $3x^2 - 10x - 8$.

a)   First look for a common factor. There is none (other than 1).

b)   Multiply the leading coefficient and the constant, 3 and -8:

$$3(-8) = -24.$$

c)   Try to factor -24 so that the sum of the factors is -10.

Pairs of factors	Sums of factors
$4, -6$	-2
$-4, 6$	2
$12, -2$	10
$-12, 2$	-10

2. $12x^2 - 17x - 5$

d)   Write $-10x$ as a sum using the results of part (c). That is, split the middle term as follows:

$$-10x = -12x + 2x.$$

e)   Factor by grouping:

$$3x^2 - 10x - 8 = 3x^2 - 12x + 2x - 8 \qquad \text{Substituting } -12x + 2x \text{ for } -10x$$
$$= 3x(x - 4) + 2(x - 4)$$
$$= (3x + 2)(x - 4).$$

DO EXERCISES 1 AND 2.

EXAMPLE 2   Factor: $8x^2 + 8x - 6$.

a)   First look for a common factor. The number 2 is common to all three terms, so we factor it out:

$$2(4x^2 + 4x - 3).$$

b)   Now factor the trinomial $4x^2 + 4x - 3$. Multiply the leading coefficient and the constant, 4 and -3:

$$4(-3) = -12.$$

c)   Try to factor -12 so that the sum of the factors is 4.

Pairs of factors	Sums of factors
$-3, 4$	1
$3, -4$	-1
$12, -1$	11
$-6, 2$	-4
$6, -2$	4

d) Split the middle term $4x$, as follows:

$$4x = 6x - 2x.$$

e) Then factor by grouping:

$$
\begin{aligned}
4x^2 + 4x - 3 &= 4x^2 + 6x - 2x - 3 \\
&= 2x(2x + 3) - (2x + 3) \\
&= 2x(2x + 3) - 1(2x + 3) \\
&= (2x - 1)(2x + 3).
\end{aligned}
$$

The factorization of $4x^2 + 4x - 3$ is $(2x - 1)(2x + 3)$. But, don't forget the common factor! We must include it to get a factorization of the original trinomial:

$$8x^2 + 8x - 6 = 2(2x - 1)(2x + 3).$$

This method of factoring is based on FOIL as shown here:

$$
\begin{aligned}
(ax + b)(cx + d) &= acx^2 + adx + bcx + bd \\
&= acx^2 + (ad + bc)x + bd
\end{aligned}
$$

$$(ac)(bd) = (ad)(bc)$$

We multiply the outside coefficients and factor them in such a way that we can write the middle term as a sum.

DO EXERCISES 3 AND 4.

Method 2

We now consider an alternative method for factoring trinomials of the type $ax^2 + bx + c$. Consider the following multiplication:

$$
\begin{array}{cccc}
 & F & O \quad I & L \\
(2x + 5)(3x + 4) &= 6x^2 & + 8x + 15x & + 20 \\
&= 6x^2 & + \quad 23x & + 20
\end{array}
$$

F	O + I	L
$2 \cdot 3$	$2 \cdot 4 + 5 \cdot 3$	$5 \cdot 4$

Now to factor $6x^2 + 23x + 20$, we do the reverse of what we just did:

$$
\begin{array}{ccc}
F & O + I & L \\
6x^2 + & 23x & + 20 \\
= (2x + 5) & (3x + 4)
\end{array}
$$

Factor.

3. $3x^2 - 19x + 20$

4. $20x^2 - 46x + 24$

Factor.

5. $6x^2 + 7x + 2$

We look for numbers (*x*-coefficients) whose product is 6—in this case, 2 and 3—and numbers whose product is 20—in this case, 4 and 5. The product of the outside terms plus the product of the inside terms must, of course, be 23.

To factor $ax^2 + bx + c$, we look for two binomials like this:

$$(\underline{}x + \underline{})(\underline{}x + \underline{}),$$

where products of numbers in the blanks are as follows.

---

1. **The numbers in the *first* blanks of each binomial have product *a*.**
2. **The *outside* product and the *inside* product add up to *b*.**
3. **The numbers in the *last* blanks of each binomial have product *c*.**

EXAMPLE 3   Factor: $3x^2 + 5x + 2$.

We look for two numbers whose product is 3. These are

$$1, 3 \quad \text{and} \quad -1, -3.$$

We have these possibilities:

$$(x + \underline{})(3x + \underline{})$$

or

$$(-x + \underline{})(-3x + \underline{}).$$

Now we look for numbers whose product is 2. These are

$$1, 2 \quad \text{and} \quad -1, -2.$$

Here are some possibilities for factorizations. There are eight possibilities, but we have not listed all of them here:

$$(x + 1)(3x + 2), \qquad (-x + 1)(-3x + 2), \qquad (x + 2)(3x + 1),$$
$$(x - 1)(3x - 2), \qquad (-x - 1)(-3x - 2), \qquad (x - 2)(3x - 1).$$

When we multiply, we must get $3x^2 + 5x + 2$. When we multiply, we find that both of the expressions in color above are factorizations. We usually choose the one in which the first coefficients are positive. Thus the factorization is

$$(x + 1)(3x + 2).$$

We always look first for a common factor. If there is one, we remove that common factor before proceeding.

DO EXERCISE 5.

EXAMPLE 4   Factor: $8x^2 + 8x - 6$.

We first look for a factor common to all three terms. The number 2 is a common factor, so we factor it out:

$$2(4x^2 + 4x - 3).$$

Now we factor the trinomial $4x^2 + 4x - 3$. We look for pairs of numbers whose product is 4. The pairs that are positive are

$$4, 1 \quad \text{and} \quad 2, 2. \qquad \text{Both positive}$$

We then have these possibilities:

$$(4x + \underline{})(x + \underline{}) \quad \text{and} \quad (2x + \underline{})(2x + \underline{}).$$

Next we look for pairs of numbers whose product is -3. They are

$$3, -1 \quad \text{and} \quad -3, 1.$$

Then we have these possibilities for factorizations:

$$(4x + 3)(x - 1), \quad (2x + 3)(2x - 1),$$
$$(4x - 1)(x + 3), \quad (2x - 3)(2x + 1).$$
$$(4x + 1)(x - 3),$$
$$(4x - 3)(x + 1),$$

We usually do not write all of these. We multiply until we find the factors that give the product $4x^2 + 4x - 3$. We find that the factorization is

$$(2x + 3)(2x - 1).$$

But don't forget the common factor. We must include it in order to get a factorization of the original polynomial:

$$8x^2 + 8x - 6 = 2(2x + 3)(2x - 1).$$

Keep in mind that no matter which of the two methods you use to factor trinomials of the type $ax^2 + bx + c$, you involve trial and error. This is the way such factoring is done. As you practice you will find that you can make better and better guesses. Don't forget: When factoring any polynomials, always look first for a common factor. Failure to do so is a common error.

DO EXERCISES 6–9.

Factor.

6. $6x^2 + 15x + 9$

7. $2y^2 + 4y - 6$

8. $4t^2 + 2t - 6$

9. $6x^2 - 5x + 1$

EXERCISE SET 4.5

i Factor.

1. $2x^2 - 7x - 4$ **2.** $3x^2 - x - 4$ **3.** $5x^2 + x - 18$ **4.** $3x^2 - 4x - 15$ **5.** $6x^2 + 23x + 7$

6. $6x^2 + 13x + 6$ **7.** $3x^2 + 4x + 1$ **8.** $7x^2 + 15x + 2$ **9.** $4x^2 + 4x - 15$ **10.** $9x^2 + 6x - 8$

11. $2x^2 - x - 1$ **12.** $15x^2 - 19x - 10$ **13.** $9x^2 + 18x - 16$ **14.** $2x^2 + 5x + 2$ **15.** $3x^2 - 5x - 2$

16. $18x^2 - 3x - 10$ **17.** $12x^2 + 31x + 20$ **18.** $15x^2 + 19x - 10$ **19.** $14x^2 + 19x - 3$ **20.** $35x^2 + 34x + 8$

21. $9x^2 + 18x + 8$ **22.** $6 - 13x + 6x^2$ **23.** $49 - 42x + 9x^2$ **24.** $25x^2 + 40x + 16$ **25.** $24x^2 + 47x - 2$

26. $16a^2 + 78a + 27$ **27.** $35x^2 - 57x - 44$ **28.** $9a^2 + 12a - 5$ **29.** $20 + 6x - 2x^2$ **30.** $15 + x - 2x^2$

31. $12x^2 + 28x - 24$ **32.** $6x^2 + 33x + 15$ **33.** $30x^2 - 24x - 54$ **34.** $20x^2 - 25x + 5$ **35.** $4x + 6x^2 - 10$

36. $-9 + 18x^2 - 21x$ **37.** $3x^2 - 4x + 1$ **38.** $6x^2 - 13x + 6$ **39.** $12x^2 - 28x - 24$

40. $6x^2 - 33x + 15$ **41.** $-1 + 2x^2 - x$ **42.** $-19x + 15x^2 + 6$ **43.** $9x^2 - 18x - 16$

44. $14x^2 + 35x + 14$ **45.** $15x^2 - 25x - 10$ **46.** $18x^2 + 3x - 10$ **47.** $12x^3 + 31x^2 + 20x$

48. $15x^3 + 19x^2 - 10x$ **49.** $14x^4 + 19x^3 - 3x^2$ **50.** $70x^4 + 68x^3 + 16x^2$ **51.** $168x^3 - 45x^2 + 3x$

Problem-solving practice

52. The earth is a sphere (or ball) that is about 40,000 km in circumference. Find the radius of the earth, in kilometers and in miles. (*Hint:* 1 km $\approx$ 0.62 mi.)

53. In an apartment, lamps, an air conditioner, and a television set are all operating at the same time. The lamps take 10 times as many watts as the television set, and the air conditioner takes 40 times as many watts as the television set. The total wattage used in the apartment is 2550 watts. How many watts are used by each appliance?

Factor, if possible.

54. $9x^4 + 18x^2 + 8$ **55.** $6 - 13x + 6x^2$ **56.** $9x^2 - 42x + 49$

57. $15x^4 - 19x^2 + 6$ **58.** $6x^3 + 4x^2 - 10x$ **59.** $18x^3 - 21x^2 - 9x$

60. $x^2 + 3x - 7$

61. $x^2 + 13x - 12$

62. $x^5 + 2x^4 + 2x + 1$

63. $27x^3 - 63x^2 - 147x + 343$

Factor.

64. $20x^{2n} + 16x^n + 3$

65. $-15x^{2m} + 26x^m - 8$

66. $3x^{6a} - 2x^{3a} - 1$

67. $x^{2n+1} - 2x^{n+1} + x$

68. $3(a+1)^{n+1}(a+3)^2 - 5(a+1)^n(a+3)^3$

OBJECTIVE

After finishing Section 4.6, you should be able to:

i Factor polynomials completely using any of the methods considered thus far in this chapter.

4.6

FACTORING: A GENERAL STRATEGY

i We now try to put all of our factoring techniques together and consider a general strategy for factoring polynomials. Here we will encounter polynomials of all the types we have considered, mixed up, so you will have to determine which method to use.

To factor a polynomial:

A. **Always look first for a common factor.**

B. **Then look at the number of terms.**

   Two terms: Determine whether you have a difference of squares. Do not try to factor a sum of squares: $A^2 + B^2$.

   Three terms: Determine whether the trinomial is a square. If so, you know how to factor. If not, try trial and error.

   Four terms: Try factoring by grouping.

C. **Always** *factor completely*. **If a factor with more than one term can still be factored, you should do so. When no factor can be factored further, you have finished.**

EXAMPLE 1 Factor: $10x^3 - 40x$.

A. We look first for a common factor:

$$10x^3 - 40x = 10x(x^2 - 4). \qquad \text{Factoring out the largest common factor}$$

B. The factor $x^2 - 4$ has only two terms. It is a difference of squares. We factor it, being careful to include the common factor:

$$10x(x - 2)(x + 2).$$

C. Have we factored completely? Yes, because no factor with more than one term can be factored further.

EXAMPLE 2 Factor: $t^4 - 16$.

A. We look for a common factor. There isn't one.

B. There are only two terms. It is a difference of squares: $(t^2)^2 - 4^2$. We factor it:

$$(t^2 + 4)(t^2 - 4).$$

We see that one of the factors is still a difference of squares. We factor it:

$$(t^2 + 4)(t - 2)(t + 2).$$

This is a sum of squares. It cannot be factored!

C. We have factored completely because no factors with more than one term can be factored further.

EXAMPLE 3 Factor: $2x^3 + 10x^2 + x + 5$.

A. We look for a common factor. There isn't one.

B. There are four terms. We try factoring by grouping:

$$2x^3 + 10x^2 + x + 5 = (2x^3 + 10x^2) + (x + 5) \qquad \text{Separating into two binomials}$$
$$= 2x^2(x + 5) + 1(x + 5) \qquad \text{Factoring each binomial}$$
$$= (2x^2 + 1)(x + 5). \qquad \text{Factoring out the common factor,} \; x + 5$$

C. No factor with more than one term can be factored further, so we have factored completely.

EXAMPLE 4 Factor: $x^5 - 2x^4 - 35x^3$.

A. We look first for a common factor. This time there is one:

$$x^5 - 2x^4 - 35x^3 = x^3(x^2 - 2x - 35).$$

B. The factor $x^2 - 2x - 35$ has three terms, but it is not a trinomial square. We factor it using trial and error:

$$x^5 - 2x^4 - 35x^3 = x^3(x^2 - 2x - 35) = x^3(x - 7)(x + 5).$$

Don't forget to include the common factor in your final answer!

C. No factor with more than one term can be factored further, so we have factored completely.

EXAMPLE 5 Factor: $x^4 - 10x^2 + 25$.

A. We look first for a common factor. There isn't one.

B. There are three terms. We see that this is a trinomial square. We factor it:

$$x^4 - 10x^2 + 25 = (x^2)^2 - 2 \cdot 5 \cdot x^2 + 5^2 = (x^2 - 5)^2.$$

C. No factor with more than one term can be factored further, so we have factored completely.

DO EXERCISES 1–5.

Factor completely.

1. $3m^4 - 3$

2. $x^6 + 8x^3 + 16$

3. $2x^4 + 8x^3 + 6x^2$

4. $3x^3 + 12x^2 - 2x - 8$

5. $8x^3 - 200x$

EXERCISE SET 4.6

i Factor.

1. $2x^2 - 128$ 2. $3t^2 - 27$ 3. $a^2 + 25 - 10a$ 4. $y^2 + 49 + 14y$

5. $2x^2 - 11x + 12$ 6. $8y^2 - 18y - 5$ 7. $x^3 + 24x^2 + 144x$ 8. $x^3 - 18x^2 + 81x$

9. $x^3 + 3x^2 - 4x - 12$ 10. $x^3 - 5x^2 - 25x + 125$ 11. $24x^2 - 54$ 12. $8x^2 - 98$

13. $20x^3 - 4x^2 - 72x$ 14. $9x^3 + 12x^2 - 45x$ 15. $x^2 + 4$ 16. $t^2 + 25$

17. $x^4 + 7x^2 - 3x^3 - 21x$ 18. $m^4 + 8m^3 + 8m^2 + 64m$ 19. $x^5 - 14x^4 + 49x^3$ 20. $2x^6 + 8x^5 + 8x^4$

21. $20 - 6x - 2x^2$ 22. $45 - 3x - 6x^2$ 23. $x^2 + 3x + 1$ 24. $x^2 + 5x + 2$

25. $4x^4 - 64$ 26. $5x^5 - 80x$ 27. $1 - y^8$ 28. $t^8 - 1$

29. $x^5 - 4x^4 + 3x^3$ 30. $x^6 - 2x^5 + 7x^4$ 31. $36a^2 - 15a + \frac{25}{16}$ 32. $\frac{1}{81}x^6 - \frac{8}{27}x^3 + \frac{16}{9}$

Factor completely.

33. $a^4 - 2a^2 + 1$ 34. $x^4 + 9$ 35. $12.25x^2 - 7x + 1$ 36. $\frac{1}{5}x^2 - x + \frac{4}{5}$

37. $5x^2 + 13x + 7.2$ 38. $x^3 - (x - 3x^2) - 3$ 39. $18 + y^3 - 9y - 2y^2$ 40. $-(x^4 - 7x^2 - 18)$

41. $a^3 + 4a^2 + a + 4$ **42.** $x^3 + x^2 - (4x + 4)$ **43.** $x^4 - 7x^2 - 18$ **44.** $3x^4 - 15x^2 + 12$

45. $x^3 - x^2 - 4x + 4$ **46.** $y^2(y + 1) - 4y(y + 1) - 21(y + 1)$ **47.** $y^2(y - 1) - 2y(y - 1) + (y - 1)$

48. Factor $x^{2k} - 2^{2k}$ when $k = 4$.   **49.** Factor $a^4 - 81$.

50. Factor $acx^{m+n} + adx^n + bcx^m + bd$, where a, b, c, and d are constants.

OBJECTIVES

After finishing Section 4.7, you should be able to:

i   Solve equations (already factored) using the principle of zero products.

ii   Solve certain equations by factoring and then using the principle of zero products.

4.7

SOLVING EQUATIONS BY FACTORING

In this section we introduce a new equation-solving method and use it along with factoring to solve certain equations.

i   The Principle of Zero Products

The product of two numbers is 0 if one of the numbers is 0. Furthermore, *if any product is 0, then a factor must be 0*. For example, if $7x = 0$, then we know that $x = 0$. If $x(2x - 9) = 0$, we know that $x = 0$ or $2x - 9 = 0$. If $(x + 3)(x - 2) = 0$, we know that $x + 3 = 0$ or $x - 2 = 0$.

EXAMPLE 1   Solve: $(x + 3)(x - 2) = 0$.

We have a product of 0. This equation will be true when either factor is 0. Hence it is true when

$$x + 3 = 0 \quad \text{or} \quad x - 2 = 0.$$

Here we have two simple equations, which we know how to solve:

$$x = -3 \quad \text{or} \quad x = 2.$$

There are two solutions, -3 and 2.

We have another principle to help in solving equations.

THE PRINCIPLE OF ZERO PRODUCTS

An equation with 0 on one side and with factors on the other can be solved by finding those numbers that make the factors 0.

EXAMPLE 2   Solve: $(5x + 1)(x - 7) = 0$.

$$
\begin{aligned}
5x + 1 &= 0 \quad &\text{or} \quad x - 7 &= 0 \quad &&\text{Using the principle of zero products} \\
5x &= -1 \quad &\text{or} \quad x &= 7 \\
x &= -\tfrac{1}{5} \quad &\text{or} \quad x &= 7 \quad &&\text{Solving the two equations separately}
\end{aligned}
$$

Check: For $-\tfrac{1}{5}$:

$$
\begin{array}{c|c}
(5x + 1)(x - 7) = 0 & \\
\hline
(5(-\tfrac{1}{5}) + 1)(-\tfrac{1}{5} - 7) & 0 \\
(-1 + 1)(-7\tfrac{1}{5}) & \\
0(-7\tfrac{1}{5}) & \\
0 &
\end{array}
$$

For 7:

$$
\begin{array}{c|c}
(5x + 1)(x - 7) = 0 & \\
\hline
(5 \cdot 7 + 1)(7 - 7) & 0 \\
(35 + 1) \cdot 0 & \\
0 &
\end{array}
$$

The solutions are $-\tfrac{1}{5}$ and 7.

The "possible solutions" we get by using the principle of zero products are actually always solutions, unless we have made an error in solving. Thus, when we use

this principle, a check is not necessary, except to detect errors. Keep in mind that you must have 0 on one side of an equation before you can apply the principle of zero products.

DO EXERCISES 1–3.

When some factors have only one term, you can still use the principle of zero products in the same way.

EXAMPLE 3 Solve: $x(2x - 9) = 0$.

$$x = 0 \quad \text{or} \quad 2x - 9 = 0 \qquad \text{Using the principle of zero products}$$
$$x = 0 \quad \text{or} \qquad 2x = 9$$
$$x = 0 \quad \text{or} \qquad x = \tfrac{9}{2}$$

The solutions are 0 and $\tfrac{9}{2}$.

DO EXERCISE 4.

ii Using Factoring to Solve Equations

Using factoring and the principle of zero products, we can solve some new kinds of equations. Thus we have extended our equation-solving abilities.

EXAMPLE 4 Solve: $x^2 + 5x + 6 = 0$.

We first factor the polynomial. Then we use the principle of zero products:

$$x^2 + 5x + 6 = 0$$
$$(x + 2)(x + 3) = 0 \qquad \text{Factoring}$$
$$x + 2 = 0 \quad \text{or} \quad x + 3 = 0 \qquad \text{Using the principle of zero products}$$
$$x = -2 \quad \text{or} \qquad x = -3.$$

Check:

$x^2 + 5x + 6 = 0$	
$(-2)^2 + 5(-2) + 6$	0
$4 - 10 + 6$	
$-6 + 6$	
0	

$x^2 + 5x + 6 = 0$	
$(-3)^2 + 5(-3) + 6$	0
$9 - 15 + 6$	
$-6 + 6$	
0	

The solutions are -2 and -3.

Again, keep in mind that you *must* have 0 on one side before you can use the principle of zero products.

DO EXERCISE 5.

EXAMPLE 5 Solve: $x^2 - 8x = -16$.

We first add 16 to get 0 on one side:

$$x^2 - 8x + 16 = 0 \qquad \text{Adding 16}$$
$$(x - 4)(x - 4) = 0 \qquad \text{Factoring}$$
$$x - 4 = 0 \quad \text{or} \quad x - 4 = 0 \qquad \text{Using the principle of zero products}$$
$$x = 4 \quad \text{or} \qquad x = 4.$$

There is only one solution, 4. The check is left to the student.

Solve using the principle of zero products.

1. $(x - 3)(x + 4) = 0$

2. $(x - 7)(x - 3) = 0$

3. $(4t + 1)(3t - 2) = 0$

Solve using the principle of zero products.

4. $y(3y - 17) = 0$

Solve.

5. $x^2 - x - 6 = 0$

POLYNOMIALS AND FACTORING

Solve.

6. $x^2 - 3x = 28$

7. $y^2 + 9 = 6y$

8. $x^2 = 4x$

9. $25t^2 = 16$

EXAMPLE 6 Solve: $x^2 + 5x = 0$.

$$x(x + 5) = 0 \qquad \text{Factoring out a common factor}$$
$$x = 0 \quad \text{or} \quad x + 5 = 0 \qquad \text{Using the principle of zero products}$$
$$x = 0 \quad \text{or} \qquad x = -5$$

The solutions are 0 and -5. The check is left to the student.

EXAMPLE 7 Solve: $4x^2 - 25 = 0$.

$$(2x - 5)(2x + 5) = 0 \qquad \text{Factoring a difference of squares}$$
$$2x - 5 = 0 \quad \text{or} \quad 2x + 5 = 0$$
$$2x = 5 \quad \text{or} \qquad 2x = -5$$
$$x = \tfrac{5}{2} \quad \text{or} \qquad x = -\tfrac{5}{2}$$

The solutions are $\tfrac{5}{2}$ and $-\tfrac{5}{2}$. The check is left to the student.

DO EXERCISES 6–9.

EXERCISE SET 4.7

i Solve.

1. $(x + 8)(x + 6) = 0$

2. $(x + 3)(x + 2) = 0$

3. $(x - 3)(x + 5) = 0$

4. $(x + 9)(x - 3) = 0$

5. $(x + 12)(x - 11) = 0$

6. $(x - 13)(x + 53) = 0$

7. $x(x + 5) = 0$

8. $y(y + 7) = 0$

9. $y(y - 13) = 0$

10. $v(v - 4) = 0$

11. $0 = y(y + 10)$

12. $0 = x(x - 21)$

13. $(2x + 5)(x + 4) = 0$

14. $(2x + 9)(x + 8) = 0$

15. $(3x - 1)(x + 2) = 0$

16. $(3x - 9)(x + 3) = 0$

17. $(5x + 1)(4x - 12) = 0$

18. $(4x + 9)(14x - 7) = 0$

19. $(7x - 28)(28x - 7) = 0$

20. $(12x - 11)(8x - 5) = 0$

21. $2x(3x - 2) = 0$

22. $75x(8x - 9) = 0$

23. $\tfrac{1}{2}x(\tfrac{2}{3}x - 12) = 0$

24. $\tfrac{5}{7}x(\tfrac{3}{4}x - 6) = 0$

25. $(\tfrac{1}{3} - 3x)(\tfrac{1}{5} - 2x) = 0$

26. $(\tfrac{1}{5} + 2x)(\tfrac{1}{9} - 3x) = 0$

27. $(\tfrac{1}{3}y - \tfrac{2}{3})(\tfrac{1}{4}y - \tfrac{3}{2}) = 0$

28. $(\tfrac{7}{4}x - \tfrac{1}{12})(\tfrac{2}{3}x - \tfrac{12}{11}) = 0$

29. $(0.3x - 0.1)(0.05x - 1) = 0$

30. $(0.1x - 0.3)(0.4x - 20) = 0$

31. $9x(3x - 2)(2x - 1) = 0$

32. $(x - 5)(x + 55)(5x - 1) = 0$

ii Solve.

33. $x^2 + 6x + 5 = 0$

34. $x^2 + 7x + 6 = 0$

35. $x^2 + 7x - 18 = 0$

36. $x^2 + 4x - 21 = 0$

37. $x^2 - 8x + 15 = 0$

38. $x^2 - 9x + 14 = 0$

39. $x^2 - 8x = 0$

40. $x^2 - 3x = 0$

41. $x^2 + 19x = 0$

42. $x^2 + 12x = 0$

43. $x^2 = 16$

44. $100 = x^2$

45. $9x^2 - 4 = 0$

46. $4x^2 - 9 = 0$

47. $0 = 6x + x^2 + 9$

48. $0 = 25 + x^2 + 10x$

49. $x^2 + 16 = 8x$

50. $1 + x^2 = 2x$

51. $5x^2 = 6x$

52. $7x^2 = 8x$

53. $6x^2 - 4x = 10$

54. $3x^2 - 7x = 20$

55. $12y^2 - 5y = 2$

56. $2y^2 + 12y = -10$

57. $x(x - 5) = 14$

58. $t(3t + 1) = 2$

59. $64m^2 = 81$

60. $100t^2 = 49$

61. $3x^2 + 8x = 9 + 2x$

62. $x^2 - 5x = 18 + 2x$

63. $10x^2 - 23x + 12 = 0$

64. $12x^2 - 17x - 5 = 0$

Solve.

65. $b(b + 9) = 4(5 + 2b)$

66. $y(y + 8) = 16(y - 1)$

67. $(t - 3)^2 = 36$

68. $(t - 5)^2 = 2(5 - t)$

69. $x^2 - \tfrac{1}{64} = 0$

70. $x^2 - \tfrac{25}{36} = 0$

71. $\tfrac{5}{16}x^2 = 5$

72. $\tfrac{27}{25}x^2 = \tfrac{1}{3}$

73. Find an equation that has the given numbers as solutions. For example, 3 and -2 are solutions to $x^2 - x - 6 = 0$.

a) $1, -3$ **b)** $3, -1$ **c)** $2, 2$ **d)** $3, 4$ **e)** $3, -4$
f) $-3, 4$ **g)** $-3, -4$ **h)** $\frac{1}{2}, \frac{1}{2}$ **i)** $5, -5$ **j)** $0, 0.1, \frac{1}{4}$

74. Check the numbers found below. What's wrong with the methods used? Try to find the solutions.

a) $(x - 3)(x + 4) = 8$
$x - 3 = 0$ or $x + 4 = 8$
$x = 3$ or $x = 4$

b) $(x - 3)(x + 4) = 8$
$x - 3 = 2$ or $x + 4 = 4$
$x = 5$ or $x = 0$

75. ▦ Solve: $(0.00005x + 0.1)(0.0097x + 0.5) = 0$.

76. For each equation on the left, find an equivalent equation on the right.

a) $3x^2 - 4x + 8 = 0$
b) $(x - 6)(x + 3) = 0$
c) $x^2 + 2x + 9 = 0$
d) $(2x - 5)(x + 4) = 0$
e) $5x^2 - 5 = 0$
f) $x^2 + 10x - 2 = 0$

g) $4x^2 + 8x + 36 = 0$
h) $(2x + 8)(2x - 5) = 0$
i) $9x^2 - 12x + 24 = 0$
j) $(x + 1)(5x - 5) = 0$
k) $x^2 - 3x - 18 = 0$
l) $2x^2 + 20x - 4 = 0$

4.8

PROBLEM SOLVING

i We can use our five-step problem-solving process and our new methods for solving equations to solve problems.

EXAMPLE 1   One more than a number times one less than a number is 8. Find the number.

1. **Familiarize**   The problem is stated explicitly enough that we can go right to the translation.

2. **Translate**   We translate as follows:

One more than a number times one less than that number is 8.

$$(x + 1) \qquad \cdot \qquad (x - 1) \qquad = 8$$

3. **Carry out**   We solve the equation as follows:

$$(x + 1)(x - 1) = 8$$
$$x^2 - 1 = 8 \qquad \text{Multiplying}$$
$$x^2 - 1 - 8 = 0 \qquad \text{Adding } -8 \text{ to get 0 on one side}$$
$$x^2 - 9 = 0$$
$$(x - 3)(x + 3) = 0 \qquad \text{Factoring}$$
$$x - 3 = 0 \quad \text{or} \quad x + 3 = 0 \qquad \text{Using the principle of zero products}$$
$$x = 3 \quad \text{or} \qquad x = -3$$

4. **Check**   One more than 3 (this is 4) times one less than 3 (this is 2) is 8. Thus, 3 checks. The check for -3 is left to the student.

5. **State**   There are two such numbers, 3 and -3.

DO EXERCISES 1–3.

1. One more than a number times one less than the number is 24.

2. Seven less than a number times eight less than the number is 0.

3. A number times one less than the number is 0.

4. The square of a number minus the number is 20.

EXAMPLE 2 The square of a number minus twice the number is 48. Find the number.

1. **Familiarize** Again, the problem is stated explicitly enough that we can go right to the translation.

2. **Translate** We translate as follows:

$$\underbrace{\text{The square of a number}}_{x^2} \underbrace{\text{minus}}_{-} \underbrace{\text{twice the number}}_{2x} \underbrace{\text{is}\ 48.}_{=\ 48}$$

3. **Carry out** We solve the equation as follows:

$$x^2 - 2x = 48$$
$$x^2 - 2x - 48 = 0 \qquad \text{Adding } -48 \text{ to get 0 on one side}$$
$$(x - 8)(x + 6) = 0$$
$$x - 8 = 0 \quad \text{or} \quad x + 6 = 0 \qquad \text{Using the principle of zero products}$$
$$x = 8 \quad \text{or} \qquad x = -6$$

4. **Check** The square of 8 is 64, and twice the number 8 is 16. Then $64 - 16$ is 48, so 8 checks. The check for -6 is left to the student.

5. **State** There are two such numbers, 8 and -6.

DO EXERCISE 4.

EXAMPLE 3 The height of a triangular sail is 7 ft more than the base. The area of the triangle is 30 ft². Find the height and the base.

1. **Familiarize** We first make a drawing. If you don't remember the formula for the area of a triangle, look it up, either in this book or a geometry book. The area is

$$\tfrac{1}{2} \cdot \text{base} \cdot \text{height.}$$

2. **Translate** It helps to reword this problem before translating:

$$\underbrace{\tfrac{1}{2}}_{\tfrac{1}{2}} \text{ times } \underbrace{\text{the base}}_{b} \text{ times } \underbrace{\text{the base plus 7}}_{(b+7)} \text{ is } \underbrace{30}_{=\ 30} \qquad \text{Rewording}$$

$$\qquad\qquad\qquad\qquad\qquad\qquad\qquad\qquad\qquad\qquad\qquad\qquad \text{Translating}$$

3. Carry out   We solve the equation as follows:

$$\tfrac{1}{2} \cdot b \cdot (b + 7) = 30$$

$$\tfrac{1}{2}(b^2 + 7b) = 30 \qquad \text{Multiplying}$$

$$b^2 + 7b = 60 \qquad \text{Multiplying by 2}$$

$$b^2 + 7b - 60 = 0 \qquad \text{Adding} -60 \text{ to get 0 on one side}$$

$$(b + 12)(b - 5) = 0 \qquad \text{Factoring}$$

$$b + 12 = 0 \qquad \text{or} \quad b - 5 = 0 \qquad \text{Using the principle of zero products}$$

$$b = -12 \quad \text{or} \qquad b = 5.$$

4. Check   The solutions of the equation are -12 and 5. The base of a triangle cannot have a negative length, so -12 cannot be a solution. Suppose the base is 5 ft. Then the height is 7 ft more than the base, so the height is 12 ft and the area is $\tfrac{1}{2}(5)(12)$ or 30 ft². These numbers check in the original problem.

5. State   The height is 12 ft and the base is 5 ft.

DO EXERCISE 5.

EXAMPLE 4   In a sports league of n teams in which each team plays every other team twice, the total number N of games to be played is given by

$$N = n^2 - n.$$

A basketball league plays a total of 240 games. How many teams are in the league?

1. Familiarize   To familiarize yourself with this problem, reread Example 4 in Section 3.1, where we first considered it.

2. Translate   We are trying to find the number of teams n in a league when 240 games are played. We substitute 240 for N in order to solve for n:

$$n^2 - n = 240. \qquad \text{Substituting 240 for } N$$

3. Carry out   We solve the equation as follows:

$$n^2 - n = 240$$

$$n^2 - n - 240 = 0 \qquad \text{Adding} -240 \text{ to get 0 on one side}$$

$$(n - 16)(n + 15) = 0 \qquad \text{Factoring}$$

$$n - 16 = 0 \quad \text{or} \quad n + 15 = 0 \qquad \text{Using the principle of zero products}$$

$$n = 16 \quad \text{or} \qquad n = -15.$$

5. The width of a rectangle is 2 cm less than the length. The area is 15 cm². Find the length and the width.

6. Use $n^2 - n$ for the following:
A slow-pitch softball league plays a total of 72 games. How many teams are in the league?

4. Check The solutions of the equation are 16 and -15. Since the number of teams cannot be negative, -15 cannot be a solution. But 16 checks, since $16^2 - 16 = 256 - 16 = 240$.

5. State There are 16 teams in the league.

DO EXERCISE 6.

EXAMPLE 5 The product of two consecutive integers is 156. Find the integers.

1. Familiarize Recall that *consecutive* integers are next to each other, such as 49 and 50, or -6 and -5. If x represents the smaller integer, then $x + 1$ represents the larger integer.

2. Translate It helps to reword the problem before translating:

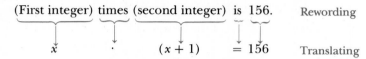

(First integer) times (second integer) is 156.   Rewording

$$x \cdot (x + 1) = 156$$ Translating

We have let x represent the first integer. Then $x + 1$ represents the second.

7. The product of two consecutive integers is 462. Find the integers.

3. Carry out We solve the equation as follows:

$$x(x + 1) = 156$$
$$x^2 + x = 156 \qquad \text{Multiplying}$$
$$x^2 + x - 156 = 0 \qquad \text{Adding } -156 \text{ to get 0 on one side}$$
$$(x - 12)(x + 13) = 0 \qquad \text{Factoring}$$
$$x - 12 = 0 \quad \text{or} \quad x + 13 = 0 \qquad \text{Using the principle of zero products}$$
$$x = 12 \quad \text{or} \qquad x = -13.$$

4. Check The solutions of the equation are 12 and -13. When x is 12, then $x + 1$ is 13, and $12 \cdot 13 = 156$. The numbers 12 and 13 are consecutive integers that are solutions to the problem. When x is -13, then $x + 1$ is -12, and $(-13)(-12) = 156$. The numbers -13 and -12 are also consecutive integers that are solutions to the problem.

5. State We have two solutions, each of which consists of a pair of numbers: 12 and 13, and -13 and -12.

DO EXERCISE 7.

EXERCISE SET 4.8

i Solve.

1. If you subtract a number from 4 times its square, the result is 3. Find the number.

2. If 7 is added to the square of a number, the result is 32. Find the number.

3. Eight more than the square of a number is 6 times the number. Find the number.

4. Fifteen more than the square of a number is 8 times the number. Find the number.

5. The product of two consecutive integers is 182. Find the integers.

6. The product of two consecutive integers is 56. Find the integers.

7. The product of two consecutive even integers is 168. Find the integers.

8. The product of two consecutive even integers is 224. Find the integers.

9. The product of two consecutive odd integers is 255. Find the integers.

10. The product of two consecutive odd integers is 143. Find the integers.

11. The length of a rectangular garden is 4 m greater than the width. The area of the rectangle is 96 m². Find the length and width.

12. The length of a rectangular calculator is 5 cm greater than the width. The area of the rectangle is 84 cm². Find the length and width.

13. The area of a square bookcase is 5 ft² more than the perimeter. Find the length of a side.

14. The perimeter of a square porch is 3 yd more than the area. Find the length of a side.

15. The base of a triangle is 10 cm greater than the height. The area is 28 cm². Find the height and base.

16. The height of a triangle is 8 m less than the base. The area is 10 m². Find the height and base.

17. If the sides of a square are lengthened by 3 m, the area becomes 81 m². Find the length of a side of the original square.

18. If the sides of a square are lengthened by 7 km, the area becomes 121 km². Find the length of a side of the original square.

19. The sum of the squares of two consecutive odd positive integers is 74. Find the integers.

20. The sum of the squares of two consecutive odd positive integers is 130. Find the integers.

Use $N = n^2 - n$ for Exercises 21–24.

21. A slow-pitch softball league has 23 teams. What is the total number of games to be played?

22. A basketball league has 14 teams. What is the total number of games to be played?

23. A slow-pitch softball league plays a total of 132 games. How many teams are in the league?

24. A basketball league plays a total of 90 games. How many teams are in the league?

The number of possible handshakes within a group of n people is given by $N = \frac{1}{2}(n^2 - n)$.

25. At a meeting there are 40 people. How many handshakes are possible?

26. At a party there are 100 people. How many handshakes are possible?

27. Everyone shook hands at a party. There were 190 handshakes in all. How many were at the party?

28. Everyone shook hands at a meeting. There were 300 handshakes in all. How many were at the meeting?

---

29. A cement walk of constant width is built around a 20-ft × 40-ft rectangular pool. The total area of the pool and walk is 1500 ft². Find the width of the walk.

30. A model rocket is launched using an engine that will generate a speed of 180 feet per second. The formula $h = rt - 16t^2$ gives the height of an object projected upward at a rate of r feet per second after t seconds. After how many seconds will the rocket reach a height of 464 feet? After how many seconds will it be at that height again?

31. When the speed of an object is measured in meters per second and distance in meters, the formula of Exercise 30 becomes $h = rt - 4.9t^2$. A baseball is thrown upward with a speed of 20.6 meters per second.

a) After how many seconds will the ball reach a height of 21.6 meters?

b) How long after it is thrown will it hit the ground?

33. The total surface area of a box is 350 m². The box is 9 m high and has a square base. Find the length of the side of the base.

34. A rectangular piece of cardboard is twice as long as it is wide. A 4-cm square is cut out of each corner, and the sides are turned up to make a box. The volume of the box is 616 cm³. Find the original dimensions of the cardboard.

32. The one's digit of a number less than 100 is 4 greater than the ten's digit. The sum of the number and the product of the digits is 58. Find the number.

35. An open rectangular gutter is made by turning up the sides of a piece of metal 20 in. wide. The area of the cross-section of the gutter is 50 in². Find the depth of the gutter.

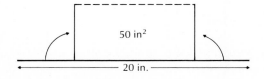

36. The length of each side of a square is increased by 5 cm to form a new square. The area of the new square is $2\frac{1}{4}$ times the area of the original square. Find the area of each square.

OBJECTIVES

After finishing Section 4.9, you should be able to:

i   Evaluate a polynomial in several variables for given values of the variables.

ii   Identify the coefficients and degrees of the terms of a polynomial and the degrees of polynomials.

iii   Collect like terms of a polynomial.

iv   Add polynomials.

v   Subtract polynomials.

vi   Multiply polynomials.

vii   Factor polynomials.

1. Evaluate the polynomial

$$4 + 3x + xy^2 + 8x^3y^3$$

for $x = 2$ and $y = -5$.

2. Evaluate the polynomial

$$8xy^2 - 2x^3z - 13x^4y^2 + 5$$

for $x = -1$, $y = 3$, and $z = 4$.

3. Given the situation below, what is the magic number for the Cincinnati Reds? Assume $G = 162$.

WESTERN DIVISION			
	W	L	Pct.	GB
Cincinnati......	77	44	.636	—
Los Angeles	65	54	.546	11
San Diego......	60	64	.484	$18\frac{1}{2}$
Houston	59	64	.480	19
Atlanta	56	65	.463	21
San Francisco ...	52	70	.426	$25\frac{1}{2}$

4.9

POLYNOMIALS IN SEVERAL VARIABLES

The polynomials we have been studying have only one variable. A *polynomial in several variables* is an expression like those you have already seen, but we allow that there can be more than one variable. Here are some examples:

$$3x + xy^2 + 5y + 4, \qquad 8xy^2z - 2x^3z - 13x^4y^2 + 15.$$

We now learn how to add, subtract, multiply, and factor polynomials in several variables.

i   Evaluating Polynomials

EXAMPLE 1   Evaluate the polynomial $4 + 3x + xy^2 + 8x^3y^3$ for $x = -2$ and $y = 5$.

We replace x by -2 and y by 5:

$$4 + 3(-2) + (-2) \cdot 5^2 + 8(-2)^3 \cdot 5^3 = 4 - 6 - 50 - 8000 = -8052.$$

DO EXERCISES 1 AND 2.

EXAMPLE 2   (*The magic number.*) The Boston Red Sox are leading the New York Yankees for the Eastern Division championship of the American League. The magic number is 8. This means that any combination of Red Sox wins and Yankee losses that totals 8 will ensure the championship for the Red Sox. The magic number is given by the polynomial

$$G - P - L + 1,$$

where G is the number of games in the season, P is the number of games the leading team has played, and L is the number of games ahead in the loss column.

Given the situation shown in the table and assuming a 162-game season, what is the magic number for the Philadelphia Phillies?

EASTERN DIVISION			
	W	L	Pct.	GB
Philadelphia........	77	40	.658	—
Pittsburgh..........	65	53	.551	$12\frac{1}{2}$
New York...........	61	60	.504	18
Chicago............	55	67	.451	$24\frac{1}{2}$
St. Louis...........	51	65	.440	$25\frac{1}{2}$
Montreal..........	41	73	.360	$34\frac{1}{2}$

We evaluate the polynomial for $G = 162$, $P = 77 + 40$, or 117, and $L = 53 - 40$, or 13:

$$162 - 117 - 13 + 1 = 33.$$

DO EXERCISE 3.

ii   Coefficients and Degrees

The *degree* of a term is the sum of the exponents of the variables. The *degree of a polynomial* is the degree of the term of highest degree.

EXAMPLE 3 Identify the coefficient and degree of each term of

$$9x^2y^3 - 14xy^2z^3 + xy + 4y + 5x^2 + 7.$$

Term	Coefficient	Degree
$9x^2y^3$	9	5
$-14xy^2z^3$	-14	6
xy	1	2
$4y$	4	1
$5x^2$	5	2
7	7	0

Think: $4y = 4y^1$

Think: $7 = 7x^0$

EXAMPLE 4 What is the degree of $5x^3y + 9xy^4 - 8x^3y^3$?

The term of highest degree is $-8x^3y^3$. Its degree is 6. Thus the degree of the polynomial is 6.

DO EXERCISES 4 AND 5.

iii Collecting Like Terms

Like terms (or *similar terms*) have exactly the same variables with exactly the same exponents. For example,

$3x^2y^3$ and $-7x^2y^3$ are like terms;

$9x^4z^7$ and $12x^4z^7$ are like terms.

But

$13xy^2$ and $-2x^2y$ are *not* like terms;

$3xyz^2$ and $4xy$ are *not* like terms.

Collecting like terms is based on the distributive law.

EXAMPLES Collect like terms.

5. $5x^2y + 3xy^2 - 5x^2y - xy^2 = (5-5)x^2y + (3-1)xy^2 = 2xy^2$

6. $3xy - 5xy^2 + 3xy^2 + 9xy = -2xy^2 + 12xy$

DO EXERCISES 6 AND 7.

iv Addition

The sum of two polynomials in several variables can be found by writing a plus sign between them and then collecting like terms.

EXAMPLE 7 Add: $-5x^3 + 3y - 5y^2$ and $8x^3 + 4x^2 + 7y^2$.

$$(-5x^3 + 3y - 5y^2) + (8x^3 + 4x^2 + 7y^2) = (-5+8)x^3 + 4x^2 + 3y + (-5+7)y^2$$
$$= 3x^3 + 4x^2 + 3y + 2y^2$$

EXAMPLE 8 Add: $(5xy^2 - 4x^2y + 5x^3 + 2) + (3xy^2 - 2x^2y + 3x^3y - 5)$.

We look for like terms. The like terms are $5xy^2$ and $3xy^2$, $-4x^2y$ and $-2x^2y$, and 2 and -5. We collect these. There are no more like terms. Thus the answer is

$$8xy^2 - 6x^2y + 5x^3 + 3x^3y - 3.$$

DO EXERCISES 8–10.

4. Identify the coefficient of each term.

$$-3xy^2 + 3x^2y - 2y^3 + xy + 2$$

5. Identify the degree of each term and the degree of the polynomial.

$$4xy^2 + 7x^2y^3z^2 - 5x + 2y + 4$$

Collect like terms.

6. $4x^2y + 3xy - 2x^2y$

7. $-3pq - 5pqr^3 + 8pq + 5pqr^3 + 4$

Add.

8. $4x^3 + 4x^2 - 8x - 3$ and $-8x^3 - 2x^2 + 4x + 5$

9. $(13x^3y + 3x^2y - 5y)$ $+ (x^3y + 4x^2y - 3xy + 3y)$

10. $(-5p^2q^4 + 2p^2q^2 + 3q)$ $+ (6pq^2 + 3p^2q + 5)$

Subtract.

11. $(-4s^4t + s^3t^2 + 2s^2t^3)$
$- (4s^4t - 5s^3t^2 + s^2t^2)$

12. $(-5p^4q + 5p^3q^2 - 3p^2q^3 - 7q^4)$
$- (4p^4q - 5p^3q^2 + p^2q^3 + 2q^4)$

Multiply.

13. $(x^2y^3 + 2x)(x^3y^2 + 3x)$

14. $(p^4q - 2p^3q^2 + 3q^3)(p + 2q)$

v ■ Subtraction

We subtract a polynomial by adding its inverse. An equivalent expression for the additive inverse of a polynomial is found by replacing each coefficient by its additive inverse, or by changing the sign of each term. For example, the additive inverse of the polynomial

$$4x^2y - 6x^3y^2 + x^2y^2 - 5y$$

can be represented by

$$-(4x^2y - 6x^3y^2 + x^2y^2 - 5y).$$

An equivalent expression can be found by replacing each coefficient by its additive inverse. Thus

$$-(4x^2y - 6x^3y^2 + x^2y^2 - 5y) = -4x^2y + 6x^3y^2 - x^2y^2 + 5y.$$

EXAMPLE 9 Subtract: $(4x^2y + x^3y^2 + 3x^2y^3 + 6y) - (4x^2y - 6x^3y^2 + x^2y^2 - 5y).$

$(4x^2y + x^3y^2 + 3x^2y^3 + 6y) - (4x^2y - 6x^3y^2 + x^2y^2 - 5y)$

$= 4x^2y + x^3y^2 + 3x^2y^3 + 6y - 4x^2y + 6x^3y^2 - x^2y^2 + 5y$ Adding the inverse

$= 7x^3y^2 + 3x^2y^3 - x^2y^2 + 11y$ Collecting like terms (Try to write just the answer!)

DO EXERCISES 11 AND 12.

vi ■ Multiplication

To multiply polynomials in several variables, we can multiply each term of one by every term of the other. Where appropriate, we use special products.

EXAMPLE 10 Multiply: $(3x^2y - 2xy + 3y)(xy + 2y).$

$$
\begin{array}{r}
3x^2y - 2xy + 3y \\
xy + 2y \\
\hline
3x^3y^2 - 2x^2y^2 + 3xy^2 \\
6x^2y^2 - 4xy^2 + 6y^2 \\
\hline
3x^3y^2 + 4x^2y^2 - xy^2 + 6y^2
\end{array}
$$

Multiplying by xy

Multiplying by $2y$

Adding

DO EXERCISES 13 AND 14.

EXAMPLES Multiply.

                       F   O   I   L

11. $(x^2y + 2x)(xy^2 + y^2) = x^3y^3 + x^2y^3 + 2x^2y^2 + 2xy^2$

12. $(p + 5q)(2p - 3q) = 2p^2 - 3pq + 10pq - 15q^2$
$= 2p^2 + 7pq - 15q^2$

$(A + B)^2 = A^2 + 2\,A\,\,B + B^2$

13. $(3x + 2y)^2 = (3x)^2 + 2(3x)(2y) + (2y)^2$
$= 9x^2 + 12xy + 4y^2$

14. $(2y^2 - 5x^2y)^2 = (2y^2)^2 - 2(2y^2)(5x^2y) + (5x^2y)^2$
$= 4y^4 - 20x^2y^3 + 25x^4y^2$

$(A + B)(A - B) = A^2 - B^2$

15. $(3x^2y + 2y)(3x^2y - 2y) = (3x^2y)^2 - (2y)^2$
$= 9x^4y^2 - 4y^2$

16. $(-2x^3y^2 + 5t)(2x^3y^2 + 5t) = (5t - 2x^3y^2)(5t + 2x^3y^2)$
$$= (5t)^2 - (2x^3y^2)^2 = 25t^2 - 4x^6y^4$$

$$(A \quad - \quad B) \quad (A \quad + \quad B) = \quad A^2 \quad - \quad B^2$$

17. $(2x + 3 - 2y)(2x + 3 + 2y) = (2x + 3)^2 - (2y)^2$
$$= 4x^2 + 12x + 9 - 4y^2$$

DO EXERCISES 15–22.

vii Factoring

To factor polynomials in several variables, we can use the same general strategy that we considered in Section 4.6, which you might review before studying the following examples.

EXAMPLE 18 Factor: $20x^3y + 12x^2y$.

A. We look first for a common factor:

$$20x^3y + 12x^2y = (4x^2y)(5x) + (4x^2y) \cdot 3$$
$$= 4x^2y(5x + 3). \qquad \text{Factoring out the largest common factor}$$

B. Then we look at the number of terms. There are only two terms, but the binomial $5x + 3$ is not a difference of squares. It cannot be factored further.

C. We have factored completely because no factors with more than one term can be factored further.

EXAMPLE 19 Factor: $6x^2y - 21x^3y^2 + 3x^2y^3$.

A. We look first for a common factor:

$$6x^2y - 21x^3y^2 + 3x^2y^3 = 3x^2y(2 - 7xy + y^2).$$

B. There are three terms in $2 - 7xy + y^2$. Determine whether the trinomial is a square. Since only y^2 is a square, we do not have a trinomial square. Can the trinomial be factored by trial? A key to the answer is that x is only in the term $-7xy$. The polynomial might be in a form like $(1 - y)(2 + y)$, but there would be no x in the middle term.

C. Have we factored completely? Yes, because no factor with more than one term can be factored further.

DO EXERCISES 23 AND 24.

EXAMPLE 20 Factor: $(p + q)(x + 2) + (p + q)(x + y)$.

A. We look first for a common factor:

$$(p + q)(x + 2) + (p + q)(x + y) = (p + q)[(x + 2) + (x + y)]$$
$$= (p + q)(2x + y + 2).$$

B. There are three terms in $2x + y + 2$, but this trinomial cannot be factored further.

C. No factor with more than one term can be factored further, so we have factored completely.

EXAMPLE 21 Factor: $px + py + qx + qy$.

A. We look first for a common factor. There isn't one.

Multiply.

15. $(3xy + 2x)(x^2 + 2xy^2)$

16. $(x - 3y)(2x - 5y)$

17. $(4x + 5y)^2$

18. $(3x^2 - 2xy^2)^2$

19. $(2xy^2 + 3x)(2xy^2 - 3x)$

20. $(3xy^2 + 4y)(-3xy^2 + 4y)$

21. $(3y + 4 - 3x)(3y + 4 + 3x)$

22. $(2a + 5b + c)(2a - 5b - c)$

Factor.

23. $x^4y^2 + 2x^3y + 3x^2y$

24. $10p^6q^2 - 4p^5q^3 + 2p^4q^4$

Factor.

25. $(a - b)(x + 5) + (a - b)(x + y^2)$

26. $ax^2 + ay + bx^2 + by$

Factor.

27. $x^4 + 2x^2y^2 + y^4$

28. $-4x^2 + 12xy - 9y^2$
(*Hint:* First factor out -1.)

B. There are four terms. We try factoring by grouping:

$$px + py + qx + qy = p(x + y) + q(x + y)$$
$$= (p + q)(x + y).$$

C. Have we factored completely? Since no factor with more than one term can be factored further, we have factored completely.

DO EXERCISES 25 AND 26.

EXAMPLE 22 Factor: $25x^2 + 20xy + 4y^2$.

A. We look first for a common factor. There isn't one.

B. There are three terms. Determine whether the trinomial is a square. The first term and the last term are squares:

$$25x^2 = (5x)^2 \quad \text{and} \quad 4y^2 = (2y)^2.$$

Twice the product of $5x$ and $2y$ should be the other term:

$$2 \cdot 5x \cdot 2y = 20xy.$$

Thus the trinomial is a perfect square.
We factor by writing the square roots of the square terms and the sign of the other term:

$$25x^2 + 20xy + 4y^2 = (5x + 2y)^2.$$

We can check by squaring $5x + 2y$.

C. No factor with more than one term can be factored further, so we have factored completely.

DO EXERCISES 27 AND 28.

EXAMPLE 23 Factor: $p^2q^2 + 7pq + 12$.

A. We look first for a common factor. There isn't one.

B. There are three terms. Determine whether the trinomial is a square. The first term is a square, but neither of the other terms is a square, so we do not have a trinomial square. We use trial and error thinking of the product pq as a single variable:

$$p^2q^2 + 7pq + 12 = (pq)^2 + (3 + 4)pq + 3 \cdot 4$$
$$= (pq + 3)(pq + 4).$$

C. No factor with more than one term can be factored further, so we have factored completely.

EXAMPLE 24 Factor: $8x^4 - 20x^2y - 12y^2$.

A. We look first for a common factor:

$$8x^4 - 20x^2y - 12y^2 = 4(2x^4 - 5x^2y - 3y^2).$$

B. There are three terms in $2x^4 - 5x^2y - 3y^2$. Determine whether the trinomial is a square. Since none of the terms is a square, we do not have a trinomial square. We use trial and error:

$$8x^4 - 20x^2y - 12y^2 = 4(2x^4 - 5x^2y - 3y^2)$$
$$= 4[(2x^2)(x^2) + (-6 + 1)x^2y + (-3y)y]$$
$$= 4(2x^2 + y)(x^2 - 3y).$$

C. No factor with more than one term can be factored further, so we have factored completely.

EXAMPLE 25 Factor: $a^4 + a^3b - 6a^2b^2$.

A. We look first for a common factor:

$$a^4 + a^3b - 6a^2b^2 = a^2(a^2 + ab - 6b^2).$$

B. There are three terms in $a^2 + ab - 6b^2$. Determine whether the trinomial is a square. Since neither ab nor $-6b^2$ is a square, we do not have a trinomial square. We use trial and error:

$$a^4 + a^3b - 6a^2b^2 = a^2(a^2 + ab - 6b^2)$$
$$= a^2(a - 2b)(a + 3b).$$

C. No factor with more than one term can be factored further, so we have factored completely.

EXAMPLE 26 Factor: $a^4 - 16b^4$.

$$a^4 - 16b^4 = (a^2 - 4b^2)(a^2 + 4b^2) = (a - 2b)(a + 2b)(a^2 + 4b^2)$$

DO EXERCISES 29–31.

Factor.

29. $x^2y^2 + 5xy + 4$

30. $2x^4y^6 + 6x^2y^3 - 20$

31. $t^5 - 4m^2t^3$

EXERCISE SET 4.9

i Evaluate each polynomial for $x = 3$ and $y = -2$.

1. $x^2 - y^2 + xy$

2. $x^2 + y^2 - xy$

Evaluate each polynomial for $x = 2$, $y = -3$, and $z = -1$.

3. $xyz^2 + z$

4. $xy - xz + yz$

An amount of money P is invested at interest rate r. In 3 years it will grow to an amount given by the polynomial

$$P + 3rP + 3r^2P + r^3P.$$

5. Evaluate the polynomial for $P = 10,000$ and $r = 0.08$ to find the amount to which \$10,000 will grow at 8% interest for 3 years.

6. Evaluate the polynomial for $P = 10,000$ and $r = 0.07$ to find the amount to which \$10,000 will grow at 7% interest for 3 years.

The area of a right circular cylinder is given by the polynomial

$$2\pi rh + 2\pi r^2,$$

where h is the height and r is the radius of the base.

7. A 12-oz beverage can has a height of 4.7 in. and a radius of 1.2 in. Evaluate the polynomial for $h = 4.7$ and $r = 1.2$ to find the area of the can. Use 3.14 for π.

8. A 16-oz beverage can has a height of 6.3 in. and a radius of 1.2 in. Evaluate the polynomial for $h = 6.3$ and $r = 1.2$ to find the area of the can. Use 3.14 for π.

ii Identify the coefficient and degree of each term of the following polynomials. Then find the degree of the polynomial.

9. $x^3y - 2xy + 3x^2 - 5$ **10.** $5y^3 - y^2 + 15y + 1$ **11.** $17x^2y^3 - 3x^3yz - 7$ **12.** $6 - xy + 8x^2y^2 - y^5$

iii Collect like terms.

13. $a + b - 2a - 3b$

14. $y^2 - 1 + y - 6 - y^2$

15. $3x^2y - 2xy^2 + x^2$

16. $m^3 + 2m^2n - 3m^2 + 3mn^2$

17. $2u^2v - 3uv^2 + 6u^2v - 2uv^2$

18. $3x^2 + 6xy + 3y^2 - 5x^2 - 10xy - 5y^2$

19. $6au + 3av - 14au + 7av$

20. $3x^2y - 2z^2y + 3xy^2 + 5z^2y$

iv Add.

21. $(2x^2 - xy + y^2) + (-x^2 - 3xy + 2y^2)$

22. $(2z - z^2 + 5) + (z^2 - 3z + 1)$

23. $(r - 2s + 3) + (2r + s) + (s + 4)$

24. $(b^3 a^2 - 2b^2 a^3 + 3ba + 4) + (b^2 a^3 - 4b^3 a^2 + 2ba - 1)$

25. $(2x^2 - 3xy + y^2) + (-4x^2 - 6xy - y^2) + (x^2 + xy - y^2)$

v Subtract.

26. $(x^3 - y^3) - (-2x^3 + x^2 y - xy^2 + 2y^3)$

27. $(xy - ab) - (xy - 3ab)$

28. $(3y^4 x^2 + 2y^3 x - 3y) - (2y^4 x^2 + 2y^3 x - 4y - 2x)$

29. $(-2a + 7b - c) - (-3b + 4c - 8d)$

30. Find the sum of $2a + b$ and $3a - 4b$. Then subtract $5a + 2b$.

vi Multiply.

31. $(3z - u)(2z + 3u)$

32. $(a - b)(a^2 + b^2 + 2ab)$

33. $(a^2 b - 2)(a^2 b - 5)$

34. $(xy + 7)(xy - 4)$

35. $(a + a^2 - 1)(a^2 + 1 - y)$

36. $(r + tx)(vx + s)$

37. $(a^3 + bc)(a^3 - bc)$

38. $(m^2 + n^2 - mn)(m^2 + mn + n^2)$

39. $(y^4 x + y^2 + 1)(y^2 + 1)$

40. $(a - b)(a^2 + ab + b^2)$

41. $(3xy - 1)(4xy + 2)$

42. $(m^3 n + 8)(m^3 n - 6)$

43. $(3 - c^2 d^2)(4 + c^2 d^2)$

44. $(6x - 2y)(5x - 3y)$

45. $(m^2 - n^2)(m + n)$

46. $(pq + 0.2)(0.4pq - 0.1)$

47. $(xy + x^5 y^5)(x^4 y^4 - xy)$

48. $(x - y^3)(2y^3 + x)$

49. $(x + h)^2$

50. $(3a + 2b)^2$

51. $(r^3 t^2 - 4)^2$

52. $(3a^2 b - b^2)^2$

53. $(p^4 + m^2 n^2)^2$

54. $(ab + cd)^2$

55. $(2a^3 - \frac{1}{2}b^3)^2$

56. $-5x(x + 3y)^2$

57. $3a(a - 2b)^2$

58. $(a^2 + b + 2)^2$

59. $(2a - b)(2a + b)$

60. $(x - y)(x + y)$

61. $(c^2 - d)(c^2 + d)$

62. $(p^3 - 5q)(p^3 + 5q)$

63. $(ab + cd^2)(ab - cd^2)$

64. $(xy + pq)(xy - pq)$

65. $(x + y - 3)(x + y + 3)$

66. $(p + q + 4)(p + q - 4)$

67. $[x + y + z][x - (y + z)]$

68. $[a + b + c][a - (b + c)]$

69. $(a + b + c)(a - b - c)$

70. $(3x + 2 - 5y)(3x + 2 + 5y)$

vii Factor.

71. $12n^2 + 24n^3$

72. $ax^2 + ay^2$

73. $9x^2 y^2 - 36xy$

74. $x^2 y - xy^2$

75. $2\pi rh + 2\pi r^2$

76. $10p^4 q^4 + 35p^3 q^3 + 10p^2 q^2$

77. $(a + b)(x - 3) + (a + b)(x + 4)$

78. $5c(a^3 + b) - (a^3 + b)$

79. $(x - 1)(x + 1) - y(x + 1)$

80. $x^2 + x + xy + y$

81. $n^2 + 2n + np + 2p$

82. $a^2 - 3a + ay - 3y$

83. $2x^2 - 4x + xz - 2z$

84. $6y^2 - 3y + 2py - p$

85. $x^2 + y^2 - 2xy$

86. $4b^2 + a^2 - 4ab$

87. $9c^2 + 6cd + d^2$

88. $16x^2 + 24xy + 9y^2$

89. $49m^4 - 112m^2 n + 64n^2$

90. $4x^2 y^2 + 12xyz + 9z^2$

91. $y^4 + 10y^2 z^2 + 25z^4$

92. $0.01x^4 - 0.1x^2 y^2 + 0.25y^4$

93. $\frac{1}{4}a^2 + \frac{1}{3}ab + \frac{1}{9}b^2$

94. $4p^2 q + pq^2 + 4p^3$

95. $a^2 - ab - 2b^2$

96. $3b^2 - 17ab - 6a^2$

97. $2mn - 360n^2 + m^2$

98. $15 + x^2 y^2 + 8xy$

99. $m^2 n^2 - 4mn - 32$

100. $p^2 q^2 + 7pq + 6$

101. $a^5 b^2 + 3a^4 b - 10a^3$

102. $m^2 n^6 + 4mn^5 - 32n^4$

103. $a^5 + 4a^4 b - 5a^3 b^2$

104. $2s^6 t^2 + 10s^3 t^3 + 12t^4$

105. $x^6 + x^3 y - 2y^2$

106. $a^4 + a^2 bc - 2b^2 c^2$

107. $x^2 - y^2$

108. $p^2 q - r^2$

109. $7p^4 - 7q^4$

110. $a^4 b^4 - 16$

111. $81a^4 - b^4$

112. $1 - 16x^{12} y^{12}$

Find a polynomial for the area of each shaded region. (Leave results in terms of π where appropriate.)

113.

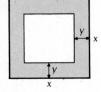

114.

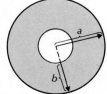

115.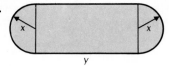

116.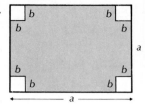

117. Find a formula for $(A + B)^3$.

Factor.

118. $6(x - 1)^2 + 7y(x - 1) - 3y^2$

119. $(y + 4)^2 + 2x(y + 4) + x^2$

120. $2(a + 3)^2 - (a + 3)(b - 2) - (b - 2)^2$

THE ASVAB TEST: ARITHMETIC REASONING   CHAPTER 4

See the discussion of the ASVAB Test at the beginning of Chapter 3.

1. The cost of a Navy A-6 Intruder attack bomber is $30 million. Eventually it must be renovated at a cost of $600,000. What percent of the original price is the cost of renovation?

- A   5%
- B   2%
- C   20%
- D   30%

2. A jeep traveling at a speed of 55 mph will reach a town 110 mi away in how many hours?

- A   2 hr
- B   3 hr
- C   1 hr
- D   4 hr

3. Sigourney bought a compact disc for $14.98 and paid for it with a $20 bill. How much change did she receive?

- A $3.98
- B $5.98
- C $4.02
- D $5.02

4. If 5 dozen eggs can be bought for $4.45, how many can you buy for $7.12?

- A   1.6 dozen
- B   8 dozen
- C   80 eggs
- D   16 eggs

5. On a map 3 in. represents 249 actual mi. A river is 4.8 in. long on the map. How long is the river in reality?

- A   398.4 mi
- B   155.6 mi
- C   57.8 mi
- D   14.4 mi

6. A soldier is off duty from 9:45 A.M. one day until 2:30 A.M. the next night. How many hours is the soldier off duty?

- A $17\frac{3}{4}$ hr
- B $16\frac{2}{3}$ hr
- C $16\frac{1}{2}$ hr
- D $16\frac{3}{4}$ hr

7. A person on a diet goes from 150 lb to 120 lb. Find the percent of decrease in weight.

- A   20%
- B   30%
- C   25%
- D   11.1%

8. A student earns $3000 one year and $4200 the next. Find the percent of increase.

- A   28.6%
- B   16.6%
- C   40%
- D   83.3%

9. A contractor sells a building for $42,525. Her costs were $19,170 for labor and $17,315 for materials. How much was her profit?

- A $36,485
- B $40,670
- C $79,010
- D $6,040

10. How many half-pint soda pop bottles can be filled from 8 gal of soda pop?

- A   64
- B   32
- C   128
- D   16

11. How many pounds of ham will it take to make 80 sandwiches if each sandwich takes $\frac{3}{5}$ lb?

    A   45

    B   48

    C $\frac{240}{5}$

    D $80\frac{3}{5}$

13. How many inches are there in $5\frac{1}{4}$ yd?

    A   63 in.

    B   15.75 in.

    C   21 in.

    D   189 in.

12. How many people can be served from 64 lb of hamburger if each person is served a "quarter pounder" ($\frac{1}{4}$ lb)?

    A   16

    B   256

    C   246

    D   17

14. On a map 1 in. represents 250 mi. How many inches does it take to represent 875 mi on the map?

    A   4 in.

    B   3.5 in.

    C   0.67 in.

    D   8 in.

OBJECTIVE

After finishing Section 4.10, you should be able to:

i **Factor sums and differences of two cubes.**

4.10

SUMS OR DIFFERENCES OF TWO CUBES

i Although a sum of two squares cannot be factored using real-number coefficients, a sum of two cubes can. In this section we develop the patterns for factoring a sum or a difference of two expressions that are cubes.

Consider the following products:

$$(A + B)(A^2 - AB + B^2) = A(A^2 - AB + B^2) + B(A^2 - AB + B^2)$$
$$= A^3 - A^2B + AB^2 + A^2B - AB^2 + B^3$$
$$= A^3 + B^3$$

and

$$(A - B)(A^2 + AB + B^2) = A(A^2 + AB + B^2) - B(A^2 + AB + B^2)$$
$$= A^3 + A^2B + AB^2 - A^2B - AB^2 - B^3$$
$$= A^3 - B^3.$$

The above equations (reversed) show how we can factor a sum or a difference of two cubes.

$$A^3 + B^3 = (A + B)(A^2 - AB + B^2)$$
$$A^3 - B^3 = (A - B)(A^2 + AB + B^2)$$

This table of cubes will help in the following examples.

N	0.2	0.1	0	1	2	3	4	5	6	7	8
N^3	0.008	0.001	0	1	8	27	64	125	216	343	512

EXAMPLE 1 Factor: $x^3 - 27$.

$$x^3 - 27 = x^3 - 3^3$$

In one set of parentheses we write the cube root of the first term, x. Then we write the cube root of the second term, -3. This gives us the expression $x - 3$:

$$(x - 3)(\qquad).$$

To get the next factor we think of $x - 3$ and do the following:

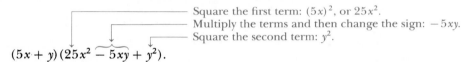

— Square the first term: x^2.
— Multiply the terms and then change the sign: $3x$.
— Square the second term: 9.

$$(x - 3)(x^2 + 3x + 9).$$

Note that we cannot factor $x^2 + 3x + 9$. (It is not a trinomial square nor can it be factored by trial).

DO EXERCISES 1 AND 2.

EXAMPLE 2 Factor: $125x^3 + y^3$.

$$125x^3 + y^3 = (5x)^3 + y^3$$

In one set of parentheses we write the cube root of the first term, then a plus sign, and then the cube root of the second term:

$$(5x + y)(\qquad).$$

To get the next factor, we think of $5x + y$ and do the following:

— Square the first term: $(5x)^2$, or $25x^2$.
— Multiply the terms and then change the sign: $-5xy$.
— Square the second term: y^2.

$$(5x + y)(25x^2 - 5xy + y^2).$$

DO EXERCISES 3 AND 4.

EXAMPLE 3 Factor: $128y^7 - 250x^6y$.

We first look for a common factor:

$$2y(64y^6 - 125x^6) = 2y[(4y^2)^3 - (5x^2)^3]$$
$$= 2y(4y^2 - 5x^2)(16y^4 + 20x^2y^2 + 25x^4).$$

EXAMPLE 4 Factor: $64a^6 - 729b^6$.

$$(8a^3 - 27b^3)(8a^3 + 27b^3) \qquad \text{Factoring a difference of squares}$$

Factor.

1. $x^3 - 8$

2. $64 - y^3$

Factor.

3. $27x^3 + y^3$

4. $8y^3 + z^3$

184

Factor.

5. $16x^7y + 54xy^7$

6. $729x^6 - 64y^6$

7. $x^3 - 0.027$

Each factor is a sum or difference of cubes. We factor:

$$(2a - 3b)(4a^2 + 6ab + 9b^2)(2a + 3b)(4a^2 - 6ab + 9b^2).$$

EXAMPLE 5 Factor: $y^3 + 0.008$.

We note that $0.008 = (0.2)^3$. Thus we have a sum of two cubes, and we factor as follows:

$$(y + 0.2)(y^2 - 0.2y + 0.04).$$

Remember the following about factoring sums or differences of squares and cubes:

Sum of cubes:	$A^3 + B^3 = (A + B)(A^2 - AB + B^2)$
Difference of cubes:	$A^3 - B^3 = (A - B)(A^2 + AB + B^2)$
Difference of squares:	$A^2 - B^2 = (A + B)(A - B)$
Sum of squares:	$A^2 + B^2$ **cannot be factored**

DO EXERCISES 5–7.

EXERCISE SET 4.10

i Factor.

1. $x^3 + 8$ **2.** $c^3 + 27$ **3.** $y^3 - 64$ **4.** $z^3 - 1$ **5.** $w^3 + 1$ **6.** $x^3 + 125$

7. $8a^3 + 1$ **8.** $27x^3 + 1$ **9.** $y^3 - 8$ **10.** $p^3 - 27$ **11.** $8 - 27b^3$ **12.** $64 - 125x^3$

13. $64y^3 + 1$ **14.** $125x^3 + 1$ **15.** $8x^3 + 27$ **16.** $27y^3 + 64$ **17.** $a^3 - b^3$ **18.** $x^3 - y^3$

19. $a^3 + \frac{1}{8}$ **20.** $b^3 + \frac{1}{27}$ **21.** $2y^3 - 128$ **22.** $3z^3 - 3$ **23.** $24a^3 + 3$ **24.** $54x^3 + 2$

25. $rs^3 + 64r$ **26.** $ab^3 + 125a$ **27.** $5x^3 - 40z^3$ **28.** $2y^3 - 54z^3$ **29.** $x^3 + 0.001$ **30.** $y^3 + 0.125$

31. $64x^6 - 8t^6$ **32.** $125c^6 - 8d^6$

Problem-solving practice

33. The University of Heidelberg was founded in 1386. This was 583 years before Neil Armstrong first stepped on the moon. When did he first step on the moon?

34. The customs duty on phonograph records is 5% of the wholesale price. You buy 350 British records for a wholesale price of $2800. How much customs duty will you pay?

Factor. Assume that variables in exponents represent natural numbers.

35. $x^{6a} + y^{3b}$ **36.** $a^3x^3 - b^3y^3$ **37.** $3x^{3a} + 24y^{3b}$

38. $\frac{8}{27}x^3 + \frac{1}{64}y^3$ **39.** $\frac{1}{24}x^3y^3 + \frac{1}{3}z^3$ **40.** $\frac{1}{16}x^{3a} + \frac{1}{2}y^{6a}z^{9b}$

41. A 4×4 ft sandbox is placed on a square lawn that is x ft on a side. Find a polynomial function giving the area of the part of the lawn not covered.

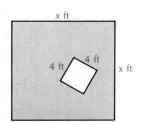

42. Find a polynomial function for the area of the color region.

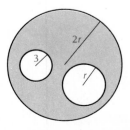

SUMMARY AND REVIEW: CHAPTER 4

The following contains a summary of what you should be able to do after completing this chapter. The review exercises are for practice. Answers are at the back of the book. If you miss an exercise, restudy the section indicated alongside the answer.

You should be able to:

Factor monomials.

Find three factorizations of each monomial.

1. $-10x^2$

2. $36x^5$

Factor polynomials when the terms have a common factor; differences of two squares; trinomial squares; trinomials of the type $x^2 + bx + c$; trinomials of the type $ax^2 + bx + c$, $a \neq 1$; and polynomials with four terms by grouping; and apply the general strategy for factoring.

Factor.

3. $5 - 20x^6$ **4.** $x^2 - 3x$ **5.** $9x^2 - 4$ **6.** $x^2 + 4x - 12$

7. $x^2 + 14x + 49$ **8.** $6x^3 + 12x^2 + 3x$ **9.** $x^3 + x^2 + 3x + 3$ **10.** $6x^2 - 5x + 1$

11. $x^4 - 81$ **12.** $9x^3 + 12x^2 - 45x$ **13.** $2x^2 - 50$ **14.** $x^4 + 4x^3 - 2x - 8$

15. $16x^4 - 1$ **16.** $8x^6 - 32x^5 + 4x^4$ **17.** $75 + 12x^2 + 60x$ **18.** $x^2 + 9$

19. $x^3 - x^2 - 30x$ **20.** $4x^2 - 25$ **21.** $9x^2 + 25 - 30x$ **22.** $6x^2 - 28x - 48$

23. $x^2 - 6x + 9$ **24.** $2x^2 - 7x - 4$ **25.** $18x^2 - 12x + 2$ **26.** $3x^2 - 27$

27. $15 - 8x + x^2$ **28.** $25x^2 - 20x + 4$

Solve equations by factoring and then by using the principle of zero products.

Solve.

29. $(x - 1)(x + 3) = 0$ **30.** $x^2 + 2x - 35 = 0$ **31.** $x^2 + x - 12 = 0$

32. $3x^2 + 2 = 5x$ **33.** $2x^2 + 5x = 12$ **34.** $16 = x(x - 6)$

Solve problems involving equations that can be factored.

35. The square of a number is 6 more than the number. Find the number.

36. The product of two consecutive even integers is 288. Find the integers.

37. The product of two consecutive odd integers is 323. Find the integers.

38. Twice the square of a number is 10 more than the number. Find the number.

Evaluate a polynomial in several variables for given values of the variables and identify the coefficients and the degrees of the terms and the degree of the polynomial. Also collect like terms of a polynomial in several variables.

39. Evaluate the polynomial $2 - 5xy + y^2 - 4xy^3 + x^6$ for $x = -1$ and $y = 2$.

Identify the coefficient and degree of each term of the following polynomials. Then find the degree of the polynomial.

40. $x^5y - 7xy + 9x^2 - 8$ **41.** $x^2y^5z^9 - y^{40} + x^{13}z^{10}$

Collect like terms.

42. $y + w - 2y + 8w - 5$ **43.** $m^6 - 2m^2n + m^2n^2 + n^2m - 6m^3 + m^2n^2 + 7n^2m$

Add, subtract, multiply, and factor polynomials in several variables.

44. Add: $(5x^2 - 7xy + y^2) + (-6x^2 - 3xy - y^2) + (x^2 + xy - 2y^2)$.

45. Subtract: $(6x^3y^2 - 4x^2y - 6x) - (-5x^3y^2 + 4x^2y + 6x^2 - 6)$.

Multiply.

46. $(p - q)(p^2 + pq + q^2)$ **47.** $(3a^4 - \frac{1}{3}b^3)^2$

Factor.

48. $x^2y^2 + xy - 12$ **49.** $12a^2 + 84ab + 147b^2$ **50.** $m^2 + 5m + mt + 5t$ **51.** $32x^4 - 128y^4z^4$

Factor sums and differences of two cubes.

Factor.

52. $27x^3 + 8$

53. $t^3 - 64m^3$

54. $0.064b^3 - 0.125c^3$

Solve.

55. The pages of a book measure 15 cm by 20 cm. Margins of equal width surround the printing on each page and constitute one half of the area of the page. Find the width of the margins.

56. The cube of a number is the same as twice the square of the number. Find the number.

57. The length of a rectangle is 2 times its width. When the length is increased by 20 and the width decreased by 1, the area is 160. Find the original length and width.

Solve.

58. $x^2 + 25 = 0$

59. $(x - 2)(x + 3)(2x - 5) = 0$

TEST: CHAPTER 4

1. Find three factorizations of $4x^3$.

Factor.

2. $x^2 - 7x + 10$

3. $x^2 + 25 - 10x$

4. $6y^2 - 8y^3 + 4y^4$

5. $x^3 + x^2 + 2x + 2$

6. $x^2 - 5x$

7. $x^3 + 2x^2 - 3x$

8. $28x - 48 + 10x^2$

9. $4x^2 - 9$

10. $x^2 - x - 12$

11. $6m^3 + 9m^2 + 3m$

12. $3w^2 - 75$

13. $60x + 45x^2 + 20$

14. $3x^4 - 48$

15. $49x^2 - 84x + 36$

16. $5x^2 - 26x + 5$

17. $x^4 + 2x^3 - 3x - 6$

18. $80 - 5x^4$

19. $4x^2 - 4x - 15$

20. $6t^3 + 9t^2 - 15t$

Solve.

21. $x^2 - x - 20 = 0$

22. $2x^2 + 7x = 15$

23. $x(x - 3) = 28$

24. The square of a number is 24 more than 5 times the number. Find the number.

25. The length of a rectangle is 6 m more than the width. The area of the rectangle is 40 m². Find the length and the width.

26. Collect like terms:

$$x^3y - y^3 + xy^3 + 8 - 6x^3y - x^2y^2 + 11.$$

27. Subtract:

$$(8a^2b^2 - ab + b^3) - (-6ab^2 - 7ab - ab^3 + 5b^3).$$

28. Multiply:

$$(3x^5 - 4y^5)(3x^5 + 4y^5).$$

29. Factor:

$$3m^2 - 9mn - 30n^2.$$

Factor.

30. $a^3 - h^3$

31. $16a^7b + 54ab^7$

32. Solve: The length of a rectangle is 5 times its width. When the length is decreased by 3 and the width is increased by 2, the area of the new rectangle is 60. Find the original length and width.

33. Factor:

$$(a + 3)^2 - 2(a + 3) - 35.$$

Slope is a number related to the slant of a line. This sign shows the grade of a road. It is a number that represents how steeply the road slants. How can we compute such numbers?

We now study the graphs of equations in two variables. These variables will be raised to the first power only and there will be no products or quotients involving variables. Such equations are called *linear*. We learn to graph such equations. The graph of an equation is a geometric picture of its solutions. The graphs of linear equations are straight lines. Later we attach numbers to the way the equations slant. Such numbers are called *slopes*.

188

OBJECTIVES

After finishing Section 5.1, you should be able to:

i Plot points associated with ordered pairs of numbers.

ii Determine the quadrant in which a point lies.

iii Find coordinates of a point on a graph.

Plot these points on the graph below.

1. (4, 5)

2. (5, 4)

3. (−2, 5)

4. (−3, −4)

5. (5, −3)

6. (−2, −1)

7. (0, −3)

8. (2, 0)

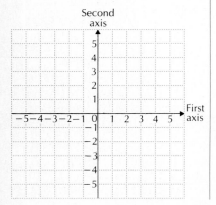

5.1

GRAPHS AND EQUATIONS

We have graphed numbers on a line. We now learn to graph number pairs on a plane, to enable us to graph an equation that contains two variables.

Points and Ordered Pairs

On a number line each point is the graph of a number. On a plane each point is the graph of a number pair. We use two perpendicular number lines called *axes*. They cross at a point called the *origin*. It has coordinates (0, 0) but is usually labeled 0. The arrows show the positive directions.

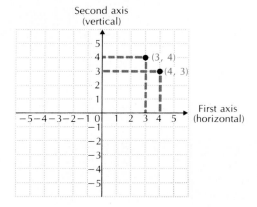

i Plotting Points

Note that (4, 3) and (3, 4) give different points. They are called *ordered pairs* of numbers because it makes a difference which number comes first.

EXAMPLE 1 Plot the point (−3, 4).

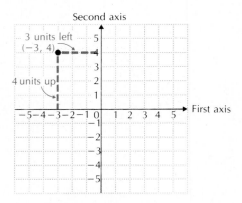

The first number, −3, is negative. We go −3 units in the first direction (3 units to the left). The second number, 4, is positive. We go 4 units in the second direction (up).

The numbers in an ordered pair are called *coordinates*. In (−3, 4), the *first coordinate* is −3 and the *second coordinate* is 4.

DO EXERCISES 1–8.

ii Quadrants

This figure shows some points and their coordinates. In region I (the first *quadrant*) both coordinates of any point are positive. In region II (the second *quadrant*) the first coordinate is negative and the second positive, and so on.

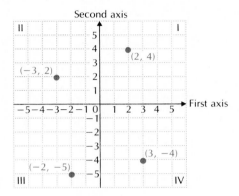

DO EXERCISES 9–14.

iii Finding Coordinates

To find coordinates of a point, we see how far to the right or left of zero it is located and how far up or down.

EXAMPLE 2 Find the coordinates of point *A*.

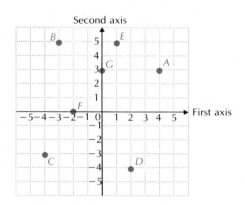

Point *A* is 4 units to the right (first direction) and 3 units up (second direction). Its coordinates are (4, 3).

DO EXERCISE 15.

9. What can you say about the coordinates of a point in the third quadrant?

10. What can you say about the coordinates of a point in the fourth quadrant?

In which quadrant is the point located?

11. (5, 3)

12. (−6, −4)

13. (10, −14)

14. (−13, 9)

15. Find the coordinates of points *B*, *C*, *D*, *E*, *F*, and *G* in the figure shown in Example 2.

EXERCISE SET 5.1

1. Plot these points.

    (2, 5) (−1, 3) (3, −2) (−2, −4)
    (0, 4) (0, −5) (5, 0) (−5, 0)

2. Plot these points.

    (4, 4) (−2, 4) (5, −3) (−5, −5)
    (0, 4) (0, −4) (3, 0) (−4, 0)

ii In which quadrant is each point located?

3. $(-5, 3)$ **4.** $(-12, 1)$ **5.** $(100, -1)$ **6.** $(35.6, -2.5)$

7. $(-6, -29)$ **8.** $(-3.6, -105.9)$ **9.** $(3.8, 9.2)$ **10.** $(1895, 1492)$

11. In quadrant III, first coordinates are always _____ and second coordinates are always _____ .

12. In quadrant II, _____ coordinates are always positive and _____ coordinates are always negative.

iii

13. Find the coordinates of points A, B, C, D, and E.

14. Find the coordinates of points A, B, C, D, and E.

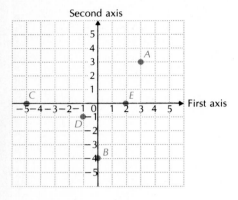

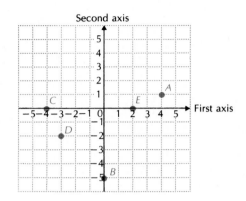

Use graph paper. Draw a first and second axis. Then plot these points.

15. $(0, -3)$, $(-1, -5)$, $(1, -1)$, $(2, 1)$

16. $(0, 1)$, $(1, 4)$, $(-1, -2)$, $(-2, -5)$

Problem-solving practice

17. The radius of Jupiter is about 11 times that of the earth. The radius of the earth is about 4030 mi. What is the radius of Jupiter?

18. The length of a rectangle is 4 in. greater than the width. The area of the rectangle is 21 in². Find the perimeter of the rectangle.

In Exercises 19–24, tell in which quadrant(s) each of the following points could be located.

19. The first coordinate is positive.

20. The second coordinate is negative.

21. The first and second coordinates are equal.

22. The first coordinate is the additive inverse of the second coordinate.

23. The points $(-1, 1)$, $(4, 1)$, and $(4, -5)$ are three vertices of a rectangle. Find the coordinates of the fourth vertex.

24. Three parallelograms share the vertices $(-2, -3)$, $(-1, 2)$, and $(4, -3)$. Find the fourth vertex of each parallelogram.

25. Graph eight points such that the sum of the coordinates is 6.

26. Graph eight points such that the first coordinate minus the second coordinate is 1.

27. Find the perimeter of a rectangle whose vertices have co-ordinates $(5, 3)$, $(5, -2)$, $(-3, -2)$, and $(-3, 3)$.

28. Find the area of a triangle whose vertices have coordinates $(0, 9)$, $(0, -4)$, and $(5, -4)$.

Coordinates on the globe. Three-dimensional objects can also be coordinatized: 0° latitude is the equator and 0° longitude is a line from the North Pole to the South Pole through France and Spain. In the drawing below, the hurricane Clara is at a point about 260 mi northwest of Bermuda near latitude 36.0 North, longitude 69.0 West.

29. Approximate the latitude and longitude of Bermuda.

30. Approximate the latitude and longitude of Lake Okeechobee.

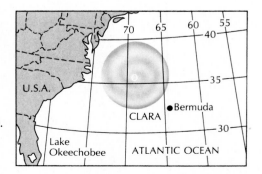

5.2

GRAPHING EQUATIONS

We now see how to graph equations in two variables on a plane.

i Solutions of Equations

An equation with two variables has *pairs* of numbers for solutions. We usually take the variables in alphabetical order. Then we get *ordered* pairs for solutions.

EXAMPLE 1 Determine whether $(3, 7)$ is a solution of $y = 2x + 1$.

$$y = 2x + 1$$

$$7 \mid \begin{array}{l} 2 \cdot 3 + 1 \\ 6 + 1 \\ 7 \end{array}$$

We substitute 3 for x and 7 for y (alphabetical order of variables).

The equation becomes true: $(3, 7)$ is a solution.

EXAMPLE 2 Determine whether $(-2, 3)$ is a solution of $2t = 4s - 8$.

$$2t = 4s - 8$$

$$2 \cdot 3 \mid \begin{array}{l} 4(-2) - 8 \\ -8 - 8 \\ -16 \end{array}$$

We substitute -2 for s and 3 for t.

6

The equation becomes false: $(-2, 3)$ is not a solution.

DO EXERCISES 1 AND 2.

ii Graphing Equations $y = mx$ and $y = mx + b$

The graph of an equation is a drawing of its solutions.

> To *graph* an equation means to make a drawing of its solutions.

If an equation has a graph that is a line, we can graph it by plotting a few points and then drawing a line through them.

EXAMPLE 3 Graph: $y = x$.

We will use alphabetical order. Thus the first axis will be the x-axis and the second axis will be the y-axis. Next, we find some solutions of the equation. In this case it is easy. Here are a few:

$$(0, 0), \quad (1, 1), \quad (5, 5), \quad (-2, -2), \quad (-4, -4).$$

Now we plot these points. We can see that if we were to plot a million solutions, the dots we draw would resemble a solid line. Once we see the pattern, we can draw the line with a ruler. The line is the graph of the equation $y = x$. We label the line $y = x$ on the graph paper.

OBJECTIVES

After finishing Section 5.2, you should be able to:

i Determine whether an ordered pair of numbers is a solution of an equation with two variables.

ii Graph equations of the type $y = mx$ and $y = mx + b$.

1. Determine whether $(2, 3)$ is a solution of $y = 2x + 3$.

2. Determine whether $(-2, 4)$ is a solution of $4q - 3p = 22$.

Graph.

3. $y = 3x$

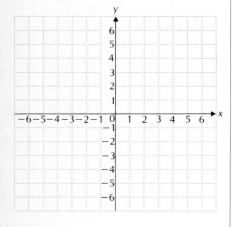

4. $y = \frac{1}{2}x$

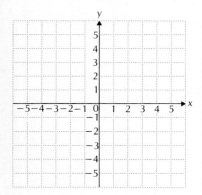

Graph.

5. $y = -x$ (or $-1 \cdot x$)

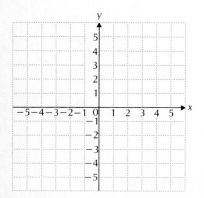

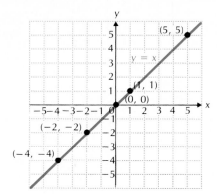

A graph of an equation is a picture of its solutions. Each point of the picture gives an ordered pair (a, b) that is a solution. No other points give solutions.

EXAMPLE 4   Graph: $y = 2x$.

We find some ordered pairs that are solutions, keeping the results in a table. We choose *any* number for x and then determine y by substitution. Suppose we choose 0 for x. Then

$$y = 2x = 2 \cdot 0 = 0.$$

We get a solution: the ordered pair $(0, 0)$. Suppose we choose 3 for x. Then

$$y = 2x = 2 \cdot 3 = 6.$$

We get a solution: the ordered pair $(3, 6)$. We make some negative choices for x as well as some positive ones. If a number takes us off the graph paper, we generally do not use it. Continuing in this manner we get a table like the one shown below. In this case, since $y = 2x$, we get y by doubling x.

Now we plot these points. If we had enough of them, they would make a line. We draw it with a ruler and label it $y = 2x$.

x	y
3	6
1	2
0	0
−2	−4
−3	−6

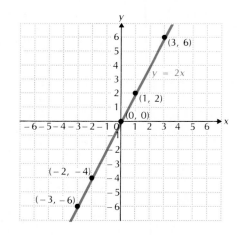

DO EXERCISES 3 AND 4.

EXAMPLE 5   Graph: $y = -3x$.

We make a table of solutions. Then we plot the points. If we had enough of them, they would make a line. We draw it with a ruler and label it $y = -3x$.

x	y
0	0
1	−3
−1	3
2	−6
−2	6

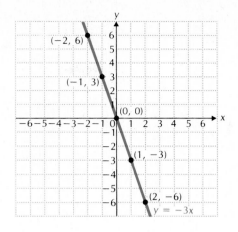

DO EXERCISES 5 AND 6. (EXERCISE 5 IS ON PRECEDING PAGE.)

EXAMPLE 6 Graph: $y = -\frac{5}{3}x$.

We make a table of solutions.

When $x = 0$, $\qquad y = -\frac{5}{3} \cdot 0 = 0$.

When $x = 3$, $\qquad y = -\frac{5}{3} \cdot 3 = -5$.

When $x = -3$, $\qquad y = -\frac{5}{3}(-3) = 5$.

When $x = 1$, $\qquad y = -\frac{5}{3} \cdot 1 = -\frac{5}{3}$.

Note that if we substitute multiples of 3, we can avoid fractions.

Next we plot the points. If we had enough of them, they would make a line.

x	y
0	0
3	−5
−3	5
1	$-\frac{5}{3}$

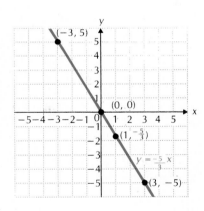

Every equation $y = mx$ has a graph that is a straight line. It contains the origin. The number m, called the *slope*, tells us how the line slants. For a positive slope a line slants up from left to right, as in Examples 3 and 4. For a negative slope a line slants down from left to right, as in Examples 5 and 6.

DO EXERCISES 7 AND 8.

We know that the graph of any equation $y = mx$ is a straight line through the origin, with slope m. What will happen if we add a number b on the right side to get an equation $y = mx + b$?

Graph.

6. $y = -2x$

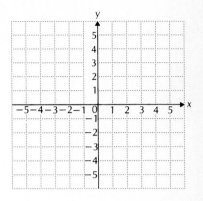

Graph.

7. $y = \frac{3}{4}x$

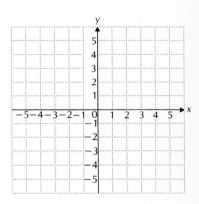

8. $y = -\frac{4}{5}x$

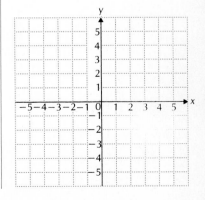

9. Graph $y = x + 3$ and compare it with $y = x$.

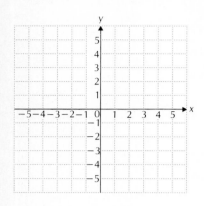

10. Graph $y = x - 1$ and compare it with $y = x$.

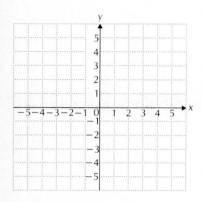

11. Graph $y = 2x + 3$ and compare it with $y = 2x$.

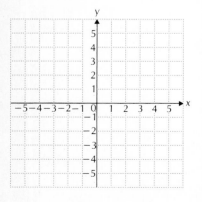

EXAMPLE 7 Graph $y = x + 2$ and compare it with $y = x$.

We first make a table of values.

x	y (or $x + 2$)
0	2
1	3
-1	1
2	4
-2	0
3	5

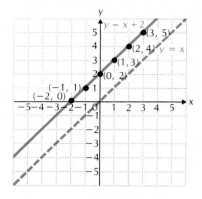

We then plot these points. If we had enough of them, they would make a line. We draw this line with a ruler and label it $y = x + 2$. The graph of $y = x$ is drawn for comparison. Note that the graph of $y = x + 2$ looks just like the graph of $y = x$, but it is moved up 2 units.

DO EXERCISES 9 AND 10.

EXAMPLE 8 Graph $y = 2x - 3$ and compare it with $y = 2x$.

We first make a table of values.

x	y (or $2x - 3$)
0	-3
1	-1
2	1
-1	-5

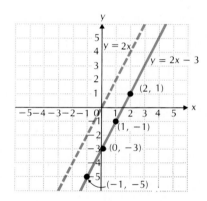

We draw the graph of $y = 2x - 3$. It looks just like the graph of $y = 2x$, but it is moved down 3 units.

DO EXERCISE 11.

The graph of $y = mx$ goes through the origin $(0, 0)$. The graph of any equation $y = mx + b$ is also a line. It is parallel to $y = mx$, but moved up or down. It goes through the point $(0, b)$. That point is called the *y-intercept*. We may also refer to the number b as the y-intercept. The number m is still called the *slope*. It tells us how steeply the line slants. We will study slope in more detail in Section 5.4.

EXAMPLE 9 Graph: $y = \frac{2}{5}x + 4$.

We first make a table of values. Using multiples of 5 avoids fractions.

When $x = 0$, $\qquad y = \frac{2}{5} \cdot 0 + 4 = 0 + 4 = 4$.

When $x = 5$, $\qquad y = \frac{2}{5} \cdot 5 + 4 = 2 + 4 = 6$.

When $x = -5$, $\qquad y = \frac{2}{5}(-5) + 4 = -2 + 4 = 2$.

Since two points determine a line, that is all you really need to graph a line, but you should always plot a third point as a check.

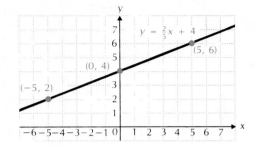

x	y
0	4
5	6
−5	2

We draw the graph of $y = \frac{2}{5}x + 4$.

EXAMPLE 10   Graph: $y = -\frac{3}{4}x - 2$.

We first make a table of values.

When $x = 0$, $y = -\frac{3}{4} \cdot 0 - 2 = 0 - 2 = -2$.

When $x = 4$, $y = -\frac{3}{4} \cdot 4 - 2 = -3 - 2 = -5$.

When $x = -4$, $y = -\frac{3}{4}(-4) - 2 = 3 - 2 = 1$.

x	y
0	−2
4	−5
−4	1

We plot these points and draw a line through them.

We plot this point for a check to see whether it is on the line.

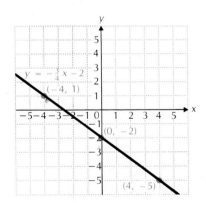

We draw the graph of $y = -\frac{3}{4}x - 2$. Every graph should be labeled.

> **Any equation $y = mx + b$ has a graph that is a straight line. It goes through the point $(0, b)$ and the y-intercept, and it has slope m.**

DO EXERCISES 12–15.

Graph.

12. $y = \frac{3}{5}x + 2$

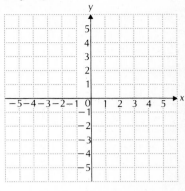

13. $y = \frac{3}{5}x - 2$

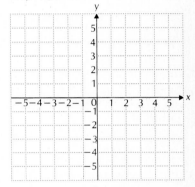

14. $y = -\frac{3}{5}x - 1$

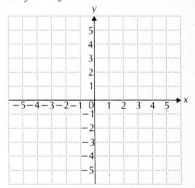

15. $y = -\frac{3}{5}x + 4$

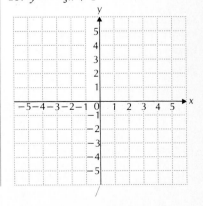

EXERCISE SET 5.2

i Determine whether the given point is a solution of the equation.

1. $(2, 5);\quad y = 3x - 1$

2. $(1, 7);\quad y = 2x + 5$

3. $(2, -3);\quad 3x - y = 4$

4. $(-1, 4);\quad 2x + y = 6$

5. $(-2, -1);\quad 2a + 2b = -7$

6. $(0, -4);\quad 4m + 2n = -9$

ii Graph.

7. $y = 4x$

8. $y = 2x$

9. $y = -2x$

10. $y = -4x$

11. $y = \frac{1}{3}x$

12. $y = \frac{1}{4}x$

13. $y = -\frac{3}{2}x$

14. $y = -\frac{5}{4}x$

15. $y = x + 1$

16. $y = -x + 1$

17. $y = 2x + 2$

18. $y = 3x - 2$

19. $y = \frac{1}{3}x - 1$

20. $y = \frac{1}{2}x + 1$

21. $y = -x - 3$

22. $y = -x - 2$

23. $y = \frac{5}{2}x + 3$

24. $y = \frac{5}{3}x - 2$

25. $y = -\frac{5}{2}x - 2$

26. $y = -\frac{5}{3}x + 3$

27. $y = x$

28. $y = -x$

29. $y = 3 - 2x$

30. $y = 7 - 5x$

31. $y = \frac{4}{3} - \frac{1}{3}x$

32. $y = -\frac{1}{4}x - \frac{1}{2}$

Problem-solving practice

33. A post is placed through some water into the mud at the bottom of the lake. Half of the post is in the mud and $\frac{1}{3}$ is in the water, and the part above water is $5\frac{1}{2}$ ft long. How long is the post?

34. The sum of two consecutive even integers is 130. Find the product of the integers.

○ ───────────────────────────

35. Complete the table for $y = x^2 + 1$. Plot the points on graph paper and draw the graph.

x	0	-1	1	-2	2	-3	3
y							

36. Find all whole-number solutions of $x + y = 6$.

37. Find all whole-number solutions of $x + 3y = 15$.

38. Translate to an equation: n nickels and d dimes total $1.95. Find three solutions.

39. Translate to an equation: n nickels and q quarters total $2.35. Find three solutions.

40. Find three solutions of $y = |x|$.

41. Find three solutions of $y = |x| + 1$.

42. Two machines A and B produce rivets. Machine A produces 68 rivets per hour, while machine B produces 76 rivets per hour. Let x represent the number of hours machine A runs and y represent the number of hours machine B runs. Translate to an equation: The combined production of A and B on a given day is 864. Find a solution to the equation. Explain your solution.

OBJECTIVES

After finishing Section 5.3, you should be able to:

i Graph using intercepts.

ii Graph equations of the type $x = a$ and $y = b$.

5.3

LINEAR EQUATIONS

We now develop faster procedures for graphing equations whose graphs are straight lines. Such equations are called *linear equations*.

i Graphing Using Intercepts

The fastest method for graphing equations whose graphs are straight lines involves the use of intercepts. Look at the graph of $y - 2x = 4$ shown below. We could graph this equation by solving for y to get $y = 2x + 4$ and proceed as before, but we want to develop a faster method.

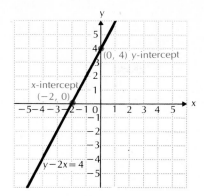

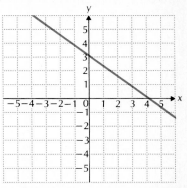

1. Look at the graph shown below.

a) Find the coordinates of the x-intercept.

b) Find the coordinates of the y-intercept.

The y-intercept is $(0, 4)$. It occurs where the line crosses the y-axis and always has 0 as the first coordinate. The x-intercept is $(-2, 0)$. It occurs where the line crosses the x-axis and always has 0 as the second coordinate.

DO EXERCISE 1.

The x-intercept is $(a, 0)$. To find a, let $y = 0$.
The y-intercept is $(0, b)$. To find b, let $x = 0$.

Now let us draw a graph using intercepts.

EXAMPLE 1   Graph: $4x + 3y = 12$.

To find the x-intercept, let $y = 0$. Then
$$4x + 3 \cdot 0 = 12$$
$$4x = 12$$
$$x = 3.$$

Thus $(3, 0)$ is the x-intercept. Note that this amounts to covering up the y-term and looking at the rest of the equation.

To find the y-intercept, let $x = 0$. Then
$$4 \cdot 0 + 3y = 12$$
$$3y = 12$$
$$y = 4.$$

Thus $(0, 4)$ is the y-intercept.

We plot these points and draw the line. A third point should be used as a check. We substitute any arbitrary value for x and solve for y.

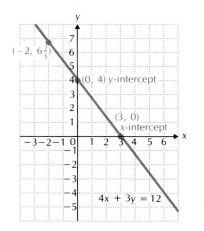

Graph using intercepts.

2. $2x + 3y = 6$

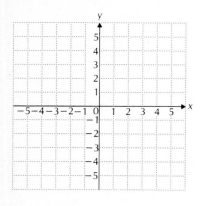

3. $3y - 4x = 12$

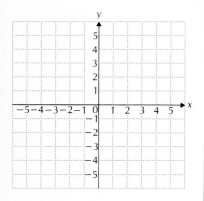

Graph.

4. $x = 5$

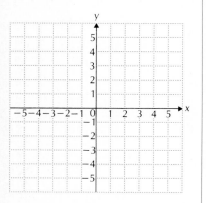

If we let $x = -2$, then

$$4(-2) + 3y = 12 \qquad \text{Substituting } -2 \text{ for } x$$
$$-8 + 3y = 12$$
$$3y = 12 + 8 = 20$$
$$y = \tfrac{20}{3}, \quad \text{or} \quad 6\tfrac{2}{3}. \qquad \text{Solving for } y$$

We see that the point $(-2, 6\tfrac{2}{3})$ is on the graph, so the graph is probably correct.

DO EXERCISES 2 AND 3.

Equations with a Missing Variable

Consider the equation $y = 3$. We can think of it as $y = 0 \cdot x + 3$. No matter what number we choose for x, we find that y is 3.

EXAMPLE 2 Graph: $y = 3$.

Any ordered pair $(x, 3)$ is a solution. So the line is parallel to the x-axis with y-intercept $(0, 3)$.

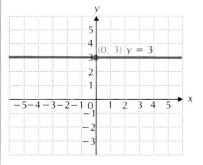

EXAMPLE 3 Graph: $x = -4$.

Any ordered pair $(-4, y)$ is a solution. So the line is parallel to the y-axis with x-intercept $(-4, 0)$.

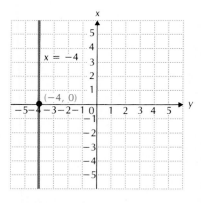

> **The graph of $y = b$ is a horizontal line. The graph of $x = a$ is a vertical line.**

DO EXERCISES 4–7. (EXERCISES 5–7 ARE ON THE FOLLOWING PAGE.)

Equations whose graphs are straight lines are linear equations. We summarize the best procedure for graphing linear equations.

TO GRAPH LINEAR EQUATIONS

1. Is the equation of the type $x = a$ or $y = b$? If so, the graph will be a line parallel to an axis.

2. If the line is not of the type $x = a$ or $y = b$, find the intercepts. Graph using the intercepts if this is feasible.

3. If the intercepts are too close together, choose another point farther from the origin.

4. In any case, use a third point as a check.

If you have trouble remembering whether a graph such as $y = 3$ or $x = -4$ is horizontal or vertical, the following may help.

Consider $y = 3$. Make up a table with all 3's in the y-column.

x	y
	3
	3
	3

Choose any numbers for x.

x	y
-2	3
0	3
4	3

(y must be 3)

Now when you plot the ordered pairs $(-2, 3)$, $(0, 3)$, and $(4, 3)$ and connect the points, you will obtain a horizontal line. Similarly, consider $x = -4$. Make up a table with all -4's in the x-column.

x	y
-4	
-4	
-4	

Choose any numbers for y.

x	y
-4	-5
-4	1
-4	3

Now when you plot the ordered pairs $(-4, -5)$, $(-4, 1)$, and $(-4, 3)$ and connect them, you will obtain a vertical line.

5. $y = -2$

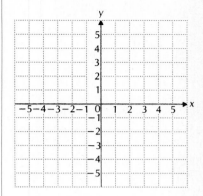

6. $x = 0$

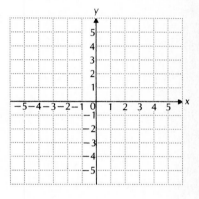

7. $x = -3$

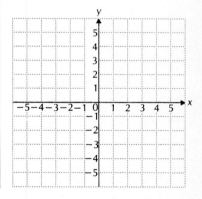

EXERCISE SET 5.3

i Find the intercepts. Then graph.

1. $x + 3y = 6$
2. $x + 2y = 8$
3. $-x + 2y = 4$
4. $-x + 3y = 9$
5. $3x + y = 9$
6. $2x + y = 6$
7. $2y - 2 = 6x$
8. $3y - 6 = 9x$
9. $3x - 9 = 3y$
10. $5x - 10 = 5y$
11. $2x - 3y = 6$
12. $2x - 5y = 10$
13. $4x + 5y = 20$
14. $2x + 6y = 12$
15. $2x + 3y = 8$
16. $x - 1 = y$
17. $x - 3 = y$
18. $2x - 1 = y$
19. $3x - 2 = y$
20. $4x - 3y = 12$
21. $6x - 2y = 18$
22. $7x + 2y = 6$
23. $3x + 4y = 5$
24. $y = -4 - 4x$
25. $y = -3 - 3x$
26. $-3x = 6y - 2$
27. $-4x = 8y - 5$

ii Graph.

28. $x = -2$

29. $x = -1$

30. $y = 2$

31. $y = 4$

32. $x = 2$

33. $x = 3$

34. $y = 0$

35. $y = -1$

36. $x = \frac{3}{2}$

37. $x = -\frac{5}{2}$

Problem-solving practice

38. A salesperson gets a weekly salary of $235 plus a $2 commission for each tire that is sold. How much did the salesperson make in a four-week period in which the salesperson sold 84 tires?

39. The base of a triangle is 5 m greater than the height. The area is 7 m². Find the base and the height.

40. Write an equation for the *y*-axis.

41. Write an equation for the *x*-axis.

42. Find the coordinates of the point of intersection of the graphs of the equations $x = -3$ and $y = 6$.

43. Write an equation of a line parallel to the *x*-axis and 5 units below it.

44. Write an equation of a line parallel to the *y*-axis and 13 units to the right of it.

45. Write an equation of a line parallel to the *x*-axis and intersecting the *y*-axis at (0, 2.8).

46. Find the value of *m* in $y = mx + 3$ so that the *x*-intercept of its graph will be (2, 0).

47. Find the value of *b* in $2y = -5x + 3b$ so that the *y*-intercept of its graph will be (0, −12).

48. *Straight-line depreciation.* A company buys a machine for $5200. The machine is expected to last for 8 years at which time its trade-in, or scrap, value will be $1300. Over its lifetime it depreciates $5200 − $1300, or $3900. If the company figures the decline in value to be the same each year—that is, $\frac{1}{8}$, or 12.5% of $3900, which is $487.50—then they are using what is called *straight-line depreciation*. We see this in the graph below. We see that the book value after one year is $5200 − $487.50, or $4712.50. After two years it is $4712.50 − $487.50, or $4225. After 3 years it is $4225 − $487.50, or $3737.50, and so on. Find the book values of the machine after each of the remaining years.

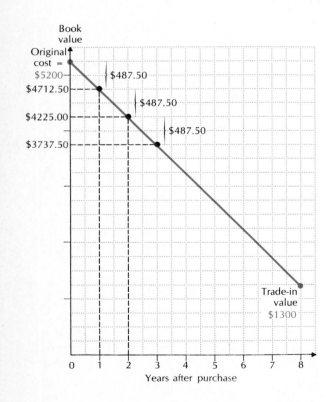

5.4

EQUATIONS OF LINES AND SLOPES

i Slope

Graphs of some linear equations slant upward from left to right. Others slant downward. Some are vertical and some are horizontal. Some slant more steeply than others. We are looking for a way to describe such possibilities with numbers.

Consider a line with two points marked P and Q. As we move from P to Q, the x-coordinate changes from 2 to 6 and the y-coordinate changes from 1 to 3. The change in x is $6 - 2$, or 4. The change in y is $3 - 1$, or 2.

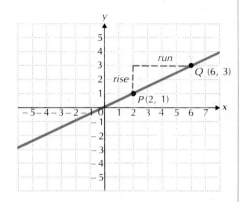

We call the change in y the *rise*, and the change in x the *run*. The ratio rise/run is the same for any two points on a line. We call this ratio the *slope*. Slope describes the slant of a line. The slope of the line in the graph is given by

$$\frac{\text{rise}}{\text{run}}, \quad \text{or } \frac{2}{4}, \quad \text{or } \frac{1}{2}.$$

The *slope* of a line containing points (x_1, y_1) and (x_2, y_2) is given by

$$m = \frac{\text{rise}}{\text{run}} = \frac{\text{the change in } y}{\text{the change in } x} = \frac{y_2 - y_1}{x_2 - x_1}.$$

EXAMPLE 1 Graph the line containing the points $(-4, 3)$ and $(2, -6)$ and find the slope.

The graph is shown below. From $(-4, 3)$ to $(2, -6)$ the change in y, or rise, is $3 - (-6)$, or 9. The change in x, or run, is $-4 - 2$, or -6.

$$\text{Slope} = \frac{\text{rise}}{\text{run}} = \frac{\text{change in } y}{\text{change in } x} = \frac{3 - (-6)}{-4 - 2}$$

$$= \frac{9}{-6} = -\frac{9}{6}, \quad \text{or } -\frac{3}{2}.$$

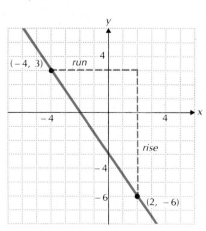

DO EXERCISES 1 AND 2.

OBJECTIVES

After finishing Section 5.4, you should be able to:

i Given the coordinates of two points, find the slope of the line containing them.

ii Determine the slope, if it exists, of a horizontal or a vertical line.

iii Find the slope of a line from an equation of the line.

iv Given an equation of a line, find the slope–intercept equation, the slope, and the y-intercept of the line.

v Find an equation of a line given a point on the line and the slope and given two points on the line.

Graph a line containing the points and find the slope.

1. $(-2, 3)$ and $(3, 5)$

2. $(0, -3)$ and $(-3, 2)$

3. Find the slope in Exercise 1, but subtract in a different order.

When we use the formula

$$m = \frac{y_2 - y_1}{x_2 - x_1},$$

we subtract in two ways. We just have to remember to subtract the y-coordinates in the same order that we subtract the x-coordinates. Let's do Example 1 again.

$$\text{Slope} = \frac{\text{change in } y}{\text{change in } x} = \frac{-6 - 3}{2 - (-4)} = \frac{-9}{6} = -\frac{3}{2}.$$

DO EXERCISES 3 AND 4.

The slope of a line tells how it slants. A line with a positive slope slants up from left to right. The larger the positive slope, the steeper the slant. A line with negative slope slants downward from left to right.

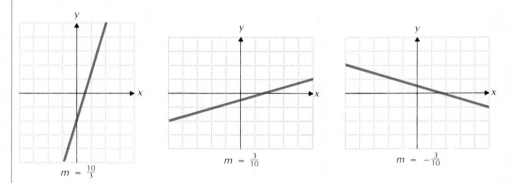

$$m = \frac{10}{3} \qquad m = \frac{3}{10} \qquad m = -\frac{3}{10}$$

4. Find the slope in Exercise 2, but subtract in a different order.

ii **Horizontal and Vertical Lines**

What about the slope of a horizontal or a vertical line?

EXAMPLE 2 Find the slope of the line $y = 4$.

Consider the points $(-3, 4)$ and $(2, 4)$, which are on the line.

The change in $y = 4 - 4$, or 0.

The change in $x = -3 - 2$, or -5.

$$m = \frac{4 - 4}{-3 - 2}$$

$$= \frac{0}{-5}$$

$$= 0$$

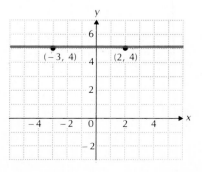

Any two points on a horizontal line have the same y-coordinate. Thus the change in y is 0, so the slope is 0.

EXAMPLE 3   Find the slope of the line $x = -3$.

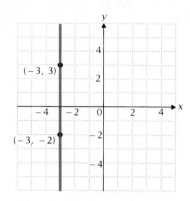

Consider the points $(-3, 3)$ and $(-3, -2)$, which are on the line.

The change in $y = 3 - (-2)$, or 5.

The change in $x = -3 - (-3)$, or 0.

$$m = \frac{3 - (-2)}{-3 - (-3)}$$

$$= \frac{5}{0}$$

Since division by 0 is not defined, this line has no slope.

> **A horizontal line has slope 0. A vertical line has *no* slope.**

DO EXERCISES 5–8.

iii Finding Slope from an Equation

It is possible to find the slope of a line from its equation. Let us consider the equation

$$y = 2x + 3.$$

Two points on the line are $(0, 3)$ and $(1, 5)$. We can find such points by picking arbitrary values for x, say 0 and 1, and substituting to find corresponding y-values. The slope of the line is found as follows.

$$m = \frac{\text{change in } y}{\text{change in } x} = \frac{5 - 3}{1 - 0} = \frac{2}{1} = 2$$

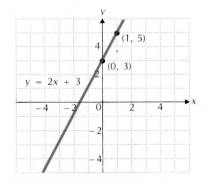

Find the slope, if it exists, of the line containing the points.

5. $(9, 7)$ and $(3, 7)$

6. $(4, -6)$ and $(4, 0)$

Find the slope of the line.

7. $x = 7$

8. $y = -5$

Find the slope of the line.

9. $y = -8x - 14$

The slope is 2. This is also the coefficient of the x-term in the equation $y = 2x + 3$.

> The slope of the line $y = mx + b$ is m. To find the slope of a nonvertical linear equation in x and y, first solve the equation for y, and get the resulting equation in the form $y = mx + b$. The coefficient of the x-term, m, is the slope of the line.

10. $4x + 5y = 7$

EXAMPLE 4 Find the slope of the line $2x + 3y = 7$.

We solve for y:

$$2x + 3y = 7$$
$$3y = -2x + 7$$
$$y = \tfrac{1}{3}(-2x + 7)$$
$$y = \tfrac{1}{3}(-2x) + \tfrac{1}{3}(7)$$
$$y = -\tfrac{2}{3}x + \tfrac{7}{3}.$$

11. $\tfrac{1}{4}x = 7 + y$

The slope is $-\tfrac{2}{3}$.

DO EXERCISES 9–12.

12. $5x - 4y = 8$

iv The Slope–Intercept Equation of a Line

In the equation $y = mx + b$, we know that m is the slope. What is the y-intercept? To find out we let $x = 0$, and solve for y:

$$y = mx + b$$
$$y = m(0) + b$$
$$y = b.$$

Thus the y-intercept is $(0, b)$.

Find the slope and y-intercept.

13. $y = 5x$

> The equation $y = mx + b$ is called the *slope–intercept* equation. The slope is m and the y-intercept is $(0, b)$.

14. $y = -\tfrac{3}{2}x - 6$

EXAMPLE 5 Find the slope and y-intercept of $y = 3x - 4$.

Since the equation is already in the form $y = mx + b$, we simply read the slope and y-intercept from the equation.

$$y = 3x - 4$$

15. $2y = 4x - 17$

The slope is 3. The y-intercept is $(0, -4)$.

16. $3x + 4y = 15$

EXAMPLE 6 Find the slope and y-intercept of $2x + 3y = 8$.

We first solve for y:

$$2x + 3y = 8$$
$$3y = -2x + 8$$
$$y = -\tfrac{2}{3}x + \tfrac{8}{3}.$$

17. $-7x - 5y = 22$

The slope is $-\tfrac{2}{3}$ and the y-intercept is $(0, \tfrac{8}{3})$.

DO EXERCISES 13–17.

▼ **The Point–Slope Equation of a Line**

Suppose we know the slope of a line and that it contains a certain point. We can use the slope–intercept equation to find an equation of the line.

EXAMPLE 7   Find an equation of the line with slope 3 that contains the point (4, 1).

Step 1.   Use the point (4, 1) and substitute 4 for x and 1 for y in $y = mx + b$. We also substitute 3 for m, the slope. Then we solve for b:

$$y = mx + b$$
$$1 = 3 \cdot 4 + b \qquad \text{Substituting}$$
$$-11 = b. \qquad \text{Solving for } b, \text{ the } y\text{-intercept}$$

Step 2.   Write the equation $y = mx + b$ by substituting 3 for m and -11 for b:

$$y = mx + b$$
$$y = 3x - 11.$$

DO EXERCISES 18 AND 19.

Now consider a line with slope 2 and containing the point (1, 3) as shown. Suppose (x, y) is any point on this line. Using the definition of slope and the two points (1, 3) and (x, y), we get

$$m = \frac{\text{change in } y}{\text{change in } x}$$
$$= \frac{y - 3}{x - 1}.$$

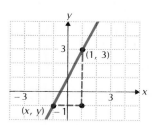

We know that the slope is 2, so

$$2 = \frac{y - 3}{x - 1}, \quad \text{or} \quad \frac{y - 3}{x - 1} = 2.$$

Multiplying on both sides by $x - 1$, we get

$$y - 3 = 2(x - 1).$$

Solving for y, we obtain

$$y - 3 = 2x - 2$$
$$y = 2x + 1.$$

THE POINT–SLOPE EQUATION

A nonvertical line that contains a point (x_1, y_1) with slope m has an equation

$$y - y_1 = m(x - x_1).$$

Find an equation of the line that contains the given point and has the given slope.

18. $(4, 2)$, $m = 5$

19. $(-2, 1)$, $m = -3$

Find an equation of the line containing the given point and with the given slope.

20. $(3, 5)$, $m = 6$

21. $(1, 4)$, $m = -\frac{2}{3}$

EXAMPLE 8 Find an equation of the line with slope 5 that contains the point $(-2, -3)$.

We use the point–slope equation. We substitute 5 for m, -2 for x_1, and -3 for y_1:

$$y - y_1 = m(x - x_1)$$
$$y - (-3) = 5[x - (-2)]$$
$$y + 3 = 5(x + 2)$$
$$y + 3 = 5x + 10$$
$$y = 5x + 7.$$

DO EXERCISES 20 AND 21.

We can also use the point–slope equation to find an equation of a line that contains two given points.

EXAMPLE 9 Find an equation of the line containing $(2, 3)$ and $(-6, 1)$.

First we find the slope:

$$m = \frac{3 - 1}{2 - (-6)}$$
$$= \frac{1}{4}.$$

Then we use the point–slope equation:

$$y - y_1 = m(x - x_1)$$
$$y - 3 = \tfrac{1}{4}(x - 2) \qquad \text{Substituting 2 for } x_1, 3 \text{ for } y_1, \text{ and } \tfrac{1}{4} \text{ for } m$$
$$y - 3 = \tfrac{1}{4}x - \tfrac{1}{2}$$
$$y = \tfrac{1}{4}x + \tfrac{5}{2}.$$

DO EXERCISES 22 AND 23.

Find an equation of the line containing the given points.

22. $(2, 4)$ and $(3, 5)$

23. $(-1, 2)$ and $(-3, -2)$

Applications of Slope

Slope has many real-world applications. For example, numbers like 2%, 3%, and 6% are often used to represent the *grade* of a road. Such a number is meant to tell how steep a road up a hill or mountain is. For example, a 3% grade means that for every horizontal distance of 100 ft, the road rises 3 ft. The concept of grade also occurs in cardiology when a person runs on a treadmill. A physician may change the steepness of the treadmill to measure its effect on heartbeat.

Another example occurs in hydrology. When a river flows, the strength or force of the river depends on how much the river falls vertically compared to how much it flows horizontally.

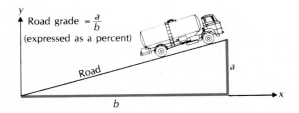

EXERCISE SET 5.4

i Find the slope, if it exists, of the line containing each pair of points.

1. $(3, 2)$ and $(-1, 2)$ **2.** $(4, 1)$ and $(-2, -3)$ **3.** $(-2, 4)$ and $(3, 0)$ **4.** $(-4, 2)$ and $(2, -3)$

5. $(4, 0)$ and $(5, 7)$ **6.** $(3, 0)$ and $(6, 2)$ **7.** $(0, 8)$ and $(-3, 10)$ **8.** $(0, 9)$ and $(4, 7)$

9. $(3, -2)$ and $(5, -6)$ **10.** $(-2, 4)$ and $(6, -7)$ **11.** $(-2, \frac{1}{2})$ and $(-5, \frac{1}{2})$ **12.** $(8, -3)$ and $(10, -3)$

13. $(9, -4)$ and $(9, -7)$ **14.** $(-10, 3)$ and $(-10, 4)$

ii Find the slope, if it exists, of each line.

15. $x = -8$ **16.** $x = -4$ **17.** $y = 2$ **18.** $y = 17$

19. $x = 9$ **20.** $x = 6$ **21.** $y = -9$ **22.** $y = -4$

iii Find the slope of each line.

23. $3x + 2y = 6$ **24.** $4x - y = 5$ **25.** $x + 4y = 8$ **26.** $x + 3y = 6$

27. $-2x + y = 4$ **28.** $-5x + y = 5$ **29.** $4x - 3y = -12$ **30.** $x - 2y = 9$

31. $x - 3y = -2$ **32.** $x + y = 7$ **33.** $-2x + 4y = 8$ **34.** $-5x + 7y = 2$

iv Find the slope and y-intercept of each line.

35. $y = -4x - 9$ **36.** $y = -3x - 5$ **37.** $y = 1.8x$ **38.** $y = -27.4x$

39. $2x + 3y = 9$ **40.** $5x + 4y = 12$ **41.** $-8x - 7y = 21$ **42.** $-2x - 9y = 13$

43. $9x = 3y + 5$ **44.** $4x = 9y + 7$ **45.** $-6x = 4y + 2$ **46.** $y = -17$

v Find an equation of the line containing the given point and having the given slope.

47. $(2, 5)$, $m = 5$ **48.** $(-3, 0)$, $m = -2$ **49.** $(2, 4)$, $m = \frac{3}{4}$ **50.** $(\frac{1}{2}, 2)$, $m = -1$

51. $(2, -6)$, $m = 1$ **52.** $(4, -2)$, $m = 6$ **53.** $(-3, 0)$, $m = -3$ **54.** $(0, 3)$, $m = -3$

55. $(5, 6)$, $m = \frac{2}{3}$ **56.** $(2, 7)$, $m = \frac{5}{6}$

Find an equation of the line that contains the given pair of points.

57. $(-6, 1)$ and $(2, 3)$ **58.** $(12, 16)$ and $(1, 5)$ **59.** $(0, 4)$ and $(4, 2)$ **60.** $(0, 0)$ and $(4, 2)$

61. $(3, 2)$ and $(1, 5)$ **62.** $(-4, 1)$ and $(-1, 4)$ **63.** $(5, 0)$ and $(0, -2)$ **64.** $(-2, -2)$ and $(1, 3)$

65. $(-2, -4)$ and $(2, -1)$ **66.** $(-3, 5)$ and $(-1, -3)$

67. Find an equation of the line that contains the point $(2, -3)$ and has the same slope as the line $3x - y + 4 = 0$.

68. Find an equation of the line that has the same y-intercept as the line $x - 3y = 6$ and contains the point $(5, -1)$.

69. Find an equation of the line with the same slope as $3x - 2y = 8$ and the same y-intercept as $2y + 3x = -4$.

70. Graph several equations that have the same slope. How are these lines related?

5.5

GRAPHING USING SLOPE AND y-INTERCEPT

i The *slope* of a line $y = mx$ or $y = mx + b$ is the number m. It tells how the line slants. A line with a positive slope slants up from left to right. The larger the positive

OBJECTIVES

After finishing Section 5.5, you should be able to:

i Given an equation $y = mx$ or $y = mx + b$, determine the slope and y-intercept and use them to graph the equation.

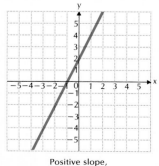

Positive slope,
$m > 0$

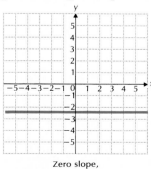

Negative slope,
$m < 0$

Zero slope,
$m = 0$

For each equation determine the slope and y-intercept.

1. $y = \frac{2}{3}x$

2. $y = -6x$

3. $y = -2x + 9$

4. $y = 0.7x - 3$

Graph.

5. $y = 4x$

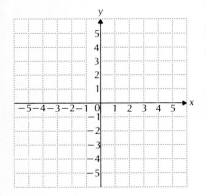

---

slope, the steeper the slant. A line with negative slope slants downward from left to right. A line with 0 slope is horizontal.

The *y-intercept* is the point $(0, b)$. It is the point at which the graph crosses the y-axis. That is, when the graph crosses the y-axis, $x = 0$, so $y = m(0) + b = b$, and the point of intersection is $(0, b)$.

EXAMPLES For each equation determine the slope and y-intercept.

1. $y = 2x$

We can think of this as

Then

$$y = 2x + \;\; 0$$
$$m = 2; \quad (0, 0)$$

The slope is 2. The y-intercept is $(0, 0)$.

2. $y = -5.3x - 12$

$$y = -5.3x - 12$$
$$m = -5.3; \; (0, -12)$$

The slope is -5.3. The y-intercept is $(0, -12)$.

DO EXERCISES 1–4.

Now we use the slope and y-intercept to graph equations $y = mx$ and $y = mx + b$. We use the following procedure.

> **To graph equations $y = mx$ and $y = mx + b$ using the slope and y-intercept:**
>
> **a)** Determine the slope and y-intercept.
> **b)** Plot the y-intercept.
> **c)** Using the slope and starting from the y-intercept, locate two other points and draw the line through the three points.

EXAMPLE 3 Graph $y = 2x$.

a) The slope is 2. The y-intercept is $(0, 0)$.

b) Plot the y-intercept.

c) The slope is 2, or $\frac{2}{1}$. It tells us that from the y-intercept the line slants *up* 2 units for every 1 unit to the *right*.

Then in this direction from the y-intercept $(0, 0)$ we get the point $(1, 2)$. Now we move to the left from the y-intercept. For every 2 units we move down, we move 1 unit to the left. This gives us the point $(-1, -2)$. We then join these three points with a line to complete the graph.

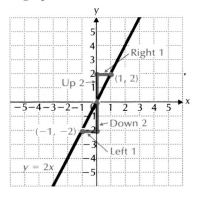

DO EXERCISE 5.

EXAMPLE 4   Graph $y = -3x$.

a) The slope is -3. The *y*-intercept is $(0, 0)$.

b) Plot the *y*-intercept.

c) The slope is -3, or $\frac{-3}{1}$. It tells us that from the *y*-intercept the line slants *down* 3 units for every 1 unit to the *right*.

Then in this direction from the *y*-intercept $(0, 0)$ we get the point $(1, -3)$. Now we move to the left from the *y*-intercept. For every 3 units we move up, we move 1 unit to the left. This gives us the point $(-1, 3)$. We then join these points with a line to complete the graph.

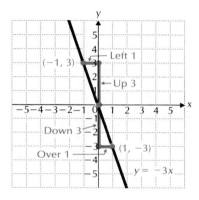

DO EXERCISE 6.

EXAMPLE 5   Graph $y = 3x - 2$.

a) The slope is 3. The *y*-intercept is $(0, -2)$.

b) Plot the *y*-intercept.

c) The slope is 3, or $\frac{3}{1}$. It tells us that from the *y*-intercept the line slants *up* 3 units for every 1 unit to the *right*.

Then in this direction from the *y*-intercept $(0, -2)$ we get the point $(1, 1)$. Now we move to the *left* from the *y*-intercept. For every 3 units we move *down*, we move 1 unit to the *left*. This gives us the point $(-1, -5)$. We then join these three points with a line to complete the graph.

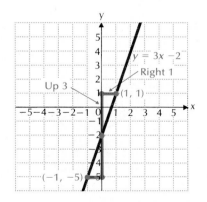

DO EXERCISES 7 AND 8.

Graph.

6. $y = -4x$

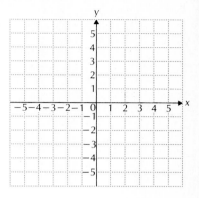

Graph.

7. $y = 2x + 1$

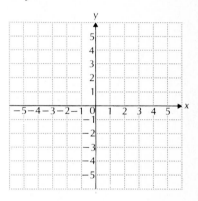

8. $y = 2x - 4$

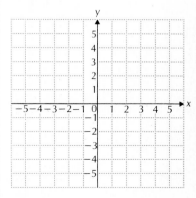

Graph.

9. $y = \frac{3}{4}x - 2$

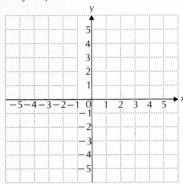

10. $y = \frac{3}{2}x + 1$

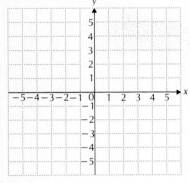

Graph.

11. $y = -\frac{3}{5}x + 5$

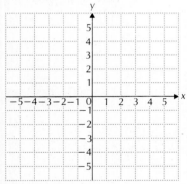

12. $y = -\frac{5}{3}x - 4$

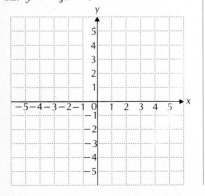

EXAMPLE 6 Graph $y = \frac{2}{3}x + 4$.

a) The slope is $\frac{2}{3}$. The y-intercept is $(0, 4)$.

b) Plot the y-intercept.

c) The slope is $\frac{2}{3}$. It tells us that from the y-intercept the line slants *up* 2 units for every 3 units to the *right*.

Then in this direction from the y-intercept $(0, 4)$ we get the point $(3, 6)$. Now we move to the left from the y-intercept. For every 2 units we move down, we move 3 units to the left. This gives us the point $(-3, 2)$. We then join these points with a line to complete the graph.

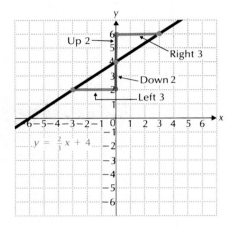

DO EXERCISES 9 AND 10.

EXAMPLE 7 Graph $y = -\frac{3}{4}x - 1$.

a) The slope is $-\frac{3}{4}$. The y-intercept is $(0, -1)$.

b) Plot the y-intercept.

c) The slope is $-\frac{3}{4}$, or $\frac{-3}{4}$. It tells us that from the y-intercept the line slants *down* 3 units for every 4 units to the *right*.

Then in this direction from the y-intercept $(0, -1)$ we get the point $(4, -4)$. Now we move to the left from the y-intercept. For every 3 units we move *up* we move 4 units to the *left*. This gives us the point $(-4, 2)$. We then join these three points with a line to complete the graph.

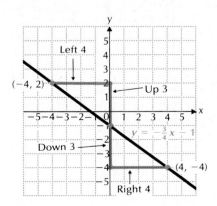

DO EXERCISES 11 AND 12.

EXERCISE SET 5.5

Graph each equation using the slope and y-intercept.

1. $y = \frac{5}{2}x + 1$ **2.** $y = \frac{2}{5}x - 4$ **3.** $y = -\frac{5}{2}x - 4$ **4.** $y = -\frac{2}{5}x + 3$

5. $y = 2x - 5$ **6.** $y = -2x + 4$ **7.** $y = \frac{1}{3}x + 6$ **8.** $y = -3x + 6$

9. $y = -0.25x + 2$ **10.** $y = 1.5x - 3$ **11.** $y = -\frac{3}{4}x$ **12.** $y = \frac{4}{5}x$

SUMMARY AND REVIEW: CHAPTER 5

The following contains a summary of what you should be able to do after completing this chapter. The review exercises are for practice. Answers are at the back of the book. If you miss an exercise, restudy the section indicated alongside the answer.

You should be able to:

Plot points associated with ordered pairs of numbers, determine the quadrant in which a point lies, and find coordinates of a point on a graph.

Use graph paper. Plot these points.

1. $(2, 5)$ **2.** $(0, -3)$ **3.** $(-4, -2)$

In which quadrant is each point located?

4. $(3, -8)$ **5.** $(-20, -14)$ **6.** $(4.9, 1.3)$

Find the coordinates of each point.

7. A

8. B

9. C

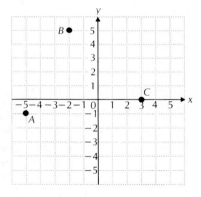

Determine whether an ordered pair of numbers is a solution of an equation in two variables.

Determine whether the given point is a solution of the equation $2y - x = 10$.

10. $(2, -6)$ **11.** $(0, 5)$

Graph linear equations.

Graph.

12. $y = 2x - 5$ **13.** $y = -\frac{3}{4}x$ **14.** $y = -x + 4$ **15.** $y = 3 - 4x$

16. $5x - 2y = 10$ **17.** $y = 3$ **18.** $x = -\frac{3}{4}$ **19.** $x - 2y = 6$

Find the slope, if it exists, of the line containing a given pair of points. Determine the slope of a horizontal or vertical line. Find the slope of a line given an equation of the line.

Find the slope, if it exists, of the line containing each pair of points.

20. $(6, 8)$ and $(-2, -4)$ **21.** $(5, 1)$ and $(-1, 1)$

22. $(-3, 0)$ and $(-3, 5)$ **23.** $(-8.3, 4.6)$ and $(-9.9, 1.4)$

Find the slope, if it exists, of each line.

24. $y = -6$ **25.** $x = 90$ **26.** $4x + 3y = -12$

Given an equation of a line, find the slope–intercept equation, the slope, and the y-intercept of the line.

Find the slope and y-intercept of each line.

27. $y = -9x + 46$ **28.** $x + y = 9$ **29.** $3x - 5y = 4$

Find an equation of a line given a point on the line and the slope and given two points on the line.

Find an equation of the line containing the given point and with the given slope.

30. $(1, 2)$, $m = 3$ **31.** $(-2, -5)$, $m = \frac{2}{3}$ **32.** $(0, -4)$, $m = -2$

Find an equation of the line that contains the given pair of points.

33. $(5, 7)$ and $(-1, 1)$ **34.** $(2, 0)$ and $(-4, -3)$

Graph linear equations using slope and y-intercept.

Graph using the slope and y-intercept. Show your work.

35. $y = -3x + 2$ **36.** $y = \frac{5}{2}x$ **37.** $y = \frac{3}{4}x - 4$

---

38. Find m in $y = mx + 3$ so that $(-2, 5)$ will be on the graph.

39. Find the value of b in $y = -5x + b$ so that $(3, 4)$ will be on the graph.

40. Find the area and the perimeter of a rectangle whose vertices are $(-2, 2)$, $(7, 2)$, $(-2, -3)$, and $(7, -3)$.

41. Find three solutions of $y = 4 - |x|$.

42. Find an equation of the line for which the second coordinate is the additive inverse of the first coordinate.

43. Find the slope and the intercepts of a line whose equation is

$$\frac{x}{a} + \frac{y}{b} = 1, \quad a \neq 0, \quad \text{and} \quad b \neq 0.$$

TEST: CHAPTER 5

In which quadrant is each point located?

1. $\left(-\frac{1}{2}, 7\right)$ **2.** $(-5, -6)$

Find the coordinates of each point.

3. A

4. B

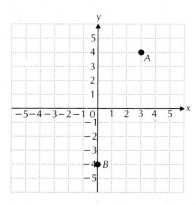

5. Determine whether the ordered pair $(2, -4)$ is a solution of the equation $y - 3x = -10$.

Graph.

6. $y = 2x - 1$ **7.** $2x - 4y = -8$ **8.** $y = 5$ **9.** $y = -\frac{3}{2}x$ **10.** $x = -4$

Find the slope, if it exists, of the line containing each pair of points.

11. $(4, 7)$ and $(4, -1)$

12. $(9, 2)$ and $(-3, -5)$

Find the slope, if it exists, of each line.

13. $y = -7$

14. $x = 6$

Find the slope and y-intercept.

15. $y = 2x - \frac{1}{4}$

16. $-4x + 3y = -6$

Find an equation of the line that contains the given point and having the given slope.

17. $(3, 5)$, $m = 1$

18. $(-2, 0)$, $m = -3$

Find an equation of the line that contains the given pair of points.

19. $(1, 1)$ and $(2, -2)$

20. $(4, -1)$ and $(-4, -3)$

Graph using the slope and y-intercept. Show your work.

21. $y = -\frac{3}{5}x$

22. $y = -2x - 5$

23. Find the area and the perimeter of a rectangle whose vertices are $(-3, 1)$, $(5, 1)$, $(5, 8)$, and $(-3, 8)$.

24. Find the slope–intercept equation of the line that contains the point $(-4, 1)$ and has the same slope as the line $2x - 3y = -6$.

THE ASVAB TEST: SAMPLE QUESTIONS   CHAPTER 5

NUMERICAL OPERATIONS   Time: 3 minutes; 50 questions

Allow yourself 3 minutes. Work as carefully as you can and try to get as many correct as possible. See the discussion of the ASVAB test at the beginning of Chapter 3.

1. $5 + 4 =$
A 11
B 9
C 1
D 20

2. $6 - 3 =$
A 3
B 4
C 9
D 2

3. $2 \times 8 =$
A 10
B 12
C 16
D 28

4. $9 + 6 =$
A 3
B 15
C 14
D 13

5. $9 \div 3 =$
A 6
B 12
C 3
D 4

6. $8 - 8 =$
A 16
B 8
C 1
D 0

7. $24 \div 4 =$
A 28
B 6
C 20
D 8

8. $5 \times 7 =$
A 30
B 35
C 37
D 40

9. $6 + 1 =$
A 5
B 6
C 7
D 8

10. $0 \times 4 =$
A 2
B 1
C 0
D 4

11. $8 - 2 =$
A 4
B 6
C 10
D 16

12. $7 + 8 =$
A 13
B 16
C 15
D 17

13. $18 \div 2 =$
A 9
B 16
C 10
D 20

14. $2 \times 4 =$
A 4
B 12
C 8
D 6

15. $6 \times 1 =$
A 0
B 1
C 7
D 6

16. $7 + 5 =$
A 2
B 12
C 11
D 13

17. $9 - 3 =$
A 3
B 7
C 6
D 12

18. $9 + 9 =$
A 18
B 1
C 16
D 0

19. $5 + 0 =$
A 0
B 1
C 4
D 5

20. $9 \times 3 =$
A 27
B 26
C 24
D 21

21. $7 \div 7 =$
A 0
B 1
C 7
D 14

22. $4 + 9 =$
A 14
B 12
C 13
D 11

23. $10 \times 4 =$
A 6
B 14
C 40
D 20

24. $6 - 1 =$
A 4
B 5
C 6
D 7

25. $6 \times 3 =$
A 12
B 9
C 3
D 18

26. $0 \div 5 =$
A 0
B 1
C 5
D 10

27. $8 + 3 =$
A 5
B 11
C 13
D 24

28. $6 + 2 =$
A 4
B 12
C 3
D 8

29. $70 \div 7 =$
A 10
B 77
C 63
D 11

30. $10 - 6 =$
A 16
B 6
C 3
D 4

31. $8 \times 8 =$
A 1
B 8
C 64
D 16

32. $63 \div 7 =$
A 70
B 8
C 56
D 9

33. $5 + 3 =$
A 8
B 2
C 9
D 7

34. $3 + 2 =$
A 1
B 2
C 5
D 6

35. $4 \times 7 =$
A 11
B 18
C 21
D 28

36. $4 - 3 =$
A 0
B 1
C 7
D 2

37. $5 \times 2 =$
A 15
B 7
C 10
D 3

38. $3 \div 1 =$
A 4
B 1
C 2
D 3

39. $9 + 2 =$
A 11
B 7
C 10
D 12

40. $9 \times 5 =$
A 4
B 14
C 40
D 45

41. $7 - 5 =$
A 1
B 2
C 3
D 4

42. $7 \times 6 =$
A 43
B 13
C 42
D 44

43. $80 \div 10 =$
A 8
B 9
C 88
D 90

44. $8 \times 4 =$
A 28
B 4
C 32
D 12

45. $9 + 8 =$
A 1
B 19
C 17
D 16

46. $7 - 0 =$
A 0
B 1
C 6
D 7

47. $5 \times 4 =$
A 20
B 9
C 11
D 22

48. $25 \div 5 =$
A 4
B 5
C 20
D 30

49. $7 + 1 =$
A 5
B 6
C 7
D 8

50. $8 \times 6 =$
A 86
B 24
C 48
D 14

6

SYSTEMS OF EQUATIONS AND PROBLEM SOLVING

How can candies of different costs be mixed in order to create a mixture that has a cost between the two costs? Systems of equations can be used to find an answer.

We now consider where two graphs of linear equations might intersect. By doing so we can solve what is called a *system of equations.* Systems of equations have extensive applications to many fields such as sociology, psychology, business, education, engineering, and science.

Richard Estes. The Candy Store. 1969. Detail. Oil and synthetic polymer. $47\frac{3}{4} \times 68\frac{3}{4}$ inches. Collection of Whitney Museum of American Art. Gift of the Friends of the Whitney Museum of American Art. Acq. # 69.21.

OBJECTIVE

After finishing Section 6.1, you should be able to:

i Translate problems to systems of equations.

Translate to a system of equations. Do not attempt to solve. Save for later use.

1. The sum of two numbers is 115. The difference is 21. Find the numbers.

6.1

TRANSLATING PROBLEMS TO EQUATIONS

i As you have probably already noted, in the five-step problem-solving process the most difficult and time-consuming part is translating to mathematical language. In this chapter we consider situations in which the translation can be done by writing more than one equation. That is usually much easier than translating to a single equation. In this section we just practice translating a problem to two equations.

EXAMPLE 1 Translate the following problem situation to mathematical language, using two equations.

The sum of two numbers is 15. One number is four times the other. Find the numbers.

2. **Translate** There are two statements in this problem. We translate the first one.

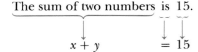

We have used x and y for the numbers. Now we translate the second statement, remembering to use x and y.

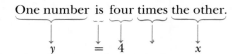

For the second statement we could have also translated to $x = 4y$. That would also have been correct. The problem has been translated to a pair or *system of equations*. We list what the variables represent and then list the equations:

$$x + y = 15,$$
$$y = 4x.$$

DO EXERCISE 1.

EXAMPLE 2 Translate the following problem situation to mathematical language, using two equations.

Badger Rent-A-Car rents compact cars at a daily rate of $43.95 plus 40¢ per mile. Thirsty Rent-A-Car rents compact cars at a daily rate of $42.95 plus 42¢ per mile. For what mileage is the cost the same?

2. **Translate** We translate the first statement, using $0.40 for 40¢.

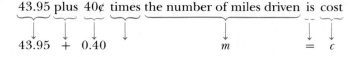

We have let m represent the mileage and c the cost. We translate the second statement, but again it helps to reword it first.

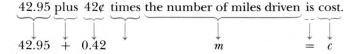

We have now translated to a system of equations:

$$43.95 + 0.40m = c,$$
$$42.95 + 0.42m = c.$$

DO EXERCISE 2.

The familiarization step often aids translating. In particular, making a drawing is often helpful.

EXAMPLE 3 Translate the following problem situation to mathematical language, using two equations.

The perimeter of a rectangle is 90 cm. The length is 20 cm greater than the width. Find the length and the width.

1. **Familiarize** We make a drawing and label it. We have called the length l and the width w.

2. **Translate** From the drawing we see that the perimeter (the distance around) of the rectangle is $l + l + w + w$, or $2l + 2w$. We translate the first statement.

The perimeter is 90 cm.

$$2l + 2w = 90$$

We translate the second statement.

The length is 20 cm greater than the width.

$$l = 20 + w$$

We have translated to a system of equations:

$$2w + 2l = 90,$$
$$l = 20 + w.$$

DO EXERCISE 3.

Translate to a system of equations. Do not attempt to solve. Save for later use.

2. Acme Rent-a-Car rents a station wagon at a daily rate of $31.95 plus 33¢ per mile. Speedo Rentzit rents a wagon for $34.95 plus 29¢ per mile. For what mileage is the cost the same?

Translate to a system of equations. Do not attempt to solve. Draw a picture if helpful. Save for later use.

3. The perimeter of a rectangle is 86 cm. The length is 19 cm more than the width. Find the length and the width.

EXERCISE SET 6.1

Translate to a system of equations. Do not attempt to solve. Save for later use.

1. The sum of two numbers is 58. The difference is 16. Find the numbers.

2. The sum of two numbers is 26.4. One number is five times the other. Find the numbers.

3. The perimeter of a rectangle is 400 m. The width is 40 m less than the length. Find the length and width.

4. The perimeter of a rectangle is 76 cm. The width is 17 cm less than the length. Find the length and width.

5. Acme Rent-A-Car rents an intermediate-size car at a daily rate of $53.95 plus 30¢ per mile. Hartz Rent-A-Car rents an intermediate-size car at a daily rate of $54.95 plus 20¢ per mile. For what mileage is the cost the same?

6. Badger rents a basic car at a daily rate of $45.95 plus 40¢ per mile. Hartz rents a basic car at a daily rate of $46.95 plus 20¢ per mile. For what mileage is the cost the same?

7. The difference between two numbers is 16. Three times the larger number is seven times the smaller. What are the numbers?

8. The difference between two numbers is 18. Twice the smaller number plus three times the larger is 74. What are the numbers?

9. Two angles are supplementary. One is 8° more than three times the other. Find the angles. (Supplementary angles are angles whose sum is 180°.)

10. Two angles are supplementary. One is 30° more than two times the other. Find the angles.

11. Two angles are complementary. Their difference is 34°. Find the angles. (Complementary angles are angles whose sum is 90°.)

12. Two angles are complementary. One angle is 42° more than $\frac{1}{2}$ the other. Find the angles.

13. In a vineyard a vintner uses 820 hectares to plant Chardonnay and Riesling grapes. The vintner knows that profits will be greatest by planting 140 hectares more of Chardonnay than Riesling. How many hectares of each grape should be planted? (A *hectare* is a unit of area, about 2.47 acres.)

14. The Hayburner Horse Farm allots 650 hectares to plant hay and oats. The owners know that their needs are best met if they plant 180 hectares more of hay than oats. How many hectares of each should they plant?

15. The difference between two numbers is 18. Twice the smaller number plus three times the larger is 74. What are the numbers?

16. The perimeter of a rectangle is 400 m. The length is 40 m more than the width. Find the length and width.

17. The perimeter of a rectangle is 76 cm. The length is 17 cm more than the width. Find the length and width.

18. The perimeter of a football field (excluding the end zones) is $306\frac{2}{3}$ yd. The length is $46\frac{2}{3}$ yd longer than the width. Find the length and width.

Translate to a system of equations. Do not attempt to solve. Save for later use.

19. Patrick's age is 20% of his father's age. Twenty years from now, Patrick's age will be 52% of his father's age. How old are Patrick and his father?

20. If 5 is added to a man's age and the total is divided by 5, the result will be his daughter's age. Five years ago the father's age was eight times his daughter's age. Find their present ages.

21. When the base of a triangle is increased by 2 ft and the height is decreased by 1 ft, the height becomes $\frac{1}{3}$ of the base, and the area becomes 24 ft^2. Find the original dimensions of the triangle.

OBJECTIVES

After finishing Section 6.2, you should be able to:

i Determine whether an ordered pair is a solution of a system of equations.

ii Solve systems of equations by graphing.

6.2

SYSTEMS OF EQUATIONS

When a problem has been translated, as in Section 6.1, we have a system of equations. We now learn how to solve such systems.

i **Identifying Solutions**

Consider the system of equations

$$x + y = 8,$$
$$2x - y = 1.$$

A *solution* of a system of two equations is an ordered pair that makes both equations true. Consider the system listed above. Look at the graphs on the following page. Recall that a graph of an equation is a picture of its solution set. Each point on a graph corresponds to a solution. Which points (ordered pairs) are solutions of *both* equations? The graph shows that there is only one. It is the point *P* where the graphs cross. This point looks as if its coordinates are (3, 5). We check:

$$\begin{array}{c|c} x + y = 8 \\ \hline 3 + 5 & 8 \\ & 8 \end{array} \qquad \begin{array}{c|c} 2x - y = 1 \\ \hline 2 \cdot 3 - 5 & 1 \\ 6 - 5 \\ 1 \end{array}$$

Determine whether the given ordered pair is a solution of the system of equations.

1. $(2, -3);$ $x = 2y + 8,$
      $2x + y = 1$

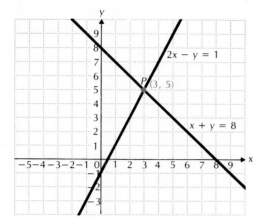

There is just one solution of the system of equations. It is $(3, 5)$. In other words, $x = 3$ and $y = 5$.

EXAMPLE 1   Determine whether $(1, 2)$ is a solution of the system

$$y = x + 1,$$
$$2x + y = 4.$$

We check:

2. $(20, 40);$ $a = \frac{1}{2}b,$
      $b - a = 60$

$$\begin{array}{c|c} y = x + 1 \\ \hline 2 & 1 + 1 \\ 2 & 2 \end{array} \qquad \begin{array}{c|c} 2x + y = 4 \\ \hline 2 \cdot 1 + 2 & 4 \\ 2 + 2 \\ 4 \end{array}$$

Thus $(1, 2)$ is a solution of the system.

EXAMPLE 2   Determine whether $(-3, 2)$ is a solution of the system

$$a + b = -1,$$
$$b + 3a = 4.$$

We check:

$$\begin{array}{c|c} a + b = -1 \\ \hline -3 + 2 & -1 \\ -1 \end{array} \qquad \begin{array}{c|c} b + 3a = 4 \\ \hline 2 + 3(-3) & 4 \\ 2 - 9 \\ -7 \end{array}$$

The point $(-3, 2)$ is not a solution of $b + 3a = 4$. Thus it is not a solution of the system.

DO EXERCISES 1 AND 2.

Solve by graphing.

3. $2x + y = 1,$
    $x = 2y + 8$

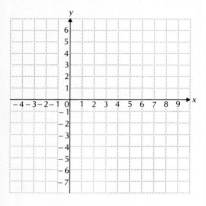

ii Solving Systems by Graphing

> To solve a system of equations by graphing, we graph both equations and find coordinates of the point(s) of intersection. Then we check. If the lines are parallel, there is no solution.

EXAMPLE 3 Solve by graphing:

$$x + 2y = 7,$$
$$x = y + 4.$$

We graph the equations. Point P looks as if it has coordinates $(5, 1)$.

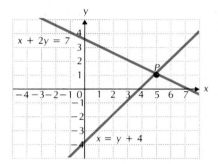

Check:

$x + 2y = 7$	$x = y + 4$
$5 + 2 \cdot 1 \mid 7$	$5 \mid 1 + 4$
$5 + 2$	$5 \mid 5$
7	

The solution is $(5, 1)$.

EXAMPLE 4 Solve by graphing:

$$y = 3x + 4,$$
$$y = 3x - 3.$$

The graphs are parallel. There is no point where they cross, so the system has no solution.

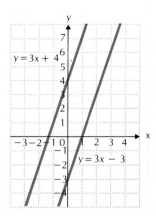

When we graph a system of two equations, one of the following three things can happen.

1. **The lines have one point of intersection, as in Example 3. The point of intersection is the only solution of the system.**
2. **The lines are parallel, as in Example 4. If so, there is no point that satisfies both equations. The system has no solution.**
3. **The lines coincide. Thus the equations have the same graph, and every solution of one equation is a solution of the other. There is an unlimited number of solutions.**

DO EXERCISES 3 AND 4. (EXERCISE 3 IS ON THE PRECEDING PAGE.)

4. $y + 4 = x,$
   $x - y = -2$

EXERCISE SET 6.2

i Determine whether the given ordered pair is a solution of the system of equations. Remember to use alphabetical order of variables.

1. $(3, 2);$ $2x + 3y = 12,$
$x - 4y = -5$

2. $(1, 5);$ $5x - 2y = -5,$
$3x - 7y = -32$

3. $(3, 2);$ $3t - 2s = 0,$
$t + 2s = 15$

4. $(2, -2);$ $b + 2a = 2,$
$b - a = -4$

5. $(15, 20);$ $3x - 2y = 5,$
$6x - 5y = -10$

6. $(-1, -3);$ $3r + s = -6,$
$2r = 1 + s$

7. $(-1, 1);$ $x = -1,$
$x - y = -2$

8. $(-3, 4);$ $2x = -y - 2,$
$y = -4$

9. $(12, 3);$ $y = \frac{1}{4}x,$
$3x - y = 33$

10. $(-3, 1);$ $y = -\frac{1}{3}x$
$3y = -5x - 12$

ii Use graph paper. Solve each system graphically and check.

11. $x + y = 3,$
$x - y = 1$

12. $x - y = 2,$
$x + y = 6$

13. $x + 2y = 10,$
$3x + 4y = 8$

14. $x - 2y = 6,$
$2x - 3y = 5$

15. $8x - y = 29,$
$2x + y = 11$

16. $4x - y = 10,$
$3x + 5y = 19$

17. $u = v,$
$4u = 2v - 6$

18. $x = 3y,$
$3y - 6 = 2x$

19. $x = -y,$
$x + y = 4$

20. $-3x = 5 - y,$
$2y = 6x + 10$

21. $a = \frac{1}{2}b + 1,$
$a - 2b = -2$

22. $x = \frac{1}{3}y + 2,$
$-2x - y = 1$

23. $y = 3,$
$x = 5$

24. $y = 3x,$
$y = -3x + 2$

25. $x + y = 9,$
$3x + 3y = 27$

26. $x + y = 4,$
$x + y = -4$

27. The solution of the following system is $(2, -3)$. Find A and B.

$$Ax - 3y = 13,$$
$$x - By = 8$$

Determine whether the given ordered pair is a solution of each system of three equations.

28. $(2, -3);$ $x + 3y = -7,$
$-x + y = -5,$
$2x - y = 1$

29. $(-1, -5);$ $4a - b = 1,$
$-a + b = -4,$
$2a + 3b = -17$

30. Describe in words the graph of the system in Exercise 29. Describe the solution.

31. Solve this system by graphing. What happens when you check your possible solution?

$$3x + 7y = 5,$$
$$6x - 7y = 1$$

OBJECTIVES

After finishing Section 6.3, you should be able to:

| i | Solve a system of two equations by the substitution method when one of the equations has a variable alone on one side. |

| ii | Solve a system of two equations by the substitution method when neither equation has a variable alone on one side. |

| iii | Solve problems involving systems of equations and their solution by the substitution method. |

Solve by the substitution method. Do not graph.

1. $x + y = 5,$
    $x = y + 1$

6.3

SOLVING BY SUBSTITUTION: PROBLEM SOLVING

Graphing helps picture the solution of a system of equations, but solving by graphing is not fast or accurate. Let us learn better ways using algebra.

i The Substitution Method

One method for solving systems is known as the *substitution method.*

EXAMPLE 1   Solve the system:

$$x + y = 6,$$
$$x = y + 2.$$

The second equation says that x and $y + 2$ name the same thing. Thus in the first equation, we can substitute $y + 2$ for x:

$$x + y = 6$$
$$(y + 2) + y = 6. \qquad \text{Substituting } y + 2 \text{ for } x$$

This last equation has only one variable. We solve it:

$$(y + 2) + y = 6$$
$$2y + 2 = 6 \qquad \text{Collecting like terms}$$
$$2y = 4$$
$$y = 2.$$

We return to the original pair of equations. We substitute 2 for y in either of them. We use the first equation:

$$x + y = 6$$
$$x + 2 = 6 \qquad \text{Substituting 2 for } y$$
$$x = 4.$$

The ordered pair $(4, 2)$ may be a solution. We check.

Check:
$$\begin{array}{c|c} x + y = 6 \\ \hline 4 + 2 & 6 \\ 6 & \end{array} \qquad \begin{array}{c|c} x = y + 2 \\ \hline 4 & 2 + 2 \\ & 4 \end{array}$$

Since $(4, 2)$ checks, we have the solution. We could also express the answer as $x = 4$, $y = 2$.

DO EXERCISE 1.

EXAMPLE 2   Solve:

$$s = 13 - 3t,$$
$$s + t = 5.$$

We substitute $13 - 3t$ for s in the second equation:

$$s + t = 5$$
$$(13 - 3t) + t = 5. \qquad \text{Substituting } 13 - 3t \text{ for } s$$

Now we solve for t:

$$13 - 2t = 5 \qquad \text{Collecting like terms}$$
$$-2t = -8 \qquad \text{Adding } -13$$
$$t = \frac{-8}{-2}, \quad \text{or } 4. \qquad \text{Multiplying by } \frac{1}{-2}$$

Next we substitute 4 for t in the second equation of the original system:

$$s + t = 5$$
$$s + 4 = 5 \qquad \text{Substituting 4 for } t$$
$$s = 1.$$

We check the ordered pair $(1, 4)$.

Check:

$$\begin{array}{c|c} s = 13 - 3t & s + t = 5 \\ \hline 1 \mid 13 - 3 \cdot 4 & 1 + 4 \mid 5 \\ 13 - 12 & 5 \\ 1 & \end{array}$$

Since $(1, 4)$ checks, it is the solution.

DO EXERCISE 2.

ii Solving for the Variable First

Sometimes neither equation of a pair has a variable alone on one side. Then we solve one equation for one of the variables and proceed as before.

EXAMPLE 3   Solve:

$$x - 2y = 6,$$
$$3x + 2y = 4.$$

We solve one equation for one variable. Since the coefficient of x is 1 in the first equation, it is easier to solve it for x:

$$x - 2y = 6$$
$$x = 6 + 2y.$$

We substitute $6 + 2y$ for x in the second equation of the original pair and solve:

$$3x + 2y = 4$$
$$3(6 + 2y) + 2y = 4 \qquad \text{Substituting } 6 + 2y \text{ for } x$$
$$18 + 6y + 2y = 4$$
$$18 + 8y = 4$$
$$8y = -14$$
$$y = \frac{-14}{8}, \quad \text{or } -\frac{7}{4}.$$

We go back to either of the original equations and substitute $-\frac{7}{4}$ for y. It will be easier to solve for x in the first equation:

$$x - 2y = 6$$
$$x - 2(-\tfrac{7}{4}) = 6$$
$$x + \tfrac{7}{2} = 6$$
$$x = 6 - \tfrac{7}{2}$$
$$x = \tfrac{5}{2}.$$

Solve by the substitution method.

2. $a - b = 4,$
$\quad b = 2 - a$

Solve.

3. $x - 2y = 8,$
   $2x + y = 8$

4. The perimeter of a rectangle is 76 cm. The length is 17 cm more than the width. Find the length and the width.

Check:

$x - 2y = 6$	
$\frac{5}{2} - 2(-\frac{7}{4})$	6
$\frac{5}{2} + \frac{7}{2}$	
$\frac{12}{2}$	
6	

$3x + 2y = 4$	
$3 \cdot \frac{5}{2} + 2(-\frac{7}{4})$	4
$\frac{15}{2} - \frac{7}{2}$	
$\frac{8}{2}$	
4	

Since $(\frac{5}{2}, -\frac{7}{4})$ checks, it is the solution.

DO EXERCISE 3.

iii Problem Solving

Now let us use the substitution method in problem solving.

EXAMPLE 4 The perimeter of a rectangle is 90 cm. The length is 20 cm greater than the width. Find the length and the width.

The Familiarize and Translate steps have been done in Example 3 of Section 6.1. The resulting system of equations is

$$2l + 2w = 90,$$
$$l = 20 + w;$$

where l represents the length and w the width.

3. Carry out We solve the system. We substitute $20 + w$ for l in the first equation and solve:

$$2(20 + w) + 2w = 90 \qquad \text{Substituting } 20 + w \text{ for } l$$
$$40 + 2w + 2w = 90$$
$$40 + 4w = 90$$
$$4w = 50$$
$$w = \frac{50}{4}, \quad \text{or } 12.5.$$

We go back to the original equations and substitute 12.5 for w in the second equation:

$$l = 20 + w$$
$$l = 20 + 12.5$$
$$l = 32.5.$$

4. Check A possible solution is a length of 32.5 cm and a width of 12.5 cm. The perimeter would be $2(32.5) + 2(12.5)$, or $65 + 25$, or 90. Also, the length is 20 cm greater than the width. These check.

5. State The length is 32.5 cm and the width is 12.5 cm.

DO EXERCISE 4.

EXERCISE SET 6.3

---

i Solve by the substitution method.

1. $x + y = 4,$
   $y = 2x + 1$

2. $x + y = 10,$
   $y = x + 8$

3. $y = x + 1,$
   $2x + y = 4$

4. $y = x - 6,$
   $x + y = -2$

5. $y = 2x - 5,$
$3y - x = 5$

6. $y = 2x + 1,$
$x + y = -2$

7. $x = -2y,$
$x + 4y = 2$

8. $r = -3s,$
$r + 4s = 10$

ii Solve by the substitution method. Get one variable alone first.

9. $s + t = -4,$
$s - t = 2$

10. $x - y = 6,$
$x + y = -2$

11. $y - 2x = -6,$
$2y - x = 5$

12. $x - y = 5,$
$x + 2y = 7$

13. $2x + 3y = -2,$
$2x - y = 9$

14. $x + 2y = 10,$
$3x + 4y = 8$

15. $x - y = -3,$
$2x + 3y = -6$

16. $3b + 2a = 2,$
$-2b + a = 8$

17. $r - 2s = 0,$
$4r - 3s = 15$

18. $y - 2x = 0,$
$3x + 7y = 17$

19. $x - 3y = 7,$
$-4x + 12y = 28$

20. $8x + 2y = 6,$
$4x = 3 - y$

iii *Problem solving*

21. The sum of two numbers is 27. One number is 3 more than the other. Find the numbers.

22. The sum of two numbers is 36. One number is 2 more than the other. Find the numbers.

23. Find two numbers whose sum is 58 and whose difference is 16.

24. Find two numbers whose sum is 66 and whose difference is 8.

25. The difference between two numbers is 16. Three times the larger number is seven times the smaller. What are the numbers?

26. The difference between two numbers is 18. Twice the smaller number plus three times the larger is 74. What are the numbers?

27. The perimeter of a rectangle is 400 m. The length is 40 m than the width. Find the length and width.

28. The perimeter of a rectangle is 76 cm. The length is 17 cm more than the width. Find the length and width.

29. The perimeter of a football field (excluding the end zones) is $306\frac{2}{3}$ yards. The length is $46\frac{2}{3}$ yards longer than the width. Find the length and width.

○ ───────────────────────────────

Solve by the substitution method.

30. ▦ $y - 2.35x = -5.97$
$2.14y - x = 4.88$

31. $\frac{1}{4}(a - b) = 2$
$\frac{1}{6}(a + b) = 1$

32. $\dfrac{x}{2} + \dfrac{3y}{2} = 2$

$\dfrac{x}{5} - \dfrac{y}{2} = 3$

33. $0.4x + 0.7y = 0.1$
$0.5x - 0.1y = 1.1$

34. Determine whether $(2, -3)$ is a solution of this system of three equations:

$$x + 3y = 7,$$
$$-x + y = -5,$$
$$2x - y = 1.$$

35. A rectangle has a perimeter of P feet. The width is 5 feet less than the length. Find the length in terms of P.

36. A rectangle has a perimeter of P meters. The length is 8 meters longer than the width. Find the width in terms of P.

Exercises 37 and 38 contain systems of three equations in three variables. A solution is an ordered triple, listed in alphabetical order. Use the substitution method to solve.

37. $x + y + z = 4,$
$x - 2y - z = 1,$
$y = -1$

38. $x + y + z = 180,$
$x = z - 70,$
$2y - z = 0$

39. Consider this system of equations:

$$3y + 3x = 14,$$
$$y = -x + 4.$$

Try to solve by the substitution method. Can you explain your results?

40. Consider this system of equations:

$$y = x + 5,$$
$$-3x + 3y = 15.$$

Try to solve by the substitution method. Can you explain your results?

41. Why is there no solution to the following system? (*Hint:* Use substitution more than once.)

$$x + y = 10,$$
$$y + z = 10,$$
$$x + z = 10,$$
$$x + y + z = 10$$

OBJECTIVES

After finishing Section 6.4, you should be able to:

i Solve a system of two equations using the addition method when no multiplication is necessary.

ii Solve a system of two equations using the addition method when the multiplication principle must be used.

iii Solve problems involving systems of equations and their solution by the addition method.

Solve using the addition method.

1. $x + y = 5$,
$2x - y = 4$

2. $3x - 3y = 6$,
$3x + 3y = 0$

6.4

THE ADDITION METHOD: PROBLEM SOLVING

The *addition method* for solving systems of equations makes use of the *addition principle*. Some systems are much easier to solve using the addition method.

i Solving by the Addition Method

Another method of solving systems of equations is called the *addition method*.

EXAMPLE 1 Solve:

$$x + y = 5,$$
$$x - y = 1.$$

We will use the addition principle for equations. According to the second equation, $x - y$ and 1 are the same thing. Thus we can add $x - y$ to the left side of the first equation and 1 to the right side:

$$x + y = 5$$
$$\underline{x - y = 1}$$
$$2x + 0y = 6.$$

We have made one variable "disappear." We have an equation with just one variable:

$$2x = 6.$$

We solve for x: $x = 3$. Next we substitute 3 for x in either of the original equations:

$$x + y = 5$$
$$3 + y = 5 \qquad \text{Substituting 3 for } x \text{ in the first equation}$$
$$y = 2. \qquad \text{Solving for } y$$

Check:

$x + y = 5$		$x - y = 1$	
$3 + 2$	5	$3 - 2$	1
5		1	

Since $(3, 2)$ checks, it is the solution.

DO EXERCISES 1 AND 2.

ii Using the Multiplication Principle First

The addition method allows us to eliminate a variable. We may need to multiply by -1 to make this happen.

EXAMPLE 2 Solve:

$$2x + 3y = 8,$$
$$x + 3y = 7.$$

If we add, we do not eliminate a variable. However, if the $3y$ were $-3y$ in one equation we could. We multiply on both sides of the second equation by -1 and then add:

$$2x + 3y = 8$$
$$\underline{-x - 3y = -7} \qquad \text{Multiplying by } -1$$
$$x = 1. \qquad \text{Adding}$$

Now we substitute 1 for x in one of the original equations:

$$x + 3y = 7$$
$$1 + 3y = 7 \qquad \text{Substituting 1 for } x \text{ in the second equation}$$
$$3y = 6$$
$$y = 2 \qquad \text{Solving for } y$$

Check:

$$
\begin{array}{c|c}
2x + 3y = 8 & \\
\hline
2 \cdot 1 + 3 \cdot 2 & 8 \\
2 + 6 & \\
8 &
\end{array}
\qquad
\begin{array}{c|c}
x + 3y = 7 & \\
\hline
1 + 3 \cdot 2 & 7 \\
1 + 6 & \\
7 &
\end{array}
$$

Since $(1, 2)$ checks, it is the solution.

DO EXERCISE 3.

In Example 2 we used the multiplication principle, multiplying by -1. We often need to multiply by something other than -1.

EXAMPLE 3 Solve:

$$3x + 6y = -6,$$
$$5x - 2y = 14.$$

This time we multiply by 3 on both sides of the second equation. Then we add:

$$
\begin{array}{rll}
3x + 6y = -6 & \\
15x - 6y = 42 & \text{Multiplying by 3} \\
\hline
18x = 36 & \text{Adding} \\
x = 2. & \text{Solving for } x
\end{array}
$$

We go back to the first equation and substitute 2 for x:

$$3 \cdot 2 + 6y = -6 \qquad \text{Substituting}$$
$$6 + 6y = -6$$
$$6y = -12$$
$$y = -2. \qquad \text{Solving for } y$$

Check:

$$
\begin{array}{c|c}
3x + 6y = -6 & \\
\hline
3 \cdot 2 + 6 \cdot (-2) & -6 \\
6 + (-12) & \\
-6 &
\end{array}
\qquad
\begin{array}{c|c}
5x - 2y = 14 & \\
\hline
5 \cdot 2 - 2 \cdot (-2) & 14 \\
10 - (-4) & \\
14 &
\end{array}
$$

Since $(2, -2)$ checks, it is the solution.

Solving a *system* of equations in two variables requires finding an ordered *pair* of numbers. Once you have solved for one variable, don't forget the other.

DO EXERCISE 4.

EXAMPLE 4 Solve:

$$3x + 5y = 6,$$
$$5x + 3y = 4.$$

3. Solve. Multiply one equation by -1 first.

$$5x + 3y = 17,$$
$$5x - 2y = -3$$

4. Solve.

$$4a + 7b = 11,$$
$$2a + 3b = 5$$

Solve.

5. $5x + 3y = 2,$
$3x + 5y = -2$

6. $2x + 3y = 1,$
$3x - 2y = 5$

7. Solve.

$2x + y = 15,$
$4x + 2y = 23$

We use the multiplication principle with both equations:

$$3x + 5y = 6,$$
$$5x + 3y = 4.$$

Thus we have

$$\begin{array}{ll} 15x + 25y = 30 & \text{Multiplying on both sides of the first equation by 5} \\ \underline{-15x - 9y = -12} & \text{Multiplying on both sides of the second equation by } -3 \\ 16y = 18 & \text{Adding} \\ y = \frac{18}{16}, \text{ or } \frac{9}{8}. \end{array}$$

We substitute $\frac{9}{8}$ for y in one of the original equations:

$$\begin{array}{ll} 3x + 5y = 6 & \\ 3x + 5 \cdot \frac{9}{8} = 6 & \text{Substituting } \frac{9}{8} \text{ for } y \text{ in the first equation} \\ 3x + \frac{45}{8} = 6 & \\ 3x = 6 - \frac{45}{8} & \\ 3x = \frac{48}{8} - \frac{45}{8} & \\ 3x = \frac{3}{8} & \\ x = \frac{3}{8} \cdot \frac{1}{3}, \text{ or } \frac{1}{8}. & \text{Solving for } x \end{array}$$

Check:

$$\begin{array}{c|c} 3x + 5y = 6 & \\ \hline 3 \cdot \frac{1}{8} + 5 \cdot \frac{9}{8} & 6 \\ \frac{3}{8} + \frac{45}{8} & \\ \frac{48}{8} & \\ 6 & \end{array}$$

$$\begin{array}{c|c} 5x + 3y = 4 & \\ \hline 5 \cdot \frac{1}{8} + 3 \cdot \frac{9}{8} & 4 \\ \frac{5}{8} + \frac{27}{8} & \\ \frac{32}{8} & \\ 4 & \end{array}$$

The solution is $(\frac{1}{8}, \frac{9}{8})$.

DO EXERCISES 5 AND 6.

Some systems have no solution.

EXAMPLE 5 Solve:

$$y - 3x = 2,$$
$$y - 3x = 1.$$

We multiply by -1 on both sides of the second equation:

$$\begin{array}{ll} y - 3x = 2 & \\ \underline{-y + 3x = -1} & \text{Multiplying by } -1 \\ 0 = -1. & \text{Adding} \end{array}$$

We obtain a false equation, $0 = 1$, so there is no solution. The graphs of the equations are parallel lines. They do not intersect.

DO EXERCISE 7.

When decimals or fractions appear, multiply to clear them. Then proceed as before.

EXAMPLE 6 Solve:

$$\frac{1}{4}x + \frac{5}{12}y = \frac{1}{2},$$
$$\frac{5}{12}x + \frac{1}{4}y = \frac{1}{3}.$$

The number 12 is a multiple of all of the denominators. We multiply on both sides of each equation by 12:

$$12(\tfrac{1}{4}x + \tfrac{5}{12}y) = 12 \cdot \tfrac{1}{2} \qquad\qquad 12(\tfrac{5}{12}x + \tfrac{1}{4}y) = 12 \cdot \tfrac{1}{3}$$
$$12 \cdot \tfrac{1}{4}x + 12 \cdot \tfrac{5}{12}y = 6 \qquad\qquad 12 \cdot \tfrac{5}{12}x + 12 \cdot \tfrac{1}{4}y = 4$$
$$3x + 5y = 6; \qquad\qquad\qquad 5x + 3y = 4.$$

The resulting system is

$$3x + 5y = 6,$$
$$5x + 3y = 4.$$

The solution of this system is given in Example 4.

DO EXERCISE 8.

ⅲ Problem Solving

Now let us use the addition method in problem solving.

EXAMPLE 7 Badger Rent-A-Car rents compact cars at a daily rate of $43.95 plus 40¢ per mile. Thirsty Rent-A-Car rents compact cars at a daily rate of $42.95 plus 42¢ per mile. For what mileage is the cost the same?

The ⟩Familiarize⟩ and ⟩Translate⟩ steps have been done in Example 2 of Section 7.1. The resulting system of equations is

$$43.95 + 0.40m = c,$$
$$42.95 + 0.42m = c;$$

where m represents the mileage and c the cost.

3. ⟩Carry out⟩ We solve the system of equations. We clear the system of decimals by multiplying on both sides by 100. Then we multiply the second equation by -1 and use the addition method.

$$
\begin{aligned}
4395 + 40m &=100c \\
-4295 - 42m &= -100c \\
\hline
100 - 2m &= 0 \\
100 &= 2m \\
50 &= m
\end{aligned}
$$

4. ⟩Check⟩ For 50 mi, the cost of the Badger car is $43.95 + 0.40(50)$, or $43.95 + 20$, or $63.95, and the cost of the other car is $42.95 + 0.42(50)$, or $42.95 + 21$, or $63.95, so the costs are the same when the mileage is 50.

5. ⟩State⟩ When the cars are driven 50 miles, the costs will be the same.

DO EXERCISE 9.

8. Solve.

$$\tfrac{1}{2}x + \tfrac{3}{10}y = \tfrac{1}{5}$$
$$\tfrac{3}{10}x + \tfrac{1}{2}y = -\tfrac{1}{5}$$

9. Acme rents a station wagon at a daily rate of $31.95 plus 33¢ per mile. Speedo Rentzit rents a wagon for $34.95 plus 29¢ per mile. For what mileages is the cost the same?

EXERCISE SET 6.4

ⅰ Solve using the addition method.

1. $x + y = 10,$
$x - y = 8$

2. $x - y = 7,$
$x + y = 3$

3. $x + y = 8,$
$-x + 2y = 7$

4. $x + y = 6,$
$-x + 3y = -2$

5. $3x - y = 9,$
$2x + y = 6$

6. $4x - y = 1,$
$3x + y = 13$

7. $4a + 3b = 7,$
$-4a + b = 5$

8. $7c + 5d = 18,$
$c - 5d = -2$

9. $8x - 5y = -9,$
$\quad 3x + 5y = -2$

10. $3a - 3b = -15,$
$\quad -3a - 3b = -3$

11. $4x - 5y = 7,$
$\quad -4x + 5y = 7$

12. $2x + 3y = 4,$
$\quad -2x - 3y = -4$

ii Solve using the multiplication principle first. Then add.

13. $-x - y = 8,$
$\quad 2x - y = -1$

14. $x + y = -7,$
$\quad 3x + y = -9$

15. $x + 3y = 19,$
$\quad x - y = -1$

16. $3x - y = 8,$
$\quad x + 2y = 5$

17. $x + y = 5,$
$\quad 5x - 3y = 17$

18. $x - y = 7,$
$\quad 4x - 5y = 25$

19. $2w - 3z = -1,$
$\quad 3w + 4z = 24$

20. $7p + 5q = 2,$
$\quad 8p - 9q = 17$

21. $2a + 3b = -1,$
$\quad 3a + 5b = -2$

22. $3x - 4y = 16,$
$\quad 5x + 6y = 14$

23. $x - 3y = 0,$
$\quad 5x - y = -14$

24. $5a - 2b = 0,$
$\quad 2a - 3b = -11$

25. $3x - 2y = 10,$
$\quad 5x + 3y = 4$

26. $2p + 5q = 9,$
$\quad 3p - 2q = 4$

27. $3x - 8y = 11,$
$\quad x + 6y - 8 = 0$

28. $m - n = 32,$
$\quad 3m - 8n - 6 = 0$

29. $0.06x + 0.05y = 0.07,$
$\quad 0.04x - 0.03y = 0.11$

30. $x - \frac{3}{2}y = 13,$
$\quad \frac{3}{2}x - y = 17$

iii *Problem solving*

Many of these problems have already been translated in Exercise Set 6.1.

31. Acme rents an intermediate-size car at a daily rate of $53.95 plus 30¢ per mile. Another company rents an intermediate-size car for $54.95 plus 20¢ per mile. For what mileage is the cost the same?

32. Badger rents a basic car at a daily rate of $45.95 plus 40¢ per mile. Another company rents a basic car for $46.95 plus 20¢ per mile. For what mileage is the cost the same?

33. Two angles are supplementary. One is 8° more than three times the other. Find the angles. (Supplementary angles are angles whose sum is 180°.)

34. Two angles are supplementary. One is 30° more than two times the other. Find the angles.

35. Two angles are complementary. Their difference is 34°. Find the angles. (Complementary angles are angles whose sum is 90°.)

36. Two angles are complementary. One angle is 42° more than $\frac{1}{2}$ the other. Find the angles.

37. In a vineyard a vintner uses 820 hectares to plant Chardonnay and Riesling grapes. The vintner knows the profits will be greatest by planting 140 hectares more of Chardonnay than Riesling. How many hectares of each grape should be planted?

38. The Hayburner Horse Farm allots 650 hectares to plant hay and oats. The owners know that their needs are best met if they plant 180 hectares more of hay than oats. How many hectares of each should they plant?

○ ──

39. Several ancient Chinese books included problems that can be solved by translating to systems of equations. *Arithmetical Rules in Nine Sections* is a book of 246 problems compiled by a Chinese mathematician, Chang Tsang, who died in 152 B.C. One of the problems is: Suppose there are a number of rabbits and pheasants confined in a cage. In all there are 35 heads and 94 feet. How many rabbits and how many pheasants are there? Solve the problem.

40–42. Solve the systems of equations you set up in Exercises 19–21 of Exercise Set 6.1.

Solve.

43. $3(x - y) = 9,$
$\quad x + y = 7$

44. $5(a - b) = 10,$
$\quad a + b = 2$

45. $2(x - y) = 3 + x,$
$\quad x = 3y + 4$

46. $2(5a - 5b) = 10,$
$\quad -5(6a + 2b) = 10$

47. $1.5x + 0.85y = 1637.5,$
$\quad 0.01(x + y) = 15.25$

48. $\frac{x}{3} + \frac{y}{2} = 1\frac{1}{3},$
$\quad x + 0.05y = 4$

Solve for x and y.

49. $y = ax + b,$
$\quad y = x + c$

50. $ax + by + c = 0,$
$\quad ax + cy + b = 0$

Solve.

51. $3(7 - a) - 2(1 + 2b) + 5 = 0,$
$\quad 3a + 2b - 18 = 0$

52. $\frac{2}{x} - \frac{3}{y} = -\frac{1}{2},$

$\quad \frac{1}{x} + \frac{2}{y} = \frac{11}{12}$

53. Suppose we can get a system into the form

$$ax + by = c$$
$$dx + ey = f,$$

where a, b, c, d, e, and f are positive or negative rational numbers.

a) Solve the system for x and y.

b) 🔲 Use the pattern of your solution to part (a) to solve the following system.

$$1.425x - 5695y = 1000,$$
$$0.875x + 275y = 2500.$$

6.5

MORE PROBLEM SOLVING

i We continue solving problems using the five-step process and our methods for solving systems of equations. The question may arise, after we have translated to a system, as to which method to use to solve the system: the substitution method or the addition method.

Although there are exceptions, the substitution method is probably better when a variable has a coefficient of 1, as in the following:

$$8x + 4y = 10, \qquad x + y = 61,$$
$$-x + y = 3; \qquad x = 2y.$$

Otherwise, the addition method is better. Both methods work. When in doubt, use the addition method.

EXAMPLE 1 Major league baseball teams play 162 games in a regular season. In a recent year, a team won 34 more games than it lost. How many games did the team win and how many did the team lose?

1. **Familiarize** Let's familiarize ourselves with the problem by making a guess. Suppose the team won 103 games. Since they play 162 games, and we know that

Number of games won plus Number of games lost = 162,

it follows that they lost $162 - 103$, or 59 games. The difference

Games won − Games lost

is $103 - 59$, or 44. The results do not quite check in the problem, but at least we have become more familiar with the problem.

2. **Translate** We let $x =$ the number of games won and $y =$ the number of games lost. Then, the first two statements of the problem can be reworded and translated as follows.

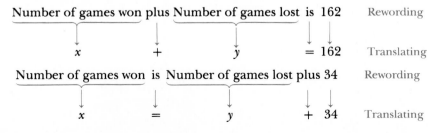

Number of games won	plus	Number of games lost	is 162	Rewording
x	$+$	y	$= 162$	Translating
Number of games won	is	Number of games lost	plus 34	Rewording
x	$=$	y	$+\ 34$	Translating

OBJECTIVE

After finishing Section 6.5, you should be able to:

i Solve problems involving systems of equations.

1. The sum of two numbers is 36 and their difference is 4. Find the numbers.

3. **Carry out** We solve the system of equations using the substitution method:

$$x + y = 162,$$
$$x = y + 34.$$

We substitute $y + 34$ for x in the first equation:

$$x + y = 162$$
$$(y + 34) + y = 162$$
$$34 + 2y = 162$$
$$2y = 128$$
$$y = 64.$$

We substitute 64 for y in the second equation and solve for x:

$$x = y + 34$$
$$x = 64 + 34$$
$$x = 98.$$

4. **Check** If the team won 98 games and lost 64, then the number of games won plus the number of games lost is $98 + 64$, or 162. Since games lost plus 34 is $64 + 34$, or 98, we know that it is true that the team won 34 more games than it lost.

5. **State** The team won 98 games and lost 64.

DO EXERCISE 1.

EXAMPLE 2 Howie is 21 years older than Izzi. In six years, Howie will be twice as old as Izzi. How old are they now?

1. **Familiarize** Let us consider some conditions of the problem. We let x represent Howie's age now and y Izzi's age now. We know that now Howie is 21 years older than Izzi. We make a table to organize our information. How do the ages relate in 6 years? In 6 years Izzi will be $y + 6$ and Howie will be $x + 6$, or $2(y + 6)$.

	Age now	Age in 6 years
Howie	x, or $y + 21$	$x + 6$, or $2(y + 6)$
Izzi	y	$y + 6$

2. **Translate** From the present ages we get the following rewording and translation.

Howie's age is 21 more than Izzi's age.   Rewording

$$x \qquad = 21 \qquad + \qquad y \qquad\qquad \text{Translating}$$

From their ages in 6 years we get the following rewording and translation.

Howie's age in six years will be twice Izzi's age in six years.   Rewording

$$x + 6 \qquad = \qquad 2 \cdot \qquad (y + 6) \qquad \text{Translating}$$

3. **Carry out** We solve the system of equations:

$$x = 21 + y,$$
$$x + 6 = 2(y + 6).$$

We use the addition method. We first add $-y$ on both sides of the first equation:

$$x - y = 21.$$

We also simplify the second equation:

$$x + 6 = 2y + 12$$
$$x - 2y = 6.$$

We solve the system

$$x - y = 21,$$
$$x - 2y = 6.$$

We multiply on both sides of the second equation by -1 and add:

$$
\begin{array}{rl}
x - y = & 21 \\
-x + 2y = & -6 \qquad \text{Multiplying by } -1 \\
\hline
y = & 15. \qquad \text{Adding}
\end{array}
$$

We find x by substituting 15 for y in $x - y = 21$:

$$x - 15 = 21$$
$$x = 36.$$

4. ▷ Check ▷ Howie's age is 36, which is 21 more than 15, Izzi's age. In six years when Howie will be 42 and Izzi 21, Howie's age will be twice Izzi's age.

5. ▷ State ▷ Howie is now 36 and Izzi is 15.

DO EXERCISE 2.

EXAMPLE 3 There were 411 people at a movie. Admission was $1.00 for adults and $0.75 for children. The receipts were $395.75. How many adults and how many children attended?

1. ▷ Familiarize ▷ There are many ways to familiarize ourselves with a problem situation. This time, let us make a guess and do some calculations. We guess:

240 adults and

180 children.

Does the guess check? Does it make sense?

  The total number of people attending was supposed to be 411, so our guess, which totaled 420, cannot be right. Let's try another guess.

240 adults and

171 children.

2. Wilma is 13 years older than Adam. In 9 years, Wilma will be twice as old as Adam. How old is Adam?

Now the total number sold is 411, so our guess is right with respect to the number of people attending.

How much money was taken in? The problem says that adults paid $1.00, so the total amount of money collected from the adults was

$$240(\$1), \quad \text{or } \$240.$$

The children paid $0.75, so the total amount of money collected from the children was

$$171(\$0.75), \quad \text{or } \$128.25.$$

This makes the total receipts

$$\$240 + \$128.25, \quad \text{or } \$368.25.$$

Our guess is not the answer to the problem because the total taken in, according to the problem, was $395.75. However, we are at least more familiar with the problem situation.

Let us list the information in a table. That usually helps a great deal in the familiarization process.

People	Paid	Number attending	Money taken in
Adults	$1.00	x	$1.00x$
Children	$0.75	y	$0.75y$
Totals		411	$395.75

2. **Translate** We let $x =$ the number of adults and $y =$ the number of children. The total number of people attending was 411, so

$$x + y = 411.$$

The amount taken in from the adults was $1.00x$, and the amount taken in from the children was $0.75y$. These amounts are in dollars. The total was $395.75, so we have

$$1.00x + 0.75y = 395.75.$$

We can multiply on both sides by 100 to clear of decimals. Thus we have the translation, a system of equations:

$$x + y = 411,$$
$$100x + 75y = 39{,}575.$$

3. **Carry out** We solve the system of equations. We use the addition method. We multiply on both sides of the first equation by -100 and then add:

$$
\begin{array}{ll}
-100x - 100y = -41{,}100 & \text{Multiplying by } -100 \\
\underline{100x + 75y = 39{,}575} & \\
-25y = -1{,}525 & \text{Adding} \\
y = \dfrac{-1525}{-25} & \text{Dividing by } -25 \\
y = 61.
\end{array}
$$

We go back to the first equation and substitute 61 for y:

$$x + y = 411$$
$$x + 61 = 411$$
$$x = 350.$$

4. **Check** We leave the check to the student. It is similar to what we did in the familiarization step.

5. **State** 350 adults and 61 children attended.

DO EXERCISE 3.

EXAMPLE 4 A chemist has one solution that is 80% acid (and the rest water) and another solution that is 30% acid. What is needed is 200 liters (L) of a solution that is 62% acid. The chemist will prepare it by mixing the two solutions on hand. How much of each should be used?

1. **Familiarize** We can draw a picture of the situation. The chemist uses x liters of the first solution and y liters of the second solution.

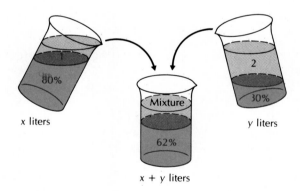

We can arrange the information in a table.

Type of solution	Amount of solution	Percent of acid	Amount of acid in solution
1	x	80%	$80\%x$
2	y	30%	$30\%y$
Mixture	200 liters	62%	$62\% \times 200$, or 124 liters

2. **Translate** The chemist uses x liters of the first solution and y liters of the second. Since the total is to be 200 liters, we have

$$x + y = 200.$$

3. There were 166 paid admissions to a game. The price was $2 for adults and $0.75 for children. The amount taken in was $293.25. How many adults and how many children attended?

4. One solution is 50% alcohol and a second is 70% alcohol. How much of each should be used to make 30 L of a solution that is 55% alcohol?

The *amount* of acid in the new mixture is to be 62% of 200 liters, or 124 liters. The amounts of acid from the two solutions are $80\%x$ and $30\%y$. Thus,

$$80\%x + 30\%y = 124,$$

or

$$0.8x + 0.3y = 124.$$

We clear the decimals by multiplying on both sides by 10:

$$10(0.8x + 0.3y) = 10 \cdot 124$$
$$8x + 3y = 1240.$$

3. Carry out ▷ We solve the system:

$$x + y = 200,$$
$$8x + 3y = 1240.$$

We use the addition method. We multiply on both sides of the first equation by -3 and then add:

$$
\begin{aligned}
-3x - 3y &= -600 \qquad &\text{Multiplying by } -3 \\
8x + 3y &= 1240 \\
\hline
5x &= 640 \qquad &\text{Adding} \\
x &= \tfrac{640}{5} \qquad &\text{Multiplying by } \tfrac{1}{5} \\
x &= 128. \qquad &\text{Solving for } x
\end{aligned}
$$

We go back to the first equation and substitute 128 for x:

$$x + y = 200$$
$$128 + y = 200$$
$$y = 72.$$

The solution is $x = 128$ and $y = 72$.

4. Check ▷ The sum of 128 and 72 is 200. Now 80% of 128 is 102.4 and 30% of 72 is 21.6. These add up to 124.

5. State ▷ The chemist should use 128 liters of the 80%-acid solution and 72 liters of the 30%-acid solution.

DO EXERCISE 4.

EXAMPLE 5 A grocer wishes to mix some candy worth 45¢ per lb and some worth 80¢ per lb to make 350 lb of a mixture worth 65¢ per lb. How much of each should be used?

1. | Familiarize ▷ Arranging information in a table will help.

	Cost of candy	Amount (lb)	Value (¢)
1	45¢	x	$45x$
2	80¢	y	$80y$
Mixture	65¢	350	65¢ (350), or 22,750

2. | Translate ▷ We use x for the amount of 45¢ candy and y for the amount of 80¢ candy. Then

$$x + y = 350.$$

Our second equation will come from the values. The value of the first candy, in cents, is $45x$ (x lb at 45¢ per lb). The value of the second is $80y$, and the value of the mixture is 350×65. Thus we have

$$45x + 80y = 350 \times 65.$$

3. | Carry out ▷ We solve the system:

$$x + y = 350,$$
$$45x + 80y = 22{,}750.$$

We use the addition method. We multiply on both sides of the first equation by -45 and then add:

$$
\begin{aligned}
-45x - 45y &= -15{,}750 \quad \text{Multiplying by } -45\\
\underline{45x + 80y} &= \underline{22{,}750}\\
35y &= 7{,}000 \quad \text{Adding}\\
y &= \tfrac{7{,}000}{35}\\
y &= 200.
\end{aligned}
$$

We go back to the first equation and substitute 200 for y:

$$x + y = 350$$
$$x + 200 = 350$$
$$x = 150.$$

4. | Check ▷ We consider $x = 150$ lb and $y = 200$ lb. The sum is 350 lb. The value of the candy is $45(150) + 80(200)$, or 22,750¢. These values check.

5. | State ▷ The grocer should mix 150 lb of the 45¢ candy with 200 lb of the 80¢ candy.

DO EXERCISE 5.

5. Grass seed A is worth $1.00 per lb and seed B is worth $1.35 per lb. How much of each would you use to make 50 lb of a mixture worth $1.14 per lb?

EXERCISE SET 6.5

i Solve.

1. A firm sells cars and trucks. There is room on its lot for 510 vehicles. From experience they know that profits will be greatest if there are 190 more cars than trucks on the lot. How many of each vehicle should the firm have for the greatest profit?

2. A family went camping at a park 45 km from town. They drove 23 km more than they walked to get to the campsite. How far did they walk?

3. Sammy is twice as old as his daughter. In four years Sammy's age will be three times what his daughter's age was six years ago. How old is each at present?

4. Ann is eighteen years older than her son. She was three times as old one year ago. How old is each?

5. Marge is twice as old as Consuelo. The sum of their ages seven years ago was 13. How old are they now?

6. Andy is four times as old as Wendy. In twelve years, Wendy's age will be half of Andy's. Find their ages now.

7. A collection of dimes and quarters is worth $15.25. There are 103 coins in all. How many of each are there?

8. A collection of quarters and nickels is worth $1.25. There are 13 coins in all. How many of each are there?

9. A collection of nickels and dimes is worth $25. There are three times as many nickels as dimes. How many of each are there?

10. A collection of nickels and dimes is worth $2.90. There are 19 more nickels than dimes. How many of each are there?

11. There were 429 people at a play. Admission was $1 for adults and 75¢ for children. The receipts were $372.50. How many adults and how many children attended?

12. The attendance at a school concert was 578. Admission was $2 for adults and $1.50 for children. The receipts were $985. How many adults and how many children attended?

13. There were 200 tickets sold for a college basketball game. Tickets for students were $0.50 and for adults were $0.75. The total amount of money collected was $132.50. How many of each type of ticket were sold?

14. There were 203 tickets sold for a wrestling match. For activity card holders the price was $1.25 and for noncard holders the price was $2. The total amount of money collected was $310. How many of each type of ticket were sold?

15. Solution A is 50% acid and solution B is 80% acid. How much of each should be used to make 100 grams of a solution that is 68% acid? (*Hint:* 68% of what is acid?)

16. Solution A is 30% alcohol and solution B is 75% alcohol. How much of each should be used to make 100 liters of a solution that is 50% alcohol?

17. Farmer Jones has 100 liters of milk that is 4.6% butterfat. How much skim milk (no butterfat) should be mixed with it to make milk that is 3.2% butterfat?

18. A tank contains 8000 liters of a solution that is 40% acid. How much water should be added to make a solution that is 30% acid?

19. A solution containing 30% insecticide is to be mixed with a solution containing 50% insecticide to make 200 liters of a solution containing 42% insecticide. How much of each solution should be used?

20. A solution containing 28% fungicide is to be mixed with a solution containing 40% fungicide to make 300 liters of a solution containing 36% fungicide. How much of each solution should be used?

21. The Nuthouse has 10 kg of mixed cashews and pecans worth $8.40 per kg. Cashews alone sell for $8.00 per kg and pecans sell for $9.00 per kg. How many kg of each are in the mixture?

22. A coffee shop mixes Brazilian coffee worth $5 per kg with Turkish coffee worth $8 per kg. The mixture is to sell for $7 per kg. How much of each type of coffee should be used to make a mixture of 300 kg?

---

23. A total of $27,000 is invested, part of it at 12% and part of it at 13%. The total yield after one year is $3385. How much was invested at each rate?

24. A two-digit number is six times the sum of its digits. The ten's digit is one more than the unit's digit. Find the number.

25. The sum of the digits of a two-digit number is 12. When the digits are reversed, the number is decreased by 18. Find the original number.

26. An automobile radiator contains 16 liters of antifreeze and water. This mixture is 30% antifreeze. How much of this mixture should be drained and replaced with pure antifreeze so that the mixture will be 50% antifreeze?

27. An employer has a daily payroll of $325 when employing some workers at $20 per day and others at $25 per day. When the number of $20 workers is increased by 50% and the number of $25 workers is decreased by $\frac{1}{3}$, the new daily payroll is $400. Find how many were originally employed at each rate.

28. A student earned $288 on investments. If $1100 was invested at one yearly rate and $1800 at a rate that was 1.5% higher, find the two rates of interest.

29. In a two-digit number, the sum of the unit's digit and the number is 43 more than five times the ten's digit. The sum of the digits is 11. Find the number.

30. The sum of the digits of a three-digit number is 9. If the digits are reversed, the number increases by 495. The sum of the ten's and hundred's digit is half the unit's digit. Find the number.

31. Together, a bat, ball, and glove cost $99.00. The bat costs $9.95 more than the ball, and the glove costs $65.45 more than the bat. How much does each cost?

32. In Lewis Carroll's "Through the Looking Glass" Tweedledum says to Tweedledee, "The sum of your weight and twice mine is 361 pounds." Then Tweedledee says to Tweedledum, "Contrariwise, the sum of your weight and twice mine is 362 pounds." Find the weight of Tweedledum and Tweedledee.

6.6

PROBLEMS INVOLVING MOTION

i   Many problems deal with distance, time, and speed. A basic formula comes from the definition of speed.

$$\text{Speed} = \frac{\text{Distance}}{\text{Time}}, \qquad r = \frac{d}{t}.$$

From $r = d/t$, we can obtain two other formulas by solving. They are

$$d = rt \qquad \text{Multiplying by } t$$

and

$$t = \frac{d}{r}. \qquad \text{Multiplying by } t \text{ and dividing by } r$$

In most problems involving motion, you will use one of these formulas. It is probably easiest to remember the definition of speed, $r = d/t$. You can easily obtain either of the other formulas as you need them.

We have a five-step process for problem solving. The following steps are also helpful when solving motion problems.

1. **Organize the information in a chart.**
2. **Look for as many things as you can that are the same, so you can write equations.**

EXAMPLE 1   A train leaves Podunk traveling east at 35 kilometers per hour (km/h). An hour later another train leaves Podunk on a parallel track at 40 km/h. How far from Podunk will the trains meet?

1. **Familiarize**   First make a drawing.

From the drawing we see that the distances are the same. Let's call the distance d. We don't know the times. Let t represent the time for the faster train. Then the time for the slower train will be $t + 1$. We can organize the information in a chart. In this case the distances are the same, so we shall use the formula $d = rt$.

	Distance	Speed	Time
Slow train	d	35	$t + 1$
Fast train	d	40	t

1. A car leaves Hereford traveling north at 56 km/h. Another car leaves Hereford one hour later traveling north at 84 km/h. How far from Hereford will the second car overtake the first? (*Hint:* The cars travel the same distance.)

2. Translate⟩ In these problems we look for things that are the same, so we can find equations. From each row of the chart we get an equation, $d = rt$. Thus we have two equations:

$$d = 35(t + 1),$$
$$d = 40t.$$

3. Carry out⟩ We solve the system using the substitution method:

$35(t + 1) = 40t$ Using the substitution method (substituting $35(t + 1)$ for d in the second equation)

$35t + 35 = 40t$ Multiplying

$35 = 5t$ Adding $-35t$

$\frac{35}{5} = t$ Multiplying by $\frac{1}{5}$

$7 = t.$

The problem asks us to find how far from Podunk the trains meet. Thus we need to find d. We can do this by substituting 7 for t in the equation $d = 40t$:

$$d = 40(7)$$
$$= 280.$$

4. Check⟩ If the time is 7 hr, then the distance the slow train travels is $35(7 + 1)$, or 280 km. The fast train travels $40(7)$, or 280 km. Since the distances are the same, we know how far from Podunk the trains will meet.

5. State⟩ The trains meet 280 km from Podunk.

DO EXERCISE 1.

EXAMPLE 2 A motorboat took 3 hr to make a downstream trip with a 6-km/h current. The return trip against the same current took 5 hr. Find the speed of the boat in still water.

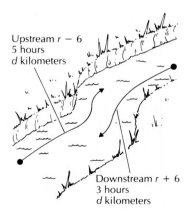

Upstream $r - 6$
5 hours
d kilometers

Downstream $r + 6$
3 hours
d kilometers

1. Familiarize⟩ We first make a drawing. From the drawing we see that the distances are the same. Let's call the distance d. Let r represent the speed of the boat in still water. Then, when the boat is traveling downstream, its speed is $r + 6$ (the current helps the boat along). When it is traveling upstream, its speed is $r - 6$ (the current holds the boat back some). We can organize the information in a chart. In this case the distances are the same, so we shall use the formula $d = rt$.

	Distance	Speed	Time
Downstream	d	$r+6$	3
Upstream	d	$r-6$	5

2. **Translate** From each row of the chart, we get an equation, $d = rt$:

$$d = (r+6)3,$$
$$d = (r-6)5.$$

3. **Carry out** We solve the system using substitution:

$(r+6)3 = (r-6)5$ Using substitution (substituting $(r+6)3$ for d in the second equation)

$3r + 18 = 5r - 30$ Multiplying

$-2r + 18 = -30$ Adding $-5r$

$-2r = -48$ Adding -18

$r = \dfrac{-48}{-2},$ or 24.   Multiplying by $-\frac{1}{2}$

4. **Check** When $r = 24$, $r + 6 = 30$, and $30 \cdot 3 = 90$, the distance. When $r = 24$, $r - 6 = 18$, and $18 \cdot 5 = 90$. In both cases we get the same distance.

5. **State** The speed in still water is 24 km/h.

DO EXERCISE 2.

EXAMPLE 3 Two cars leave town at the same time going in opposite directions. One of them travels 60 mph and the other 30 mph. In how many hours will they be 150 miles apart?

1. **Familiarize** We first make a drawing.

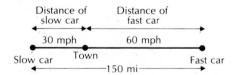

From the wording of the problem and the drawing, we see that the distances may *not* be the same. But the times the cars travel are the same, so we can just use t for time. We can organize the information in a chart.

	Distance	Speed	Time
Fast car	Distance of fast car	60	t
Slow car	Distance of slow car	30	t

2. **Translate** From the drawing we see that

(Distance of fast car) + (Distance of slow car) = 150.

Then using $d = rt$ in each row of the table we get

$$60t + 30t = 150.$$

2. An airplane flew for 5 hr with a 25-km/h tail wind. The return flight against the same wind took 6 hr. Find the speed of the airplane in still air. (*Hint:* The distance is the same both ways. The speeds are $r + 25$ and $r - 25$, where r is the speed in still air.)

3. Two cars leave town at the same time in opposite directions. One travels 48 mph and the other 60 mph. How far apart will they be 3 hr later? (*Hint:* The times are the same. Be *sure* to make a drawing.)

4. Two cars leave town at the same time in the same direction. One travels 35 mph and the other 40 mph. In how many hours will they be 15 mi apart? (*Hint:* The times are the same. Be *sure* to make a drawing.)

3. [Carry out] We solve the equation:

$$90t = 150 \qquad \text{Collecting like terms}$$
$$t = \tfrac{150}{90}, \quad \text{or} \quad \tfrac{5}{3}, \quad \text{or} \quad 1\tfrac{2}{3} \text{ hours.} \qquad \text{Multiplying by } \tfrac{1}{90}$$

4. [Check] When $t = \tfrac{5}{3}$ hr,

$$(\text{Distance of fast car}) + (\text{Distance of slow car}) = 60\left(\tfrac{5}{3}\right) + 30\left(\tfrac{5}{3}\right)$$
$$= 100 + 50, \quad \text{or } 150 \text{ miles.}$$

Thus the time of $\tfrac{5}{3}$ hr, or $1\tfrac{2}{3}$ hr checks.

5. [State] In $1\tfrac{2}{3}$ hr the cars will be 150 miles apart.

DO EXERCISES 3 AND 4. (EXERCISE 3 IS ON THE PRECEDING PAGE.)

EXERCISE SET 6.6

i Solve.

1. Two cars leave town at the same time going in opposite directions. One travels 55 mph and the other travels 48 mph. In how many hours will they be 206 miles apart?

2. Two cars leave town at the same time going in opposite directions. One travels 44 mph and the other travels 55 mph. In how many hours will they be 297 miles apart?

3. Two cars leave town at the same time going in the same direction. One travels 30 mph and the other travels 46 mph. In how many hours will they be 72 miles apart?

4. Two cars leave town at the same time going in the same direction. One travels 32 mph and the other travels 47 mph. In how many hours will they be 69 miles apart?

5. A train leaves a station and travels east at 72 km/h. Three hours later a second train leaves on a parallel track and travels east at 120 km/h. When will it overtake the first train?

6. A private airplane leaves an airport and flies due south at 192 km/h. Two hours later a jet leaves the same airport and flies due south at 960 km/h. When will the jet overtake the plane?

7. A canoeist paddled for 4 hours with a 6-km/h current to reach a campsite. The return trip against the same current took 10 hours. Find the speed of the canoe in still water.

8. An airplane flew for 4 hours with a 20-km/h tail wind. The return flight against the same wind took 5 hours. Find the speed of the plane in still air.

9. It takes a passenger train 2 hours less time than it takes a freight train to make the trip from Central City to Clear Creek. The passenger train averages 96 km/h, whereas the freight train averages 64 km/h. How far is it from Central City to Clear Creek?

10. It takes a small jet 4 hours less time than it takes a propeller-driven plane to travel from Glen Rock to Oakville. The jet averages 637 km/h, whereas the propeller plane averages 273 km/h. How far is it from Glen Rock to Oakville?

11. An airplane took 2 hours to fly 600 km against a head wind. The return trip with the wind took $1\tfrac{2}{3}$ hours. Find the speed of the plane in still air.

12. It took 3 hours to row a boat 18 km against the current. The return trip with the current took $1\tfrac{1}{2}$ hours. Find the speed of the rowboat in still water.

13. A motorcycle breaks down and the rider has to walk the rest of the way to work. The motorcycle was being driven at 45 mph and the rider walks at a speed of 6 mph. The distance from home to work is 25 miles and the total time for the trip was 2 hours. How far did the motorcycle go before it broke down?

14. A student walks and jogs to college each day. The student averages 5 km/h walking and 9 km/h jogging. The distance from home to college is 8 km and the student makes the trip in 1 hour. How far does the student jog?

15. An airplane flew for 4.23 hours with a 25.5-km/h tail wind. The return flight against the same wind took 4.97 hours. Find the speed of the plane in still air.

16. An airplane took $2\tfrac{1}{2}$ hours to fly 625 miles with the wind. It took 4 hours and 10 minutes to make the return trip against the same wind. Find the wind speed and the speed of the plane in still air.

17. To deliver a package, a messenger must travel at a speed of 60 mph on land and then use a motorboat whose speed is 20 mph in still water. While delivering the package, the messenger goes by land to a dock and then travels on a river against a current of 4 mph. The messenger reaches the destination in 4.5 hours and then returns to the starting point in 3.5 hours. How far did the messenger travel by land and how far by water?

18. Against a headwind, Gary computes his flight time for a trip of 2900 miles at 5 hours. The flight would take 4 hours and 50 minutes if the headwind were half as much. Find the headwind and the plane's air speed.

19. A car travels from one town to another at a speed of 32 mph. If it had gone 4 mph faster, it could have made the trip in $\frac{1}{2}$ hour less time. How far apart are the towns?

20. Two airplanes start at the same time and fly toward each other from points 1000 km apart at rates of 420 km/h and 330 km/h. When will they meet?

21. A truck and a car leave a service station at the same time and travel in the same direction. The truck travels at 55 mph and the car at 40 mph. They can maintain CB ratio contact within a range of 10 miles. When will they lose contact?

22. Charles Lindbergh flew the Spirit of St. Louis in 1927 from New York to Paris at an average speed of 107.4 mph. Eleven years later, Howard Hughes flew the same route, averaged 217.1 mph and took 16 hours, 57 minutes less time. Find the length of their route.

SUMMARY AND REVIEW: CHAPTER 6

The following contains a summary of what you should be able to do after completing this chapter. The review exercises are for practice. Answers are at the back of the book. If you miss an exercise, restudy the section indicated alongside the answer.

You should be able to:

Determine whether an ordered pair is a solution of a system of equations.

Determine whether the given ordered pair is a solution of the system of equations.

1. $(6, -1)$; $x - y = 3$,
$2x + 5y = 6$

2. $(2, -3)$; $2x + y = 1$,
$x - y = 5$

3. $(-2, 1)$; $x + 3y = 1$,
$2x - y = -5$

4. $(-4, -1)$; $x - y = 3$,
$x + y = -5$

Solve systems of equations using graphing and the substitution and addition methods.

Use graph paper. Solve each system graphically.

5. $x + y = 4$,
$x - y = 8$

6. $x + 3y = 12$,
$2x - 4y = 4$

7. $y = 5 - x$,
$3x - 4y = -20$

8. $3x - 2y = -4$,
$2y - 3x = -2$

Solve by the substitution method.

9. $y = 5 - x$,
$3x - 4y = -20$

10. $x + 2y = 6$,
$2x + 3y = 8$

11. $3x + y = 1$,
$x - 2y = 5$

12. $x + y = 6$,
$y = 3 - 2x$

13. $s + t = 5$,
$s = 13 - 3t$

14. $x - y = 4$,
$y = 2 - x$

Solve by the addition method.

15. $x + y = 4$
$2x - y = 5$

16. $x + 2y = 9$,
$3x - 2y = -5$

17. $x - y = 8$,
$2x + y = 7$

18. $\frac{2}{3}x + y = -\frac{5}{3}$,
$x - \frac{1}{3}y = -\frac{13}{3}$

19. $2x + 3y = 8$,
$5x + 2y = -2$

20. $5x - 2y = 2$,
$3x - 7y = 36$

21. $-x - y = -5$,
$2x - y = 4$

22. $6x + 2y = 4$,
$10x + 7y = -8$

Solve problems using the five-step process and the substitution and addition methods for solving systems of equations.

23. The sum of two numbers is 8. Their difference is 12. Find the numbers.

24. The sum of two numbers is 27. One half of the first number plus one third of the second number is 11. Find the numbers.

25. The perimeter of a rectangle is 96 cm. The length is 27 cm more than the width. Find the length and the width.

26. An airplane flew for 4 hr with a 15-km/h tailwind. The return flight against the wind took 5 hr. Find the speed of the airplane in still air.

27. There were 508 people at an organ recital. Orchestra seats cost $5.00 per person and balcony seats cost $3.00. The total receipts were $2118. Find the number of orchestra seats and the number of balcony seats sold.

28. Solution A is 30% alcohol and solution B is 60% alcohol. How much of each is needed to make 80 L of a solution that is 45% alcohol?

29. Jeff is three times as old as his son. In nine years, Jeff will be twice as old as his son. How old is each now?

30. The solution of the following system is $(6, 2)$. Find C and D.

$$2x - Dy = 6,$$
$$Cx + 4y = 14$$

31. Solve using the substitution method.

$$x - y + 2z = -3,$$
$$2x + y - 3z = 11,$$
$$z = -2$$

32. Solve: $3(x - y) = 4 + x,$

$\qquad x = 5y + 2.$

33. For a two-digit number, the sum of the unit's digit and the ten's digit is 6. When the digits are reversed, the new number is 18 more than the original number. Find the original number.

34. A stable boy agreed to work for one year. At the end of that time he was to receive $240 and one horse. After 7 months the boy quit the job, but still received the horse and $100. What was the value of the horse?

TEST: CHAPTER 6

1. Determine whether the given ordered pair is a solution of the system of equations.

$$(-2, -1); \quad x = 4 + 2y,$$
$$2y - 3x = 4$$

2. Use graph paper. Solve graphically.

$$x - y = 3,$$
$$x - 2y = 4$$

Solve by the substitution method.

3. $y = 6 - x,$
$\quad 2x - 3y = 22$

4. $x + 2y = 5,$
$\quad x + y = 2$

Solve by the addition method.

5. $x - y = 6,$
$\quad 3x + y = -2$

6. $\frac{1}{2}x - \frac{1}{3}y = 8,$
$\quad \frac{2}{3}x + \frac{1}{2}y = 5$

7. $4x + 5y = 5,$
$\quad 6x + 7y = 7$

8. $2x + 3y = 13,$
$\quad 3x - 5y = 10$

9. The difference of two numbers is 12. One fourth of the larger number plus one half of the smaller is 9. Find the numbers.

10. A motorboat traveled for 2 hr with an 8-km/h current. The return trip against the same current took 3 hr. Find the speed of the motorboat in still water.

11. Solution A is 25% acid and solution B is 40% acid. How much of each is needed to make 60 L of a solution that is 30% acid?

12. Find the numbers C and D such that $(-2, 3)$ is a solution of the system.

$$Cx - 4y = 7,$$
$$3x + Dy = 8$$

13. You are in line at a ticket window. There are two more people ahead of you in line than there are behind you. In the entire line there are three times as many people as there are behind you. How many are ahead of you in line?

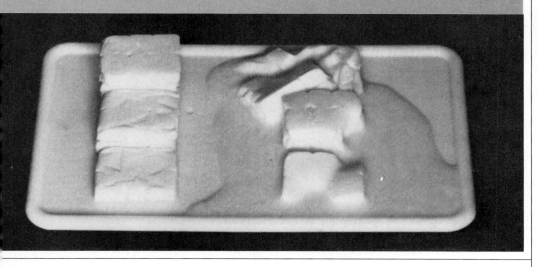

7

INEQUALITIES AND SETS

How can we find those Celsius temperatures for which butter stays solid? Inequalities can be a help in solving such a problem.

An inequality is a sentence like $2x + 3 < -9$. There are principles for solving inequalities very much like the addition and multiplication principles that we have used to solve equations. We learn how to solve inequalities in one variable and how to graph them. Then we use our new skills for problem solving.

Another goal of the chapter is to graph inequalities in two variables in the plane.

OBJECTIVES

After finishing Section 7.1, you should be able to:

i Determine whether a given number is a solution of an inequality.

ii Solve inequalities using the addition principle.

Determine whether the given number is a solution of the inequality.

1. $x < 3$

   a) 2

   b) 0

   c) -5

   d) 15

   e) 3

2. $x \geqslant 6$

   a) 6

   b) 0

   c) -4

   d) 25

   e) -6

7.1

USING THE ADDITION PRINCIPLE

In this section we learn an addition principle for solving inequalities that is similar to the one we learned for equations.

i Solutions of Inequalities

In Section 1.1 we defined the symbols $>$ (greater than) and $<$ (less than). The symbol $\geqslant$ means *greater than or equal to*. The symbol $\leqslant$ means *less than or equal to*. For example, $3 \leqslant 4$ and $3 \leqslant 3$ are both true, but $-3 \leqslant -4$ and $0 \geqslant 2$ are both false. An *inequality* is a number sentence with $>$, $<$, $\geqslant$, or $\leqslant$ for its verb—for example,

$$-4 < 5, \qquad x < 3, \qquad 2x + 5 \geqslant 0, \quad \text{and} \quad -3y + 7 \leqslant -8.$$

Some replacements for a variable in an inequality make it true and some make it false.

EXAMPLE 1   Determine whether each number is a solution of $x < 3$.

   -2 Since $-2 < 3$ is true, -2 is a solution.

   5   Since $5 < 3$ is false, 5 is not a solution.

   $\frac{1}{4}$ Since $\frac{1}{4} < 3$ is true, $\frac{1}{4}$ is a solution.

When we have found *all* the numbers that make an inequality true, we have *solved* the inequality.

> The replacements that make an inequality true are called its *solutions*. To *solve* an inequality means to find all of its solutions.

DO EXERCISES 1 AND 2.

ii The Addition Principle

Consider the true inequality

$$3 < 7.$$

Add 2 on both sides and we get another true inequality:

$$3 + 2 < 7 + 2 \quad \text{or} \quad 5 < 9.$$

Similarly, if we add -3 to both numbers we get another true inequality:

$$3 + (-3) < 7 + (-3) \quad \text{or} \quad 0 < 4.$$

> **THE ADDITION PRINCIPLE FOR INEQUALITIES**
>
> **If any number is added on both sides of a true inequality, we get another true inequality.**

The addition principle holds whether the inequality contains $<$, $>$, $\leqslant$, or $\geqslant$. Let's see how we use the addition principle to solve inequalities.

EXAMPLE 2   Solve: $x + 2 > 8$.

We use the addition principle, adding -2:

$$x + 2 + (-2) > 8 + (-2)$$
$$x > 6.$$

By using the addition principle we get an inequality for which we can determine the solutions easily.

Any number greater than 6 makes the last sentence true, hence is a solution of that sentence. Any such number is also a solution of the original sentence. Thus we have it solved.

We cannot check all the solutions of an inequality by substitution, as we can check solutions of equations. There are too many of them.

A partial check could be done by substituting a number greater than 6, say, 7, into the original inequality:

$$\begin{array}{c|c} x + 2 > 8 \\ \hline 7 + 2 & 8 \\ 9 & \end{array}$$

Since $9 > 8$ is true, 7 is a solution.

However, we don't really need to check. Let us see why. Consider the first and last inequalities

$$x + 2 > 8 \quad \text{and} \quad x > 6.$$

Any number that makes the first one true must make the last one true. We know this by the addition principle. Now the question is, will any number that makes the last one true also be a solution of the first one? Let us use the addition principle again, adding 2:

$$x > 6$$
$$x + 2 > 6 + 2$$
$$x + 2 > 8.$$

Now we know that any number that makes $x > 6$ true also makes $x + 2 > 8$ true. Therefore the sentences $x > 6$ and $x + 2 > 8$ have the same solutions. When two equations or inequalities have the same solutions, we say that they are *equivalent*. Any time we use the addition principle a similar thing happens. Thus, whenever we use the principle with inequalities, the first and last sentences will be equivalent.

EXAMPLE 3   Solve: $3x + 1 < 2x - 3$.

$$3x + 1 < 2x - 3$$
$$3x + 1 - 1 < 2x - 3 - 1 \qquad \text{Adding } -1$$
$$3x < 2x - 4 \qquad \text{Simplifying}$$
$$3x - 2x < 2x - 4 - 2x \qquad \text{Adding } -2x$$
$$x < -4 \qquad \text{Simplifying}$$

Any number less than -4 is a solution. The following are some of the solutions:

$$-5, \qquad -6, \qquad -4.1, \qquad -2045, \qquad -18\pi.$$

To describe all the solutions, we use the set notation

$$\{x \,|\, x < -4\},$$

Solve. Write set notation for your answer.

3. $x + 3 > 5$

4. $x - 5 < 8$

5. $5x + 1 < 4x - 2$

Solve.

6. $x + \frac{2}{3} \leqslant \frac{4}{5}$

7. $5y + 2 \leqslant -1 + 4y$

which is read:

> The set of all x such that x is less than -4.

DO EXERCISES 3–5.

EXAMPLE 4 Solve: $x + \frac{1}{3} \geqslant \frac{5}{4}$.

$$x + \tfrac{1}{3} \geqslant \tfrac{5}{4}$$
$$x + \tfrac{1}{3} - \tfrac{1}{3} \geqslant \tfrac{5}{4} - \tfrac{1}{3} \qquad \text{Adding } -\tfrac{1}{3}$$
$$x \geqslant \tfrac{5}{4} \cdot \tfrac{3}{3} - \tfrac{1}{3} \cdot \tfrac{4}{4} \qquad \text{Finding a common denominator}$$
$$x \geqslant \tfrac{15}{12} - \tfrac{4}{12}$$
$$x \geqslant \tfrac{11}{12}$$

Any number greater than or equal to $\frac{11}{12}$ is a solution. The *solution set* is

$$\{x \,|\, x \geqslant \tfrac{11}{12}\},$$

which is read:

> The set of all x such that x is greater than or equal to $\frac{11}{12}$.

DO EXERCISES 6 AND 7.

EXERCISE SET 7.1

i Determine whether each number is a solution of the given inequality.

1. $x > 4$
   a) 4
   b) 0
   c) -4
   d) 6

2. $y < 5$
   a) 0
   b) 5
   c) -1
   d) -5

3. $x \geqslant 6$
   a) -6
   b) 0
   c) 6
   d) 8

4. $x \leqslant 10$
   a) 4
   b) -10
   c) 0
   d) 11

5. $x < -8$
   a) 0
   b) -8
   c) -9
   d) -7

6. $x \geqslant 0$
   a) 2
   b) -3
   c) 0
   d) 3

7. $y \geqslant -5$
   a) 0
   b) -4
   c) -5
   d) -6

8. $y \leqslant -\frac{1}{2}$
   a) -1
   b) $-\frac{2}{3}$
   c) 0
   d) -0.5

ii Solve using the addition principle. Write set notation for answers.

9. $x + 7 > 2$

10. $x + 6 > 3$

11. $y + 5 > 8$

12. $y + 7 > 9$

13. $x + 8 \leqslant -10$

14. $x + 9 \leqslant -12$

15. $a + 12 < 6$

16. $a + 20 < 8$

17. $x - 7 \leqslant 9$

18. $x - 3 \leqslant 14$

19. $x - 6 > 2$

20. $x - 9 > 4$

21. $y - 7 > -12$

22. $y - 10 > -16$

23. $2x + 3 > x + 5$

24. $2x + 4 > x + 7$

25. $3x + 9 \leqslant 2x + 6$

26. $3x + 10 \leqslant 2x + 8$

27. $3x - 6 \geqslant 2x + 7$

28. $3x - 9 \geqslant 2x + 11$

29. $5x - 6 < 4x - 2$

30. $6x - 8 < 5x - 9$

31. $3y + 4 \geqslant 2y - 7$

32. $4y + 5 \leqslant 3y - 8$

33. $7 + c > 7$

34. $-9 + c > 9$

35. $y + \frac{1}{4} \leqslant \frac{1}{2}$

36. $y + \frac{1}{3} \leqslant \frac{5}{6}$

37. $x - \frac{1}{3} > \frac{1}{4}$

38. $x - \frac{1}{8} > \frac{1}{2}$

39. $-14x + 21 > 21 - 15x$

40. $-10x + 15 > 18 - 11x$

Problem-solving practice

41. When the sides of a square are lengthened by 0.2 km, the area becomes 0.64 km². Find the length of a side of the original square.

42. In one year 43,500 hunting and fishing licenses were sold in the United States. There were 5500 more fishing licenses sold than hunting licenses. How many of each were sold that year?

Solve.

43. $3(r + 2) < 2r + 4$ **44.** $4(r + 5) \geqslant 3r + 7$ **45.** $0.8x + 5 \geqslant 6 - 0.2x$ **46.** $0.7x + 6 \leqslant 7 - 0.3x$

47. $2x + 2.4 > x - 9.4$ **48.** $5x + 2.5 > 4x - 1.5$ **49.** $12x + 1.2 \leqslant 11x$ **50.** $x + 0.8 \leqslant 7.8 - 6$

51. 🖩 Solve: $17x + 9,479,756 \leqslant 16x - 8,579,243$.

52. Suppose that $2x - 5 \geqslant 9$ is true for some value of x. Determine whether $2x - 5 \geqslant 8$ is true for that same value of x.

7.2

USING THE MULTIPLICATION PRINCIPLE

There is a multiplication principle for inequalities similar to that for equations, but it must be qualified when multiplying on both sides by a negative number. We consider the multiplication principle and then use it with the addition principle to solve inequalities.

i The Multiplication Principle

Consider the true inequality

$$3 < 7.$$

Multiply both numbers by 2 and we get another true inequality:

$$3 \cdot 2 < 7 \cdot 2 \quad \text{or } 6 < 14.$$

Multiply both numbers by -3 and we get the false inequality:

$$3 \cdot (-3) < 7 \cdot (-3)$$

or

$$-9 < -21. \quad \text{False}$$

However, if we reverse the inequality symbol we get a true inequality:

$$-9 > -21. \quad \text{True}$$

THE MULTIPLICATION PRINCIPLE FOR INEQUALITIES

If we multiply on both sides of a true inequality by a positive number, we get another true inequality. If we multiply by a negative number and the inequality symbol is reversed, we get another true inequality.

The multiplication principle holds whether the inequality contains $<$, $>$, $\geqslant$, or $\leqslant$.

EXAMPLE 1 Solve: $4x < 28$.

$$\frac{1}{4} \cdot 4x < \frac{1}{4} \cdot 28 \qquad \text{Multiplying by } \frac{1}{4}$$

The symbol stays the same.

$$x < 7 \qquad \text{Simplifying}$$

The solution set is $\{x \mid x < 7\}$.

DO EXERCISES 1 AND 2.

OBJECTIVES

After finishing Section 7.2, you should be able to:

i Solve inequalities using the multiplication principle.

ii Solve inequalities using the addition and multiplication principles together.

Solve.

1. $8x < 64$

2. $5y \geqslant 160$

Solve.

3. $-4x \leqslant 24$

4. $-5y > 13$.

5. Solve.

$$7 - 4x < 8$$

6. Solve.

$$13x + 5 \leqslant 12x + 4$$

7. Solve.

$$24 - 7y \geqslant 11y - 14$$

EXAMPLE 2   Solve: $-2y < 18$.

$$-\tfrac{1}{2}(-2y) > -\tfrac{1}{2} \cdot 18 \qquad \text{Multiplying by } -\tfrac{1}{2}$$

             — The symbol has to be reversed!

$$y > -9 \qquad \text{Simplifying}$$

The solution set is $\{y \mid y > -9\}$.

DO EXERCISES 3 AND 4.

ⅱ Using the Principles Together

We use the addition and multiplication principles together in solving inequalities in much the same way as in solving equations. We generally use the addition principle first.

EXAMPLE 3   Solve: $6 - 5y > 7$.

$$-6 + 6 - 5y > -6 + 7 \qquad \text{Adding } -6$$
$$-5y > 1 \qquad \text{Simplifying}$$
$$-\tfrac{1}{5} \cdot (-5y) < -\tfrac{1}{5} \cdot 1 \qquad \text{Multiplying by } -\tfrac{1}{5}$$

            — The symbol has to be reversed!

$$y < -\tfrac{1}{5}$$

The solution set is $\{y \mid y < -\tfrac{1}{5}\}$.

DO EXERCISE 5.

EXAMPLE 4   Solve: $5x + 9 \leqslant 4x + 3$.

$$5x + 9 - 9 \leqslant 4x + 3 - 9 \qquad \text{Adding } -9$$
$$5x \leqslant 4x - 6 \qquad \text{Simplifying}$$
$$5x - 4x \leqslant 4x - 6 - 4x \qquad \text{Adding } -4x$$
$$x \leqslant -6 \qquad \text{Simplifying}$$

The solution set is $\{x \mid x \leqslant -6\}$.

DO EXERCISE 6.

EXAMPLE 5   Solve: $8y - 5 > 17 - 5y$.

$$-17 + 8y - 5 > -17 + 17 - 5y \qquad \text{Adding } -17$$
$$8y - 22 > -5y \qquad \text{Simplifying}$$
$$-8y + 8y - 22 > -8y - 5y \qquad \text{Adding } -8y$$
$$-22 > -13y \qquad \text{Simplifying}$$
$$-\tfrac{1}{13} \cdot (-22) < -\tfrac{1}{13} \cdot (-13y) \qquad \text{Multiplying by } -\tfrac{1}{13}$$

            — The symbol has to be reversed.

$$\tfrac{22}{13} < y$$

The solution set is $\{y \mid \tfrac{22}{13} < y\}$. Since $\tfrac{22}{13} < y$ has the same meaning as $y > \tfrac{22}{13}$, we can also describe the solution set as $\{y \mid y > \tfrac{22}{13}\}$.

DO EXERCISE 7.

EXERCISE SET 7.2

i Solve using the multiplication principle.

1. $5x < 35$ **2.** $8x \geqslant 32$ **3.** $9y \leqslant 81$ **4.** $10x > 240$ **5.** $7x < 13$

6. $8y < 17$ **7.** $12x > -36$ **8.** $16x < -64$ **9.** $5y \geqslant -2$ **10.** $7x > -4$

11. $-2x \leqslant 12$ **12.** $-3y \leqslant 15$ **13.** $-4y \geqslant -16$ **14.** $-7x < -21$ **15.** $-3x < -17$

16. $-5y > -23$ **17.** $-2y > \frac{1}{7}$ **18.** $-4x \leqslant \frac{1}{9}$ **19.** $-\frac{6}{5} \leqslant -4x$ **20.** $-\frac{7}{8} > -56t$

ii Solve using the addition and multiplication principles.

21. $4 + 3x < 28$ **22.** $5 + 4y < 37$ **23.** $6 + 5y \geqslant 36$ **24.** $7 + 8x \geqslant 71$

25. $3x - 5 \leqslant 13$ **26.** $5y - 9 \leqslant 21$ **27.** $10y - 9 > 31$ **28.** $12y - 6 > 42$

29. $13x - 7 < -46$ **30.** $8y - 4 < -52$ **31.** $5x + 3 \geqslant -7$ **32.** $7y + 4 \geqslant -10$

33. $4 - 3y > 13$ **34.** $6 - 8x > 22$ **35.** $3 - 9x < 30$ **36.** $5 - 7y < 40$

37. $3 - 6y > 23$ **38.** $8 - 2y > 14$ **39.** $4x + 2 - 3x \leqslant 9$ **40.** $15x + 3 - 14x \leqslant 7$

41. $8x + 7 - 7x > -3$ **42.** $9x + 8 - 8x > -5$ **43.** $6 - 4y > 4 - 3y$ **44.** $7 - 8y > 5 - 7y$

45. $5 - 9y \leqslant 2 - 8y$ **46.** $6 - 13y \leqslant 4 - 12y$ **47.** $19 - 7y - 3y < 39$ **48.** $18 - 6y - 9y < 63$

49. $21 - 8y < 6y + 49$ **50.** $33 - 12x < 4x + 97$ **51.** $14 - 5y - 2y \geqslant -19$ **52.** $17 - 6y - 7y \leqslant -13$

53. $27 - 11x > 14x - 18$ **54.** $42 - 13y > 15y - 19$

Problem-solving practice

55. The first angle of a triangle is five times as large as the second. The measure of the third angle is $16°$ less than that of the second. How large are the angles?

56. A number subtracted from twice its square is six. Find the number.

Solve.

57. $5(12 - 3t) \geqslant 15(t + 4)$ **58.** $6(z - 5) < 5(7 - 2z)$

59. $4(0.5 - y) + y > 4y - 0.2$ **60.** $3 + 3(0.6 + y) > 2y + 6.6$

61. $\frac{x}{3} - 2 \leqslant 1$ **62.** $\frac{2}{3} - \frac{x}{5} < \frac{4}{15}$ **63.** $\frac{y}{5} + 1 \leqslant \frac{2}{5}$

64. $\frac{3x}{5} \geqslant -15$ **65.** $\frac{-x}{4} - \frac{3x}{8} + 2 > 3 - x$ **66.** $11 - x > 5 + \frac{2x}{5}$

Solve for x.

67. $-(x + 5) \geqslant 4a - 5$ **68.** $\frac{1}{2}(2x + 2b) > \frac{1}{3}(21 + 3b)$ **69.** $-6(x + 3) \leqslant -9(y + 2)$

70. $y < ax + b$ **71.** $x^2 > 0$ **72.** $x^2 + 1 > 0$

Determine whether each of the following statements is true for all rational numbers a, b, and c.

73. If $a + c < b + c$, then $a < b$.   **74.** If $a - c \leqslant b - c$, then $a \leqslant b$.

75. If $ac \leqslant bc$, then $a \leqslant b$.   **76.** If $a^2 < b^2$, then $a < b$.

77. Suppose we are considering *only* integer solutions to $x > 5$. Find an equivalent inequality involving $\geqslant$.

Translate to an inequality.

78. a is at least b **79.** x is at most y

7.3

PROBLEM SOLVING USING INEQUALITIES

i We can use inequalities to solve certain kinds of problems. We continue to use our five-step problem-solving process, but we translate to inequalities rather than equations.

OBJECTIVE

After finishing Section 7.3, you should be able to:

i Solve problems involving the solution of inequalities.

1. The perimeter of a rectangular swimming pool is not to exceed 70 ft. The length is to be twice the width. What widths will meet these conditions?

EXAMPLE 1 A student is taking an introductory algebra course in which four tests are to be given. To get an A, the student must average at least 90 on the four tests. The student got scores of 91, 86, and 89 on the first three tests. Determine (in terms of an inequality) what scores on the last test will allow the student to get an A.

1. **Familiarize** Suppose the student gets a 96 on the last test. The average of the four scores is their sum divided by the number of tests, 4, and is given by

$$\frac{91 + 86 + 89 + 96}{4}.$$

For this average to be *at least* 90 means it must be greater than or equal to 90. Thus we must have

$$\frac{91 + 86 + 89 + 96}{4} \geq 90.$$

Since the average is $\frac{362}{4}$, or 90.5, the test score of 96 will give the student an A. But there are other possible scores. To find all of them, we translate to an inequality and solve.

2. **Translate** We let x represent the student's score on the last test. The average of the four scores is given by

$$\frac{91 + 86 + 89 + x}{4}.$$

Since this average must be *at least* 90, this means that it must be greater than or equal to 90. Thus we can translate the problem to the inequality

$$\frac{91 + 86 + 89 + x}{4} \geq 90.$$

3. **Carry out** We solve the inequality. We first multiply by 4 to clear of fractions.

$$\frac{91 + 86 + 89 + x}{4} \geq 90$$

$$4\left(\frac{91 + 86 + 89 + x}{4}\right) \geq 4 \cdot 90 \qquad \text{Multiplying by 4}$$

$$91 + 86 + 89 + x \geq 360$$

$$266 + x \geq 360 \qquad \text{Collecting like terms}$$

$$x \geq 94$$

4. **Check** Suppose x is a score greater than or equal to 94. Then by successively adding 91, 86, and 89 on both sides of the inequality we get

$$91 + 86 + 89 + x \geq 360,$$

so

$$\frac{91 + 86 + 89 + x}{4} \geq \frac{360}{4}, \quad \text{or } 90.$$

From what we did in the familiarization, we know that 96 is a score that checks. This score is greater than or equal to 94. This also gives us a partial check.

5. **State** Any score that is at least 94 will give the student an A in the course.

DO EXERCISE 1.

EXAMPLE 2 A formula for converting Celsius temperatures C to Fahrenheit temperatures F is

$$F = \tfrac{9}{5}C + 32.$$

Butter stays solid at Fahrenheit temperatures below 88°. Determine (in terms of an inequality) those Celsius temperatures for which butter stays solid.

2. The formula

$$R = -0.028t + 20.8$$

can be used to predict the world record in the 200-m dash t years after 1920. Determine (in terms of an inequality) those years for which the world record will be less than 19.0 sec.

1. Familiarize We can draw a picture of the situation as follows.

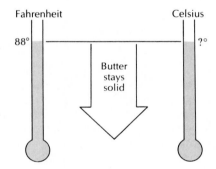

2. Translate The Fahrenheit temperature F is to be less than 88. We have the inequality

$$F < 88.$$

To find Celsius temperatures C that satisfy this condition, we substitute $\tfrac{9}{5}C + 32$ for F:

$$\tfrac{9}{5}C + 32 < 88.$$

Thus we have translated the problem to an inequality.

3. Carry out We solve the inequality:

$$\begin{aligned} \tfrac{9}{5}C + 32 &< 88 \\ \tfrac{9}{5}C &< 56 \qquad \text{Adding} -32 \\ C &< \tfrac{5}{9} \cdot 56 \qquad \text{Multiplying by } \tfrac{5}{9} \\ C &< 31.1. \qquad \text{Rounding to the nearest tenth} \end{aligned}$$

4. Check The check is left to the student.

5. State Butter stays solid at Celsius temperatures below 31.1°.

DO EXERCISE 2.

EXERCISE SET 7.3

1. Your quiz grades are 73, 75, 89, and 91. Determine (in terms of an inequality) those scores you can obtain on the last quiz in order to receive an average quiz grade of 85.

3. The formula

$$R = -0.075t + 3.85$$

can be used to predict the world record in the 1500-meter run *t* years after 1930. Determine (in terms of an inequality) those years for which the world record will be less than 3.5 minutes.

5. Find all numbers such that three times the number minus ten times the number is greater than or equal to eight times the number.

7. Atlas rents an intermediate-size car at a daily rate of $44.95 plus $0.39 per mile. A businessperson is not to exceed a daily car rental budget of $250. Determine (in terms of an inequality) those mileages that will allow the businessperson to stay within budget. Round to the nearest tenth of a mile.

9. The width of a rectangle is 16 yd. Determine (in terms of an inequality) those lengths for which the area will be greater than or equal to 264 yd^2.

11. The width of a rectangle is 8 ft. What lengths will make the perimeter at least 200 ft? at most 200 ft?

13. A salesperson made 18 customer calls last week and 22 calls this week. How many calls must be made next week to maintain an average of at least 20 for the three-week period?

15. A student is shopping for a new pair of jeans and a sweater. The student is determined to spend no more than $40.00 for the outfit. The student buys jeans for $21.95. What is the most the student can spend for the sweater?

17. The height of a triangle is 20 cm. What lengths of the base will make the area at most 40 cm^2?

18. The area of a square can be no more than 64 cm^2. What lengths of a side will allow this?

2. A human body is considered to be fevered when its temperature is higher than 98.6°F. Using the formula in Example 2, determine (in terms of an inequality) those Celsius temperatures for which the body is fevered.

4. Find all numbers such that the sum of the number and 15 is less than four times the number.

6. Acme rents station wagons at a daily rate of $42.95 plus $0.46 per mile. A family wants to rent a wagon one day while on vacation, but must stay within a budget of $200. Determine (in terms of an inequality) those mileages that will allow the family to stay within budget. Round to the nearest tenth of a mile.

8. The length of a rectangle is 4 cm. Determine (in terms of an inequality) those widths for which the area will be less than 86 cm^2.

10. The length of a rectangle is 26 cm. What widths will make the perimeter greater than 80 cm?

12. One side of a triangle is 2 cm shorter than the base. The other side is 3 cm longer than the base. What lengths of the base will allow the perimeter to be greater than 19 cm?

14. George and Joan do volunteer work at a hospital. Joan worked 3 more hours than George and together they worked more than 27 hours. What possible hours did each work?

16. The width of a rectangle is 32 km. What lengths will make the area at least 2048 km^2?

19. The sum of two consecutive odd integers is less than 100. What is the largest possible pair of such integers?

20. The sum of three consecutive odd integers is less than 30. What are the three largest of such integers?

21. A salesperson can choose to be paid in one of two ways

Plan A.   A salary of \$600 per month, plus a commission of 4% of gross sales.

Plan B: A salary of \$800 per month, plus a commission of 6% of gross sales over \$10,000.

For what gross sales is plan A better than plan B, assuming that gross sales are always more than \$10,000?

22. A mason can be paid in one of two ways.

Plan A: \$3000 plus \$3.00 per hour.

Plan B: \$8.50 per hour.

If a job takes n hours, for what values of n is plan B better for the mason?

7.4

GRAPHS OF INEQUALITIES

A *graph* of an inequality is a drawing of its solutions. An inequality in one variable can be graphed on a number line. An inequality in two variables can be graphed on a coordinate plane.

i   Inequalities in One Variable

We graph inequalities in one variable on a number line.

EXAMPLE 1   Graph: $x < 2$.

The solutions of $x < 2$ are those numbers less than 2. They are shown on the graph by shading all points to the left of 2. The open circle at 2 indicates that 2 is not part of the graph.

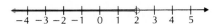

EXAMPLE 2   Graph: $y \geqslant -3$.

The solutions of $y \geqslant -3$ are shown by shading the point for -3 and all points to the right of -3. The closed circle at -3 *is* part of the graph.

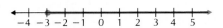

DO EXERCISES 1 AND 2.

EXAMPLE 3   Graph: $3x + 2 < 5x - 1$.

We first solve:

$$3x + 2 < 5x - 1$$
$$2 < 2x - 1 \qquad \text{Adding } -3x$$
$$3 < 2x \qquad \text{Adding 1}$$
$$\tfrac{3}{2} < x. \qquad \text{Multiplying by } \tfrac{1}{2}$$

OBJECTIVES

After finishing Section 7.4, you should be able to:

i   Graph inequalities in one variable on a number line.

ii   Graph inequalities that contain absolute value on a number line.

iii   Graph linear inequalities in two variables on a plane.

Graph on a number line.

1. $x < 4$

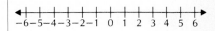

2. $y \geqslant -5$

Graph on a number line.

3. $x + 2 > 1$

We shade all points to the right of $\frac{3}{2}$.

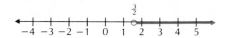

DO EXERCISES 3 AND 4.

ii Graphing Inequalities with Absolute Value

EXAMPLE 4 Graph: $|x| < 3$.

The absolute value of a number is its distance from 0 on a number line. For the absolute value of a number to be less than 3 it must be between 3 and -3. Therefore we use open circles at 3 and -3 and shade the points between these two numbers.

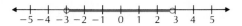

4. $4x + 6 < 7x - 3$

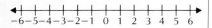

DO EXERCISES 5 AND 6.

EXAMPLE 5 Graph: $|x| \geqslant 2$.

For the absolute value of a number to be greater than or equal to 2 its distance from 0 must be 2 or more. Thus the number must be 2 or greater, or it must be less than or equal to -2. Therefore we shade the point for 2 and all points to its right. We also shade the point for -2 and all points to its left.

Graph on a number line.

5. $|x| < 5$

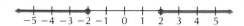

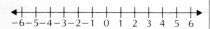

DO EXERCISES 7 AND 8.

6. $|x| \leqslant 4$

iii Inequalities in Two Variables

The solutions of inequalities in two variables are ordered pairs.

EXAMPLE 6 Determine whether $(-3, 2)$ is a solution of $5x - 4y < 13$.

We use alphabetical order of variables. We replace x by -3 and y by 2.

$$
\begin{array}{c|c}
\multicolumn{2}{c}{5x - 4y < 13} \\
\hline
5(-3) - 4 \cdot 2 & 13 \\
-15 - 8 & \\
-23 &
\end{array}
$$

Since $-23 < 13$ is true, $(-3, 2)$ is a solution.

Graph on a number line.

7. $|x| \geqslant 3$

DO EXERCISE 9.

8. $|y| > 5$

EXAMPLE 7 Graph: $y > x$.

We first graph the line $y = x$ for comparison. Every solution of $y = x$ is an ordered pair such as $(3, 3)$. The first and second coordinates are the same. The graph of $y = x$ is shown to the left on the following page.

9. Determine whether $(4, 3)$ is a solution of $3x - 2y < 1$.

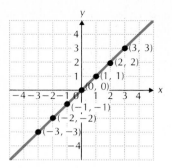

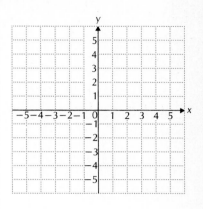

10. Graph $y < x$.

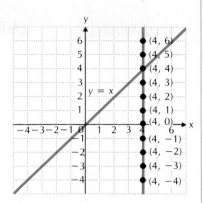

Now look at the graph to the right above. We consider a vertical line and ordered pairs on it. For all points above $y = x$, the second coordinate is greater than the first, $y > x$. For all points below the line, $y < x$. The same thing happens for any vertical line. Then for all points above $y = x$, the ordered pairs are solutions. We shade the half-plane above $y = x$. This is the graph of $y > x$. Points on $y = x$ are not in the graph, so we draw it dashed.

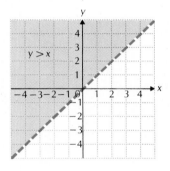

DO EXERCISE 10.

EXAMPLE 8 Graph: $y \leqslant x - 1$.

First we sketch the line $y = x - 1$. Points on the line $y = x - 1$ are also in the graph of $y \leqslant x - 1$, so we draw the line solid. For points above the line, $y > x - 1$. These points are not in the graph. For points below the line, $y < x - 1$. These are in the graph, so we shade the lower half-plane.

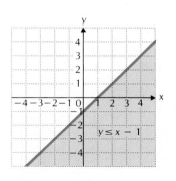

11. Graph $y \geqslant x + 2$.

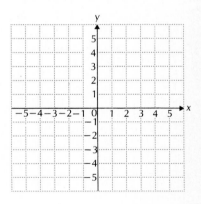

DO EXERCISE 11.

12. Graph $2x + 4y < 8$.

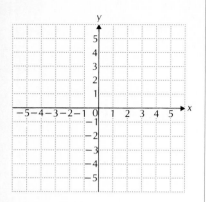

A *linear inequality* is one that we can get from a linear equation by changing the equals symbol to an inequality symbol. Every linear equation has a graph that is a straight line. The graph of a linear inequality is a half-plane, sometimes including the line along the edge. In the following example we show a different method of graphing. We graph the line using intercepts. Then we determine which side to shade by substituting a point from either half-plane.

EXAMPLE 9 Graph: $6x - 2y < 10$.

We first graph the line $y = 3x - 5$. The intercepts are $(0, -5)$ and $(\frac{5}{3}, 0)$. The point $(3, 4)$ is also on the graph. This line forms the boundary of the solutions of the inequality. In this case points on the line are not solutions of the inequality. We shade the half-plane above the line. We know to do this by picking a test point on the line. We try $(3, -2)$ and substitute:

$$6(3) - 2(-2) < 10, \quad \text{or} \quad 22 < 10.$$

Since this inequality is *false*, the point $(3, -2)$ is *not* a solution; no point in the half-plane containing $(3, -2)$ is a solution. Thus the points in the opposite half-plane are solutions.

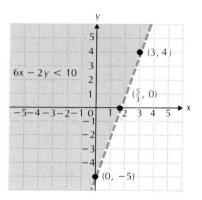

13. Graph $3x - 5y < 15$.

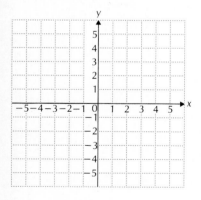

DO EXERCISE 12.

EXAMPLE 10 Graph: $2x + 3y < 6$.

a) First we graph the line $2x + 3y = 6$. The intercepts are $(3, 0)$ and $(0, 2)$. We use a dashed line for the graph since we have $<$.

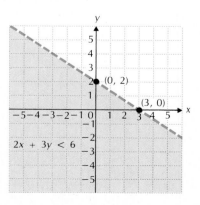

b) Let's pick a point that does not belong to the line. Substitute to determine whether this point is a solution. The origin $(0, 0)$ is usually an easy one to use: $2 \cdot 0 + 3 \cdot 0 < 6$ is true, so the origin is a solution. This means we shade the lower half-plane. Had the substitution given us a false inequality, we would have shaded the other half-plane.

If the line goes through the origin, we must test some other point not on the line. The point $(1, 1)$ is often a good one to try.

DO EXERCISES 13 AND 14. (EXERCISE 13 IS ON THE PRECEDING PAGE.)

14. Graph $2x + 3y \geqslant 12$.

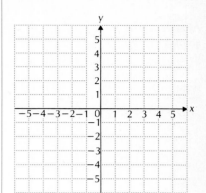

EXERCISE SET 7.4

i Graph on a number line.

1. $x < 5$	**2.** $y < 0$	**3.** $t < -3$	**4.** $y > 5$
5. $x \geqslant 6$	**6.** $y \geqslant -4$	**7.** $x + 2 > 7$	**8.** $x + 3 > 9$
9. $z - 3 < 4$	**10.** $x - 2 < 6$	**11.** $t - 3 \leqslant -7$	**12.** $x - 4 \leqslant -8$
13. $x - 8 \geqslant 0$	**14.** $2x + 6 < 14$	**15.** $3y + 5 < 26$	**16.** $4x - 8 \geqslant 12$
17. $5y - 4 > 11$	**18.** $3x + 7 < 8x - 3$	**19.** $4y + 9 > 11y - 12$	**20.** $5t + 8 \geqslant 12t - 27$

21. $6x + 11 \leqslant 14x + 7$

ii Graph on a number line.

22. $|x| < 2$	**23.** $|y| < 6$	**24.** $|t| \leqslant 1$	**25.** $|x| \leqslant 7$
26. $|y| > 3$	**27.** $|t| > 4$	**28.** $|x| \geqslant 7$	**29.** $|y| \geqslant 9$

30. Determine whether $(-3, -5)$ is a solution of
$$-x - 3y < 18.$$

31. Determine whether $(5, -3)$ is a solution of
$$-2x + 4y \leqslant -2.$$

32. Determine whether $(\frac{1}{2}, -\frac{1}{4})$ is a solution of
$$7y - 9x > -3.$$

33. Determine whether $(-6, 5)$ is a solution of
$$x + 0 \cdot y < 3.$$

iii Graph on a plane.

34. $x > 2y$	**35.** $x > 3y$	**36.** $y \leqslant x - 3$	**37.** $y \leqslant x - 5$	**38.** $y < x + 1$
39. $y < x + 4$	**40.** $y \geqslant x - 2$	**41.** $y \geqslant x - 1$	**42.** $y \leqslant 2x - 1$	**43.** $y \leqslant 3x + 2$
44. $x + y \leqslant 3$	**45.** $x + y \leqslant 4$	**46.** $x - y > 7$	**47.** $x - y > -2$	**48.** $x - 3y < 6$
49. $x - y < -10$	**50.** $2x + 3y \leqslant 12$	**51.** $5x + 4y \geqslant 20$	**52.** $y \geqslant 1 - 2x$	**53.** $y - 2x \leqslant -1$
54. $y + 4x > 0$	**55.** $y - x < 0$	**56.** $y > -3x$	**57.** $y < -5x$	

○ ────────────────────────────────────

Graph on a plane.

58. $y > 2$ **59.** $x \geqslant 3$ **60.** $x > 0$ **61.** $y \leqslant 0$
(*Hint:* Think of this as $0 \cdot x + y > 2$.)

Graph the following pair of inequalities using the same set of axes. Then determine whether each point satisfies *both* inequalities:
$$x + y \leqslant 1, \qquad x - y < 1.$$

62. $(0, 0)$	**63.** $(1, 1)$	**64.** $(1, 0)$	**65.** $(0, 1)$
66. $(-1, -1)$	**67.** $(1, -3)$	**68.** $(5, 2)$	**69.** $(-4, 2)$

OBJECTIVES

After finishing Section 7.5, you should be able to:

i Graph a system of inequalities in two variables.

ii Solve linear programming problems.

1. Find the intersection of the solution sets of each equation:

$$2x - y = 1$$
$$x + y = 8.$$

That is, solve the system.

7.5

LINEAR PROGRAMMING

i Systems of Inequalities

The common part of two sets is called their *intersection*. When we solve a system of equations such as

$$y - x = 1$$
$$x + y = 3,$$

we are finding the intersection of the solution sets of each equation. The solution set of $y - x = 1$ is all points on the line in the graph shown here. The solution set of $x + y = 3$ is also a line, as shown on the graph. The point $(1, 2)$ is the part common to both lines and is the intersection of the solution sets of each line.

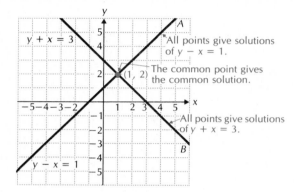

DO EXERCISE 1.

A *system of inequalities* generally has a very large solution set. It is the intersection of the solution sets of each inequality. We can find the solution set of a system of inequalities in much the same way that we find the solution of a system of equations graphically. We graph each inequality separately and look for the intersection.

EXAMPLE 1 Graph the system of inequalities:

$$6x - 2y \leqslant 12,$$
$$y - 3 \leqslant 0,$$
$$x + y \geqslant 0.$$

We graph the individual inequalities. The arrows indicate the half-planes in the solution sets. To graph $y - 3 \leqslant 0$, consider $y \leqslant 3$. The graph consists of all ordered pairs in which the second coordinate is less than or equal to 3 and is the line $y = 3$ and all points below.

a)
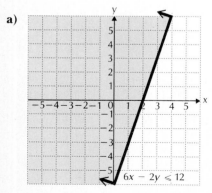

b)
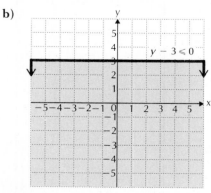

c)
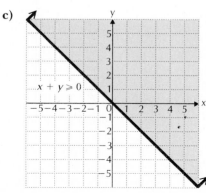

We now find the intersection of the three graphs.

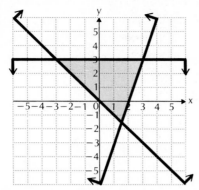

We now have a picture, or graph, of the solution set of the system of inequalities. In making such graphs you should generally use the same set of axes.

DO EXERCISES 2 AND 3.

ii Linear Programming

There are many situations in real life in which we want to find a greatest value (a maximum) or a least value (a minimum). For example, if you are in business you would like to know how to make the *most* profit. Or, you might like to know how to make your expenses the *least* possible. Some such problems can be solved using systems of inequalities.

We will be considering *linear expressions* of the type

$$F = ax + by + c,$$

where values of F are found by substituting the numbers x and y from an ordered pair (x, y). These ordered pairs will come from a region and its interior like the following:

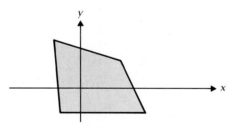

On such a region, F has both a maximum value and a minimum value and those values occur at the vertices of the polygon. The branch of mathematics that uses this approach is called *linear programming*. It was developed during World War II in connection with the shipments of supplies to Europe.

To solve problems using linear programming, we often take advantage of the results that mathematicians have proved. That is the case here. The five-step procedure is still appropriate, but we now have some new methods to use in translating and in carrying out a mathematical procedure.

EXAMPLE 2 You are taking a test in which items of type A are worth 10 points and items of type B are worth 15 points. It takes you 3 min to answer each item of type A and 6 min to answer each item of type B. The total time allowed is 60 min and you are not allowed to answer more than 16 questions. Assuming that all of your answers are correct, how many items of each type should you answer in order to get the best score?

Graph each system of inequalities.

2. $x + y \geqslant 1$,
   $y - x \geqslant 2$

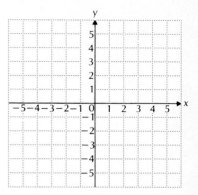

3. $5x + 6y \leqslant 30$,
   $y \geqslant 0$,
   $y \leqslant 3$,
   $x \geqslant 0$,
   $x \leqslant 4$

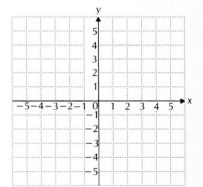

1. **Familiarize**   Tabulating information will help us to see the picture.

Type	Number of points for each	Time required for each	Number answered
A	10	3 min	x
B	15	6 min	y
Total time: 60 minutes			
Total number of items: 16 or fewer			

Note that we have used x to represent the number of items of type A and y to represent the number of items of type B.

2. **Translate**   In this case it will help to extend the table.

Type	Number of points for each	Time required for each	Number answered	Total time for type	Total points for type
A	10	3 min	x	$3x$	$10x$
B	15	6 min	y	$6y$	$15y$
Total			$x + y \leqslant 16$	$3x + 6y \leqslant 60$	$10x + 15y$

                                              ↑   ↑   ↑

                   Because no   Because the   This is the
                   more than   time cannot   total score
                   16 items can be more than on the test
                   be answered 60 min

Suppose that the total score on the test is T. We have a formula for T expressed in terms of x and y as

$$T = 10x + 15y.$$

We know the following facts, called *constraints*, about the variables x and y:

$$x + y \leqslant 16,$$
$$3x + 6y \leqslant 60,$$
$$\left. \begin{array}{c} x \geqslant 0, \\ y \geqslant 0. \end{array} \right\} \quad \text{Because the number of items answered} \\ \text{cannot be negative}$$

This system of inequalities describes a region over which we will consider values of T.

3. **Carry out**   The mathematical manipulation we do here consists of graphing the system of inequalities and then computing values of T at the corners, or *vertices*. The graph of the system of inequalities is as shown here. The arrows indicate the half-planes that are the solution sets of each inquality.

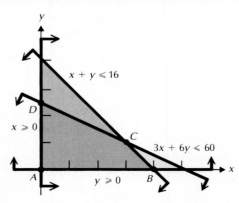

We need to find the coordinates of each vertex. Note that each vertex is the intersection of two lines. Thus we can find each one by solving a system of two linear equations. The coordinates of point A are obviously $(0, 0)$. Let us find the coordinates of point C. We solve the system

$$3x + 6y = 60,$$
$$x + y = 16$$

as follows:

$$
\begin{array}{rl}
3x + 6y = & 60 \\
-3x - 3y = & -48 \qquad \text{Multiplying by } -3 \\
\hline
3y = & 12 \qquad \text{Adding} \\
y = & 4
\end{array}
$$

Then we find that $x = 12$. Thus the coordinates of vertex C are $(12, 4)$.

Continuing to find the coordinates of the vertices and computing the test score for each ordered pair, we obtain the following:

Vertices (x, y)	Score $T = 10x + 15y$
$A\ (0, 0)$	0
$B\ (16, 0)$	160
$C\ (12, 4)$	180
$D\ (0, 10)$	150

The greatest score in the table is 180, obtained when 12 items of type A are answered and 4 items of type B are answered.

4. **Check** ▷ We can go back to the original conditions of the problem and calculate the scores, using the ordered pairs in the table above. We can also check our work, the algebra and the arithmetic. In this case there is no further checking that we can do without an undue amount of work.

5. **State** ▷ The answer is that you should answer 12 items of type A and 4 items of type B.

DO EXERCISE 4.

4. Find the maximum and minimum values of

$$H = 2x - 4y + 7,$$

subject to

$$5x + 6y \leqslant 30,$$
$$y \geqslant 0,$$
$$y \leqslant 3,$$
$$x \geqslant 0,$$
$$x \leqslant 4.$$

EXERCISE SET 7.5

i Graph these systems of inequalities.

1. $x + y \leqslant 6,$
  $y \leqslant 5,$
  $x \geqslant 0,$
  $y \geqslant 0$

2. $x + 2y \leqslant 8,$
  $x \leqslant 4,$
  $x \geqslant 0,$
  $y \geqslant 0$

3. $3x + 2y \leqslant 12,$
  $x + y \leqslant 5,$
  $x \geqslant 0,$
  $y \geqslant 0$

4. $y \leqslant x,$
  $y \geqslant 3 - x$

5. $y \geqslant x,$
  $y \leqslant 1 - x$

6. $y \geqslant -2,$
  $x \leqslant 1$

7. $x \geqslant -3,$
  $y \leqslant 4$

8. $x + y \leqslant 1,$
  $x - y \leqslant 2$

ii Find the maximum and minimum values of the expression and the values of x and y where they occur.

9. $F = x + 2y$, subject to
  $x + y \leqslant 6,$
  $y \leqslant 5,$
  $x \geqslant 0,$
  $y \geqslant 0$

10. $T = x - y$, subject to
  $x + 2y \leqslant 8,$
  $x \leqslant 4,$
  $x \geqslant 0,$
  $y \geqslant 0$

11. $G = 5x + 4y$, subject to
$$3x + 2y \leqslant 12,$$
$$x + y \leqslant 5,$$
$$x \geqslant 0,$$
$$y \geqslant 0$$

12. $P = 2x + y$, subject to
$$3x + 5y \leqslant 15,$$
$$3x + 2y \leqslant 12,$$
$$x \geqslant 0,$$
$$y \geqslant 0$$

Solve.

13. *Allocation of resources in a manufacturing process.* A clothier makes suits and dresses. Each suit requires 1 yd of polyester and 4 yd of wool. Each dress requires 2 yd of polyester and 3 yd of wool. The clothier has in stock 60 yd of polyester and 120 yd of wool. A suit sells for $120 and a dress sells for $75. How many of each should the clothier make in order to maximize income?

14. *Allocation of resources in a manufacturing process.* A manufacturer of hi-fi speakers makes two speaker assemblies. The inexpensive speaker assembly consists of one midrange speaker and one tweeter. It sells for $25. The expensive speaker assembly consists of one woofer, one midrange speaker, and two tweeters. It sells for $150. The manufacturer has in stock 22 12-in. woofers, 30 5-in. midrange speakers, and 45 $1\frac{1}{2}$-in. tweeters. How many of each type of speaker assembly should be made in order to maximize income?

15. A company manufactures motorcycles and bicycles. To stay in business it must produce at least 10 motorcycles each month, but it does not have facilities to produce more than 60 motorcycles. It also does not have facilities to produce more than 120 bicycles. The total production of motorcycles and bicycles cannot exceed 160. The profit on a motorcycle is $134 and on a bicycle is $20. Find the number of each that should be manufactured in order to maximize profit.

16. A college snack bar cooks and sells hamburgers and hot dogs during the lunch hour. To stay in business it must sell at least 10 hamburgers but cannot cook more than 40. It must also sell at least 30 hot dogs but cannot cook more than 70. It cannot cook more than 90 sandwiches altogether. The profit on a hamburger is $.33 and on a hot dog it is $.21. How many of each kind of sandwich should they sell in order to make the maximum profit? What is the maximum profit?

17. *Ecology: Maximizing animal support in a forest.* A certain area of forest is populated by two species of animals, A1 and A2. The forest supplies three kinds of food, F1, F2, and F3. For one year, species A1 requires 1 unit of food F1, 2 units of food F2, and 2 units of food F3, whereas species A2 requires 1.2 units of food F1, 1.8 units of food F2, and 0.6 unit of food F3. The forest can normally supply at most 600 units of food F1, 960 units of food F2, and 720 units of food F3 per year. What are the maximum numbers of these animals that the forest can support?

18. *Time utilization.* You are about to take a test that contains questions of type A worth 4 points and questions of type B worth 7 points. You must do at least 5 questions of type A, but time restricts doing more than 10. You must do at least 3 questions of type B but time restricts doing more than 10. In total, you can do no more than 18 questions. How many of each type of question must you do in order to maximize your score? What is this maximum score?

19. *Time utilization.* You are about to take a test that contains questions of type A worth 10 points and questions of type B worth 25 points. You must do at least 3 questions of type A, but time restricts doing more than 12. You must do at least 4 questions of type B, but time restricts doing more than 15. In total, you can do no more than 20 questions. How many of each type of question must you do in order to maximize your score? What is this maximum score?

SUMMARY AND REVIEW: CHAPTER 7

The following contains a summary of what you should be able to do after completing this chapter. The review exercises are for practice. Answers are at the back of the book. If you miss an exercise, restudy the section indicated alongside the answer.

You should be able to:

Determine whether a number is a solution of an inequality.

Determine whether the given number is a solution of the inequality $x \leqslant 4$.

1. -3 **2.** 7 **3.** 4

Solve inequalities using the addition principle, the multiplication principle, and both principles together.

Solve. Write set notation for the answers.

4. $y + \frac{2}{3} \geqslant \frac{1}{6}$ **5.** $9x \geqslant 63$ **6.** $2 + 6y > 14$ **7.** $7 - 3y \geqslant 27 + 2y$

8. $3x + 5 < 2x - 6$ **9.** $-4y < 28$ **10.** $3 - 4x < 27$ **11.** $4 - 8x < 13 + 3x$

12. $-3y \geqslant -21$ **13.** $-4x \leqslant \frac{1}{3}$

Solve problems involving inequalities using the five-step problem-solving process.

14. Your quiz grades are 71, 75, 82, and 86. What is the lowest grade you can get on the last quiz and still have an average of at least 80?

15. The length of a rectangle is 43 cm. What widths will make the perimeter greater than 120 cm?

Determine whether a point is a solution of an inequality.

Determine whether the given point is a solution of the inequality $x - 2y > 1$.

16. $(0, 0)$

17. $(1, 3)$

18. $(4, -1)$

Graph inequalities on a number line and a plane.

Graph on a number line.

19. $y \leqslant 9$

20. $6x - 3 < x + 2$

21. $|x| \leqslant 2$

Graph on a plane.

22. $x < y$

23. $x + 2y \geqslant 4$

Graph systems of inequalities.

Graph.

24. $y \geqslant -3,$
$\quad x \geqslant 2$

25. $x + 3y \geqslant -3,$
$\quad x + 3y \leqslant 6$

26. $x - 3y \leqslant 3,$
$\quad x + 3y \geqslant 9,$
$\quad y \leqslant 6$

Solve linear programming problems.

Solve.

27. Find the maximum and minimum values of

$$P = 2x - 3y + 24, \text{ subject to}$$
$$x + y \leqslant 7,$$
$$x + y \geqslant 3,$$
$$x \geqslant 0,$$
$$y \geqslant 0,$$

Solve.

28. $3[2 - 4(y - 3)] < 6(y - 1)$

29. $-\dfrac{x}{5} - \dfrac{4x}{15} + 6 \leqslant 10 - x$

30. The sum of three consecutive odd integers is less than 198. What are the largest three such integers?

TEST: CHAPTER 7

Solve. Write set notation for the answers.

1. $x + 6 \leqslant 2$

2. $14x + 9 > 13x - 4$

3. $12x \leqslant 60$

4. $-2y \geqslant 26$

5. $-4y \leqslant -32$

6. $-5x \geqslant \frac{1}{4}$

7. $4 - 6x > 40$

8. $5 - 9x \geqslant 19 + 5x$

Problem solving

9. The sum of three consecutive integers is greater than 29. What are the least possible values of these integers?

10. Find all numbers such that six times the number is greater than the number plus 30.

Graph on a number line.

11. $4x - 6 < x + 3$

12. $|x| \geqslant 5$

Graph on a plane.

13. $y > x - 1$

14. $2x - y \leqslant 4$

Determine whether the given point is a solution of the inequality $3y - 2x < -2$.

15. $(0, 0)$

16. $(-4, -10)$

17. Graph

$$y - x \geqslant 2,$$
$$y - x \leqslant 5,$$
$$x \leqslant 2,$$
$$x \geqslant -3$$

18. Find the maximum and minimum values of

$$T = 3x + 10y, \text{ subject to}$$
$$y - x \geqslant 2,$$
$$y - x \leqslant 5,$$
$$x \leqslant 8,$$
$$x \geqslant 3.$$

---

19. Give all integer solutions for $|2x| \leqslant 5$.

20. Solve: $x^2 + 7 > 0$.

21. Graph on a plane: $x < -4$.

MAN AND THE MANLIKE APES

How can the length of the human bone be used to find the height of the entire body? A linear function can give the answer.

8

FUNCTIONS AND VARIATION

A function is a special kind of correspondence between sets. The notion of function is basic to most of mathematics. This chapter contains a brief introduction to functions. Our goal is to give you an idea of what a function is and to extend your graphing skills. If time is short, Sections 8.1 and 8.2 can be omitted.

The last two sections of this chapter cover two classes of functions—those involving direct and inverse variation. These have applications in problem solving, as we shall see.

OBJECTIVES

After finishing Section 8.1, you should be able to:

i Determine whether a given correspondence is a function.

ii Find function outputs by substituting into formulas, and find function values using function notation $f(x)$ and a formula.

8.1

FUNCTIONS

i **Identifying Functions**

Consider a set of years. To each year there corresponds the cost of a first-class postage stamp that year.

Year	Cost of first-class stamp
1948	3¢
1958	4¢
1964	5¢
1974	10¢
1978	15¢
1983	20¢
1984	20¢
1985	22¢

To each year there corresponds *exactly* one cost. Such a correspondence is called a *function*.

> A *function* is a correspondence (or rule) that assigns to each member of some set (called the *domain*) exactly one member of a set (called the *range*).

The members of the domain are also called *inputs* and the members of the range are called *outputs*. Arrows can be used to describe functions.

Domain (set of inputs)	Range (set of outputs)
1948 ——————→	3
1958 ——————→	4
1964 ——————→	5
1974 ——————→	10
1978 ——————→	15
1983 ——————→	20
1984 —————↗	
1985 ——————→	22

Note that each input has exactly one output, even though in the case of 1983 and 1984 those outputs are the same.

EXAMPLE 1 Determine whether or not each of the following correspondences is a function.

f:
Domain   Range
a ——————→ 4
b ——————→ 0
c ——————↗

g:
Domain   Range
3 ——————→ 5
4 ——————→ 9
5 ——————→ −7
6 ——————↗

Domain   Range   Domain   Range
4 ———→ 0   Cheese pizza ——→ $9.75
h: 6   p: Tomato pizza ✕ $7.25
2   Meat pizza $8.50

The correspondence f is a function because each member of the domain is matched to only one member of the range. The correspondence g is not a function because the member 4 of the domain is matched to more than one member of the range. The correspondence h is a function. The correspondence p is not a function because the member cheese pizza of the domain is matched to two members of the range. In a function, a member of the domain can be matched with only one member of the range.

For functions we refer to a member of the domain as an *input* and the corresponding member of the range as its *output*.

DO EXERCISES 1–3.

ii Formulas for Functions

Functions are also described by formulas. Such formulas are "recipes" for finding outputs.

EXAMPLE 2   You see a flash of lightning. After a few seconds you hear the thunder associated with that flash. How far away was the lightning?

Your distance from the storm, M (in miles), is a function of n, the number of seconds it takes the sound of the thunder to reach you. We can approximate M by the formula

$$M = \tfrac{1}{5} n.$$

Complete the following table for this function.

n (sec)	0	1	2	3	4	5	6	10
M (mi)	0	$\tfrac{1}{5}$						

To complete the table we successively substitute values of n, and compute M:

For $n = 2$: $M = \tfrac{1}{5}(2) = \tfrac{2}{5}$;   For $n = 3$: $M = \tfrac{1}{5}(3) = \tfrac{3}{5}$;

For $n = 4$: $M = \tfrac{1}{5}(4) = \tfrac{4}{5}$;   For $n = 5$: $M = \tfrac{1}{5}(5) = 1$;

For $n = 6$: $M = \tfrac{1}{5}(6) = \tfrac{6}{5}$, or $1\tfrac{1}{5}$;   For $n = 10$: $M = \tfrac{1}{5}(10) = 2.$

DO EXERCISE 4.

Determine whether the correspondence is a function.

1.

Year (domain)	Alcohol-related fatalities (range)
1980 ———→	51,091
1981 ———→	49,301
1982 ———→	43,145
1983 ———→	42,589
1984 ———→	45,800

2. Domain   Range
1 ———→ 4
1 ———→ 1
−1
2 ———→ 3

3. Domain   Range
14 ⟶ 545
     ↘ 636
15 ⟶ 782
     ↘ 643
16 ———→ 700

4. For the function of Example 2, complete the following table.

n (sec)	15	7	8	10
M (mi)				

Function Notation

Finding outputs can be thought of in terms of a "function machine." Inputs are entered into the machine. The machine then gives the outputs.

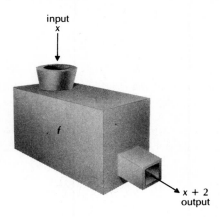

The symbol $f(x)$, read "f of x," denotes the number assigned to x by the correspondence. The preceding function f assigns to each input x the output $x + 2$. We describe this with the formula $f(x) = x + 2$, and then we can calculate as follows:

For the input 8, the output is 10:

$$f(8) = 8 + 2, \quad \text{or } 10.$$

For the input -3, the output is -1:

$$f(-3) = -3 + 2, \quad \text{or } -1.$$

For the input 0, the output is 2:

$$f(0) = 0 + 2, \quad \text{or } 2.$$

For the input 5, the output is 7:

$$f(5) = 5 + 2, \quad \text{or } 7.$$

EXAMPLE 3   A function H assigns to each input x the output 5. Find (a) $H(x)$, (b) $H(0)$, and (c) $H(2)$.

a) $H(x) = 5$

b) $H(0) = 5$

c) $H(2) = 5$

A function such as H is known as a *constant* function.

EXAMPLE 4   For the function $f(t) = 2t^2 + 5$, find (a) $f(-2)$, (b) $f(0)$, and (c) $f(3)$.

a) $f(-2) = 2(-2)^2 + 5$
$$= 2 \cdot 4 + 5$$
$$= 13$$

b) $f(0) = 2(0)^2 + 5$
$$= 5$$

c) $f(3) = 2(3)^2 + 5$
$\qquad = 2 \cdot 9 + 5$
$\qquad = 23$

DO EXERCISES 5–8.

Outputs are also called *function values*. In Example 4, $f(-2) = 13$. We could say that the "function value at -2 is 13," or "when x is -2, the value of the function is 13."

EXAMPLE 5 (*An application: Predicting heights in anthropology*). An anthropologist can use certain functions to estimate the height of a male or female, given the lengths of certain bones. A *humerus* is the bone from the elbow to the shoulder. Let $x =$ the length of the humerus in cm. Then the height, in cm, of a male with a humerus of length x is given by

$$M(x) = 2.89x + 70.64.$$

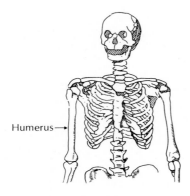

Humerus→

The height, in cm, of a female with a humerus of length x is given by

$$F(x) = 2.75x + 71.48.$$

A 45-cm humerus was uncovered in a ruin.

a) If it was from a male, how tall was he?

b) If it was from a female, how tall was she?

a) We find $M(45)$:

$$M(45) = 2.89(45) + 70.64$$
$$= 130.05 + 70.64$$
$$= 200.69.$$

The male's height was about 200.69 cm.

b) We find $F(45)$:

$$F(45) = 2.75(45) + 71.48$$
$$= 123.75 + 71.48$$
$$= 195.23.$$

The female's height was about 195.23 cm.

DO EXERCISE 9.

5. For the function machine shown below, find $f(5)$, $f(-8)$, and $f(-2)$.

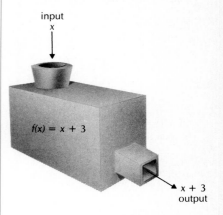

input
x

$f(x) = x + 3$

x + 3
output

6. For the function
$$G(x) = 3x - x^2,$$
find $G(0)$, $G(-2)$, and $G(1)$.

7. For the function
$$f(y) = 8y^2 + 3,$$
find $f(-1)$, $f(2)$, and $f(\frac{1}{2})$.

8. For the function
$$p(x) = 2x^2 + x - 1,$$
find $p(0)$, $p(-2)$, and $p(3)$.

9. Consider the functions given in Example 5. A 48-cm humerus was uncovered in a ruin.

a) If it was a male, how tall was he?

b) If it was a female, how tall was she?

EXERCISE SET 8.1

Determine whether each of the following correspondences is a function.

1. Domain   Range

2 ⟶ 9
5 ⟶ 8
19

2. Domain   Range

5 ⟶ 3
−3 ⟶ 7
7
−7

3. Domain   Range

−5 ⟶ 1
5
8

4. Domain   Range

6 ⟶ −6
7 ⟶ −7
3 ⟶ −3

5.   Domain   Range

Los Angeles ⟶ Mets
New York ⟶ Lakers
⟶ Dodgers
⟶ Yankees

6.   Domain   Range

(3, 4) ⟶ 12
(8, 10) ⟶ −11
(4, −2) ⟶ 18
(−3, −8) ⟶ 2

Find the indicated outputs.

7.

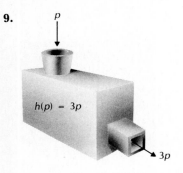

$f(x) = x + 5$

Find $f(3)$, $f(7)$, and $f(-9)$.

8.

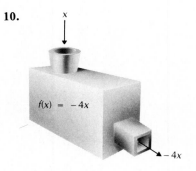

$g(t) = t - 6$

Find $g(0)$, $g(6)$, and $g(18)$.

9.

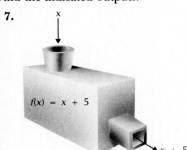

$h(p) = 3p$

Find $h(-2)$, $h(5)$, and $h(24)$.

10.

$f(x) = -4x$

Find $f(6)$, $f(-\frac{1}{2})$, and $f(20)$.

Find the indicated outputs.

11. $g(s) = 2s + 4$; find $g(1)$, $g(-7)$, and $g(6)$.

12. $h(x) = 19$; find $h(4)$, $h(-6)$, and $h(12)$.

13. $F(x) = 2x^2 - 3x + 2$; find $F(0)$, $F(-1)$, and $F(2)$.

14. $P(x) = 3x^2 - 2x + 5$; find $P(0)$, $P(-2)$, and $P(3)$.

15. $h(x) = |x|$; find $h(-4)$, $h(5)$, and $h(-3)$.

16. $f(t) = |t| + 1$; find $f(-5)$, $f(0)$, and $f(-9)$.

17. $f(x) = |x| - 2$; find $f(-3)$, $f(93)$, and $f(-100)$.

18. $g(t) = t^3 + 3$; find $g(1)$, $g(-5)$, and $g(0)$.

19. $h(x) = x^4 - 3$; find $h(0)$, $h(-1)$, and $h(3)$.

20. $f(x) = 2/x$; find $f(-3), f(-2), f(-1), f(\frac{1}{2}), f(1), f(2), f(3)$, and $f(10)$.

21. The function $P(d) = 1 + (d/33)$ gives the pressure, in *atmospheres* (atm), at a depth d in the sea (d is in feet). Note that $P(0) = 1$ atm, $P(33) = 2$ atm, and so on. Find the pressure at 20 ft, 30 ft, and 100 ft.

22. The function $R(t) = 33\frac{1}{3}t$ gives the number of revolutions of a $33\frac{1}{3}$ RPM record as a function of t, the time (in minutes) it turns. Find the number of revolutions at 5 min, 20 min, and 25 min.

23. The function $T(d) = 10d + 20$ gives the temperature (in degrees Celsius) inside the earth as a function of d, the depth in kilometers. Find the temperature at 5 km, 20 km, 1000 km.

24. Snow is melted to water in a cylindrical tube. The function $W(d) = 0.112d$ approximates the amount (in centimeters) of water W that will melt from snow that is d cm deep. Find the amount of water that results from snow melting from depths of 16 cm, 25 cm, and 100 cm.

In many physical situations we speak of one quantity being "a function of" another. For instance, the cost of replacing a defective tire is a function of the tread depth. For Exercises 25–28, the following chart gives a rule for this function for a tire with original tread depth of 9 millimeters (mm).

Tread depth (mm)	% of regular price charged
9	No charge
8	20%
7	30%
6	40%
5	55%
4	70%
3	80%
2	90%
1	100%

25. Find the cost of replacing a tire whose regular price is $64.50 and whose tread depth is 4 mm.

26. Find the cost of replacing a tire whose regular price is $78.50 and whose tread depth is 7 mm.

27. Find the cost of replacing a tire whose regular price is $67.80 and whose tread depth is 3 mm.

28. Find the cost of replacing a tire whose regular price is $72.40 and whose tread depth is 5 mm.

Find the range of each function for the given domain.

29. $f(x) = 3x + 5$, when the domain is the set of whole numbers less than 4

30. $g(t) = t^2 - 5$, when the domain is the set of integers between -4 and 2

31. $h(x) = |x| - x$, when the domain is the set of integers between -2 and 20

32. $f(m) = m^3 + 1$, when the domain is the set of integers between -3 and 3

Suppose that $f(x) = 3x$ and $g(x) = -4x^2$. Find each of the following.

33. $f(8) - g(2)$

34. $f(0) - g(-5)$

35. $2f(1) + 3g(4)$

36. $g(-3) \cdot f(-8) + 16$

37. If $f(-1) = -7$ and $f(3) = 8$, find a linear equation for $f(x)$.

38. $H(x - 1) = 5x$; find $H(6)$.

39. When you flip a coin n times, determine whether or not the number of "heads" you get is a function of the number (n) of flips.

OBJECTIVES

After finishing Section 8.2, you should be able to:

i Graph a function.

ii Determine whether a graph is of a function.

1. Graph $f(x) = 4 - x$.

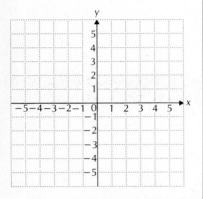

2. Graph $h(x) = |x| - 2$.

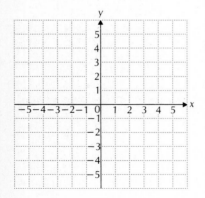

8.2

FUNCTIONS AND GRAPHS

i **Graphs of Functions**

To graph a function, such as $f(x) = x + 2$, we do just what we do to graph $y = x + 2$. We find ordered pairs (x, y), or $(x, f(x))$; plot them; and connect the points.

EXAMPLE 1 Graph: $f(x) = x + 2$.

A list of function values is shown in this table.

x	$f(x)$
-4	-2
-3	-1
-2	0
-1	1
0	2
1	3
2	4
3	5
4	6

We plot the points and connect them. The graph is a straight line.

EXAMPLE 2 Graph: $g(x) = |x|$.

A list of function values is shown in this table.

x	$g(x)$
-3	3
-2	2
-1	1
0	0
1	1
2	2
3	3

We plot the points and connect them. The graph is V-shaped. Note that as x increases through positive values, the absolute value of x increases. As x decreases through negative values, the absolute value of x increases. Thus the graph is a V-shaped curve that rises on either side of the vertical axis.

DO EXERCISES 1 AND 2.

EXAMPLE 3   Graph: $h(x) = \dfrac{4}{x}$.

A list of function values is shown in the table.

x	$h(x)$
4	1
3	$1\frac{1}{3}$
2	2
1	4
-1	-4
-2	-2
-3	$-1\frac{1}{3}$
-4	-1

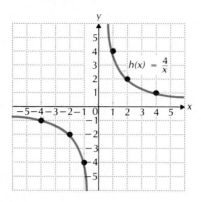

$h(x) = \dfrac{4}{x}$

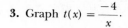

3. Graph $t(x) = \dfrac{-4}{x}$.

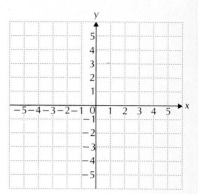

Note that 0 is not in the domain of this function because we cannot substitute 0 into the formula. It would result in division by 0. We plot the points. We connect them in the first quadrant, and we connect them in the third quadrant. There are two parts to the graph, but they are not connected because there is no function value at 0.

DO EXERCISE 3.

ii Recognizing Graphs of Functions

Consider the function f described by $f(x) = x - 2$. Its graph is shown below. It is also the graph of the equation $y = x - 2$.

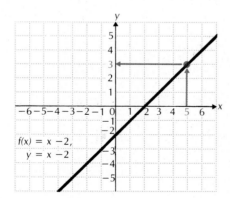

$f(x) = x - 2$,
$y = x - 2$

The domain, the set of all possible inputs, is represented by a set of points on the horizontal axis. The range, the set of all possible outputs, is represented by a set of points on the vertical axis. To find a function value, we locate the input on the horizontal axis, move as shown to the graph of the function, and move horizontally to find the output on the vertical axis.

Some graphs are graphs of functions and some are not. If a vertical line drawn anywhere on the graph can intersect the graph at more than one point, there will be

Is the graph that of a function?

4.

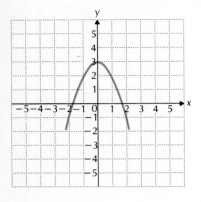

5.

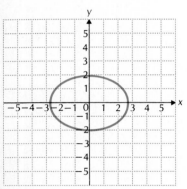

6.

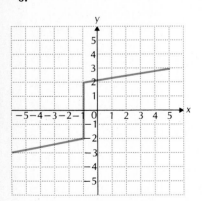

more than one number in the range corresponding to a member of the domain. This graph is not the graph of a function.

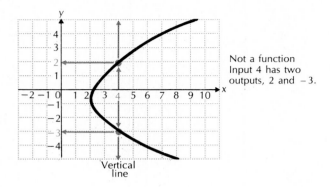

Not a function
Input 4 has two
outputs, 2 and −3.

Vertical
line

To recognize a graph of a function we can use a vertical-line test. If it is possible for any vertical line to intersect the graph at more than one point, the graph is not the graph of a function.

EXAMPLE 3 Which of the following are graphs of functions?

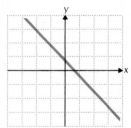

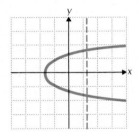

A function. No vertical line
could cross the graph at
more than one point.

Not a function. A vertical
line does cross the graph at
more than one point.

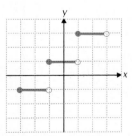

A function. No vertical line
can cross the graph at more
than one point.

DO EXERCISES 4–6.

EXERCISE SET 8.2

i Graph each function.

1. $f(x) = x + 4$

2. $g(x) = x + 3$

3. $h(x) = 2x - 3$

4. $f(x) = 3x - 1$

5. $g(x) = x - 6$

6. $h(x) = x - 5$

7. $f(x) = 2x - 7$

8. $g(x) = 4x - 13$

9. $f(x) = \frac{1}{2}x + 1$

10. $f(x) = -\frac{3}{4}x - 2$

11. $g(x) = 2|x|$

12. $h(x) = -|x|$

13. $g(x) = x^2$

14. $f(x) = x^2 - 1$

15. $f(x) = \dfrac{2}{x}$

16. $f(x) = -\dfrac{1}{x}$

Which of the following are graphs of functions?

17.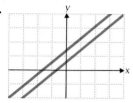

18.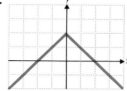

19.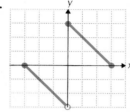

20.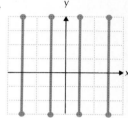

Problem-solving practice

21. A student is 150 km from home. The student drives farther away from home in a straight line at a speed of 60 km/h. How far is the student from home after 4 hr?

22. The length of a rectangle is 2 m greater than the width. The area is 35 m². Find the perimeter.

23. Sketch a graph that is not a function.

24. Draw the graph of $|y| = x$. Is this the graph of a function?

25. Draw the graph of $y^2 = x$. Is this the graph of a function?

8.3

DIRECT VARIATION

i **Equations of Direct Variation**

A bicycle is traveling at 10 km/h. In 1 hr it goes 10 km. In 2 hr it goes 20 km. In 3 hr it goes 30 km, and so on. We will use the number of hours as the first coordinate and the number of kilometers traveled as the second coordinate $(1, 10)$, $(2, 20)$, $(3, 30)$, $(4, 40)$, and so on. Note that as the first number gets larger, so does the second. Note also that the ratio of distance to time for each of these ordered pairs is $\frac{10}{1}$, or 10.

OBJECTIVES

After finishing Section 8.3, you should be able to:

i Find an equation of direct variation given a pair of values of the variables.

ii Solve problems involving direct variation.

Find an equation of variation where y varies directly as x.

1. $y = 84$ when $x = 12$

2. $y = 50$ when $x = 80$

Whenever a situation produces pairs of numbers in which the ratio is constant, we say that there is *direct variation*. Here the distance varies directly as the time:

$$\frac{d}{t} = 10 \text{ (a constant)}, \quad \text{or } d = 10t.$$

If a situation translates to a function described by $y = kx$, where k is a positive constant, $y = kx$ is called an *equation of direct variation*, and k is called the *variation constant.*

The graph of $y = kx$, $k > 0$, always goes through the origin and rises from left to right. Note that as x increases, y increases.

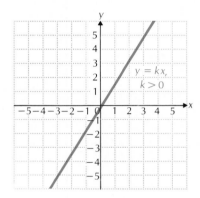

When there is direct variation $y = kx$, the variation constant can be found if one pair of values of x and y is known. Then other values can be found.

EXAMPLE 1 Find an equation of variation where y varies directly as x, and $y = 7$ when $x = 25$.

We substitute to find k:

$$y = kx$$
$$7 = k \cdot 25$$
$$\tfrac{7}{25} = k, \quad \text{or } k = 0.28.$$

Then the equation of variation is $y = 0.28x$. Note that the answer is an *equation.*

DO EXERCISES 1 AND 2.

ii Problem Solving with Direct Variation

EXAMPLE 2 It is known that the karat rating K of a gold object varies directly as the actual percentage of gold in the object. A 14-karat gold object is 58.25% gold. What is the percentage of gold in a 24-karat object?

1. **Familiarize** The problem states that we have direct variation between the variables K and P. Thus an equation $K = kP$, $k > 0$, applies. As the percentage of gold increases, the karat rating increases.

2. **Translate** We write an equation of variation:

$$K = kP.$$

Karat rating varies directly as percentage of gold.

3. **Carry out** The mathematical manipulation has two steps: First, find the equation of variation. Second, compute the percentage of gold in a 24-karat object.

   a) First, find an equation of variation:

   $$K = kP$$
   $$14 = k \cdot 0.5825 \qquad \text{Substituting 14 for } K \text{ and 0.5825 for } P$$
   $$\frac{14}{0.5825} = k$$
   $$24.03 \approx k. \qquad \text{Dividing and rounding to the nearest hundredth}$$

   The equation of variation is $K = 24.03P$.

   b) Use the equation to find the percentage of gold in a 24-karat object:

   $$K = 24.03P$$
   $$24 = 24.03P$$
   $$\frac{24}{24.03} = P$$
   $$0.999 \approx P$$
   $$99.9\% \approx P.$$

4. **Check** The check might be done by repeating the computations. You might also do some reasoning about the answer. The karat rating increased from 14 to 24. Similarly, the percentage increased from 58% to 99.9%.

5. **State** A 24-karat object is 99.9% gold.

DO EXERCISES 3 AND 4.

3. The cost c of operating a TV varies directly as the number of hours n it is in operation. It costs $14.00 to operate a standard-size color TV continuously for 30 days. At this rate, how much would it cost to operate the TV for 1 day? for 1 hour?

4. The weight V of an object on Venus varies directly as its weight E on earth. A person weighing 75 kg on earth would weigh 66 kg on Venus. How much would a person weighing 90 kg on earth weigh on Venus?

EXERCISE SET 8.3

i Find an equation of variation where y varies directly as x and the following are true.

1. $y = 28$, when $x = 7$
2. $y = 30$, when $x = 8$
3. $y = 0.7$, when $x = 0.4$
4. $y = 0.8$, when $x = 0.5$
5. $y = 400$, when $x = 125$
6. $y = 630$, when $x = 175$
7. $y = 200$, when $x = 300$
8. $y = 500$, when $x = 60$

ii **Problem solving**

9. A person's paycheck P varies directly as the number H of hours worked. For working 15 hours, the pay is $78.75. Find the pay for 35 hours of work.

10. The number B of bolts a machine can make varies directly as the time it operates. It can make 6578 bolts in 2 hours. How many can it make in 5 hours?

11. The number of servings S of meat that can be obtained from a turkey varies directly as its weight W. From a turkey weighing 14 kg one can get 40 servings of meat. How many servings can be obtained from an 8-kg turkey?

12. The number of servings S of meat that can be obtained from round steak varies directly as the weight W. From 9 kg of round steak one can get 70 servings of meat. How many servings can one get from 12 kg of round steak?

13. The weight M of an object on the moon varies directly as its weight E on earth. A person who weighs 78 kg on earth weighs 13 kg on the moon. How much would a 100-kg person weigh on the moon?

14. The weight M of an object on Mars varies directly as its weight E on earth. A person who weighs 95 kg on earth weighs 36.1 kg on Mars. How much would an 80-kg person weigh on Mars?

15. The number of kilograms of water W in a human body varies directly as the total body weight B. A person weighing 75 kg contains 54 kg of water. How many kilograms of water are in a person weighing 95 kg?

16. The amount C that a family spends on car expenses varies directly as its income I. A family making $21,760 a year will spend $3264 a year for car expenses. How much will a family making $30,000 a year spend for car expenses?

Write an equation of direct variation for each situation in Exercises 17–20. If possible, give a value for k and graph the equation.

17. The perimeter P of an equilateral polygon varies directly as the length S of a side.

18. The circumference C of a circle varies directly as the radius r.

19. The number of bags B of peanuts sold at a baseball game varies directly as the number N of people in attendance.

20. The cost C of building a new house varies directly as the area A of the floor space of the house.

21. Show that if p varies directly as q, then q varies directly as p.

22. The area of a circle varies directly as the square of the length of the radius. What is the variation constant?

Write an equation of variation to describe these situations.

23. In a stream, the amount S of salt carried varies directly as the sixth power of the speed V of the stream.

24. The square of the pitch P of a vibrating string varies directly as the tension t on the string.

The volume of a box varies directly as its length. It also varies directly as the height. We then say that the volume varies *jointly* as the length and the height. An equation of variation can be written using the *product* of variables $V = k \cdot l \cdot h$. Write an equation of variation for the following.

25. The power P required in an electric circuit varies jointly as the resistance R and the square of the current I.

OBJECTIVES

After finishing Section 8.4, you should be able to:

i Find an equation of inverse variation given a pair of values of the variables.

ii Solve problems involving inverse variation.

8.4

INVERSE VARIATION

i Equations of Inverse Variation

A car is traveling a distance of 10 km. At a speed of 10 km/h it will take 1 hr. At 20 km/h, it will take $\frac{1}{2}$ hr. At 30 km/h it will take $\frac{1}{3}$ hr, and so on. This determines a set of pairs of numbers, all having the same product:

$$(10, 1), \quad (20, \tfrac{1}{2}), \quad (30, \tfrac{1}{3}), \quad (40, \tfrac{1}{4}), \quad \text{and so on.}$$

Note that as the first number gets larger, the second number gets smaller. Whenever a situation produces pairs of numbers whose product is constant, we say that there is *inverse variation*. Here the time varies inversely as the speed:

$$rt = 10 \text{ (a constant)}, \quad \text{or } t = \frac{10}{r}.$$

If a situation translates to a function described by $y = k/x$, where k is a positive constant, $y = k/x$ is called an *equation of inverse variation*. We say that y varies inversely as x.

The graph of $y = k/x$, $k > 0$, is shaped like the following for positive values of x. Note that as x increases, y decreases; and as x decreases, y increases.

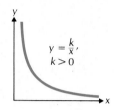

EXAMPLE 1   Find an equation of variation where y varies inversely as x, and $y = 145$ when $x = 0.8$.

We substitute to find k:

$$y = \frac{k}{x}$$

$$145 = \frac{k}{0.8}$$

$$(0.8)145 = k$$

$$116 = k.$$

The equation of variation is $y = 116/x$.

DO EXERCISES 1 AND 2.

ii   Problem Solving with Inverse Variation

EXAMPLE 2   The time T to do a certain job varies inversely as the number N of people working (assuming all work at the same rate). It takes 4 hr for 20 people to wash and wax the floors in a building. How long would it then take 25 people to do the job?

1. **Familiarize**   The problem states that we have inverse variation between the variables T and N. Thus an equation $T = k/N$, $k > 0$, applies. As the number of people increases, the time it takes to do the job decreases.

Find an equation of variation where y varies directly as x.

1. $y = 105$ when $x = 0.6$

2. $y = 45$ when $x = 20$

3. In Example 2, how long would it take 10 people to do the job?

2. Translate ▷ We write an equation of variation:

$$T = \frac{k}{N}.$$

Time varies inversely as the number of people.

3. Carry out ▷ The mathematical manipulation has two steps: First, find the equation of variation. Second, compute the amount of time it would take 25 people to do the job.

   a) First, find an equation of variation:

$$T = \frac{k}{N}$$

$$4 = \frac{k}{20} \qquad \text{Substituting 4 for } T \text{ and 20 for } N$$

$$20 \cdot 4 = k$$

$$80 = k.$$

The equation of variation is $T = 80/N$.

4. The time required to drive a fixed distance varies inversely as the speed r. It takes 5 hr at 60 km/h to drive a fixed distance. How long would it take at 40 km/h?

   b) Use the equation to find the amount of time that it takes 25 people to do the job:

$$T = \frac{80}{N}$$

$$T = \frac{80}{25}$$

$$= 3.2.$$

4. Check ▷ The check might be done by repeating the computations. We might also analyze the results. The number of people increased from 20 to 25. Did the time decrease? It did, and this confirms what we expect with inverse variation.

5. State ▷ It takes 3.2 hr for 25 people to do the job.

DO EXERCISES 3 AND 4.

EXERCISE SET 8.4

---

i Find an equation of variation where y varies inversely as x and the following are true.

1. $y = 25$, when $x = 3$

2. $y = 45$, when $x = 2$

3. $y = 8$, when $x = 10$

4. $y = 7$, when $x = 10$

5. $y = 0.125$, when $x = 8$

6. $y = 6.25$, when $x = 0.16$

7. $y = 42$, when $x = 25$

8. $y = 42$, when $x = 50$

9. $y = 0.2$, when $x = 0.3$

10. $y = 0.4$, when $x = 0.6$

ii *Problem solving*

11. It takes 16 hr for 2 people to resurface a gym floor. How long would it take 6 people to do the job?

12. It takes 4 hr for 9 cooks to prepare a school lunch. How long would it take 8 cooks to prepare the lunch?

13. The volume V of a gas varies inversely as the pressure P on it. The volume of a gas is 200 cubic centimeters (cm^3) under a pressure of 32 kg/cm^2. What will be its volume under a pressure of 20 kg/cm^2?

14. The current I in an electrical conductor varies inversely as the resistance R of the conductor. The current is 2 amperes when the resistance is 960 ohms. What is the current when the resistance is 540 ohms?

15. The time t required to empty a tank varies inversely as the rate r of pumping. A pump can empty a tank in 90 minutes at the rate of 1200 L/min. How long will it take the pump to empty the tank at 2000 L/min?

16. The height H of triangles of fixed area varies inversely as the base B. Suppose the height is 50 cm when the base is 40 cm. Find the height when the base is 8 cm. What is the fixed area?

17. The pitch P of a musical tone varies inversely as its wavelength W. One tone has a pitch of 660 vibrations per second and a wavelength of 1.6 feet. Find the wavelength of another tone that has a pitch of 440 vibrations per second.

18. The time t required to drive a fixed distance varies inversely as the speed r. It takes 5 hr at 60 km/h to drive a fixed distance. How long would it take at 40 km/h?

Write an equation of inverse variation for each situation.

19. The cost C per person of chartering a fishing boat varies inversely as the number N of persons sharing the cost.

20. The number N of revolutions of a tire rolling over a given distance varies inversely as the circumference C of the tire.

21. The amount of current I flowing in an electrical circuit varies inversely with the resistance R of the circuit.

22. The density D of a given mass varies inversely as its volume V.

23. The intensity of illumination I from a light source varies inversely as the square of the distance d from the source.

Which of the following vary inversely?

24. The cost of mailing a letter in the United States and the distance it travels.

25. A runner's speed in a race and the time it takes to run it.

26. The number of plays to go 80 yards for a touchdown and the average gain per play.

27. The weight of a turkey and the cooking time.

28. Graph the equation of inverse variation $y = 6/x$.

The time it takes n people to do s jobs varies directly as the number of jobs and inversely as the number of people. An equation of variation is

$$T = ks \cdot \frac{1}{n}.$$

This is *combined variation*. Write an equation of variation for each of the following.

29. The force F needed to keep a car from skidding on a curve varies directly as the square of the car's speed S and its mass m and inversely as the radius of the curve r.

30. For a horizontal beam supported at both ends, the maximum safe load L varies directly as its width w and the square of its thickness t and inversely as the distance d between the supports.

SUMMARY AND REVIEW: CHAPTER 8

The following contains a summary of what you should be able to do after completing this chapter. The review exercises are for practice. Answers are at the back of the book. If you miss an exercise, restudy the section indicated alongside the answer.

You should be able to:

Determine whether a correspondence is a function.

Determine whether each of the following correspondences is a function.

1. *Domain*   *Range*
    $-1 \longleftrightarrow 3$
    $0 \longrightarrow 4$
    $1 \longrightarrow 5$

2. *Domain*   *Range*
    $-2 \longrightarrow 0$
    $5 \longrightarrow 1$
    $7 \longrightarrow 4$

Given a function defined by a formula, find function values (outputs) for specified inputs.

Find the indicated outputs.

3. $f(x) = 3x - 4$; find $f(2)$, $f(0)$, and $f(-1)$.

4. $g(t) = |t| - 3$; find $g(3)$, $g(-5)$, and $g(0)$.

5. $h(x) = x^3 + 1$; find $h(-2)$, $h(0)$, and $h(1)$.

6. If you are moderately active, you need each day about 15 calories per pound of body weight. The function $C(p) = 15p$ approximates the number of calories C that are needed to maintain body weight, p, in lb. How many calories would be needed to maintain a body weight of 180 lb?

Graph a function and recognize the graph of a function.

Graph.

7. $g(x) = x + 7$

8. $f(x) = x^2 - 3$

9. $h(x) = 3|x|$

Which of the following are graphs of functions?

10.
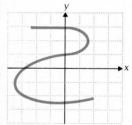

11.
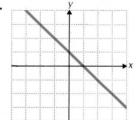

Find an equation of variation where y varies directly as x and y varies inversely as x.

Find an equation of variation where y varies directly as x, and the following are true.

12. $y = 12$, when $x = 4$

13. $y = 4$, when $x = 8$

14. $y = 0.4$, when $x = 0.5$

Find an equation of variation where y varies inversely as x and the following are true.

15. $y = 5$, when $x = 6$

16. $y = 0.5$, when $x = 2$

17. $y = 1.3$, when $x = 0.5$

Solve a problem involving variation using the five-step problem-solving process.

18. A person's paycheck P varies directly as the number H of hours worked. The pay is $165.00 for working 20 hr. Find the pay for 30 hr of work.

19. It takes 5 hr for 2 washing machines to wash a fixed amount. How long would it take 10 washing machines? (The number of hours varies inversely as the number of washing machines.)

20. The stopping distance (at some fixed speed) of regular tires on glare ice is given by a linear function of the air temperature F,

$$D(F) = 2F + 115,$$

where $D(F) =$ the stopping distance, in feet, when the air temperature is F, in degrees Fahrenheit.

a) Find the stopping distance when the air temperature is $0°F$.

b) Find the stopping distance when the air temperature is $32°F$.

21. Suppose that $f(x) = 2x$ and $g(x) = -3x^2$. Find $3f(2) - f(-1) \cdot g(4)$.

22. Draw the graph of $y^2 + 1 = x$. Is this the graph of a function?

TEST: CHAPTER 8

1. Determine whether the following correspondence is a function.

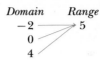

Domain   Range
$-2 \longrightarrow 5$
0
4

Find the indicated outputs.

2. $f(x) = \frac{1}{2}x + 1$; find $f(0)$, $f(1)$, and $f(2)$.

3. $g(t) = -2|t| + 3$; find $t(-1)$, $t(0)$, and $t(3)$.

4. The world record for the 10,000-m run has been decreasing steadily since 1940. The record is 30.18 min minus 0.12 times the number of years since 1940. The function $R(t) = 30.18 - 0.12t$ gives the record R in min as a function of t, the time in yr since 1940. What will the record be in 1986?

Graph each function.

5. $h(x) = x - 4$

6. $g(x) = x^2 - 4$

Which of the following are graphs of functions?

7.

8.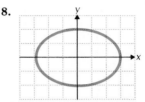

Find an equation of variation where y varies directly as x and the following are true.

9. $y = 6$, when $x = 3$

10. $y = 1.5$, when $x = 3$

Find an equation of variation where y varies inversely as x and the following are true.

11. $y = 6$, when $x = 2$

12. $y = \frac{1}{3}$, when $x = 3$

13. The distance d traveled by a train varies directly as the time t it travels. The train travels 60 km in $\frac{1}{2}$ hr. How far will it travel in 2 hr?

14. It takes 3 hr for 2 cement mixers to mix a certain amount. The number of hours varies inversely as the number of cement mixers. How long would it take 5 cement mixers to do the job?

---

15. The population of a small town, in thousands, is estimated by a linear function

$$P(t) = 1.25t + 15,$$

where $P(t)$ = the population the tth year after 1940; and $P(0)$ is the population in 1940, $P(45)$ is the population in 1985, and so on.

a) Find the population in 1940.
b) Find the population in 1985.

16. Suppose that $f(x) = x - 4$ and $g(x) = -2x^2$. Find $f(-2) - f(0) \cdot g(3) + g(-1)$.

THE ASVAB TEST: MATHEMATICS KNOWLEDGE

CHAPTER 8

See the discussion of the ASVAB test at the beginning of Chapter 3.

1. 60% of what is 96?

A   150
B   1.60
C   0.625
D   160

2. What percent of 80 is 32?

A   10%
B   25%
C   0.4%
D   40%

3. 57% of 45 is what?

  A   15.65

  B   25.65

  C   78.95

  D   2565

4. If $-5x + 7 = 12$, then $x = ?$

  A   1

  B $-19/5$

  C -1

  D $5/19$

5. The temperature of an airfield in Iceland is 13°. During the night the temperature drops 37°. What is the resulting temperature?

  A   50°

  B $-24°$

  C $-50°$

  D $-23°$

6. A student has $2000 in a checking account and spends N dollars. How much does the student have left?

  A $N + 2000$

  B $N - 2000$

  C $2000 - N$

  D $2000 + N$

7. $0.0053 \times 8.5 = ?$

  A   0.04505

  B   0.4405

  C   8.5053

  D   13.8

8. $20.16 \div 48 = ?$

  A   0.43

  B   28.16

  C   0.42

  D   41

9. 25 to the third power is?

  A $3\sqrt{25}$

  B 25×3

  C $25 \times 25 \times 25$

  D 3^{25}

10. If $0.625t = 1$, then $t = ?$

  A   0.625

  B   62.5

  C   16

  D   1.6

11. If t in. of rain falls in 30 min, how many inches would fall in D days?

  A $48tD$

  B $48t/D$

  C $2D/t$

  D $tD + 30$

12. The geometric figure shown here is what?

  A   an octagon

  B   a pentagon

  C   a hexagon

  D   a nonagon

13. If $A = 4(-7) - 8n + 15/n$ and $n = -3$, then $A = ?$

  A   1

  B -1

  C -9

  D -57

14. If $A = P(1 + rt)$, $P = \$1000$, $t = 2$, and $r = 8\%$, then $A = ?$

  A $1160

  B $1080

  C $1016

  D $1200

15. The sum of one third and one fourth is?

  A   One seventh

  B   Seven twelfths

  C   Twelve sevenths

  D   Two sevenths

16. If 20 is multiplied by $x + 3$, the result is 67. Find x.

  A $\frac{20}{7}$

  B $\frac{16}{5}$

  C   44

  D $\frac{7}{20}$

9

RATIONAL EXPRESSIONS, EQUATIONS, AND FUNCTIONS

How can you make up a number "trick" to amaze your friends? One way is to use rational expressions.

A fractional expression is an expression that indicates division, as the fractional symbols in arithmetic do. In this chapter we consider algebraic fractional expressions, also called *rational expressions*. You will learn, or review, how to add them, multiply them, simplify them, and use them in equations or in describing functions. Then, of course, you will use the expressions, equations, and functions in problem solving. As you will see, there are problems for which you need to use fractional expression in the translation.

OBJECTIVES

After finishing Section 9.1, you should be able to:

i Find function values for rational functions.

ii Multiply fractional expressions.

iii Multiply a fractional expression by 1, using an expression like A/A.

iv Simplify fractional expressions.

v Multiply fractional expressions and simplify.

vi Divide fractional expressions and simplify.

1. For the rational function
$$f(x) = \frac{2x^2 - 5x + 1}{2x},$$
find each of the following.
   a) $f(1)$
   b) $f(-1)$
   c) $f(3)$
   d) $f(0)$

9.1

RATIONAL EXPRESSIONS: MULTIPLYING AND DIVIDING

An expression that is a quotient of two expressions is called a fractional expression, or a *rational expression*. The term *rational* comes from the word *ratio*. The following are rational expressions:

$$\frac{7}{8}, \quad \frac{a}{b}, \quad \frac{8}{y+5}, \quad \frac{x^2 + 7xy - 4}{x^3 - y^3}, \quad \frac{1 + \dfrac{3}{x}}{y - \dfrac{y}{3}}.$$

In most of the expressions that we will consider, the numerator and denominator will be polynomials.

Rational expressions indicate division. For example,

$$\frac{7}{8} \quad \text{means} \quad 7 \div 8 \quad \text{and} \quad \frac{2x + y}{x - 5} \quad \text{means} \quad (2x + y) \div (x - 5).$$

Certain substitutions are not sensible in rational expressions, since division by zero is not defined. For example, in the expression on the right above, 5 is not a sensible replacement because it results in division by 0.

i Rational Functions

Like polynomials, rational expressions can be used to define or describe functions. A function that can be described by a rational expression is known as a *rational function*.

EXAMPLE 1 The rational function

$$v(t) = \frac{3t^2 + 2t + 5}{t}$$

gives the velocity of a car, in ft/sec, after t sec have gone by. Find the following function values (velocities): **(a)** $v(1)$; **(b)** $v(3)$; **(c)** $v(5)$.

a) $v(1) = \dfrac{3(1)^2 + 2(1) + 5}{1} = 10$ ft/sec

b) $v(3) = \dfrac{3(3)^2 + 2(3) + 5}{3} = \dfrac{38}{3}$ ft/sec

c) $v(5) = \dfrac{3(5)^2 + 2(5) + 5}{5} = 18$ ft/sec

DO EXERCISE 1.

ii Multiplying

Most of the calculations we do with rational expressions are very much like the calculations we do with fractional notation in arithmetic. By proceeding as in arithmetic, we will know that when we obtain one rational expression from another, the two expressions will be equivalent, that is, they will name the same number for

all sensible replacements. The sensible replacements are, of course, those numbers that are in the domains of both functions.

Multiply.

2. $\dfrac{x-2}{5} \cdot \dfrac{x+2}{x+4}$

> For any rational expressions A/B and C/D,
>
> $$\frac{A}{B} \cdot \frac{C}{D} = \frac{A \cdot C}{B \cdot D}.$$
>
> (To multiply two rational expressions, multiply numerators and multiply denominators.)

EXAMPLE 2 Multiply.

$$\frac{x+3}{y-4} \cdot \frac{x^3}{y+5} = \frac{(x+3)x^3}{(y-4)(y+5)} \qquad \text{Multiplying numerators and multiplying denominators}$$

3. $\dfrac{x+y}{x+3} \cdot \dfrac{x+y}{x-3}$

Usually we do not carry out the multiplications because it is easier to simplify if we do not.

DO EXERCISES 2 AND 3.

iii Multiplying by 1. Any number multiplied by 1 is that same number. Any expression with the same numerator and denominator names the number 1:

Multiply.

$$\frac{y+5}{y+5}, \qquad \frac{4x^2-5}{4x^2-5}, \qquad \frac{-1}{-1} \qquad \text{All name the number 1 for all sensible replacements.}$$

4. $\dfrac{3x+2y}{5x+4y} \cdot \dfrac{x}{x}$

We can multiply by 1 to get equivalent expressions. For example, let us multiply $(x+y)/5$ by 1:

$$\frac{x+y}{5} \cdot \frac{x-y}{x-y} = \frac{(x+y)(x-y)}{5(x-y)} \qquad \text{Multiplying by } \frac{x-y}{x-y}, \text{ which is 1}$$

We know that

$$\frac{x+y}{5} \quad \text{and} \quad \frac{(x+y)(x-y)}{5(x-y)}$$

5. $\dfrac{2x^2-y}{3x+4} \cdot \dfrac{3x+2}{3x+2}$

are equivalent. That means they will name the same number for all replacements that do not make a denominator 0.

EXAMPLES Multiply to obtain equivalent expressions.

3. $\dfrac{x^2+3}{x-1} \cdot \dfrac{x+1}{x+1} = \dfrac{(x^2+3)(x+1)}{(x-1)(x+1)}$

4. $\dfrac{-1}{-1} \cdot \dfrac{x-4}{x-y} = \dfrac{-1 \cdot (x-4)}{-1 \cdot (x-y)}$

6. $\dfrac{-1}{-1} \cdot \dfrac{2a-5}{a-b}$

DO EXERCISES 4–6.

iv Simplifying

We can simplify rational expressions by reversing the procedure of multiplying by 1. We "remove" a factor of 1. We first factor the numerator and denominator, and then factor the rational expression, finding a factor that is equal to 1.

Simplify by removing factors of 1.

7. $\dfrac{7x^2}{x}$

8. $\dfrac{6a+9}{3}$

EXAMPLES Simplify by removing factors equal to 1.

5. $\dfrac{5x^2}{x} = \dfrac{5x \cdot x}{1 \cdot x}$ Factoring numerator and denominator

$\qquad = \dfrac{5x}{1} \cdot \dfrac{x}{x}$ Factoring the fractional expression

$\qquad = 5x \cdot 1$ $\dfrac{x}{x} = 1$ for all sensible replacements

$\qquad = 5x$ "Removing" a factor of 1

In this example we supplied a 1 in the denominator. This can always be done, if necessary.

6. $\dfrac{4a+8}{2} = \dfrac{2(2a+4)}{2 \cdot 1}$ Factoring numerator and denominator

$\qquad = \dfrac{2}{2} \cdot \dfrac{2a+4}{1}$ Factoring the fractional expression

$\qquad = \dfrac{2a+4}{1}$ "Removing" a factor of 1

$\qquad = 2a + 4$

A common error in Example 6 is to get $2a + 8$. That happens if you take an incorrect shortcut.

DO EXERCISES 7 AND 8.

Simplify by removing factors of 1.

9. $\dfrac{6x^2 + 4x}{4x^2 + 8x}$

10. $\dfrac{2y^2 + 6y + 4}{y^2 - 1}$

7. $\dfrac{2x^2 + 4x}{6x^2 + 2x} = \dfrac{2x(x+2)}{2x(3x+1)}$ Factoring numerator and denominator

$\qquad = \dfrac{2x}{2x} \cdot \dfrac{x+2}{3x+1}$ Factoring the fractional expression

$\qquad = \dfrac{x+2}{3x+1}$ "Removing" a factor of 1

8. $\dfrac{x^2 - 1}{2x^2 - x - 1} = \dfrac{(x-1)(x+1)}{(2x+1)(x-1)}$ Factoring numerator and denominator

$\qquad = \dfrac{x-1}{x-1} \cdot \dfrac{x+1}{2x+1}$ Factoring the fractional expression

$\qquad = \dfrac{x+1}{2x+1}$ "Removing" a factor of 1

9. $\dfrac{9x^2 + 6xy - 3y^2}{12x^2 - 12y^2} = \dfrac{3(x+y)(3x-y)}{12(x+y)(x-y)}$ Factoring numerator and denominator

$\qquad = \dfrac{3(x+y)}{3(x+y)} \cdot \dfrac{3x-y}{4(x-y)}$ Factoring the fractional expression

$\qquad = \dfrac{3x-y}{4(x-y)}$ "Removing" a factor of 1

DO EXERCISES 9 AND 10.

v Multiplying and Simplifying

After multiplying we ordinarily simplify, if possible. That is why we do not multiply out the numerator and denominator too soon. If we did that, we might need to factor them again anyway, in order to simplify.

EXAMPLES   Multiply. Then simplify by removing factors of 1.

10. $\dfrac{x+2}{x-3} \cdot \dfrac{x^2-4}{x^2+x-2} = \dfrac{(x+2)(x^2-4)}{(x-3)(x^2+x-2)}$ Multiplying numerators and also denominators

$\qquad = \dfrac{(x+2)(x-2)(x+2)}{(x-3)(x+2)(x-1)}$ Factoring numerators and denominators

$\qquad = \dfrac{x+2}{x+2} \cdot \dfrac{(x+2)(x-2)}{(x-3)(x-1)}$ Factoring the fractional expression

$\qquad = \dfrac{(x+2)(x-2)}{(x-3)(x-1)},$ "Removing" a factor of 1

$\qquad$ or $\dfrac{x^2-4}{x^2-4x+3}$ Multiplying out numerator and denominator

We may not always multiply out numerator and denominator.

11. $\dfrac{a^3-b^3}{a^2-b^2} \cdot \dfrac{a^2+2ab+b^2}{a^2+ab+b^2} = \dfrac{(a^3-b^3)(a^2+2ab+b^2)}{(a^2-b^2)(a^2+ab+b^2)}$

$\qquad = \dfrac{(a-b)(a^2+ab+b^2)(a+b)(a+b)}{(a-b)(a+b)(a^2+ab+b^2)}$

$\qquad = \dfrac{(a-b)(a^2+ab+b^2)(a+b)}{(a-b)(a^2+ab+b^2)(a+b)} \cdot \dfrac{a+b}{1}$

$\qquad = a+b$

DO EXERCISES 11 AND 12.

vi Dividing and Simplifying

Two expressions are reciprocals of each other if their product is 1. As in arithmetic, to find the reciprocal of a rational expression, we interchange numerator and denominator.

$\qquad$ The reciprocal of $y-8$ is $\dfrac{1}{y-8}.$

$\qquad$ The reciprocal of $\dfrac{x}{x^2+3}$ is $\dfrac{x^2+3}{x}.$

DO EXERCISES 13–15.

It should be noted that any replacement that makes a numerator zero will not be a sensible replacement in the expression for the reciprocal, because it will make the denominator zero.

Multiply and simplify.

11. $\dfrac{(x-y)^3}{x+y} \cdot \dfrac{3x+3y}{x^2-y^2}$

12. $\dfrac{a^3+b^3}{a^2-b^2} \cdot \dfrac{a^2-2ab+b^2}{a^2-ab+b^2}$

Find the reciprocal.

13. $\dfrac{x+3}{x-5}$

14. $x+7$

15. $\dfrac{1}{y^3-9}$

Divide. Simplify by removing a factor of 1 if possible. Then multiply out the numerator and denominator.

16. $\dfrac{x^2 + 7x + 10}{2x - 4} \div \dfrac{x^2 - 3x - 10}{x - 2}$

17. $\dfrac{a^2 - b^2}{ab} \div \dfrac{a^2 - 2ab + b^2}{2a^2 b^2}$

> For any rational expressions A/B and C/D,
> $$\frac{A}{B} \div \frac{C}{D} = \frac{A}{B} \cdot \frac{D}{C}.$$
> (To divide two rational expressions, multiply by the reciprocal of the divisor, that is, invert and multiply.)

EXAMPLES Divide. Simplify by removing a factor of 1 if possible.

12. $\dfrac{x - 2}{x + 1} \div \dfrac{x + 5}{x - 3} = \dfrac{x - 2}{x + 1} \cdot \dfrac{x - 3}{x + 5}$ Multiplying by the reciprocal

$= \dfrac{(x - 2)(x - 3)}{(x + 1)(x + 5)}$ Multiplying numerators and denominators

or $\dfrac{x^2 - 5x + 6}{x^2 + 6x + 5}$

13. $\dfrac{a^2 - 1}{a - 1} \div \dfrac{a^2 - 2a + 1}{a + 1} = \dfrac{a^2 - 1}{a - 1} \cdot \dfrac{a + 1}{a^2 - 2a + 1}$ Multiplying by the reciprocal

$= \dfrac{(a^2 - 1)(a + 1)}{(a - 1)(a^2 - 2a + 1)}$ Multiplying numerators and denominators

$= \dfrac{(a + 1)(a - 1)(a + 1)}{(a - 1)(a - 1)(a - 1)}$ Factoring numerator and denominator

$= \dfrac{a - 1}{a - 1} \cdot \dfrac{(a + 1)(a + 1)}{(a - 1)(a - 1)}$ Factoring the fractional expression

$= \dfrac{(a + 1)(a + 1)}{(a - 1)(a - 1)},$ "Removing" a factor of 1

or $\dfrac{a^2 + 2a + 1}{a^2 - 2a + 1}$ Multiplying out numerator and denominator

DO EXERCISES 16 AND 17.

EXERCISE SET 9.1

i For each rational function, find the function values indicated.

1. $v(t) = \dfrac{4t^2 - 5t + 2}{t + 3}$;

$v(0), \ v(3), \ v(7)$

2. $s(x) = \dfrac{5x^2 + 4x - 12}{6 - x}$;

$s(4), \ s(-1), \ s(3)$

3. $r(y) = \dfrac{3y^3 - 2y}{y - 5}$;

$r(0), \ r(4), \ r(5)$

4. $f(r) = \dfrac{\pi r^2 + 2\pi r h}{r - 1}$; (use 3.14 for π)

$f(2), \ f(5), \ f(1)$

ii , **iii** Multiply to obtain equivalent expressions. Do not simplify.

5. $\dfrac{3x}{3x} \cdot \dfrac{x + 1}{x + 3}$

6. $\dfrac{4 - y^2}{6 - y} \cdot \dfrac{-1}{-1}$

7. $\dfrac{t - 3}{t + 2} \cdot \dfrac{t + 3}{t + 3}$

8. $\dfrac{p - 4}{p - 5} \cdot \dfrac{p + 5}{p + 5}$

iv Simplify by removing a factor of 1.

9. $\dfrac{9y^2}{15y}$

10. $\dfrac{6x^3}{18x^2}$

11. $\dfrac{2a - 6}{2}$

12. $\dfrac{3a - 6}{3}$

13. $\dfrac{4y - 12}{4y + 12}$

14. $\dfrac{8x + 16}{8x - 16}$

15. $\dfrac{t^2 - 16}{t^2 - 8t + 16}$

16. $\dfrac{p^2 - 25}{p^2 + 10p + 25}$

v Multiply. Simplify by removing a factor of 1 if possible.

17. $\dfrac{x^2 - 16}{x^2} \cdot \dfrac{x^2 - 4x}{x^2 - x - 12}$

18. $\dfrac{y^2 + 10y + 25}{y^2 - 9} \cdot \dfrac{y + 3}{y + 5}$

19. $\dfrac{y^2 - 16}{2y + 6} \cdot \dfrac{y + 3}{y - 4}$

20. $\dfrac{m^2 - n^2}{4m + 4n} \cdot \dfrac{m + n}{m - n}$

21. $\dfrac{x^2 - 2x - 35}{2x^3 - 3x^2} \cdot \dfrac{4x^3 - 9x}{7x - 49}$

22. $\dfrac{y^2 - 10y + 9}{y^2 - 1} \cdot \dfrac{y + 4}{y^2 - 5y - 36}$

23. $\dfrac{c^3 + 8}{c^2 - 4} \cdot \dfrac{c^2 - 4c + 4}{c^2 - 2c + 4}$

24. $\dfrac{x^3 - 27}{x^2 - 9} \cdot \dfrac{x^2 - 6x + 9}{x^2 + 3x + 9}$

25. $\dfrac{x^2 - y^2}{x^3 - y^3} \cdot \dfrac{x^2 + xy + y^2}{x^2 + 2xy + y^2}$

26. $\dfrac{4x^2 - 9y^2}{8x^3 - 27y^3} \cdot \dfrac{4x^2 + 6xy + 9y^2}{4x^2 + 12xy + 9y^2}$

vi Divide. Simplify by removing a factor of 1 if possible.

27. $\dfrac{3y + 15}{y} \div \dfrac{y + 5}{y}$

28. $\dfrac{6x + 12}{x} \div \dfrac{x + 2}{x^3}$

29. $\dfrac{y^2 - 9}{y} \div \dfrac{y + 3}{y + 2}$

30. $\dfrac{x^2 - 4}{x} \div \dfrac{x - 2}{x + 4}$

31. $\dfrac{4a^2 - 1}{a^2 - 4} \div \dfrac{2a - 1}{a - 2}$

32. $\dfrac{25x^2 - 4}{x^2 - 9} \div \dfrac{5x - 2}{x + 3}$

33. $\dfrac{x^2 - 16}{x^2 - 10x + 25} \div \dfrac{3x - 12}{x^2 - 3x - 10}$

34. $\dfrac{y^2 - 36}{y^2 - 8y + 16} \div \dfrac{3y - 18}{y^2 - y - 12}$

35. $\dfrac{y^3 + 3y}{y^2 - 9} \div \dfrac{y^2 + 5y - 14}{y^2 + 4y - 21}$

36. $\dfrac{a^3 + 4a}{a^2 - 16} \div \dfrac{a^2 + 8a + 15}{a^2 + a - 20}$

37. $\dfrac{x^3 - 64}{x^3 + 64} \div \dfrac{x^2 - 16}{x^2 - 4x + 16}$

38. $\dfrac{8y^3 - 27}{64y^3 - 1} \div \dfrac{4y^2 - 9}{16y^2 + 4y + 1}$

Perform the indicated operations and simplify.

39. $\left[\dfrac{r^2 - 4s^2}{r + 2s} \div (r + 2s) \right] \cdot \dfrac{2s}{r - 2s}$

40. $\left[\dfrac{d^2 - d}{d^2 - 6d + 8} \cdot \dfrac{d - 2}{d^2 + 5d} \right] \div \dfrac{5d}{d^2 - 9d + 20}$

Multiply or divide.

41. $\dfrac{834x}{y - 427.2} \cdot \dfrac{26.3x}{y + 427.2}$

42. $\dfrac{527}{x + 93.87} \div \dfrac{x - 93.87}{468}$

43. $\dfrac{0.0049t}{t + 0.007} \cdot \dfrac{27{,}000t}{t - 0.007}$

44. $\dfrac{y + 924.6}{0.003} \div \dfrac{0.421}{y - 924.6}$

Simplify.

45. $\dfrac{x(x + 1) - 2(x + 3)}{(x + 1)(x + 2)(x + 3)}$

46. $\dfrac{2x - 5(x + 2) - (x - 2)}{x^2 - 4}$

47. $\dfrac{m^2 - t^2}{m^2 + t^2 + m + t + 2mt}$

48. $\dfrac{a^3 - 2a^2 + 2a - 4}{a^3 - 2a^2 - 3a + 6}$

49. $\dfrac{x^3 + x^2 - y^3 - y^2}{x^2 - 2xy + y^2}$

50. $\dfrac{u^6 + v^6 + 2u^3v^3}{u^3 - v^3 + u^2v - uv^2}$

51. $\dfrac{x^5 - x^3 + x^2 - 1 - (x^3 - 1)(x + 1)^2}{(x^2 - 1)^2}$

52. Here is a number "trick." You tell a friend to think of a number, then add 3, double the result, then subtract 2, and finally divide by 2. Your friend tells you the answer. You then subtract 2 from that result and announce the original number.

How does this work? It depends on equivalent rational expressions. Here is what you had your friend do.

$$\text{Original number}$$
$$\text{Double} \quad \text{Add 3} \quad \text{Subtract 2}$$
$$\dfrac{2(x + 3) - 2}{2}$$
$$\text{Divide by 2}$$

This expression can be simplified to $x + 2$, so the expressions represent the same number (have the same function value) for **every** x. You knew the simplified expression, so you simply subtracted 2 to get x, the original number.

Make up two or three other such number "tricks."

OBJECTIVES

After finishing Section 9.2, you should be able to:

i Find the LCM of several numbers by factoring.

ii Add numbers (fractional notation) by first finding a common denominator.

iii Find the LCM of several algebraic expressions by factoring.

Find the LCM by factoring.

1. 18, 30

2. 12, 18, 24

9.2

LEAST COMMON MULTIPLES

i To add rational expressions when denominators are different, we first find a common denominator. Let us review the procedure in arithmetic first. To do the addition

$$\frac{5}{42} + \frac{7}{12}$$

we first find a common denominator. We look for a number that is a multiple of both 42 and 12 (a common multiple). Usually we try to get the smallest such number, or the *least common multiple* (LCM). To find the LCM we factor both numbers completely (into primes).

$$42 = 2 \cdot 3 \cdot 7 \longleftarrow \text{Any multiple of 42 has these factors.}$$
$$12 = 2 \cdot 2 \cdot 3 \longleftarrow \text{Any multiple of 12 has these factors.}$$

The LCM is the number that has 2 as a factor twice, 3 as a factor once, and 7 as a factor once. The LCM is $2 \cdot 2 \cdot 3 \cdot 7$, or 84.

> **To obtain the LCM we use each factor the greatest number of times it occurs in any one factorization.**

EXAMPLE 1 Find the LCM of 18 and 24.

$$\begin{array}{l} 18 = 3 \cdot 3 \cdot 2 \\ 24 = 2 \cdot 2 \cdot 2 \cdot 3 \end{array} \Big\} \qquad \text{The LCM is } 3 \cdot 3 \cdot 2 \cdot 2 \cdot 2, \quad \text{or} \quad 72.$$

DO EXERCISES 1 AND 2.

ii Now let us return to adding $\frac{5}{42}$ and $\frac{7}{12}$:

$$\frac{5}{42} + \frac{7}{12} = \frac{5}{2 \cdot 3 \cdot 7} + \frac{7}{2 \cdot 2 \cdot 3}. \qquad \text{Factoring denominators}$$

The LCM is $2 \cdot 2 \cdot 3 \cdot 7$. To get this LCM in the first denominator we need a 2. In the second denominator we need a 7. We multiply by 1, as follows:

$$\frac{5}{2 \cdot 3 \cdot 7} \cdot \frac{2}{2} + \frac{7}{2 \cdot 2 \cdot 3} \cdot \frac{7}{7} = \frac{10}{2 \cdot 2 \cdot 3 \cdot 7} + \frac{49}{2 \cdot 2 \cdot 3 \cdot 7}$$

$$= \frac{59}{2 \cdot 2 \cdot 3 \cdot 7}$$

$$= \frac{59}{84}$$

Multiplying the first fraction by $\frac{2}{2}$ gave us a denominator that is the LCM. Multiplying the second fraction by $\frac{7}{7}$ also gave us a denominator that is the LCM. When we have a common denominator, we add the numerators.

DO EXERCISES 3 AND 4.

iii LCMs of Algebraic Expressions

To find the LCM of several algebraic expressions we proceed as in arithmetic, beginning by factoring the expressions.

EXAMPLE 2   Find the LCM of $12xy^2$ and $15x^3y$.

We factor each expression completely.

$$12xy^2 = 2 \cdot 2 \cdot 3 \cdot x \cdot y \cdot y$$

Remember: To find the LCM, use each factor the greatest number of times it occurs in any one factorization.

$$15x^3y = 3 \cdot 5 \cdot x \cdot x \cdot x \cdot y$$

$$\text{LCM} = 2 \cdot 2 \cdot 3 \cdot 5 \cdot x \cdot x \cdot x \cdot y \cdot y = 60x^3y^2$$

EXAMPLE 3   Find the LCM of $x^2 + 2x + 1$, $5x^2 - 5x$, and $x^2 - 1$.

$$x^2 + 2x + 1 = (x + 1)(x + 1)$$
$$5x^2 - 5x = 5x(x - 1) \qquad \text{Factoring}$$
$$x^2 - 1 = (x + 1)(x - 1)$$
$$\text{LCM} = 5x(x + 1)(x + 1)(x - 1)$$

EXAMPLE 4   Find the LCM of $x^2 - y^2$, $x^3 + y^3$, and $x^2 + 2xy + y^2$.

$$x^2 - y^2 = (x - y)(x + y)$$
$$x^3 + y^3 = (x + y)(x^2 - xy + y^2)$$
$$x^2 + 2xy + y^2 = (x + y)(x + y)$$
$$\text{LCM} = (x - y)(x + y)(x + y)(x^2 - xy + y^2)$$

DO EXERCISES 5–7.

The additive inverse of an LCM is also an LCM. For example, if $(x + 2)(x - 3)$ is an LCM, then $-(x + 2)(x - 3)$ is also an LCM. We can name the latter

$$(x + 2)(-1)(x - 3), \quad \text{or} \quad (x + 2)(3 - x).$$

If, when we are finding LCMs, factors that are additive inverses occur, we do not use them both. For example, if $(a - b)$ occurs in one factorization and $(b - a)$ occurs in another, we do not use them both since they are additive inverses.

EXAMPLE 5   Find the LCM of $x^2 - y^2$ and $3y - 3x$.

$$x^2 - y^2 = (x + y)(x - y) \longleftarrow \text{We can use } (x - y) \text{ or } (y - x) \text{ but we do not use both.}$$
$$3y - 3x = 3(y - x) \longleftarrow$$
$$\text{LCM} = 3(x + y)(x - y), \quad \text{or} \quad 3(x + y)(y - x)$$

DO EXERCISE 8.

Add, first finding the LCM of the denominators.

3. $\dfrac{5}{12} + \dfrac{11}{30}$

4. $\dfrac{7}{12} + \dfrac{13}{18} + \dfrac{1}{24}$

Find the LCM.

5. $a^2b^2, \quad 5a^3b$

6. $y^2 + 7y + 12, \quad y^2 + 8y + 16, \quad y + 4$

7. $x^2 - 9, \quad x^3 - x^2 - 6x, \quad 2x^2$

Find the LCM.

8. $a^2 - b^2, \quad 2b - 2a$

EXERCISE SET 9.2

i Find the LCM by factoring.

1. 12, 18   **2.** 15, 20   **3.** 18, 48   **4.** 45, 54   **5.** 24, 36

6. 30, 75   **7.** 9, 15, 5   **8.** 27, 35, 63   **9.** 24, 36, 42   **10.** 24, 42, 60

ii Add, first finding the LCM of the denominators.

11. $\dfrac{5}{6} + \dfrac{4}{15}$ **12.** $\dfrac{5}{36} + \dfrac{5}{24}$ **13.** $\dfrac{7}{12} + \dfrac{11}{18}$ **14.** $\dfrac{11}{30} + \dfrac{19}{75}$ **15.** $\dfrac{3}{4} + \dfrac{2}{5} + \dfrac{1}{6}$ **16.** $\dfrac{5}{8} + \dfrac{7}{12} + \dfrac{11}{40}$

iii Find the LCM.

17. $8x^2,\ 12x^3$ **18.** $4y^2,\ 24y^3$ **19.** $12x^2y,\ 4xy$

20. $18r^2s,\ 12rs^3$ **21.** $15ab^2,\ 3ab,\ 10a^3b$ **22.** $6x^2y^2,\ 9x^3y,\ 15y^3$

23. $a + b,\ a - b$ **24.** $x - 4,\ x + 4$ **25.** $3(y - 2),\ 6(2 - y)$

26. $5(y - 1),\ 10(1 - y)$ **27.** $y^2 - 9,\ 3y + 9$ **28.** $a^2 - b^2,\ ab + b^2$

29. $5y - 15,\ y^2 - 6y + 9$ **30.** $4x - 16,\ x^2 - 8x + 16$ **31.** $(a + 1),\ (a - 1)^2,\ a^2 - 1$

32. $(x - 2),\ (x + 2)^2,\ x^2 - 4$ **33.** $x^2 - 4,\ 2 - x$ **34.** $y^2 - 9,\ 3 - y$

35. $x^2 + 10x + 25,\ x^2 + 2x - 15$ **36.** $y^2 + 8x + 16,\ y^2 - 3y - 28$ **37.** $2r^2 - 5r - 12,\ 3r^2 - 13r + 4$

38. $3x^2 - 4x - 4,\ 4x^2 - 5x - 6$

Problem-solving practice

39. A kitchen floor is 3 ft longer than it is wide. The area of the floor is 180 ft^2. Find the dimensions of the kitchen.

40. The sum of the squares of three consecutive even integers is equal to eight more than three times the square of the second number. Find the numbers.

Find the LCM.

41. $2x^2 - 5x - 3,\ 2x^2 - x - 1,\ x^2 - 6x + 9$ **42.** $3x^2 + 4x - 4,\ 2x^2 + 7x + 6,\ x^2 - 4x + 4$

43. $x^8 - x^4,\ x^5 - x^2,\ x^5 - x^3,\ x^5 + x^2$

44. The LCM of two expressions is $8a^4b^7$. One of the expressions is $2a^3b^7$. List all the possibilities for the other expression.

OBJECTIVES

After finishing Section 9.3, you should be able to:

i Add and subtract fractional expressions having the same denominator.

ii Add and subtract fractional expressions whose denominators are additive inverses of each other.

iii Add and subtract fractional expressions having different denominators.

9.3

ADDITION AND SUBTRACTION

Addition and subtraction of rational expressions are done with the same methods that we used for fractional notation in arithmetic.

i **When Denominators Are the Same**

If the denominators of two or more rational expressions are the same, we keep that denominator and add the numerators. That procedure works for rational numbers. It also works for other real-number replacements in rational expressions, and it will always give us equivalent expressions.

> When denominators are the same, we add or subtract the numerators and keep the same denominator.

EXAMPLE 1   Add: $\dfrac{3+x}{x}+\dfrac{4}{x}$.

$$\dfrac{3+x}{x}+\dfrac{4}{x}=\dfrac{7+x}{x}\longleftarrow \text{This expression does \textit{not} simplify to 7.}$$

Example 1 shows that

$$\dfrac{3+x}{x}+\dfrac{4}{x}\quad\text{and}\quad\dfrac{7+x}{x}$$

are equivalent expressions. They name the same number for all replacements except 0. As functions

$$\dfrac{3+x}{x}+\dfrac{4}{x}\quad\text{and}\quad\dfrac{7+x}{x}$$

are the same function.

EXAMPLE 2   Add: $\dfrac{4x^2-5xy}{x^2-y^2}+\dfrac{2xy-y^2}{x^2-y^2}$.

$$\dfrac{4x^2-5xy}{x^2-y^2}+\dfrac{2xy-y^2}{x^2-y^2}=\dfrac{4x^2-3xy-y^2}{x^2-y^2}\quad\text{Adding numerators}$$

$$=\dfrac{(4x+y)(x-y)}{(x+y)(x-y)}\quad\text{Factoring numerator and denominator}$$

$$=\dfrac{x-y}{x-y}\cdot\dfrac{4x+y}{x+y}\quad\text{Factoring the fractional expression}$$

$$=\dfrac{4x+y}{x+y}\quad\text{``Removing'' a factor of 1}$$

DO EXERCISES 1 AND 2.

EXAMPLE 3   Subtract: $\dfrac{4x+5}{x+3}-\dfrac{x-2}{x+3}$.

$$\dfrac{4x+5}{x+3}-\dfrac{x-2}{x+3}=\dfrac{4x+5-(x-2)}{x+3}\quad\text{Subtracting numerators}$$

$$=\dfrac{4x+5-x+2}{x+3}$$

$$=\dfrac{3x+7}{x+3}$$

The parentheses in the numerator in Example 3 are important. If they are forgotten, you may incorrectly subtract only *part* of the numerator $x-2$.

DO EXERCISES 3 AND 4.

ii   When Denominators Are Additive Inverses of Each Other

When one denominator is the additive inverse of the other, we multiply one expression by $-1/-1$. That gives us a common denominator.

Add.

1. $\dfrac{5+y}{y}+\dfrac{7}{y}$

2. $\dfrac{2x^2+5x-9}{x-5}+\dfrac{x^2-19x+4}{x-5}$

Subtract.

3. $\dfrac{a}{b+2}-\dfrac{b}{b+2}$

4. $\dfrac{4y+7}{x^2+y^2}-\dfrac{3y-5}{x^2+y^2}$

Add.

5. $\dfrac{b}{3b} + \dfrac{b^3}{-3b}$

6. $\dfrac{3x^2 + 4}{x - 5} + \dfrac{x^2 - 7}{5 - x}$

Subtract.

7. $\dfrac{3}{4y} - \dfrac{7x}{-4y}$

8. $\dfrac{4x^2}{2x - y} - \dfrac{7x^2}{y - 2x}$

EXAMPLE 4 Add: $\dfrac{a}{2a} + \dfrac{a^3}{-2a}$.

$$\dfrac{a}{2a} + \dfrac{a^3}{-2a} = \dfrac{a}{2a} + \dfrac{-1}{-1} \cdot \dfrac{a^3}{-2a} \qquad \text{Multiplying by } \dfrac{-1}{-1}, \text{ which is 1}$$

$$= \dfrac{a}{2a} + \dfrac{-a^3}{2a}$$

$$= \dfrac{a - a^3}{2a} \qquad \text{Adding numerators}$$

$$= \dfrac{a(1 - a^2)}{2a} \qquad \text{Factoring numerator and denominator}$$

$$= \dfrac{a}{a} \cdot \dfrac{1 - a^2}{2} \qquad \text{Factoring the expression}$$

$$= \dfrac{1 - a^2}{2} \qquad \text{"Removing" a factor of 1}$$

DO EXERCISES 5 AND 6.

EXAMPLE 5 Subtract: $\dfrac{5x}{x - 2y} - \dfrac{3y - 7}{2y - x}$.

$$\dfrac{5x}{x - 2y} - \dfrac{3y - 7}{2y - x} = \dfrac{5x}{x - 2y} - \dfrac{-1}{-1} \cdot \dfrac{3y - 7}{2y - x}$$

$$= \dfrac{5x}{x - 2y} - \dfrac{7 - 3y}{x - 2y}$$

$$= \dfrac{5x - (7 - 3y)}{x - 2y} \qquad \text{Subtracting numerators}$$

$$= \dfrac{5x - 7 + 3y}{x - 2y}$$

DO EXERCISES 7 AND 8.

iii When Denominators Are Different

When denominators are different, but not additive inverses of each other, we first find a common denominator, using the LCM, and then add or subtract numerators.

EXAMPLE 6 Add: $\dfrac{2a}{5} + \dfrac{3b}{2a}$.

First find the LCM of the denominators.

$$\begin{array}{c} 5 \\ 2a \end{array} \qquad \text{The LCM is } 5 \cdot 2a, \quad \text{or} \quad 10a.$$

Now we multiply each expression by 1. We choose whatever symbol for 1 will give us the LCM in each denominator. In this case we use $2a/2a$ and $5/5$:

$$\dfrac{2a}{5} \cdot \dfrac{2a}{2a} + \dfrac{3b}{2a} \cdot \dfrac{5}{5} = \dfrac{4a^2}{10a} + \dfrac{15b}{10a} = \dfrac{4a^2 + 15b}{10a}.$$

Multiplying by $2a/2a$ in the first term gave us a denominator of $10a$. Multiplying by $5/5$ in the second term also gave us a denominator of $10a$.

DO EXERCISE 9.

EXAMPLE 7   Add: $\dfrac{3x^2 + 3xy}{x^2 - y^2} + \dfrac{2 - 3x}{x - y}$.

We first find the LCM of the denominators.

$$\left. \begin{array}{l} x^2 - y^2 = (x + y)(x - y) \\[4pt] x - y = x - y \end{array} \right\} \quad \text{The LCM is } (x + y)(x - y).$$

We now multiply by 1 to get the LCM in the second expression. Then we add and simplify if possible.

$$\dfrac{3x^2 + 3xy}{(x + y)(x - y)} + \dfrac{2 - 3x}{x - y} \cdot \dfrac{x + y}{x + y} \qquad \text{Multiplying by 1 to get the LCM}$$

$$= \dfrac{3x^2 + 3xy}{(x + y)(x - y)} + \dfrac{(2 - 3x)(x + y)}{(x - y)(x + y)}$$

$$= \dfrac{3x^2 + 3xy}{(x + y)(x - y)} + \dfrac{2x + 2y - 3x^2 - 3xy}{(x - y)(x + y)} \qquad \text{Multiplying out the numerator}$$

$$= \dfrac{3x^2 + 3xy + 2x + 2y - 3x^2 - 3xy}{(x + y)(x - y)} \qquad \text{Adding numerators}$$

$$= \dfrac{2x + 2y}{(x + y)(x - y)} = \dfrac{2(x + y)}{(x + y)(x - y)} \qquad \begin{array}{l}\text{Combining like terms and}\\\text{then factoring the numerator}\end{array}$$

$$= \dfrac{x + y}{x + y} \cdot \dfrac{2}{x - y} = \dfrac{2}{x - y} \qquad \text{Simplifying by ``removing'' a factor of 1}$$

DO EXERCISES 10 AND 11.

EXAMPLE 8   Subtract: $\dfrac{2y + 1}{y^2 - 7y + 6} - \dfrac{y + 3}{y^2 - 5y - 6}$.

$$\dfrac{2y + 1}{y^2 - 7y + 6} - \dfrac{y + 3}{y^2 - 5y - 6}$$

$$= \dfrac{2y + 1}{(y - 6)(y - 1)} - \dfrac{y + 3}{(y - 6)(y + 1)} \qquad \text{The LCM is } (y - 6)(y - 1)(y + 1).$$

$$= \dfrac{2y + 1}{(y - 6)(y - 1)} \cdot \dfrac{y + 1}{y + 1} - \dfrac{y + 3}{(y - 6)(y + 1)} \cdot \dfrac{y - 1}{y - 1}$$

$$= \dfrac{(2y + 1)(y + 1) - (y + 3)(y - 1)}{(y - 6)(y - 1)(y + 1)}$$

$$= \dfrac{2y^2 + 3y + 1 - (y^2 + 2y - 3)}{(y - 6)(y - 1)(y + 1)}$$

$$= \dfrac{2y^2 + 3y + 1 - y^2 - 2y + 3}{(y - 6)(y - 1)(y + 1)} = \dfrac{y^2 + y + 4}{(y - 6)(y - 1)(y + 1)}$$

Add.

9. $\dfrac{3x}{7} + \dfrac{4y}{3x}$

Add.

10. $\dfrac{2xy - 2x^2}{x^2 - y^2} + \dfrac{2x + 3}{x + y}$

11. $\dfrac{1}{3y} + \dfrac{4y}{y^2 - 1} + \dfrac{7}{y - 1}$

Subtract.

12. $\dfrac{u}{a+3} - \dfrac{a-4}{a}$

13. $\dfrac{4y-5}{y^2-7y+12} - \dfrac{y+7}{y^2+2y-15}$

Do this calculation.

14. $\dfrac{8x}{x^2-1} + \dfrac{2}{1-x} - \dfrac{4}{x+1}$

We may not always multiply out a numerator or denominator. In fact, it will be helpful when we solve equations not to multiply out denominators.

DO EXERCISES 12 AND 13.

EXAMPLE 9 Calculate: $\dfrac{2x}{x^2-4} + \dfrac{5}{2-x} - \dfrac{1}{2+x}$.

$$\dfrac{2x}{x^2-4} + \dfrac{5}{2-x} - \dfrac{1}{2+x} = \dfrac{2x}{(x-2)(x+2)} + \dfrac{5}{2-x} - \dfrac{1}{2+x}$$

$$= \dfrac{2x}{(x-2)(x+2)} + \dfrac{-1}{-1} \cdot \dfrac{5}{(2-x)} - \dfrac{1}{x+2}$$

$$= \dfrac{2x}{(x-2)(x+2)} + \dfrac{-5}{x-2} - \dfrac{1}{x+2} \qquad \text{The LCM is } (x-2)(x+2).$$

$$= \dfrac{2x}{(x-2)(x+2)} + \dfrac{-5}{x-2} \cdot \dfrac{x+2}{x+2} - \dfrac{1}{x+2} \cdot \dfrac{x-2}{x-2}$$

$$= \dfrac{2x - 5(x+2) - (x-2)}{(x-2)(x+2)}$$

$$= \dfrac{2x - 5x - 10 - x + 2}{(x-2)(x+2)}$$

$$= \dfrac{-4x - 8}{(x-2)(x+2)} = \dfrac{-4(x+2)}{(x-2)(x+2)}$$

$$= \dfrac{-4}{x-2} \cdot \dfrac{x+2}{x+2} = \dfrac{-4}{x-2}$$

DO EXERCISE 14.

EXERCISE SET 9.3

i , **ii** , **iii** Perform the indicated operations. Simplify when possible. If a denominator has three or more factors (other than monomials), leave it factored.

1. $\dfrac{a-3b}{a+b} + \dfrac{a+5b}{a+b}$

2. $\dfrac{x-5y}{x+y} + \dfrac{x+7y}{x+y}$

3. $\dfrac{4y+3}{y-2} - \dfrac{y-2}{y-2}$

4. $\dfrac{3t+2}{t-4} - \dfrac{t-4}{t-4}$

5. $\dfrac{a^2}{a-b} + \dfrac{b^2}{b-a}$

6. $\dfrac{s^2}{r-s} + \dfrac{r^2}{s-r}$

7. $\dfrac{3}{x} - \dfrac{8}{-x}$

8. $\dfrac{2}{a} - \dfrac{5}{-a}$

9. $\dfrac{2x-10}{x^2-25} - \dfrac{5-x}{25-x^2}$

10. $\dfrac{y-9}{y^2-16} - \dfrac{7-y}{16-y^2}$

11. $\dfrac{y-2}{y+4} + \dfrac{y+3}{y-5}$

12. $\dfrac{x-2}{x+3} + \dfrac{x+2}{x-4}$

13. $\dfrac{4xy}{x^2-y^2} + \dfrac{x-y}{x+y}$

14. $\dfrac{5ab}{a^2-b^2} + \dfrac{a+b}{a-b}$

15. $\dfrac{9x+2}{3x^2-2x-8} + \dfrac{7}{3x^2+x-4}$

16. $\dfrac{3y+2}{2y^2-y-10} + \dfrac{8}{2y^2-7y+5}$

17. $\dfrac{4}{x+1} + \dfrac{x+2}{x^2-1} + \dfrac{3}{x-1}$

18. $\dfrac{-2}{y+2} + \dfrac{5}{y-2} + \dfrac{y+3}{y^2-4}$

19. $\dfrac{x-1}{3x+15} - \dfrac{x+3}{5x+25}$

20. $\dfrac{y-2}{4y+8} - \dfrac{y+6}{5y+10}$

21. $\dfrac{5ab}{a^2-b^2} - \dfrac{a-b}{a+b}$

22. $\dfrac{6xy}{x^2 - y^2} - \dfrac{x + y}{x - y}$

23. $\dfrac{3y}{y^2 - 7y + 10} - \dfrac{2y}{y^2 - 8y + 15}$

24. $\dfrac{5x}{x^2 - 6x + 8} - \dfrac{3x}{x^2 - x - 12}$

25. $\dfrac{y}{y^2 - y - 20} + \dfrac{2}{y + 4}$

26. $\dfrac{6}{y^2 + 6y + 9} + \dfrac{5}{y^2 - 9}$

27. $\dfrac{3y + 2}{y^2 + 5y - 24} + \dfrac{7}{y^2 + 4y - 32}$

28. $\dfrac{3y + 2}{y^2 - 7y + 10} + \dfrac{2y}{y^2 - 8y + 15}$

29. $\dfrac{3x - 1}{x^2 + 2x - 3} - \dfrac{x + 4}{x^2 - 9}$

30. $\dfrac{3p - 2}{p^2 + 2p - 24} - \dfrac{p - 3}{p^2 - 16}$

31. $\dfrac{1}{x + 1} - \dfrac{x}{x - 2} + \dfrac{x^2 + 2}{x^2 - x - 2}$

32. $\dfrac{2}{y + 3} - \dfrac{y}{y - 1} + \dfrac{y^2 + 2}{y^2 + 2y - 3}$

33. $\dfrac{4x}{x^2 - 1} + \dfrac{3x}{1 - x} - \dfrac{4}{x - 1}$

34. $\dfrac{5y}{1 - 2y} - \dfrac{2y}{2y + 1} + \dfrac{3}{4y^2 - 1}$

Problem-solving practice

35. The cost c of an insurance policy varies directly as the age a of the insured. A 32-year-old person pays an annual premium of \$152. What is the age of a person who pays \$270?

36. Debbie wants to buy tapes to record her favorite music. She needs some 30-min tapes and some 60-min tapes. If she buys 12 tapes with a total recording time of 10 hr, how many of each length tape did she buy?

Perform the indicated operations and simplify.

37. $2x^{-2} + 3x^{-2}y^{-2} - 7xy^{-1}$

38. $5(x - 3)^{-1} + 4(x + 3)^{-1} - 2(x + 3)^{-2}$

39. $4(y - 1)(2y - 5)^{-1} + 5(2y + 3)(5 - 2y)^{-1} + (y - 4)(2y - 5)^{-1}$

Simplify, using $A = x + y$ and $B = x - y$.

40. $\dfrac{A + B}{A - B} - \dfrac{A - B}{A + B}$

41. $\left(\dfrac{1}{A} + \dfrac{x}{B} \right) \div \left(\dfrac{1}{B} - \dfrac{x}{A} \right)$

9.4

COMPLEX FRACTIONAL EXPRESSIONS

OBJECTIVE

After finishing Section 9.4, you should be able to:

i Simplify complex fractional expressions.

i A *complex fractional expression* is one that has one or more fractional expressions somewhere in its numerator or denominator. The following are examples of complex fractional expressions:

$$\dfrac{x}{x - \dfrac{1}{3}}, \quad \dfrac{2x - \dfrac{4x}{3y}}{\dfrac{5x^2 + 2x}{6y^2}}, \quad \dfrac{\dfrac{1}{a} + \dfrac{1}{b}}{\dfrac{1}{a} - \dfrac{1}{b}}, \quad \dfrac{\dfrac{5}{x}}{\dfrac{x}{y}}.$$

To simplify a complex fractional expression:

1. **Add or subtract, as necessary, to get a single fractional expression in the numerator.**

2. **Add or subtract, as necessary, to get a single fractional expression in the denominator.**

3. **Divide the numerator by the denominator.**

4. **If possible, simplify by removing a factor of 1.**

EXAMPLE 1 Simplify: $\dfrac{x + \dfrac{1}{5}}{x - \dfrac{1}{3}}.$

We have

$$\frac{x + \dfrac{1}{5}}{x - \dfrac{1}{3}} = \frac{x \cdot \dfrac{5}{5} + \dfrac{1}{5}}{x - \dfrac{1}{3}}$$

Finding the LCM in the numerator, multiplying by 1, and adding

$$= \frac{\dfrac{5x + 1}{5}}{x - \dfrac{1}{3}}$$

$$= \frac{\dfrac{5x + 1}{5}}{x \cdot \dfrac{3}{3} - \dfrac{1}{3}}$$

Finding the LCM in the denominator, multiplying by 1, and subtracting

$$= \frac{\dfrac{5x + 1}{5}}{\dfrac{3x - 1}{3}}$$

$$= \frac{5x + 1}{5} \cdot \frac{3}{3x - 1}$$

Multiplying by the reciprocal of the denominator

$$= \frac{15x + 3}{15x - 5}$$

Multiplying numerators and denominators

Simplifying by removing a factor of 1 is not possible in this case.

EXAMPLE 2 Simplify: $\dfrac{1 + \dfrac{1}{x}}{1 - \dfrac{1}{x^2}}.$

$$\frac{1 + \dfrac{1}{x}}{1 - \dfrac{1}{x^2}} = \frac{\dfrac{x}{x} + \dfrac{1}{x}}{\dfrac{x^2}{x^2} - \dfrac{1}{x^2}}$$

Finding the LCM and multiplying by 1

$$= \frac{\dfrac{x + 1}{x}}{\dfrac{x^2 - 1}{x^2}}$$

Adding in the numerator and subtracting in the denominator

$$= \frac{x + 1}{x} \cdot \frac{x^2}{x^2 - 1}$$

Multiplying by the reciprocal of the denominator

Then

$$\frac{x + 1}{x} \cdot \frac{x^2}{x^2 - 1} = \frac{(x + 1) \cdot x^2}{x(x^2 - 1)}$$

$$= \frac{x(x + 1)}{x(x + 1)} \cdot \frac{x}{x - 1}$$

Factoring the fractional expression

$$= \frac{x}{x - 1}$$

Removing a factor of 1

Example 2 shows that the expressions

$$\frac{1 + \dfrac{1}{x}}{1 - \dfrac{1}{x^2}} \quad \text{and} \quad \frac{x}{x - 1}$$

are equivalent. They name the same number for all replacements except 0, 1, and −1. As functions, they are also the same except that the domain does not include 0, 1, or −1.

DO EXERCISES 1 AND 2.

EXAMPLE 3 Simplify: $\dfrac{\dfrac{1}{a} + \dfrac{1}{b}}{\dfrac{1}{a^3} + \dfrac{1}{b^3}}.$

$$\frac{\dfrac{1}{a} + \dfrac{1}{b}}{\dfrac{1}{a^3} + \dfrac{1}{b^3}} = \frac{\dfrac{1}{a} \cdot \dfrac{b}{b} + \dfrac{1}{b} \cdot \dfrac{a}{a}}{\dfrac{1}{a^3} \cdot \dfrac{b^3}{b^3} + \dfrac{1}{b^3} \cdot \dfrac{a^3}{a^3}} = \frac{\dfrac{b}{ab} + \dfrac{a}{ab}}{\dfrac{b^3}{a^3 b^3} + \dfrac{a^3}{a^3 b^3}}$$

$$= \frac{\dfrac{b + a}{ab}}{\dfrac{b^3 + a^3}{a^3 b^3}} \qquad \text{Adding in the numerator and denominator}$$

$$= \frac{b + a}{ab} \cdot \frac{a^3 b^3}{b^3 + a^3} \qquad \text{Multiplying by the reciprocal of the denominator}$$

$$= \frac{(b + a) a^3 b^3}{ab(b^3 + a^3)} = \frac{(b + a) \cdot ab \cdot a^2 b^2}{ab(b + a)(b^2 - ab + a^2)}$$

$$= \frac{(b + a) ab}{(b + a) ab} \cdot \frac{a^2 b^2}{b^2 - ab + a^2}$$

$$= \frac{a^2 b^2}{b^2 - ab + a^2}$$

DO EXERCISES 3 AND 4.

Simplify.

1. $\dfrac{y + \dfrac{1}{2}}{y - \dfrac{1}{7}}$

2. $\dfrac{1 - \dfrac{1}{x}}{1 - \dfrac{1}{x^2}}$

Simplify.

3. $\dfrac{\dfrac{1}{a} + \dfrac{1}{b}}{\dfrac{1}{a} - \dfrac{1}{b}}$

4. $\dfrac{\dfrac{1}{a} - \dfrac{1}{b}}{\dfrac{1}{a^3} - \dfrac{1}{b^3}}$

EXERCISE SET 9.4

i Simplify.

1. $\dfrac{\dfrac{1}{x} + 4}{\dfrac{1}{x} - 3}$

2. $\dfrac{\dfrac{1}{y} + 7}{\dfrac{1}{y} - 5}$

3. $\dfrac{x - \dfrac{1}{x}}{x + \dfrac{1}{x}}$

4. $\dfrac{y + \dfrac{1}{y}}{y - \dfrac{1}{y}}$

5. $\dfrac{\dfrac{3}{x} + \dfrac{4}{y}}{\dfrac{4}{x} - \dfrac{3}{y}}$

6. $\dfrac{\dfrac{2}{y}+\dfrac{5}{z}}{\dfrac{1}{y}-\dfrac{4}{z}}$

7. $\dfrac{\dfrac{x^2-y^2}{xy}}{\dfrac{x-y}{y}}$

8. $\dfrac{\dfrac{a^2-b^2}{ab}}{\dfrac{a-b}{b}}$

9. $\dfrac{a-\dfrac{3a}{b}}{b-\dfrac{b}{a}}$

10. $\dfrac{1-\dfrac{2}{3x}}{x-\dfrac{4}{9x}}$

11. $\dfrac{\dfrac{1}{a}+\dfrac{1}{b}}{\dfrac{a^2-b^2}{ab}}$

12. $\dfrac{\dfrac{x^2-y^2}{xy}}{\dfrac{1}{x}+\dfrac{1}{y}}$

13. $\dfrac{\dfrac{1}{x+h}-\dfrac{1}{x}}{h}$

14. $\dfrac{\dfrac{1}{a-h}-\dfrac{1}{a}}{h}$

15. $\dfrac{\dfrac{y^2-y-6}{y^2-5y-14}}{\dfrac{y^2+6y+5}{y^2-6y-7}}$

16. $\dfrac{\dfrac{x^2-x-12}{x^2-2x-15}}{\dfrac{x^2+8x+12}{x^2-5x-14}}$

Problem-solving practice

17. The length of one rectangle is 3 less than the length of a second rectangle. The width of the first rectangle is 4 less than the width of the second. If the perimeter of the second rectangle is 1 less than twice the perimeter of the first, what are their respective perimeters?

18. A waitress received \$3.50 in tips on Monday, \$2.75 in tips on Tuesday, and \$4.50 in tips on Wednesday. How much will she have to earn in tips on Thursday if her average tips for the four days is to be \$3.75?

○ ───────────────────────────────────

Simplify.

19. $\dfrac{5x^{-1}-5y^{-1}+10x^{-1}y^{-1}}{6x^{-1}-6y^{-1}+12x^{-1}y^{-1}}$

20. $\dfrac{\dfrac{4a}{2a^2-a-1}-\dfrac{4}{a-1}}{\dfrac{1}{a-1}+\dfrac{6}{2a+1}}$

21. $2+\dfrac{2}{2+\dfrac{2}{2+\dfrac{2}{2+\dfrac{2}{x}}}}$

22. $\left[\dfrac{\dfrac{x+3}{x-3}+1}{\dfrac{x+3}{x-3}-1}\right]^4$

23. $(a^2-ab+b^2)^{-1}(a^2b^{-1}+b^2a^{-1})(a^{-2}-b^{-2})(a^{-2}+2a^{-1}b^{-1}+b^{-2})^{-1}$

Find the reciprocal and simplify.

24. $x^2-\dfrac{1}{x}$

25. $\dfrac{1-\dfrac{1}{a}}{a-1}$

26. $\dfrac{a^3+b^3}{a+b}$

27. $1+\dfrac{1}{1+\dfrac{1}{1+\dfrac{1}{1+\dfrac{1}{x}}}}$

28. For $f(x)=\dfrac{1}{1-x}$, find $f(f(x))$ and $f(f(f(x)))$.

Find and simplify $\dfrac{f(x+h)-f(x)}{h}$ for each rational function f.

29. $f(x)=\dfrac{3}{x^2}$

30. $f(x)=\dfrac{5}{x}$

31. $f(x)=\dfrac{1}{1-x}$

32. $f(x)=\dfrac{x}{1+x}$

9.5

SOLVING FRACTIONAL EQUATIONS

i A *fractional equation* is an equation that contains one or more fractional expressions. Here are some examples:

$$\frac{2}{3} - \frac{5}{6} = \frac{1}{x}, \qquad \frac{x-1}{x-5} = \frac{4}{x-5}.$$

Solve.

1. $\dfrac{2}{3} + \dfrac{5}{6} = \dfrac{1}{x}$

> To solve a fractional equation, we multiply on both sides by the LCM of all the denominators. This is called *clearing of fractions.*

EXAMPLE 1 Solve: $\dfrac{2}{3} - \dfrac{5}{6} = \dfrac{1}{x}$.

The LCM of all denominators is $6x$, or $2 \cdot 3 \cdot x$. We multiply on both sides of the equation by the LCM:

$$(2 \cdot 3 \cdot x) \cdot \left(\frac{2}{3} - \frac{5}{6}\right) = (2 \cdot 3 \cdot x) \cdot \frac{1}{x} \qquad \text{Multiplying by the LCM}$$

$$2 \cdot 3 \cdot x \cdot \frac{2}{3} - 2 \cdot 3 \cdot x \cdot \frac{5}{6} = 2 \cdot 3 \cdot x \cdot \frac{1}{x} \qquad \text{Multiplying to remove parentheses}$$

$$\frac{2 \cdot 3 \cdot x \cdot 2}{3} - \frac{2 \cdot 3 \cdot x \cdot 5}{6} = \frac{2 \cdot 3 \cdot x}{x}$$

$$4x - 5x = 6 \qquad \text{Simplifying}$$

$$-x = 6$$

$$-1 \cdot x = 6$$

$$x = -6.$$

Check:

$$\frac{2}{3} - \frac{5}{6} = \frac{1}{x}$$

$$\begin{array}{c|c} \dfrac{2}{3} - \dfrac{5}{6} & \dfrac{1}{-6} \\[2ex] \dfrac{4}{6} - \dfrac{5}{6} & -\dfrac{1}{6} \\[2ex] -\dfrac{1}{6} & \end{array}$$

When clearing of fractions, as in Example 1, be sure to multiply *every* term in the equation by the LCM.

Note that when we *clear of fractions*, all the denominators disappear. Then we have an equation without fractional expressions, which we know how to solve. Note too that we have introduced a new use of the LCM here. Before, we used the LCM in adding or subtracting. Now we are working with equations. We clear of fractions by multiplying on both sides by the LCM.

DO EXERCISE 1.

Solve.

2. $\dfrac{y-4}{5} - \dfrac{y+7}{2} = 5$

Solve.

3. $\dfrac{x-7}{x-9} = \dfrac{2}{x-9}$

EXAMPLE 2 Solve: $\dfrac{x+1}{2} - \dfrac{x-3}{3} = 3$.

The LCM of all the denominators is $2 \cdot 3$, or 6. We multiply on both sides of the equation by the LCM:

$$2 \cdot 3 \cdot \left(\dfrac{x+1}{2} - \dfrac{x-3}{3} \right) = 2 \cdot 3 \cdot 3 \qquad \text{Multiplying on both sides by the LCM}$$

$$2 \cdot 3 \cdot \dfrac{x+1}{2} - 2 \cdot 3 \cdot \dfrac{x-3}{3} = 2 \cdot 3 \cdot 3 \qquad \text{Multiplying to remove parentheses}$$

$$\dfrac{2 \cdot 3(x+1)}{2} - \dfrac{2 \cdot 3(x-3)}{3} = 18$$

$$3(x+1) - 2(x-3) = 18$$

$$\left.\begin{array}{r} 3x + 3 - 2x + 6 = 18 \\ x + 9 = 18 \end{array}\right\} \quad \text{Multiplying and collecting like terms}$$

$$x = 9.$$

Check:

$$\begin{array}{c|c} \dfrac{x+1}{2} - \dfrac{x-3}{3} = 3 \\ \hline \dfrac{9+1}{2} - \dfrac{9-3}{3} & 3 \\ 5 - 2 & \\ 3 & \end{array}$$

In solving fractional equations, we often multiply on both sides by an expression containing a variable. When we do so, we may obtain an equation that is not equivalent to the first one. Thus it is extremely important to check possible solutions in the original equation. They may not check, even though we have made no error.

DO EXERCISE 2.

EXAMPLE 3 Solve: $\dfrac{x-1}{x-5} = \dfrac{4}{x-5}$.

The LCM of the denominators is $x - 5$. We multiply by $x - 5$:

$$(x-5) \cdot \dfrac{x-1}{x-5} = (x-5) \cdot \dfrac{4}{x-5}$$

$$x - 1 = 4$$

$$x = 5.$$

Check:

$$\begin{array}{c|c} \dfrac{x-1}{x-5} = \dfrac{4}{x-5} \\ \hline \dfrac{5-1}{5-5} & \dfrac{4}{5-5} \\ \dfrac{4}{0} & \dfrac{4}{0} \end{array}$$

We know that 5 is not a solution of the original equation because it results in division by 0. In fact, the equation has no solution.

DO EXERCISE 3.

In solving an equation, when we multiply by an expression containing a variable, we may get an equation having solutions that are not solutions of the original equation.

EXAMPLE 4   Solve: $\dfrac{x^2}{x-2} = \dfrac{4}{x-2}$.

The LCM of the denominators is $x - 2$. We multiply by $x - 2$:

$$(x-2)\cdot\frac{x^2}{x-2} = (x-2)\cdot\frac{4}{x-2}$$

$$x^2 = 4$$

$$x^2 - 4 = 0$$

$$(x+2)(x-2) = 0$$

$$x = -2 \quad \text{or} \quad x = 2 \qquad \text{Using the principle of zero products}$$

Check:

For 2: $\dfrac{x^2}{x-2} = \dfrac{4}{x-2}$

$$\frac{2^2}{2-2} \;\bigg|\; \frac{4}{2-2}$$

$$\frac{4}{0} \;\bigg|\; \frac{4}{0}$$

For -2: $\dfrac{x^2}{x-2} = \dfrac{4}{x-2}$

$$\frac{(-2)^2}{-2-2} \;\bigg|\; \frac{4}{-2-2}$$

$$\frac{4}{-4} \;\bigg|\; \frac{4}{-4}$$

The number -2 is a solution, but 2 is not (it results in division by 0).

DO EXERCISE 4.

EXAMPLE 5   Solve: $x + \dfrac{6}{x} = 5$.

The LCM of the denominators is x. We multiply on both sides by x:

$$x\left(x + \frac{6}{x}\right) = 5\cdot x \qquad \text{Multiplying on both sides by } x$$

$$x^2 + x\cdot\frac{6}{x} = 5x$$

$$x^2 + 6 = 5x \qquad \text{Simplifying}$$

$$x^2 - 5x + 6 = 0 \qquad \text{Getting 0 on one side}$$

$$(x-3)(x-2) = 0 \qquad \text{Factoring}$$

$$x = 3 \quad \text{or} \quad x = 2. \qquad \text{Using the principle of zero products}$$

Check:

For 3: $x + \dfrac{6}{x} = 5$

$$3 + \frac{6}{3} \;\bigg|\; 5$$

$$3 + 2$$

$$5$$

For 2: $x + \dfrac{6}{x} = 5$

$$2 + \frac{6}{2} \;\bigg|\; 5$$

$$2 + 3$$

$$5$$

The solutions are 2 and 3.

DO EXERCISES 5 AND 6.

Solve.

4. $\dfrac{x^2}{x+3} = \dfrac{9}{x+3}$

Solve.

5. $\dfrac{x^2}{x-3} = \dfrac{9}{x-3}$

6. $x - \dfrac{12}{x} = 1$

Solve.

7. $\dfrac{2}{x-1} = \dfrac{3}{x+2}$

EXAMPLE 6 Solve: $\dfrac{2}{x-1} = \dfrac{3}{x+1}$.

The LCM of the denominators is $(x-1)(x+1)$.

$$(x-1)(x+1) \cdot \frac{2}{(x-1)} = (x-1)(x+1) \cdot \frac{3}{x+1} \qquad \text{Multiplying}$$

$$2(x+1) = 3(x-1) \qquad \text{Simplifying}$$

$$2x + 2 = 3x - 3$$

$$5 = x$$

Check:

$$\frac{2}{x-1} = \frac{3}{x+1}$$

$$\begin{array}{c|c} \dfrac{2}{5-1} & \dfrac{3}{5+1} \\[2mm] \hline \dfrac{2}{4} & \dfrac{3}{6} \\[2mm] \dfrac{1}{2} & \dfrac{1}{2} \end{array}$$

The number 5 is the solution.

EXAMPLE 7 Solve: $\dfrac{2}{x+5} + \dfrac{1}{x-5} = \dfrac{16}{x^2-25}$.

The LCM is $(x+5)(x-5)$. We multiply by $(x+5)(x-5)$:

$$(x+5)(x-5) \cdot \left[\frac{2}{x+5} + \frac{1}{x-5} \right] = (x+5)(x-5) \cdot \frac{16}{x^2-25}$$

8. $\dfrac{2}{x^2-9} + \dfrac{5}{x-3} = \dfrac{3}{x+3}$

$$(x+5)(x-5) \cdot \frac{2}{x+5} + (x+5)(x-5) \cdot \frac{1}{x-5} = (x+5)(x-5) \cdot \frac{16}{x^2-25}$$

$$2(x-5) + (x+5) = 16$$

$$2x - 10 + x + 5 = 16$$

$$3x - 5 = 16$$

$$3x = 21$$

$$x = 7$$

Check:

$$\frac{2}{x+5} + \frac{1}{x-5} = \frac{16}{x^2-25}$$

$$\begin{array}{c|c} \dfrac{2}{7+5} + \dfrac{1}{7-5} & \dfrac{16}{7^2-25} \\[2mm] \hline \dfrac{2}{12} + \dfrac{1}{2} & \dfrac{16}{49-25} \\[2mm] \dfrac{8}{12} & \dfrac{16}{24} \\[2mm] \dfrac{2}{3} & \dfrac{2}{3} \end{array}$$

The solution is 7.

DO EXERCISES 7 AND 8.

EXERCISE SET 9.5

i Solve.

1. $\dfrac{2}{5} + \dfrac{7}{8} = \dfrac{y}{20}$

2. $\dfrac{4}{5} + \dfrac{1}{3} = \dfrac{t}{9}$

3. $\dfrac{1}{3} - \dfrac{5}{6} = \dfrac{1}{x}$

4. $\dfrac{5}{8} - \dfrac{2}{5} = \dfrac{1}{y}$

5. $\dfrac{x}{3} - \dfrac{x}{4} = 12$

6. $\dfrac{y}{5} - \dfrac{y}{3} = 15$

7. $y + \dfrac{5}{y} = -6$

8. $x + \dfrac{4}{x} = -5$

9. $\dfrac{4}{z} + \dfrac{2}{z} = 3$

10. $\dfrac{4}{3y} - \dfrac{3}{y} = \dfrac{10}{3}$

11. $\dfrac{x-3}{x+2} = \dfrac{1}{5}$

12. $\dfrac{y-5}{y+1} = \dfrac{3}{5}$

13. $\dfrac{3}{y+1} = \dfrac{2}{y-3}$

14. $\dfrac{4}{x-1} = \dfrac{3}{x+2}$

15. $\dfrac{y-1}{y-3} = \dfrac{2}{y-3}$

16. $\dfrac{x-2}{x-4} = \dfrac{2}{x-4}$

17. $\dfrac{x+1}{x} = \dfrac{3}{2}$

18. $\dfrac{y+2}{y} = \dfrac{5}{3}$

19. $\dfrac{2}{x} - \dfrac{3}{x} + \dfrac{4}{x} = 5$

20. $\dfrac{4}{y} - \dfrac{6}{y} + \dfrac{8}{y} = 8$

21. $\dfrac{1}{2} - \dfrac{4}{9x} = \dfrac{4}{9} - \dfrac{1}{6x}$

22. $-\dfrac{1}{3} - \dfrac{5}{4y} = \dfrac{3}{4} - \dfrac{1}{6y}$

23. $\dfrac{60}{x} - \dfrac{60}{x-5} = \dfrac{2}{x}$

24. $\dfrac{50}{y} - \dfrac{50}{y-2} = \dfrac{4}{y}$

25. $\dfrac{7}{5x-2} = \dfrac{5}{4x}$

26. $\dfrac{5}{y+4} = \dfrac{3}{y-2}$

27. $\dfrac{x}{x-2} + \dfrac{x}{x^2-4} = \dfrac{x+3}{x+2}$

28. $\dfrac{3}{y-2} + \dfrac{2y}{4-y^2} = \dfrac{5}{y+2}$

29. $\dfrac{a}{2a-6} - \dfrac{3}{a^2-6a+9} = \dfrac{a-2}{3a-9}$

30. $\dfrac{2}{x+4} + \dfrac{2x-1}{x^2+2x-8} = \dfrac{1}{x-2}$

31. $\dfrac{2x+3}{x-1} = \dfrac{10}{x^2-1} + \dfrac{2x-3}{x+1}$

32. $\dfrac{y}{y+1} + \dfrac{3y+5}{y^2+4y+3} = \dfrac{2}{y+3}$

Problem-solving practice

33. Beaker A contains a solution that is 40% ammonia. Beaker B contains a solution that is 60% ammonia. How much of each should be added together to obtain 35 L of a solution that is 55% ammonia?

34. Find two consecutive even numbers whose product is 288.

Solve.

35. $\left(\dfrac{1}{1+x} + \dfrac{x}{1-x} \right) \div \left(\dfrac{x}{1+x} - \dfrac{1}{1-x} \right) = -1$

36. $\dfrac{x+3}{x+2} - \dfrac{x+4}{x+3} = \dfrac{x+5}{x+4} - \dfrac{x+6}{x+5}$

37. $\dfrac{2.315}{y} - \dfrac{12.6}{17.4} = \dfrac{6.71}{7} + 0.763$

38. $\dfrac{6.034}{x} - 43.17 = \dfrac{0.793}{x} + 18.15$

39. $\dfrac{x^3+8}{x+2} = x^2 - 2x + 4$

40. $\dfrac{(x-3)^2}{x-3} = x - 3$

Equations that are true for all sensible replacements of the variables are called *identities*. Determine which equations are identities.

41. $\dfrac{x^2+6x-16}{x-2} = x + 8$

42. $\dfrac{x^3+8}{x^2-4} = \dfrac{x^2-2x+4}{x-2}$

43. $(x+3)^2 = x^2 + 9$

44. $(y-5)^2 = y^2 - 25$

45. $(x+3)^2 = x^2 + 6x + 9$

46. $(y-5)^2 = y^2 - 10y + 25$

9.6

SOLVING PROBLEMS

Now that we have considered equations containing fractional expressions, we can use that skill in solving certain kinds of problems that we could not have handled before. The problem-solving steps and strategy are the same.

i Problems Using Fractional Notation

EXAMPLE 1 If a certain number is added to 5 times the reciprocal of 2 more than that number, the result is 4. Find the number.

OBJECTIVES

After finishing Section 9.6, you should be able to:

i Solve problems using fractional equations.

ii Solve certain problems involving amounts of work.

iii Solve certain problems involving motion.

1. When 12 is divided by 5 more than a certain number, the result is 3 more than 2 less than the number. Find the number.

This problem is stated in such a way that we can translate directly to an equation.

2. **Translate**

A certain number added to five times the reciprocal of 2 more than the number is 4

$$y + 5 \cdot \frac{1}{y + 2} = 4$$

3. **Carry out** We solve as follows:

$$y + 5 \cdot \frac{1}{y + 2} = 4$$

$$(y + 2) \cdot y + (y + 2) \cdot 5 \cdot \frac{1}{y + 2} = (y + 2) \cdot 4 \qquad \text{Multiplying by the LCM}$$

$$y^2 + 2y + 5 = 4y + 8 \qquad \text{Simplifying}$$

$$y^2 - 2y - 3 = 0 \qquad \text{Collecting like terms all on one side}$$

$$(y - 3)(y + 1) = 0 \qquad \text{Factoring}$$

$$y - 3 = 0 \quad \text{or} \quad y + 1 = 0 \qquad \text{Principle of zero products}$$

$$y = 3 \quad \text{or} \qquad y = -1$$

4. **Check** The possible solutions are 3 and -1. We check 3 in the conditions of the problem.

Number:	3
2 more than the number:	5
Reciprocal of 2 more than the number:	$\frac{1}{5}$
5 times the reciprocal of 2 more than the number:	$5 \cdot \frac{1}{5} = 1$
Sum of number and 5 times the reciprocal of 2 more than the number:	$3 + 1 = 4$

The number 3 checks. So does the number -1, but we leave that check to you.

5. **State** The answer is that there are two numbers satisfying the conditions of the problem. They are 3 and -1.

DO EXERCISE 1.

ii Problems Involving Work

EXAMPLE 2 Tom can mow a lawn in 4 hours. Penny can mow the same lawn in 5 hours. How long would it take them working together, with 2 mowers, to mow the lawn?

1. **Familiarize** We list the facts:

Tom: Can mow in 4 hours;

Penny: Can mow in 5 hours.

We want to know how long it will take them working together. Let's let t be that number of hours.

We might try a guess. Suppose we guess 9 hours (we get that by adding 4 and 5). It's a good guess, but we are way off. Either one of them can do it alone in less than 9 hours.

Let's try making some kind of table. First of all, we know that Tom should be able to mow $\frac{1}{4}$ of the lawn in 1 hour, since he can do it all in 4 hours.

Number of hours	1	2	3	4	5	$\frac{1}{4}$ in 1 hour
Tom	$\frac{1}{4}$	$\frac{2}{4}$	$\frac{3}{4}$	$\frac{4}{4}$		— All in 4 hours
Penny	$\frac{1}{5}$	$\frac{2}{5}$	$\frac{3}{5}$	$\frac{4}{5}$	$\frac{5}{5}$	— All in 5 hours

$\frac{1}{5}$ in 1 hour

In two hours, Tom can do $\frac{2}{4}$ of the lawn. In 4 hours, he does it all, or $\frac{4}{4}$ of the lawn. Similarly, Penny can do $\frac{1}{5}$ of the lawn in 1 hour, $\frac{2}{5}$ in two hours, and so on. She can do $\frac{5}{5}$, or all of it, in 5 hours.

Since Tom can do $\frac{1}{4}$ of the work in 1 hour and Penny can do $\frac{1}{5}$ of it in 1 hour, then working together they should do

$$\frac{1}{4} + \frac{1}{5}$$

of the lawn in 1 hour. We are supposing that it takes them t hours working together, so they should do $1/t$ of the lawn in 1 hour.

2. **Translate** They can do $1/t$ of the lawn in 1 hour. They can also do $\frac{1}{4} + \frac{1}{5}$ of it in one hour. That gives us an equation:

$$\frac{1}{4} + \frac{1}{5} = \frac{1}{t}.$$

3. **Carry out** We solve the equation:

$$20t\left(\frac{1}{4} + \frac{1}{5}\right) = 20t \cdot \frac{1}{t}$$

$$\frac{20t}{4} + \frac{20t}{5} = \frac{20t}{t}$$

Multiplying on both sides by the LCM of the denominators and clearing of fractions

$$5t + 4t = 20$$

$$9t = 20$$

$$t = \frac{20}{9}$$

4. **Check** The possible solution is $\frac{20}{9}$, or $2\frac{2}{9}$ hours. If Tom works $\frac{20}{9}$ hours, he will do $\frac{1}{4} \cdot \frac{20}{9}$ of the work, or $\frac{5}{9}$ of it. If Penny works $\frac{20}{9}$ hours, she will do $\frac{1}{5} \cdot \frac{20}{9}$ of the work, or $\frac{4}{9}$ of it. Altogether, they will do $\frac{5}{9} + \frac{4}{9}$ of the work, or all of it.

5. **State** The answer is that it will take $\frac{20}{9}$, or $2\frac{2}{9}$ hours for Tom and Penny to mow the lawn when they work together.

DO EXERCISE 2.

EXAMPLE 3 At a factory, smokestack A pollutes the air twice as fast as smokestack B. When the stacks operate together, they yield a certain amount of pollution in

2. Fred does a certain typing job in 6 hours. Fran can do the same job in 4 hours. How long would it take them to do the same amount of typing working together?

15 hours. Find the time it would take each to yield that same amount of pollution operating individually.

1. **Familiarize** Smokestack A takes half as much time as smokestack B to pollute the same amount. Let's let t represent the time that it will take smokestack A. Then $2t$ represents the time it takes smokestack B. We list what we know.

Time for smokestack A: t hours;

Time for smokestack B: twice that long, or $2t$ hours;

Time together: 15 hours

Thus in one hour,

A does $\dfrac{1}{t}$ of the total;

B does $\dfrac{1}{2t}$ of the total;

Together they do $\dfrac{1}{15}$ of the total.

2. **Translate** In 1 hour stacks A and B will yield $\frac{1}{15}$ of the total. They will also yield

$$\frac{1}{t} + \frac{1}{2t}$$

of the total. We now have an equation:

$$\frac{1}{t} + \frac{1}{2t} = \frac{1}{15}.$$

3. **Carry out** We solve the equation:

$$30t\left(\frac{1}{t} + \frac{1}{2t}\right) = 30t \cdot \frac{1}{15}$$

$$\frac{30t}{t} + \frac{30t}{2t} = \frac{30t}{15}$$

$$30 + 15 = 2t$$

$$45 = 2t.$$

Multiplying on both sides by the LCM of the denominators and clearing of fractions

$$t = \frac{45}{2}, \quad \text{or} \quad 22\frac{1}{2}.$$

4. | Check ▷ The solution of the equation is $\frac{45}{2}$, the time in hours for smokestack A. If it takes smokestack B twice as long, then its time will be 45. Let's try these numbers. In one hour, A will do $\frac{1}{45/2}$ of the total and B will do $\frac{1}{45}$ of the total. Together, they will do

$$\frac{1}{45/2} + \frac{1}{45}$$

of the total. Adding, we get

$$\frac{2}{45} + \frac{1}{45} = \frac{3}{45} = \frac{1}{15}.$$

The answers check.

5. | State ▷ The answer is that it would take smokestack A $\frac{45}{2}$, or $22\frac{1}{2}$ hours and smokestack B 45 hours.

DO EXERCISE 3.

iii Problems Involving Motion

When a problem deals with speed, distance, and time, we can expect to use the definition of speed, or something equivalent to it.

DEFINITION OF SPEED

$$\textbf{Speed} = \frac{\textbf{distance}}{\textbf{time}}, \quad \textbf{or} \quad r = \frac{d}{t}$$

The equation

$$r = \frac{d}{t}$$

can be solved for either d or t. That gives us two other equations we can use:

$$d = rt \quad \text{or} \quad t = \frac{d}{r}.$$

You can remember $d = rt$, as many do, or you can remember the definition of speed. In any case, you should be able to obtain any of the three equations as needed in a problem situation.

EXAMPLE 4   A train leaves Sioux City traveling east at 30 mph. Two hours later another train leaves Sioux City in the same direction on a parallel track at 45 mph. How far from Sioux City will the faster train catch the slower one?

3. Two pipes carry water to the same tank. Pipe A, working alone, can fill the tank three times as fast as pipe B. Together the pipes can fill the tank in 24 hours. Find the time each would take to fill the tank alone.

4. A train leaves Gilville traveling east at 35 mph. One hour later a faster train leaves Gilville, also traveling east on a parallel track at 40 mph. How far from Gilville will the faster train catch the slower one?

1. | Familiarize ▷ We make a drawing and label it.

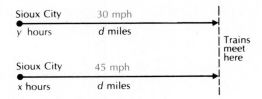

We have used x and y for the times and d for the distance. From the figure we see that the distance is the same for both trains. We can organize the information in a table.

	Distance	Speed	Time
Slow train	d	30	y (2 hours more than the fast train)
Fast train	d	45	x

2. | Translate ▷ The table suggests that we use the relation $d = rt$. If we do, we get two equations:

$$d = 30y,$$
$$d = 45x.$$

Then we have

$$45x = 30y.$$

From the table, we also see that $y = x + 2$. Thus we have a system of equations:

$$45x = 30y,$$
$$y = x + 2.$$

3. | Carry out ▷ We solve the system of equations. When we do we find that $x = 4$ and $y = 6$.

4. | Check ▷ Since we have used x and y for the times, we know that the time for the slower train is 6 hours and the time for the faster train is 4 hours. Let's compute the distance in two cases:

Slow train: $d = rt = 30 \cdot 6 = 180$ miles;

Fast train: $d = rt = 45 \cdot 4 = 180$ miles.

We have an answer.

5. | State ▷ The trains meet at a distance of 180 miles from Sioux City.

DO EXERCISE 4.

EXAMPLE 5 An airplane flies 1062 miles with the wind. In the same amount of time, it can fly 738 miles against the wind. The speed of the plane in still air is 200 mph. Find the speed of the wind.

1. **Familiarize** We know that the speed of the plane is 200, but we don't know the speed of the wind. Let's call it w. Now we make a drawing and label it.

```
1062 mi                    t hours
→
200 + w          (The wind increases the
                  speed over the ground.)

        738 mi             t hours
←
        200 − w  (The wind decreases the
                  speed over the ground.)
```

We can also keep the data in a table.

	Distance	Speed	Time
With the wind	1062	$200 + w$	t
Against the wind	738	$200 - w$	t

2. **Translate** The table suggests that we use

$$t = \frac{d}{r}.$$

Thus we get two equations:

$$t = \frac{1062}{200 + w} \quad \text{and} \quad t = \frac{738}{200 - w}.$$

3. **Carry out** We solve:

$$\frac{1062}{200 + w} = \frac{738}{200 - w} \qquad \text{Using substitution}$$

$$(200 + w)(200 - w)\left(\frac{1062}{200 + w}\right) = (200 + w)(200 - w)\left(\frac{738}{200 - w}\right) \qquad \begin{array}{l}\text{Multiplying}\\\text{by the LCM}\\\text{to clear of}\\\text{fractions}\end{array}$$

$$(200 - w)1062 = (200 + w)738$$

$$212{,}400 - 1062w = 147{,}600 + 738w$$

$$64{,}800 = 1800w$$

$$w = 36.$$

5. A boat travels 246 miles downstream in the same time that it takes to travel 180 miles upstream. The speed of the current in the stream is 5.5 mph. Find the speed of the boat in still water.

4. | Check | If it checks, the speed of the wind is 36 mph. Let's calculate:

With the wind: Speed of the plane is $200 + 36 = 236$ mph;

Against the wind: Speed of the plane is $200 - 36 = 164$ mph.

Now we calculate time, using

$$t = \frac{d}{r}.$$

236 mph: $t = \dfrac{1062}{236} = 4.5$ hours;

164 mph: $t = \dfrac{738}{164} = 4.5$ hours.

We have an answer.

5. | State | The answer is that the speed of the wind is 36 mph.

DO EXERCISE 5.

EXERCISE SET 9.6

Solve.

i

1. The reciprocal of 5 plus the reciprocal of 7 is the reciprocal of what number?

2. The reciprocal of 3 plus the reciprocal of 6 is the reciprocal of what number?

3. The sum of a number and 6 times its reciprocal is -5. Find the number.

4. The sum of a number and 21 times its reciprocal is -10. Find the number.

ii

5. Sam, an experienced shipping clerk, can fill a certain order in 5 hours. Willy, a new clerk, needs 9 hours to do the same job. Working together, how long will it take them to fill the order?

6. Paul can paint a room in 4 hours. Sally can paint the same room in 3 hours. Working together, how long will it take them to paint the room?

7. A swimming pool can be filled in 12 hours if water enters through a pipe alone, or in 30 hours if water enters through a hose alone. If water is entering through both the pipe and the hose, how long will it take to fill the pool?

8. A tank can be filled in 18 hours by pipe A alone and in 22 hours by pipe B alone. How long will it take to fill the tank if both pipes are working?

9. Bill can clear a lot in 5.5 hours. His partner can do the same job in 7.5 hours. How long will it take them to clear the lot working together?

10. One printing press can print an order of booklets in 4.5 hours. Another press can do the same job in 5.5 hours. How long will it take if both presses are used?

11. A can paint the neighbor's house 4 times as fast as B. The year they worked together it took them 8 days. How long would it take each to paint the house alone?

12. A can deliver papers 3 times as fast as B. If they work together, it takes them 1 hour. How long would it take each to deliver the papers alone?

iii

13. The speed of a stream is 3 mph. A boat travels 4 mi upstream in the same time it takes to travel 10 mi downstream. What is the speed of the boat in still water?

14. The speed of a stream is 4 mph. A boat travels 6 mi upstream in the same time it takes to travel 12 mi downstream. What is the speed of the boat in still water?

15. A train leaves a station and travels north at 75 km/h. Two hours later a second train leaves on a parallel track traveling north at 125 km/h. How far from the station will the second train overtake the first train?

16. A private airplane leaves an airport and flies due east at 180 km/h. Two hours later a jet leaves the same airport and flies due east at 900 km/h. How far from the airport will the jet overtake the private plane?

17. The speed of train A is 12 mph slower than the speed of train B. Train A travels 230 mi in the same time it takes train B to travel 290 mi. Find the speed of each train.

18. The speed of a passenger train is 14 mph faster than the speed of a freight train. The passenger train travels 400 mi in the same time it takes the freight train to travel 330 mi. Find the speed of each train.

19. Suzie has a boat that can move at a speed of 15 km/h in still water. She rides 140 km downstream in a river in the same time it takes to ride 35 km upstream. What is the speed of the river?

20. A paddleboat can move at a speed of 2 km/h in still water. The boat is paddled 4 km downstream in a river in the same time it takes to go 1 km upstream. What is the speed of the river?

21. At what time after 4:00 will the minute hand and the hour hand of a clock first be in the same position?

22. At what time after 10:30 will the hands of a clock first be perpendicular?

23. A boat travels 96 km downstream in 4 hr. It travels 28 km upstream in 7 hr. Find the speed of the boat and the speed of the stream.

24. An airplane carries enough fuel for 6 hours of flight time, and its speed in still air is 240 mph. It leaves an airport against a wind of 40 mph and returns to the same airport with a wind of 40 mph. How far can it fly under those conditions without refueling?

25. A motor boat travels 3 times as fast as the current. A trip up the river and back takes 10 hours, and the total distance of the trip is 100 km. Find the speed of the current.

26. An employee drives to work at 50 mph and arrives one minute late. The employee drives to work at 60 mph and arrives 5 minutes early. How far does the employee live from work?

Average speed is defined as *total distance divided by total time.*

27. A driver went 200 km. For the first half of the trip the driver traveled at a speed of 40 km/h. For the second half of the trip the driver traveled at a speed of 60 km/h. What was the average speed for half of the trip? (It is *not* 50 km/h.)

28. For half of the time of a trip a driver travels at 40 mph. What speed would the driver have to travel for the last half of the trip so that the average speed for the entire trip would be 45 mph?

29. Three trucks A, B, and C, working together, can move a load of sand in t hours. When working alone, it takes A, 1 extra hour to move the sand; B, 6 extra hours; and C, t extra hours. Find t.

30. An employee drives to work at 45 mph and arrives one minute early. The employee drives to work at 40 mph and arrives one minute late. How far does the employee live from work?

9.7

DIVISION OF POLYNOMIALS

Remember that a rational expression indicates division. In some cases, it is useful to actually carry out that division. We now consider division of polynomials.

i Divisor a Monomial

To divide a polynomial by a monomial, we can first write a fractional expression, as in the following example.

EXAMPLE 1 Divide $12x^3 + 8x^2 + x + 4$ by $4x$.

$$\frac{12x^3 + 8x^2 + x + 4}{4x} \quad \text{Writing a fractional expression}$$

$$= \frac{12x^3}{4x} + \frac{8x^2}{4x} + \frac{x}{4x} + \frac{4}{4x} \quad \text{Doing the reverse of adding}$$

$$= 3x^2 + 2x + \frac{1}{4} + \frac{1}{x} \quad \text{Doing the four indicated divisions}$$

DO EXERCISE 1.

OBJECTIVES

After finishing Section 9.7, you should be able to:

i Divide a polynomial by a monomial.

ii Divide a polynomial by a divisor that is not a monomial, and if there is a remainder, express the result in two ways.

Divide.

1. $\dfrac{x^3 + 16x^2 + 6x}{2x}$

Divide.

2. $(15y^5 - 6y^4 + 18y^3) \div 3y^2$

3. $(x^4 + 10x^3 + 16x^2) \div 2x^2$

Divide and check.

4. $x - 2 \overline{)\, x^2 + 3x - 10}$

EXAMPLE 2   **Divide:** $(8x^4 - 3x^3 + 5x^2) \div x^2$.

$$\frac{8x^4 - 3x^3 + 5x^2}{x^2} = \frac{8x^4}{x^2} - \frac{3x^3}{x^2} + \frac{5x^2}{x^2}$$

$$= 8x^2 - 3x + 5$$

It is not necessary to write out all of the steps shown in these examples. You should try to write only the answer.

To divide a polynomial by a monomial, we can divide each term by the monomial.

DO EXERCISES 2 AND 3.

ii   Divisor Not a Monomial

When the divisor is not a monomial, we use a procedure very much like long division in arithmetic.

EXAMPLE 3   Divide $x^2 + 5x + 8$ by $x + 3$.

$$
\begin{array}{r}
x \\
x + 3 \overline{)\, x^2 + 5x + 8} \\
x^2 + 3x \\
\hline
2x
\end{array}
$$

← Divide the first term of the dividend by the first term of the divisor: $x^2/x = x$.

← Multiply x above by divisor.

← Subtract.

The subtraction we have done is $(x^2 + 5x) - (x^2 + 3x)$. Remember: To subtract, add the inverse (change the sign of every term, then add).

We now "bring down" the other terms of the dividend—in this case, 8:

$$
\begin{array}{r}
x + 2 \\
x + 3 \overline{)\, x^2 + 5x + 8} \\
x^2 + 3x \\
\hline
2x + 8 \\
2x + 6 \\
\hline
2
\end{array}
$$

← Divide the first term by the first term: $2x/x = 2$.

← The 8 has been "brought down."

← Multiply 2 by the divisor.

← Subtract: $(2x + 8) - (2x + 6)$.

Answer: Quotient $x + 2$, remainder 2.

To check, we multiply the quotient by the divisor and add the remainder, to see if we get the dividend.

$$
\overbrace{(x + 3)}^{\text{Divisor}} \cdot \overbrace{(x + 2)}^{\text{Quotient}} + \overbrace{2}^{\text{Remainder}} = (x^2 + 5x + 6) + 2
$$

$$
= \underbrace{x^2 + 5x + 8}_{\text{Dividend}}
$$

The answer checks.

DO EXERCISE 4.

You should remember the following.

---

1. **Arrange polynomials in descending order.**
2. **If there are missing terms in the dividend, either write them with zero coefficients or leave space for them.**

EXAMPLE 4 Divide: $(125y^3 - 8) \div (5y - 2)$.

a)
$$
\begin{array}{r}
25y^2 + 10y + 4 \\
5y - 2 \,)\, \overline{125y^3 + 0y^2 + 0y - 8} \\
\end{array}
$$

$125y^3 - 50y^2$

$\quad\quad 50y^2 + 0y - 8$

$\quad\quad 50y^2 - 20y$

$\quad\quad\quad\quad 20y - 8$

$\quad\quad\quad\quad 20y - 8$

$\quad\quad\quad\quad\quad\quad 0$

This subtraction is $125y^3 - (125y^3 - 50y^2)$. We get $50y^2$

b)
$$
\begin{array}{r}
25y^2 + 10y + 4 \\
5y - 2 \,)\, \overline{125y^3 \quad\quad\quad - 8} \\
\end{array}
$$

$125y^3 - 50y^2$

$\quad\quad 50y^2 \quad\quad - 8$

$\quad\quad 50y^2 - 20y$

$\quad\quad\quad\quad 20y - 8$

$\quad\quad\quad\quad 20y - 8$

$\quad\quad\quad\quad\quad\quad 0$

This subtraction is $50y^2 - (50y^2 - 20y)$. We get $20y$.

DO EXERCISE 5.

EXAMPLE 5 Divide: $(x^4 - 9x^2 - 5) \div (x - 2)$.

$$
\begin{array}{r}
x^3 + 2x^2 - 5x - 10 \\
x - 2 \,)\, \overline{x^4 \quad\quad - 9x^2 \quad\quad - 5} \\
\end{array}
$$

$x^4 - 2x^3$

The first subtraction is $x^4 - (x^4 - 2x^3)$.

$\quad 2x^3 - 9x^2 \quad\quad - 5$

$\quad 2x^3 - 4x^2$

The second subtraction is $(2x^3 - 9x^2) - (2x^3 - 4x^2)$.

$\quad\quad\quad - 5x^2 \quad\quad - 5$

$\quad\quad\quad - 5x^2 + 10x$

$\quad\quad\quad\quad\quad - 10x - 5$

$\quad\quad\quad\quad\quad - 10x + 20$

$\quad\quad\quad\quad\quad\quad\quad - 25$

The answer is $x^3 + 2x^2 - 5x - 10$, with $R = -25$, or

$$ x^3 + 2x^2 - 5x - 10 + \frac{-25}{x - 2}. $$

DO EXERCISES 6 AND 7.

When dividing, we may "come out even" (have a remainder of 0), or we may not. If not, how long should we keep working? We continue until the degree of the remainder is less than the degree of the divisor, as in the next example. The answer can be given by writing the quotient and remainder, or by using a fractional expression, as you saw in Example 5 and as you will see in the next example.

Divide and check.

5. $(9y^4 + 14y^2 - 8) \div (3y + 2)$

Divide and check.

6. $(y^3 - 11y^2 + 6) \div (y - 3)$

7. $(x^3 + 9x^2 - 5) \div (x - 1)$

Divide and check.

8. $(y^3 - 11y^2 + 6) \div (y^2 - 3)$

EXAMPLE 6   Divide: $(x^3 + 9x^2 - 5) \div (x^2 - 1)$.

$$
\begin{array}{r}
x + 9 \\
x^2 - 1 \overline{)\, x^3 + 9x^2 + 0x - 5} \\
\underline{x^3 \qquad\quad - x} \\
9x^2 + x - 5 \\
\underline{9x^2 \qquad - 9} \\
x + 4
\end{array}
$$

The degree of the remainder is less than the degree of the divisor, so we are finished. The answer is $x + 9$, with R $= x + 4$, or

$$x + 9 + \frac{x + 4}{x^2 - 1}.$$

The fractional expression is the remainder over the divisor.

DO EXERCISE 8.

EXERCISE SET 9.7

i Divide.

1. $\dfrac{30x^8 - 15x^6 + 40x^4}{5x^4}$

2. $\dfrac{24y^6 + 18y^5 - 36y^2}{6y^2}$

3. $\dfrac{-14a^3 + 28a^2 - 21a}{7a}$

4. $\dfrac{-32x^4 - 24x^3 - 12x^2}{4x}$

5. $(9y^4 - 18y^3 + 27y^2) \div 9y$

6. $(24a^3 + 28a^2 - 20a) \div 2a$

7. $(36x^6 - 18x^4 - 12x^2) \div -6x$

8. $(18y^7 - 27y^4 - 3y^2) \div -3y^2$

9. $(a^2b - a^3b^3 - a^5b^5) \div a^2b$

10. $(x^3y^2 - x^3y^3 - x^4y^2) \div x^2y^2$

11. $(6p^2q^2 - 9p^2q + 12pq^2) \div -3pq$

12. $(16y^4z^2 - 8y^6z^4 + 12y^8z^3) \div 4y^4z$

ii Divide and check.

13. $(x^2 + 10x + 21) \div (x + 3)$

14. $(y^2 - 8y + 16) \div (y - 4)$

15. $(a^2 - 8a - 16) \div (a + 4)$

16. $(y^2 - 10y - 25) \div (y - 5)$

17. $(y^2 - 25) \div (y + 5)$

18. $(a^2 - 81) \div (a - 9)$

19. $(y^3 - 4y^2 + 3y - 6) \div (y - 2)$

20. $(x^3 - 5x^2 + 4x - 7) \div (x - 3)$

21. $(a^3 - a + 12) \div (a - 4)$

22. $(x^3 - x + 6) \div (x + 2)$

23. $(8x^3 + 27) \div (2x + 3)$

24. $(64y^3 - 8) \div (4y - 2)$

25. $(x^4 - x^2 - 42) \div (x^2 - 7)$

26. $(y^4 - y^2 - 54) \div (y^2 - 3)$

27. $(x^4 - x^2 - x + 2) \div (x - 1)$

28. $(y^4 - y^2 - y + 3) \div (y + 1)$

29. $(10y^3 + 6y^2 - 9y + 10) \div (5y - 2)$

30. $(6x^3 - 11x^2 + 11x - 2) \div (2x - 3)$

31. $(2x^4 - x^3 - 5x^2 + x - 6) \div (x^2 + 2)$

32. $(3x^4 + 2x^3 - 11x^2 - 2x + 5) \div (x^2 - 2)$

Problem-solving practice

33. Find three consecutive positive integers such that the product of the first and second integers is 26 less than the product of the second and third integers.

34. For a certain fractional expression, the value of the numerator is 4 less than twice the value of the denominator. The reciprocal of the expression has a value $\frac{5}{6}$ greater than that of the original expression. What is the value of the original expression?

Divide.

35. $(x^4 - x^3y + x^2y^2 + 2x^2y - 2xy^2 + 2y^3) \div (x^2 - xy + y^2)$

36. $(4a^3b + 5a^2b^2 + a^4 + 2ab^3) \div (a^2 + 2b^2 + 3ab)$

37. $(x^4 - y^4) \div (x - y)$

38. $(a^7 + b^7) \div (a + b)$

Solve.

39. Find k so that when $x^3 - kx^2 + 3x + 7k$ is divided by $x + 2$, the remainder will be 0.

40. When $x^2 - 3x + 2k$ is divided by $x + 2$, the remainder is 7. Find k.

9.8

FORMULAS AND PROBLEM SOLVING

i To solve a formula, we use the same methods we would to solve any other equation. Many formulas contain fractional expressions, and to solve them we proceed as in solving any equations with fractional expressions, as follows.

OBJECTIVE

After finishing Section 9.8, you should be able to:

i Solve a formula containing a fractional expression for a specified letter, and use that skill in solving problems.

TO SOLVE FRACTIONAL FORMULAS

1. **Multiply by the LCM to clear of fractions.**
2. **Multiply to remove parentheses, if necessary.**
3. **Get all terms with the unknown on one side of the equation.**
4. **Factor out the unknown.**
5. **Use the multiplication principle to get the unknown alone on one side.**

Remember: To solve a problem, part of the familiarization process may be to look up a formula and perhaps some definitions of the terms used in the problem. You may then have to solve a formula for a certain unknown.

EXAMPLE 1 A pitcher's earned run average is 3, and the pitcher has given up 12 earned runs. How many innings have been pitched?

1. **Familiarize** If we do not already know what "earned run average" means or what "earned runs" are, we must look up those terms. When we do, we find that an "earned run" is one that does not result from an error committed by someone on the defensive team, and that "earned run average" is defined by the formula

$$A = \frac{9R}{I},$$

where A is the earned run average (ERA), R is the number of earned runs, and I is the number of innings pitched.

1. The formula

$$\frac{PV}{T} = k$$

relates the pressure, volume, and temperature of a gas. Solve it for T. (*Hint:* Begin by clearing of fractions.)

In this problem, $R = 12$ and $A = 3$. We want to find I, the number of innings pitched.

2. Translate> The translation in this case is the formula

$$A = \frac{9R}{I}.$$

3. Carry out> We solve the formula for I:

$$A = \frac{9R}{I}$$

$$I \cdot A = I \cdot \frac{9R}{I} \qquad \text{Multiplying by the LCM to clear of fractions}$$

$$IA = 9R$$

$$IA \cdot \frac{1}{A} = 9R \cdot \frac{1}{A} \qquad \text{Using the multiplication principle to isolate } I$$

$$I = \frac{9R}{A}. \qquad \text{Simplifying}$$

We now substitute and calculate:

$$I = \frac{9 \cdot 12}{3}$$

$$I = 36.$$

4. Check> The possible answer is 36. If there are 12 earned runs and the number of innings is 36, then the ERA will be

$$\frac{9 \cdot 12}{36}.$$

This simplifies to 3, so the answer checks.

5. State> The pitcher has pitched 36 innings.

DO EXERCISE 1.

As Example 1 indicates, it is important to be able to solve a formula for a particular letter. In the following example, we illustrate that skill.

EXAMPLE 2 Solve the formula

$$I = \frac{pT}{M + pn}$$

for p.

We first clear of fractions by multiplying by the LCM, $M + pn$:

$$(M + pn)I = (M + pn)\frac{pT}{M + pn}$$

$$IM + Ipn = pT.$$

We now get all terms containing p alone on one side:

$$IM = pT - Ipn \qquad \text{Adding } - Ipn$$

$$IM = p(T - In) \qquad \text{Factoring out the unknown } p$$

$$\frac{IM}{T - In} = p \qquad \text{Multiplying by } \frac{1}{T - In}$$

We now have p alone on one side, and p does *not* appear on the other side, so we have solved the formula for p.

DO EXERCISE 2.

EXAMPLE 3 Two electrical resistors are connected in parallel. The resistance of one of them is 6 ohms and the resistance of the combination is 3.75 ohms. What is the resistance of the other resistor?

1. **Familiarize** In a book on electricity, radio, or physics we will find that resistors in parallel are diagrammed as shown here:

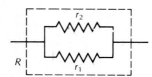

The rule, or law, for finding the resistance of a combination is given by the formula

$$\frac{1}{R} = \frac{1}{r_1} + \frac{1}{r_2},$$

where r_1 and r_2 are the resistances of the individual resistors and R is the resistance of the combination.

In this problem, we have the following:

$$R = 3.75 \text{ ohms};$$

$$r_1 = 6 \text{ ohms};$$

$$r_2 = ?$$

We could assign a letter to the unknown resistance, but we already have the letter r_2, so we will use it.

2. **Translate** The translation in this problem is the formula

$$\frac{1}{R} = \frac{1}{r_1} + \frac{1}{r_2}.$$

3. **Carry out** We solve the formula for the unknown resistance r_2. We multiply by the LCM, which is Rr_1r_2:

$$Rr_1r_2 \cdot \frac{1}{R} = Rr_1r_2 \cdot \left[\frac{1}{r_1} + \frac{1}{r_2} \right] \qquad \text{Multiplying by the LCM}$$

$$Rr_1r_2 \cdot \frac{1}{R} = Rr_1r_2 \cdot \frac{1}{r_1} + Rr_1r_2 \cdot \frac{1}{r_2} \qquad \text{Multiplying to remove parentheses}$$

$$\left. \begin{array}{l} \dfrac{R}{R} \cdot r_1r_2 = \dfrac{r_1}{r_1} \cdot Rr_2 + \dfrac{r_2}{r_2} \cdot Rr_1 \\[2mm] r_1r_2 = Rr_2 + Rr_1. \end{array} \right\} \quad \begin{array}{l} \text{Simplifying by} \\ \text{removing factors of 1} \end{array}$$

2. Solve $V = \dfrac{A}{n(T - t)}$ for T.

3. The formula

$$\frac{1}{p} + \frac{1}{q} = \frac{1}{f}$$

applies to a lens. The letters p and q represent the distances from an object to the lens and from the image to the lens, respectively. The letter f represents the focal length of the lens. Find the focal length of a lens that forms an image 15 cm from the lens when an object is placed 10 cm from the lens.

You might be tempted at this point to multiply by $1/r_1$ to get r_2 alone on the left, BUT note that there is an r_2 on the right. We must get all the terms involving r_2 on the *same side* of the equation:

$$r_1 r_2 - R r_2 = R r_1 \qquad \text{Adding } -R r_2$$

$$r_2 (r_1 - R) = R r_1 \qquad \text{Factoring out } r_2$$

$$r_2 = \frac{R r_1}{r_1 - R}. \qquad \text{Multiplying by } \frac{1}{r_1 - R} \text{ to get } r_2 \text{ alone}$$

We now substitute into the formula, as we have solved it, and compute:

$$r_2 = \frac{3.75 \cdot 6}{6 - 3.75}$$

$$r_2 = 10.$$

4. Check The possible answer is 10 ohms. We substitute into the original formula, as in checking an equation.

$$\frac{\dfrac{1}{R} = \dfrac{1}{r_1} + \dfrac{1}{r_2}}{\begin{array}{c|c} \dfrac{1}{3.75} & \dfrac{1}{6} + \dfrac{1}{10} \\[2mm] 0.26666\ldots & \dfrac{5}{30} + \dfrac{3}{30} \\[2mm] & \dfrac{8}{30} \\[2mm] & 0.266666\ldots \end{array}}$$

The answer checks.

5. State The answer is that the resistor has a resistance of 10 ohms.

In the preceding example, you should note carefully that R, r_1, and r_2 are all different variables (or constants). We commonly use subscripts as in r_1 and r_2 to distinguish variables. It is important that you write the subscripts correctly and that you never write a capital letter in place of a lower-case letter.

DO EXERCISE 3.

EXERCISE SET 9.8

i Solve each formula for the specified letter.

1. $\dfrac{W_1}{W_2} = \dfrac{d_1}{d_2}; \quad d_1$

2. $\dfrac{W_1}{W_2} = \dfrac{d_1}{d_2}; \quad W_2$

3. $s = \dfrac{(v_1 + v_2)t}{2}; \quad t$

4. $s = \dfrac{(v_1 + v_2)t}{2}; \quad v_1$

5. $\dfrac{1}{R} = \dfrac{1}{r_1} + \dfrac{1}{r_2}; \quad r_1$

6. $\dfrac{1}{R} = \dfrac{1}{r_1} + \dfrac{1}{r_2}; \quad R$

7. $R = \dfrac{gs}{g + s}; \quad s$

8. $R = \dfrac{gs}{g + s}; \quad g$

9. $I = \dfrac{2V}{R + 2r}; \quad r$

10. $I = \dfrac{2V}{R + 2r}; \quad R$

11. $\dfrac{1}{p} + \dfrac{1}{q} = \dfrac{1}{f}; \quad f$

12. $\dfrac{1}{p} + \dfrac{1}{q} = \dfrac{1}{f}; \quad p$

13. $I = \dfrac{nE}{R + nr}$; r

14. $I = \dfrac{nE}{R + nr}$; n

15. $S = \dfrac{H}{m(t_1 - t_2)}$; H

16. $S = \dfrac{H}{m(t_1 - t_2)}$; t_1

17. $\dfrac{E}{e} = \dfrac{R + r}{r}$; e

18. $\dfrac{E}{e} = \dfrac{R + r}{r}$; r

19. $S = \dfrac{a - ar^n}{1 - r}$; a

20. $S = \dfrac{a}{1 - r}$; r

Problem-solving practice

21. A pitcher's earned run average is 2.4 and he has pitched 45 innings. How many earned runs were given up?

22. Two resistors are connected in parallel. Their resistances are, respectively, 8 ohms and 15 ohms. What is the resistance of the combination?

23. A resistor has a resistance of 50 ohms. What size resistor should be put with it, in parallel, in order to obtain a resistance of 5 ohms?

24. The area of a certain trapezoid is 25 cm². Its height is 5 cm and the length of one base is 4 cm. Find the length of the other base.

25. Pam can do a certain job in a hours working alone. Elaine, working alone, can do the same job in b hours. Working together it takes them t hours to do the job.

a) Find a formula for t.
b) Solve the formula for a.
c) Solve the formula for b.

26. The *harmonic mean* of two numbers a and b is a number M such that the reciprocal of M is the average of the reciprocals of a and b. Find a formula for the harmonic mean.

SUMMARY AND REVIEW: CHAPTER 9

The following contains a summary of what you should be able to do after completing this chapter. The review exercises are for practice. Answers are at the back of the book. If you miss an exercise, restudy the section indicated alongside the answer.

You should be able to:

Simplify rational expressions by multiplying.

Simplify.

1. $\dfrac{y^2 - 64}{2y + 10} \cdot \dfrac{y + 5}{y + 8}$

2. $\dfrac{x^3 - 8}{x^2 - 25} \cdot \dfrac{x^2 + 10x + 25}{x^2 + 2x + 4}$

Simplify rational expressions by dividing.

Simplify.

3. $\dfrac{9a^2 - 1}{a^2 - 9} \div \dfrac{3a + 1}{a + 3}$

4. $\dfrac{x^3 - 64}{x^2 - 36} \div \dfrac{x^2 + 5x + 6}{x^2 - 3x - 18}$

Find the least common multiple of several polynomials.

Find the LCM.

5. $3y, 5y^2$

6. $6x^3, 16x^2$

7. $x^2 - 49, 3x + 21$

8. $x^2 + x - 20, x^2 + 3x - 10$

Add and subtract rational expressions, and simplify when possible.

Perform the indicated operation and simplify if possible.

9. $\dfrac{9xy}{x^2 - y^2} + \dfrac{x + y}{x - y}$

10. $\dfrac{2x^2}{x - y} + \dfrac{2y^2}{x + y}$

11. $\dfrac{3}{y + 4} - \dfrac{y}{y - 1} + \dfrac{y^2 + 3}{y^2 + 3y - 4}$

Simplify complex fractional expressions.

Simplify.

12. $\dfrac{\dfrac{1}{x} - x}{\dfrac{1}{x} + x}$

13. $\dfrac{\dfrac{1}{a^3} + \dfrac{1}{b^3}}{\dfrac{1}{a^2} - \dfrac{1}{b^2}}$

14. $\dfrac{\dfrac{y^2 - 5y - 6}{y^2 - 7y - 18}}{\dfrac{y^2 + 3y + 2}{y^2 + 4y + 4}}$

Solve fractional equations.

Solve.

15. $\dfrac{6}{x} + \dfrac{4}{x} = 5$

16. $\dfrac{x}{7} + \dfrac{x}{4} = 1$

17. $\dfrac{5}{3x+2} = \dfrac{3}{2x}$

Solve problems involving fractional equations.

Solve.

18. Dave can paint the outside of his house in 12 hours. Bill can paint the same house in 9 hours. How long would it take them to paint the house together?

19. A car gets on the turnpike traveling at a speed of 45 mph. Three hours later another car gets on at the same place traveling in the same direction at 55 mph. How far from the entrance to the turnpike will the second car overtake the first?

Divide polynomials.

Divide.

20. $(20r^2s^3 - 15r^2s^2 - 10r^3s^3) \div 5r^2s$

21. $(y^2 - 8y + 15) \div (y - 5)$

22. $(y^3 - 27) \div (y - 3)$

23. $(4x^3 + 3x^2 - 5x - 2) \div (x^2 + 1)$

24. $(x^3 + 3x^2 + 2x - 6) \div (x - 3)$

25. $(4x^3 + 6x^2 - 5) \div (x + 3)$

Solve formulas for specified variables.

Solve for the specified letter.

26. $R = \dfrac{gs}{g+s};\quad g$

27. $S = \dfrac{H}{m(t_1 - t_2)};\quad m$

Simplify.

28. $\dfrac{\dfrac{1}{a-b} + \dfrac{1}{a+b}}{1 + \dfrac{b^2}{a^2 - b^2}}$

29. $\dfrac{2-x}{x^2 - 2x + 1} - \dfrac{2x-3}{x^2 - 3x + 2}$

TEST: CHAPTER 9

Simplify.

1. $\dfrac{4y^2 - 4}{3y + 9} \cdot \dfrac{y+3}{y+1}$

2. $\dfrac{x^3 + 27}{x^2 - 16} \div \dfrac{x^2 + 8x + 15}{x^2 + x - 20}$

Find the LCM.

3. $x^2 - 16,\ x^3 - 64$

4. $(x^2 + 8x - 33),\ (x^2 - 12x + 27)$

Perform the indicated operation and simplify.

5. $\dfrac{4ab}{a^2 - b^2} + \dfrac{a^2 + b^2}{a+b}$

6. $\dfrac{6}{x^3 - 64} - \dfrac{4}{x^2 - 16}$

7. $\dfrac{4}{y+3} - \dfrac{y}{y-2} + \dfrac{y^2 + 4}{y^2 + y - 6}$

Simplify.

8. $\dfrac{3 + \dfrac{3}{y}}{4 + \dfrac{4}{y}}$

9. $\dfrac{\dfrac{y^2 + 4y - 77}{y^2 - 10y + 25}}{\dfrac{y^2 - 5y - 14}{y^2 - 25}}$

Solve.

10. $\dfrac{1}{x} + \dfrac{3}{x} = 4$

11. $\dfrac{6}{5a+3} = \dfrac{4}{2a-5}$

12. Tom can mow the yard in 3.5 hours. Larry can mow the yard in 4.5 hours. How long will it take them to mow it together?

Divide.

13. $(16ab^3c - 10ab^2c^2 + 12a^2b^2c) \div (4a^2b)$

14. $(y^2 - 20y + 64) \div (y - 6)$

15. $(6x^4 + 3x^2 + 5x + 4) \div (x^2 + 2)$

16. $(x^3 + 5x^2 + 4x - 7) \div (x - 4)$

17. Solve the formula for the specified letter:

$$A = \frac{h(b_1 + b_2)}{2}, \quad \text{for } b_1.$$

Simplify.

18. $\dfrac{3x^7 - x^6 + 31x^4 + 21x + 5}{x + 2}$

SOMETHING EXTRA: METRIC–AMERICAN CONVERSION

In many situations—for example, chemistry and medical applications—it is important to be able to convert between metric units of measure and American (also called British–American) units of measure. The following table is useful and should be memorized.

Metric units		British–American units	
1 m	(meter)	39.37 in.	(inches)
1 m		3.3 ft	(feet)
2.54 cm	(centimeters)	1 in.	
1 km	(kilometer)	0.621 mi	(miles)
1.609 km		1 mi	
1 kg	(kilogram)	2.2 lb	(pounds)
1 L	(liter)	1.057 qt	(quarts)

We will now use the table to make conversions between the systems. It is assumed that you will be using a calculator when doing the computations. Sometimes conversions can be made by simply making substitutions.

EXAMPLE 1 Complete: 4 m = _____ in.

$$4 \text{ m} = 4(1 \text{ m}) \qquad \text{We think of 4 m as } 4 \cdot (1 \text{ m}), \text{ or } 4(1 \text{ m}).$$
$$\approx 4(39.37 \text{ in.}) \qquad \text{Substituting 39.37 in. for 1 m}$$
$$= 157.48 \text{ in.} \qquad \text{Multiplying}$$

EXAMPLE 2 Complete: 7 ft = _____ cm.

$$7 \text{ ft} = 7(1 \text{ ft})$$
$$= 7(12 \text{ in.}) \qquad \text{Substituting 12 in. for 1 ft}$$
$$= 84 \text{ in.} \qquad \text{Multiplying}$$
$$= 84(1 \text{ in.})$$
$$\approx 84(2.54 \text{ cm}) \qquad \text{Substituting 2.54 cm for 1 in.}$$
$$= 213.36 \text{ cm}$$

Sometimes it helps to use multiplying by 1 in making conversions. For example, 12 in. = 1 ft, so

$$\frac{12 \text{ in.}}{1 \text{ ft}} = 1 \quad \text{and} \quad \frac{1 \text{ ft}}{12 \text{ in.}} = 1.$$

EXAMPLE 3 Complete: 255 in. = _____ m.

$$255 \text{ in.} \approx 255 \text{ in.} \cdot \frac{1 \text{ m}}{39.37 \text{ in.}}$$

This is *approximately* equal to 1.

$$= \frac{255}{39.37} \cdot \frac{\text{in.}}{\text{in.}} \cdot \text{m}$$

$$\approx 6.48 \text{ m} \qquad \text{Rounding to the nearest hundredth of a meter}$$

The rounding would depend on the precision desired. In Example 3 we used the following symbol for 1:

"m" in the numerator is the unit we are changing *to*.

$$\frac{1 \text{ m}}{39.37 \text{ in.}}$$

"in." in the denominator is the unit we are changing *from*.

When more than one conversion is involved, we may need to multiply by 1 more than once.

EXAMPLE 4 Complete: 26 gal = _____ L.

$$26 \text{ gal} \approx 26 \text{ gal} \cdot \frac{4 \text{ qt}}{1 \text{ gal}} \cdot \frac{1 \text{ L}}{1.057 \text{ qt}} \approx \frac{26 \cdot 4}{1.057} \cdot \frac{\text{gal}}{\text{gal}} \cdot \frac{\text{qt}}{\text{qt}} \cdot \text{L} \approx 98.39 \text{ L}$$

EXERCISES

Complete. Use a calculator. Give answers to two decimal places.

1. 18 m = _____ in.

2. 456 m = _____ ft

3. 300 in. = _____ cm

4. 500 km = _____ mi

5. 100 kg = _____ lb

6. 400 mi = _____ km

7. 324 L = _____ qt

8. 1 yd = _____ m

9. 700 in. = _____ m

10. 1000 lb = _____ kg

11. 1 L = _____ oz

12. 45 yd = _____ m

13. 1 ft = _____ cm

14. 1 yd = _____ cm

15. 10 L = _____ oz

16. 12 oz = _____ mL; 1000 mL = 1 L.
(The volume of a soda pop bottle)

17. 16 oz = _____ L

18. 200 lb = _____ kg

19. 120 kg = _____ lb

20. 100 m = _____ yd
(The length of a dash in track)

21. 100 yd = _____ m
(The length of a football field)

22. 330 ft = _____ m
(The length of most baseball foul lines)

23. 500 mi = _____ km
(The Indianapolis 500-mile race)

24. 26.2 mi = _____ km
(The length of the Olympic marathon)

25. 237,087 mi = _____ km
(The distance from the earth to the moon)

Miles per hour = mph; kilometers per hour = km/h

26. 55 mph = _____ km/h
   (A speed limit)

27. 240 km/h = _____ mph
   (A speed on the Autobahn)

Monetary Conversion: $1 = 1 U.S. dollar = 2.03 Deutsche Marks (MK),
$1 = 1 U.S. dollar = 0.69 British Pounds (BP),
$1 = 1 U.S. dollar = 625 Turkish Liras (TL).

Complete.

28. $345 = _____ MK

29. 180 MK = $_____

30. $345 = _____ BP

31. 200 BP = $_____

32. 180 MK = _____ BP

33. 550 BP = _____ MK

34. 1 TL = $_____

35. $7400 = _____ TL

36. 550 TL = _____ BP

10

EXPONENTIAL AND RADICAL SYMBOLISM

Exponential and radical expressions and equations are used by surveyors to solve problems.

In this chapter we will be concerned with square roots, cube roots, and the like. To name such numbers, we use a radical symbol $\sqrt{}$, and we also use those symbols in algebraic expressions known as *radical expressions*. Curiously, certain radical expressions are equivalent to expressions with exponents. In this chapter we will study that connection. Of course, you will again have the opportunity to apply your knowledge of this kind of algebraic symbolism to the solving of problems.

OBJECTIVES

After finishing Section 10.1, you should be able to:

i Find the square roots of whole numbers from 0^2 to 25^2.

ii Identify rational numbers and irrational numbers when named with radical notation.

iii Identify rational numbers and irrational numbers when named by fractional notation and decimal notation.

iv Approximate square roots using a calculator or a table.

v Solve problems involving square roots.

Find the following.

1. The square roots of 169

2. $\sqrt{256}$

3. $-\sqrt{100}$

10.1

SQUARE ROOTS AND REAL NUMBERS

In this section we study square roots, irrational numbers, and real numbers.

i Square Roots

When we raise a number to the second power, we have *squared* the number. Sometimes we may need to find the number that was squared. We call this process finding a square root of a number.

> **The number c is a *square root* of a if $c^2 = a$.**

Every positive number has two square roots. For example, the square roots of 25 are 5 and -5 because $5^2 = 25$ and $(-5)^2 = 25$. The positive square root is also called the *principal square root*. The symbol $\sqrt{\ }$ is called a *radical* symbol. The radical symbol refers only to the principal root. Thus $\sqrt{25} = 5$. To name the negative square root of a number, we use $-\sqrt{\ }$. The number 0 has only one square root, 0.

EXAMPLES

1. Find the square roots of 81.

$$9 \text{ and } -9 \text{ are square roots.}$$

2. Find $\sqrt{225}$.

$$\sqrt{225} = 15, \quad \text{taking the principal root.}$$

3. Find $-\sqrt{64}$.

$$\sqrt{64} = 8, \quad \text{so } -\sqrt{64} = -8.$$

DO EXERCISES 1–3.

ii Irrational Numbers

Recall that all rational numbers can be named by fractional notation a/b, where a and b are integers and $b \neq 0$. Rational numbers can be named in other ways, such as with decimal notation, but they can all be named with fractional notation. Suppose we try to find a rational number for $\sqrt{2}$. We look for a number a/b for which $(a/b)^2 = 2$.

We can find rational numbers whose squares are quite close to 2:

$$\left(\frac{14}{10}\right)^2 = (1.4)^2 = 1.96,$$

$$\left(\frac{141}{100}\right)^2 = (1.41)^2 = 1.9881,$$

$$\left(\frac{1414}{1000}\right)^2 = (1.414)^2 = 1.999396,$$

$$\left(\frac{14142}{10000}\right)^2 = (1.4142)^2 = 1.99996164,$$

$$\left(\frac{141421}{100000}\right)^2 = (1.41421)^2 = 1.99998992.$$

Actually, we can never find one whose square is exactly 2. That can be proved, but we will not do so here. Since $\sqrt{2}$ is not a rational number, we call it an *irrational number*.

> An *irrational number* is a number that cannot be named by fractional notation a/b, where a and b are integers and $b \neq 0$.

The square roots of most whole numbers are irrational. Only the perfect squares 1, 4, 9, 16, 25, 36, 49, 64, 81, 100, etc., have rational square roots.

EXAMPLES Identify the rational numbers and the irrational numbers.

4. $\sqrt{3}$ $\sqrt{3}$ is irrational, since 3 is not a perfect square.

5. $\sqrt{25}$ $\sqrt{25}$ is rational, since 25 is a perfect square.

6. $\sqrt{35}$ $\sqrt{35}$ is irrational, since 35 is not a perfect square.

DO EXERCISES 4–7.

ⅲ Real Numbers

The rational numbers are very close together on the number line. Yet no matter how close together two rational numbers are, we can find many rational numbers between them. By averaging, we can find the number *halfway* between. This process can be repeated indefinitely. For instance, the number halfway between $\frac{1}{32}$ and $\frac{2}{32}$ is $\frac{3}{64}$, the average of $\frac{1}{32}$ and $\frac{2}{32}$. In turn, $\frac{5}{128}$ is halfway between $\frac{1}{32}$ and $\frac{3}{64}$. It looks as if the rational numbers fill up the number line, but they do not. There are many points on the line for which there are no rational numbers. These points correspond to irrational numbers.

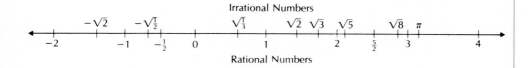

> The *real numbers* consist of the rational numbers and the irrational numbers. There is a real number for each point on a number line.

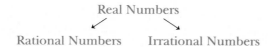

We know that decimal notation for a rational number either ends or repeats a group of digits. For example,

$\frac{1}{4} = 0.25$ The decimal ends.

$\frac{1}{3} = 0.3333\ldots$ The 3 repeats.

$\frac{5}{11} = 0.45\overline{45}$ The bar indicates that "45" repeats.

Decimal notation for an irrational number never ends and does not repeat. The number π is an example:

$$\pi = 3.1415926535\ldots.$$

Identify the rational numbers and the irrational numbers.

4. $\sqrt{5}$

5. $-\sqrt{36}$

6. $-\sqrt{32}$

7. $\sqrt{101}$

Identify the rational numbers and the irrational numbers.

8. $-\frac{95}{37}$

9. 612

10. 0.353535. . .
(numeral repeats)

11. 3.010010001000010000001. . .
(numeral does not repeat)

Decimal notation for π never ends and never repeats. The numbers 3.1416, 3.14, or $\frac{22}{7}$ are only rational number approximations for π. Decimal notation for $\frac{22}{7}$ is 3.142857142857. . . . It repeats.

Here are some other examples of irrational numbers:

2.818118111811118111118. . .   No group of digits repeats.

0.0350355035550355550. . .   No group of digits repeats.

> **Decimal notation for an irrational number is nonrepeating and nonending.**

EXAMPLES   Identify the rational numbers and the irrational numbers.

7. $\frac{18}{37}$ Rational, since it is the ratio of two integers.

8. 2.565656. . .   (Numeral repeats). Rational, since the digits "56" repeat.

9. 4.020020002. . .   (Numeral does not repeat). Irrational, since the decimal notation neither ends nor repeats.

DO EXERCISES 8–11.

iv Approximating Square Roots

We often need to use rational numbers to approximate square roots that are irrational. Such approximations can be found using a table such as Table 2 at the back of the book. They can also be found on a calculator with a square root key [$\sqrt{}$].

Before we use either of these we will use a calculator, but not the square root key, and a "guess-and-multiply" method in order to get an understanding of how rational numbers can be used to approximate square roots.

EXAMPLE 10   Use a calculator and "guess and multiply," but not the square root key, to approximate $\sqrt{10}$ to two decimal places.

Step 1.   We are looking for a number whose square is 10. Since

and
$$3^2 = 9 \quad\longleftarrow 10$$
$$4^2 = 16,$$

it seems reasonable that whatever the approximation is, it is between 3 and 4. Thus the ones digit of the approximation is 3.

Step 2.   We move on to the *tenths* place. We guess some numbers and square them:
$$3.1^2 = 9.61, \quad\longleftarrow 10$$
$$3.2^2 = 10.24,$$
$$3.3^2 = 10.89.$$

Since 10 is between the square of 3.1 and 3.2, it seems reasonable that the tenths digit of the approximation is 1. Thus we have improved our approximation of $\sqrt{10}$ to 3.1.

Step 3.   We move on to the *hundredths* place. We guess some numbers and square them:
$$3.11^2 = 9.6721,$$
$$3.12^2 = 9.7344,$$
$$3.13^2 = 9.7969,$$
$$3.14^2 = 9.8596,$$
$$3.15^2 = 9.9225,$$
$$3.16^2 = 9.9856, \quad\longleftarrow 10$$
$$3.17^2 = 10.0489.$$

Since 10 is between the square of 3.16 and 3.17, it seems reasonable that the hundredths digit of the approximation is 6. Thus we have improved our approximation of $\sqrt{10}$ to 3.16.

This is the place at which we are directed to stop. We could continue as far as we wish.

DO EXERCISE 12.

EXAMPLE 11 Use your calculator and the square root key or Table 2 to approximate $\sqrt{10}$. Round to three decimal places.

You will need to consult the instruction manual to find square roots on your calculator. Calculators vary in their methods of operation.

$$\sqrt{10} \approx 3.162277660 \qquad \text{Using a calculator with a 10-digit readout}$$

Different calculators give different numbers of digits in their readouts. This may cause some variance in answers. We round to the third decimal place. Then

$$\sqrt{10} \approx 3.162. \qquad \text{This can also be found in Table 2.}$$

The symbol $\approx$ means "is approximately equal to."

It would be helpful before beginning the exercise set or continuing this chapter to list the squares of numbers from 0 to 20, and to memorize the list.

DO EXERCISES 13 AND 14.

v **Problem Solving**

We now consider an application involving a formula with a square root. Since it just involves evaluating an expression, we need not use the entire problem-solving process.

EXAMPLE 12 (*An application: Parking lot arrival spaces*). A parking lot has attendants to park cars, and it uses spaces for cars to be left before they are taken to permanent parking stalls. The number N of such spaces needed is approximated by the formula

$$N = 2.5\sqrt{A},$$

where A is the average number of arrivals in peak hours. Find the number of spaces needed when the average number of arrivals in peak hours is 49 and 77.

12. Use your calculator, but not the square root key, to approximate $\sqrt{7}$ to three decimal places, using "guess and multiply."

Use your calculator and the square root key or Table 2 to approximate the square root to three decimal places.

13. $\sqrt{7}$

14. $\sqrt{72}$

15. In reference to Example 12, find the number of spaces needed when the average number of arrivals in peak hours is 64; when it is 39.

We substitute 49 into the formula:

$$N = 2.5\sqrt{49} = 2.5(7) = 17.5 \approx 18.$$

Note that we round up to 18 spaces because 17.5 spaces would give us a half space, which we could not use. To ensure that we have enough, we need 18. We substitute 77 into the formula. We use a calculator or Table 2 to find an approximation:

$$N = 2.5\sqrt{77} \approx 2.5(8.775) = 21.938 \approx 22.$$

When the average number of arrivals is 49, about 18 spaces are needed. When the average number of arrivals is 77, about 22 spaces are needed.

DO EXERCISE 15.

EXERCISE SET 10.1

i Find the square roots of each number.

1. 1   **2.** 4   **3.** 16   **4.** 9

5. 100   **6.** 121   **7.** 169   **8.** 144

Simplify.

9. $\sqrt{4}$ **10.** $\sqrt{1}$ **11.** $-\sqrt{9}$ **12.** $-\sqrt{25}$ **13.** $-\sqrt{64}$

14. $-\sqrt{81}$ **15.** $-\sqrt{225}$ **16.** $\sqrt{400}$ **17.** $\sqrt{361}$ **18.** $\sqrt{441}$

ii Identify each square root as rational or irrational.

19. $\sqrt{2}$ **20.** $\sqrt{6}$ **21.** $\sqrt{8}$ **22.** $\sqrt{10}$ **23.** $\sqrt{49}$ **24.** $\sqrt{100}$

25. $\sqrt{98}$ **26.** $\sqrt{75}$ **27.** $-\sqrt{4}$ **28.** $-\sqrt{1}$ **29.** $-\sqrt{12}$ **30.** $-\sqrt{14}$

iii Identify each number as rational or irrational.

31. 4.23   **32.** 0.03   **33.** 23   **34.** -19

35. $-\frac{2}{3}$ **36.** 0   **37.** $\frac{2.3}{0.01}$

38. 0.424242...
(numeral repeats)
     39. 0.156156156...
(numeral repeats)
     40. 4.282282228...
(numeral does not repeat)

41. 7.767767776...
(numeral does not repeat)
     42. $-34.69191191119...$
(numeral does not repeat)
     43. $-63.030030003...$
(numeral does not repeat)

iv Use your calculator or Table 2 to approximate these square roots. Round to three decimal places.

44. $\sqrt{5}$ **45.** $\sqrt{6}$ **46.** $\sqrt{17}$ **47.** $\sqrt{19}$

48. $\sqrt{93}$ **49.** $\sqrt{43}$ **50.** $\sqrt{40}$

v *Problem solving*

Use the formula $N = 2.5\sqrt{A}$ of Example 11.

51. Find the number of spaces needed when the average number of arrivals is 25 and 89.

52. Find the number of spaces needed when the average number of arrivals is 62 and 100.

53. Simplify $\sqrt{\sqrt{16}}$.

54. Simplify $\sqrt{3^2 + 4^2}$.

55. Between what two consecutive integers is $-\sqrt{33}$?

Use a calculator to approximate these square roots. Round to three decimal places.

56. $\sqrt{12.8}$ **57.** $\sqrt{198}$ **58.** $\sqrt{930}$ **59.** $\sqrt{4932}$ **60.** $\sqrt{1043.89}$

61. What number is halfway between x and y?

62. Find a number that is the square of an integer and the cube of a different integer.

63. Find the number one-third of the way from $2\frac{3}{4}$ to $4\frac{5}{8}$.

64. Find the number one-fifth of the way from -1 to $-\frac{4}{3}$.

65. What number is one-fourth of the way between z and w if $z < w$?

Find the following square roots.

66. $-\sqrt{\dfrac{49}{36}}$ **67.** $-\sqrt{\dfrac{361}{9}}$ **68.** $\sqrt{196}$ **69.** $\sqrt{441}$ **70.** $-\sqrt{\dfrac{16}{81}}$

71. $-\sqrt{\dfrac{81}{144}}$ **72.** $\sqrt{0.09}$ **73.** $\sqrt{0.36}$ **74.** $-\sqrt{0.0049}$ **75.** $\sqrt{0.0144}$

10.2

RADICAL EXPRESSIONS

In this section we consider radical notation for square roots and the ways in which we can manipulate such notation to get equivalent expressions. Such manipulation can be important in problem solving.

i Radicands

When an expression is written under a radical, we have a *radical expression*. Here are some examples:

$$\sqrt{14}, \qquad \sqrt{x}, \qquad \sqrt{x^2 + 4}, \qquad \sqrt{\dfrac{x^2 - 5}{2}}.$$

The expression written under the radical is called the *radicand*.

EXAMPLES Identify the radicand in each expression.

1. $\sqrt{x}$ The radicand is x.
2. $\sqrt{y^2 - 5}$ The radicand is $y^2 - 5$.

DO EXERCISES 1 AND 2.

ii Meaningless Expressions

The square of any number is always positive. For example, $8^2 = 64$ and $(-11)^2 = 121$. There are no real numbers that can be squared to get negative numbers.

> Radical expressions with negative radicands have no meaning in the real-number system.

Thus the following expressions do not represent real numbers:

$$\sqrt{-100}, \qquad \sqrt{-49}, \qquad -\sqrt{-3}.$$

Later in your study of mathematics, you may encounter a number system in which negative numbers have square roots.

EXAMPLE 3 Determine whether 6 is a sensible replacement in the expression $\sqrt{1 - y}$.

If we replace y by 6, we get $\sqrt{1 - 6} = \sqrt{-5}$, which has no meaning because the radicand is negative.

OBJECTIVES

After finishing Section 10.2, you should be able to:

i Identify radicands.

ii Identify meaningless radical expressions, determine whether a given number is a sensible replacement in a radical expression, and determine the sensible replacements in a radical expression.

iii Simplify radical expressions with a perfect-square radicand.

iv Find cube roots, simplifying certain expressions.

v Simplify expressions involving odd and even roots.

Identify the radicand.

1. $\sqrt{45 + x}$

2. $\sqrt{\dfrac{x}{x + 2}}$

Is the expression meaningless? Write "yes" or "no."

3. $-\sqrt{25}$

4. $\sqrt{-25}$

5. $-\sqrt{-36}$

6. $-\sqrt{36}$

7. Determine whether 8 is a sensible replacement in $\sqrt{x}$.

8. Determine whether 10 is a sensible replacement in $\sqrt{4-x}$.

Determine the sensible replacements in the expression.

9. $\sqrt{a}$

10. $\sqrt{x-3}$

11. $\sqrt{2x-5}$

12. $\sqrt{2x^2+3}$

Simplify.

13. $\sqrt{(xy)^2}$

14. $\sqrt{x^2y^2}$

15. $\sqrt{(x-1)^2}$

16. $\sqrt{x^2+8x+16}$

17. $\sqrt{25y^2}$

18. $\sqrt{\frac{1}{4}t^2}$

EXAMPLES Determine the sensible replacements in each expression.

4. $\sqrt{x}$ Any number greater than or equal to 0 is sensible.

5. $\sqrt{x+2}$ We solve the inequality $x+2 \geqslant 0$. Any number greater than or equal to -2 is sensible.

6. $\sqrt{x^2}$ Squares of numbers are never negative. All replacements are sensible.

7. $\sqrt{x^2+1}$ Since x^2 is never negative, x^2+1 is never negative. All real-number replacements are sensible.

DO EXERCISES 3–12.

iii Perfect-Square Radicands

We now consider radical expressions with a perfect-square radicand, $\sqrt{x^2}$. If x represents a nonnegative number, $\sqrt{x^2}$ simplifies to x. If x represents a negative number, $\sqrt{x^2}$ simplifies to $-x$ (the additive inverse of x). That is because $\sqrt{a}$ denotes the *principal* square root of a.

Suppose $x = 3$. Then $\sqrt{x^2} = \sqrt{3^2}$, which is $\sqrt{9}$, or 3.
Suppose $x = -3$. Then $\sqrt{x^2} = \sqrt{(-3)^2}$, which is $\sqrt{9}$ or 3, the absolute value of -3.

In either case we have $\sqrt{x^2} = |x|$.

> Any radical expression $\sqrt{A^2}$ can be simplified to $|A|$.

EXAMPLES Simplify.

8. $\sqrt{(3x)^2} = |3x|$ Absolute-value notation is necessary.
9. $\sqrt{a^2b^2} = \sqrt{(ab)^2} = |ab|$
10. $\sqrt{x^2+2x+1} = \sqrt{(x+1)^2} = |x+1|$

We can sometimes simplify absolute-value notation. In Example 8, $|3x|$ simplifies to $|3| \cdot |x|$, or $3|x|$. In Example 9, we can change $|ab|$ to $|a| \cdot |b|$ if we wish. The absolute value of a product is always the product of the absolute values.

> For any real numbers a and b, $|a \cdot b| = |a| \cdot |b|$.

DO EXERCISES 13–18.

iv Cube Roots

A number c is the *cube root* of a if its third power is a, that is, if $c^3 = a$. For example,

2 is the cube root of 8 because $2^3 = 2 \cdot 2 \cdot 2 = 8$;
-4 is the cube root of -64 because $(-4)^3 = (-4)(-4)(-4) = -64$.

In the real-number system, every number has exactly one cube root. The cube root of a positive number is positive, and the cube root of a negative number is negative. In simplifying expressions involving cube roots, we need not use absolute-value signs.

We use the symbol $\sqrt[3]{a}$ to name the cube root of a.

EXAMPLE 11 Simplify: $\sqrt[3]{8}$.

Since $2 \cdot 2 \cdot 2 = 2^3 = 8$, it follows that $\sqrt[3]{8} = 2$.

EXAMPLE 12 Simplify: $\sqrt[3]{-27}$.

Since $(-3)(-3)(-3) = -27$, it follows that $\sqrt[3]{-27} = -3$.

EXAMPLE 13 Simplify: $\sqrt[3]{-8y^3}$.

$$\sqrt[3]{-8y^3} = -2y$$

(No absolute-value signs are needed.)

DO EXERCISES 19–22.

ⅴ Odd and Even *k*th Roots

The fifth root of a number a is the number c for which $c^5 = a$. There are also 7th roots, 9th roots, and so on. Whenever the number k in $\sqrt[k]{}$ is an odd number, we say that we are taking an *odd* root. The number k is called the *index*. When the *index* is 2 we do not write it.

Every number has just one root for an odd k. If the number is positive, the root is positive. If the number is negative, the root is negative.

EXAMPLES Find each of the following.

14. $\sqrt[5]{32} = 2$
15. $\sqrt[5]{-32} = -2$
16. $-\sqrt[5]{32} = -2$
17. $-\sqrt[5]{-32} = -(-2) = 2$
18. $\sqrt[7]{x^7} = x$
19. $\sqrt[9]{(x-1)^9} = x - 1$

Absolute-value signs are never needed when finding odd roots.

When the index k in $\sqrt[k]{}$ is an even number, we say that we are taking an *even* root. Every positive real number has two kth roots when k is even. One of those roots is positive and one is negative. Negative real numbers do not have kth roots when k is even. When finding even kth roots, absolute-value signs are sometimes necessary, as with square roots.

EXAMPLES Find the following. Assume that variables can represent any real numbers.

20. $\sqrt[4]{16} = 2$
21. $-\sqrt[4]{16} = -2$
22. $\sqrt[4]{-16}$ No real-number fourth roots
23. $\sqrt[4]{81x^4} = 3|x|$
24. $\sqrt[6]{(y+7)^6} = |y+7|$

For any real number a:

a) $\sqrt[k]{a^k} = |a|$ when k is even. We use absolute value when k is even unless a is nonnegative.

b) $\sqrt[k]{a^k} = a$ when k is odd. We do not use absolute value when k is odd.

DO EXERCISES 23–28.

Simplify.

19. $\sqrt[3]{64}$

20. $\sqrt[3]{-125}$

21. $\sqrt[3]{8(x+2)^3}$

22. $\sqrt[3]{-27(y-4)^3}$

Find the following. Assume that letters can represent any real number.

23. $\sqrt[5]{-32x^5}$

24. $\sqrt[7]{(3x+2)^7}$

25. a) $\sqrt[4]{81}$

    b) $-\sqrt[4]{81}$

    c) $\sqrt[4]{-81}$

26. $\sqrt[4]{16(x-2)^4}$

27. $\sqrt[6]{x^6}$

28. $\sqrt[8]{(x+3)^8}$

EXERCISE SET 10.2

i Identify the radicand in each expression.

1. $\sqrt{a-4}$ **2.** $\sqrt{t+3}$ **3.** $5\sqrt{t^2+1}$ **4.** $8\sqrt{x^2+5}$ **5.** $x^2y\sqrt{\dfrac{3}{x+2}}$ **6.** $ab^2\sqrt{\dfrac{a}{a-b}}$

ii Which of these expressions are meaningless? Write "yes" or "no."

7. $\sqrt{-16}$ **8.** $\sqrt{-81}$ **9.** $-\sqrt{81}$ **10.** $-\sqrt{64}$

Determine the sensible replacements in each expression.

11. $\sqrt{5x}$ **12.** $\sqrt{3y}$ **13.** $\sqrt{t-5}$ **14.** $\sqrt{y-8}$ **15.** $\sqrt{y+8}$ **16.** $\sqrt{x+6}$
17. $\sqrt{x+20}$ **18.** $\sqrt{m-18}$ **19.** $\sqrt{2y-7}$ **20.** $\sqrt{3x+8}$ **21.** $\sqrt{t^2+5}$ **22.** $\sqrt{y^2+1}$

iii Simplify.

23. $\sqrt{t^2}$ **24.** $\sqrt{x^2}$ **25.** $\sqrt{9x^2}$ **26.** $\sqrt{4a^2}$
27. $\sqrt{(-7)^2}$ **28.** $\sqrt{(-5)^2}$ **29.** $\sqrt{(-4d)^2}$ **30.** $\sqrt{(-3b)^2}$
31. $\sqrt{(x+3)^2}$ **32.** $\sqrt{(x-7)^2}$ **33.** $\sqrt{a^2-10a+25}$ **34.** $\sqrt{x^2+2x+1}$

iv Simplify.

35. $\sqrt[3]{27}$ **36.** $-\sqrt[3]{64}$ **37.** $\sqrt[3]{-64x^3}$ **38.** $\sqrt[3]{-125y^3}$
39. $\sqrt[3]{-216}$ **40.** $-\sqrt[3]{-1000}$ **41.** $\sqrt[3]{0.343(x+1)^3}$ **42.** $\sqrt[3]{0.000008(y-2)^3}$

v Find each of the following. Do not assume that variables represent nonnegative numbers.

43. $\sqrt[4]{625}$ **44.** $-\sqrt[4]{256}$ **45.** $\sqrt[5]{-1}$ **46.** $-\sqrt[5]{-32}$ **47.** $\sqrt[5]{-\dfrac{32}{243}}$

48. $\sqrt[5]{-\dfrac{1}{32}}$ **49.** $\sqrt[6]{x^6}$ **50.** $\sqrt[8]{y^8}$ **51.** $\sqrt[4]{(5a)^4}$ **52.** $\sqrt[4]{(7b)^4}$

53. $\sqrt[10]{(-6)^{10}}$ **54.** $\sqrt[12]{(-10)^{12}}$ **55.** $\sqrt[414]{(a+b)^{414}}$ **56.** $\sqrt[1976]{(2a+b)^{1976}}$ **57.** $\sqrt[7]{y^7}$

58. $\sqrt[3]{(-6)^3}$ **59.** $\sqrt[5]{(x-2)^5}$ **60.** $\sqrt[9]{(2xy)^9}$

Problem-solving practice

61. The amount F that a family spends on food varies directly as its income I. A family making $19,600 a year will spend $5096 on food. At this rate, how much would a family making $20,500 spend on food?

62. A collection of dimes and nickels is worth $15.25. There are 157 coins in all. How many of each kind are there?

Solve.

63. $\sqrt{x^2}=6$ **64.** $\sqrt{y^2}=-7$ **65.** $-\sqrt{x^2}=-3$ **66.** $t^2=49$

Simplify.

67. $\sqrt{(3a)^2}$ **68.** $(\sqrt{3a})^2$ **69.** $\sqrt{\dfrac{144x^8}{36y^6}}$ **70.** $\sqrt{\dfrac{y^{12}}{8100}}$ **71.** $\sqrt{\dfrac{169}{m^{16}}}$ **72.** $\sqrt{\dfrac{p^2}{3600}}$

Determine the sensible replacements in each expression.

73. $\sqrt{m(m+3)}$ **74.** $\sqrt{x^2(x-3)}$ **75.** $\sqrt{(x+3)(x-2)}$
76. $\sqrt{x^2+7x+12}$ **77.** $\sqrt{x^2-4}$ **78.** $\sqrt{4x^2-1}$

For each function, determine whether the specified inputs are members of the domain. If so, find the function values.

79. $f(y)=\sqrt{5y-10}$;
Inputs: 6, 2, 0, −3

80. $g(x)=\sqrt{x^2-25}$;
Inputs: −5, 5, 0, 6, −6

81. $t(x)=-\sqrt{2x+1}$;
Inputs: 4, −4, 0, 12

82. $p(z)=\sqrt{2z^2-20}$;
Inputs: 0, 5, −5, 10, −10

83. $f(t)=\sqrt{t^2+1}$;
Inputs: 5, −5, 0, 10, −10

84. $g(x)=-\sqrt{(x+1)^2}$;
Inputs: 1, −1, 3, −3, 5, −5

85. $f(x)=\sqrt[3]{x+1}$;
Inputs: 7, 26, −9, −65

86. $g(x)=-\sqrt[3]{2x-1}$;
Inputs: 0, −62, −13, 63

87. Graph the function $y=\sqrt{x}$.

88. Graph the function $y=-\sqrt{x}$.

THE ASVAB TEST: MATHEMATICS KNOWLEDGE   CHAPTER 10

See the discussion of the ASVAB Test at the beginning of Chapter 3.

1. Solve for C: $F = \frac{9}{5}C + 32$.

   A $\frac{9}{5}(F - 32)$

   B $\frac{5}{9}F - 32$

   C $\frac{9}{5}F + 32$

   D $\frac{5}{9}(F - 32)$

2. Solve the following inequality: $x - 11 > 2$.

   A $x > 13$

   B $x < -9$

   C $x > -9$

   D $x < 13$

3. What is the reciprocal of $2\frac{1}{2}$?

   A $\frac{2}{5}$

   B $-\frac{2}{5}$

   C $-\frac{5}{2}$

   D $\frac{2}{3}$

4. What is the value of $7 \times 3 - (5 - 9)$?

   A 17

   B 25

   C 49

   D 7

5. In a class, $\frac{1}{5}$ of the students received A's, $\frac{1}{2}$ received B's, $\frac{1}{4}$ received C's, and the remaining 3 people failed. How many people were in the class?

   A 40

   B 20

   C 30

   D 60

6. If one person is picked at random from a group of 28 men and 7 women, what is the probability that a woman will be chosen?

   A $\frac{4}{5}$

   B $\frac{1}{7}$

   C $\frac{1}{5}$

   D $\frac{1}{4}$

7. Divide 20.442 by 0.06.

   A 340.7

   B 3407

   C 3.407

   D 34,070

8. It takes 3 qt of paint to cover a section of fence 24 ft long. How many gallons are needed for a fence that surrounds a lot measuring 120 ft × 150 ft?

   A 562.5

   B 16.875

   C 1000

   D 250

9. What is the area A of a circle in terms of its circumference C?

   A $A = C^2\pi/2$

   B $A = C^2/2$

   C $A = 4\pi/C^2$

   D $A = C^2/4\pi$

10. If a real estate salesman's commission is $3\frac{1}{2}\%$, how much does he earn selling a \$67,000 house?

   A \$1914.29

   B \$234.50

   C \$2345.00

   D \$2043.50

11. 36 is what percent of 16?

   A 22.5

   B 2250

   C 225

   D 2.25

12. A circle has the same perimeter as an equilateral triangle. If the radius of the circle is 6 in., what is the length of the base of the triangle? Use 3.14 for π.

   A 12.56 in.

   B 6.28 in.

   C 37.68 in.

   D 9.42 in.

13. The opposite corners of a square are 8 cm apart. What is the area of the square?

   A 64 cm^2

   B 4 cm^2

   C 32 cm^2

   D 1024 cm^2

14. The length of each side of an octagon is $4\frac{1}{6}$ in. What is the perimeter?

   A $66\frac{2}{3}$ in.

   B 25 in.

   C 50 in.

   D $33\frac{1}{3}$ in.

15. An inch is 2.54 cm. How many centimeters are in $1\frac{1}{2}$ ft?

   A 45.72

   B 7.09

   C 50.8

   D 38.1

16. A magazine devotes $\frac{9}{16}$ of its pages to advertising. How many pages of a 216-page magazine are not ads?

   A 121.5

   B 108

   C 135

   D 94.5

17. City B is 40 mi due north of city A and 30 mi due east of city C. How far apart are A and C?

    A   25 mi
    B   50 mi
    C   70 mi
    D   35 mi

18. The area of the triangle shown is 32 cm². What is the area of the circle? Express your answer in terms of π.

    A 16π cm²
    B 64π cm²
    C 32π cm²
    D 128π cm²

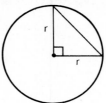

OBJECTIVES

After finishing Section 10.3, you should be able to:

  i Multiply with radical notation.

  ii Simplify radical expressions by factoring, assuming that all expressions represent nonnegative numbers.

  iii Multiply and simplify radical expressions.

  iv Identify rational and irrational numbers.

  v Approximate square roots using a table or calculator.

Multiply.

1. $\sqrt{19}\sqrt{7}$

2. $\sqrt{x+2y}\sqrt{x-2y}$

3. $\sqrt[4]{403}\sqrt[4]{7}$

4. $\sqrt[3]{8x}\sqrt[3]{x^4+5}$

10.3

MULTIPLYING AND SIMPLIFYING WITH RADICAL NOTATION

In this section, we look at multiplying and factoring with radical notation, making sure, as usual, that we obtain equivalent expressions.

i   Multiplying

Notice that

$$\sqrt{4}\sqrt{25} = 2\cdot 5 = 10 \qquad \text{and} \qquad \sqrt{4\cdot 25} = \sqrt{100} = 10.$$

Likewise,

$$\sqrt[3]{27}\sqrt[3]{8} = 3\cdot 2 = 6 \qquad \text{and} \qquad \sqrt[3]{27\cdot 8} = \sqrt[3]{216} = 6.$$

These examples suggest the following.

> For any nonnegative real numbers a and b and any index k,
>
> $$\sqrt[k]{a}\cdot\sqrt[k]{b} = \sqrt[k]{a\cdot b}.$$
>
> **(To multiply, we multiply the radicands.)**

EXAMPLES Multiply.

1. $\sqrt{3}\cdot\sqrt{5} = \sqrt{3\cdot 5} = \sqrt{15}$

2. $\sqrt{x+2}\sqrt{x-2} = \sqrt{(x+2)(x-2)} = \sqrt{x^2-4}$

3. $\sqrt[3]{4}\,\sqrt[3]{5} = \sqrt[3]{4\cdot 5} = \sqrt[3]{20}$

4. $\sqrt[4]{\dfrac{y}{5}}\sqrt[4]{\dfrac{7}{x}} = \sqrt[4]{\dfrac{y}{5}\cdot\dfrac{7}{x}} = \sqrt[4]{\dfrac{7y}{5x}}$

DO EXERCISES 1–4.

ii   Simplifying by Factoring

Reversing the above statement, we have $\sqrt[k]{a\cdot b} = \sqrt[k]{a}\cdot\sqrt[k]{b}$. This shows a way to factor and thus simplify radical expressions. Consider $\sqrt{20}$. The number 20 has the factor 4, which is a perfect square. Therefore,

$$\sqrt{20} = \sqrt{4\cdot 5} \qquad \text{Factoring the radicand (4 is a perfect square)}$$
$$= \sqrt{4}\cdot\sqrt{5} \qquad \text{Factoring into two radicals}$$
$$= 2\sqrt{5}. \qquad \text{Taking the square root of 4}$$

> To simplify a radical expression by factoring, look for factors of the radicand that are perfect kth powers (where k is the index). Then take the kth root of those factors.

EXAMPLES Simplify by factoring.

5. $\sqrt{50} = \sqrt{25 \cdot 2} = \sqrt{25} \cdot \sqrt{2} = 5\sqrt{2}$ ⌐ This factor is a perfect square.

6. $\sqrt[3]{32} = \sqrt[3]{8 \cdot 4} = \sqrt[3]{8} \cdot \sqrt[3]{4} = 2\sqrt[3]{4}$ ⌐ This factor is a perfect cube (third power).

DO EXERCISES 5 AND 6.

In many situations, expressions under radicals never represent negative numbers. In such cases, absolute-value notation is not necessary.

EXAMPLES Simplify by factoring. Assume that all expressions under radicals represent nonnegative numbers.

7. $\sqrt{5x^2} = \sqrt{x^2 \cdot 5}$ Factoring the radicand

$\quad\quad = \sqrt{x^2} \cdot \sqrt{5}$ Factoring into two radicals

$\quad\quad = x \cdot \sqrt{5}$ Taking the square root of x^2

8. $\sqrt{2x^2 - 4x + 2} = \sqrt{2(x-1)^2}$

$\quad\quad\quad\quad\quad\quad = \sqrt{(x-1)^2} \cdot \sqrt{2}$

$\quad\quad\quad\quad\quad\quad = (x-1) \cdot \sqrt{2}$

9. $\sqrt{216x^5y^3} = \sqrt{36 \cdot 6 \cdot x^4 \cdot x \cdot y^2 \cdot y}$ ⎱

$\quad\quad\quad\quad = \sqrt{36 \cdot x^4 \cdot y^2 \cdot 6 \cdot x \cdot y}$ ⎰ Factoring the radicand

$\quad\quad\quad\quad = \sqrt{36}\sqrt{x^4}\sqrt{y^2}\sqrt{6xy}$ Factoring into several radicals

$\quad\quad\quad\quad = 6x^2y\sqrt{6xy}$ Taking square roots

Note: Had we not seen that $216 = 36 \cdot 6$, where 36 is the largest square factor of 216, we could have found the prime factorization $2 \cdot 2 \cdot 2 \cdot 3 \cdot 3 \cdot 3$. Each pair of factors makes a square, so

$$\sqrt{2 \cdot 2 \cdot 2 \cdot 3 \cdot 3 \cdot 3} = \sqrt{2^2 \cdot 3^2 \cdot 2 \cdot 3} = 2 \cdot 3\sqrt{2 \cdot 3}.$$

DO EXERCISES 7–13.

▓ Multiplying and Simplifying

Sometimes after we multiply we can then simplify by factoring.

EXAMPLES Multiply and then simplify by factoring. Assume that all expressions under radicals represent nonnegative numbers.

10. $\sqrt{15}\sqrt{6} = \sqrt{15 \cdot 6} = \sqrt{90} = \sqrt{9 \cdot 10} = 3\sqrt{10}$

11. $3\sqrt[3]{25} \cdot 2\sqrt[3]{5} = 6 \cdot \sqrt[3]{25 \cdot 5}$ ⎱

$\quad\quad\quad\quad\quad = 6 \cdot \sqrt[3]{125}$ ⎰ Multiplying radicands

$\quad\quad\quad\quad\quad = 6 \cdot 5$, or 30 Taking the cube root of 125

12. $\sqrt[3]{18y^3}\sqrt[3]{4x^2} = \sqrt[3]{18y^3 \cdot 4x^2}$

$\quad\quad\quad\quad\quad = \sqrt[3]{72y^3x^2}$ Multiplying radicands

$\quad\quad\quad\quad\quad = \sqrt[3]{8y^3 \cdot 9x^2}$ Factoring the radicand

$\quad\quad\quad\quad\quad = \sqrt[3]{8y^3}\sqrt[3]{9x^2}$ Factoring into two radicals

$\quad\quad\quad\quad\quad = 2y\sqrt[3]{9x^2}$ Taking the cube root

DO EXERCISES 14–18.

Simplify by factoring.

5. $\sqrt{32}$

6. $\sqrt[3]{80}$

Simplify by factoring. Assume that all expressions represent nonnegative numbers. Hence no absolute-value signs will be needed.

7. $\sqrt{300}$

8. $\sqrt{36y^2}$

9. $\sqrt{3x^2 + 12x + 12}$

10. $\sqrt{12ab^3c^2}$

11. $\sqrt[3]{16}$

12. $\sqrt[3]{81x^4y^8}$

13. $\sqrt[3]{(a+b)^4}$

Multiply and then simplify by factoring. Assume that all expressions under radicals represent nonnegative numbers.

14. $\sqrt{3}\sqrt{6}$

15. $\sqrt{18y}\sqrt{14y}$

16. $\sqrt[3]{3x^2y}\sqrt[3]{36x}$

17. $\sqrt{7a}\sqrt{21b}$

18. $\sqrt[3]{2(y+5)}\sqrt[3]{4(y+5)^4}$

Which of the following are rational?
Which are irrational?

19. $\dfrac{-4}{5}$

20. $\dfrac{59}{37}$

21. 7.42

22. 0.47474747...
(Numeral repeats)

23. 2.57340046631...
(Numeral does not repeat)

24. $\sqrt{7}$

25. $\sqrt{16}$

iv **Rational and Irrational Numbers**

The real numbers consist of the rational numbers and the irrational numbers. We can describe these kinds of numbers in several ways.

> **Rational numbers are:**
>
> 1. Those numbers that can be named with fractional notation a/b, where a and b are integers and $b \neq 0$. (Definition)
> 2. Those numbers for which decimal notation either ends or repeats.

EXAMPLES

13. $\frac{5}{16} = 0.3125$ Ending (terminating) decimal

14. $\frac{8}{7} = 1.142857142857\ldots = 1.\overline{142857}$ Repeating decimal

15. $\frac{3}{11} = 0.2727\ldots = 0.\overline{27}$ Repeating decimal

We can also describe the irrational numbers in several ways.

> **Irrational numbers are:**
>
> 1. Those real numbers that are not rational.
> 2. Those real numbers that cannot be named with fractional notation a/b, where a and b are integers and $b \neq 0$.
> 3. Those real numbers for which decimal notation does not end and does not repeat.

There are many irrational numbers. For example, $\sqrt{2}$ is irrational. We can find rational numbers a/b for which $(a/b) \cdot (a/b)$ is close to 2, but we cannot find such a number a/b for which $(a/b) \cdot (a/b)$ is *exactly* 2.

Unless a whole number is a perfect square, its square root is irrational. For example, $\sqrt{9}$ and $\sqrt{25}$ are rational, but all of the following are irrational:

$$\sqrt{3}, \qquad -\sqrt{14}, \qquad \sqrt{45}.$$

There are also many irrational numbers that we do not obtain by taking square roots. The number π is an example.* Decimal notation for π does not end and does not repeat.

EXAMPLES   All of these are irrational.

16. $\pi = 3.1415926535\ldots$ Numeral does not repeat.

17. $1.101001000100001000001\ldots$ Numeral does not repeat.

18. $\sqrt{2} = 1.41421356\ldots$ Numeral does not repeat.

In Example 17 there is a pattern, but it is not a repeating pattern.

DO EXERCISES 19–25.

v **Finding Square Roots**

Not all square roots can be found directly from Table 2 or by using a calculator. In that event we proceed by factoring.

---

* 3.14 and $\frac{22}{7}$ are only rational approximations to π.

EXAMPLE 19 Use Table 2 to approximate $\sqrt{275}$ to the nearest tenth.

$$\sqrt{275} = \sqrt{25 \cdot 11} \qquad \text{Factoring the radicand (note that 25 is a perfect square)}$$
$$= \sqrt{25} \cdot \sqrt{11} \qquad \text{Factoring the expression}$$
$$= 5\sqrt{11}$$
$$\approx 5 \times 3.317 \qquad \text{Using Table 2}$$
$$\approx 16.6 \qquad \text{Multiplying and rounding}$$

DO EXERCISES 26 AND 27.

EXAMPLE 20 Use a calculator to approximate $\sqrt{1,350,000,000}$.

If the number does not fit into a calculator, we factor the radicand, as follows:

$$\sqrt{1,350,000,000} = \sqrt{1350 \cdot 10^6} \qquad \begin{array}{l}\text{Factoring the radicand (note} \\ \text{that the exponent, 6, is even)}\end{array}$$
$$= \sqrt{1350} \cdot \sqrt{10^6} \qquad \text{Factoring the expression}$$
$$\approx 36.74 \cdot 10^3 \qquad \begin{array}{l}\text{Approximating } \sqrt{1350} \text{ with a} \\ \text{calculator and finding } \sqrt{10^6}\end{array}$$
$$\approx 36,740. \qquad \text{Multiplying}$$

EXAMPLE 21 Use a calculator to approximate $\sqrt{0.000000005768}$.

If the number will not fit into a calculator, we factor:

$$\sqrt{0.000000005768} = \sqrt{57.68 \times 10^{-10}} \qquad \begin{array}{l}\text{Factoring the radicand} \\ \text{(note that the exponent, } -10, \text{ is even)}\end{array}$$
$$= \sqrt{57.68} \times \sqrt{10^{-10}} \qquad \text{Factoring the expression}$$
$$\approx 7.595 \times 10^{-5} \qquad \begin{array}{l}\text{Approximating } \sqrt{57.68} \text{ with a calculator and} \\ \text{finding } \sqrt{10^{-10}}\end{array}$$
$$\approx 0.00007595.$$

DO EXERCISES 28 AND 29.

Use Table 2 to approximate to the nearest tenth.

26. $\sqrt{252}$

27. $\sqrt{405}$

Use a calculator to approximate.

28. $\sqrt{473,100,000,000,000}$

29. $\sqrt{0.00000000000003142}$

EXERCISE SET 10.3

i Multiply.

1. $\sqrt{3}\sqrt{2}$
2. $\sqrt{5}\sqrt{7}$
3. $\sqrt[3]{2}\sqrt[3]{5}$
4. $\sqrt[3]{7}\sqrt[3]{2}$

5. $\sqrt[4]{8}\sqrt[4]{9}$
6. $\sqrt[4]{6}\sqrt[4]{3}$
7. $\sqrt{3a}\sqrt{10b}$
8. $\sqrt{2x}\sqrt{13y}$

9. $\sqrt[5]{9t^2}\sqrt[5]{2t}$
10. $\sqrt[5]{8y^3}\sqrt[5]{10y}$
11. $\sqrt{x-a}\sqrt{x+a}$
12. $\sqrt{y-b}\sqrt{y+b}$

13. $\sqrt[3]{0.3x}\sqrt[3]{0.2x}$
14. $\sqrt[3]{0.7y}\sqrt[3]{0.3y}$
15. $\sqrt[4]{x-1}\sqrt[4]{x^2+x+1}$
16. $\sqrt[5]{x-2}\sqrt[5]{(x-2)^2}$

17. $\sqrt{\dfrac{6}{x}}\sqrt{\dfrac{y}{5}}$
18. $\sqrt{\dfrac{7}{t}}\sqrt{\dfrac{s}{11}}$

ii Simplify by factoring. Assume that all variables and radicands represent nonnegative numbers.

19. $\sqrt{8}$
20. $\sqrt{18}$
21. $\sqrt{24}$
22. $\sqrt{20}$
23. $\sqrt{180x^4}$
24. $\sqrt{175y^6}$

25. $\sqrt[3]{54x^8}$
26. $\sqrt[3]{40y^3}$
27. $\sqrt[3]{80x^8}$
28. $\sqrt[3]{108m^5}$
29. $\sqrt[4]{32}$
30. $\sqrt[4]{80}$

31. $\sqrt[4]{162c^4d^6}$
32. $\sqrt[4]{243x^8y^{10}}$
33. $\sqrt[3]{(x+y)^4}$

iii Multiply and simplify by factoring. Assume that all variables and radicands represent nonnegative numbers.

34. $\sqrt{3}\sqrt{6}$
35. $\sqrt{5}\sqrt{10}$
36. $\sqrt{15}\sqrt{12}$
37. $\sqrt{2}\sqrt{32}$

38. $\sqrt{6}\sqrt{8}$
39. $\sqrt{18}\sqrt{14}$
40. $\sqrt[3]{3}\sqrt[3]{18}$
41. $\sqrt{45}\sqrt{60}$

42. $\sqrt{5b^3}\sqrt{10c^4}$ **43.** $\sqrt{2x^3y}\sqrt{12xy}$ **44.** $\sqrt[3]{y^4}\sqrt[3]{16y^5}$ **45.** $\sqrt[3]{5^2t^4}\sqrt[3]{5^4t^6}$

46. $\sqrt[3]{(b+3)^4}\sqrt[3]{(b+3)^2}$ **47.** $\sqrt[3]{(x+y)^3}\sqrt[3]{(x+y)^5}$ **48.** $\sqrt{12a^3b}\sqrt{8a^4b^2}$

iv   Which of the following are irrational numbers? Which are rational?

49. $-\frac{5}{6}$ **50.** 8.93 **51.** $8.23\overline{23}$ **52.** $\sqrt{5}$

53. $\sqrt{36}$ **54.** $-\sqrt{15}$ **55.** $2.101001\ldots$
(Numeral does not repeat.)   **56.** $3.1415926\ldots$
(Numeral does not repeat.)

v   Use Table 2 to approximate to the nearest tenth.

57. $\sqrt{252}$ **58.** $\sqrt{140}$ **59.** $\sqrt{189}$ **60.** $\sqrt{490}$ **61.** $\sqrt{350}$

62. $\sqrt{320}$ **63.** $\sqrt{891}$ **64.** $\sqrt{376}$ **65.** $\sqrt{240}$ **66.** $\sqrt{369}$

▦ Use a calculator to approximate each of the following.

67. $\sqrt{24,500,000,000}$ **68.** $\sqrt{16,500,000,000}$ **69.** $\sqrt{468,200,000,000}$ **70.** $\sqrt{99,400,000,000}$

71. $\sqrt{175,420,000,000}$ **72.** $\sqrt{0.000004631}$ **73.** $\sqrt{0.0000000395}$ **74.** $\sqrt{0.0000001543}$

75. $\sqrt{0.0000005001}$ **76.** $\sqrt{0.000010101}$

Problem-solving practice

77. During a one-hour television show there were 12 commercials. Some of the commercials were 30 seconds long and the others were 60 seconds long. If the number of 30-second commercials was six less than the total number of minutes of commercial time during the show, how many 60-second commercials were used during the hour?

78. The length of a rectangle is one-fourth of the width. The area of the rectangle is twice the perimeter. Find the dimensions of the rectangle.

79. (*An application: Speed of a skidding car*). After an accident, police can estimate the speed at which a car was traveling by measuring its skid marks. The formula

$$r = 2\sqrt{5L}$$

can be used, where r is the speed in mph and L is the length of the skid marks in feet. Estimate the speed of a car that left skid marks (a) 20 ft long; (b) 70 ft long; (c) 90 ft long.

80. ▦ (*An application: Wind chill temperature*). In cold weather we feel colder if there is wind than if there is not. *Wind chill temperature* is the temperature at which, without wind, we would feel as cold in an actual situation with wind. Here is a formula for finding wind chill temperature:

$$T_w = 33 - \frac{(10.45 + 10\sqrt{v} - v)(33 - T)}{22},$$

where T is the actual temperature given in degrees Celsius and v is the wind speed in m/sec. Find the wind chill temperature for the given actual temperatures and wind speeds.

a) $T = 7°C$, $v = 8$ m/sec
b) $T = 0°C$, $v = 12$ m/sec
c) $T = -5°C$, $v = 14$ m/sec
d) $T = -23°C$, $v = 15$ m/sec

OBJECTIVES

After finishing Section 10.4, you should be able to:

i   Simplify radical expressions having a quotient for a radicand.

ii   Divide and simplify radical expressions.

iii   Calculate combinations of roots and powers.

10.4

DIVIDING AND SIMPLIFYING WITH RADICAL NOTATION

In this section we will manipulate radical expressions using division rather than multiplication.

i   **Roots of Quotients**

Notice that

$$\sqrt[3]{\frac{27}{8}} = \frac{3}{2} \quad \text{and} \quad \frac{\sqrt[3]{27}}{\sqrt[3]{8}} = \frac{3}{2}.$$

This example suggests that we can take the root of a quotient by taking the root of the numerator and denominator separately. This is true:

> **Rule A:** For any nonnegative number a and any positive number b, and any index k,
>
> $$\sqrt[k]{\frac{a}{b}} = \frac{\sqrt[k]{a}}{\sqrt[k]{b}}.$$
>
> **(We can take the kth root of the numerator and denominator separately.)**

EXAMPLES Simplify by taking roots of the numerator and denominator. Assume that all expressions under radicals represent positive numbers.

1. $\sqrt[3]{\dfrac{27}{125}} = \dfrac{\sqrt[3]{27}}{\sqrt[3]{125}} = \dfrac{3}{5}$ Taking the cube root of the numerator and denominator

2. $\sqrt{\dfrac{25}{y^2}} = \dfrac{\sqrt{25}}{\sqrt{y^2}} = \dfrac{5}{y}$ Taking the square root of the numerator and denominator

3. $\sqrt{\dfrac{16x^3}{y^4}} = \dfrac{\sqrt{16x^3}}{\sqrt{y^4}} = \dfrac{\sqrt{16x^2 \cdot x}}{\sqrt{y^4}} = \dfrac{4x\sqrt{x}}{y^2}$

4. $\sqrt[3]{\dfrac{27y^5}{343x^3}} = \dfrac{\sqrt[3]{27y^5}}{\sqrt[3]{343x^3}} = \dfrac{\sqrt[3]{27y^3 \cdot y^2}}{\sqrt[3]{343x^3}} = \dfrac{\sqrt[3]{27y^3} \cdot \sqrt[3]{y^2}}{\sqrt[3]{343x^3}} = \dfrac{3y\sqrt[3]{y^2}}{7x}$

We are assuming that no expression represents zero or a negative number. Thus we need not be concerned about zero denominators or absolute-value signs.

DO EXERCISES 1–3.

ii   Dividing Radical Expressions

Reversing Rule A shows a way to divide radical expressions as follows.

> **Rule B:**
>
> $$\frac{\sqrt[k]{a}}{\sqrt[k]{b}} = \sqrt[k]{\frac{a}{b}}$$
>
> **To divide, we divide the radicands. After doing so we can sometimes simplify by taking roots.**

EXAMPLES Divide. Then simplify by taking roots if possible. Assume that all expressions under radicals represent positive numbers.

5. $\dfrac{\sqrt{80}}{\sqrt{5}} = \sqrt{\dfrac{80}{5}} = \sqrt{16} = 4$

6. $\dfrac{3\sqrt{2}}{5\sqrt{3}} = \dfrac{3}{5} \cdot \dfrac{\sqrt{2}}{\sqrt{3}} = \dfrac{3}{5} \cdot \sqrt{\dfrac{2}{3}}$

7. $\dfrac{5\sqrt[3]{32}}{\sqrt[3]{2}} = 5\sqrt[3]{\dfrac{32}{2}} = 5\sqrt[3]{16} = 5\sqrt[3]{8 \cdot 2} = 5\sqrt[3]{8}\sqrt[3]{2} = 5 \cdot 2\sqrt[3]{2} = 10\sqrt[3]{2}$

Simplify by taking roots of numerator and denominator. Assume that all expressions under radicals represent positive numbers.

1. $\sqrt{\dfrac{25}{36}}$

2. $\sqrt{\dfrac{x^2}{100}}$

3. $\sqrt[3]{\dfrac{54x^5}{125}}$

Divide. Then simplify by taking roots if possible. Assume that all expressions under radicals represent positive numbers.

4. $\dfrac{\sqrt{75}}{\sqrt{3}}$

5. $\dfrac{14\sqrt{128xy}}{2\sqrt{2}}$

6. $\dfrac{\sqrt{50a^3}}{\sqrt{2a}}$

7. $\dfrac{4\sqrt[3]{250}}{7\sqrt[3]{2}}$

8. $\dfrac{\sqrt[3]{8a^3b}}{\sqrt[3]{27b^{-2}}}$

Calculate as shown. Then use Rule C to calculate another way.

9. $(\sqrt[3]{125})^2$

10. $(\sqrt{6y})^3$

8. $\dfrac{\sqrt{72xy}}{2\sqrt{2}} = \dfrac{1}{2}\dfrac{\sqrt{72xy}}{\sqrt{2}} = \dfrac{1}{2}\sqrt{\dfrac{72xy}{2}} = \dfrac{1}{2}\sqrt{36xy} = \dfrac{1}{2}\sqrt{36}\sqrt{xy}$

$\qquad = \dfrac{1}{2}\cdot 6\sqrt{xy} = 3\sqrt{xy}$

9. $\dfrac{\sqrt[4]{33a^5b^3}}{\sqrt[4]{2b^{-1}}} = \sqrt[4]{\dfrac{33a^5b^3}{2b^{-1}}} = \sqrt[4]{\dfrac{33}{2}a^5b^4} = \sqrt[4]{a^4b^4}\,\sqrt[4]{\dfrac{33}{2}}\,a = ab\sqrt[4]{\dfrac{33}{2}}\,a$

DO EXERCISES 4–8.

iii Powers and Roots Combined

Consider the following:

$$\sqrt[3]{8^2} = \sqrt[3]{64} = 4,$$
$$(\sqrt[3]{8})^2 = (2)^2 = 4.$$

This suggests another important property of radical expressions.

> **Rule C:** **For any nonnegative number a, any index k, and any natural number m,**
>
> $$\sqrt[k]{a^m} = (\sqrt[k]{a})^m.$$
>
> **(We can raise to a power and then take a root, or we can take a root and then raise to a power.)**

In some cases one way of calculating is easier than the other.

EXAMPLES Calculate as shown. Then use Rule C to calculate in another way. Assume that all expressions under radicals are positive.

10. a) $\sqrt[3]{27^2} = \sqrt[3]{729} = 9$ Finding 27^2 and then taking the cube root

   b) $(\sqrt[3]{27})^2 = (3)^2 = 9$ Taking the cube root and then squaring

11. a) $\sqrt[3]{2^6} = \sqrt[3]{64} = 4$

   b) $(\sqrt[3]{2})^6 = \sqrt[3]{2}\,\sqrt[3]{2}\,\sqrt[3]{2}\,\sqrt[3]{2}\,\sqrt[3]{2}\,\sqrt[3]{2} = 2\cdot 2 = 4$

12. a) $(\sqrt{5x})^3 = \sqrt{5x}\,\sqrt{5x}\,\sqrt{5x} = 5x\sqrt{5x}$

   b) $\sqrt{(5x)^3} = \sqrt{5^3x^3} = \sqrt{5^2x^2}\,\sqrt{5x} = 5x\sqrt{5x}$

DO EXERCISES 9 AND 10.

EXERCISE SET 10.4

---

i Simplify by taking roots of the numerator and denominator. Assume that all expressions under radical signs represent positive numbers.

1. $\sqrt{\dfrac{16}{25}}$ **2.** $\sqrt{\dfrac{100}{81}}$ **3.** $\sqrt[3]{\dfrac{64}{27}}$ **4.** $\sqrt[3]{\dfrac{343}{512}}$ **5.** $\sqrt{\dfrac{49}{y^2}}$

6. $\sqrt{\dfrac{121}{x^2}}$ **7.** $\sqrt{\dfrac{25y^3}{x^4}}$ **8.** $\sqrt{\dfrac{36a^5}{b^6}}$ **9.** $\sqrt[3]{\dfrac{8x^5}{27y^3}}$ **10.** $\sqrt[3]{\dfrac{64x^7}{216y^6}}$

ii Divide. Then simplify by taking roots if possible. Assume that all expressions under radical signs represent positive numbers.

11. $\dfrac{\sqrt{21a}}{\sqrt{3a}}$

12. $\dfrac{\sqrt{28y}}{\sqrt{4y}}$

13. $\dfrac{\sqrt[3]{54}}{\sqrt[3]{2}}$

14. $\dfrac{\sqrt[3]{40}}{\sqrt[3]{5}}$

15. $\dfrac{\sqrt{40xy^3}}{\sqrt{8x}}$

16. $\dfrac{\sqrt{56ab^3}}{\sqrt{7a}}$

17. $\dfrac{\sqrt[3]{96a^4b^2}}{\sqrt[3]{12a^2b}}$

18. $\dfrac{\sqrt[3]{189x^5y^7}}{\sqrt[3]{7x^2y^2}}$

19. $\dfrac{\sqrt{72xy}}{2\sqrt{2}}$

20. $\dfrac{\sqrt{75ab}}{3\sqrt{3}}$

21. $\dfrac{\sqrt{x^3-y^3}}{\sqrt{x-y}}$

22. $\dfrac{\sqrt{r^3+s^3}}{\sqrt{r+s}}$

iii Calculate as shown. Then use Rule C to calculate in another way. Assume that all expressions under radicals represent positive numbers.

23. $\sqrt{(6a)^3}$

24. $\sqrt{(7y)^3}$

25. $(\sqrt[3]{16b^2})^2$

26. $(\sqrt[3]{25r^2})^2$

27. $\sqrt{(18a^2b)^3}$

28. $\sqrt{(12x^2y)^3}$

29. $(\sqrt[3]{12c^2d})^2$

30. $(\sqrt[3]{9x^2y})^2$

Problem-solving practice

31. Bob's salary increased 10% after his first year on the job. At the end of the second year he received another 10% raise. If his salary at the beginning of the third year was $6.05 per hour, what did he earn per hour during his first year?

32. A fractional expression has a value of $\frac{3}{5}$. The value of the numerator is two less than three times a certain number, and the value of the denominator is one more than six times the same number. Find the values of the numerator and the denominator.

33. (*An application: Pendulums*). The *period* of a pendulum is the time it takes to complete one cycle, swinging to and fro. If a pendulum consists of a ball on a string, the period T is given by the following formula:

$$T = 2\pi \sqrt{\dfrac{L}{980}},$$

where T is in seconds and L is the length of the pendulum in centimeters. Find the period of a pendulum of the given lengths. Use 3.14 for π.

a) 65 cm

b) 98 cm

c) 120 cm

Divide and simplify.

34. $\dfrac{7\sqrt{a^2b}\sqrt{25xy}}{5\sqrt{a^{-4}b^{-1}}\sqrt{49x^{-1}y^{-3}}}$

35. $\dfrac{(\sqrt[3]{81mn^2})^2}{(\sqrt[3]{mn})^2}$

36. $\dfrac{\sqrt{44x^2y^9z}\sqrt{22y^9z^6}}{(\sqrt{11xy^8z^2})^2}$

10.5

ADDITION AND SUBTRACTION INVOLVING RADICALS

i Any two real numbers can be added. For instance, the sum of 7 and $\sqrt{3}$ can be named

$$7 + \sqrt{3}.$$

We cannot simplify this name for the sum. However, when we have *like radicals* (radicals having the same index and radicand), we can use the distributive laws to simplify by collecting like radical terms.

EXAMPLES   Add or subtract. Simplify by collecting like radical terms if possible.

1. $6\sqrt{7} + 4\sqrt{7} = (6+4)\sqrt{7}$ Using the distributive law (factoring out $\sqrt{7}$)

$$= 10\sqrt{7}$$

2. $8\sqrt[3]{2} - 7x\sqrt[3]{2} + 5\sqrt[3]{2} = (8 - 7x + 5)\sqrt[3]{2}$ Factoring out $\sqrt[3]{2}$

$$= (13 - 7x)\sqrt[3]{2}$$

3. $6\sqrt[5]{4x} + 4\sqrt[5]{4x} - \sqrt[3]{4x} = (6+4)\sqrt[5]{4x} - \sqrt[3]{4x}$

$$= 10\sqrt[5]{4x} - \sqrt[3]{4x}$$

DO EXERCISES 1 AND 2.

OBJECTIVE

After finishing Section 10.5, you should be able to:

i Add or subtract with radical notation and simplify.

Add or subtract. Simplify by collecting like radical terms if possible.

1. $5\sqrt{2} + 8\sqrt{2}$

2. $7\sqrt[4]{5x} + 3\sqrt[4]{5x} - \sqrt{7}$

Add or subtract. Simplify by collecting like radical terms if possible.

3. $7\sqrt{45} - 2\sqrt{5}$

Sometimes we need to factor in order to have terms with like radicals.

EXAMPLES Add or subtract. Simplify by collecting like radical terms if possible.

4. $3\sqrt{8} - 5\sqrt{2} = 3\sqrt{4 \cdot 2} - 5\sqrt{2}$ Factoring 8

$\qquad\qquad = 3\sqrt{4} \cdot \sqrt{2} - 5\sqrt{2}$ Factoring $\sqrt{4 \cdot 2}$ into two radicals

$\qquad\qquad = 3 \cdot 2\sqrt{2} - 5\sqrt{2}$ Taking the square root of 4

$\qquad\qquad = 6\sqrt{2} - 5\sqrt{2}$

$\qquad\qquad = (6 - 5)\sqrt{2}$

$\qquad\qquad = \sqrt{2}$ Collecting like radical terms

4. $3\sqrt[3]{y^5} + 4\sqrt[3]{y^2} + \sqrt[3]{8y^6}$

5. $5\sqrt{2} - 4\sqrt{3}$ No simplification possible

6. $5\sqrt[3]{16y^4} + 7\sqrt[3]{2y} = 5\sqrt[3]{8y^3 \cdot 2y} + 7\sqrt[3]{2y}$ $\Big\}$ Factoring the first radical

$\qquad\qquad = 5\sqrt[3]{8y^3} \cdot \sqrt[3]{2y} + 7\sqrt[3]{2y}\Big\}$

5. $\sqrt{25x - 25} - \sqrt{9x - 9}$

$\qquad\qquad = 5 \cdot 2y \cdot \sqrt[3]{2y} + 7\sqrt[3]{2y}$ Taking the cube root

$\qquad\qquad = 10y\sqrt[3]{2y} + 7\sqrt[3]{2y}$

$\qquad\qquad = (10y + 7)\sqrt[3]{2y}$ Collecting like radical terms

DO EXERCISES 3–5.

EXERCISE SET 10.5

i   Add or subtract. Simplify by collecting like radical terms if possible, assuming that all expressions under radicals represent nonnegative numbers.

1. $6\sqrt{3} + 2\sqrt{3}$

2. $8\sqrt{5} + 9\sqrt{5}$

3. $9\sqrt[3]{5} - 6\sqrt[3]{5}$

4. $14\sqrt[5]{2} - 6\sqrt[5]{2}$

5. $4\sqrt[3]{y} + 9\sqrt[3]{y}$

6. $6\sqrt[4]{t} - 3\sqrt[4]{t}$

7. $8\sqrt{2} - 6\sqrt{2} + 5\sqrt{2}$

8. $2\sqrt{6} + 8\sqrt{6} - 3\sqrt{6}$

9. $4\sqrt[3]{3} - \sqrt{5} + 2\sqrt[3]{3} + \sqrt{5}$

10. $5\sqrt{7} - 8\sqrt[4]{11} + \sqrt{7} + 9\sqrt[4]{11}$

11. $8\sqrt{27} - 3\sqrt{3}$

12. $9\sqrt{50} - 4\sqrt{2}$

13. $8\sqrt{45} + 7\sqrt{20}$

14. $9\sqrt{12} + 16\sqrt{27}$

15. $18\sqrt{72} + 2\sqrt{98}$

16. $12\sqrt{45} - 8\sqrt{80}$

17. $3\sqrt[3]{16} + \sqrt[3]{54}$

18. $\sqrt[3]{27} - 5\sqrt[3]{8}$

19. $2\sqrt{128} - \sqrt{18} + 4\sqrt{32}$

20. $5\sqrt{50} - 2\sqrt{18} + 9\sqrt{32}$

21. $\sqrt{5a} + 2\sqrt{45a^3}$

22. $4\sqrt{3x^3} - \sqrt{12x}$

23. $\sqrt[3]{24x} - \sqrt[3]{3x^4}$

24. $\sqrt[3]{54x} - \sqrt[3]{2x^4}$

25. $\sqrt{8y - 8} + \sqrt{2y - 2}$

26. $\sqrt{12t + 12} + \sqrt{3t + 3}$

27. $\sqrt{x^3 - x^2} + \sqrt{9x - 9}$

28. $\sqrt{4x - 4} - \sqrt{x^3 - x^2}$

29. $6\sqrt{8} + 3\sqrt{18} - 4\sqrt{32}$

30. $4\sqrt{27} - 6\sqrt{75} + 6\sqrt{48}$

31. $5\sqrt[3]{32} - \sqrt[3]{108} + 2\sqrt[3]{256}$

32. $3\sqrt[3]{8x} - 4\sqrt[3]{27x} + 2\sqrt[3]{64x}$

33. $9\sqrt{45} - 10\sqrt{20} - 4\sqrt{80}$

34. $12\sqrt{44} + 4\sqrt{28} - 8\sqrt{99}$

35. $\sqrt{x^3 + x^2} + \sqrt{4x^3 + 4x^2} - \sqrt{9x^3 + 9x^2}$

36. $\sqrt{5x^2 + 4} - 5\sqrt{45x^2 + 36} + 3\sqrt{20x^2 + 16}$

Problem-solving practice

37. The perimeter of a hexagon with all six sides congruent is the same as the perimeter of a square. One side of the hexagon is three less than a side of the square. Find the perimeter of each polygon.

38. John is one-third as old as his father, Bill. Bill is three-fifths as old as his father, Sam. The sum of the three ages is 144. What are the ages of John, Bill, and Sam?

39. Find as many pairs of numbers, a and b, as you can for which

$$\sqrt{a} + \sqrt{b} = \sqrt{a + b}.$$

Find several pairs of numbers for which the above equation is false.

Add or subtract by collecting like terms.

40. $\sqrt{432} - \sqrt{6125} + \sqrt{845} - \sqrt{4800}$

41. $\sqrt{1250x^3y} - \sqrt{1800xy^3} - \sqrt{162x^3y^3}$

42. $\frac{1}{2}\sqrt{36a^5bc^4} - \frac{1}{2}\sqrt[3]{64a^4bc^6} + \frac{1}{6}\sqrt{144a^3bc^2}$

43. $7x\sqrt{(x + y)^3} - 5xy\sqrt{x + y} - 2y\sqrt{(x + y)^3}$

10.6

MORE ABOUT MULTIPLICATION WITH RADICALS

i We now consider multiplication of expressions in which some factors contain more than one term. To do that we use the procedures for multiplying polynomials, which are based on the distributive laws.

EXAMPLES Multiply.

1. $\sqrt{3}(x - \sqrt{5}) = \sqrt{3} \cdot x - \sqrt{3} \cdot \sqrt{5}$ Using the distributive law
$= x\sqrt{3} - \sqrt{15}$ Multiplying radicals

2. $\sqrt[3]{y}(\sqrt[3]{y^2} + \sqrt[3]{2}) = \sqrt[3]{y} \cdot \sqrt[3]{y^2} + \sqrt[3]{y} \cdot \sqrt[3]{2}$ Using the distributive law
$= \sqrt[3]{y^3} + \sqrt[3]{2y}$ Multiplying radicals
$= y + \sqrt[3]{2y}$ Simplifying $\sqrt[3]{y^3}$

DO EXERCISES 1 AND 2.

EXAMPLE 3 Multiply: $(4\sqrt{3} + \sqrt{2})(\sqrt{3} - 5\sqrt{2})$.

$$\begin{array}{cccc} & \text{F} & \text{O} & \text{I} & \text{L} \end{array}$$
$$(4\sqrt{3} + \sqrt{2})(\sqrt{3} - 5\sqrt{2}) = 4(\sqrt{3})^2 - 20\sqrt{3} \cdot \sqrt{2} + \sqrt{2} \cdot \sqrt{3} - 5(\sqrt{2})^2$$
$$= 4 \cdot 3 - 20\sqrt{6} + \sqrt{6} - 5 \cdot 2$$
$$= 12 - 20\sqrt{6} + \sqrt{6} - 10 = 2 - 19\sqrt{6}$$

EXAMPLE 4 Multiply: $(\sqrt{a} + \sqrt{3})(\sqrt{b} + \sqrt{3})$. Assume that all expressions under radicals represent positive numbers.

$$(\sqrt{a} + \sqrt{3})(\sqrt{b} + \sqrt{3}) = \sqrt{a}\sqrt{b} + \sqrt{a}\sqrt{3} + \sqrt{3}\sqrt{b} + \sqrt{3}\sqrt{3}$$
$$= \sqrt{ab} + \sqrt{3a} + \sqrt{3b} + 3$$

DO EXERCISES 3 AND 4.

EXAMPLE 5 Multiply: $(\sqrt{5} + \sqrt{7})(\sqrt{5} - \sqrt{7})$.
$$(\sqrt{5} + \sqrt{7})(\sqrt{5} - \sqrt{7}) = (\sqrt{5})^2 - (\sqrt{7})^2$$ This is now a difference of two squares.
$$= 5 - 7$$
$$= -2$$

DO EXERCISE 5.

EXAMPLE 6 Multiply: $(\sqrt{3} + x)^2$.
$$(\sqrt{3} + x)^2 = (\sqrt{3})^2 + 2x\sqrt{3} + x^2$$ Squaring a binomial
$$= 3 + 2x\sqrt{3} + x^2$$

DO EXERCISES 6 AND 7.

OBJECTIVE

After finishing Section 10.6, you should be able to:

i Multiply expressions involving radicals, where some of the expressions contain more than one term.

Multiply.

1. $\sqrt{2}(5\sqrt{3} + 3\sqrt{7})$

2. $\sqrt[3]{a^2}(\sqrt[3]{3a} - \sqrt[3]{2})$

Multiply.

3. $(\sqrt{3} - 5\sqrt{2})(2\sqrt{3} + \sqrt{2})$

4. $(\sqrt{a} + 2\sqrt{3})(3\sqrt{b} - 4\sqrt{3})$

Multiply.

5. $(\sqrt{2} + \sqrt{5})(\sqrt{2} - \sqrt{5})$

Multiply.

6. $(2\sqrt{5} - y)^2$

7. $(3\sqrt{6} + 2)^2$

EXERCISE SET 10.6

i Multiply.

1. $\sqrt{6}(2 - 3\sqrt{6})$

2. $\sqrt{3}(4 + \sqrt{3})$

3. $\sqrt{2}(\sqrt{3} - \sqrt{5})$

4. $\sqrt{5}(\sqrt{5} - \sqrt{2})$

5. $\sqrt{3}(2\sqrt{5} - 3\sqrt{4})$

6. $\sqrt{2}(3\sqrt{10} - 2\sqrt{2})$

7. $\sqrt[3]{2}\,(\sqrt[3]{4} - 2\sqrt[3]{32})$

8. $\sqrt[3]{3}\,(\sqrt[3]{9} - 4\sqrt[3]{21})$

9. $\sqrt[3]{a}\,(\sqrt[3]{2a^2} + \sqrt[3]{16a^2})$

10. $\sqrt[3]{x}\,(\sqrt[3]{3x^2} - \sqrt[3]{81x^2})$

11. $(\sqrt{3} - \sqrt{2})(\sqrt{3} + \sqrt{2})$

12. $(\sqrt{5} + \sqrt{6})(\sqrt{5} - \sqrt{6})$

13. $(\sqrt{8} + 2\sqrt{5})(\sqrt{8} - 2\sqrt{5})$

14. $(\sqrt{18} + 3\sqrt{7})(\sqrt{18} - 3\sqrt{7})$.

Multiply. Assume that all expressions under radicals represent positive numbers.

15. $(\sqrt{a} + \sqrt{b})(\sqrt{a} - \sqrt{b})$

16. $(\sqrt{x} - \sqrt{y})(\sqrt{x} + \sqrt{y})$

17. $(3 - \sqrt{5})(2 + \sqrt{5})$

18. $(2 + \sqrt{6})(4 - \sqrt{6})$

19. $(\sqrt{3} + 1)(2\sqrt{3} + 1)$

20. $(4\sqrt{3} + 5)(\sqrt{3} - 2)$

21. $(2\sqrt{7} - 4\sqrt{2})(3\sqrt{7} + 6\sqrt{2})$

22. $(4\sqrt{5} + 3\sqrt{3})(3\sqrt{5} - 4\sqrt{3})$

23. $(\sqrt{a} + \sqrt{2})(\sqrt{a} + \sqrt{3})$

24. $(2 - \sqrt{x})(1 - \sqrt{x})$

25. $(2\sqrt[3]{3} + \sqrt[3]{2})(\sqrt[3]{3} - 2\sqrt[3]{2})$

26. $(3\sqrt[4]{7} + \sqrt[4]{6})(2\sqrt[4]{9} - 3\sqrt[4]{6})$

27. $(2 + \sqrt{3})^2$

28. $(\sqrt{5} + 1)^2$

Problem-solving practice

29. The sum of the digits of a two-digit number is 10. If the digits are reversed, the new number is two less than three times the original number. Find the numbers.

30. Harry can fill his pool with a garden hose in 12 hours. If he also uses his neighbor's larger hose, it takes only 3 hours to fill the pool. How long would it take to fill the pool using only the neighbor's hose?

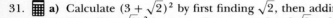

31. **a)** Calculate $(3 + \sqrt{2})^2$ by first finding $\sqrt{2}$, then adding 3, and then squaring.

   b) Find $(3 + \sqrt{2})^2$ by expanding, to get $9 + 6\sqrt{2} + 2$, and then evaluating. Compare your answers.

Multiply and simplify.

32. $\sqrt{9 + 3\sqrt{5}}\,\sqrt{9 - 3\sqrt{5}}$

33. $(\sqrt{x + 2} - \sqrt{x - 2})^2$

34. $(\sqrt{3} + \sqrt{5} - \sqrt{6})^2$

35. $\sqrt[3]{y}\,(1 - \sqrt[3]{y})(1 + \sqrt[3]{y})$

36. $(\sqrt[3]{9} - 2)(\sqrt[3]{9} + 4)$

37. $[\sqrt{3 + \sqrt{2 + \sqrt{1}}}]^4$

38. $(\sqrt{y + 10} - \sqrt{10})(\sqrt{y + 10} + \sqrt{10})$

39. $(8\sqrt{a + 3} - a\sqrt{3})(2\sqrt{a + 3} + 5a\sqrt{3})$

40. $(8\sqrt{3} + 5\sqrt{2})(6\sqrt{7} - 2\sqrt{5})$

41. $(\sqrt{17} + \sqrt{2} - \sqrt{19})(\sqrt{17} + \sqrt{2} + \sqrt{19})$

42. Evaluate: $(\sqrt{7 + 4\sqrt{3}} - \sqrt{7 - 4\sqrt{3}})^2$.

OBJECTIVES

After finishing Section 10.7, you should be able to:

i Rationalize the denominator of a radical expression.

ii Rationalize the numerator of a radical expression.

iii Rationalize denominators or numerators having two terms.

10.7

RATIONALIZING NUMERATORS OR DENOMINATORS

When a fractional expression contains a radical in its denominator, we often want to find an equivalent expression that does not have a radical in the denominator. When we do that, we say that we have *rationalized the denominator*. Similarly, if we find an equivalent expression without a radical sign in a numerator, we say that we have *rationalized the numerator*.

i **Rationalizing Denominators**

Let us consider several ways of finding $\sqrt{\dfrac{1}{2}}$.

1. $\sqrt{\dfrac{1}{2}} = \dfrac{\sqrt{1}}{\sqrt{2}} = \dfrac{1}{\sqrt{2}}$

We find an approximation to $\sqrt{2}$ to be 1.414. Then

$$\frac{1}{\sqrt{2}} \approx \frac{1}{1.414}.$$

By long division we can then find decimal notation for the answer. The division would be rather lengthy without a calculator.

2. $\sqrt{\dfrac{1}{2}} = \sqrt{\dfrac{1}{2} \cdot \dfrac{2}{2}}$ Multiplying by $\dfrac{2}{2}$

$= \sqrt{\dfrac{2}{4}} = \dfrac{\sqrt{2}}{\sqrt{4}} = \dfrac{\sqrt{2}}{2}$

We find an approximation to $\sqrt{2}$ and then compute:

$$\frac{\sqrt{2}}{2} \approx \frac{1.414}{2} = 0.707.$$

This time the division is easy, even without a calculator.

3. $\sqrt{\dfrac{1}{2}} = \sqrt{0.5}$ Dividing

≈ 0.707 Using a calculator

Before the days of calculators, method 2 was almost always used. Today, with calculators it doesn't matter much which of the three methods we use for computing. It is still more or less standard, however, to write radical expressions with the denominator rationalized. The ability to rationalize a numerator or a denominator is a skill that is important in a number of situations.

EXAMPLE 1   Rationalize the denominator: $\sqrt[3]{\dfrac{7}{9}}$.

$\sqrt[3]{\dfrac{7}{9}} = \sqrt[3]{\dfrac{7}{3 \cdot 3} \cdot \dfrac{3}{3}}$ Multiplying by $\dfrac{3}{3}$ to make the denominator a perfect cube

$= \sqrt[3]{\dfrac{21}{3 \cdot 3 \cdot 3}} = \dfrac{\sqrt[3]{21}}{\sqrt[3]{3^3}} = \dfrac{\sqrt[3]{21}}{3}$

DO EXERCISES 1–3.

The idea in rationalizing a denominator is to multiply by 1 in order to make the denominator a perfect power. In Example 1 we multiplied by 1 under the radical. We can also multiply by 1 outside the radical, as in Example 2.

EXAMPLE 2   Rationalize the denominator: $\sqrt{\dfrac{2a}{5b}}$. Assume that all expressions under radicals represent positive numbers.

$\sqrt{\dfrac{2a}{5b}} = \dfrac{\sqrt{2a}}{\sqrt{5b}}$ Converting to a quotient of radicals

$= \dfrac{\sqrt{2a}}{\sqrt{5b}} \cdot \dfrac{\sqrt{5b}}{\sqrt{5b}}$ Multiplying by 1

$= \dfrac{\sqrt{10ab}}{\sqrt{25b^2}}$ The radicand in the denominator is a perfect square.

$= \dfrac{\sqrt{10ab}}{5b}$ Taking the square root of $25b^2$

DO EXERCISES 4 AND 5.

Rationalize the denominator.

1. $\sqrt{\dfrac{2}{3}}$

2. $\sqrt{\dfrac{20}{7}}$

3. $\sqrt[3]{\dfrac{3}{6}}$

Rationalize the denominator. Assume that all expressions under radicals represent positive numbers.

4. $\sqrt{\dfrac{4a}{3b}}$

5. $\dfrac{\sqrt{4x^5}}{\sqrt{3y^3}}$

Rationalize the denominator.

6. $\dfrac{\sqrt[3]{7}}{\sqrt[3]{2}}$

7. $\sqrt[3]{\dfrac{3x^5}{2y}}$

Rationalize the numerator.

8. $\dfrac{\sqrt{11}}{\sqrt{6}}$

9. $\dfrac{\sqrt[3]{7}}{\sqrt[3]{4}}$

EXAMPLE 3 Rationalize the denominator: $\dfrac{\sqrt[3]{a}}{\sqrt[3]{9x}}$.

$$\frac{\sqrt[3]{a}}{\sqrt[3]{9x}} = \frac{\sqrt[3]{a}}{\sqrt[3]{9x}} \cdot \frac{\sqrt[3]{3x^2}}{\sqrt[3]{3x^2}} \qquad \text{Multiplying by 1}$$

To choose the symbol for 1, we look at the radicand $9x$. This is $3 \cdot 3 \cdot x$. To make it a cube we need another 3 and two more x's. Thus we multiply by $\sqrt[3]{3x^2}/\sqrt[3]{3x^2}$. We have

$$\frac{\sqrt[3]{3ax^2}}{\sqrt[3]{27x^3}},$$

which simplifies to

$$\frac{\sqrt[3]{3ax^2}}{3x}.$$

DO EXERCISES 6 AND 7.

ⅱ Rationalizing Numerators

Sometimes in advanced mathematics it is necessary to rationalize a numerator. We use a similar procedure.

EXAMPLE 4 Rationalize the numerator: $\dfrac{\sqrt{7}}{\sqrt{5}}$.

$$\frac{\sqrt{7}}{\sqrt{5}} = \frac{\sqrt{7}}{\sqrt{5}} \cdot \frac{\sqrt{7}}{\sqrt{7}} \qquad \text{Multiplying by 1}$$

$$= \frac{\sqrt{49}}{\sqrt{35}} \qquad \text{The radicand in the numerator is a perfect square.}$$

$$= \frac{7}{\sqrt{35}}$$

EXAMPLE 5 Rationalize the numerator: $\dfrac{\sqrt[3]{5}}{\sqrt[3]{3}}$.

$$\frac{\sqrt[3]{5}}{\sqrt[3]{3}} = \frac{\sqrt[3]{5}}{\sqrt[3]{3}} \cdot \frac{\sqrt[3]{5^2}}{\sqrt[3]{5^2}} \qquad \text{Multiplying by 1}$$

$$= \frac{\sqrt[3]{5^3}}{\sqrt[3]{3 \cdot 5^2}}$$

$$= \frac{\sqrt[3]{5^3}}{\sqrt[3]{75}}$$

$$= \frac{5}{\sqrt[3]{75}}$$

DO EXERCISES 8 AND 9.

iii Rationalizing When There Are Two Terms

When the denominator to be rationalized has two terms, choose a symbol for 1 as illustrated in the following example.

EXAMPLE 6 Rationalize the denominator: $\dfrac{4 + \sqrt{2}}{\sqrt{5} - \sqrt{2}}$.

$$\frac{4 + \sqrt{2}}{\sqrt{5} - \sqrt{2}} = \frac{4 + \sqrt{2}}{\sqrt{5} - \sqrt{2}} \cdot \frac{\sqrt{5} + \sqrt{2}}{\sqrt{5} + \sqrt{2}} \qquad \text{Multiplying by 1}$$

$$= \frac{(4 + \sqrt{2})(\sqrt{5} + \sqrt{2})}{(\sqrt{5} - \sqrt{2})(\sqrt{5} + \sqrt{2})} \qquad \begin{array}{l}\text{Multiplying numerators}\\\text{and denominators}\end{array}$$

$$= \frac{4\sqrt{5} + 4\sqrt{2} + \sqrt{2}\sqrt{5} + (\sqrt{2})^2}{(\sqrt{5})^2 - (\sqrt{2})^2} \qquad \text{Using FOIL}$$

$$= \frac{4\sqrt{5} + 4\sqrt{2} + \sqrt{10} + 2}{5 - 2} \qquad \text{Squaring in the denominator}$$

$$= \frac{4\sqrt{5} + 4\sqrt{2} + \sqrt{10} + 2}{3}$$

Note that the denominator in this example was $\sqrt{5} - \sqrt{2}$. We chose a symbol for 1 that had $\sqrt{5} + \sqrt{2}$ in the numerator and denominator. If the denominator had been $\sqrt{5} + \sqrt{2}$ we would have chosen

$$\frac{\sqrt{5} - \sqrt{2}}{\sqrt{5} - \sqrt{2}}$$

for 1.

EXAMPLES What symbol for 1 would you use to rationalize the denominator?

Expression	*Symbol for 1*
7. $\dfrac{3}{x + \sqrt{7}}$	$\dfrac{x - \sqrt{7}}{x - \sqrt{7}}$
8. $\dfrac{\sqrt{7} + 4}{3 - 2\sqrt{5}}$	$\dfrac{3 + 2\sqrt{5}}{3 + 2\sqrt{5}}$

DO EXERCISES 10 AND 11.

EXAMPLE 9 Rationalize the denominator: $\dfrac{4}{\sqrt{3} + x}$.

We have

$$\frac{4}{\sqrt{3} + x} = \frac{4}{\sqrt{3} + x} \cdot \frac{\sqrt{3} - x}{\sqrt{3} - x}$$

$$= \frac{4(\sqrt{3} - x)}{(\sqrt{3} + x)(\sqrt{3} - x)} = \frac{4(\sqrt{3} - x)}{(\sqrt{3})^2 - x^2} = \frac{4(\sqrt{3} - x)}{3 - x^2}$$

DO EXERCISE 12.

What symbol for 1 would you use to rationalize the denominator?

10. $\dfrac{\sqrt{5} + 1}{\sqrt{3} - y}$

11. $\dfrac{1}{\sqrt{2} + \sqrt{3}}$

Rationalize the denominator.

12. $\dfrac{5}{1 - \sqrt{2}}$

What symbol for 1 would you use to rationalize the numerator?

13. $\dfrac{5 - \sqrt{7}}{\sqrt{6} - 3}$

We can also rationalize numerators. Of course we choose our symbol for 1 a little differently.

EXAMPLES   What symbol for 1 would you use to rationalize the numerator?

	Expression	*Symbol for 1*
10.	$\dfrac{\sqrt{5} + \sqrt{2}}{1 - \sqrt{3}}$	$\dfrac{\sqrt{5} - \sqrt{2}}{\sqrt{5} - \sqrt{2}}$
11.	$\dfrac{x - \sqrt{7}}{3}$	$\dfrac{x + \sqrt{7}}{x + \sqrt{7}}$
12.	$\dfrac{3 - 2\sqrt{5}}{\sqrt{7} + 4}$	$\dfrac{3 + 2\sqrt{5}}{3 + 2\sqrt{5}}$

DO EXERCISE 13.

Rationalize the numerator.

14. $\dfrac{3 + \sqrt{5}}{\sqrt{2} - \sqrt{6}}$

EXAMPLE 13   Rationalize the numerator: $\dfrac{4 + \sqrt{2}}{\sqrt{5} - \sqrt{2}}$.

$$\frac{4 + \sqrt{2}}{\sqrt{5} - \sqrt{2}} = \frac{4 + \sqrt{2}}{\sqrt{5} - \sqrt{2}} \cdot \frac{4 - \sqrt{2}}{4 - \sqrt{2}}$$

$$= \frac{16 - (\sqrt{2})^2}{4\sqrt{5} - \sqrt{5}\sqrt{2} - 4\sqrt{2} + (\sqrt{2})^2}$$

$$= \frac{14}{4\sqrt{5} - \sqrt{10} - 4\sqrt{2} + 2}$$

DO EXERCISE 14.

EXERCISE SET 10.7

i Rationalize the denominator.

1. $\sqrt{\dfrac{6}{5}}$ **2.** $\sqrt{\dfrac{11}{6}}$ **3.** $\sqrt{\dfrac{10}{7}}$ **4.** $\sqrt{\dfrac{22}{3}}$ **5.** $\dfrac{6\sqrt{5}}{5\sqrt{3}}$

6. $\dfrac{2\sqrt{3}}{5\sqrt{2}}$ **7.** $\sqrt[3]{\dfrac{16}{9}}$ **8.** $\sqrt[3]{\dfrac{3}{9}}$ **9.** $\dfrac{\sqrt[3]{3a}}{\sqrt[3]{5c}}$ **10.** $\dfrac{\sqrt[3]{7x}}{\sqrt[3]{3y}}$

11. $\dfrac{\sqrt[3]{2y^4}}{\sqrt[3]{6x^4}}$ **12.** $\dfrac{\sqrt[3]{3a^4}}{\sqrt[3]{7b^2}}$ **13.** $\dfrac{1}{\sqrt[3]{xy}}$ **14.** $\dfrac{1}{\sqrt[3]{ab}}$

ii Rationalize the numerator.

15. $\dfrac{\sqrt{7}}{\sqrt{3x}}$ **16.** $\dfrac{\sqrt{6}}{\sqrt{5x}}$ **17.** $\sqrt{\dfrac{14}{21}}$ **18.** $\sqrt{\dfrac{12}{15}}$ **19.** $\dfrac{4\sqrt{13}}{3\sqrt{7}}$

20. $\dfrac{5\sqrt{21}}{2\sqrt{6}}$ **21.** $\dfrac{\sqrt[3]{7}}{\sqrt[3]{2}}$ **22.** $\dfrac{\sqrt[3]{5}}{\sqrt[3]{4}}$ **23.** $\sqrt{\dfrac{7x}{3y}}$ **24.** $\sqrt{\dfrac{6a}{2b}}$

25. $\dfrac{\sqrt[3]{5y^4}}{\sqrt[3]{6x^5}}$ **26.** $\dfrac{\sqrt[3]{3a^5}}{\sqrt[3]{7b^2}}$ **27.** $\dfrac{\sqrt{ab}}{3}$ **28.** $\dfrac{\sqrt{xy}}{5}$

iii Rationalize the denominator. Assume that all expressions under radicals represent positive numbers.

29. $\dfrac{5}{8 - \sqrt{6}}$ **30.** $\dfrac{7}{9 + \sqrt{10}}$ **31.** $\dfrac{-4\sqrt{7}}{\sqrt{5} - \sqrt{3}}$ **32.** $\dfrac{-3\sqrt{2}}{\sqrt{3} - \sqrt{5}}$ **33.** $\dfrac{\sqrt{5} - 2\sqrt{6}}{\sqrt{3} - 4\sqrt{5}}$

34. $\dfrac{\sqrt{6} - 3\sqrt{5}}{\sqrt{3} - 2\sqrt{7}}$ **35.** $\dfrac{\sqrt{x} - \sqrt{y}}{\sqrt{x} + \sqrt{y}}$ **36.** $\dfrac{\sqrt{a} + \sqrt{b}}{\sqrt{a} - \sqrt{b}}$ **37.** $\dfrac{5\sqrt{3} - 3\sqrt{2}}{3\sqrt{2} - 2\sqrt{3}}$ **38.** $\dfrac{7\sqrt{2} + 4\sqrt{3}}{4\sqrt{3} - 3\sqrt{2}}$

Rationalize the numerator. Assume that all expressions under radicals represent positive numbers.

39. $\dfrac{\sqrt{3} + 5}{8}$ **40.** $\dfrac{3 - \sqrt{2}}{5}$ **41.** $\dfrac{\sqrt{3} - 5}{\sqrt{2} + 5}$ **42.** $\dfrac{\sqrt{6} - 3}{\sqrt{3} + 7}$

43. $\dfrac{\sqrt{x} - \sqrt{y}}{\sqrt{x} + \sqrt{y}}$ **44.** $\dfrac{\sqrt{x} + \sqrt{y}}{\sqrt{x} - \sqrt{y}}$ **45.** $\dfrac{4\sqrt{6} - 5\sqrt{3}}{2\sqrt{3} + 7\sqrt{6}}$ **46.** $\dfrac{8\sqrt{2} + 5\sqrt{3}}{5\sqrt{3} - 7\sqrt{2}}$

Rationalize the denominator.

47. $\dfrac{\sqrt{5} + \sqrt{10} - \sqrt{6}}{\sqrt{50}}$ **48.** $\dfrac{3\sqrt{y} + 4\sqrt{yz}}{5\sqrt{y} - 2\sqrt{z}}$ **49.** $\dfrac{b + \sqrt{b}}{1 + b + \sqrt{b}}$ **50.** $\dfrac{12}{3 - \sqrt{3 - y}}$ **51.** $\dfrac{36a^2 b}{\sqrt[3]{6a^2 b}}$

Rationalize the numerator.

52. $\dfrac{\sqrt{y + 18} - \sqrt{y}}{18}$ **53.** $\dfrac{\sqrt{x + 6} - 5}{\sqrt{x + 6} + 5}$

Simplify. (*Hint:* Rationalize the denominator.)

54. $\sqrt{a^2 - 3} - \dfrac{a^2}{\sqrt{a^2 - 3}}$ **55.** $5\sqrt{\dfrac{x}{y}} + 4\sqrt{\dfrac{y}{x}} - \dfrac{3}{\sqrt{xy}}$

56. $\dfrac{\dfrac{1}{\sqrt{w}} - \sqrt{w}}{\dfrac{\sqrt{w} + 1}{\sqrt{w}}}$ **57.** $\dfrac{1}{4 + \sqrt{3}} + \dfrac{1}{\sqrt{3}} + \dfrac{1}{\sqrt{3} - 4}$

10.8

RATIONAL NUMBERS AS EXPONENTS

i Fractional Exponents

Expressions like $a^{1/2}$, $5^{-1/4}$, and $(2y)^{4/5}$ have not yet been defined. We shall define such expressions in such a way that the usual properties of exponents hold.

Consider $a^{1/2} \cdot a^{1/2}$. If we still want to multiply by adding exponents, it must follow that $a^{1/2} \cdot a^{1/2} = a^{1/2 + 1/2}$ or a^1. Thus we should define $a^{1/2}$ to be a square root of a. Similarly, $a^{1/3} \cdot a^{1/3} \cdot a^{1/3} = a^{1/3 + 1/3 + 1/3}$ or a^1, so $a^{1/3}$ should be defined to mean $\sqrt[3]{a}$.

> For any nonnegative number a and any index n, $a^{1/n}$ means $\sqrt[n]{a}$ (the nonnegative nth root of a).

Whenever we use fractional exponents we assume that the bases are nonnegative.

EXAMPLES Rewrite without fractional exponents.

1. $x^{1/2} = \sqrt{x}$
2. $27^{1/3} = \sqrt[3]{27}$, or 3
3. $(abc)^{1/5} = \sqrt[5]{abc}$

DO EXERCISES 1–5.

OBJECTIVES

After finishing Section 10.8, you should be able to:

i Write expressions with or without fractional exponents.

ii Write expressions without negative exponents.

iii Use the laws of exponents with fractional exponents.

iv Use fractional exponents to simplify radical expressions.

Rewrite without fractional exponents.

1. $y^{1/4}$

2. $(3a)^{1/2}$

3. $16^{1/4}$

4. $(125)^{1/3}$

5. $(a^3 b^2 c)^{1/5}$

Rewrite with fractional exponents.

6. $\sqrt[3]{19}$

7. $\sqrt{abc}$

8. $\sqrt[5]{\dfrac{x^2 y}{16}}$

Rewrite without fractional exponents.

9. $x^{3/2}$

10. $8^{2/3}$

11. $4^{5/2}$

Rewrite with fractional exponents.

12. $(\sqrt[3]{7abc})^4$

13. $\sqrt[5]{6^7}$

Rewrite with positive exponents.

14. $16^{-1/4}$

15. $(3xy)^{-7/8}$

EXAMPLES Rewrite with fractional exponents.

4. $\sqrt[5]{7xy} = (7xy)^{1/5}$

5. $\sqrt[7]{\dfrac{x^3 y}{9}} = \left(\dfrac{x^3 y}{9}\right)^{1/7}$

DO EXERCISES 6–8.

How should we define $a^{2/3}$? If the usual properties of exponents are to hold, we have $a^{2/3} = (a^{1/3})^2$, or $(\sqrt[3]{a})^2$, or $\sqrt[3]{a^2}$. We make our definition accordingly.

For any natural numbers m and n ($n \neq 1$) and any nonnegative number a,

$$a^{m/n} \quad \text{means} \quad \sqrt[n]{a^m}, \quad \text{or} \quad (\sqrt[n]{a})^m.$$

EXAMPLES Rewrite without fractional exponents.

6. $(27)^{2/3} = \sqrt[3]{27^2}, \quad \text{or} \quad (\sqrt[3]{27})^2$
$= 3^2, \quad \text{or } 9$

7. $4^{3/2} = \sqrt[2]{4^3}, \quad \text{or} \quad (\sqrt[2]{4})^3$
$= 2^3, \quad \text{or } 8$

DO EXERCISES 9–11.

EXAMPLES Rewrite with fractional exponents.

8. $\sqrt[3]{9^4} = 9^{4/3}$

9. $(\sqrt[4]{7xy})^5 = (7xy)^{5/4}$

DO EXERCISES 12 AND 13.

ii Negative Rational Exponents

Negative fractional exponents have a meaning similar to that of negative integer exponents. Changing the sign of an exponent amounts to finding a reciprocal.

For any rational number m/n and any positive real number a, $a^{-m/n}$ means $\dfrac{1}{a^{m/n}}$ ($a^{m/n}$ and $a^{-m/n}$ are reciprocals).

EXAMPLES Rewrite with positive exponents.

10. $4^{-1/2} = \dfrac{1}{4^{1/2}}$ $4^{-1/2}$ is the reciprocal of $4^{1/2}$.

Since $4^{1/2} = \sqrt{4} = 2$, the answer simplifies to $\frac{1}{2}$.

11. $(5xy)^{-4/5} = \dfrac{1}{(5xy)^{4/5}}$ $(5xy)^{-4/5}$ is the reciprocal of $(5xy)^{4/5}$.

DO EXERCISES 14 AND 15.

iii Laws of Exponents

The same laws hold for rational-number exponents as for integer exponents. We list them for review.

> **For any real number a and any rational exponents m and n:**
>
> 1. $a^m \cdot a^n = a^{m+n}$ In multiplying, we can add exponents if the bases are the same.
> 2. $\dfrac{a^m}{a^n} = a^{m-n}$ In dividing, we can subtract exponents if the bases are the same.
> 3. $(a^m)^n = a^{m \cdot n}$ To raise a power to a power, we can multiply the exponents.

EXAMPLES Use the laws of exponents to simplify.

12. $3^{1/5} \cdot 3^{3/5} = 3^{1/5 + 3/5} = 3^{4/5}$ Adding exponents

13. $5^{2/3} \cdot 5^{1/4} = 5^{2/3 + 1/4} = 5^{8/12 + 3/12} = 5^{11/12}$ Adding exponents

14. $\dfrac{7^{1/4}}{7^{1/2}} = 7^{1/4 - 1/2} = 7^{1/4 - 2/4} = 7^{-1/4}$ Subtracting exponents

15. $(7.2^{2/3})^{3/4} = 7.2^{2/3 \cdot 3/4} = 7.2^{6/12}$ Multiplying exponents
$= 7.2^{1/2}$

DO EXERCISES 16–18.

iv Simplifying Radical Expressions

Fractional exponents can be used to simplify some radical expressions. The procedure is as follows.

> 1. Convert radical expressions to exponential expressions.
> 2. Use arithmetic and the laws of exponents to simplify.
> 3. Convert back to radical notation when appropriate.
>
> *Important:* This works only when all expressions under radicals are nonnegative, since fractional exponents are not defined otherwise. No absolute-value signs will be needed.

EXAMPLE 16 Use fractional exponents to simplify.

$$\sqrt[6]{x^3} = x^{3/6}$$ Converting to an exponential expression
$$= x^{1/2}$$ Using arithmetic to simplify the exponent
$$= \sqrt{x}$$ Converting back to radical notation

EXAMPLE 17 Use fractional exponents to simplify.

$$\sqrt[8]{a^2 b^4} = (a^2 b^4)^{1/8}$$ Converting to exponential notation
$$= a^{2/8} \cdot b^{4/8}$$ Using the third law of exponents
$$= a^{1/4} \cdot b^{2/4}$$ Using arithmetic to simplify the exponents
$$= (ab^2)^{1/4}$$ Using the third law of exponents (in reverse)
$$= \sqrt[4]{ab^2}$$ Converting back to radical notation

DO EXERCISES 19–24.

Use the laws of exponents to simplify.

16. $7^{1/3} \cdot 7^{3/5}$

17. $\dfrac{5^{7/6}}{5^{5/6}}$

18. $(9^{3/5})^{2/3}$

Use fractional exponents to simplify.

19. $\sqrt[4]{a^2}$

20. $\sqrt[4]{x^4}$

21. $\sqrt[6]{8}$

Use fractional exponents to simplify.

22. $\sqrt[5]{a^5 b^{10}}$

23. $\sqrt[4]{x^4 y^{12}}$

24. $\sqrt[12]{x^3 y^6}$

Use fractional exponents to write a single radical expression.

25. $\sqrt[4]{7} \cdot \sqrt{3}$

We can use properties of fractional exponents to write a single radical expression for a product or quotient.

EXAMPLES Use fractional exponents to write a single radical expression.

18. $\sqrt[3]{5} \cdot \sqrt{2} = 5^{1/3} \cdot 2^{1/2}$ Converting to exponential notation

$\qquad = 5^{2/6} \cdot 2^{3/6}$ Rewriting so that exponents have a common denominator

$\qquad = (5^2 \cdot 2^3)^{1/6}$ Using the third law of exponents

$\qquad = \sqrt[6]{5^2 \cdot 2^3}$ Converting back to radical notation

$\qquad = \sqrt[6]{200}$ Multiplying under the radical

26. $\dfrac{\sqrt[4]{(a-b)^5}}{(a-b)}$

19. $\dfrac{\sqrt[4]{(x+y)^3}}{\sqrt{x+y}} = \dfrac{(x+y)^{3/4}}{(x+y)^{1/2}}$ Converting to exponential notation

$\qquad = (x+y)^{3/4-1/2}$ Using the second law of exponents

$\qquad = (x+y)^{1/4}$ Subtracting exponents

$\qquad = \sqrt[4]{x+y}$ Converting back to radical notation

DO EXERCISES 25 AND 26.

EXAMPLE 20 Write a single radical expression.

$a^{1/2}b^{-1/2}c^{5/6} = a^{3/6}b^{-3/6}c^{5/6}$ Rewriting exponents with a common denominator

$\qquad = (a^3 b^{-3} c^5)^{1/6}$ Using the third law of exponents

$\qquad = \sqrt[6]{a^3 b^{-3} c^5}$ Converting to radical notation

DO EXERCISES 27 AND 28.

Write a single radical expression.

27. $x^{2/3}y^{1/2}z^{5/6}$

We have now seen several different methods of simplifying radical expressions. We list them.

28. $\dfrac{a^{1/2}b^{3/8}}{a^{1/4}b^{1/8}}$

SOME WAYS TO SIMPLIFY RADICAL EXPRESSIONS

1. *Simplifying by factoring.* We factor the radicand, looking for factors that are perfect powers.

   Example: $\sqrt[3]{16} = \sqrt[3]{8}\,\sqrt[3]{2}$
   $\qquad\qquad\quad = 2\sqrt[3]{2}$

2. *Rationalizing denominators.* Radical expressions are usually considered simpler if there are no radicals in the denominator.

   Example: $\dfrac{1}{\sqrt{2}} = \dfrac{\sqrt{2}}{2}$

3. *Collecting like radical terms.*

   Example: $\sqrt{8} + 3\sqrt{2} = \sqrt{4}\cdot\sqrt{2} + 3\sqrt{2}$
   $\qquad\qquad\qquad\quad = 2\sqrt{2} + 3\sqrt{2}$
   $\qquad\qquad\qquad\quad = 5\sqrt{2}$

4. *Using fractional exponents to simplify.* We convert to exponential notation and then use arithmetic and the laws of exponents to simplify the exponents. Then we convert back to radical notation. *Caution:* This works only when there are only non-negative expressions under radical signs.

EXERCISE SET 10.8

i Rewrite without fractional exponents.

1. $x^{1/4}$ **2.** $y^{1/5}$ **3.** $(8)^{1/3}$ **4.** $(16)^{1/2}$ **5.** $(a^2 b^2)^{1/5}$

6. $(x^3 y^3)^{1/4}$ **7.** $a^{2/3}$ **8.** $b^{3/2}$ **9.** $16^{3/4}$ **10.** $4^{7/2}$

Rewrite with fractional exponents.

11. $\sqrt[3]{20}$ **12.** $\sqrt[3]{19}$ **13.** $\sqrt{17}$ **14.** $\sqrt{6}$ **15.** $\sqrt[4]{cd}$ **16.** $\sqrt[5]{xy}$

17. $\sqrt[5]{xy^2 z}$ **18.** $\sqrt[7]{x^3 y^2 z^2}$ **19.** $(\sqrt{3mn})^3$ **20.** $(\sqrt[3]{7xy})^4$ **21.** $(\sqrt[7]{8x^2 y})^5$ **22.** $(\sqrt[6]{2a^5 b})^7$

ii Rewrite with positive exponents.

23. $x^{-1/3}$ **24.** $y^{-1/4}$ **25.** $(2rs)^{-3/4}$ **26.** $(5xy)^{-5/6}$

27. $\left(\dfrac{1}{10}\right)^{-2/3}$ **28.** $\left(\dfrac{1}{8}\right)^{-3/4}$ **29.** $\dfrac{1}{x^{-2/3}}$ **30.** $\dfrac{1}{x^{-5/6}}$

iii Use the properties of exponents to simplify.

31. $5^{3/4} \cdot 5^{1/8}$ **32.** $11^{2/3} \cdot 11^{1/2}$ **33.** $\dfrac{7^{5/8}}{7^{3/8}}$ **34.** $\dfrac{9^{9/11}}{9^{7/11}}$

35. $\dfrac{8.3^{3/4}}{8.3^{2/5}}$ **36.** $\dfrac{3.9^{3/5}}{3.9^{1/4}}$ **37.** $(10^{3/5})^{2/5}$ **38.** $(5^{5/4})^{3/7}$

iv Use fractional exponents to simplify.

39. $\sqrt[6]{a^4}$ **40.** $\sqrt[6]{y^2}$ **41.** $\sqrt[3]{8y^6}$ **42.** $\sqrt{x^4 y^6}$ **43.** $\sqrt[4]{32}$ **44.** $\sqrt[8]{81}$

45. $\sqrt[6]{4x^2}$ **46.** $\sqrt[4]{16x^4 y^2}$ **47.** $\sqrt[5]{32c^{10} d^{15}}$ **48.** $\sqrt[4]{16x^{12} y^{16}}$ **49.** $\sqrt[6]{\dfrac{m^{12} n^{24}}{64}}$ **50.** $\sqrt[5]{\dfrac{x^{15} y^{20}}{32}}$

51. $\sqrt[8]{r^4 s^2}$ **52.** $\sqrt[12]{64t^6 s^6}$ **53.** $\sqrt[3]{27a^3 b^9}$ **54.** $\sqrt[4]{81x^8 y^8}$

Use fractional exponents to write a single radical expression.

55. $\sqrt[3]{7} \cdot \sqrt{2}$ **56.** $\sqrt[3]{7} \cdot \sqrt[5]{5}$ **57.** $\sqrt{x} \sqrt[3]{2x}$ **58.** $\sqrt[3]{y} \sqrt[5]{3y}$ **59.** $\sqrt{x} \sqrt[3]{x-2}$ **60.** $\sqrt[4]{3x} \sqrt{y+4}$

61. $\dfrac{\sqrt[3]{(a+b)^2}}{\sqrt{(a+b)}}$ **62.** $\dfrac{\sqrt[3]{(x+y)^2}}{\sqrt[4]{(x+y)^3}}$ **63.** $a^{2/3} \cdot b^{3/4}$ **64.** $x^{1/3} \cdot y^{1/4} \cdot z^{1/6}$ **65.** $\dfrac{s^{7/12} \cdot t^{5/6}}{s^{1/3} \cdot t^{-1/6}}$ **66.** $\dfrac{x^{8/15} \cdot y^{4/5}}{x^{1/3} \cdot y^{-1/5}}$

Use fractional exponents to write a single radical expression and simplify.

67. $\sqrt[5]{yx^2} \sqrt{xy}$ **68.** $\sqrt{x^5 \sqrt[3]{x^4}}$ **69.** $\dfrac{\sqrt{(a+b)^3} \sqrt[3]{(a+b)^2}}{\sqrt[4]{a+b}}$ **70.** $\sqrt[4]{\sqrt[3]{8x^3 y^6}}$

Simplify.

71. $(-\sqrt[4]{7} \sqrt[3]{w})^{12}$ **72.** $\sqrt[12]{p^2 + 2pq + q^2}$ **73.** $\dfrac{1}{\sqrt[3]{3} - \sqrt[3]{2}}$

74. $[\sqrt[10]{\sqrt[5]{x^{15}}}]^5 [\sqrt[5]{\sqrt[10]{x^{15}}}]^5$ **75.** $\sqrt[p]{x^{5p} y^{7p+1} z^{p+3}}$

76. ▦ (*An application: Road pavement messages*). In a psychological study it was determined that the proper length L of the letters of a word printed on pavement is given by

$$L = \frac{(0.00252) d^{2.27}}{h},$$

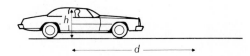

where d is the distance of a car from the lettering and h is the height of the eye above the surface of the road. All units are in meters. This formula says that if a person is h meters above the surface of the road and is to be able to recognize a message d meters away, that message will be the most recognizable if the length of the letters is L. Find L, given d and h.

a) $h = 1$ m, $d = 60$ m   **b)** $h = 0.9906$ m, $d = 75$ m

c) $h = 2.4$ m, $d = 80$ m   **d)** $h = 1.1$ m, $d = 100$ m

OBJECTIVES

After finishing Section 10.9, you should be able to:

i Solve radical equations with one radical term.

ii Solve radical equations with two radical terms.

iii Use the Pythagorean property of right triangles to find lengths of sides of triangles and to solve problems.

Solve.

1. $\sqrt{x} - 7 = 3$

2. $\sqrt{x} = -2$

10.9

SOLVING RADICAL EQUATIONS

i A radical equation is an equation in which variables occur in one or more radicands. Here are some examples:

$$\sqrt[3]{2x} + 1 = 5, \qquad \sqrt{x} + \sqrt{4x - 2} = 7.$$

To solve such equations we need a new principle for equations. Suppose that the equation $a = b$ is true. When we square both sides we get another true equation, $a^2 = b^2$. Generalizing, we have the principle of powers.

THE PRINCIPLE OF POWERS

If an equation $a = b$ is true, then $a^n = b^n$ is true for any natural number n.

If an equation $a^n = b^n$ is true, it *may not* be true that $a = b$. For example, $3^2 = (-3)^2$ is true, but $3 = -3$ is not true.

EXAMPLE 1 Solve: $\sqrt{x} - 3 = 4$.

$$\sqrt{x} - 3 = 4$$
$$\sqrt{x} = 7 \qquad \text{Adding to isolate the radical}$$
$$(\sqrt{x})^2 = 7^2 \qquad \text{Using the principle of powers}$$
$$x = 49$$

Check:

$$\begin{array}{c|c} \sqrt{x} - 3 = 4 \\ \hline \sqrt{49} - 3 & 4 \\ 7 - 3 & \\ 4 & \end{array}$$

The solution is 49.

The principle of powers does not always produce equivalent equations.

EXAMPLE 2 Solve: $\sqrt{x} = -3$.

We might observe at the outset that this equation has no solution because the principal square root of a number is never negative. Let us continue as above, for comparison.

$$x = (-3)^2, \quad \text{or } 9 \qquad \text{Principle of powers (squaring)}$$

Check:

$$\begin{array}{c|c} \sqrt{x} = -3 \\ \hline \sqrt{9} & -3 \\ 3 & \end{array}$$

The number 9 does not check. Hence the equation has no solution.

> In solving radical equations, possible solutions found using the principle of powers *must* be checked!

DO EXERCISES 1 AND 2.

To solve an equation with a radical term, we first isolate the radical term on one side of the equation. Then we use the principle of powers.

EXAMPLE 3 Solve: $x = \sqrt{x + 7} + 5$.

$$x = \sqrt{x + 7} + 5$$
$$x - 5 = \sqrt{x + 7} \qquad \text{Adding } -5 \text{ to isolate the radical term}$$
$$(x - 5)^2 = (\sqrt{x + 7})^2 \qquad \text{Principle of powers: squaring both sides}$$
$$x^2 - 10x + 25 = x + 7$$
$$x^2 - 11x + 18 = 0$$
$$(x - 9)(x - 2) = 0 \qquad \text{Factoring}$$
$$x = 9 \quad \text{or} \quad x = 2 \qquad \text{Using the principle of zero products}$$

The possible solutions are 9 and 2. Let us check.

For 9:
$$\begin{array}{c|c} x = \sqrt{x + 7} + 5 \\ \hline 9 & \sqrt{9 + 7} + 5 \\ & 9 \end{array}$$

For 2:
$$\begin{array}{c|c} x = \sqrt{x + 7} + 5 \\ \hline 2 & \sqrt{2 + 7} + 5 \\ & 8 \end{array}$$

Since 9 checks but 2 does not, the solution is 9.

EXAMPLE 4 Solve: $\sqrt[3]{2x + 1} + 5 = 0$.

$$\sqrt[3]{2x + 1} + 5 = 0$$
$$\sqrt[3]{2x + 1} = -5 \qquad \text{Adding } -5; \text{ this isolates the radical term}$$
$$2x + 1 = (-5)^3, \quad \text{or } -125 \qquad \text{Principle of powers: raising to the third power}$$
$$2x = -126 \qquad \text{Adding } -1$$
$$x = -63$$

Check:
$$\begin{array}{c|c} \sqrt[3]{2x + 1} + 5 = 0 \\ \hline \sqrt[3]{2 \cdot (-63) + 1} + 5 & 0 \\ \sqrt[3]{-125} + 5 & \\ -5 + 5 & \\ 0 & \end{array}$$

The solution is -63.

DO EXERCISES 3 AND 4.

ii Equations With Two Radical Terms

A general strategy for solving equations with two radical terms is as follows.

---
1. **Isolate one of the radical terms.**
2. **Use the principle of powers.**
3. **If a radical remains, perform steps 1 and 2 again.**
4. **Check possible solutions.**

Solve.

3. $x - 1 = \sqrt{x + 5}$

4. $\sqrt[4]{x - 1} - 2 = 0$

Solve.

5. $\sqrt{x} - \sqrt{x-5} = 1$

EXAMPLE 5 Solve: $\sqrt{x-3} + \sqrt{x+5} = 4$.

$$\sqrt{x-3} + \sqrt{x+5} = 4$$

$$\sqrt{x-3} = 4 - \sqrt{x+5} \qquad \text{Adding } -\sqrt{x+5}; \text{ this isolates one of the radical terms}$$

$$(\sqrt{x-3})^2 = (4 - \sqrt{x+5})^2 \qquad \text{Principle of powers: squaring both sides}$$

In the above step 2, we are squaring a binomial. We square 4, then find twice the product of 4 and $\sqrt{x+5}$, and then the square of $\sqrt{x+5}$.

$$x - 3 = 16 - 8\sqrt{x+5} + (x+5)$$

$$-24 = -8\sqrt{x+5} \qquad \text{Isolating the remaining radical term}$$

$$3 = \sqrt{x+5}$$

$$3^2 = x + 5 \qquad \text{Squaring}$$

$$4 = x$$

The number 4 checks, so it is the solution.

EXAMPLE 6 Solve: $\sqrt{2x-5} = 1 + \sqrt{x-3}$.

$$\sqrt{2x-5} = 1 + \sqrt{x-3}$$

$$(\sqrt{2x-5})^2 = (1 + \sqrt{x-3})^2 \qquad \text{One radical is already isolated; we square both sides.}$$

$$2x - 5 = 1 + 2\sqrt{x-3} + (x-3)$$

$$x - 3 = 2\sqrt{x-3} \qquad \text{Isolating the remaining radical term}$$

$$(x-3)^2 = (2\sqrt{x-3})^2 \qquad \text{Squaring both sides}$$

$$x^2 - 6x + 9 = 4(x-3)$$

$$x^2 - 6x + 9 = 4x - 12$$

$$x^2 - 10x + 21 = 0$$

$$(x-7)(x-3) = 0 \qquad \text{Factoring}$$

$$x = 7 \quad \text{or} \quad x = 3 \qquad \text{Using the principle of zero products}$$

The numbers 7 and 3 check, so they are the solutions.

6. $\sqrt{3x+1} = 1 + \sqrt{x+4}$

DO EXERCISES 5 AND 6.

ⅲ The Pythagorean Property of Right Triangles

A famous property of right triangles, relating the lengths of the sides, is named for Pythagoras, a mathematician of ancient Greece. His theorem is very useful in problem solving. As we shall see, the use of the property in problem solving involves the use of exponential and radical symbolism.

In a right triangle, the longest side is called the *hypotenuse,* and the other two sides are called the *legs.*

> **THE PYTHAGOREAN THEOREM**
>
> **In a right triangle, if c is the length of the hypotenuse and a and b are the lengths of the legs, then**
>
> $$a^2 + b^2 = c^2.$$

If we know the lengths of any two sides of a right triangle, we can find the length of the other side.

EXAMPLE 7 Find the length of the hypotenuse of the right triangle shown here. Approximate your answer to three decimal places.

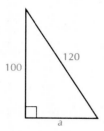

$$a^2 + b^2 = c^2 \qquad \text{The Pythagorean equation}$$
$$7^2 + 5^2 = c^2 \qquad \text{Substituting}$$
$$49 + 25 = c^2$$
$$74 = c^2$$
$$\sqrt{74} = c \qquad \text{Taking the principal square root}$$
$$8.602 \approx c \qquad \text{Using Table 2 or a calculator}$$

EXAMPLE 8 Find the length of the leg of the right triangle shown here. Approximate your answer to two decimal places.

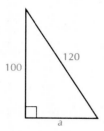

We use the Pythagorean equation
$$a^2 + b^2 = c^2.$$

We solve the equation for a:
$$a^2 = c^2 - b^2$$
$$a = \sqrt{c^2 - b^2}.$$

Now we substitute:
$$a = \sqrt{120^2 - 100^2}$$
$$= \sqrt{14{,}400 - 10{,}000}$$
$$= \sqrt{4400}, \quad \text{or } 20\sqrt{11}.$$

Using Table 2 or a calculator, we find that $a \approx 66.33$.

DO EXERCISES 7 AND 8.

Problem Solving

Right triangles occur in many problems. When that happens, we should always think of trying to use the Pythagorean theorem.

7. Find the length of the hypotenuse of this right triangle. Approximate your answer to three decimal places.

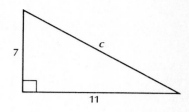

8. Find the length of the leg of this right triangle. Approximate your answer to two decimal places.

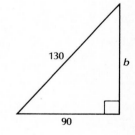

9. Hometown is 13 miles east of Georgetown. Northwood is 18 miles north of Georgetown. How far is it from Hometown to Northwood?

EXAMPLE 9 How much wire will be needed to make a guy wire for a telephone pole if the wire is to be fastened 20 ft above the ground and is to be anchored in the ground 9 ft from the base of the pole?

1. **Familiarize** We draw a picture and label it. In the picture we see a right triangle, so we should see if the Pythagorean theorem can be applied. We have called the length of the wire d.

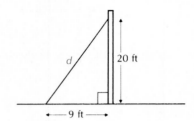

2. **Translate** From the right triangle in our drawing, we have

$$d^2 = 9^2 + 20^2. \qquad \text{Substituting in the Pythagorean equation}$$

3. **Carry out** We solve the equation for d:

$$d = \sqrt{9^2 + 20^2} \qquad \text{Taking the principal square root on both sides}$$
$$= \sqrt{481}$$
$$\approx 21.93. \qquad \text{Using Table 2 or a calculator}$$

4. **Check** We square the lengths 9 and 20, obtaining $81 + 400$, or 481. We then square the length 21.93:

$$21.93 \times 21.93 = 480.92.$$

This is very close to 481, so the answer checks.

5. **State** The answer is that the guy wire will be 21.93 ft long, plus an amount that will be needed for fastening it to the pole and the ground.

DO EXERCISE 9.

EXERCISE SET 10.9

i Solve.

1. $\sqrt{2x - 3} = 1$
2. $\sqrt{x + 3} = 6$
3. $\sqrt{y + 1} - 5 = 8$
4. $\sqrt{x - 2} - 7 = -4$
5. $\sqrt[3]{x + 5} = 2$
6. $\sqrt[3]{x - 2} = 3$
7. $\sqrt[4]{y - 3} = 2$
8. $\sqrt[4]{x + 3} = 3$
9. $\sqrt{3y + 1} = 9$
10. $\sqrt{2y + 1} = 13$
11. $3\sqrt{x} = 6$
12. $8\sqrt{y} = 2$
13. $\sqrt[3]{x} = -3$
14. $\sqrt[3]{y} = -4$
15. $\sqrt{y + 3} - 20 = 0$
16. $\sqrt{x + 4} - 11 = 0$
17. $\sqrt{x + 2} = -4$
18. $\sqrt{y - 3} = -2$
19. $8 = \dfrac{1}{\sqrt{x}}$
20. $3 = \dfrac{1}{\sqrt{y}}$
21. $\sqrt[3]{6x + 9} + 8 = 5$
22. $\sqrt[3]{3y + 6} + 2 = 3$

ii Solve.

23. $\sqrt{3y + 1} = \sqrt{2y + 6}$
24. $\sqrt{5x - 3} = \sqrt{2x + 3}$
25. $2\sqrt{1 - x} = \sqrt{5}$
26. $2\sqrt{2y - 3} = \sqrt{4y}$
27. $2\sqrt{t - 1} = \sqrt{3t - 1}$
28. $\sqrt{y + 10} = 3\sqrt{2y + 3}$
29. $\sqrt{y - 5} + \sqrt{y} = 5$
30. $\sqrt{x - 9} + \sqrt{x} = 1$
31. $3 + \sqrt{z - 6} = \sqrt{z + 9}$

32. $\sqrt{4x - 3} = 2 + \sqrt{2x - 5}$

33. $\sqrt{20 - x} + 8 = \sqrt{9 - x} + 11$

34. $4 + \sqrt{10 - x} = 6 + \sqrt{4 - x}$

35. $\sqrt{x + 2} + \sqrt{3x + 4} = 2$

36. $\sqrt{6x + 7} - \sqrt{3x + 3} = 1$

37. $\sqrt{4y + 1} - \sqrt{y - 2} = 3$

38. $\sqrt{y + 15} - \sqrt{2y + 7} = 1$

39. $\sqrt{3x - 5} + \sqrt{2x + 3} + 1 = 0$

40. $\sqrt{2m - 3} = \sqrt{m + 7} - 2$

iii In each exercise, the lengths of two sides of a right triangle are given. Find the length of the other side (c is always the length of the hypotenuse). Use Table 2 or a calculator to find an approximation.

41. $a = 3$, $b = 4$

42. $b = 8$, $c = 10$

43. $a = 16$, $c = 20$

44. $a = 5$, $b = 12$

45. $a = 6$, $b = 8$

46. $b = 20$, $c = 25$

47. $a = 3$, $b = 7$

48. $b = 11$, $c = 12$

49. $a = 1$, $b = 1$

50. $a = 4$, $b = 7$

51. $a = 2$, $b = 2\sqrt{2}$

52. $a = 3$, $b = 3\sqrt{6}$

Problem solving

53. Triangle ABC has sides that are of lengths 25 ft, 25 ft, and 30 ft. Triangle PQR has sides that are of lengths 25 ft, 25 ft, and 40 ft. Which triangle has the greater area and by how much?

54. During the summer heat a 2-mile bridge expands 2 ft in length. Assuming the bulge occurs straight up from the middle, how high is the bulge?

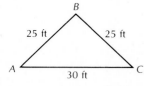 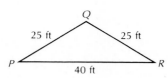

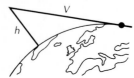

55. Two sides of a triangle have length s, and the height is h. Find a formula for the length of the third side.

56. Each side of a regular octagon has length s. Find the distance d between the parallel sides of the octagon.

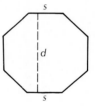

(*An application: Sighting to the horizon*). The formula $V = 1.2\sqrt{h}$ can be used to approximate the distance V, in miles, that a person can see to the horizon from a height h, in feet. Use a calculator for Exercises 57 and 58.

57. How far can you see to the horizon through an airplane window at a height of 30,000 ft?

58. How far can a sailor see to the horizon from the top of a 72-ft mast?

Solve.

59. $\dfrac{x + \sqrt{x + 1}}{x - \sqrt{x + 1}} = \dfrac{5}{11}$

60. $\sqrt[3]{\dfrac{z}{4}} - 10 = 2$

61. $\sqrt[4]{z^2 + 17} = 3$

62. $\sqrt{\sqrt{y} + 49} = 7$

63. $\sqrt[3]{x^2 + x + 15} - 3 = 0$

64. $x^2 - 5x - \sqrt{x^2 - 5x - 2} = 4$

65. $\sqrt{8 - b} = b\sqrt{8 - b}$

66. $\sqrt{x - 2} - \sqrt{x + 2} + 2 = 0$

67. $6\sqrt{y} + 6y^{-1/2} = 37$

68. $\sqrt{a^2 + 30a} = a + \sqrt{5a}$

69. $\sqrt{\sqrt{x} + 4} = \sqrt{x} - 2$

70. $\dfrac{x - 1}{\sqrt{x^2 + 3x + 6}} = \dfrac{1}{4}$

71. $\sqrt{x + 1} - \dfrac{2}{\sqrt{x + 1}} = 1$

72. $\sqrt{11x + \sqrt{6 + x}} = 6$

73. $\sqrt{2x - 2} + \sqrt{7x + 4} = \sqrt{13x + 10}$

(*The Pythagorean formula in three dimensions*). The length d of a diagonal of a rectangular box is given by $d = \sqrt{a^2 + b^2 + c^2}$, where a, b, and c are the lengths of the sides.

74. ▦ Find the length of a diagonal of a box whose sides have lengths 2, 4, and 5.

75. ▦ Will a baton 57 cm long fit inside an attache case that measures 12.5 cm by 45 cm by 31.5 cm?

EXTRA PRACTICE ON SOLVING RADICAL EQUATIONS

Solve.

1. $\sqrt{x} = 5$

2. $\sqrt{x} = 7$

3. $\sqrt{x} = 6.2$

4. $\sqrt{x} = 4.3$

5. $\sqrt{x + 3} = 20$

6. $\sqrt{x + 4} = 11$

7. $\sqrt{2x + 4} = 25$

8. $\sqrt{2x + 1} = 13$

9. $3 + \sqrt{x - 1} = 5$

10. $4 + \sqrt{y - 3} = 11$

11. $6 - 2\sqrt{3n} = 0$

12. $8 - 4\sqrt{5n} = 0$

13. $\sqrt{5x - 7} = \sqrt{x + 10}$

14. $\sqrt{4x - 5} = \sqrt{x + 9}$

15. $\sqrt{x} = -7$

16. $\sqrt{x} = -5$

17. $\sqrt{2y + 6} = \sqrt{2y - 5}$

18. $2\sqrt{3x - 2} = \sqrt{2x - 3}$

19. $\sqrt{5x^2 + 5} = 5$

20. $\sqrt{x} = -x$

21. $x - 1 = \sqrt{x + 5}$

22. $\sqrt{y^2 + 6} + y - 3 = 0$

23. $\sqrt{x - 5} + \sqrt{x} = 5$

24. $\sqrt{3x + 1} = 1 + \sqrt{x + 4}$

25. $4 + \sqrt{10 - x} = 6 + \sqrt{4 - x}$

26. $x = (x - 2)\sqrt{x}$

SUMMARY AND REVIEW: CHAPTER 10

The following contains a summary of what you should be able to do after completing this chapter. The review exercises are for practice. Answers are at the back of the book. If you miss an exercise, restudy the section indicated alongside the answer.

You should be able to:

Simplify radical expressions.

Simplify. Assume that expressions in radicands represent *any* real numbers.

1. $\sqrt{(-48)^2}$

2. $\sqrt{25x^2}$

3. $\sqrt[3]{\dfrac{-27}{64x^3}}$

4. $\sqrt[5]{-32}$

5. $\sqrt[4]{(-5x)^4}$

Multiply and simplify with radical notation.

Multiply. Assume that expressions in radicands are positive for the remainder of this review.

6. $\sqrt{15x}\sqrt{30x}$

7. $\sqrt[3]{m^4n}\sqrt[3]{m^2n^8}$

8. $\sqrt[3]{9a^3b^4}\sqrt[3]{6ab^2}$

Approximate square roots to the nearest tenth.

Approximate to the nearest tenth.

9. $\sqrt{245}$

10. $\sqrt{640}$

Approximate.

11. $\sqrt{63,000,000,000}$

12. $\sqrt{0.000003033}$

Divide and simplify with radical notation.

Divide and simplify.

13. $\dfrac{\sqrt[3]{54}}{\sqrt[3]{2}}$

14. $\sqrt{\dfrac{20x^4}{y^6}}$

15. $\sqrt{\dfrac{72a^7}{40a^4}}$

Add and subtract with radical notation.

Add and subtract.

16. $3\sqrt{242} - \sqrt{32} + \sqrt{98}$

17. $\sqrt[3]{128} - \sqrt[3]{54}$

18. $2\sqrt[3]{4x^4} - \sqrt[3]{32x}$

Multiply radical expressions whose factors contain more than one term.

Multiply.

19. $(3 + 5\sqrt{2})(3 - 5\sqrt{2})$

20. $(4\sqrt{6} - 3)^2$

21. $(3 + \sqrt{x})(2 + \sqrt{x})$

Rationalize the denominator in fractional expressions containing radicals.

Rationalize the denominator.

22. $\dfrac{\sqrt{7x}}{\sqrt{12x}}$

23. $\dfrac{8}{6 + \sqrt{5}}$

24. $\dfrac{\sqrt{2} - 5}{\sqrt{5} - 2}$

Simplify radical expressions using fractional exponents.

Simplify.

25. $\sqrt[8]{x^5 y^4 z^3}$

26. $(\sqrt[3]{25})^2$

27. $\sqrt[10]{36}$

28. $\sqrt[4]{81 x^{16} y^{20}}$

Use fractional exponents to write a single radical expression.

Write as a single radical expression.

29. $x^{1/4} y^{1/2}$

30. $\sqrt[3]{x}\sqrt{x}$

31. $\dfrac{\sqrt[4]{(a-b)^3}}{\sqrt[5]{(a-b)^2}}$

Solve radical equations.

Solve.

32. $\sqrt{4x-5}=3$

33. $\sqrt{5x+1}-\sqrt{x}=5$

Identify rational and irrational numbers.

Identify as rational or irrational.

34. $\sqrt{7}$

35. $16.79\overline{79}$

36. 16.79

37. $-\frac{4}{5}$

38. π

39. $4.191991999199991\ldots$
(Numeral does not repeat)

40. $\sqrt{0.4}$

○ _____

41. Multiply.

$$(\sqrt{x}-2\sqrt{y})(2\sqrt{x}+\sqrt{y})$$

42. Rationalize the denominator.

$$\dfrac{\sqrt{3xy}}{\sqrt{2x}-\sqrt{3y}}$$

43. Write fractional exponent notation with no denominator.

$$(4\sqrt{x})(3\sqrt{xy^{-1}})$$

TEST: CHAPTER 10

Simplify. Assume that expressions in radicands represent *any* real number.

1. $-\sqrt{144 x^2}$

2. $\sqrt[3]{-216}$

3. $\sqrt{24x}\sqrt{48x}$

4. $\sqrt[3]{x^3 y^5}\sqrt[3]{x^2 y^2}$

5. $\dfrac{\sqrt{21 x^7}}{\sqrt{28 x^2}}$

6. $\dfrac{\sqrt[3]{810}}{\sqrt[3]{15}}$

Add or subtract.

7. $\sqrt{56}+\sqrt{126}$

8. $\sqrt[3]{120x}-\sqrt[3]{15x}$

Multiply. Assume that expressions in radicands are positive for the remainder of this test.

9. $(4+3\sqrt{2})(4-3\sqrt{2})$

10. $(\sqrt{4x}+\sqrt{9y})(\sqrt{4x}-\sqrt{9y})$

Rationalize the denominator.

11. $\dfrac{\sqrt{3}}{\sqrt{11}}$

12. $\dfrac{\sqrt[3]{3a^4}}{\sqrt[3]{6b^4}}$

Simplify.

13. $\sqrt[3]{6x^4 y^2}$

14. $\sqrt[6]{49 a^9 b^9}$

15. $\sqrt[6]{25 m^{18} n^{24}}$

Solve.

16. $\sqrt{a+5}=a+3$

17. $\sqrt{x-2}+1=\sqrt{2x-6}$

Identify as rational or irrational.

18. $12.34344344434444\ldots$
(Numeral does not repeat)

19. $\sqrt{10}$

20. $\frac{2}{3}$

21. $3.\overline{452}$

○ _____

Solve.

22. $(\sqrt{x-1}-3)(\sqrt{x-1}+3)=14$

11

QUADRATIC EQUATIONS AND FUNCTIONS

How should you cut out corners to make the biggest possible box? Quadratic functions can be used to solve such problems.

When you translate a problem situation to mathematical language, you may obtain a function or equation containing a polynomial of second degree. Such equations and functions are called *quadratic*. In this chapter you will consider quadratic equations and learn to solve them. Then you will have the opportunity to solve problems whose translation contains quadratic equations or functions.

OBJECTIVES

After finishing Section 11.1, you should be able to:

i Write a quadratic equation in standard form $ax^2 + bx + c = 0$, $a > 0$, and determine the coefficients a, b, and c.

ii Solve equations of the type $ax^2 + c = 0$, $a \neq 0$.

iii Solve problems involving quadratic equations of the type $ax^2 + c = 0$.

Write standard form and determine a, b, and c.

1. $x^2 = 7x$

2. $3 - x^2 = 9x$

3. $3x + 5x^2 = x^2 - 4 + x$

11.1

INTRODUCTION TO QUADRATIC EQUATIONS

The following are *quadratic equations*. They contain polynomials of second degree.

$$x^2 + 7x - 5 = 0, \qquad 3t^2 - \tfrac{1}{2}t = 9, \qquad 5y^2 = -6y, \qquad 3m^2 = 0$$

In this section we learn about the standard form of a quadratic equation. We also learn to solve quadratic equations of the type $ax^2 + c = 0$ and to use them for problem solving.

i Standard Form

The quadratic equation

$$4x^2 + 7x - 5 = 0$$

is said to be in *standard form*. The quadratic equation

$$4x^2 = 5 - 7x$$

is equivalent to the preceding equation, but it is not in standard form.

> An equation of the type $ax^2 + bx + c = 0$, where a, b, and c are real-number constants and $a > 0$, is called the *standard form of a quadratic equation.*

To write standard form for a quadratic equation, we find an equivalent equation that is in standard form.

EXAMPLES Write standard form and determine a, b, and c.

1. $4x^2 + 7x - 5 = 0$ The equation is already in standard form.

$a = 4$; $b = 7$; $c = -5$

2. $3x^2 - 0.5x = 9$

$3x^2 - 0.5x - 9 = 0$ Adding -9. This is standard form.

$a = 3$; $b = -0.5$; $c = -9$

3. $-4y^2 = 5y$

$-4y^2 - 5y = 0$ Adding $-5y$

Not positive!

$4y^2 + 5y = 0$ Multiplying by -1. This is standard form.

$a = 4$; $b = 5$; $c = 0$

DO EXERCISES 1–3.

ii Solving Equations of the Type $ax^2 + c = 0$

When b is 0, we solve for x^2 and take the principal square root.

EXAMPLE 4 Solve: $5x^2 = 15$.

$$5x^2 = 15$$
$$x^2 = 3 \qquad \text{Solving for } x^2 \text{; multiplying by } \tfrac{1}{5}$$
$$|x| = \sqrt{3}$$

Since $|x|$ is either x or $-x$,

$$x = \sqrt{3} \quad \text{or} \quad -x = \sqrt{3}$$
$$x = \sqrt{3} \quad \text{or} \quad x = -\sqrt{3}.$$

Check: For $\sqrt{3}$:

$$\frac{5x^2 = 15}{\begin{array}{c|c} 5(\sqrt{3})^2 & 15 \\ 5 \cdot 3 & \\ 15 & \end{array}}$$

For $-\sqrt{3}$:

$$\frac{5x^2 = 15}{\begin{array}{c|c} 5(-\sqrt{3})^2 & 15 \\ 5 \cdot 3 & \\ 15 & \end{array}}$$

The solutions are $\sqrt{3}$ and $-\sqrt{3}$.

DO EXERCISE 4.

EXAMPLE 5 Solve: $\frac{1}{3}x^2 = 0$.

$$\frac{1}{3}x^2 = 0$$
$$x^2 = 0 \qquad \text{Multiplying by 3}$$
$$|x| = \sqrt{0} \qquad \text{Taking the principal square root}$$
$$|x| = 0$$

The only number with absolute value 0 is 0. It checks, so it is the solution.

DO EXERCISE 5.

EXAMPLE 6 Solve: $-3x^2 + 7 = 0$.

$$-3x^2 + 7 = 0$$
$$-3x^2 = -7 \qquad \text{Adding } -7$$
$$x^2 = \frac{-7}{-3} \qquad \text{Multiplying by } -\frac{1}{3}$$
$$x^2 = \frac{7}{3}$$
$$|x| = \sqrt{\frac{7}{3}} \qquad \text{Taking the principal square root}$$
$$x = \sqrt{\frac{7}{3}} \quad \text{or} \quad x = -\sqrt{\frac{7}{3}}$$
$$x = \sqrt{\frac{7}{3} \cdot \frac{3}{3}} \quad \text{or} \quad x = -\sqrt{\frac{7}{3} \cdot \frac{3}{3}} \qquad \text{Rationalizing the denominators}$$
$$x = \frac{\sqrt{21}}{3} \quad \text{or} \quad x = -\frac{\sqrt{21}}{3}$$

Check:

$$\frac{-3x^2 + 7 = 0}{\begin{array}{c|c} -3\left(\dfrac{\sqrt{21}}{3}\right)^2 + 7 & 0 \\ -3 \cdot \dfrac{21}{9} + 7 & \\ -7 + 7 & \\ 0 & \end{array}}$$

$$\frac{-3x^2 + 7 = 0}{\begin{array}{c|c} -3\left(-\dfrac{\sqrt{21}}{3}\right)^2 + 7 & 0 \\ -3 \cdot \dfrac{21}{9} + 7 & \\ -7 + 7 & \\ 0 & \end{array}}$$

The solutions are $\dfrac{\sqrt{21}}{3}$ and $-\dfrac{\sqrt{21}}{3}$.

DO EXERCISES 6 AND 7.

Solve.

4. $4x^2 = 20$

Solve.

5. $2x^2 = 0$

Solve.

6. $2x^2 - 3 = 0$

7. $4x^2 - 9 = 0$

iii Problem Solving

Quadratic equations can be used for problem solving.

EXAMPLE 7 The Sears Tower in Chicago is 1451 ft tall. How long would it take an object to fall from the top?

1. **Familiarize** If we did not know anything about this problem, we might consider looking up a formula in a mathematics or physics book. Actually, we studied such a formula earlier. A formula that fits this situation is

$$s = 16t^2,$$

where s is the distance, in feet, traveled by a body falling freely from rest in t seconds. This formula is actually an approximation in that it does not account for air resistance. In this problem we know the distance s to be 1451. We want the time t.

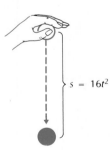

$s = 16t^2$

2. **Translate** We know the distance of 1451 and need to solve for t. We substitute 1451 for s:

$$1451 = 16t^2.$$

This gives us a translation.

3. **Carry out** We solve the equation:

$$1451 = 16t^2$$

$$\frac{1451}{16} = t^2 \qquad \text{Solving for } t^2$$

Then

$$90.6875 = t^2 \quad \text{Dividing}$$

$$\sqrt{90.6875} = |t| \quad \text{Taking the principal square root}$$

$$9.5 \approx |t| \quad \begin{array}{l}\text{Using a calculator to find the square root}\\ \text{and rounding to the nearest tenth}\end{array}$$

$$9.5 \approx t \quad \text{or} \quad -9.5 \approx t.$$

4. Check⟩ The number -9.5 cannot be a solution because time cannot be negative in this situation. We substitute 9.5 in the original equation:

$$s = 16(9.5)^2 = 16(90.25) = 1444.$$

This is close. Remember we approximated a solution. Thus we have a check.

5. State⟩ It takes about 9.5 sec for the object to fall.

DO EXERCISE 8.

8. The height of the World Trade Center in New York is 1377 ft (excluding TV towers and antennas). How long would it take an object to fall from the top?

EXERCISE SET 11.1

i Write standard form and determine a, b, and c.

1. $x^2 - 3x + 2 = 0$

2. $x^2 - 8x - 5 = 0$

3. $2x^2 = 3$

4. $5x^2 = 9$

5. $7x^2 = 4x - 3$

6. $9x^2 = x + 5$

7. $5 = -2x^2 + 3x$

8. $2x = x^2 - 5$

9. $2x - 1 = 3x^2 + 7$

ii Solve.

10. $x^2 = 121$

11. $x^2 = 10$

12. $5x^2 = 35$

13. $3x^2 = 30$

14. $5x^2 = 3$

15. $2x^2 = 5$

16. $4x^2 - 25 = 0$

17. $9x^2 - 4 = 0$

18. $3x^2 - 49 = 0$

19. $5x^2 - 16 = 0$

20. $4y^2 - 3 = 9$

21. $49y^2 - 16 = 0$

22. $25y^2 - 36 = 0$

23. $5x^2 - 100 = 0$

24. $100x^2 - 5 = 0$

iii *Problem solving*

Use the formula $s = 16t^2$.

25. A body falls 1000 ft. How many seconds does this take?

26. A body falls 2496 ft. How many seconds does this take?

27. The world record for free-fall, to the ground without a parachute, by a woman is 175 ft and is held by Kitty O'Neill. Approximately how long did the fall take?

28. The world record for free-fall to the ground, without a parachute, by a man is 311 ft and is held by Dar Robinson. Approximately how long did the fall take?

Stuntwoman Kitty O'Neill, leaping 127 feet from the top of the Valley Hilton Hotel in Los Angeles during the filming of a television series. She established a new world record for a high fall for women.

Solve.

29. $4.82x^2 = 12,000$

30. $\dfrac{x}{4} = \dfrac{9}{x}$

31. $1 = \dfrac{1}{3}x^2$

32. $\dfrac{x}{9} = \dfrac{36}{4x}$

33. $\dfrac{4}{m^2 - 7} = 1$

Solve for x.

34. $\dfrac{1}{4}x^2 + \dfrac{1}{6} = \dfrac{2}{3}x^2$

35. $\dfrac{4}{x^2 - 3} = \dfrac{6}{x^2}$

36. $3ax^2 - 9b = 3b^2$

37. $x^2 + 9a^2 = 9 + ax^2$

OBJECTIVES

After finishing Section 11.2, you should be able to:

i Solve quadratic equations of the type $ax^2 + bx = 0$, $a \neq 0$, $b \neq 0$, by factoring.

ii Solve quadratic equations of the type $ax^2 + bx + c = 0$, $a \neq 0$, $b \neq 0$, $c \neq 0$, by factoring.

iii Solve problems involving quadratic equations.

11.2

SOLVING BY FACTORING

Sometimes we can use factoring and the principle of zero products to solve quadratic equations.

i Equations of the Type $ax^2 + bx = 0$

When c is 0 (and $b \neq 0$), we can factor and use the principle of zero products.

EXAMPLE 1 Solve: $7x^2 + 2x = 0$.

$$7x^2 + 2x = 0$$
$$x(7x + 2) = 0 \qquad \text{Factoring}$$
$$x = 0 \quad \text{or} \quad 7x + 2 = 0 \qquad \text{Principle of zero products}$$
$$x = 0 \quad \text{or} \quad 7x = -2$$
$$x = 0 \quad \text{or} \quad x = -\tfrac{2}{7}$$

Check:

$$\begin{array}{c|c} 7x^2 + 2x = 0 \\ \hline 7 \cdot 0^2 + 2 \cdot 0 & 0 \\ 0 & \end{array}$$

$$\begin{array}{c|c} 7x^2 + 2x = 0 \\ \hline 7(-\tfrac{2}{7})^2 + 2(-\tfrac{2}{7}) & 0 \\ 7(\tfrac{4}{49}) - \tfrac{4}{7} & \\ \tfrac{4}{7} - \tfrac{4}{7} & \\ 0 & \end{array}$$

The solutions are 0 and $-\tfrac{2}{7}$.

When we use the principle of zero products, we need not check except to detect errors in solving.

EXAMPLE 2 Solve: $20x^2 - 15x = 0$.

$$20x^2 - 15x = 0$$
$$5x(4x - 3) = 0 \qquad \text{Factoring}$$
$$5x = 0 \quad \text{or} \quad 4x - 3 = 0 \qquad \text{Principle of zero products}$$
$$x = 0 \quad \text{or} \quad 4x = 3$$
$$x = 0 \quad \text{or} \quad x = \tfrac{3}{4}$$

The solutions are 0 and $\tfrac{3}{4}$.

A quadratic equation of this type will always have 0 as one solution and a non-zero number as the other solution.

DO EXERCISES 1 AND 2.

ii Equations of the Type $ax^2 + bx + c = 0$

When neither b nor c is 0, we can sometimes solve by factoring.

EXAMPLE 3 Solve: $5x^2 - 8x + 3 = 0$.

$$5x^2 - 8x + 3 = 0$$
$$(5x - 3)(x - 1) = 0 \qquad \text{Factoring}$$
$$5x - 3 = 0 \quad \text{or} \quad x - 1 = 0$$
$$5x = 3 \quad \text{or} \qquad x = 1$$
$$x = \tfrac{3}{5} \quad \text{or} \qquad x = 1$$

The solutions are $\tfrac{3}{5}$ and 1.

EXAMPLE 4 Solve: $(y - 3)(y - 2) = 6(y - 3)$.

We write standard form and then try to factor:

$$y^2 - 5y + 6 = 6y - 18 \qquad \text{Multiplying}$$
$$y^2 - 11y + 24 = 0 \qquad \text{Standard form}$$
$$(y - 8)(y - 3) = 0$$
$$y - 8 = 0 \quad \text{or} \quad y - 3 = 0$$
$$y = 8 \quad \text{or} \qquad y = 3.$$

The solutions are 8 and 3.

DO EXERCISES 3 AND 4.

iii Problem Solving

EXAMPLE 5 The number of diagonals d of a polygon of n sides is given by the formula

$$d = \frac{n^2 - 3n}{2}.$$

If a polygon has 27 diagonals, how many sides does it have?

1. **Familiarize** We can make a drawing to become familiar with the problem. We draw an octagon (8 sides). We count the diagonals and see that there are 20. Let us check this in the formula. We evaluate the formula for $n = 8$:

$$d = \frac{8^2 - 3(8)}{2} = \frac{64 - 24}{2} = \frac{40}{2} = 20.$$

Solve.

1. $3x^2 + 5x = 0$

2. $10x^2 - 6x = 0$

Solve.

3. $3x^2 + x - 2 = 0$

4. $(x - 1)(x + 1) = 5(x - 1)$

5. Use $d = \dfrac{n^2 - 3n}{2}$.

a) A heptagon has 7 sides. How many diagonals does it have?

b) A polygon has 44 diagonals. How many sides does it have?

2. Translate > We know that the number of diagonals is 27. We substitute 27 for d:

$$27 = \frac{n^2 - 3n}{2}.$$

This gives us a translation.

3. Carry out > We solve the equation for n. We first reverse the equation for convenience.

$$\frac{n^2 - 3n}{2} = 27$$

$$n^2 - 3n = 54 \qquad \text{Multiplying by 2 to clear of fractions}$$
$$n^2 - 3n - 54 = 0$$
$$(n - 9)(n + 6) = 0$$
$$n - 9 = 0 \quad \text{or} \quad n + 6 = 0$$
$$n = 9 \quad \text{or} \qquad n = -6$$

4. Check > Since the number of sides cannot be negative, -6 cannot be a solution. We leave it to the student to show that 9 checks by substitution.

5. State > The polygon has 9 sides (it is a nonagon).

DO EXERCISE 5.

EXERCISE SET 11.2

i Solve.

1. $x^2 + 7x = 0$

2. $x^2 + 5x = 0$

3. $3x^2 + 6x = 0$

4. $4x^2 + 8x = 0$

5. $5x^2 = 2x$

6. $7x = 3x^2$

7. $4x^2 + 4x = 0$

8. $2x^2 - 2x = 0$

9. $0 = 10x^2 - 30x$

10. $0 = 10x^2 - 50x$

11. $11x = 55x^2$

12. $33x^2 = -11x$

13. $14t^2 = 3t$

14. $8m = 17m^2$

15. $5y^2 - 3y^2 = 72y + 9y$

ii Solve.

16. $x^2 - 16x + 48 = 0$

17. $x^2 + 8x - 48 = 0$

18. $x^2 + 6 + 7x = 0$

19. $5 + 6x + x^2 = 0$

20. $t^2 + 4t = 21$

21. $18 = 7p + p^2$

22. $m^2 + 14 = 9m$

23. $-15 = -8y + y^2$

24. $x^2 + 10x + 25 = 0$

25. $x^2 + 6x + 9 = 0$

26. $x^2 + 1 = 2x$

27. $r^2 = 8r - 16$

28. $2x^2 - 13x + 15 = 0$

29. $6x^2 + x - 2 = 0$

30. $3a^2 = 10a + 8$

31. $15b - 9b^2 = 4$

32. $3x^2 - 7x = 20$

33. $6x^2 - 4x = 10$

34. $2t^2 + 12t = -10$

35. $12w^2 - 5w = 2$

36. $6z^2 + z - 1 = 0$

37. $t(t - 5) = 14$

38. $3y^2 + 8y = 12y + 15$

39. $t(9 + t) = 4(2t + 5)$

40. $(2x - 3)(x + 1) = 4(2x - 3)$

41. $16(p - 1) = p(p + 8)$

42. $(t - 1)(t + 3) = t - 1$

43. $(x - 2)(x + 2) = x + 2$

44. $m(3m + 1) = 2$

Problem solving

Use $d = \dfrac{n^2 - 3n}{2}$ for Exercises 45–48.

45. A hexagon is a figure with 6 sides. How many diagonals does a hexagon have?

46. A decagon is a figure with 10 sides. How many diagonals does a decagon have?

47. A polygon has 14 diagonals. How many sides does it have?

48. A polygon has 9 diagonals. How many sides does it have?

Solve.

49. $4m^2 - (m + 1)^2 = 0$

50. $x^2 + \sqrt{3}\,x = 0$

51. $\sqrt{5}\,x^2 - x = 0$

52. $\sqrt{7}\,x^2 + \sqrt{3}\,x = 0$

53. $\dfrac{5}{y + 4} - \dfrac{3}{y - 2} = 4$

54. $\dfrac{2z + 11}{2z + 8} = \dfrac{3z - 1}{z - 1}$

55. Solve for x: $ax^2 + bx = 0$.

56. 🖩 Solve: $0.0025x^2 + 70,400x = 0$.

Solve.

57. $y^4 - 4y^2 + 4 = 0$ (*Hint:* Let $x = y^2$. Then $x^2 = y^4$. Write a quadratic equation in x and solve. Remember to solve for y after finding x.)

58. $z - 10\sqrt{z} + 9 = 0$ (Let $x = \sqrt{z}$.)

11.3

COMPLETING THE SQUARE

The equation $(x - 5)^2 = 9$ can be solved by taking the square root on both sides. We will see that other equations can be made to look like this one.

i Solving Equations of the Type $(x + k)^2 = d$

In equations of the type $(x + k)^2 = d$, we have the square of a binomial equal to a constant.

EXAMPLE 1 Solve: $(x - 5)^2 = 9$.

$$(x - 5)^2 = 9$$
$$|x - 5| = \sqrt{9} \qquad \text{Taking the principal square root}$$
$$|x - 5| = 3$$
$$x - 5 = 3 \quad \text{or} \quad x - 5 = -3$$
$$x = 8 \quad \text{or} \quad x = 2$$

The solutions are 8 and 2.

EXAMPLE 2 Solve: $(x + 2)^2 = 7$.

$$(x + 2)^2 = 7$$
$$|x + 2| = \sqrt{7}$$
$$x + 2 = \sqrt{7} \qquad \text{or} \quad x + 2 = -\sqrt{7}$$
$$x = -2 + \sqrt{7} \quad \text{or} \qquad x = -2 - \sqrt{7}$$

The solutions are $-2 + \sqrt{7}$ and $-2 - \sqrt{7}$, or simply $-2 \pm \sqrt{7}$ (read "-2 plus or minus $\sqrt{7}$").

DO EXERCISES 1–3.

ii Completing the Square

The following is the square of a binomial:

$$x^2 + 10x + 25.$$

An equivalent expression is $(x + 5)^2$. We could find the 25 from $10x$, by taking half the coefficient of x and squaring it.

OBJECTIVES

After finishing Section 11.3, you should be able to:

i Solve equations of the type $(x + k)^2 = d$.

ii Complete the square for a binomial $x^2 + bx$.

iii Solve certain compound-interest problems using quadratic equations.

Solve.

1. $(x - 3)^2 = 16$

2. $(x + 3)^2 = 10$

3. $(x - 1)^2 = 5$

Complete the square.

4. $x^2 - 8x$

We can make $x^2 + 10x$ the square of a binomial by adding the proper number to it. This is called *completing the square*.

$$x^2 + 10x$$
$$\downarrow$$
$$\frac{10}{2} = 5 \qquad \text{Taking half the } x\text{-coefficient}$$
$$\downarrow$$
$$5^2 = 25 \qquad \text{Squaring}$$
$$\downarrow$$
$$x^2 + 10x + 25 \qquad \text{Adding}$$

The trinomial $x^2 + 10x + 25$ is the square of $x + 5$.

EXAMPLES   Complete the square.

5. $x^2 + 20x$

3. $x^2 - 12x$

$$\left(\frac{-12}{2}\right)^2 = (-6)^2 = 36 \qquad \text{Taking half the } x\text{-coefficient and squaring}$$
$$\downarrow$$
$$x^2 - 12x + 36$$

The trinomial $x^2 - 12x + 36$ is the square of $x - 6$.

4. $x^2 - 5x$

$$\left(\frac{-5}{2}\right)^2 = \frac{25}{4}$$
$$\downarrow$$
$$x^2 - 5x + \frac{25}{4}$$

6. $x^2 + 7x$

The trinomial $x^2 - 5x + \frac{25}{4}$ is the square of $x - \frac{5}{2}$.

DO EXERCISES 4–7.

iii Problem Solving: Compound Interest

7. $x^2 - 3x$

We studied compound interest in Section 4.1. Recall that if you put money in a savings account, the bank will pay you interest. At the end of a year, the bank will start paying you interest on both the original amount and the interest. This is called *compounding interest annually*.

> If an amount of money P is invested at interest rate r, compounded annually, then in t years it will grow to the amount A given by
> $$A = P(1 + r)^t.$$

We can use quadratic equations to solve certain interest problems. Before we do that let us review the use of the preceding formula.

EXAMPLE 5 $1000 invested at 16% for 2 years compounded annually will grow to what amount?

We have

$$A = P(1 + r)^t$$
$$A = 1000(1 + 0.16)^2 \quad \text{Substituting into the formula}$$
$$A = 1000(1.16)^2$$
$$A = 1000(1.3456)$$
$$A = 1345.60. \quad \text{Computing}$$

The amount is \$1345.60.

DO EXERCISE 8.

EXAMPLE 6 \$2560 is invested at interest rate r compounded annually. In 2 years it grows to \$2890. What is the interest rate?

1. **Familiarize** We know that \$2560 is originally invested. Thus P is \$2560. That amount grows to \$2890 in 2 years. Thus A is \$2890 and t is 2.

2. **Translate** We substitute 2560 for P, 2890 for A, and 2 for t in the formula, and obtain a translation:

$$A = P(1 + r)^t$$
$$2890 = 2560(1 + r)^2.$$

3. **Carry out** We solve the equation:

$$2890 = 2560(1 + r)^2$$

$$\frac{2890}{2560} = (1 + r)^2$$

$$\frac{289}{256} = (1 + r)^2$$

$$\sqrt{\frac{289}{256}} = |1 + r| \quad \text{Taking the principal square root}$$

$$\frac{17}{16} = |1 + r|.$$

We then have

$$\frac{17}{16} = 1 + r \quad \text{or} \quad -\frac{17}{16} = 1 + r$$

$$-\frac{16}{16} + \frac{17}{16} = r \quad \text{or} \quad -\frac{16}{16} - \frac{17}{16} = r$$

$$\frac{1}{16} = r \quad \text{or} \quad -\frac{33}{16} = r.$$

4. **Check** Since the interest rate cannot be negative,

$$\frac{1}{16} = r$$

$$0.0625 = r$$

$$\text{or} \quad 6.25\% = r. \quad \text{This checks in the formula.}$$

5. **State** The interest rate must be 6.25% in order for \$2560 to grow to \$2890 in 2 years.

DO EXERCISE 9.

Solve.

8. \$1000 invested at 14% compounded annually for 2 years will grow to what amount?

9. Suppose \$2560 is invested at interest rate r compounded annually, and grows to \$3240 in 2 years. What is the interest rate?

EXERCISE SET 11.3

i Solve.

1. $(x - 2)^2 = 49$
2. $(x + 1)^2 = 6$
3. $(x + 3)^2 = 21$
4. $(x - 3)^2 = 6$

5. $(x + 13)^2 = 8$
6. $(x - 13)^2 = 64$
7. $(x - 7)^2 = 12$
8. $(x + 1)^2 = 14$

9. $(x + 9)^2 = 34$
10. $(t + 2)^2 = 25$
11. $(x + \frac{3}{2})^2 = \frac{7}{2}$
12. $(y - \frac{3}{4})^2 = \frac{17}{16}$

ii Complete the square.

13. $x^2 - 2x$
14. $x^2 - 4x$
15. $x^2 + 18x$
16. $x^2 + 22x$

17. $x^2 - x$
18. $x^2 + x$
19. $t^2 + 5t$
20. $y^2 - 9y$

21. $x^2 - \frac{3}{2}x$
22. $x^2 + \frac{4}{3}x$
23. $m^2 + \frac{9}{2}m$
24. $r^2 - \frac{2}{5}r$

iii *Problem solving*

Use $A = P(1 + r)^t$ for Exercises 25–32. What is the interest rate?

25. $1000 grows to $1210 in 2 years
26. $1000 grows to $1440 in 2 years
27. $2560 grows to $3610 in 2 years

28. $4000 grows to $4410 in 2 years
29. $6250 grows to $7290 in 2 years
30. $6250 grows to $6760 in 2 years

31. $2500 grows to $3600 in 2 years
32. $1600 grows to $2500 in 2 years

Factor the left side of the equation. Then solve.

33. $x^2 + 2x + 1 = 81$
34. $x^2 - 2x + 1 = 16$
35. $x^2 + 4x + 4 = 29$

36. $y^2 + 16y + 64 = 15$
37. $t^2 + 3t + \frac{9}{4} = \frac{49}{4}$
38. $m^2 - \frac{3}{2}m + \frac{9}{16} = \frac{17}{16}$

39. $9x^2 - 24x + 16 = 2$
40. $0.81x^2 + 0.36x + 0.04 = 5.76$
41. $64y^2 + 48y + 9 = 100$

42. In order for $2000 to double itself in 2 years, what would the interest rate have to be?

43. $1000 is invested at interest rate r. In 2 years it grows to $1267.88. What is the interest rate?

44. $4000 is invested at interest rate r. In 2 years it grows to $5267.03. What is the interest rate?

45. In two years you want to have $3000. How much do you need to invest now if you can get an interest rate of 15.75% compounded annually?

Solve (for x or y).

46. $\dfrac{x - 1}{9} = \dfrac{1}{x - 1}$
47. $(y - b)^2 = 4b^2$
48. $2(3x + 1)^2 = 8$
49. $5(5x - 2)^2 - 7 = 13$

Complete the square.

50. $x^2 - ax$
51. $x^2 + (2b - 4)x$
52. $ax^2 + bx$

OBJECTIVE

11.4

After finishing Section 11.4, you should be able to:

i Solve quadratic equations by completing the square.

SOLVING BY COMPLETING THE SQUARE

i We have seen that a quadratic equation $(x + k)^2 = d$ can be solved by taking the principal square root on both sides. An equation such as $x^2 + 6x + 8 = 0$ can be put in this form by completing the square. Then we can solve as before.

EXAMPLE 1 Solve: $x^2 + 6x + 8 = 0$.

$$x^2 + 6x + 8 = 0$$
$$x^2 + 6x = -8 \qquad \text{Adding } -8$$

We take half of 6 and square it, to get 9. Then we add 9 on *both* sides of the

equation. This makes the left side the square of a binomial. We have *completed the square*.

$$x^2 + 6x + 9 = -8 + 9$$

$$(x + 3)^2 = 1$$

$$|x + 3| = \sqrt{1} \qquad \text{Taking the principal square root}$$

$$|x + 3| = 1$$

$$x + 3 = 1 \quad \text{or} \quad x + 3 = -1$$

$$x = -2 \quad \text{or} \quad x = -4$$

The solutions are -2 and -4.

This method of solving is called *completing the square*.

EXAMPLE 2   Solve $x^2 - 4x - 7 = 0$ by completing the square.

$$x^2 - 4x - 7 = 0$$

$$x^2 - 4x = 7 \qquad \text{Adding 7}$$

$$x^2 - 4x + 4 = 7 + 4 \qquad \text{Adding 4: } (\tfrac{-4}{2})^2 = (-2)^2 = 4$$

$$(x - 2)^2 = 11$$

$$|x - 2| = \sqrt{11}$$

$$x - 2 = \sqrt{11} \quad \text{or} \quad x - 2 = -\sqrt{11}$$

$$x = 2 + \sqrt{11} \quad \text{or} \quad x = 2 - \sqrt{11}$$

The solutions are $2 \pm \sqrt{11}$.

DO EXERCISES 1–3.

EXAMPLE 3   Solve $x^2 + 3x - 10 = 0$ by completing the square.

$$x^2 + 3x - 10 = 0$$

$$x^2 + 3x = 10$$

$$x^2 + 3x + \frac{9}{4} = 10 + \frac{9}{4} \qquad \text{Adding } \frac{9}{4}: \left(\frac{3}{2}\right)^2 = \frac{9}{4}$$

$$\left(x + \frac{3}{2}\right)^2 = \frac{40}{4} + \frac{9}{4} = \frac{49}{4}$$

$$\left|x + \frac{3}{2}\right| = \sqrt{\frac{49}{4}} = \frac{7}{2}$$

We then have

$$x + \frac{3}{2} = \frac{7}{2} \quad \text{or} \quad x + \frac{3}{2} = -\frac{7}{2}$$

$$x = \frac{4}{2} \quad \text{or} \quad x = -\frac{10}{2}$$

$$x = 2 \quad \text{or} \quad x = -5.$$

The solutions are 2 and -5.

DO EXERCISES 4 AND 5.

When the coefficient of x^2 is not 1, we can make it 1.

Solve by completing the square.

1. $x^2 + 8x + 12 = 0$

2. $x^2 - 10x + 22 = 0$

3. $x^2 + 6x - 1 = 0$

Solve by completing the square.

4. $x^2 - 3x - 10 = 0$

5. $x^2 + 5x - 14 = 0$

Solve by completing the square.

6. $2x^2 + 3x - 3 = 0$

7. $3x^2 - 2x - 3 = 0$

EXAMPLE 4 Solve $2x^2 - 3x - 1 = 0$ by completing the square.

$$2x^2 - 3x - 1 = 0$$

$$\frac{1}{2}(2x^2 - 3x - 1) = \frac{1}{2} \cdot 0 \qquad \text{Multiplying by } \frac{1}{2} \text{ to make the } x^2\text{-coefficient 1}$$

$$x^2 - \frac{3}{2}x - \frac{1}{2} = 0$$

$$x^2 - \frac{3}{2}x = \frac{1}{2}$$

$$x^2 - \frac{3}{2}x + \frac{9}{16} = \frac{1}{2} + \frac{9}{16} \qquad \text{Adding } \frac{9}{16}: \left[\frac{1}{2}\left(-\frac{3}{2}\right)\right]^2 = \left[-\frac{3}{4}\right]^2 = \frac{9}{16}$$

$$\left(x - \frac{3}{4}\right)^2 = \frac{8}{16} + \frac{9}{16}$$

$$\left(x - \frac{3}{4}\right)^2 = \frac{17}{16}$$

$$\left|x - \frac{3}{4}\right| = \sqrt{\frac{17}{16}}$$

$$\left|x - \frac{3}{4}\right| = \frac{\sqrt{17}}{4}$$

$$x - \frac{3}{4} = \frac{\sqrt{17}}{4} \qquad \text{or} \quad x - \frac{3}{4} = -\frac{\sqrt{17}}{4}$$

$$x = \frac{3}{4} + \frac{\sqrt{17}}{4} \quad \text{or} \qquad x = \frac{3}{4} - \frac{\sqrt{17}}{4}$$

The solutions are $\dfrac{3 \pm \sqrt{17}}{4}$.

DO EXERCISES 6 AND 7.

EXERCISE SET 11.4

i Solve by completing the square. Show your work.

1. $x^2 - 6x - 16 = 0$ **2.** $x^2 + 8x + 15 = 0$ **3.** $x^2 + 22x + 21 = 0$ **4.** $x^2 + 14x - 15 = 0$

5. $x^2 - 2x - 5 = 0$ **6.** $x^2 - 4x - 11 = 0$ **7.** $x^2 - 22x + 102 = 0$ **8.** $x^2 - 18x + 74 = 0$

9. $x^2 + 10x - 4 = 0$ **10.** $x^2 - 10x - 4 = 0$ **11.** $x^2 - 7x - 2 = 0$ **12.** $x^2 + 7x - 2 = 0$

13. $x^2 + 3x - 28 = 0$ **14.** $x^2 - 3x - 28 = 0$ **15.** $x^2 + \frac{3}{2}x - \frac{1}{2} = 0$ **16.** $x^2 - \frac{3}{2}x - 2 = 0$

17. $2x^2 + 3x - 17 = 0$ **18.** $2x^2 - 3x - 1 = 0$ **19.** $3x^2 + 4x - 1 = 0$ **20.** $3x^2 - 4x - 3 = 0$

21. $2x^2 - 9x - 5 = 0$ **22.** $2x^2 - 5x - 12 = 0$ **23.** $4x^2 + 12x - 7 = 0$ **24.** $6x^2 + 11x - 10 = 0$

Problem-solving practice

25. There were 12,000 people at a rock concert. A ticket cost $7.00 at the door and $6.50 if it was bought in advance. Total receipts were $81,165. How many people bought their tickets in advance?

26. It is known that 100 g of a certain kind of milk contains 3.5 g of protein. How many grams of protein are contained in 450 g of milk?

Find b such that each trinomial is a square.

27. $x^2 + bx + 36$

28. $x^2 + bx + 55$

29. $x^2 + bx + 128$

30. $4x^2 + bx + 16$

31. $x^2 + bx + c$

32. $ax^2 + bx + c$

Solve for x by completing the square.

33. $x^2 - ax - 6a^2 = 0$

34. $x^2 + 4bx + 2b = 0$

35. $x^2 - x - c^2 - c = 0$

36. $3x^2 - bx + 1 = 0$

37. $ax^2 + 4x + 3 = 0$

38. $4x^2 + 4x + c = 0$

39. $kx^2 + mx + n = 0$

40. $b^2x^2 - 2bx + c^2 = 0$

11.5

THE QUADRATIC FORMULA

There are at least two reasons for learning to complete the square. One is to enhance your ability to graph certain second-degree equations, which you will encounter later in mathematics. The other is to prove a general formula that can be used to solve quadratic equations.

i Solving Using the Quadratic Formula

Each time you solve by completing the square, you do about the same thing. In situations like this in mathematics, when we do about the same kind of computation many times, we look for a formula so we can speed up our work. Consider any quadratic equation in standard form:

$$ax^2 + bx + c = 0, \qquad a > 0.$$

Let's solve by completing the square.

$$x^2 + \frac{b}{a}x + \frac{c}{a} = 0 \qquad \text{Multiplying by } \frac{1}{a}$$

$$x^2 + \frac{b}{a}x = -\frac{c}{a} \qquad \text{Adding } -\frac{c}{a}$$

Half of $\dfrac{b}{a}$ is $\dfrac{b}{2a}$. The square is $\dfrac{b^2}{4a^2}$. We add $\dfrac{b^2}{4a^2}$ on both sides.

$$x^2 + \frac{b}{a}x + \frac{b^2}{4a^2} = -\frac{c}{a} + \frac{b^2}{4a^2}$$

$$\left(x + \frac{b}{2a}\right)^2 = -\frac{4ac}{4a^2} + \frac{b^2}{4a^2}$$

$$\left(x + \frac{b}{2a}\right)^2 = \frac{b^2 - 4ac}{4a^2}$$

$$\left|x + \frac{b}{2a}\right| = \sqrt{\frac{b^2 - 4ac}{4a^2}} \qquad \text{Taking the principal square root}$$

$$x + \frac{b}{2a} = \sqrt{\frac{b^2 - 4ac}{4a^2}} \quad \text{or} \quad x + \frac{b}{2a} = -\sqrt{\frac{b^2 - 4ac}{4a^2}}$$

OBJECTIVES

After finishing Section 11.5, you should be able to:

i Solve quadratic equations using the quadratic formula.

ii Find approximate solutions using a calculator or a square-root table.

Since $a > 0$, $|a| = a$. Then

$$x + \frac{b}{2a} = \frac{\sqrt{b^2 - 4ac}}{2a} \quad \text{or} \quad x + \frac{b}{2a} = -\frac{\sqrt{b^2 - 4ac}}{2a}.$$

Thus,

$$x + \frac{b}{2a} = \pm\frac{\sqrt{b^2 - 4ac}}{2a},$$

so

$$x = -\frac{b}{2a} + \frac{\sqrt{b^2 - 4ac}}{2a} \quad \text{or} \quad x = -\frac{b}{2a} - \frac{\sqrt{b^2 - 4ac}}{2a}.$$

The solutions are given by the following.

THE QUADRATIC FORMULA

$$x = \frac{-b \pm \sqrt{b^2 - 4ac}}{2a}$$

EXAMPLE 1 Solve $5x^2 - 8x = -3$ using the quadratic formula.

First find standard form and determine a, b, and c:

$$5x^2 - 8x + 3 = 0$$
$$a = 5, \quad b = -8, \quad c = 3.$$

Then use the quadratic formula:

$$x = \frac{-b \pm \sqrt{b^2 - 4ac}}{2a}$$

$$x = \frac{-(-8) \pm \sqrt{(-8)^2 - 4 \cdot 5 \cdot 3}}{2 \cdot 5} \qquad \text{Substituting}$$

$$x = \frac{8 \pm \sqrt{64 - 60}}{10}$$

$$x = \frac{8 \pm \sqrt{4}}{10}$$

$$x = \frac{8 \pm 2}{10}$$

$$x = \frac{8 + 2}{10} \quad \text{or} \quad x = \frac{8 - 2}{10}$$

$$x = \frac{10}{10} \quad \text{or} \quad x = \frac{6}{10}$$

$$x = 1 \quad \text{or} \quad x = \frac{3}{5}.$$

The solutions are 1 and $\frac{3}{5}$.

It turns out that the equation in Example 1 could have been solved by factoring, which would actually have been easier.

To solve a quadratic equation:

1. **Try factoring.**

2. **If it is not possible to factor or if factoring seems difficult, use the quadratic formula.**

The solutions of a quadratic equation can always be found using the quadratic formula. They cannot always be found by factoring. When $b^2 - 4ac \geq 0$, the equation has solutions. When $b^2 - 4ac < 0$, the equation has no real-number solutions. The expression $b^2 - 4ac$ is called the *discriminant*.

DO EXERCISE 1.

When using the quadratic formula, it is wise to compute the discriminant first. If it is negative, there are no real-number solutions.

EXAMPLE 2 Solve $3x^2 = 7 - 2x$ using the quadratic formula.

Find standard form and determine a, b, and c:

$$3x^2 + 2x - 7 = 0$$
$$a = 3, \quad b = 2, \quad c = -7.$$

We compute the discriminant:

$$b^2 - 4ac = 2^2 - 4 \cdot 3 \cdot (-7) = 4 + 84 = 88.$$

This is positive, so there are solutions. They are given by

$$x = \frac{-2 \pm \sqrt{88}}{6} \qquad \text{Substituting into the quadratic formula}$$

$$x = \frac{-2 \pm \sqrt{4 \cdot 22}}{6} = \frac{-2 \pm 2\sqrt{22}}{6}$$

$$x = \frac{2(-1 \pm \sqrt{22})}{2 \cdot 3} \qquad \text{Factoring out 2 in the numerator and denominator}$$

$$x = \frac{-1 \pm \sqrt{22}}{3}.$$

The solutions are $\dfrac{-1 + \sqrt{22}}{3}$ and $\dfrac{-1 - \sqrt{22}}{3}$.

DO EXERCISE 2.

▐ii▐ Approximating Solutions

A calculator or Table 2 can be used to approximate solutions.

EXAMPLE 3 Use a calculator or Table 2 to approximate to the nearest tenth the solutions to the equation in Example 2.

Solve using the quadratic formula.

1. $2x^2 = 4 - 7x$

Solve using the quadratic formula.

2. $5x^2 - 8x = 3$

3. Approximate the solutions to the equation in Exercise 2 above. Round to the nearest tenth.

Using a calculator or Table 2, we see that $\sqrt{22} \approx 4.690$:

$$\frac{-1 + \sqrt{22}}{3} \approx \frac{-1 + 4.690}{3} \qquad\qquad \frac{-1 - \sqrt{22}}{3} \approx \frac{-1 - 4.690}{3}$$

$$\approx \frac{3.69}{3} \qquad\qquad\qquad \approx \frac{-5.69}{3}$$

$$\approx 1.2 \quad \text{to the} \qquad\qquad \approx -1.9 \quad \text{to the}$$
$$\text{nearest tenth;} \qquad\qquad\qquad \text{nearest tenth.}$$

DO EXERCISE 3.

EXERCISE SET 11.5

i Solve. Try factoring first. If factoring is not possible or is difficult, use the quadratic formula.

1. $x^2 - 4x = 21$

2. $x^2 + 7x = 18$

3. $x^2 = 6x - 9$

4. $x^2 = 8x - 16$

5. $3y^2 - 2y - 8 = 0$

6. $3y^2 - 7y + 4 = 0$

7. $4x^2 + 12x = 7$

8. $4x^2 + 4x = 15$

9. $x^2 - 9 = 0$

10. $x^2 - 4 = 0$

11. $x^2 - 2x - 2 = 0$

12. $x^2 - 4x - 7 = 0$

13. $y^2 - 10y + 22 = 0$

14. $y^2 + 6y - 1 = 0$

15. $x^2 + 4x + 4 = 7$

16. $x^2 - 2x + 1 = 5$

17. $3x^2 + 8x + 2 = 0$

18. $3x^2 - 4x - 2 = 0$

19. $2x^2 - 5x = 1$

20. $2x^2 + 2x = 3$

21. $4y^2 - 4y - 1 = 0$

22. $4y^2 + 4y - 1 = 0$

23. $3x^2 + 5x = 0$

24. $5x^2 - 2x = 0$

25. $2t^2 + 6t + 5 = 0$

26. $4y^2 + 3y + 2 = 0$

27. $4x^2 = 100$

28. $5t^2 = 80$

29. $3x^2 = 5x + 4$

30. $2x^2 + 3x = 1$

31. $2y^2 - 6y = 10$

32. $5m^2 = 3 + 11m$

33. $3p^2 + 2p = 3$

ii Solve using the quadratic formula. Use a calculator or Table 2 to approximate the solutions to the nearest tenth.

34. $x^2 - 4x - 7 = 0$

35. $x^2 + 2x - 2 = 0$

36. $y^2 - 6y - 1 = 0$

37. $y^2 + 10y + 22 = 0$

38. $4x^2 + 4x = 1$

39. $4x^2 = 4x + 1$

40. $3x^2 + 4x - 2 = 0$

41. $3x^2 - 8x + 2 = 0$

42. $2y^2 + 2y - 3 = 0$

Solve for x or y.

43. $5x + x(x - 7) = 0$

44. $x(3x + 7) - 3x = 0$

45. $3 - x(x - 3) = 4$

46. $x(5x - 7) = 1$

47. $(y + 4)(y + 3) = 15$

48. $(y + 5)(y - 1) = 27$

49. $x^2 + (x + 2)^2 = 7$

50. $x^2 + (x + 1)^2 = 5$

51. $(x + 2)^2 + (x + 1)^2 = 0$

52. $(x + 3)^2 + (x + 1)^2 = 0$

53. $ax^2 + 2x = 3$

54. $2bx^2 - 5x + 3b = 0$

55. $4x^2 - 4cx + c^2 - 3d^2 = 0$

56. $0.8x^2 + 0.16x - 0.09 = 0$

57. $bdx^2 + bcx - ac = adx$

58. $\frac{1}{2}x^2 + bx + (b - \frac{1}{2}) = 0$

59. a) In $ax^2 + bx + c = 0$, $b^2 > 4ac$. Will the equation have real-number solutions? Does it make any difference whether b is positive, negative, or zero?

b) In $ax^2 + bx + c = 0$, $ac < 0$. Will the equation have real-number solutions? Does it make any difference whether b is positive, negative, or zero?

c) In $ax^2 + bx + c = 0$, a and c are both positive. Under what conditions will the equation have real-number solutions?

60. Use the two roots given by the quadratic formula to find a formula for the sum of the solutions of any quadratic equation. What is the product of the solutions? Without solving, find the sum and product of the solutions of $2x^2 + 5x - 3 = 0$.

61. One solution of the equation $2x^2 + bx - 3 = 0$ is known to be -5. Use the results of the preceding exercise to find the other solution.

THE ASVAB TEST: ARITHMETIC REASONING   CHAPTER 11

See the discussion of the ASVAB Test at the beginning of Chapter 3.

1. The sales tax rate is 5%. What is the total cost of a television set priced at $673?

  A $674.35
  B $716.65
  C $676.37
  D $706.65

2. What is the percent of increase in the speed of a turntable when it is changed from $33\frac{1}{3}$ rpm (revolutions per minute) to 45 rpm?

  A 135
  B 74
  C $11\frac{2}{3}$
  D 35

3. If a board that is 10 ft long is cut in half, one of those pieces is cut in thirds, and one of those pieces is cut in fourths, how long is each of the smaller pieces?

  A 10 in.
  B 5 in.
  C 4 in.
  D 6 in.

4. It took 4 hr to drive 180 mi. How long would it take to make the same trip driving 5 mph slower?

  A 3.6 hr
  B 5.4 hr
  C 5 hr
  D 4.5 hr

5. A single piece of floor tile is 9 in. × 9 in. How many such tiles does it take to cover a floor that is 12 ft × 18 ft?

  A 216
  B 384
  C 288
  D 267

6. Compute: $\frac{12}{32} \div \frac{16}{24}$.

  A $\frac{9}{16}$
  B $\frac{1}{4}$
  C 1
  D $\frac{16}{9}$

7. A checking account had a balance of $437.32. Two checks were then written for $68.29 and $183.47. What is the new balance?

  A $251.76
  B $689.08
  C $322.14
  D $185.56

8. If $1\frac{1}{2}$ lb of cheese costs $3.00, how much is the cheese in cents per ounce?

  A 8
  B 20
  C 12.5
  D 0.08

9. A marathon footrace is 26 mi, 385 yd. How many feet is it?

  A 138,435
  B 15,382
  C 137,280
  D 137,665

10. A 12 ft × 30 ft swimming pool is built in a yard that is 90 ft × 60 ft. How much yard is left over?

  A 1800 sq ft
  B 5040 sq ft
  C 2340 sq ft
  D 5367 sq ft

11. A 20-in. pipe is cut into 5 equal pieces but $\frac{1}{4}$ in. is lost with each cut due to the width of the saw blade. How long is each piece?

  A 3.5 in.
  B 4.75 in.
  C 3.8 in.
  D 3.75 in.

12. If $\frac{5}{8}$ of a class are present, what percent are not present?

  A 0.375
  B $\frac{3}{8}$
  C 25
  D 37.5%

13. A baby's diaper is changed every 3 hr. How many diapers are changed in 4 weeks?

  A 112
  B 84
  C 224
  D 168

OBJECTIVES

After finishing Section 11.6, you should be able to:

i Solve a formula for a specified letter.

ii Solve problems in which the solving of a formula is needed.

1. Solve $A = \sqrt{\dfrac{w_1}{w_2}}$ for w_2.

2. Solve $V = \pi r^2 h$ for r.

3. Solve $Ls^2 + Rs = -\dfrac{1}{C}$ for s.

11.6

FORMULAS AND PROBLEM SOLVING

We can now apply our knowledge and skill with quadratic equations to solving formulas and solving problems.

i **Solving Formulas**

Recall that to solve a formula for a certain letter, we use the principles for solving equations to get that letter alone on one side. When square roots appear, we can usually eliminate the radical signs by squaring both sides.

EXAMPLE 1 (*A pendulum formula*). Solve $T = 2\pi \sqrt{\dfrac{l}{g}}$ for l.

$$T = 2\pi \sqrt{\frac{l}{g}}$$

$$T^2 = \left(2\pi \sqrt{\frac{l}{g}}\right)^2 \qquad \text{Principle of powers (squaring)}$$

$$T^2 = 2^2\pi^2 \frac{l}{g}$$

$$gT^2 = 4\pi^2 l \qquad \text{Clearing of fractions}$$

$$\frac{gT^2}{4\pi^2} = l \qquad \text{Multiplying by } \frac{1}{4\pi^2}$$

We now have l alone on one side and l does not appear on the other side, so the formula is solved for l.

In most formulas the letters represent nonnegative numbers, so we do not need to use absolute-value signs when taking square roots.

DO EXERCISE 1.

EXAMPLE 2 (*A motion formula*). Solve $s = gt + 16t^2$ for t.

This time we use the quadratic formula to get t alone on one side of the equation:

$$16t^2 + gt - s = 0 \qquad \text{Writing standard form}$$

$$a = 16, \quad b = g, \quad \text{and} \quad c = -s$$

$$t = \frac{-g \pm \sqrt{g^2 - 4 \cdot 16 \cdot (-s)}}{2 \cdot 16} \qquad \text{Using the quadratic formula}$$

Since taking the negative square root would result in a negative answer, we take the positive one:

$$t = \frac{-g + \sqrt{g^2 + 64s}}{32}.$$

DO EXERCISES 2 AND 3.

The following list of steps should help you when solving formulas. Remember that when solving formulas you do the same things you would do to solve any equation.

TO SOLVE A FORMULA FOR A LETTER, SAY b:

1. **Clear of fractions** (if needed).
2. **Use the principle of powers until radicals are eliminated.**
   (*Note:* You may sometimes do step 2 before step 1.)
3. **Collect all terms with b^2 in them. Also collect all terms with b in them.**
4. **If b^2 does not appear, finish by using the addition and multiplication principles.**
5. **If b^2 appears but b does not, solve for b^2 and take the square root.**
6. **If b^2 and b both appear, write standard form and factor or use the quadratic formula.**

---

4. Solve $\dfrac{b}{\sqrt{a^2 - b^2}} = p$ for b.

EXAMPLE 3 Solve the following formula for a: $q = \dfrac{a}{\sqrt{a^2 + b^2}}.$

In this case, we could either clear of fractions first or use the principle of powers first. Let us clear of fractions. We then have

$$q\sqrt{a^2 + b^2} = a.$$

Now we square both sides and then continue:

$$\begin{aligned}
(q\sqrt{a^2 + b^2})^2 &= a^2 && \text{Squaring} \\
q^2(a^2 + b^2) &= a^2 \\
q^2 a^2 + q^2 b^2 &= a^2 \\
q^2 b^2 &= a^2 - q^2 a^2 && \text{Getting all } a^2 \text{ terms together} \\
q^2 b^2 &= a^2(1 - q^2) && \text{Factoring out } a^2 \\
\frac{q^2 b^2}{1 - q^2} &= a^2 && \text{Multiplying by } \frac{1}{1 - q^2} \\
\sqrt{\frac{q^2 b^2}{1 - q^2}} &= a && \text{Taking the square root}
\end{aligned}$$

Since we have a alone on one side and a does not appear on the other side, we have solved the formula for a.

DO EXERCISE 4.

ii Solving Problems

In problem solving we often need to use a formula, and it is not uncommon for us to have to solve a formula for a certain letter. Let's look at an example.

A function that gives the distance, in meters, that an object will fall in t seconds is as follows:

$$s(t) = 4.9t^2 + v_0 t,$$

where v_0 is the initial speed, or velocity, and is hence a constant. This function gives us a formula that we can solve for t if necessary.

EXAMPLE 4 An object is dropped from the top of the Gateway Arch in St. Louis, which is 195 meters high. How long will it take to reach the ground?

5. The lateral area of a cone is 157 cm² , and the slant height is 10 cm. What is the radius of the base?

[*Hint:* A formula for the lateral area of a cone is $A = \pi rs$, where r is the radius of the base and s is the slant height. Use 3.14 for π.]

1. **Familiarize** We make a sketch, and label it.

195 m

$v_0 = 0$, since the object was just dropped (not thrown)

Time to fall 195 m = ?

Part of the familiarization consists of finding the proper formula to use. If it had not been given to us in the above discussion, we would need to look it up somewhere, probably in an encyclopedia or physics book. The formula is $s = 4.9t^2 + v_0t$.

2. **Translate** We have the formula $s = 4.9t^2 + v_0t$ and we know that $v_0 = 0$. Substituting, we have the translation

$$s = 4.9t^2 + 0 \cdot t, \quad \text{or} \quad 4.9t^2.$$

3. **Carry out** We solve the formula for t:

$$s = 4.9t^2$$

$$\frac{s}{4.9} = t^2 \qquad \text{Multiplying by } \frac{1}{4.9}$$

$$\sqrt{\frac{s}{4.9}} = t \qquad \text{Taking the principal square root}$$

Next, we substitute our value of s and calculate:

$$t = \sqrt{\frac{195}{4.9}} \approx 6.31.$$

4. **Check** We go back to the original formula, or function, and substitute 6.31:

$$s = 4.9t^2 + v_0t = 4.9t^2 + 0 \cdot t$$
$$= 4.9(6.31)^2$$
$$= 195.1.$$

Our answer is of course approximate, but it checks.

5. **State** The answer is that it will take approximately 6.31 seconds for the object to reach the ground.

DO EXERCISE 5.

EXERCISE SET 11.6

---

i Solve each formula for the indicated letter. Assume that all variables represent nonnegative numbers.

1. $A = 6s^2$, for s
(Area of a cube)

2. $A = 4\pi r^2$, for r
(Area of a sphere)

3. $F = \dfrac{Gm_1 m_2}{r^2}$, for r
(Law of gravity)

4. $N = \dfrac{kQ_1 Q_2}{s^2}$, for s
(Number of phone calls between two cities)

5. $E = mc^2$, for c
(Energy–mass relationship)

6. $A = \pi r^2$, for r
(Area of a circle)

7. $a^2 + b^2 = c^2$, for b
(Pythagorean formula in two dimensions)

8. $a^2 + b^2 + c^2 = d^2$, for c
(Pythagorean formula in three dimensions)

9. $N = \dfrac{k^2 - 3k}{2}$, for k
(Number of diagonals of a polygon)

10. $s = v_0 t + \dfrac{gt^2}{2}$, for t
(A motion formula)

11. $A = 2\pi r^2 + 2\pi rh$, for r
(Area of a cylinder)

12. $A = \pi r^2 + \pi rs$, for r
(Area of a cone)

13. $T = 2\pi \sqrt{\dfrac{l}{g}}$, for g
(A pendulum formula)

14. $W = \sqrt{\dfrac{l}{LC}}$, for L
(An electricity formula)

15. $P_1 - P_2 = \dfrac{32LV}{gD^2}$, for D

16. $N + p = \dfrac{6.2A^2}{pR^2}$, for R

17. $m = \dfrac{m_0}{\sqrt{1 - \dfrac{v^2}{c^2}}}$, for v
(A relativity formula)

18. Solve the formula given in Exercise 17 for c.

ii Solve.

19. a) An object is dropped 75 m from an airplane. How long does it take to reach the ground?
b) An object is thrown downward 75 m from the plane at an initial velocity of 30 m/sec. How long does it take to reach the ground?
c) How far will an object fall in 2 sec, thrown downward at an initial velocity of 30 m/sec?

20. a) An object is dropped 500 m from an airplane. How long does it take to reach the ground?
b) An object is thrown downward 500 m from the plane at an initial velocity of 30 m/sec. How long does it take to reach the ground?
c) How far will an object fall in 5 sec, thrown downward at an initial velocity of 30 m/sec?

21. An amount of money P is invested at interest rate r. In t years it will grow to the amount A given by $A = P(1 + r)^t$, where interest is compounded annually. For the following situations, find the interest rate for interest compounded annually.

a) $2560 grows to $3610 in 2 years
b) $1000 grows to $1210 in 2 years
c) $8000 grows to $9856.80 in 2 years (Use a calculator.)
d) $1000 grows to $1271.26 in 2 years (Use a calculator.)

22. A ladder 10 ft long leans against a wall. The bottom of the ladder is 6 ft from the wall. How much would the lower end of the ladder have to be pulled away so that the top end would be pulled down the same amount?

23. A ladder 13 ft long leans against a wall. The bottom of the ladder is 5 ft from the wall. How much would the lower end of the ladder have to be pulled away so that the top end would be pulled down the same amount?

24. The area of a triangle is 18 cm^2. The base is 3 cm longer than the height. Find the height.

25. A baseball diamond is a square 90 ft on a side. How far is it directly from second base to home?

26. Trains A and B leave the same city at right angles at the same time. Train B travels 5 mph faster than train A. After 2 hr they are 50 mi apart. Find the speed of each train.

11.7

EQUATIONS REDUCIBLE TO QUADRATIC

i Certain equations that are not really quadratic can be thought of in such a way that they can be solved as quadratic. For example, consider this fourth-degree equation.

OBJECTIVE

After finishing Section 11.7, you should be able to:

i Solve equations that are reducible to quadratic.

394

Solve.

1. $x^4 - 10x^2 + 9 = 0$

$$
\begin{array}{c}
x^4 \quad - \quad 9x^2 \quad + 8 = 0 \\
\downarrow \qquad \downarrow \qquad \downarrow \quad \downarrow \\
(x^2)^2 - 9(x^2) + 8 = 0 \qquad \text{Thinking of } x^4 \text{ as } (x^2)^2 \\
\downarrow \qquad \downarrow \qquad \downarrow \quad \downarrow \\
u^2 \quad - \quad 9u \quad + 8 = 0 \qquad \text{To make this clearer, write } u \text{ instead of } x^2.
\end{array}
$$

The equation $u^2 - 9u + 8 = 0$ can be solved by factoring or the quadratic formula. After that, we can find x by remembering that $x^2 = u$. Equations that can be solved like this are said to be *reducible to quadratic*.

EXAMPLE 1 Solve: $x^4 - 9x^2 + 8 = 0$.

Let $u = x^2$. Then we solve the equation found by substituting u for x^2:

$$
\begin{aligned}
u^2 - 9u + 8 &= 0 \\
(u - 8)(u - 1) &= 0 \qquad \text{Factoring} \\
u - 8 = 0 \quad \text{or} \quad u - 1 &= 0 \qquad \text{Principle of zero products} \\
u = 8 \quad \text{or} \qquad u &= 1
\end{aligned}
$$

Now we substitute x^2 for u and solve these equations:

$$
\begin{aligned}
x^2 = 8 & \quad \text{or} \quad x^2 = 1 \\
x = \pm\sqrt{8} & \quad \text{or} \quad x = \pm 1 \\
x = \pm 2\sqrt{2} & \quad \text{or} \quad x = \pm 1
\end{aligned}
$$

To check first note that when $x = 2\sqrt{2}$, $x^2 = 8$ and $x^4 = 64$. Also, when $x = -2\sqrt{2}$, $x^2 = 8$ and $x^4 = 64$. Similarly, when $x = 1$, $x^2 = 1$, and $x^4 = 1$, and when $x = -1$, $x^2 = 1$, and $x^4 = 1$. Thus instead of making four checks we need make only two.

Check: For $\pm 2\sqrt{2}$:

$$
\begin{array}{c|c}
x^4 - 9x^2 + 8 = 0 & \\
\hline
(\pm 2\sqrt{2})^4 - 9(\pm 2\sqrt{2})^2 + 8 & 0 \\
64 - 9 \cdot 8 + 8 & \\
& 0
\end{array}
$$

For ± 1:

$$
\begin{array}{c|c}
x^4 - 9x^2 + 8 = 0 & \\
\hline
(\pm 1)^4 - 9(\pm 1)^2 + 8 & 0 \\
1 - 9 + 8 & \\
& 0
\end{array}
$$

The solutions are 1, -1, $2\sqrt{2}$, and $-2\sqrt{2}$.

DO EXERCISE 1.

A check by substituting is necessary when solving equations reducible to quadratic. Reducing the equation to quadratic can sometimes introduce numbers that are not solutions of the original equation.

EXAMPLE 2 Solve: $x - 3\sqrt{x} - 4 = 0$.

Let $u = \sqrt{x}$. Then we solve the equation found by substituting u for $\sqrt{x}$ (and, of course, u^2 for x):

$$
\begin{aligned}
u^2 - 3u - 4 &= 0 \\
(u - 4)(u + 1) &= 0 \\
u = 4 \quad \text{or} \quad u &= -1.
\end{aligned}
$$

Now we substitute $\sqrt{x}$ for u and solve these equations:

$$
\sqrt{x} = 4 \quad \text{or} \quad \sqrt{x} = -1.
$$

Squaring the first equation we get $x = 16$. The second equation has no real solution since principal square roots are never negative.

The number 16 checks, so it is the solution.

DO EXERCISE 2.

EXAMPLE 3 Solve: $(x^2 - 1)^2 - (x^2 - 1) - 2 = 0$.

Let $u = x^2 - 1$. Then we solve the equation found by substituting u for $x^2 - 1$:

$$u^2 - u - 2 = 0$$
$$(u - 2)(u + 1) = 0$$
$$u = 2 \quad \text{or} \quad u = -1.$$

Now we substitute $x^2 - 1$ for u and solve these equations:

$$x^2 - 1 = 2 \qquad \text{or} \quad x^2 - 1 = -1$$
$$x^2 = 3 \qquad \text{or} \qquad x^2 = 0$$
$$x = \pm\sqrt{3} \quad \text{or} \qquad x = 0.$$

The numbers $\sqrt{3}$, $-\sqrt{3}$, and 0 check. They are the solutions.

DO EXERCISE 3.

EXAMPLE 4 Solve: $y^{-2} - y^{-1} - 2 = 0$.

Let $u = y^{-1}$. Then we solve the equation found by substituting u for y^{-1} and u^2 for y^{-2}:

$$u^2 - u - 2 = 0$$
$$(u - 2)(u + 1) = 0$$
$$u = 2 \quad \text{or} \quad u = -1.$$

Now we substitute y^{-1} or $1/y$ for u and solve these equations:

$$\frac{1}{y} = 2 \quad \text{or} \quad \frac{1}{y} = -1.$$

Solving, we get

$$y = \frac{1}{2} \quad \text{or} \quad y = \frac{1}{(-1)} = -1.$$

The numbers $\frac{1}{2}$ and -1 both check. They are the solutions.

DO EXERCISE 4.

2. Solve $x + 3\sqrt{x} - 10 = 0$. Be sure to check.

Solve.

3. $(x^2 - x)^2 - 14(x^2 - x) + 24 = 0$

Solve.

4. $x^{-2} + x^{-1} - 6 = 0$

EXERCISE SET 11.7

i Solve.

1. $x - 10\sqrt{x} + 9 = 0$

2. $2x - 9\sqrt{x} + 4 = 0$

3. $x^4 - 10x^2 + 25 = 0$

4. $x^4 - 3x^2 + 2 = 0$

5. $t^{2/3} + t^{1/3} - 6 = 0$

6. $w^{2/3} - 2w^{1/3} - 8 = 0$

7. $z^{1/2} - z^{1/4} - 2 = 0$

8. $m^{1/3} - m^{1/6} - 6 = 0$

9. $(x^2 - 6x)^2 - 2(x^2 - 6x) - 35 = 0$

10. $(1 + \sqrt{x})^2 + (1 + \sqrt{x}) - 6 = 0$

11. $(y^2 - 5y)^2 - 2(y^2 - 5y) - 24 = 0$

12. $(2t^2 + t)^2 - 4(2t^2 + t) + 3 = 0$

13. $w^4 - 4w^2 - 2 = 0$

14. $t^4 - 5t^2 + 5 = 0$

15. $x^{-2} - x^{-1} - 6 = 0$

16. $4x^{-2} - x^{-1} - 5 = 0$

17. $2x^{-2} + x^{-1} - 1 = 0$

18. $m^{-2} + 9m^{-1} - 10 = 0$

Problem-solving practice

19. A department store having a sale reduced all their appliances by 20% and all their hardware by 25%. After the discount, the price of a drill was the same as the price of a toaster. If the original price of the drill was $1.50 more than the toaster, what does each appliance cost during the sale?

20. The average of two positive integers is 171. If one of the numbers is the square root of the other, find the numbers.

Solve. Check possible solutions by substituting into the original equation.

21. $6.75x - 35\sqrt{x} - 5.36 = 0$

22. $\pi x^4 - \pi^2 x^2 - \sqrt{99.3} = 0$

Solve.

23. $\left(\dfrac{y^2 - 1}{y}\right)^2 - 4\left(\dfrac{y^2 - 1}{y}\right) - 12 = 0$

24. $\left(\sqrt{\dfrac{x}{x - 3}}\right)^2 - 24 = 10\sqrt{\dfrac{x}{x - 3}}$

25. $\left(\dfrac{x^2 - 1}{x}\right)^2 - \left(\dfrac{x^2 - 1}{x}\right) - 2 = 0$

26. $\left(\dfrac{x^2 - 2}{x}\right)^2 - 7\left(\dfrac{x^2 - 2}{x}\right) - 18 = 0$

27. $\left(\dfrac{x^2 + 1}{x}\right)^2 - 8\left(\dfrac{x^2 + 1}{x}\right) + 15 = 0$

28. $\dfrac{x}{x - 1} - 6\sqrt{\dfrac{x}{x - 1}} - 40 = 0$

29. $\left(\dfrac{x + 1}{x - 1}\right)^2 + \left(\dfrac{x + 1}{x - 1}\right) - 2 = 0$

30. $5\left(\dfrac{x + 2}{x - 2}\right)^2 - 3\left(\dfrac{x + 2}{x - 2}\right) - 2 = 0$

31. $9x^{3/2} - 8 = x^3$

32. $\sqrt[3]{2x + 3} = \sqrt[6]{2x + 3}$

33. $\sqrt{x - 3} - \sqrt[4]{x - 3} = 2$

34. $a^3 - 26a^{3/2} - 27 = 0$

35. $\dfrac{2x + 1}{x} = 3 + 7\sqrt{\dfrac{2x + 1}{x}}$

OBJECTIVES

After finishing Section 11.8, you should be able to:

i Given a situation of variation, write an equation of variation or a proportion.

ii Solve problems involving variation.

11.8

OTHER KINDS OF VARIATION

We have already studied direct variation and inverse variation. We now consider variation in which the equations involved are of second degree or greater. In some cases, we also consider equations with variables in the denominators.

Consider the equation for the area of a circle, where A and r are variables and π is a constant:

$$A = \pi r^2.$$

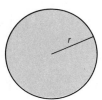

We say that the area varies directly as the square of the radius. When (r_1, A_1) and (r_2, A_2) are solutions of the equation, we have $A_1 = \pi r_1^2$ and $A_2 = \pi r_2^2$. Then dividing we get

$$\frac{A_1}{A_2} = \frac{\pi r_1^2}{\pi r_2^2} = \frac{r_1^2}{r_2^2}.$$

A proportion like

$$\frac{A_1}{A_2} = \frac{r_1^2}{r_2^2}$$

can be helpful in solving problems.

y varies directly as the square of x if there is some positive constant k such that $y = kx^2$.

EXAMPLE 1 Find an equation of variation where y varies directly as the square of x, and $y = 12$ when $x = 2$.

We write an equation of variation and find k:

$$y = kx^2, \quad \text{so } 12 = k \cdot 2^2 \quad \text{and} \quad 3 = k.$$

Thus $y = 3x^2$.

DO EXERCISE 1.

From the law of gravity, we know that the weight W of an object varies inversely as the square of its distance d from the center of the earth:

$$W = \frac{k}{d^2}.$$

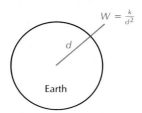

When (d_1, W_1) and (d_2, W_2) are solutions of the equation we have $W_1 = k/d_1^2$ and $W_2 = k/d_2^2$. Then

$$\frac{W_1}{W_2} = \frac{\dfrac{k}{d_1^2}}{\dfrac{k}{d_2^2}} = \frac{k}{d_1^2} \cdot \frac{d_2^2}{k} = \frac{d_2^2}{d_1^2}.$$

A proportion like

$$\frac{W_1}{W_2} = \frac{d_2^2}{d_1^2}$$

can be helpful in solving problems.

y varies inversely as the square of x if there is some positive constant k such that $y = k/x^2$.

EXAMPLE 2 Find an equation of variation where W varies inversely as the square of d and $W = 3$ when $d = 5$.

$$W = \frac{k}{d^2}, \quad \text{so} \quad 3 = \frac{k}{5^2} \quad \text{and} \quad k = 75$$

Thus

$$W = \frac{75}{d^2}.$$

DO EXERCISE 2.

1. Find an equation of variation where y varies directly as the square of x and $y = 175$ when $x = 5$.

2. Find an equation of variation where y varies inversely as the square of x and $y = \frac{1}{4}$ when $x = 6$.

3. Find an equation of variation where y varies jointly as x and z and $y = 65$ when $x = 10$ and $z = 13$.

Consider the equation for the area A of a triangle with height h and base b:

$$A = \tfrac{1}{2}bh.$$

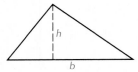

We say that the area varies *jointly* as the height and the base. From two solutions of the equation we get the proportion

$$\frac{A_1}{A_2} = \frac{\tfrac{1}{2}b_1 h_1}{\tfrac{1}{2}b_2 h_2} = \frac{b_1 h_1}{b_2 h_2}.$$

> y **varies jointly as x and z if there is some positive constant k such that $y = kxz$.**

EXAMPLE 3 Find an equation of variation where y varies jointly as x and z, and $y = 42$ when $x = 2$ and $z = 3$.

$$y = kxz, \qquad \text{so} \quad 42 = k \cdot 2 \cdot 3 \quad \text{and} \quad k = 7.$$

Thus $y = 7xz$.

DO EXERCISE 3.

4. Find an equation of variation where y varies jointly as x and the square of z and inversely as w and $y = 80$ when $x = 4$, $z = 10$, and $w = 25$.

The equation

$$y = k \cdot \frac{xz^2}{w}$$

asserts that y varies jointly as x and the square of z, and inversely as w.

EXAMPLE 4 Find an equation of variation where y varies jointly as x and z and inversely as the square of w, and $y = 105$ when $x = 3$, $z = 20$, and $w = 2$.

$$y = k \cdot \frac{xz}{w^2}, \qquad \text{so} \quad 105 = k \cdot \frac{3 \cdot 20}{2^2} \quad \text{and} \quad k = 7.$$

Thus

$$y = 7 \cdot \frac{xz}{w^2}.$$

DO EXERCISE 4.

ii Problem Solving

There are many problem situations that translate to mathematical language in the form of equations of variation. The following is an example.

EXAMPLE 5 The volume of wood V in a tree trunk varies jointly as the height h and the square of the girth g (girth is distance around). If the volume of a redwood tree

is 216 m³ when the height is 30 m and the girth is 1.5 m, what is the height of a tree whose volume is 960 m³ and girth is 2 m?

1. **Familiarize** We'll make a table, including the data from the problem and the data we need to find.

Volume of wood	Height of tree	Girth of tree
216 m³, or V_1	30 m, or h_1	1.5 m, or g_1
960 m³, or V_2	h_2	2 m, or g_2

We have of course used h_2 to represent the height of the tree in question, a quantity that we do not know and wish to determine.

We know from the statement of the problem that in this situation the volume varies jointly as the height and the square of the girth.

2. **Translate** We will do this twice, once with an equation of variation and once with a proportion.

Method 1. First we find k using the first set of data. Then we solve for h using the second set of data:

$$V = khg^2$$
$$216 = k \cdot 30 \cdot 1.5^2$$
$$3.2 = k. \quad \text{This is the variation constant.}$$

The equation of variation is $V = 3.2hg^2$.

Method 2. We first write a proportion:

$$\frac{V_1}{V_2} = \frac{h_1 \cdot g_1^2}{h_2 \cdot g_2^2}.$$

This is the translation.

3. **Carry out**

Method 1. The translation is $V = 3.2hg^2$. We substitute and solve for h:

$$960 = 3.2 \cdot h \cdot 2^2$$
$$75 = h.$$

5. The distance s that an object falls when dropped from some point above the ground varies directly as the square of the time t it falls. If the object falls 19.6 m in 2 sec, how far will the object fall in 10 sec?

6. At a fixed temperature, the resistance R of a wire varies directly as the length l and inversely as the square of its diameter d. If the resistance is 0.1 ohm when the diameter is 1 mm and the length is 50 cm, what is the resistance when the length is 2000 cm and the diameter is 2 mm?

Method 2. The translation is

$$\frac{V_1}{V_2} = \frac{h_1 \cdot g_1^2}{h_2 \cdot g_2^2}.$$

We substitute and solve for h_2:

$$\frac{216}{960} = \frac{30 \cdot 1.5^2}{h_2 \cdot 2^2}$$

$$h_2 = \frac{960 \cdot 67.5}{4 \cdot 216}$$

$$h_2 = 75.$$

4. ▷Check◁ We already have a pretty good check, since we have done the translating and calculating in two different ways. Had we not done that, we would need to recheck all calculations and perhaps do an estimate to see whether our answer is reasonable.

5. ▷State◁ The answer is that the height of the tree is 75 meters.

DO EXERCISES 5 AND 6.

EXERCISE SET 11.8

i Find an equation of variation in which:

1. y varies directly as the square of x and $y = 0.15$ when $x = 0.1$.

2. y varies directly as the square of x and $y = 6$ when $x = 3$.

3. y varies inversely as the square of x and $y = 0.15$ when $x = 0.1$.

4. y varies inversely as the square of x and $y = 6$ when $x = 3$.

5. y varies jointly as x and z and $y = 56$ when $x = 7$ and $z = 8$.

6. y varies directly as x and inversely as z and $y = 4$ when $x = 12$ and $z = 15$.

7. y varies jointly as x and the square of z and $y = 105$ when $x = 14$ and $z = 5$.

8. y varies jointly as x and z and inversely as w and $y = \frac{3}{2}$ when $x = 2$, $z = 3$, and $w = 4$.

9. y varies jointly as x and z and inversely as the product of w and p, and $y = \frac{3}{28}$ when $x = 3$, $z = 10$, $w = 7$, and $p = 8$.

10. y varies jointly as x and z and inversely as the square of w, and $y = \frac{12}{5}$ when $x = 16$, $z = 3$, and $w = 5$.

ii *Problem solving*

11. (*Stopping distance of a car*). The stopping distance d of a car after the brakes have been applied varies directly as the square of the speed r. If a car traveling 60 mph can stop in 200 ft, how many feet will it take the same car to stop when it is traveling 80 mph?

12. (*Area of a cube*). The area of a cube varies directly as the square of the length of a side. If a cube has an area 168.54 in^2 when the length of a side is 5.3 in., what will the area be when the length of a side is 10.2 in.?

13. (*Weight of an astronaut*). The weight W of an object varies inversely as the square of the distance d from the center of the earth. At sea level (6400 km from the center of the earth) an astronaut weighs 100 kg. Find his weight when he is 200 km above the surface of the earth and the spacecraft is not in motion.

14. (*Intensity of light*). The intensity I of light from a light bulb varies inversely as the square of the distance d from the bulb. Suppose I is 90 W/m^2 when the distance is 5 m. Find the intensity at a distance of 10 m.

15. (*Earned run average*). A pitcher's earned run average A varies directly as the number R of earned runs allowed and inversely as the number I of innings pitched. In a recent year Tom Seaver had an earned run average of 2.92. He gave up 85 earned runs in 262 innings. How many earned runs would he have given up had he pitched 300 innings with the same average? Round to the nearest whole number.

16. (*Volume of a gas*). The volume V of a given mass of a gas varies directly as the temperature T and inversely as the pressure P. If $V = 231$ cm^3 when $T = 42°$ and $P = 20$ kg/cm^2, what is the volume when $T = 30°$ and $P = 15$ kg/cm^2?

17. Suppose y varies directly as x and x is tripled. What is the effect on y?

18. Suppose y varies inversely as x and x is tripled. What is the effect on y?

19. Suppose y varies inversely as the square of x and x is multiplied by n. What is the effect on y?

20. Suppose y varies directly as the square of x and x is multiplied by n. What is the effect on y?

21. The area of a circle varies directly as the square of the length of a diameter. What is the variation constant?

22. A peanut butter jar in the shape of a right circular cylinder is 4 in. high and 3 in. in diameter and sells for $1.20. Assuming the same ratio of volume of peanut butter to cost, how much should a jar 6 in. high and 6 in. in diameter cost?

23. Suppose y varies inversely as the cube of x and x is multiplied by 0.5. What is the effect on y?

24. (*An application: The gravity model*). It has been determined that the average number of telephone calls in a day N, between two cities, is directly proportional to the populations P_1 and P_2 of the cities and inversely proportional to the square of the distance between the cities. That is, $N = kP_1P_2/d^2$. This model is called the "gravity model" because the equation of variation resembles the equation that applies to Newton's law of gravity.

a) The population of Indianapolis is 744,624 and the population of Cincinnati is 452,524, and the distance between the cities is 174 km. The average number of daily phone calls between the two cities is 11,153. Find the value k and write the equation of variation.

b) The average number of daily phone calls between Indianapolis and New York is 4270 and the population of New York is 7,895,563. Find the distance between Indianapolis and New York.

THE ASVAB TEST: MATHEMATICS KNOWLEDGE   CHAPTER 11

See the discussion of the ASVAB Test at the beginning of Chapter 3.

1. If the length of a side of a square inscribed in a circle is 7 cm, what is the area of the circle? Use $\frac{22}{7}$ for π.

   A   154 cm^2
   B   308 cm^2
   C $\frac{22}{7}$ cm^2
   D   77 cm^2

2. Compute: $(-0.5)^3$.

   A   0.0125
   B -12.5
   C -0.125
   D   0.125

3. The measure of the first angle of a triangle is x degrees and the second angle is 3 times the first. What is the measure of the third angle?

   A $180 - 4x$
   B $90 - 4x$
   C $360 - 3x$
   D $180 - 3x$

4. Compute: $\sqrt{8^2 + 6^2}$.

   A   12
   B   10
   C   50
   D $\sqrt{24}$

5. Subtract: $-7 - (-16)$.

   A   9
   B -23
   C -9
   D   23

6. If $m\angle A = 60°$, what is $m\angle B$?

   A   120°
   B   60°
   C   30°
   D   45°

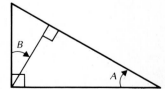

7. One factor of $6x^2 - 14x + 8$ is $2x - 2$. What is the other factor?

   A $3x - 4$
   B $3x + 4$
   C $7x - 4$
   D $3x - 7$

8. If 3 times the quantity $a + 4$ is $5a$, what is the value of a?

   A $\frac{3}{2}$
   B $\frac{1}{4}$
   C   6
   D   2

9. The temperature is 14°, but the wind makes it feel 27° colder. How cold does it feel?

A 41°
B 13°
C $-13°$
D $-41°$

10. If $x > y$ and $y < z$, then which of the following is true?

A $x + z > y + y$
B $x > z$
C $x + y > y + z$
D $x = z$

11. If 2 is 20% of M, then M is 20% of what?

A 2
B 8
C 50
D 0.08

12. What is the value of $4x + 6y$ when $x = \frac{3}{2}$ and $y = \frac{3}{4}$?

A 10.5
B 12
C 15
D 7.5

13. Find the value of $(x - 3)(x + 4)(x + 1)$ when $x = -2$.

A 10
B -10
C 18
D -18

14. Solve: $8x - 3 = 4x + 9$.

A 1
B 3
C $\frac{3}{2}$
D 4

15. It takes $2\frac{1}{2}$ hr to get from A to B flying at 400 mph. How much faster must the plane fly in order to make the trip in only 2 hr?

A 500 mph
B 100 mph
C 40 mph
D 50 mph

16. Compute: $4! - 3!$.

A 1
B 4
C 16
D 18

17. Subtract $3x^3 - 4x^2 + 3x - 8$ from $2x^3 - 6x^2 + 2x + 9$.

A $-x^3 - 10x^2 - x + 1$
B $-x^3 + 2x^2 + x - 17$
C $-x^3 - 2x^2 - x + 17$
D $-x^3 + 2x^2 - x - 1$

18. Solve for R: $S = \dfrac{4Q}{RT}$.

A $R = \dfrac{ST}{4Q}$

B $R = 4QST$

C $R = \dfrac{4Q}{ST}$

D $R = 4Q - ST$

19. Compute: $8^2 - 2^6$.

A 0
B 4
C 32
D 2

20. Solve: $x^2 + 2x = -3x$.

A 1, 0
B 5, -1
C -1, 5
D 0, -5

OBJECTIVES

After finishing Section 11.9, you should be able to:

i Without graphing, tell whether the graph of an equation equivalent to $y = ax^2 + bx + c$ opens upward or downward. Then graph the equation.

ii Approximate the solutions of $ax^2 + bx + c = 0$ by graphing.

11.9

GRAPHS OF QUADRATIC EQUATIONS AND FUNCTIONS

In this section you will learn to graph equations of the form

$$y = ax^2 + bx + c, \qquad a \neq 0.$$

The polynomial on the right is of second degree, or *quadratic*. If you studied functions in Chapter 8, you will recognize that the equation is that of a quadratic function

$$f(x) = ax^2 + bx + c, \qquad a \neq 0.$$

Examples of the types of equations we are going to graph are

$$y = x^2 \quad \text{and} \quad y = x^2 + 2x - 3.$$

i Graphing Quadratic Equations, $y = ax^2 + bx + c$

Graphs of quadratic equations, $y = ax^2 + bx + c$ (where $a \neq 0$), are always cup-shaped. They all have a *line of symmetry* like the dashed line shown in the figure. If you fold on this line, the two halves will match exactly. The curve goes on forever.

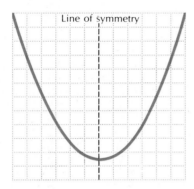

These curves are called *parabolas*. Some parabolas are thin and others are flat, but they all have the same general shape.

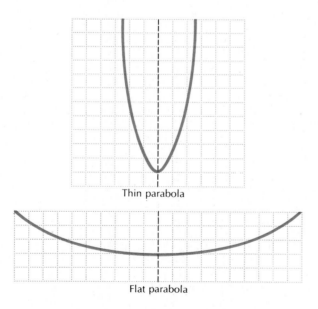

Thin parabola

Flat parabola

To graph a quadratic equation, we begin by choosing some numbers for x and computing the corresponding values of y.

EXAMPLE 1 Graph: $y = x^2$.

We plot the ordered pairs resulting from the computations shown in the table and connect the points with a smooth curve.

1. a) Without graphing tell whether the graph of

$$y = x^2 - 3$$

opens upward or downward.

b) Graph the equation.

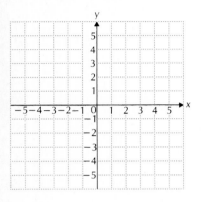

x	y
−2	4
−1	1
0	0
1	1
2	4

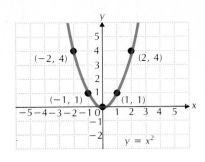

EXAMPLE 2 Graph: $y = x^2 + 2x - 3$.

x	y
1	0
0	−3
−1	−4
−2	−3
−3	0
−4	5
2	5

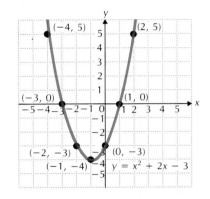

2. a) Without graphing tell whether the graph of

$$y = x^2 + 6x + 9$$

opens upward or downward.

b) Graph the equation.

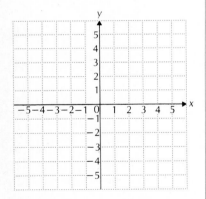

EXAMPLE 3 Graph: $y = -2x^2 + 3$.

x	y
0	3
1	1
−1	1
2	−5
−2	−5

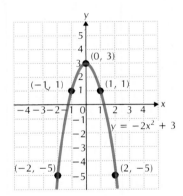

The graphs in Examples 1 and 2 open upward and the coefficients of x^2 are both 1, which is positive. The graph in Example 3 opens downward and the coefficient of x^2 is -2, which is negative.

> Graphs of quadratic equations $y = ax^2 + bx + c$ are all parabolas. They are *smooth* cup-shaped symmetric curves, with no sharp points or kinks in them.
>
> The graph of $y = ax^2 + bx + c$ opens upward if $a > 0$. It opens downward if $a < 0$.
>
> In drawing parabolas, be sure to plot enough points to see the general shape of each graph.

If your graphs look like any of the following, they are incorrect.

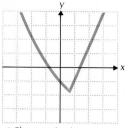

a) Sharp point is wrong.

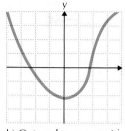

b) Outward nonsymmetric curve is wrong.

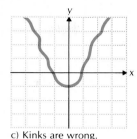

c) Kinks are wrong.

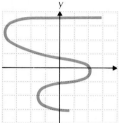

d) S-shaped curve is wrong.

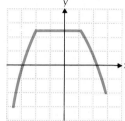

e) Flat nose is wrong.

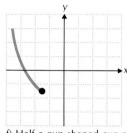

f) Half a cup-shaped curve is wrong.

DO EXERCISES 1–3. (EXERCISES 1 AND 2 ARE ON THE PRECEDING PAGE.)

ii Approximating Solutions of $ax^2 + bx + c = 0$

We can use graphing to approximate the solutions of quadratic equations, $0 = ax^2 + bx + c$. We graph the equation $y = ax^2 + bx + c$. If the graph crosses the x-axis, the points of crossing will give us solutions.

EXAMPLE 4 Approximate the solutions of

$$-2x^2 + 3 = 0$$

by graphing.

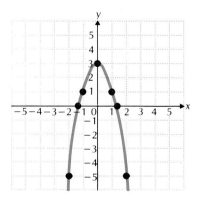

The graph was found in Example 3. The graph crosses the x-axis at about $(-1.2, 0)$ and $(1.2, 0)$. So the solutions are about -1.2 and 1.2.

DO EXERCISES 4 AND 5.

3. a) Without graphing tell whether the graph of

$$y = -3x^2 + 6x$$

opens upward or downward.

b) Graph the equation.

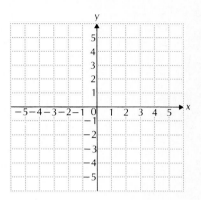

Approximate the solutions by graphing.

4. $x^2 - 4x + 4 = 0$

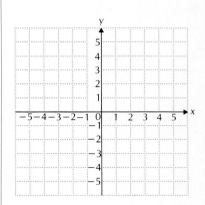

5. $-2x^2 - 4x + 1 = 0$

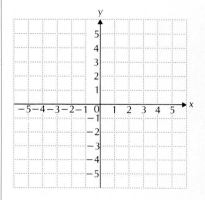

EXERCISE SET 11.9

i Without graphing, tell whether the graph of the equation opens upward or downward. Then graph the equation using graph paper.

1. $y = x^2 + 1$ **2.** $y = 2x^2$ **3.** $y = -1 \cdot x^2$ **4.** $y = x^2 - 1$

5. $y = -x^2 + 2x$ **6.** $y = x^2 + x - 6$ **7.** $y = 8 - x - x^2$ **8.** $y = x^2 + 2x + 1$

9. $y = x^2 - 2x + 1$ **10.** $y = -\frac{1}{2}x^2$ **11.** $y = -x^2 + 2x + 3$ **12.** $y = -x^2 - 2x + 3$

13. $y = -2x^2 - 4x + 1$ **14.** $y = 2x^2 + 4x - 1$ **15.** $y = \frac{1}{4}x^2$ **16.** $y = -0.1x^2$

17. $y = 3 - x^2$ **18.** $y = x^2 + 3$ **19.** $y = -x^2 + x - 1$ **20.** $y = x^2 + 2x$

21. $y = -2x^2$ **22.** $y = -x^2 - 1$ **23.** $y = x^2 - x - 6$ **24.** $y = 8 + x - x^2$

ii Approximate the solutions by graphing.

25. $x^2 - 5 = 0$ **26.** $x^2 - 3 = 0$ **27.** $x^2 + 2x = 0$ **28.** $x^2 - 2x = 0$

29. $8 - x - x^2 = 0$ **30.** $8 + x - x^2 = 0$ **31.** $x^2 + 10x + 25 = 0$ **32.** $x^2 - 8x + 16 = 0$

33. $-2x^2 - 4x + 1 = 0$ **34.** $2x^2 + 4x - 1 = 0$ **35.** $x^2 + 5 = 0$ **36.** $x^2 + 3 = 0$

37. Graph the equation $y = x^2 - x - 6$. Use your graph to approximate the solutions of the following equations.

a) $x^2 - x - 6 = 2$ (*Hint:* Graph $y = 2$ on the same set of axes as your graph of $y = x^2 - x - 6$.)

b) $x^2 - x - 6 = -3$

38. Graph $y = 2x^2 - 4x + 7$. Use the graph to approximate the solutions of $2x^2 - 4x + 7 = 0$. Solve $2x^2 - 4x + 7 = 0$ using the quadratic formula. What might you guess is true of any quadratic equation whose graph does not cross the x-axis?

39. The y-intercept of the linear equation $y = mx + b$ is $(0, b)$. What is the y-intercept of the quadratic equation $y = ax^2 + bx + c$?

40. The graph of a quadratic equation has either a high point (maximum) or a low point (minimum). Examine the graphs in Exercises 1–24. What characteristic of each equation tells whether the graph has a maximum or minimum?

41. Find an equation of the line that is the axis of symmetry of each graph in Exercises 1–10.

42. Using the same set of axes, graph $y = x^2$, $y = 2x^2$, $y = 3x^2$, $y = \frac{1}{2}x^2$, $y = -x^2$, $y = -2x^2$, $y = -3x^2$, and $y = -\frac{1}{2}x^2$. Describe the change in the graph of a quadratic equation $y = ax^2$ as $|a|$ increases.

43. What is the largest rectangular area that can be enclosed with 16 feet of fence?

$$2w + 2l = 16$$
$$w + l = 8$$
$$w = 8 - l$$

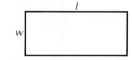

(*Hint:* Find $A = lw$ in terms of l. Graph the resulting quadratic equation and find its maximum.)

OBJECTIVES

After finishing Section 11.10, you should be able to:

i Graph functions like $y = ax^2$ or $f(x) = ax^2$.

ii Graph functions like $f(x) = a(x - h)^2$.

iii Graph functions like $f(x) = a(x - h)^2 + k$ and determine the vertex, line of symmetry, and the maximum or minimum function values.

iv Use translations to graph functions or to determine properties of the function without graphing.

11.10

QUADRATIC FUNCTIONS

A *quadratic function* is a function that can be described by a quadratic polynomial, as follows:

$$f(x) = ax^2 + bx + c.$$

The constants b and c may be 0, but of course a cannot be 0, or else the polynomial will not be quadratic. In this section we will look at the graphs of quadratic functions.

i Graphs of $f(x) = ax^2$

We first consider graphs of quadratic functions in which b and c are both 0. In fact, let's start with a very simple one, a function for which $a = 1$.

EXAMPLE 1   Graph: $f(x) = x^2$.

We choose some numbers for x, some positive and some negative, and for each number compute $f(x)$.

x	0	1	2	3	-1	-2	-3
$f(x)$	0	1	4	9	1	4	9

We plot these ordered pairs and connect the points with a smooth curve.

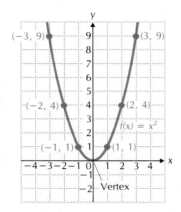

All quadratic functions have graphs similar to the one in Example 1. Such curves are called *parabolas*. They are smooth, cup-shaped curves, with a line of symmetry. In the graph above, the vertical axis is the *line of symmetry*. If the paper were folded on this line, the two halves of the curve would match. The point $(0, 0)$ is known as the *vertex* of the parabola.

Some common errors in drawing graphs of quadratic functions are shown in the figure below.

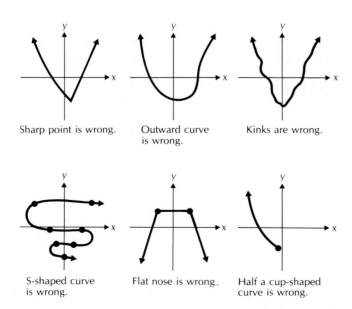

Sharp point is wrong.

Outward curve is wrong.

Kinks are wrong.

S-shaped curve is wrong.

Flat nose is wrong.

Half a cup-shaped curve is wrong.

DO EXERCISE 1.

1. Graph $y = \dfrac{1}{2}x^2$.

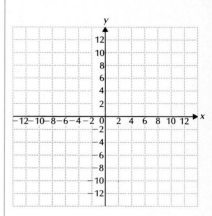

2. Graph $y = 3x^2$.

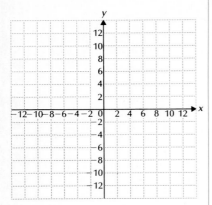

The graph of $f(x) = \frac{1}{2}x^2$ is a flatter parabola than the graph of $y = x^2$. The graph of $f(x) = 2x^2$ is narrower, as shown here, but the vertex and line of symmetry have not changed.

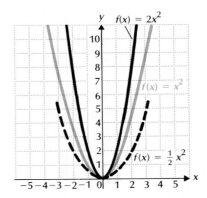

> The graph of $y = ax^2$ is a parabola with the vertical axis as its line of symmetry and its vertex at the origin.
>
> If $|a|$ is greater than 1, the parabola is narrower than $y = x^2$.
>
> If $|a|$ is between 0 and 1, the parabola is flatter than $y = x^2$.
>
> If a is positive, the parabola opens upward; if a is negative, the parabola opens downward.

3. Graph $y = -2x^2$.

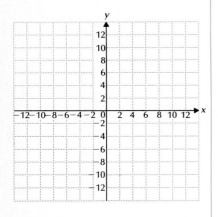

DO EXERCISES 2 AND 3.

ii Graphs of $f(x) = a(x - h)^2$

Why not now consider graphs of $f(x) = ax^2 + bx + c$, where b and c are not both 0? In effect, we shall do that, but in a disguised form. It turns out to be convenient to consider functions $f(x) = a(x - h)^2$, that is, where we start with ax^2 but then replace x by $x - h$, where h is some constant.

EXAMPLE 2 Graph: $f(x) = 2(x - 3)^2$.

We choose some values of x and compute $f(x)$.

x	0	1	2	3	4	5	6
$f(x)$, or $2(x - 3)^2$	18	8	2	0	2	8	18

Now we plot the points and draw the curve (see the following page).

The curve in Example 2 looks just like the graph of $f(x) = 2x^2$, except that it is moved three units to the right. The line of symmetry is now the line $x = 3$ and the vertex is the point $(3, 0)$.

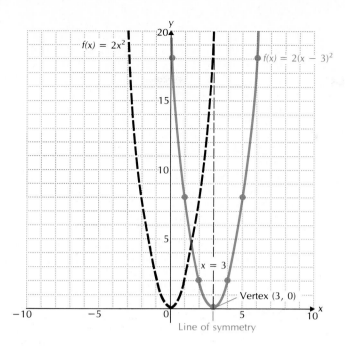

The graph of $y = a(x - h)^2$ looks just like the graph of $y = ax^2$, except that it is moved to the right or left. If h is positive, it is moved to the right. If h is negative, it is moved to the left. The vertex is now $(h, 0)$ and the line of symmetry is $x = h$.

EXAMPLE 3   Graph: $f(x) = 2(x + 3)^2$.

We know that the graph looks like that of $f(x) = 2x^2$, but moved right or left. Think of $x + 3$ as $x - (-3)$. The number we are *subtracting* is negative, so we move *left*. We can draw $f(x) = 2x^2$ lightly, then move it three units to the left. See the following graph. We should compute a few values as a check.

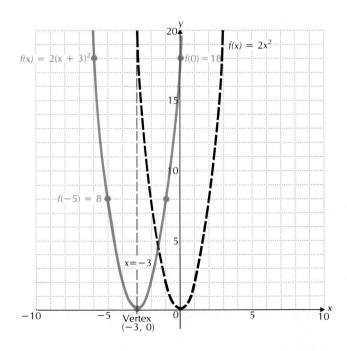

4. a) Graph $y = \frac{1}{2}(x - 4)^2$. Label the vertex and draw the line of symmetry.

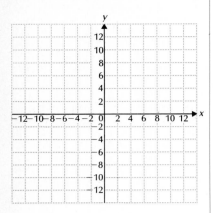

b) Graph $y = -\frac{1}{2}(x - 4)^2$.

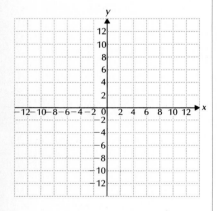

EXAMPLE 4 Graph: $y = -2(x + 2)^2$.

We know that the graph looks like that of $y = 2x^2$, but moved to the left two units, and it will also open downward, because of the negative coefficient, -2. See the graph below.

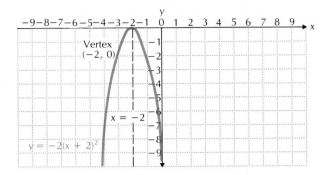

DO EXERCISE 4.

iii Graphs of $f(x) = a(x - h)^2 + k$

Given a graph of $f(x) = a(x - h)^2$, what happens to it if we add a constant k? Suppose we add 2. This increases each function value $f(x)$ by 2, so the curve is moved up. If we add a negative number, the curve is moved down. The vertex of the parabola will be at the point (h, k) and the line of symmetry will be $x = h$.

Note that if a parabola opens upward ($a > 0$), the function value, or y-value, at the vertex is a least, or *minimum* value. That is, it is less than the y-value at any other point. If the parabola opens downward ($a < 0$), the function value at the vertex will be a greatest, or *maximum* value.

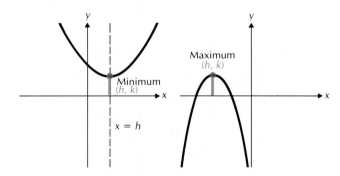

The graph of $f(x) = a(x - h)^2 + k$ looks just like the graph of $f(x) = a(x - h)^2$, except that it is moved up or down. If k is positive, the curve is moved up k units. If k is negative, the curve is moved down. The vertex is at (h, k) and the line of symmetry is $x = h$. If $a > 0$, then k is the minimum function value. If $a < 0$, then k is the maximum function value.

EXAMPLE 5 Graph $f(x) = 2(x + 3)^2 - 5$ and find the minimum function value.

We know that the graph looks like that of $f(x) = 2(x + 3)^2$ (this is shown in Example 3) but moved down five units. The vertex is now $(-3, -5)$, and the minimum function value is -5.

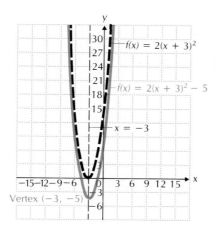

EXAMPLE 6 Graph $f(x) = \frac{1}{2}(x - 3)^2 + 5$ and find the minimum function value.

We know that the graph looks just like that of $f(x) = \frac{1}{2}x^2$, but moved to the right three units and up five units. The vertex is $(3, 5)$ and the line of symmetry is the line $x = 3$. We draw $f(x) = \frac{1}{2}x^2$ and then move the curve over and up. We plot a few points as a check. The minimum function value is 5.

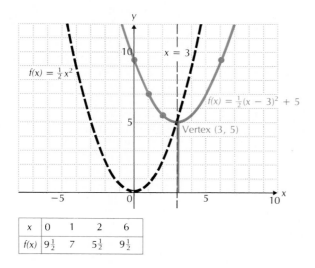

x	0	1	2	6
f(x)	$9\frac{1}{2}$	7	$5\frac{1}{2}$	$9\frac{1}{2}$

EXAMPLE 7 Graph $y = -2(x + 3)^2 + 5$ and find the maximum function value.

We know that the graph looks like that of $f(x) = 2x^2$, but moved to the left three units and up five units, except that it now opens downward. The vertex is $(-3, 5)$. This time there is a *greatest* (maximum) function value, 5.

5. a) Graph $y = \dfrac{1}{2}(x + 2)^2 - 4$.

b) Label the vertex and draw the line of symmetry.

c) Find the minimum function value.

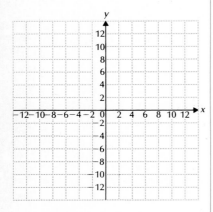

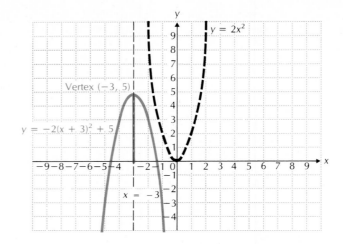

DO EXERCISES 5 AND 6.

6. a) Graph
$f(x) = -2(x - 5)^2 + 3$.

b) Label the vertex and draw the line of symmetry.

c) Find the maximum function value.

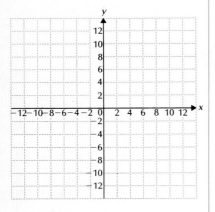

iv Translations

In this section we have seen a comparison of the graphs of

$$f(x) = ax^2 \quad \text{and} \quad f(x) = a(x - h)^2.$$

The graphs are alike, except that the second one has been moved horizontally. We have also seen a comparison of

$$f(x) = a(x - h)^2 \quad \text{and} \quad f(x) = a(x - h)^2 + k.$$

Again, the graphs look alike, but in this case the second one has been moved vertically.

A motion of a graph like those we are considering, in which we do not change the size or shape of a graph and in which there is no rotation, is called a *translation*. What we have seen with parabolas holds for any other function. It also holds for other graphs, whether or not they are graphs of functions. We summarize as follows. Although the properties are stated for functions, they hold in general.

> For a function $y = f(x)$, if we replace x by $x - h$, where h is a constant, the graph will be moved horizontally. If h is positive, the graph will be moved in the positive x-direction. If h is negative, the graph will be moved in the negative x-direction.

EXAMPLE 8 The following figure is a graph of a function $y = g(x)$. Draw graphs for (a) $y = g(x - 4)$; (b) $y = g(x + 3)$.

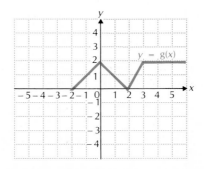

a) The graph of $y = g(x - 4)$ looks just like the graph of $y = g(x)$, but it will be moved four units in the positive x-direction (since we are subtracting a positive constant from x, in $x - 4$).

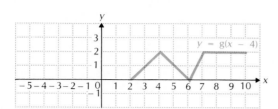

b) We can think of the graph of $y = g(x + 3)$ as the graph of $y = g(x - (-3))$. Since we are subtracting a negative constant from x, the graph will be moved in the negative x-direction.

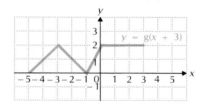

DO EXERCISE 7.

What we have said about replacing x by $x - h$ also holds for replacing y by $y - k$, or replacing $f(x)$ by $f(x) - k$, where k is some constant.

> **For a function $y = f(x)$, if we replace y by $y - k$, where k is a constant, the graph will be moved vertically. If k is positive, the graph will be moved in the positive y-direction. If k is negative, the graph will be moved in the negative y-direction.**

For functions, we are most likely to write $y = f(x) + k$ than $y - k = f(x)$. The rule is easier to remember, however, if you write, or at least think of writing, $y - k = f(x)$. In equations that do not represent functions, it will often be *necessary* to write $y - k$ in place of y.

EXAMPLE 9 Use the function $y = g(x)$ in Example 8 to draw graphs for (a) $y - 2 = g(x)$, or $y = g(x) + 2$; (b) $y + 4 = g(x)$, or $y = g(x) - 4$.

a) The graph of $y - 2 = g(x)$ looks just like the graph of $y = g(x)$, but it will be moved two units in the positive y-direction (since we are subtracting a positive constant from y, in $y - 2$).

7. This is the graph of a function $y = f(x)$.

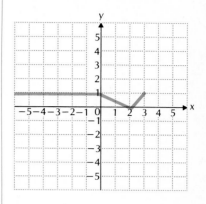

Draw the graph of each of the following.

a) $y = f(x + 2)$

b) $y = f(x - 3)$

8. Use the function of Exercise 7. Draw the graph of each of the following.

a) $y + 3 = f(x)$, or $y = f(x) - 3$

b) $y - 2 = f(x)$, or $y = f(x) + 2$

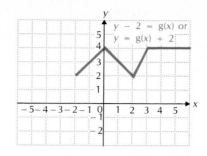

b) The graph of $y + 4 = g(x)$ can be thought of as the graph of $y - (-4) = g(x)$. Since we are subtracting a negative constant from y, the graph will be moved in the negative y-direction.

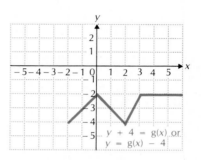

DO EXERCISE 8.

EXERCISE SET 11.10

i , **ii** For each of the following, graph the function, find the vertex, and find the line of symmetry.

1. $f(x) = x^2$ **2.** $f(x) = -x^2$ **3.** $f(x) = -4x^2$ **4.** $f(x) = 2x^2$

5. $f(x) = (x - 3)^2$ **6.** $f(x) = (x - 7)^2$ **7.** $f(x) = -(x + 4)^2$ **8.** $f(x) = -(x - 2)^2$

9. $f(x) = 2(x - 3)^2$ **10.** $f(x) = -4(x - 7)^2$ **11.** $f(x) = -2(x + 9)^2$ **12.** $f(x) = 2(x + 7)^2$

13. $f(x) = 3(x - 1)^2$ **14.** $f(x) = -4(x - 2)^2$ **15.** $f(x) = -3(x - \frac{1}{2})^2$ **16.** $f(x) = -2(x + \frac{1}{2})^2$

17. $f(x) = \frac{1}{2}(x + 1)^2$ **18.** $f(x) = \frac{1}{3}(x - 2)^2$

iii For each of the following, graph the function, find the vertex, find the line of symmetry, and find the maximum value or the minimum value.

19. $f(x) = (x - 3)^2 + 1$ **20.** $f(x) = (x + 2)^2 - 3$ **21.** $f(x) = (x + 1)^2 - 2$ **22.** $f(x) = (x - 1)^2 + 2$

23. $f(x) = 2(x - 1)^2 - 3$ **24.** $f(x) = 2(x + 1)^2 + 4$ **25.** $f(x) = -3(x + 4)^2 + 1$ **26.** $f(x) = -2(x - 5)^2 - 3$

Without graphing, find the vertex, find the line of symmetry, and find the maximum value or the minimum value.

27. $f(x) = 8(x - 9)^2 + 5$ **28.** $f(x) = 10(x + 5)^2 - 8$ **29.** $f(x) = 5(x + \frac{1}{4})^2 - 13$

30. $f(x) = 6(x - \frac{1}{4})^2 + 19$ **31.** $f(x) = -7(x - 10)^2 - 20$ **32.** $f(x) = -9(x + 12)^2 + 23$

33. $f(x) = \sqrt{2}\,(x + 4.58)^2 + 65\pi$ **34.** $f(x) = 4\pi(x - 38.2)^2 - \sqrt{34}$

iv   Use the graph of the function $y = h(x)$ below for Exercises 35–40.

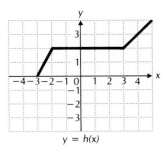

$y = h(x)$

Draw a graph of each of the following.

35. $y = h(x + 2)$

36. $y = h(x - 3)$

37. $y - 3 = h(x)$

38. $y = h(x) + 4$

39. $y + 2 = h(x - 4)$

40. $y = h(x - 2) + 3$

Problem-solving practice

41. A restaurant sells 12-oz sodas for 45¢ and 16-oz sodas for 80¢. In one day they sold a total of 36 sodas for $23.20. If the restaurant sold 4 gal of soda that day, how much of each size drink did they sell?

42. Given four consecutive odd numbers, the product of the first and third is 56 less than the product of the second and fourth. Find the numbers.

For each of the following, write the equation of the parabola that is a translation of $f(x) = 2x^2$ or $f(x) = -2x^2$ and has a maximum or minimum value at the specific point.

43. Maximum: $(0, 4)$

44. Minimum: $(2, 0)$

45. Minimum: $(6, 0)$

46. Maximum: $(0, 3)$

47. Maximum: $(3, 8)$

48. Minimum: $(-2, 3)$

49. Minimum: $(-3, 6)$

50. Maximum: $(-4, -3)$

51. Minimum: $(2, -3)$

For each of the following, write an equation of the parabola.

52. The parabola has a minimum value at the same point as $f(x) = 3(x - 4)^2$, but for all x in the domain except 4 the function values are doubled.

53. The parabola is a translation of

$$f(x) = -\tfrac{1}{2}(x - 2)^2 + 4$$

and has a maximum value at the same point as $f(x) = -2(x - 1)^2 - 6$.

11.11

MORE ABOUT GRAPHING QUADRATIC FUNCTIONS

Now that we know how to graph $f(x) = a(x - h)^2 + k$, we are in a position to graph any quadratic function

$$f(x) = ax^2 + bx + c.$$

All we have to do is find a way to rewrite a quadratic polynomial as $a(x - h)^2 + k$.

i   Completing the Square

Actually, we already know a way to rewrite a quadratic function as described: The method is that of *completing the square*.

OBJECTIVES

After finishing Section 11.11, you should be able to:

i   Given a quadratic function $f(x) = ax^2 + bx + c$, complete the square to put it into the form

$$f(x) = a(x - h)^2 + k,$$

and then graph the function.

ii   Find the x-intercepts of a quadratic function.

1. For $f(x) = x^2 - 4x + 7$:

a) find an equation of the type $f(x) = a(x - h)^2 + k$;

b) find the vertex and line of symmetry;

c) graph the function.

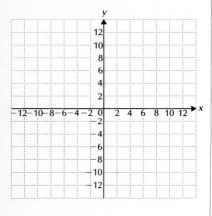

EXAMPLE 1 Graph: $f(x) = x^2 - 6x + 4$.

$$f(x) = x^2 - 6x + 4$$
$$= (x^2 - 6x) + 4$$

We complete the square inside the parentheses. We take half the x-coefficient:

$$\tfrac{1}{2} \cdot (-6) = -3.$$

We square it:

$$(-3)^2 = 9.$$

Then we add $9 - 9$ inside the parentheses.

$$f(x) = (x^2 - 6x + 9 - 9) + 4$$
$$= (x^2 - 6x + 9) + (-9 + 4) \qquad \text{Rearranging terms}$$
$$= 1 \cdot (x - 3)^2 - 5$$

The vertex is $(3, -5)$. The line of symmetry is $x = 3$.

We plot a few points as a check, and draw the curve.

x	$f(x)$
0	4
1	-1
5	-1
6	4

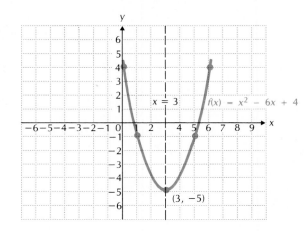

DO EXERCISE 1.

EXAMPLE 2 Graph: $f(x) = -2x^2 + 10x - 7$.

We first factor the expression $-2x^2 + 10x$ by "removing" -2 from the first two terms. This makes the coefficient of x^2 inside the parentheses 1. Then we proceed as before.

$$f(x) = -2x^2 + 10x - 7$$

and

$$f(x) = -2(x^2 - 5x) - 7.$$

We take half the x-coefficient and square it, to get $\frac{25}{4}$. Then we add $\frac{25}{4} - \frac{25}{4}$ inside the parentheses:

$$f(x) = -2(x^2 - 5x + \tfrac{25}{4} - \tfrac{25}{4}) - 7$$
$$= -2(x^2 - 5x + \tfrac{25}{4}) + 2(\tfrac{25}{4}) - 7 \qquad \text{Multiplying by } -2, \text{ using the distributive law, and rearranging terms}$$
$$= -2(x - \tfrac{5}{2})^2 + \tfrac{11}{2}$$

The vertex is $(\frac{5}{2}, \frac{11}{2})$. The line of symmetry is $x = \frac{5}{2}$. The coefficient -2 is negative, so the graph opens downward.

We plot a few points as a check and draw the curve.

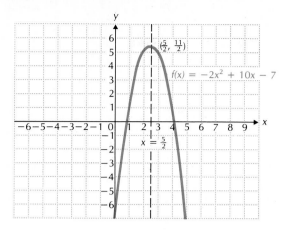

DO EXERCISE 2.

ii **Intercepts**

The points at which a graph crosses the *x*-axis are called its *x-intercepts*. These are, of course, the points at which $y = 0$.

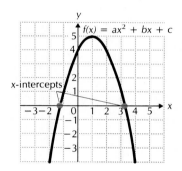

To find the *x*-intercepts of a quadratic function $f(x) = ax^2 + bx + c$ we solve the equation

$$0 = ax^2 + bx + c.$$

EXAMPLE 3 Find the *x*-intercepts of the graph of $f(x) = x^2 - 2x - 2$.

We solve the equation

$$0 = x^2 - 2x - 2.$$

The equation is difficult to factor, so we use the quadratic formula and get $x = 1 \pm \sqrt{3}$. Thus the *x*-intercepts are $(1 - \sqrt{3}, 0)$ and $(1 + \sqrt{3}, 0)$.

For plotting, we approximate, to get $(-0.7, 0)$ and $(2.7, 0)$. We sometimes refer to the *x*-coordinates as *intercepts*. It is useful to have these points when graphing a function.

DO EXERCISE 3.

2. For $f(x) = -4x^2 + 12x - 5$:

   a) find an equation of the type $f(x) = a(x - h)^2 + k$;

   b) find the vertex and line of symmetry;

   c) graph the function.

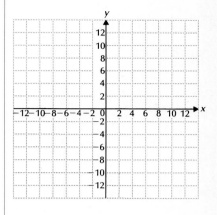

Find the *x*-intercepts.

3. $f(x) = x^2 - 2x - 5$

Find the x-intercepts if they exist.

4. $f(x) = x^2 - 2x - 3$

5. $f(x) = x^2 + 8x + 16$

6. $f(x) = -2x^2 - 4x - 3$

The discriminant, $b^2 - 4ac$, tells us how many real-number solutions the equation $0 = ax^2 + bx + c$ has, so it also indicates how many intercepts there are. Compare.

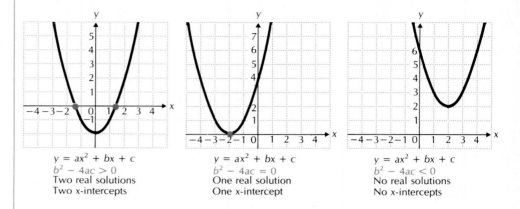

$y = ax^2 + bx + c$
$b^2 - 4ac > 0$
Two real solutions
Two x-intercepts

$y = ax^2 + bx + c$
$b^2 - 4ac = 0$
One real solution
One x-intercept

$y = ax^2 + bx + c$
$b^2 - 4ac < 0$
No real solutions
No x-intercepts

DO EXERCISES 4–6.

EXERCISE SET 11.11

i Complete the square. Find the vertex, the line of symmetry, and the maximum or minimum function value.

1. $f(x) = x^2 - 2x - 3$ **2.** $f(x) = x^2 + 2x - 5$ **3.** $f(x) = -x^2 + 4x + 6$ **4.** $f(x) = -x^2 - 4x + 3$

5. $f(x) = x^2 + 3x - 10$ **6.** $f(x) = x^2 + 5x + 4$ **7.** $f(x) = x^2 - 9x$ **8.** $f(x) = x^2 + x$

9. $f(x) = 3x^2 - 24x + 50$ **10.** $f(x) = 4x^2 + 8x - 3$ **11.** $f(x) = \frac{3}{4}x^2 + 9x$ **12.** $f(x) = -2x^2 + 2x + 1$

ii Find the x-intercepts.

13. $f(x) = x^2 - 4x + 1$ **14.** $f(x) = x^2 + 6x + 10$ **15.** $f(x) = -x^2 + 2x + 3$ **16.** $f(x) = x^2 - 2x - 5$

17. $f(x) = x^2 - 3x - 4$ **18.** $f(x) = x^2 - 8x + 5$ **19.** $f(x) = -x^2 + 3x + 4$ **20.** $f(x) = 2x^2 - 4x + 6$

21. $f(x) = 2x^2 + 4x - 1$ **22.** $f(x) = x^2 - x + 2$ **23.** $f(x) = x^2 - x + 1$ **24.** $f(x) = 4x^2 + 12x + 9$

25. $f(x) = -x^2 - 3x - 3$ **26.** $f(x) = -5x^2 + 6x - 5$

Problem-solving practice

27. The length of a diagonal of a rectangle is one more than the length of the rectangle and three more than twice the width of the rectangle. Find the dimensions of the rectangle.

28. Bill leaves the hotel parking lot going due north at 30 mph. Mary leaves the same lot going due east at 40 mph. How long will they have to travel before they are 240 mi apart?

Find the maximum or minimum value.

29. ▦ $f(x) = 2.31x^2 - 3.135x - 5.89$

30. ▦ $f(x) = -18.8x^2 + 7.92x + 6.18$

Find the x-intercepts.

31. ▦ $f(x) = 0.05x^2 - 4.735x + 100.23$

32. ▦ $f(x) = 1.13x^2 + 2.809x - 7.114$

33. ▦ $f(x) = 2.12x^2 + 3.21x + 9.73$

34. ▦ $f(x) = 0.13x^2 - 0.071x - 0.12$

35. Graph the function $f(x) = x^2 - x - 6$. Then use your graph to approximate the solutions to the following equations.

a) $x^2 - x - 6 = 2$

b) $x^2 - x - 6 = -3$

36. Graph the function

$$f(x) = \frac{x^2}{8} + \frac{x}{4} - \frac{3}{8}.$$

Use your graph to approximate solutions to the following equations.

a) $\frac{x^2}{8} + \frac{x}{4} - \frac{3}{8} = 0$ **b)** $\frac{x^2}{8} + \frac{x}{4} - \frac{3}{8} = 1$ **c)** $\frac{x^2}{8} + \frac{x}{4} - \frac{3}{8} = 2$

Find an equation of the type $f(x) = a(x - h)^2 + k$.

37. $f(x) = ax^2 + bx + c$ **38.** $f(x) = 3x^2 + mx + m^2$

Graph.

39. $f(x) = |x^2 - 1|$ **40.** $f(x) = |3 - 2x - x^2|$

Graph these quadratic inequalities.

41. $y < x^2 - 4x - 1$ **42.** $y \geqslant x^2 + 3x - 4$ **43.** $y \leqslant x^2 + 5x + 6$ **44.** $y < -x^2 - 2x + 3$

45. $y > 3x^2 + 6x + 2$ **46.** $y > 2x^2 + 4x - 2$ **47.** $y \geqslant 4x^2 + 8x + 3$ **48.** $y < 2x^2 - 4x + 2$

11.12

PROBLEM SOLVING AND QUADRATIC FUNCTIONS

There are many ways in which quadratic functions are used in problem solving. In this section we will consider two of those methods.

i Maximum and Minimum Problems

For a quadratic function, the value $f(x)$ at the vertex will either be greater than any other $f(x)$ or less than all other $f(x)$. If the graph opens upward, $f(x)$ will be a minimum. If the graph opens downward, $f(x)$ will be a maximum. In certain problems we want to find a maximum or a minimum. If a problem situation can be translated to a quadratic function, we can solve by finding $f(x)$ at the vertex.

OBJECTIVES

After finishing Section 11.12, you should be able to:

i Solve maximum and minimum problems.

ii Solve problems by fitting a quadratic function to three data points.

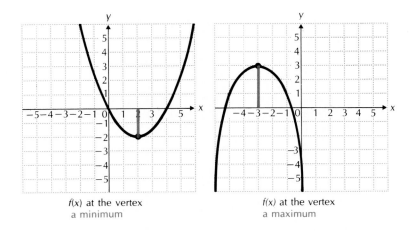

$f(x)$ at the vertex
a minimum

$f(x)$ at the vertex
a maximum

EXAMPLE 1 What are the dimensions of the largest rectangular pen that can be enclosed with 64 m of fence?

1. What is the maximum product of two numbers whose sum is 30?

1. ⟩ **Familiarize** ⟩ We make a drawing and label it.

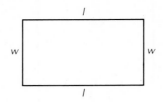

Perimeter: $2w + 2l = 64$ m

Area: $A = l \cdot w$

We may need to look up definitions or formulas in step 1. In addition, we may be able to get a better idea of the problem by looking at some possible dimensions for a rectangular pen that can be enclosed with 64 m of fence.

l	w	A
10	22	220
12	20	240
14	18	252
16	16	256
17	15	255
16.4	15.6	255.84

2. What are the dimensions of the largest rectangular pen that can be enclosed with 100 m of fence?

2. ⟩ **Translate** ⟩ We have a system of equations

$$2w + 2l = 64,$$
$$A = l \cdot w.$$

3. ⟩ **Carry out** ⟩ We will use substitution. Solving the first equation for l, we get $l = 32 - w$. Substituting for l in the second equation we get a quadratic function, $A(w)$, or just A:

$$A = (32 - w)w$$
$$A = -w^2 + 32w.$$

Completing the square, we get

$$A = -(w - 16)^2 + 256.$$

The maximum function value is 256.

4. ⟩ **Check** ⟩ The maximum value of 256 occurs when $w = 16$. We can check a function value for x less than 16 and for x greater than 16. That is not a 100% check, but it is fairly close:

$$A(15) = -(15)^2 + 32 \cdot 15 = 255,$$
$$A(17) = -(17)^2 + 32 \cdot 17 = 255.$$

Since 256 is greater than these numbers, it looks as though we have a maximum.

5. ⟩ **State** ⟩ The answer is that the dimensions of the largest pen that can be enclosed are 16 m by 16 m (the pen will be square).

DO EXERCISES 1 AND 2.

ii Fitting Quadratic Functions to Data

In many problem situations, we can make the translation by finding a quadratic function that describes how the variables behave. If we know that a quadratic function is appropriate in a certain situation, then we can proceed to find that function. If we know three inputs and their outputs, we can do so. Each such ordered pair is called a *data point*.

The following example provides a good illustration.

EXAMPLE 2 The instruction booklet for a video cassette recorder (VCR) includes a table relating the counter reading to the time the tape has run.

Time tape has run (hr)	Counter reading
0	000
1	300
2	500
3	675
4	800

Find a quadratic function that fits the data. Then find the counter reading after the tape has run for $1\frac{1}{2}$ hr.

We look for a function

$$N = at^2 + bt + c,$$

where N is the number of revolutions shown on the counter and t is the time in hours. We substitute some values of t and N:

$$0 = a(0)^2 + b(0) + c,$$
$$300 = a(1)^2 + b(1) + c,$$
$$500 = a(2)^2 + b(2) + c.$$

That gives us a system of three equations in the unknowns a, b, and c. When we solve that system, we find that

$$a = -50, \quad b = 350, \quad \text{and} \quad c = 0.$$

Therefore, the function we are looking for is

$$N = -50t^2 + 350t.$$

To find the counter reading after $1\frac{1}{2}$ hr, we substitute:

$$N = -50(1.5)^2 + 350(1.5)$$
$$= 412.5.$$

Thus after $1\frac{1}{2}$ hr the counter should read about 412.

DO EXERCISES 3 AND 4.

3. Find the quadratic function that fits the data points $(1, 0)$, $(-1, 4)$, and $(2, 1)$.

4. The following table shows the accident records in a city. It has values that a quadratic function will fit.

Age of driver	Number of accidents (in a year)
20	400
40	150
60	400

a) Assuming that a quadratic function will describe the situation, find the number of accidents as a function of age.

b) Use the function to calculate the number of accidents in which a typical 16-year-old is involved.

EXERCISE SET 11.12

i Solve.

1. A rancher is fencing off a rectangular field with a fixed perimeter of 76 ft. What dimensions will yield the maximum area? What is the maximum area?

2. A carpenter is building a rectangular room with a fixed perimeter of 68 ft. What dimensions will yield the maximum area? What is the maximum area?

3. What is the maximum product of two numbers whose sum is 22? What numbers yield this product?

4. What is the maximum product of two numbers whose sum is 45? What numbers yield this product?

5. What is the minimum product of two numbers whose difference is 4? What are the numbers?

6. What is the minimum product of two numbers whose difference is 6? What are the numbers?

7. What is the minimum product of two numbers whose difference is 5? What are the numbers?

8. What is the minimum product of two numbers whose difference is 7? What are the numbers?

ii Find the quadratic function that fits each set of data points.

9. $(1, 4)$, $(-1, -2)$, $(2, 13)$

10. $(1, 4)$, $(-1, 6)$, $(-2, 16)$

11. $(1, 5)$, $(2, 9)$, $(3, 7)$

12. $(1, -4)$, $(2, -6)$, $(3, -6)$

13. (*Predicting earnings*). A business earns $38 in the first week, $66 in the second week, and $86 in the third week. The manager graphs the points $(1, 38)$, $(2, 66)$, and $(3, 86)$ and uses a quadratic function to describe the situation.

a) Find a quadratic function that fits the data.
b) Using the function, predict the earnings for the fourth week.

14. (*Predicting earnings*). A business earns $1000 in its first month, $2000 in the second month, and $8000 in the third month. The manager plots the points $(1, 1000)$, $(2, 2000)$, and $(3, 8000)$ and uses a quadratic function to describe the situation.

a) Find a quadratic function that fits the data.
b) Using the function, predict the earnings for the fourth month.

15. a) Find a quadratic function that fits the following data.

Travel speed in km/h	Number of daytime accidents (for every 200 million km)
60	100
80	130
100	200

16. a) Find a quadratic function that fits the following data.

Travel speed in km/h	Number of nighttime accidents (for every 200 million km)
60	250
80	250
100	400

b) Use the function to calculate the number of nighttime accidents that occur at 50 km/h.

b) Use the function to calculate the number of daytime accidents that occur at 50 km/h.

Find the maximum or minimum value for each function.

17. $f(x) = 2.31x^2 - 3.105x - 5.98$

18. $f(x) = -17.7x^2 + 6.29x + 6.08$

Find the quadratic function that fits each set of data points.

19. $(20.34, -5.86)$, $(34.67, -6.02)$, $(28.55, -8.46)$

20. $(0.789, 245)$, $(0.988, 350)$, $(1.233, 404)$

21. The sum of the base and the height of a triangle is 38 cm. Find the dimensions for which the area is a maximum, and find the maximum area.

22. The perimeter of a rectangle is 44 ft. Find the least possible length of a diagonal.

23. A horticulturist has 180 ft of fencing with which to form a rectangular garden. A greenhouse will provide one side of the garden, and the fencing will be used for the other three sides. What is the area of the largest region that can be enclosed?

24. (*Maximizing yield*). An orange grower finds that she gets an average yield of 40 bu per tree when she plants 20 trees on an acre of ground. Each time she adds a tree to an acre the yield per tree decreases by 1 bu, due to congestion. How many trees per acre should she plant for maximum yield?

25. (*Maximizing revenue*). When a theater owner charges $2 for admission he averages 100 people attending. For each 10¢ increase in admission price the average number attending decreases by 1. What should he charge in order to make the most money?

26. A farmer wants to build a rectangular fence near a river, and will use 120 ft of fencing. What is the area of the largest region that can be enclosed? (The side next to the river is not fenced.)

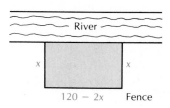

SUMMARY AND REVIEW: CHAPTER 11

The following contains a summary of what you should be able to do after completing this chapter. The review exercises are for practice. Answers are at the back of the book. If you miss an exercise, restudy the section indicated alongside the answer.

You should be able to:

Solve a quadratic equation by factoring or by the quadratic formula.

Solve.

1. $8x^2 = 24$

2. $5x^2 - 8x + 3 = 0$

3. $x^2 - 2x - 10 = 0$

4. $3y^2 + 5y = 2$

5. $(x + 8)^2 = 13$

6. $9x^2 = 0$

7. $5t^2 - 7t = 0$

8. $9x^2 - 6x - 9 = 0$

9. $x^2 + 6x = 9$

10. $1 + 4x^2 = 8x$

11. $6 + 3y = y^2$

12. $3m = 4 + 5m^2$

13. $3x^2 = 4x$

14. $40 = 5y^2$

Solve a quadratic equation by completing the square.

Solve by completing the square. Show your work.

15. $3x^2 - 2x - 5 = 0$

16. $x^2 - 5x + 2 = 0$

Use a calculator or Table 2 to approximate the solutions of a quadratic equation.

Approximate the solutions to the nearest tenth.

17. $x^2 - 5x + 2 = 0$

18. $4y^2 + 8y + 1 = 0$

Without graphing, tell whether the graph of an equation of the type $y = ax^2 + bx + c$ opens upward or downward. Then graph the equation. Approximate the solutions of $0 = ax^2 + bx + c$ by graphing.

Without graphing, tell whether the graph of the equation opens upward or downward. Then graph the equation.

19. $y = 2 - x^2$

20. $y = x^2 - 4x - 2$

21. Approximate the solutions of $0 = x^2 - 4x - 2$ by graphing.

Solve problems using the five-step process and the solution of quadratic equations.

22. The hypotenuse of a right triangle is 5 m long. One leg is 3 m longer than the other. Find the lengths of the legs. Round to the nearest tenth.

23. $1000 is invested at interest rate r, compounded annually. In 2 years it grows to $1690. What is the interest rate?

24. The length of a rectangle is 3 m greater than the width. The area is 70 m². Find the length and width.

Solve a formula for a given letter. Solve problems using formulas.

25. Solve $N = at^2 - bt$ for t.

26. Solve $N = 3\pi \sqrt{\dfrac{1}{p}}$ for p.

27. Solve $2A = \dfrac{3B}{T^2}$ for T.

28. a) The area of a right circular cone is given by $A = \frac{1}{3}\pi r^2 h$. Solve for r.
  b) The area of a right circular cone with height 8 cm is 45 cm². Find the radius of the cone.

Solve equations that are reducible to quadratic.

29. $x^4 - 13x^2 + 36 = 0$

30. $15x^{-2} - 2x^{-1} - 1 = 0$

31. $(x^2 - 4)^2 - (x^2 - 4) - 6 = 0$

32. $x - 4\sqrt{x} - 5 = 0$

Solve problems involving variation.

33. The power P expended by heat in an electric circuit of fixed resistance varies directly as the square of the current C in the circuit. A circuit expends 180 watts when a current of 6 amperes is flowing. What is the heat expended when the current is 10 amperes?

34. A warning dye is used by people in lifeboats to aid searching airplanes. The radius r of the circle formed by the dye varies directly as the square root of the volume V. It is found that 4 L of dye will spread to a circle of radius 5 m. To what radius will 9 L spread?

Graph functions like $f(x) = a(x - h)^2 + k$ and determine the vertex, the line of symmetry, and the maximum or minimum function values. Given a quadratic function $f(x) = ax^2 + bx + c$, complete the square to put it in the form $f(x) = a(x - h)^2 + k$. Find the x-intercepts of a quadratic equation.

35. a) Graph $f(x) = -3(x + 2)^2 + 4$.
  b) Find the vertex.
  c) Find the line of symmetry.
  d) Find the maximum or minimum function value.

36. Complete the square: $f(x) = 2x^2 - 12x + 23$.

37. Find the x-intercepts: $f(x) = x^2 - 9x + 14$.

Solve maximum and minimum problems. Solve problems by fitting a quadratic function to three data points.

38. What is the minimum product of two numbers whose difference is 22? What numbers yield this product?

39. Find the quadratic function that fits these data points.
$$(0, -2), \quad (1, 3), \quad (3, 7)$$

40. Two consecutive integers have squares that differ by 63. Find the integers.

41. Find b such that $x^2 + bx + 49$ is a square.

42. A square with sides of length s has the same area as a circle with radius 5 in. Find s.

TEST: CHAPTER 11

Solve.

1. $7x^2 = 35$

2. $7x^2 + 8x = 0$

3. $48 = t^2 + 2t$

4. $3y^2 + 5y = 2$

5. $(x + 8)^2 = 13$

6. $x^2 = x + 3$

7. $m^2 - 3m = 7$

8. $10 = 4x + x^2$

9. $3x^2 - 7x + 1 = 0$

10. Solve $x^2 - 4x - 10 = 0$ by completing the square. Show your work.

11. Approximate the solutions of $x^2 - 4x - 10 = 0$ to the nearest tenth.

12. Solve for n: $d = an^2 + bn$.

13. a) Without graphing, tell whether the graph of the equation
$$y = -x^2 + x + 5$$
opens upward or downward.
  b) Graph the equation. Use any method you studied.

14. Approximate the solutions of $0 = -x^2 + x + 5$ by graphing.

Solve.

15. $x^4 - 8x^2 + 15 = 0$

16. $(x^2 - 1)^2 - 4(x^2 - 1) - 5 = 0$

Solve.

17. 4000 is invested at interest rate r, compounded annually. In 2 years it grows to $6250. What is the interest rate?

18. The width of a rectangle is 4 m less than the length. The area is 16.25 m². Find the length and the width.

19. A ladder 25 ft long leans against a wall. The bottom of the ladder is 7 ft from the wall. How much would the lower end have to be pulled away so that the top end would be pulled down the same amount?

20. The area of a balloon varies directly as the square of its radius. The area is 3.4 cm^2 when the radius is 5 cm. What is the area when the radius is 7 cm?

21. a) Graph $f(x) = 4(x - 3)^2 + 5$.
    b) Find the vertex.
    c) Find the line of symmetry.
    d) Find the maximum or minimum function value.

22. Complete the square: $f(x) = 2x^2 + 4x - 6$. Show your work.

23. Find the x-intercepts: $f(x) = x^2 - x - 6$.

24. What is the minimum product of two numbers having a difference of 8?

25. Find the quadratic function that fits these data points.

$$(0, 0), \quad (3, 0), \quad (5, 2)$$

26. Find the side of a square whose diagonal is 5 ft longer than a side.

27. Solve this system for x. Use the substitution method.

$$x - y = 2,$$
$$xy = 4$$

APPENDIX A

SYSTEMS OF EQUATIONS IN THREE VARIABLES

Some problem situations naturally give rise to a translation to two equations. Others more naturally give rise to a translation to three equations, or more. In this appendix we consider how to solve systems of three linear equations and we will use such systems to solve problems.

i Identifying Solutions

A solution of a system of three equations in three variables is an ordered triple that makes *all three* equations true.

EXAMPLE 1 Determine whether $(\frac{3}{2}, -4, 3)$ is a solution of the following system.

$$4x - 2y - 3z = 5$$
$$-8x - y + z = -5$$
$$2x + y + 2z = 5$$

We substitute $(\frac{3}{2}, -4, 3)$ into the three equations, using alphabetical order.

$$
\begin{array}{c|c}
4x - 2y - 3z = 5 & \\
\hline
4 \cdot \frac{3}{2} - 2(-4) - 3 \cdot 3 & 5 \\
6 + 8 - 9 & \\
5 &
\end{array}
\qquad
\begin{array}{c|c}
-8x - y + z = -5 & \\
\hline
-8 \cdot \frac{3}{2} - (-4) + 3 & -5 \\
-12 + 4 + 3 & \\
-5 &
\end{array}
$$

$$
\begin{array}{c|c}
2x + y + 2z = 5 & \\
\hline
2 \cdot \frac{3}{2} + (-4) + 2 \cdot 3 & 5 \\
3 - 4 + 6 & \\
5 &
\end{array}
$$

The triple makes all three equations true, so it is a solution.

DO EXERCISE 1.

OBJECTIVES

After finishing Appendix A, you should be able to:

i Determine whether an ordered triple is a solution of a system of three equations in three variables.

ii Solve systems of three equations in three variables.

iii Solve problems by translating to a system of three equations in three variables.

---

1. Consider the system

$$4x + 2y + 5z = 6$$
$$2x - y + z = 5$$
$$x + 2y - z = 0.$$

a) Determine whether $(1, 2, 3)$ is a solution.

b) Determine whether $(2, -1, 0)$ is a solution.

ⅱ Solving

Graphical methods for solving linear equations in three variables are unsatisfactory, because a three-dimensional coordinate system is required. The substitution method becomes cumbersome for most systems of more than two equations. Therefore, we will use the addition method. It is essentially the same for systems of three equations as for systems of two equations.

EXAMPLE 2   Solve the following system of equations.

$$\begin{array}{ll} x + y + z = 4 & \text{①} \\ x - 2y - z = 1 & \text{②} \\ 2x - y - 2z = -1 & \text{③} \end{array}$$

These numbers indicate the equations in the first, second, and third positions, respectively.

We begin by multiplying ① by -1, and adding it to ② to eliminate x from the second equation. Equations ① and ③ are unchanged by this step.

$$\begin{array}{ll} x + y + z = 4, & \text{①} \\ -3y - 2z = -3, & \text{②} \\ 2x - y - 2z = -1. & \text{③} \end{array}$$

$$\begin{array}{ll} -x - y - z = -4 & \text{(Multiplying ① by } -1) \\ \underline{x - 2y - z = 1} & \text{②} \\ -3y - 2z = -3 & \text{(Adding)} \end{array}$$

To eliminate x from the third equation we multiply ① by -2 and add it to ③:

$$\begin{array}{ll} x + y + z = 4, & \text{①} \\ -3y - 2z = -3, & \text{②} \\ -3y - 4z = -9. & \text{③} \end{array}$$

$$\begin{array}{ll} -2x - 2y - 2z = -8 & \text{(Multiplying ① by } -2) \\ \underline{2x - y - 2z = -1} & \text{③} \\ -3y - 4z = -9 & \text{(Adding)} \end{array}$$

Next we eliminate y from the third equation by multiplying ② by -1 and adding the result to ③:

$$\begin{array}{ll} x + y + z = 4 & \text{①} \\ -3y - 2z = -3 & \text{②} \\ -2z = -6 & \text{③} \end{array}$$

$$\begin{array}{ll} 3y + 2z = 3 & \text{(Multiplying ② by } -1) \\ \underline{-3y - 4z = -9} & \text{③} \\ -2z = -6 & \text{(Adding)} \end{array}$$

When the system is in this triangular form, we can easily solve for the three variables.
First we solve ③ for z:

$$-2z = -6$$
$$z = 3. \qquad \text{(Multiplying ③ by } -\tfrac{1}{2})$$

Next we substitute 3 for z in ② and solve for y:

$$-3y - 2(3) = -3$$
$$-3y - 6 = -3$$
$$-3y = 3$$
$$y = -1.$$

Finally, we substitute -1 for y and 3 for z in ①:

$$x + (-1) + 3 = 4$$
$$x + 2 = 4$$
$$x = 2.$$

We have obtained the triple $(2, -1, 3)$. We check, in *all three* equations:

$$\frac{x + y + z = 4}{2 + (-1) + 3 \;\big|\; 4}$$
$$4 \;\big|$$

$$\frac{x - 2y - z = 1}{2 - 2(-1) - 3 \;\big|\; 1}$$
$$1 \;\big|$$

$$\frac{2x - y - 2z = -1}{2 \cdot 2 - (-1) - 2 \cdot 3 \;\big|\; -1}$$
$$-1 \;\big|$$

The solution is $(2, -1, 3)$.

In Example 2, we obtained a system of equations equivalent to the original, but in a *triangular form*. After that, finding the solution was not difficult. The following is a description of the procedure. It should be followed for systems of three or more variables, except in very simple cases.

TO SOLVE A SYSTEM OF THREE LINEAR EQUATIONS

1. **Make each x-coefficient a multiple of the first. Do so by multiplying or interchanging equations, or both.**
2. **Eliminate the x-terms from the second and third equations.**
3. **Make the y-coefficient of the third equation a multiple of the second. Do so by multiplying or interchanging equations.**
4. **Eliminate the y-term from the third equation.**
5. **Solve the third equation for z. Then substitute upward to find y and x.**

DO EXERCISE 2.

EXAMPLE 3 Solve the following system of equations.

$$
\begin{array}{ll}
2x - 4y + 6z = 22 & \text{①} \\
4x + 2y - 3z = 4 & \text{②} \\
3x + 3y - z = 4 & \text{③}
\end{array}
$$

We begin by multiplying ③ by 2, to make each x-coefficient a multiple of the first:

$$
\begin{array}{l}
2x - 4y + 6z = 22, \\
4x + 2y - 3z = 4, \\
6x + 6y - 2z = 8.
\end{array}
$$

Next, we multiply ① by -2 and add it to ②. We also multiply ① by -3 and add it to ③:

$$
\begin{array}{l}
2x - 4y + 6z = 22 \\
10y - 15z = -40 \\
18y - 20z = -58.
\end{array}
$$

Now we multiply ③ by -5 to make the y-coefficient a multiple of the y-coefficient in ②:

$$
\begin{array}{l}
2x - 4y + 6z = 22 \\
10y - 15z = -40 \\
-90y + 100z = 290.
\end{array}
$$

2. Solve. Don't forget to check.

$$
\begin{array}{rcrcrcr}
4x & - & y & + & z & = & 6 \\
-3x & + & 2y & - & z & = & -3 \\
2x & + & y & + & 2z & = & 3
\end{array}
$$

3. Solve. Don't forget to check.

$$2x + y - 4z = 0$$
$$x - y + 2z = 5$$
$$3x + 2y + 2z = 3$$

Next, we multiply ② by 9 and add it to ③:

$$2x - 4y + 6z = 22$$
$$10y - 15z = -40$$
$$-35z = -70.$$

Now we solve ③ for z:

$$-35z = -70$$
$$z = 2.$$

Next we substitute 2 for z in ②, and solve for y:

$$10y - 15(2) = -40$$
$$10y - 30 = -40$$
$$y = -1.$$

Finally, we substitute -1 for y and 2 for z in ①:

$$2x - 4(-1) + 6(2) = 22$$
$$2x + 4 + 12 = 22$$
$$x = 3.$$

We have obtained the triple $(3, -1, 2)$. It checks, but the check is left to the reader.

DO EXERCISE 3.

EXAMPLE 4   Solve:

$$x + y + z = 180 \quad ①$$
$$x \quad - z = -70 \quad ②$$
$$2y - z = 0. \quad ③$$

Note that there is no y in equation (2).

We use equations ① and ③ and eliminate y.

$$x + y + z = 180 \quad ①$$
$$2y - z = 0 \quad ③$$
$$-2x - 2y - 2z = -360 \quad \text{Multiplying by } -2$$
$$\underline{2y - z = 0}$$
$$-2x \quad - 3z = -360 \quad ④$$

4. Solve. Don't forget to check.

$$x + y + z = 100$$
$$x - y \quad = -10$$
$$x \quad - z = -30$$

Now we solve the resulting system of equations ② and ④.

$$x - z = -70 \quad ②$$
$$-2x - 3z = -360 \quad ④$$
$$2x - 2z = -140$$
$$\underline{-2x - 3z = -360}$$
$$-5z = -500$$
$$z = 100$$

Continuing as we did in the previous examples, we get the solution $(30, 50, 100)$.

DO EXERCISE 4.

iii Solving Problems with Three Equations

Solving systems of three or more equations is important in many applications. Systems of equations arise very often in the use of statistics, for example, in such fields as the social sciences. They also come up in problems of business, science, and engineering.

EXAMPLE 5   The sum of three numbers is 4. The first number minus twice the second minus the third is 1. Twice the first number minus the second minus twice the third is −1. Find the numbers.

1. **Familiarize**   In this case there are three obvious statements in the problem. The translation looks as though it can be made directly from these statements, as soon as we decide what letters to assign to the unknown numbers.

2. **Translate**   Let us call the three numbers x, y, and z. Then we can translate directly, from the words of the problem, as follows.

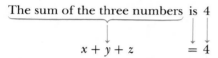

The sum of the three numbers   is   4

$$x + y + z = 4$$

The first number minus twice the second minus the third is 1.

$$x - 2y - z = 1$$

Twice the first number minus the second minus twice the third is −1.

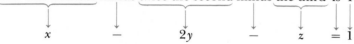

$$2x - y - 2z = -1$$

We now have a system of three equations:

$$x + y + z = 4$$
$$x - 2y - z = 1$$
$$2x - y - 2z = -1.$$

3. **Carry out**   We need to solve the system of equations. Note that we have already solved the system, as Example 2. We obtained $(2, -1, 3)$ for the solution.

4. **Check**   We go to the original problem. The first statement says that the sum of the three numbers is 4. That checks.

   The second statement says that the first number minus twice the second minus the third is 1. We calculate:

$$x - 2y - z = 1$$
$$2 - 2(-1) - 3 = 1$$
$$2 + 2 - 3 = 1$$
$$1 = 1.$$

That checks.

   We leave the check of the third statement to the reader.

5. **State**   The answer is that the three numbers are 2, −1, and 3.

DO EXERCISE 5.

5. The sum of three numbers is 3. The first number minus twice the second minus the third is 4. Twice the first number plus the other two is 5. Find the numbers.

6. The first angle of a triangle is twice as large as a second angle. The remaining angle is 20° greater than the first angle. Find the measures of the angles.

EXAMPLE 6   In a triangle the largest angle is 70° greater than the smallest angle. The largest angle is twice as large as the remaining angle. Find the measure of each angle.

1. ⟩Familiarize⟩   The first thing to do in a case like this is to make a drawing. Since we don't know the size of any angle, we have used x, y, and z for the measures of the angles. Here we use x for the smallest angle, z for the largest angle, and y for the remaining angle.

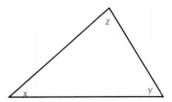

A geometric fact will be needed here, the fact that the measures of the angles of a triangle add up to 180°.

2. ⟩Translate⟩   This geometric fact about triangles gives us one equation:

$$x + y + z = 180.$$

There are two statements in the problem that we can translate almost directly.

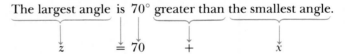

The largest angle   is   70° greater than the smallest angle.

$$z \qquad = \qquad 70 \qquad + \qquad x$$

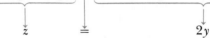

The largest angle   is   twice as large as the remaining angle.

$$z \qquad = \qquad 2y$$

We now have a system of three equations:

$$
\begin{aligned}
x + y + z &= 180 & \qquad x + y + z &= 180 \\
x + 70 &= z & \text{or} \qquad x \qquad - z &= -70 \\
2y &= z; & 2y - z &= 0.
\end{aligned}
$$

3. ⟩Carry out⟩   We solved this system in Example 4. The solution is $(30, 50, 100)$.

4. ⟩Check⟩   The sum of the numbers is 180, so that checks.
   The largest angle measures 100° and the smallest measures 30°. The largest angle is 70° greater than the smallest.
   The remaining angle measures 50°. The largest angle measures 100°, so it is twice as large. We do have an answer to the problem.

5. ⟩State⟩   The measures of the angles of the triangle are 30°, 50°, and 100°.

DO EXERCISE 6.

EXAMPLE 7   In a factory there are three machines, A, B, and C. When all three are running, they produce 222 suitcases per day. If A and B work, but C does not, they produce 159 suitcases per day. If B and C work, but A does not, they produce 147 suitcase per day. What is the daily production of each machine?

1. **Familiarize** Keeping information in a table will be of great help in this problem.

Machines working	A	B	C	A and B	B and C	A, B, and C
Daily production	x	y	z	159	147	222

2. **Translate** From the table, it is apparent that we have decided to use x, y, and z for the daily productions of the individual machines. Also from the table, we can readily obtain three equations:

$$x + y + z = 222 \qquad \text{(All three machines work)}$$
$$x + y \phantom{{}+z} = 159 \qquad \text{(A and B work)}$$
$$\phantom{x + {}} y + z = 147. \qquad \text{(B and C work)}$$

3. **Carry out** We solve the system of equations. The details are left to the reader, but the solution is $(75, 84, 63)$.

4. **Check** The numbers in the triple are supposed to be the daily numbers of suitcases produced. We add the three numbers, and obtain 222. That checks, since the problem states that the total number produced by all three machines is 222.

When A and B are working, the numbers are 75 and 84. They add up to 159. That checks.

When B and C are working, the numbers are 84 and 63. They add up to 147. That checks.

5. **State** The answer is that machine A produces 75 suitcases per day, machine B produces 84 suitcases per day, and machine C produces 63 suitcases per day.

DO EXERCISE 7.

7. There are three machines, A, B, and C, in a factory. When all three work, they produce 287 screws per hour. When only A and C work, they produce 197 screws per hour. When only A and B work, they produce 202 screws per hour. How many screws per hour can each produce alone?

EXERCISES: APPENDIX A

i

1. Determine whether $(30, 50, 100)$ is a solution of the system.

$$x + y + z = 180$$
$$x - z = -70$$
$$2y - z = 0$$

2. Determine whether $(2, -1, -2)$ is a solution of the system.

$$x + y - 2z = 5$$
$$2x - y - z = 7$$
$$-x - 2y + 3z = 6$$

ii Solve.

3. $x + y + z = 6$
$2x - y + 3z = 9$
$-x + 2y + 2z = 9$

4. $2x - y + z = 10$
$4x + 2y - 3z = 10$
$x - 3y + 2z = 8$

5. $2x - y - 3z = -1$
$2x - y + z = -9$
$x + 2y - 4z = 17$

6. $x - y + z = 6$
$2x + 3y + 2z = 2$
$3x + 5y + 4z = 4$

7. $2x - 3y + z = 5$
$x + 3y + 8z = 22$
$3x - y + 2z = 12$

8. $6x - 4y + 5z = 31$
$5x + 2y + 2z = 13$
$x + y + z = 2$

9. $3a - 2b + 7c = 13$
$a + 8b - 6c = -47$
$7a - 9b - 9c = -3$

10. $x + y + z = 0$
$2x + 3y + 2z = -3$
$-x + 2y - 3z = -1$

11. $2x + 3y + z = 17$
$x - 3y + 2z = -8$
$5x - 2y + 3z = 5$

12. $2x + y - 3z = -4$
$4x - 2y + z = 9$
$3x + 5y - 2z = 5$

13. $2x + y + z = -2$
$2x - y + 3z = 6$
$3x - 5y + 4z = 7$

14. $2x + y + 2z = 11$
$3x + 2y + 2z = 8$
$x + 4y + 3z = 0$

15. $x - y + z = 4$
$5x + 2y - 3z = 2$
$3x - 7y + 4z = 8$

16. $2x + y + 2z = 3$
$x + 6y + 3z = 4$
$3x - 2y + z = 0$

17. $4x - y - z = 4$
$2x + y + z = -1$
$6x - 3y - 2z = 3$

18. $a + 2b + c = 1$
$7a + 3b - c = -2$
$a + 5b + 3c = 2$

19. $2r + 3s + 12t = 4$
$4r - 6s + 6t = 1$
$r + s + t = 1$

20. $10x + 6y + z = 7$
$5x - 9y - 2z = 3$
$15x - 12y + 2z = -5$

21. $4a + 9b = 8$
$8a + 6c = -1$
$6b + 6c = -1$

22. $3p + 2r = 11$
$q - 7r = 4$
$p - 6q = 1$

iii Solve.

23. The sum of three numbers is 105. The third is 11 less than 10 times the second. Twice the first is 7 more than 3 times the second. Find the numbers.

24. The sum of three numbers is 57. The second is 3 more than the first. The third is 6 more than the first. Find the numbers.

25. The sum of three numbers is 5. The first number minus the second plus the third is 1. The first minus the third is 3 more than the second. Find the numbers.

26. The sum of three numbers is 26. Twice the first minus the second is 2 less than the third. The third is the second minus three times the first. Find the numbers.

27. In triangle ABC, the measure of angle B is 2° more than three times the measure of angle A. The measure of angle C is 8° more than the measure of angle A. Find the angle measures.

28. In triangle ABC, the measure of angle B is three times the measure of angle A. The measure of angle C is 30° greater than the measure of angle A. Find the angle measures.

29. In triangle ABC, the measure of angle B is twice the measure of angle A. The measure of angle C is 80° more than that of angle A. Find the angle measures.

30. In triangle ABC, the measure of angle B is three times that of angle A. The measure of angle C is 20° more than that of angle A. Find the angle measures.

31. Gina sells magazines part time. On Thursday, Friday, and Saturday, she sold $66 worth. On Thursday she sold $3 more than on Friday. On Saturday she sold $6 more than on Thursday. How much did she take in each day?

32. Pat picked strawberries on three days. He picked a total of 87 quarts. On Tuesday he picked 15 quarts more than on Monday. On Wednesday he picked 3 quarts fewer than on Tuesday. How many quarts did he pick each day?

33. Linda has a total of 225 on three tests. The sum of the scores on the first and second tests exceeds her third score by 61. Her first score exceeds her second by 6. Find the three scores.

34. Fred, Jane, and Mary made a total bowling score of 575 Fred's score was 15 more than Jane's. Mary's was 20 more than Jane's. Find the scores.

Solve.

35. $\dfrac{x+2}{3} - \dfrac{y+4}{2} + \dfrac{z+1}{6} = 0$

$\dfrac{x-4}{3} + \dfrac{y+1}{4} - \dfrac{z-2}{2} = -1$

$\dfrac{x+1}{2} + \dfrac{y}{2} + \dfrac{z-1}{4} = \dfrac{3}{4}$

36. $0.2x + 0.3y + 1.1z = 1.6$
$0.5x - 0.2y + 0.4z = 0.7$
$-1.2x + y - 0.7z = -0.9$

37. $w + x + y + z = 2$
$w + 2x + 2y + 4z = 1$
$w - x + y + z = 6$
$w - 3x - y + z = 2$

38. $w + x - y + z = 0$
$w - 2x - 2y - z = -5$
$w - 3x - y + z = 4$
$2w - x - y + 3z = 7$

For Exercises 39 and 40, let u represent $1/x$, v represent $1/y$, and w represent $1/z$. Solve for u, v, and w, and then for x, y, and z.

39. $\dfrac{2}{x} - \dfrac{1}{y} - \dfrac{3}{z} = -1$

$\dfrac{2}{x} - \dfrac{1}{y} + \dfrac{1}{z} = -9$

$\dfrac{1}{x} + \dfrac{2}{y} - \dfrac{4}{z} = 17$

40. $\dfrac{2}{x} + \dfrac{2}{y} - \dfrac{3}{z} = 3$

$\dfrac{1}{x} - \dfrac{2}{y} - \dfrac{3}{z} = 9$

$\dfrac{7}{x} - \dfrac{2}{y} + \dfrac{9}{z} = -39$

41. Determine a, b, and c such that $(2, 3, -4)$ is a solution of the system.

$$ax + by + cz = -11$$
$$bx - cy + az = -19$$
$$ax + cy - bz = 9$$

In each case three solutions of an equation are given. Find the equation using a system of equations.

42. $Ax + By + Cz = 12$;
$(1, \frac{3}{4}, 3)$, $(\frac{4}{3}, 1, 2)$, and $(2, 1, 1)$

43. $z = b - mx - ny$;
$(1, 1, 2)$, $(3, 2, -6)$, and $(\frac{3}{2}, 1, 1)$

44. In a factory there are three polishing machines, A, B, and C. When all three of them are working, 5700 lenses can be polished in one week. When only A and B are working, 3400 lenses can be polished in one week. When only B and C are working, 4200 lenses can be polished in one week. How many lenses can be polished in a week by each machine?

46. When three pumps, A, B, and C, are running together, they can pump 3700 gallons per hour. When only A and B are running, 2200 gallons per hour can be pumped. When only A and C are running, 2400 gallons per hour can be pumped. What is the pumping capacity of each pump?

48. Tammy's age is the sum of the ages of Carmen and Dennis. Carmen's age is 2 more than the sum of the ages of Dennis and Mark. Dennis's age is four times Mark's age. The sum of all four age is 42. How old is Tammy?

50. Find the sum of the angle measures at the tips of the star in this figure.

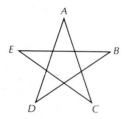

52. Find the year in which the first U.S. transcontinental railroad was completed. The following are some facts about the number. The sum of the digits in the year is 24. The one's digit is one more than the hundred's digit. Both the ten's and one's digits are multiples of three.

54. At a county fair, adults' tickets sold for $5.50, senior citizens' tickets sold for $4.00, and children's tickets sold for $1.50. On the opening day the number of children's and senior citizens' tickets sold was 30 more than half the number of adults' tickets sold. The number of senior citizens' tickets sold was 5 more than 4 times the number of children's tickets. How many of each type of ticket were sold if the total receipts from the ticket sales were $14,970?

45. Sawmills A, B, and C can produce 7400 board-feet of lumber per day. Mills A and B together can produce 4700 board-feet, while mills B and C together can produce 5200 board-feet. How many board-feet can each mill produce by itself?

47. Three welders, A, B, and C, can weld 37 linear feet per hour when working together. Welders A and B together can weld 22 linear feet per hour, while A and C together can weld 25 linear feet per hour. How many linear feet per hour can each weld alone?

49. Find a three-digit positive integer such that the sum of all three digits is 14, the ten's digit is 2 more than the unit's digit, and if the digits are reversed the number is unchanged.

51. A theater audience of 100 people consists of men, women, and children. The ticket prices are $10 for men, $3 for women, and 50¢ for children. The total amount of money taken in is $100. How many men, women, and children are in attendance? Does there seem to be some information missing? Do some more careful reasoning.

53. Hal gives Tom as many raffle tickets as Tom has and Gary as many as Gary has. In like manner, Tom then gives Hal and Gary as many tickets as each then has. Similarly, Gary gives Hal and Tom as many tickets as each then has. If each finally has 40 tickets, with how many tickets does Tom begin?

55. The graph of $y = ax^2 + bx + c$ contains the three points $(4, 2)$, $(2, 0)$, and $(1, 2)$. Find a, b, and c.

APPENDIX B

INCONSISTENT AND DEPENDENT EQUATIONS

Some systems of equations have no solution. Others may have more than one solution. In this appendix, we consider such systems.

i Inconsistent Equations

If a system of equations has a solution, we say that it is *consistent*. If a system does not have a solution, we say that it is *inconsistent*.

OBJECTIVES

After finishing Appendix B, you should be able to:

i   Determine whether a system of two or three linear equations is consistent or inconsistent.

ii   Determine whether a system of two or three linear equations is dependent or independent.

1. Determine whether these systems are consistent or inconsistent.

   a) $3x - y = 2$
   $6x - 2y = 3$

   b) $x + 4y = 2$
   $2x - y = 1$

EXAMPLE 1   Determine whether this system is consistent or inconsistent:

$$x - 3y = 1$$
$$-2x + 6y = 5.$$

We attempt to find a solution:

$$2x - 6y = 2 \qquad \text{Multiplying the first equation by 2}$$
$$\underline{-2x + 6y = 5}$$
$$0 = 7 \qquad \text{Adding}$$

The last equation says that $0 \cdot x + 0 \cdot y = 7$. There are no numbers x and y for which this is true, so there is no solution. The system is inconsistent.

Whenever we obtain a statement such as $0 = 7$, which is clearly false, we know that the system we are trying to solve is inconsistent.

DO EXERCISE 1.

EXAMPLE 2   Determine whether this system is consistent or inconsistent:

$$y + 3z = 4, \qquad (1)$$
$$-x - y + 2z = 0, \qquad (2)$$
$$x + 2y + z = 1. \qquad (3)$$

We attempt to find a solution. If we add equations (2) and (3), we get

$$y + 3z = 1.$$

If we multiply this by -1 and then add the result to equation (1), we get

$$0 = 3.$$

The system is inconsistent.

DO EXERCISE 2.

2. Determine whether these systems are consistent or inconsistent.

   a) $x + 2y + z = 1$
   $3x + 3y + z = 2$
   $2x + y = 2$

   b) $x + z = 1$
   $y + z = 1$
   $x + y = 1$

ⅱ   Dependent Equations

Consider this system:

$$2x + 3y = 1$$
$$4x + 6y = 2.$$

If we multiply the first equation by 2, we get the second equation, so the two equations have the same solutions (are equivalent). The system of two equations is equivalent to only one of them. We call them *dependent*.

> If a system of n linear equations is equivalent to a system of fewer than n of them, we say that the system is *dependent*. If such is not the case, we call the system *independent*.

EXAMPLE 3   Determine whether this system is dependent or independent:

$$2x + 3y = 1$$
$$4x + 6y = 2.$$

We attempt to solve:

$$-4x - 6y = -2 \quad \text{Multiplying the first equation by } -2$$
$$\underline{4x + 6y = 2}$$
$$0 = 0 \quad \text{Adding}$$

The last equation says that $0 \cdot x + 0 \cdot y = 0$. This is true for all numbers x and y. The system is dependent.

Whenever we get an obviously true sentence like this, we know that the system we are trying to solve is dependent.

DO EXERCISE 3.

EXAMPLE 4   Determine whether this system is dependent or independent:

$$x + 2y + z = 1$$
$$x - y + z = 1$$
$$2x + y + 2z = 2.$$

We attempt to solve:

$$x + 2y + z = 1$$
$$-3y = 0 \quad \text{Multiplying the first equation by } -1 \text{ and adding it to the second}$$
$$-3y = 0. \quad \text{Multiplying the first equation by } -2 \text{ and adding it to the third}$$

It is now obvious that our system of three equations is equivalent to a system of two equations. The system is dependent.

When we attempt to solve a system and obtain two equations that are identical, or obviously equivalent, we know that the system is dependent. Or, if we obtain a statement that is obviously true, such as $3 = 3$, we know that the system is dependent.

DO EXERCISE 4.

3. Determine whether these systems are dependent or independent.

a) $3x - 2y = 1$
   $-6x + 4y = -2$

b) $x - y = 2$
   $x + y = 4$

4. Determine whether these systems are dependent or independent.

a) $x + y + 2z = 1$
   $x - y + z = 1$
   $2x + 3z = 2$

b) $x + y + 2z = 1$
   $x - y + z = 1$
   $x + 2y + z = 2$

EXERCISES: APPENDIX B

i Determine whether these systems are consistent or inconsistent.

1. $x + 2y = 6$
   $2x = 8 - 4y$

2. $y - 2x = 1$
   $2x - 3 = y$

3. $y - x = 4$
   $x + 2y = 2$

4. $y + x = 5$
   $y = x - 3$

5. $x - 3 = y$
   $2x - 2y = 6$

6. $3y = x - 2$
   $3x = 6 + 9y$

7. $x + z = 0$
   $x + y + 2z = 3$
   $y + z = 2$

8. $x + y = 0$
   $x + z = 1$
   $2x + y + z = 2$

9. $x + z = 0$
   $x + y = 1$
   $y + z = 1$

10. $2x + y = 1$
   $x + 2y + z = 0$
   $x + z = 1$

ii Determine whether these systems are dependent or independent.

11. $x - 3 = y$
   $2x - 2y = 6$

12. $3y = x - 2$
   $3x = 6 + 9y$

13. $y - x = 4$
   $x + 2y = 2$

14. $y + x = 5$
   $y = x - 3$

15. $2x + 3y = 1$
   $x + 1.5y = 0.5$

16. $15x + 6y = 20$
   $7.5x - 10 = -3y$

17. $x + z = 0$
   $x + y = 1$
   $y + z = 1$

18. $2x + y = 1$
   $x + 2y + z = 0$
   $x + z = 1$

19. $x + y + z = 1$
   $-x + 2y + z = 2$
   $2x - y = -1$

20. $y + z = 1$
   $x + y + z = 1$
   $x + 2y + 2z = 2$

Problem-solving practice

21. The difference of two numbers is 4. The second number is 6 less than the first number. What are the numbers?

22. The graph of the equation $y = ax^2 + bx$ contains the points $(2, 6)$ and $(1, 3)$. Find a and b.

23. Two years ago, Roland was 5 times as old as Jennifer. Ten years from now, Roland's age will be 6 more than twice Jennifer's age. How old are they now?

24. The perimeter of a rectangle is 192. The width is 12 less than the length, and the sum of the length and 3 times the width is 140. Find the dimensions of the rectangle.

25. A hockey team won 15 more games than it tied, lost 10 more games than it won, and lost 7 more games than it tied. How many games did it win?

26. The sum of three numbers is 71. The first number is 12 more than twice the third number. The sum of the second number and 3 times the third number is 59. What are the numbers?

Solve. If a system has more than one solution, list three of them.

27. $9x - 3y = 15$
$6x - 2y = 10$

28. $2s - 3t = 9$
$4s - 6t = 9$

29. $x + 2y - z = -8$
$2x - y + z = 4$
$8x + y + z = 2$

30. $2x + y + z = 0$
$x + y - z = 0$
$x + 2y + 2z = 0$

31. $2x + 4y + 8z = 5$
$x + 2y + 4z = 13$
$4x + 8y + 16z = 10$

32. $x + y + z = 3$
$5x + 5y + 5z = 15$
$2x + 2y + 2z = 6$

33. Classify each of the systems in Exercises 27, 29, and 31 as consistent or inconsistent, dependent or independent.

34. Classify each of the systems in Exercises 28, 30, and 32 as consistent or inconsistent, dependent or independent.

Determine the constant k such that each system is dependent.

35. $6x - 9y = -3$
$-4x + 6y = k$

36. $8x - 16y = 20$
$10x - 20y = k$

Consider the following dependent systems. For each system, find an ordered pair in terms of y that describes the entire solution set of the system.

37. $2x + 3y = 1$
$4x + 6y = 2$

38. $-6x + 4y = 10$
$3x - 2y = -5$

TABLE 4 PHYSICAL CONSTANTS

SI PREFIXES

Prefix	Connotation
Pico-	$0.000\ 000\ 000\ 001 \times$ (or $10^{-12} \times$)
Nano-	$0.000\ 000\ 001 \times$ (or $10^{-9} \times$)
Micro-	$0.000\ 001 \times$ (or $10^{-6} \times$)
Milli-	$0.001 \times$ (or $10^{-3} \times$)
Centi-	$0.01 \times$ (or $10^{-2} \times$)
Deci-	$0.1 \times$ (or $10^{-1} \times$)
Deka-	$10 \times$ (or $10^{1} \times$)
Hecto-	$100 \times$ (or $10^{2} \times$)
Kilo-	$1\ 000 \times$ (or $10^{3} \times$)
Mega-	$1\ 000\ 000 \times$ (or $10^{6} \times$)
Giga-	$1\ 000\ 000\ 000 \times$ (or $10^{9} \times$)
Tera-	$1\ 000\ 000\ 000\ 000 \times$ (or $10^{12} \times$)

PHYSICAL AND CHEMICAL CONSTANTS

Speed of light in vacuum	c	3.00×10^{8} m/s
Charge on electron	e	1.60×10^{-19} C
Gravitational constant	G	6.67×10^{-11} N·m/kg²
Acceleration of gravity at earth's surface	g	9.81 m/s² 32.2 ft/s²
Planck's constant	h	6.63×10^{-34} J·s
Electrostatic constant	k	8.99×10^{9} N·m/C²
Electron rest mass	m_e	9.11×10^{-31} kg
Neutron rest mass	m_n	1.675×10^{-27} kg
Proton rest mass	m_p	1.673×10^{-27} kg
Avogadro's number	N_o	6.02×10^{23} formula units/mole
Boltzmann's constant	k	1.381×10^{-23} J·K^{-1}
Gas constant	R	8.314 J·mol^{-1}·K^{-1}
Permittivity of free space	ϵ_0	8.854×10^{-12} C²·N^{-1}·m^{-2}
	$\frac{1}{4}\pi\epsilon_0$	8.987×10^{9} N·m²·C^{-2}
Permeability of free space	μ_0	$4\pi \times 10^{-7}$ Wb·A^{-1}·m^{-1}
Faraday constant	F	9.65×10^{4} C/equivalent
Rydberg constant	R_{∞}	1.097×10^{7} m^{-1}

ASTRONOMICAL CONSTANTS

Mass of earth	$= 5.976 \times 10^{24}$ kg
Equatorial radius of earth	$= 6378.164$ km
Mass of sun ($M_\odot$)	$= 1.989 \times 10^{30}$ kg
Radius of sun ($R_\odot$)	$= 6.9599 \times 10^{8}$ m
Solar luminosity ($L_\odot$)	$= 3.826 \times 10^{26}$ J/sec
Mass of moon	$= 7.36 \times 10^{22}$ kg
Radius of moon	$= 1.74 \times 10^{6}$ m
Radius of earth orbit	$= 1.49 \times 10^{11}$ m
Period of earth orbit	$= 365.3$ d
Radius of moon orbit	$= 3.8 \times 10^{8}$ m
Period of moon orbit	$= 27.3$ days (d)

Distance

1 meter (m) = 100 cm = 39.4 in. = 3.28 ft
1 centimeter (cm) = 10 millimeters (mm) = 0.394 in.
1 kilometer (km) = 1000 m = 0.621 mi
1 foot (ft) = 12 in. = 0.305 m
1 inch (in.) = 0.0833 ft = 2.54 cm
1 mile (mi) = 5280 ft = 1.61 km

Time

1 day = 86,400 seconds (s) = 2.74×10^{-3} year
1 year = 3.16×10^7 s = 365 days

Speed

1 m/s = 3.28 ft/s = 2.24 mi/h = 3.60 km/h
1 ft/s = 0.305 m/s = 0.682 mi/h = 1.10 km/h
(Note: 88 ft/s = 60 mi/h)
1 mi/h = 1.47 ft/s = 0.447 m/s = 1.61 km/h

Mass

1 kilogram (kg) = 1000 grams (g) = 0.0685 slug
(Note: kg corresponds to 2.21 lb in the sense that the weight
of 1 kg is 2.21 lb.)
1 slug = 14.6 kg
(Note: 1 slug corresponds to 32.2 lb in the sense that the
weight of 1 slug is 32.2 lb.)
1 atomic mass unit (u) = 1.66×10^{-27} kg
$\qquad = 1.49 \times 10^{-10}$ J
$\qquad = 931$ MeV
1 metric ton = 1000 kg

Weight

1 pound (lb) = 454 grams (g)
1 ounce (oz) = 28.4 grams (g)
1 pound (lb) = 0.454 kilogram (kg)
1 grain (gr) = 0.0648 gram (g)
1 carat (car) = 200 milligrams (mg)

Temperature

	Kelvin K	Centigrade °C	Fahrenheit °F
Absolute zero	0 K	−273°C	−459°F
Freezing point of water	273 K	0°C	32°F
Boiling point of water	373 K	100°C	212°F

Conversions

$K = {}^{\circ}C + 273$
${}^{\circ}C = \frac{5}{9}({}^{\circ}F - 32)$
${}^{\circ}F = \frac{9}{5}{}^{\circ}C + 32$

Force

1 newton (N) = 0.225 lb
1 pound (lb) = 4.45 N

Work/Energy/Power

1 joule (J) = 0.738 ft·lb = 2.39×10^{-4} kcal
$\qquad = 6.24 \times 10^{18}$ eV
1 kilocalorie (kcal) = 4185 J = 3089 ft·lb
1 foot-pound (ft-lb) = 1.36 J = 3.25×10^{-4} kcal
1 electron volt (eV) = 10^{-6} MeV = 10^{-9} GeV
$\qquad = 1.60 \times 10^{-19}$ J
$\qquad = 1.18 \times 10^{-19}$ ft·lb = 3.83×10^{-23} kcal
1 watt (W) = 1 J/s = 0.73 ft·lb/s = 1.34×10^{-3} hp
1 kilowatt (kW) = 1000 W = 738 ft·lb/s = 1.34 hp
1 horsepower (hp) = 550 ft·lb/s = 746 W
1 calorie (cal) = 4.184 joules (J)
1 British thermal unit (Btu) = 1053 joules (J)
$\qquad = 252$ calories (cal)
1 food "Calorie" = 1 kilocalorie (kcal) = 1000 calories (cal)
$\qquad = 4184$ joules (J)

Pressure

1 atmosphere of pressure (atm) = 1.013×10^5 N/m^2
$\qquad = 14.7$ lb/in.^{2}
1 bar = 10^5 N/m^2 = 14.5 lb/in.^{2}
1 lb/in.^{2} = 6.9×10^3 N/m^2
1 millimeter of mercury (mm Hg) = 1 torr
1 atmosphere (atm) = 760 millimeters of mercury (mm Hg)
$\qquad = 760$ torr
1 atmosphere (atm) = 29.92 inches of mercury (in. Hg)
$\qquad = 14.7$ pounds per square inch (psi)
$\qquad = 101.325$ kPa

Volume

1 liter (L) = 0.2642 gallon (gal)
1 U.S. quart (qt) = 0.946 liter (L)
1 U.S. pint (pt) = 0.473 L
1 fluid ounce (fl oz) = 29.6 milliliters (mL)
1 gallon (gal) = 3.78 L
1 tablespoon (tbsp) = 14.79 mL
1 mL = 0.03381 fl oz = 0.2029 teaspoon (tsp)
$\qquad = 0.06763$ tbsp
$\qquad = 1$ cubic centimeter (cc)

Astronomy

1 Ångstrom (Å) = 10^{-8} cm
$\qquad = 10^{-10}$ m
1 astronomical unit (AU) = 1.495979×10^{11} m
$\qquad = 92.95582 \times 10^6$ miles
1 light-year (ly) = 6.3240×10^4 AU
$\qquad = 9.46053 \times 10^{15}$ m
$\qquad = 5.9 \times 10^{12}$ miles
1 parsec (pc) = 206265 AU
$\qquad = 3.085678 \times 10^{16}$ m
$\qquad = 3.261633$ ly
1 kiloparsec (kpc) = 1000 pc
1 megaparsec (Mpc) = 1,000,000 pc

ANSWERS

ANSWERS

The following are answers to certain exercises. ALL answers to the Margin Exercises and Summary-Reviews appear here. For the Exercise Sets:
1) All the answers to the odd-numbered exercises appear here.
2) There are worked out solutions to every third odd-numbered exercise starting with the first. These are found in the next part of this back matter. You will know that an exercise has a worked-out solution, when its number is underlined.
3) The even-numbered exercise answer following one which has a worked-out solution is NOT included in the text.   Thus,

$\underline{1}$, $\underline{7}$, $\underline{13}$, $\underline{19}$, $\underline{25}$, $\underline{31}$, and so on, have worked out solutions, and

2, 8, 14, 20, 26, 32, and so on, do not have answers in the text.

CHAPTER 1

Margin Exercises, Section 1.1

1. -12, 235   2. 8, -5   3. -10, 63   4. 14 > 7
5. 11 > -2   6. -15 < -5   7. -3 > -19   8. 0 > -4
9. -45 < -13   10. 14 > -5   11. 17   12. 8   13. 14
14. 0   15. $-\frac{3}{4}$, -4.8, 0, $\frac{12}{5}$, 17; answers may vary

16.

17.

18.

19.

20. 0.875   21. $0.\overline{63}$ 22. $\frac{59}{10}$ 23. $\frac{125}{1000}$

24. 4.62 > 4.26   25. 3.11 > -9.56   26. $\frac{6}{7} < \frac{13}{15}$

27. $-\frac{3}{5} < -\frac{5}{9}$ 28. 4.1   29. $\frac{8}{3}$ 30. 3.5

31.   32.

33.

Exercise Set 1.1

$\underline{1}$. 5, -12   3. -17, 12   4. 1200, -560   5. -1286, 29,028   6. Team A: -34, Team B: 0; Team A: 0, Team B: 34
$\underline{7}$. 750, -125   9. 20, -150, 300   10. -3, 128
11. 5 > 0   12. 9 > 0 $\underline{13}$. -9 < 5   15. -6 < 6
16. 0 > -7   17. -8 < -5   18. -4 < -3 $\underline{19}$. -5 > -11
21. -6 < -5   22. -10 > -14   23. -5 > -8
24. -7 < -1 $\underline{25}$. 120 > -20   27. -500,000 < 1,000,000
28. 60 > 20 $\underline{29}$. -2 > -10   30. 3 $\underline{31}$. 7   33. 11
34. 0   35. 4   36. 24 $\underline{37}$. 325   39. 3   40. 224
41. $-\frac{100}{7}$, -8.6, -5, $-\frac{15}{4}$, -1.3, 0, $\frac{17}{2}$, 10.86, 14,

29.6; answers may vary   42. $-\frac{135}{6}$, -12.4, -4.38, $-\frac{10}{3}$,
-0.1, 1.2, $\frac{38}{5}$, $\frac{147}{9}$, 29.804, 100.01; answers may vary

---

43.

45.

46.

47.   48.

49.   51.

52.   53. 0.375   54. 0.125

$\underline{55}$. $1.\overline{6}$ 57. $1.1\overline{6}$ 58. $0.41\overline{6}$ 59. $0.\overline{6}$ 60. 0.25

$\underline{61}$. 0.5   63. 0.1   64. 0.35   65. $\frac{27}{10}$ 66. $\frac{1391}{100}$

$\underline{67}$. $\frac{145}{1000}$ 69. $\frac{34}{100}$ 70. $\frac{2}{100}$ 71. $\frac{14}{10}$ 72. $\frac{2347}{1000}$

$\underline{73}$. 2.14 > 1.24   75. -14.5 < 0.011   76. 17.2 > -1.67

77. -12.88 < -6.45   78. -14.34 > -17.88 $\underline{79}$. $\frac{5}{12} < \frac{11}{25}$

81. $\frac{2}{3}$ 82. $\frac{10}{7}$ 83. 0   84. 14.8 $\underline{85}$. 3   87. $\frac{7}{8}$

88. 0   89. -17, -12, 5, 13   90. -23, -17, 0, 4
$\underline{91}$. The negative number indicates a score less than par or a weight less than a certain weight, which is better than scoring over par or being overweight.
93. 2 < |-4| 94. 0 = |0| 95. |-5| > |-2|
96. |4| < |-7| $\underline{97}$. |-8| = |8| 99. $-1\frac{2}{3}$, answers may

vary   100. $\frac{3}{4}$, answers may vary   101. a) 1   b) $\frac{1}{9}$

Margin Exercises, Section 1.2

1. 1·9, or 3·3   2. 1·16, or 2·8, or 4·4, or 2·2·2·2, or 2·2·4   3. 1·18, 2·9, 3·6, 2·3·3   4. 1·20, 2·10, 4·5, 2·2·5   5. 13   6. 2·2·2·2·3   7. 2·5·5
8. 2·5·7·11   9. 15, 30, 45, 60, . . .   10. 40
11. 120   12. 48   13. 48   14. 80   15. 40   16. 360
17. 2520   18. 18   19. 24   20. 36   21. 210   22. $\frac{96}{60}$
23. $\frac{28}{161}$ 24. $\frac{2}{3}$ 25. $\frac{8}{3}$ 26. $\frac{8}{7}$ 27. ½   28. 4
29. $\frac{5}{2}$ 30. $\frac{35}{16}$ 31. $\frac{23}{15}$ 32. $\frac{7}{12}$ 33. $\frac{2}{15}$ 34. $\frac{7}{36}$
35. $\frac{21}{20}$ 36. $\frac{5}{6}$ 37. $\frac{45}{28}$ 38. $\frac{8}{21}$ 39. $\frac{12}{35}$ 40. $\frac{14}{45}$
41. $\frac{5}{94}$ 42. $\frac{106}{65}$ 43. 0.462   44. 1   45. $\frac{67}{100}$
46. $\frac{456}{1000}$, or $\frac{57}{125}$ 47. $\frac{25}{10,000}$, or $\frac{1}{400}$ 48. 677%
49. 99.44%   50. 25%   51. $\frac{300}{8}$%, or 37.5%
52. $\frac{200}{3}$%, or $66\frac{2}{3}$%, or $66.\overline{6}$%   53. 1100   54. 25
55. 4   56. 11

Exercise Set 1.2

$\underline{1}$. 1·21, 3·7   3. 1·49, 7·7   4. 1·28, 2·14, 4·7, 2·2·7
5. 1·76, 2·38, 4·19, 2·2·19   6. 1·56, 2·28, 4·14, 8·7, 2·2·2·7 $\underline{7}$. 1·93, 3·31   9. 2·7   10. 3·11   11. 3·3
12. 7·7 $\underline{13}$. 2·3·3   15. 2·3·3·5   16. 2·5·3·7
17. 36   18. 90 $\underline{19}$. 360   21. 150   22. 72   23. 120
24. 299 $\underline{25}$. 72   27. 315   28. 420   29. 30
30. 105 $\underline{31}$. 72   33. 60   34. 360   35. 36   36. 360
$\underline{37}$. $\frac{2}{5}$ 39. $\frac{7}{2}$ 40. $\frac{8}{3}$ 41. $\frac{1}{7}$ 42. $\frac{1}{8}$ $\underline{43}$. 8
45. $\frac{27}{10}$ 46. $\frac{4}{3}$ 47. $\frac{14}{9}$ 48. $\frac{25}{9}$ $\underline{49}$. $\frac{9}{4}$ 51. 144
52. 315   53. $\frac{31}{100}$ 54. $\frac{93}{100}$ $\underline{55}$. $\frac{41}{60}$ 57. $\frac{26}{75}$ 58. $\frac{106}{175}$

59. $\frac{9}{100}$ 60. $\frac{23}{100}$ <u>61.</u> $14\frac{7}{12}$ 63. $12\frac{1}{10}$ 64. $9\frac{1}{5}$

65. $21\frac{17}{24}$ 66. $6\frac{9}{16}$ <u>67.</u> $12\frac{1}{4}$ 69. $22\frac{2}{3}$ 70. $18\frac{3}{4}$

71. $9\frac{31}{40}$ 72. $22\frac{1}{20}$ <u>73.</u> $24\frac{91}{100}$ 75. $6\frac{1}{4}$ 76. $1\frac{1}{5}$

77. $3\frac{9}{16}$ 78. $2\frac{8}{25}$ <u>79.</u> $1\frac{1}{8}$ 81. $\frac{1}{8}$

82. $\frac{44}{25}$ 83. $\frac{51}{8}$ 84. 1 <u>85.</u> 1 87. $\frac{7}{6}$ 88. $\frac{4}{3}$

89. $\frac{5}{6}$ 90. $\frac{41}{24}$ <u>91.</u> $\frac{1}{2}$ 93. $\frac{5}{18}$ 94. $\frac{31}{45}$ 95. $\frac{31}{60}$

96. $\frac{13}{48}$ <u>97.</u> $\frac{35}{18}$ 99. $\frac{10}{3}$ 100. $\frac{7}{4}$ 101. ½ 102. ½

<u>103.</u> $\frac{5}{36}$ 105. 500 106. 468 107. $\frac{3}{40}$ 108. $\frac{1}{18}$

<u>109.</u> $\frac{4}{25}$ 111. $\frac{28}{11}$ 112. $\frac{99}{8}$ 113. 3 114. $\frac{35}{9}$

<u>115.</u> $\frac{56}{39}$ 117. 0.76 118. 0.54 119. 0.547

120. 0.962 <u>121.</u> 1 123. 0.0061 124. 1.25

125. $\frac{20}{100}$, or $\frac{1}{5}$ 126. $\frac{80}{100}$, or $\frac{4}{5}$ <u>127.</u> $\frac{786}{1000}$, or $\frac{393}{500}$

129. $\frac{125}{1000}$, or $\frac{1}{8}$ 130. $\frac{120}{100}$, or $\frac{6}{5}$ 131. $\frac{42}{100,000}$, or $\frac{21}{50,000}$ 132. $\frac{68}{10,000}$, or $\frac{17}{2500}$ <u>133.</u> 454% 135. 99.8%

136. 73% 137. 200% 138. 0.57% <u>139.</u> 7.2%

141. $\frac{100}{8}$%, or 12.5% 142. $\frac{100}{3}$%, or $33\frac{1}{3}$% 143. 68%

144. 55% <u>145.</u> 17% 147. 70% 148. 80% 149. 60%

150. 34% <u>151.</u> $\frac{200}{3}$%, or $66\frac{2}{3}$% 153. $\frac{700}{4}$%, or 175%

154. $\frac{700}{8}$%, or 87.5% 155. 75% 156. 99.4%

<u>157.</u> 20, answers may vary 159. 8, answers may vary
<u>160.</u> 0.02, answers may vary 161. 1, answers may vary
162. 37, answers may vary <u>163.</u> 1000, answers may vary
165. 6000, answers may vary 166. 6000, answers may vary 167. 25, answers may vary 168. 0.2, answers may vary <u>169.</u> a) No; b) no; c) no; d) yes. The greatest number of times 2 occurs in the prime factorizations of 8 and 12 is 3 times. The greatest number of times 3 occurs is once. Thus the LCM must contain exactly three factors of 2 and one factor of 3. 171. 70,200 172. 24 in. 173. Every 60 yr

174. Every 420 yr <u>175.</u> Every 420 yr 177. $\frac{2072}{15,045}$
178. $\frac{69}{40}$ 179. $\frac{39}{20}$

180.

Product	56	63	36	72	140	96	48	168	110	90	432	63
Factor	7	7	2	2	10	8	6	21	10	9	24	3
Factor	8	9	18	36	14	12	8	8	11	10	18	21
Sum	15	16	20	38	24	20	14	29	21	19	42	24

Margin Exercises, Section 1.3

1. -8 2. -3 3. -8 4. 4 5. 0 6. -2 7. -11
8. -12 9. 2 10. -4 11. -2 12. -34 13. -22
14. 3 15. 0.53 16. 2.3 17. -7.7 18. -6.2
19. $-\frac{2}{9}$ 20. $-\frac{19}{20}$ 21. -58 22. -56 23. -14
24. -12 25. 4 26. -8.7 27. 7.74 28. $\frac{8}{9}$ 29. 0
30. -12 31. -14, 14 32. -1, 1 33. 19, -19
34. 1.6, -1.6 35. $-\frac{2}{3}, \frac{2}{3}$ 36. $\frac{9}{8}, -\frac{9}{8}$ 37. 4
38. 13.4 39. 0 40. -¼ 41. 510.8 mL 42. -10
43. 3 44. -5 45. -2 46. -11 47. 4 48. -2

49. -5 50. 10 51. -21 52. 9 53. -5.2 54. $-\frac{3}{10}$
55. -9 56. $-\frac{1}{5}$ 57. -7.5 58. $-\frac{1}{24}$ 59. -9
60. 21.6 61. 11.2 62. -$216.78

Exercise Set 1.3

1. -7 3. -4 4. 5 5. 0 6. 0 <u>7.</u> -8 9. -7
10. -13 11. -27 12. -35 <u>13.</u> 0 15. -42 16. -41
17. 0 18. 0 <u>19.</u> 0 21. 3 22. 1 23. -9 24. -2
<u>25.</u> 7 27. 2 <u>28.</u> -33 29. -26 30. 0 <u>31.</u> -22
<u>33.</u> 32 34. -32 35. 0 36. 0 <u>37.</u> 45 <u>39.</u> -1.8
40. -1.7 41. -8.1 42. -14.4 <u>43.</u> $-\frac{1}{5}$ 45. $-\frac{8}{7}$
46. $-\frac{10}{9}$ 47. $-\frac{3}{8}$ 48. $-\frac{1}{6}$ <u>49.</u> $-\frac{29}{35}$ 51. $-\frac{11}{15}$
52. $-\frac{11}{18}$ 53. -6.3 54. -14.7 <u>55.</u> -4.3 57. $\frac{7}{16}$
58. $\frac{1}{84}$ 59. 39 60. -62 <u>61.</u> 50 63. -1093
64. -1021 65. -24 66. 64 <u>67.</u> 9 69. 26.9
70. -48.2 71. -9 72. 26 <u>73.</u> $\frac{14}{3}$ 75. -0.101
76. 0 77. -65 78. 29 <u>79.</u> $\frac{5}{3}$ 81. 1 82. 7
83. -7 84. -10 <u>85.</u> 14 87. 0 88. $\frac{7}{8}$ 89. 8 yd
90. $77,320 <u>91.</u> 999 mb 93. -4 94. -5 95. -7
96. -10 <u>97.</u> -6 99. 0 100. 0 101. 0 102. 0
<u>103.</u> 14 105. 11 106. -11 107. -14 108. 16
<u>109.</u> 5 111. -7 112. -6 113. -1 114. -1
<u>115.</u> 18 117. -5 118. -6 119. -3 120. -2
<u>121.</u> -21 123. 5 124. 1 125. -8 126. -9
<u>127.</u> 12 129. -23 130. -45 131. -68 132. -81
<u>133.</u> -73 135. 116 136. 121 137. 0 138. 0
<u>139.</u> -¼ 141. $\frac{1}{12}$ 142. $-\frac{1}{8}$ 143. $-\frac{17}{12}$ 144. $-\frac{11}{8}$
<u>145.</u> $\frac{1}{8}$ 147. 19.9 148. 5 149. -9 150. -8.6
<u>151.</u> -0.01 153. -193 154. -413 155. 500
156. 1500 <u>157.</u> -2.8 159. -3.53 160. 17.3
161. -½ 162. $\frac{1}{8}$ <u>163.</u> $\frac{6}{7}$ 165. $-\frac{41}{30}$ 166. 0
167. $-\frac{1}{156}$ 168. $\frac{1}{42}$ <u>169.</u> 37 171. -62 172. 22
173. -139 174. 5 <u>175.</u> 6 177. -$330.54 178. $264
179. $665.50 180. 7°C <u>181.</u> 116 m

Margin Exercises, Section 1.4

1. 20, 10, 0, -10, -20, -30, -40 2. -18 3. -100
4. -90 5. $-\frac{3}{2}$ 6. -10, 0, 10, 20, 30 7. 20
8. 72 9. 12.6 10. $-\frac{3}{56}$ 11. -140 12. -90 13. 120
14. -30.75 15. 120 16. $-\frac{5}{3}$ 17. -3 18. -5
19. 5 20. -8 21. -6 22. Not possible 23. 0
24. $-\frac{24}{5}$ 25. $\frac{3}{2}$ 26. $-\frac{4}{5}$ 27. $-\frac{1}{3}$ 28. -5 29. $-\frac{8}{7}$
30.

Number	Additive Inverse	Reciprocal
$\frac{2}{3}$	$-\frac{2}{3}$	$\frac{3}{2}$
$-\frac{5}{4}$	$\frac{5}{4}$	$-\frac{4}{5}$
0	0	Not possible
1	-1	1
-4.5	4.5	$-\frac{1}{4.5}$

31. -6·5
32. -5 · $\frac{1}{7}$
33. 13·$\left(\frac{3}{-2}\right)$
34. $-\frac{4}{7} \cdot \left(-\frac{5}{3}\right)$ 35. $\frac{11}{20}$ 36. $-\frac{12}{5}$ 37. -16.2
38. 26.2

Exercise Set 1.4

1. -16 3. -42 4. -18 5. -24 6. -45 7. -72
9. 16 10. 10 11. 42 12. 18 13. -120 15. -238
16. 195 17. 1200 18. -1677 19. 98 21. -72
22. -63 23. -12.4 24. -6.6 25. 24 27. 21.7
28. 12.8 29. $-\frac{2}{5}$ 30. $-\frac{10}{21}$ 31. $\frac{1}{12}$ 33. -17.01
34. -38.95 35. $-\frac{5}{12}$ 36. -6 37. 420 39. $\frac{2}{7}$
40. $-\frac{3}{160}$ 41. -60 42. -70 43. 150 45. $-\frac{2}{45}$
46. $\frac{6}{35}$ 47. 1911 48. -5712 49. 50.4 51. $\frac{10}{189}$
52. -1 53. -960 54. 120 55. 17.64 57. $-\frac{5}{784}$
58. $\frac{4}{7}$ 59. 0 60. 0 61. -720
63. $-30{,}240$ 64. $151{,}200$ 65. -6 66. -4 67. -13
69. -2 70. 11 71. 4 72. 7 73. -8 75. 2
76. -25 77. -12 78. $\frac{64}{7}$ 79. -8 81. Not possible
82. 0 83. $-\frac{88}{9}$ 84. 29 85. $-\frac{7}{15}$ 87. $-\frac{13}{47}$
88. $-\frac{12}{31}$ 89. $\frac{1}{13}$ 90. $-\frac{1}{10}$ 91. $\frac{1}{4.3}$ 93. -7.1
94. -4.9 95. $3 \cdot \frac{1}{19}$ 96. $4 \cdot \left(-\frac{1}{9}\right)$ 97. $6 \cdot \left(-\frac{1}{13}\right)$
99. $13.9 \cdot \left(-\frac{1}{1.5}\right)$ 100. $-47.3 \cdot \left(\frac{1}{21.4}\right)$ 101. $-\frac{9}{8}$
102. $-\frac{7}{4}$ 103. $\frac{5}{3}$ 105. $\frac{9}{14}$ 106. $\frac{24}{25}$ 107. $\frac{9}{64}$
108. $\frac{25}{48}$ 109. -2 111. $\frac{11}{13}$ 112. $-\frac{1.9}{20}$, or $-\frac{19}{200}$
113. -16.2 114. $-\frac{17.8}{3.2}$, or $-\frac{89}{16}$ 115. -60
117. $-\frac{161}{12}$ 118. $-\frac{5}{2}$ 119. $\frac{1}{2}$ 120. 6 121. $\frac{20}{9}$
123. 0 124. 2 125. -9 126. $\frac{69}{25}$ 127. $-\frac{17}{2}$
129. $-\frac{40}{7}$ 130. 1 131. 72 132. -49 133. -6
135. 0 136. 0 137. $2\sqrt{3}$ 138. 1 139. 4π

Margin Exercises, Section 1.5

1. $2128 + x = 2866$ 2. 64 3. 28 4. 25 5. 60
6. 192 7. 16 8. $n - 12$ 9. $x + 12$, or $12 + x$
10. $y - 4$ 11. $\frac{1}{2}x$, or $\frac{x}{2}$ 12. $8t + 6$, or $6 + 8t$
13. $m - n$, or $n - m$ 14. $59\%y$, or $0.59y$ 15. $xy - 200$
16. $a + b$, or $b + a$ 17. 10^4 18. $5 \cdot 5 \cdot 5 \cdot 5$
19. $x \cdot x \cdot x \cdot x \cdot x$ 20. (a) 25; (b) -25 21. 4 22. 1
23. 8490.56 cm²   24. 119

Exercise Set 1.5

1. $10, 25, 48$ 3. 72 yd²   4. 100.1 cm²   5. 56
6. 54 7. 3 9. 2 10. 6 11. 20 12. 4
13. $m + 6$, or $6 + m$ 15. $c - 9$ 16. $d - 4$
17. $q + 6$, or $6 + q$ 18. $11 + z$, or $z + 11$
19. $a + b$, or $b + a$ 21. $y - x$ 22. $b - c$
23. $98\%x$, or $0.98x$ 24. $45\%m$, or $0.45m$
25. $r + s$, or $s + r$ 27. $2x$ 28. $4p$ 29. $5t$ 30. $9d$
31. $3 - b$, or $b - 3$ 33. $n + 6$, or $6 + n$
34. $y + 1$, or $1 + y$ 35. $m - 4$ 36. $t - 43$
37. $x + 3y$, or $3y + x$ 39. $2 \cdot 2 \cdot 2 \cdot 2$ 40. $5 \cdot 5 \cdot 5$
41. $(1.4) \cdot (1.4) \cdot (1.4) \cdot (1.4) \cdot (1.4)$ 42. $t \cdot t \cdot t \cdot t$ 43. n
45. $7p \cdot 7p$ 46. $11c \cdot 11c \cdot 11c$ 47. $19k \cdot 19k \cdot 19k \cdot 19k$
48. $104d \cdot 104d \cdot 104d \cdot 104d \cdot 104d$ 49. 1 51. 10^6 52. 6^4
53. x^7 54. y^3 55. $(3y)^4$ 57. 27 58. 64 59. 9
60. -9 61. 256 63. 1 64. 93 65. 37
66. 3629.84 ft²   67. 576 m²   69. $y + 2x$, or $2x + y$
70. $a + 2 + b$ 71. $2x - 3$ 72. $a + 5$, or $5 + a$
73. $b - 2$ 75. $98\%x$, or $0.98x$ 76. $s + s + s + s$, or $4s$
77. 6 78. 5 79. 6 81. 9 82. $w + 4$
83. d 84. $t + 3$, or $t - 3$ 85. $2v$ 87. $n + 0.1n$
88. 10^2 89. 10^5 90. 5^2 91. 2^0, or 1 93. 111
94. 225 95. $216; 24$ 96. $729; 81$ 97. $-1728; -192$

99. $400; 80$ 100. $1225; 245$ 101. $3025; 605$
102. $16{,}900; 3380$ 103. x^3y^3 105. 0 106. 9
107. 127 108. 81 109. Answers may vary. For a
10-digit readout, 6^{13} and larger will be too large
111. Yes   112. No   113. Yes   114. No   115. No

Margin Exercises, Section 1.6

1. $9 + x$ 2. qp 3. $yx + t$, $t + xy$, or $t + yx$ 4. 11
5. 27 6. 6 7. 225 8. 75 9. 16 10. 36 11. 16
12. 24 13. 139 14. 120 15. $\frac{1}{5}$ 16. 8 17. 63
18. 20 19. 72 20. $(a + b) + 2$ 21. $(3 \cdot v) \cdot w$
22. $(t \cdot 4) \cdot u$; $(u \cdot t) \cdot 4$; $u \cdot (4 \cdot t)$. There are other correct answers.   23. $(2 + r) + s$; $(s + 2) + r$; $s + (r + 2)$. There are other correct answers.   24. 28 25. 84
26. 54 27. $5x, -4y, 3$ 28. $-4y, -2x, 3z$
29. $3x - 15$ 30. $5x - 5y + 20$ 31. $-2x + 6$
32. $bx - 2by + 4bz$ 33. $\frac{5}{6}x - \frac{5}{6}y - \frac{5}{6}z$ 34. $4(x - 2)$
35. $3(x - 2y - 5)$ 36. $2(-y + 4z - 1)$
37. $4(3z - 4x - 1)$ 38. $b(x - y + z)$
39. $\frac{1}{4}(m + 3n - 5)$ 40. $3x$ 41. $6y$ 42. $0.59m$
43. $3x + 3y$ 44. $-4x - 5y - 15$

Exercise Set 1.6

1. $8 + y$ 3. nm 4. ba 5. $xy + 9$, $9 + yx$, $yx + 9$
6. $11 + ba$, $ab + 11$, $ba + 11$ 7. $c + ab$, $c + ba$, $ba + c$ 9. 19 10. 27 11. 86 12. 51 13. 7
15. 5 16. 28 17. 12 18. 400 19. 324 21. 100
22. 100 23. 512 24. 11 25. 22 27. 1 28. 1
29. 4 30. 28 31. 1500 33. 76 34. 66 35. 60
36. 343 37. 125 39. 96 40. 925 41. 500 42. $\frac{4}{5}$
43. $\frac{11}{5}$ 45. 3 46. 80 47. 32 48. $a + (b + 3)$
49. $5 + (x + y)$ 51. $6 \cdot (x \cdot y)$ 52. $2 + (b + a)$; $(2 + b) + a$; $b + (a + 2)$. There are other correct answers.   53. $y + (x + 3)$; $y + (3 + x)$; $x + (3 + y)$. There are other correct answers.   54. $(5 + w) + v$; $v + (5 + w)$; $(v + 5) + w$. There are other correct answers.   55. $(6 + y) + x$; $(x + 6) + y$; $y + (6 + x)$. There are other correct answers.   57. $a \cdot (5 \cdot b)$; $(5 \cdot b) \cdot a$; $b \cdot (a \cdot 5)$. There are other correct answers.
58. $(a \cdot 7) \cdot b$; $b \cdot (a \cdot 7) \cdot a$. There are other correct answers.   59. $(x \cdot 5) \cdot y$; $y \cdot (x \cdot 5)$; $(5 \cdot y) \cdot x$. There are other correct answers.   60. $d \cdot (c \cdot 2)$; $(2 \cdot d) \cdot c$; $(d \cdot c) \cdot 2$. There are other correct answers.
61. $4x, 3z$ 63. $7x, 8y, -9z$ 64. $8a, 10b, -18c$
65. $12x, -13.2y, \frac{5}{8}z, -4.5$ 66. $-7.8a, -3.4y, -8.7z, -12.4$ 67. 7 69. 12 70. 3.48 71. -12.71 72. $\frac{8}{9}$
73. $7x - 14$ 75. $-7y + 14$ 76. $-9y + 63$
77. $45x + 54y - 72$ 78. $14x + 35y - 63$
79. $-4x + 12y + 8z$ 81. $-3.72x + 9.92y - 3.41$
82. $8.82x + 9.03y + 4.62$ 83. $8(x - 3)$ 84. $10(x - 5)$
85. $4(8 - y)$ 87. $2(4x + 5y - 11)$ 88. $3(3a + 2b - 5)$
89. $a(x - 7)$ 90. $b(y - 9)$ 91. $a(x - y - z)$ 93. $8x$
94. $-8t$ 95. $5n$ 96. $-16y$ 97. $4x + 2y$ 99. $7x + y$
100. $11a + 5b$ 101. $0.8x + 0.5y$ 102. $2.6a + 1.4b$
103. $\frac{3}{5}x + \frac{3}{5}y$ 105. $9a$ 106. $12x$ 107. $-3b$ 108. $-3c$
109. $15y$ 111. $11a$ 112. $14x$ 113. $-8t$ 114. $-5x$
115. $10x$ 117. $8x - 8y$ 118. $13a - 10b$
119. $2c + 10d$ 120. $7a + 9b$ 121. $22x + 18$
123. $2x - 33y$ 124. $5a - 21b$ 125. No   126. No
127. Yes; commutative law of addition   129. Yes; commutative law of multiplication   130. No   131. No
132. Yes; commutative law of multiplication
133. Any number except 0   135. Any number except 2
136. Any nonzero number except 1   137. $n^2 + 7$
138. $n + 7^2$, or $n + 49$ 139. $(7 + x)^2$ 141. $\frac{y + 3}{(y + 3)^2}$
142. $m(m + 5)$ 143. $2(x - 1)$
144. No; $(12 - 5) - 6 \neq 12 - (5 - 6)$ 145. $\pi r(2 + s)$
147. $1.08P$ 148. $1.06P$ 149. $8(x - y)$, or $8(y - x)$

150. $9y - 6z$, or $12z - 9y$ **151.** $-4a + 3b$

153. $2500(x + y)$ **154.** $5420\left[41\frac{1}{8} - 37\frac{3}{4}\right]$, or $5420\left[41\frac{1}{8}\right] -$ $5420\left[37\frac{3}{4}\right]$; The loss is \$18,292.50.   **155.** $\frac{2}{3}$ **156.** $\frac{p}{t}$

157. $\frac{3sb}{2}$ **159.** $\frac{r}{g}$ **160.** $\frac{5}{2}$ **161.** No; $12 \div 4 \neq 4 \div 12$

162. $\frac{yz}{10z}$ **163.** $\frac{st}{20t}$

Margin Exercises, Section 1.7

1. $-x - 2$ **2.** $-5x - 2y - 8$ **3.** $-6 + t$ **4.** $-x + y$, or $y - x$ **5.** $4a - 3t + 10$ **6.** $-18 + m + 2n - 4z$
7. $2x - 9$ **8.** $6y + 2$ **9.** $8x - 4y + 4$ **10.** $-9x - 8y$
11. $-16a + 18$ **12.** 6 **13.** -1 **14.** 4 **15.** $5x - y - 8$
16. -1237 **17.** 381 **18.** -135 **19.** $(1/(4*5))^2$
20. $A^2 + B^2 - 2*A*B$ **21.** $X/Y - T/S$ **22.** $\frac{2}{a + 3}$
23. $a^2 - 2ab + b^2$

Exercise Set 1.7

1. $-2x - 7$ **3.** $-5x + 8$ **4.** $-6x + 7$ **5.** $-4a + 3b - 7c$
6. $-5x + 2y + 3z$ **7.** $-6x + 8y - 5$ **9.** $-3x + 5y + 6$
10. $-6a + 4b + 7$ **11.** $8x + 6y + 43$ **12.** $2a - 9b + 5c$
13. $5x - 3$ **15.** $-3a + 9$ **16.** $8n + 7$ **17.** $5x - 6$
18. $a - 7$ **19.** $-19x + 2y$ **21.** $9y - 25z$
22. $-16a + 27b - 32c$ **23.** 7 **24.** -14 **25.** -40
27. 19 **28.** 39 **29.** $12x + 30$ **30.** $13x - 1$
31. $3x + 30$ **33.** $9x - 18$ **34.** $-16x + 44$ **35.** -16
36. 8 **37.** -334 **39.** 14 **40.** 23 **41.** 1880 **42.** 305
43. 4682.688 **45.** 12 **46.** 8 **47.** 8 **48.** 8 **49.** 16
51. $A^2 + 2*A*B + B^2$ **52.** $A^3 - B^3$ **53.** $2*(3 - B)/C$
54. $(A + B)/(C - D)$ **55.** $A/B - C/D$ **57.** $2a + 7$
58. $a + \frac{1}{b}$ **59.** $3a^2 - 5$ **60.** $\frac{a + 4}{2a}$ **61.** $(a + b)^2$
63. $6y - (-2x + 3a - c)$ **64.** $x - (y + a + b)$
65. $6m - (-3n + 5m - 4b)$ **66.** 0; $a + b + (-a) + (-b) =$ $a + (-a) + b + (-b) = 0 + 0 = 0$ **67.** $-4z$ **69.** $x - 3$
70. False   **71.** False   **72.** True   **73.** False
75. False   **76.** True   **77.** True   **78.** True.
$(-x)^2 = (-x)(-x) = (-1)(x)(-1)(x) = (-1)(-1)(x)(x) =$ $1 \cdot x^2 = x^2$.   **79.** True. $(-a)(-b) = (-1)(a)(-1)(b) =$ $(-1)(-1)(a)(b) = 1 \cdot ab = ab$.   **81.** 4096 **82.** $34,992$

Margin Exercises, Section 1.8

1. $\frac{1}{4^3}$, or $\frac{1}{4 \cdot 4 \cdot 4}$, or $\frac{1}{64}$ **2.** $\frac{1}{5^2}$, or $\frac{1}{5 \cdot 5}$, or $\frac{1}{25}$

3. $\frac{1}{2^4}$, or $\frac{1}{2 \cdot 2 \cdot 2 \cdot 2}$, or $\frac{1}{16}$ **4.** 3^{-2} **5.** 5^{-4} **6.** 7^{-3}

7. $\frac{1}{5^3}$ **8.** $\frac{1}{7^5}$ **9.** $\frac{1}{10^4}$ **10.** 3^8 **11.** 5^2 **12.** 6^{-7}
13. x^{-4} **14.** y^{-2} **15.** x^{-8} **16.** 4^3 **17.** 7^{-5} **18.** a^7
19. b **20.** x^4 **21.** x^7 **22.** 3^{20} **23.** x^{-12} **24.** y^{15}
25. x^{-32} **26.** $16x^{20}y^{-12}$ **27.** $25x^{10}y^{-12}z^{-6}$
28. $27y^{-6}x^{-15}z^{24}$ **29.** \$3121.79

Exercise Set 1.8

1. $\frac{1}{3^2}$, or $\frac{1}{3 \cdot 3}$, or $\frac{1}{9}$ **3.** $\frac{1}{10^4}$, or $\frac{1}{10 \cdot 10 \cdot 10 \cdot 10}$, or $\frac{1}{10,000}$ **4.** $\frac{1}{5^6}$, or $\frac{1}{5 \cdot 5 \cdot 5 \cdot 5 \cdot 5 \cdot 5}$, or $\frac{1}{15,625}$ **5.** 4^{-3}
6. 5^{-2} **7.** x^{-3} **9.** a^{-4} **10.** t^{-5} **11.** p^{-n} **12.** m^{-n}
13. $\frac{1}{7^3}$ **15.** $\frac{1}{a^3}$ **16.** $\frac{1}{x^2}$ **17.** $\frac{1}{y^4}$ **18.** $\frac{1}{t^7}$ **19.** $\frac{1}{2^n}$
21. 2^7 **22.** 3^7 **23.** 3^3 **24.** 5 **25.** x^{-1}, or $\frac{1}{x}$
27. x^7 **28.** x^{13} **29.** x^{-13}, or $\frac{1}{x^{13}}$ **30.** y^{-13}, or $\frac{1}{y^{13}}$
31. 1 **33.** 7^3 **34.** 4^4 **35.** x^2 **36.** x^5 **37.** x^9
39. z^{-4}, or $\frac{1}{z^4}$ **40.** y^{-4}, or $\frac{1}{y^4}$
41. x^3 **42.** y^5 **43.** 1 **45.** 2^6 **46.** 3^{12} **47.** 5^{-6}, or $\frac{1}{5^6}$ **48.** 9^{-12}, or $\frac{1}{9^{12}}$ **49.** x^{12} **51.** $x^{-12}y^{-15}$, or $\frac{1}{x^{12}y^{15}}$ **52.** $t^{-20}x^{-12}$, or $\frac{1}{t^{20}x^{12}}$
53. $x^{24}y^8$ **54.** $x^{10}y^{35}$ **55.** $9x^6y^{-16}z^{-6}$, or $\frac{9x^6}{y^{16}z^6}$
57. \$2508.80   **58.** \$3041.75   **59.** \$22,318.40
60. \$63,677.44   **61.** \$15,819.02   **63.** 2^{13} **64.** 4^{13}

65. No   **66.** y^{5x}
67. 5^{-1} **69.** 3^{11} **70.** 1 **71.** 1 **72.** 7 **73.** $(\frac{1}{2})^{-1}$, or 2 **75.** $3^{-2} = 3^{0-2} = \frac{3^0}{3^2} = \frac{1}{3^2}$ **76.** No   **77.** No
78. 25 **79.** 81 **81.** $<$ **82.** $<$ **83.** $<$ **84.** False
85. True   **87.** False   **88.** True   **89.** False   **90.** 0.01
91. 0.72 **93.** $-\frac{7a^{-4}b^2}{4}$ **94.** $\frac{3x^3y^{-2}}{2}$ **95.** 11^{-2b+5}
96. $3a^{-x-4}$ **97.** $-3x^{2a-1}$ **99.** $-4b^{10}y^8$

100. $\frac{625x^{-20}y^{24}}{256}$ **101.** 7^{2ab} **102.** 3^{a^2+2a}
103. 12^{6b-2ab} **105.** $5^{2c}x^{2ac-2c}y^{2bc+2c}$, or $25^c x^{2ac-2c}y^{2bc+2c}$ **106.** $4^{5c}x^{15ac}y^{10bc}$ **107.** $2x^{a+2}y^{b-2}$
108. $-5x^{2b}y^{-2a}$ **109.** $25\frac{1}{4}$ **111.** 7 **112.** $\frac{328}{3}$
113. $2^{-2a-2b+ab}$ **114.** 8^{-2abc} **115.** $\frac{64x^{32a}y^{-68b}}{81}$

Margin Exercises, Section 1.9

1. 4.6×10^{11} **2.** 9.3×10^7 **3.** 1.235×10^{-8}
4. 1.7×10^{-24} **5.** $789,300,000,000$ **6.** 0.0000567
7. 7.462×10^{-13} **8.** 5.6×10^{-15} **9.** 2×10^3
10. 5.5×10^2 **11.** 3×10^{-4}; answers may vary
12. 7 or 8 depending on how you estimate.   **13.** -7
14. The mass of an electron is a number between 1 and 10, times 10^{-28}.   **15.** 4.15668×10^{-5} m^3 **16.** No. The correct answer should be less than 1.

Exercise Set 1.9

1. 4.7×10^{10} **3.** 8.63×10^{17} **4.** 9.57×10^{17}
5. 1.6×10^{-8} **6.** 2.63×10^{-7} **7.** 7×10^{-11}
9. 4.07×10^{11} **10.** 3.09×10^{12} **11.** 0.0004
12. 0.00005 **13.** $673,000,000$ **15.** 0.0000000008923
16. 0.07034 **17.** $90,300,000,000$
18. $1,010,000,000,000$ **19.** 9.66×10^{-5}
21. 1.3338×10^{-11} **22.** 2.6732×10^{-11}
23. 8.32×10^{10} **24.** 3.1411×10^{16} **25.** 1.9565×10^3
27. 2.5×10^3 **28.** 1.5×10^3 **29.** 5×10^{-4}
30. 3×10^{-5} **31.** 3×10^{11} **32.** 2×10^{10}
34. 3×10^{-22} **35.** 4.2×10^2 **36.** 1.6×10^{-17}
37. 1.1×10^{11} **39.** 6×10 **40.** 10^3 **41.** 1 **42.** 4
43. 1 **45.** 3 **46.** 4 **47.** -3 **48.** -3
49. 2.4×10^{-2}m^3 **51.** 5.8404×10^8 miles
52. 6.7×10^4, or \$67,000   **53.** 1.22×10^{-5}
54. 1.095×10^9 gallons   **55.** 1.512×10^{10} cubic feet
57. No. The order of magnitude should be 9.   **58.** No. The order of magnitude should be 9.   **59.** No. The order of magnitude should be 9.   **60.** Yes. The order of magnitude is 9.   **61.** ≈ 4 **63.** $8 \cdot 10^{-90}$; $7.1 \cdot 10^{-90}$

larger   **64.** $\frac{1}{3}$ **65.** 8 **66.** 7
67. $c = 2 - \frac{4}{a}$ **69.** $\frac{2}{27}$ **70.** $\frac{3^{12S+4}}{625}$

Summary and Review - Chapter 1

1. [1.5] 15 **2.** [1.5] 6 **3.** [1.5] 5 **4.** [1.5] 4
5. [1.5] 9 **6.** [1.5] 1 **7.** [1.5] 25 **8.** [1.5] 32
9. [1.6] 119 **10.** [1.7] 29 **11.** [1.7] 7
12. [1.5] $z - 8$ **13.** [1.5] $3x$ **14.** [1.5] $19\%x$, or $0.19x$ **15.** [1.5] $x - 1$ **16.** [1.5] $2m \cdot 2m \cdot 2m$
17. [1.5] $(6z)^2$ **18.** [1.6] $y + 4$ **19.** [1.6] ba
20. [1.6] $2 + pq$, or $qp + 2$, or $2 + qp$
21. [1.6] $3 + (x + 1)$ **22.** [1.6] $(m \cdot 4) \cdot n$
23. [1.6] $1 + (m + n)$; $1 + (n + m)$; $(1 + n) + m$
24. [1.6] $(4 \cdot x) \cdot y$; $y \cdot (4 \cdot x)$; $4 \cdot (y \cdot x)$
25. [1.2] $2 \cdot 2 \cdot 23$ **26.** [1.2] $2 \cdot 2 \cdot 2 \cdot 5 \cdot 5 \cdot 7$
27. [1.2] 96 **28.** [1.2] 90 **29.** [1.2] $\frac{5}{12}$
30. [1.2] $\frac{5}{9}$ **31.** [1.2] $\frac{31}{36}$ **32.** [1.2] $\frac{1}{4}$
33. [1.2] $\frac{3}{5}$ **34.** [1.2] $\frac{72}{25}$ **35.** [1.2] 0.047
36. [1.2] $\frac{60}{100}$, or $\frac{3}{5}$ **37.** [1.1, 1.2] 88.6%, $\frac{886}{1000}$ or $\frac{443}{500}$

38. [1.1, 1.2] 0.625; 62.5% 39. [1.1, 1.2] 1.16; 116%
40. [1.1] -45, 72 41. [1.1] 38 42. [1.1] 7
43. [1.1] $\frac{5}{2}$ 44. [1.1] (number line, -2.5 marked between -3 and -2)
45. [1.1] (number line, $\frac{8}{9}$ marked between 0 and 1) 46. [1.1] < 47. [1.1] >
48. [1.1] > 49. [1.1] < 50. [1.3] -3.8
51. [1.3] $\frac{3}{4}$ 52. [1.3] 34 53. [1.3] 5 54. [1.4] $\frac{8}{3}$
55. [1.4] $-\frac{1}{7}$ 56. [1.4] $-\frac{8}{9}$ 57. [1.3] -3
58. [1.3] $-\frac{7}{12}$ 59. [1.3] -4 60. [1.3] -5
61. [1.3] 4 62. [1.3] $-\frac{7}{5}$ 63. [1.3] -7.9
64. [1.4] 54 65. [1.4] -9.18 66. [1.4] $-\frac{2}{7}$
67. [1.4] -210 68. [1.4] -7 69. [1.4] -3
70. [1.4] $\frac{3}{4}$ 71. [1.3] 8-yd gain 72. [1.3] -$130
73. [1.6] 15x - 35 74. [1.6] -8x + 10
75. [1.6] 4x + 15 76. [1.6] -24 + 48x
77. [1.6] 2(x - 7) 78. [1.6] 6(x - 1)
79. [1.6] 5(x + 2) 80. [1.6] 3(4 - x)
81. [1.6] 7a - 3b 82. [1.6] -2x + 5y
83. [1.6] 5x - y 84. [1.6] -a + 8b 85. [1.7] 19
86. [1.7] 8 87. [1.7] 6x + 3 88. [1.7] 11y - 16
89. [1.7] -3a + 9 90. [1.7] -2b + 21 91. [1.7] 6
92. [1.7] 12y - 34 93. [1.7] 5x + 24
94. [1.7] -15x + 25 95. [1.8] y^{-4} 96. [1.8] $\frac{1}{5^3}$
97. [1.8] x^{-2}, or $\frac{1}{x^2}$ 98. [1.8] t^9 99. [1.8] 7^{-10}, or $\frac{1}{7^{10}}$ 100. [1.8] 4^{-15}, or $\frac{1}{4^{15}}$ 101. [1.8] 8^9
102. [1.8] $81a^{-24}$, or $\frac{81}{a^{24}}$ 103. [1.8] $x^{10}y^{-5}z^{-35}$, or $\frac{x^{10}}{y^5 z^{35}}$ 104. [1.8] $5180.12
105. [1.9] 1.08×10^{-2} 106. [1.9] 2.6378×10^{-2}
107. [1.9] 5×10^{-12} 108. [1.9] 4×10
109. [1.9] 4×10^{-17} 110. [1.9] 1×10
111. [1.9] 3.0144×10^9 miles 112. [1.1], [1.3] $-\frac{5}{8}$
113. [1.4] a) $\frac{3}{11}$ b) $\frac{10}{11}$ 114. [1.5] 25,281
115. [1.1], [1.5], [1.7] -2.1 116. [1.8] $-4b^{x-5}$
117. [1.8] $5^{4c}x^{8ac}y^{4bc-4c}$, or $625^{c}x^{8ac}y^{4bc-4c}$

CHAPTER 2

Margin Exercises, Section 2.1
1. 536.25 = 8.25t 2. 2,130,000 = x + 40,000
3. False 4. True 5. Neither 6. Neither
7. 1, 2, or 3; answers can vary 8. 7 9. 6 10. 4
11. 5 12. -5 13. 13.2 14. -2 15. 10.8
16. -3, 3 17. -27, 27 18. $-\frac{5}{4}, \frac{5}{4}$ 19. 5 20. $-\frac{7}{4}$
21. 14 22. $-\frac{5}{4}$ 23. -6 24. -50 25. 28 26. -3
27. -3, 3 28. $-\frac{16}{25}, \frac{16}{25}$

Exercise Set 2.1
1. 4 3. -20 4. 34 5. -14 6. -21 7. -18
9. 15 10. 13 11. -14 12. -11 13. 2 15. 20
16. 24 17. -6 18. -15 19. $\frac{7}{3}$ 21. $-\frac{7}{4}$ 22. $-\frac{3}{2}$
23. $\frac{41}{24}$ 24. $\frac{19}{12}$ 25. $-\frac{1}{20}$ 27. 5.1 28. 4.7
29. 12.4 30. 17.8 31. -5 33. $\frac{11}{6}$ 34. $\frac{7}{12}$
35. $-\frac{10}{21}$ 36. $123\frac{1}{8}$ 37. 6 39. 9 40. 8 41. 12
42. 7 43. -40 45. 1 46. 68 47. -7 48. -4
49. -6 51. 6 52. 8 53. -63 54. -88 55. 36
57. -21 58. -54 59. $-\frac{3}{5}$ 60. $-\frac{7}{9}$ 61. $\frac{3}{2}$ 63. $\frac{9}{2}$

64. -1 65. 7 66. 20 67. -7 69. 8 70. 8
71. 15.9 72. -9.38 73. -5, 5 75. -11, 11
76. -8.3, 8.3 77. -389, 389 78. -433, 433
79. -8, 8 81. -7, 7 82. -9, 9 83. 342.246
84. -4 85. -10 87. No solution 88. No solution
89. x = b + 3 90. x = 1 - c - a
91. x = a + 4 93. $-\frac{5}{17}$ 94. 11,074 95. Subtracting
c is the same as adding its inverse -c
96. No solution; the absolute value of a number cannot
be negative. 97. -8655 99. All real numbers
100. 5 101. $x = \frac{b}{3a}$ 102. $x = \frac{a^2 + 1}{c}$ 103. $x = \frac{4b}{a}$

105. No solution 106. False. x = y is not equivalent
to $x^2 = y^2$. The solution of x = 1 is 1, but the
solutions of $x^2 = 1$ are -1 and 1. But, if x = y is
true, then $x^2 = y^2$ is true. 107. 250 108. $\frac{13}{20}$
109. $-\frac{26}{15}$

Margin Exercises, Section 2.2
1. 5 2. -4 3. 4 4. 39 5. $-\frac{3}{2}$ 6. -4.3 7. -3
8. 800 9. 1 10. 2 11. 2 12. $\frac{17}{2}$ 13. $\frac{8}{3}$
14. -4.3 15. -3, $\frac{7}{5}$ 16. -6, $\frac{22}{3}$

Exercise Set 2.2
1. 5 3. 8 4. 9 5. 10 6. 3 7. 14 9. -8
10. -9 11. -8 12. -11 13. -7 15. 15 16. 19
17. 18 18. 21 19. 6 21. 4 22. 8 23. 6
24. 4 25. 5 27. -3 28. -4 29. 1 30. 3
31. -20 33. 6 34. 8 35. 7 36. -3 37. 7
39. 2 40. 5 41. 5 42. 2 43. 2 45. 10 46. 10
47. 4 48. 5 49. 8 51. -1 52. $\frac{1}{2}$ 53. $-\frac{4}{3}$
54. $-\frac{2}{3}$ 55. $\frac{2}{5}$ 57. -2 58. -3 59. -4 60. $\frac{5}{3}$, or $1.\overline{6}$ 61. 0.8 63. 3 64. 70 65. 28 66. $\frac{32}{7}$
67. 4 69. $-\frac{11}{2}$, 7 70. $-\frac{25}{7}$, 3 71. 4, $-\frac{12}{5}$
72. $-\frac{51}{4}, \frac{49}{4}$ 73. 4.42 75. $x = -\frac{1}{3}$, y = -5
76. x = 0.13, y = -0.324 77. 2 78. $y = \frac{a}{1 - 3a}$
79. $-\frac{7}{2}$ 81. -2 82. 10 83. 0 84. 0 85. $\frac{52}{45}$

Margin Exercises, Section 2.3
1. 2 2. 3 3. -2 4. $-\frac{1}{2}$

Exercise Set 2.3
1. 6 3. 2 4. 1 5. 6 6. 9 7. 8 9. 4 10. 5
11. 1 12. 2 13. 17 15. -8 16. -13 17. $-\frac{5}{3}$
18. -8 19. -3 21. 2 22. 1 23. 5 24. 6
25. $\frac{4}{7}$ 27. 8 28. 7 29. $\frac{13}{2}$ 30. $-\frac{405}{28}$ 31. $\frac{11}{18}$
33. $-\frac{51}{31}$ 34. $\frac{39}{14}$ 35. 2 36. -7.4 37. $\frac{837,353}{1929}$
39. -2 40. All real numbers 41. x = a + 4
42. About -0.000036365 43. No 45. Yes 46. No

Margin Exercises, Section 2.4
1. 28 ft 2. (a) $20\pi \approx 62.8$ m; (b) $D = \frac{C}{\pi}$; (c) $D = \frac{400}{\pi}$
3. $I = \frac{E}{R}$ 4. 4A - a - b - d = c 5. $I = \frac{9R}{A}$

Exercise Set 2.4
1. (a) 68 ft²; (b) $w = \frac{A}{\ell}$; (c) 8 cm 3. (a) $17\frac{3}{16}$ m²;
(b) $h = \frac{2A}{b}$; (c) 84 yd 4. (a) 138 cm; (b) $\ell = \frac{P - 2w}{2}$,
or $\frac{1}{2}P - w$; (c) 111 mi 5. $b = \frac{A}{h}$ 6. $h = \frac{A}{b}$ 7. $r = \frac{d}{t}$
9. $P = \frac{I}{rt}$ 10. $r = \frac{I}{Pt}$ 11. $a = \frac{F}{m}$ 12. $m = \frac{F}{a}$
13. $w = \frac{P - 2\ell}{2}$, or $\frac{P}{2} - \ell$ 15. $r^2 = \frac{A}{\pi}$ 16. $\pi = \frac{A}{r^2}$

17. $m = \frac{E}{c^2}$

18. $c^2 = \frac{E}{m}$ 19. $b = 3A - a - c$ 21. $t = \frac{3k}{v}$

22. $c = \frac{ab}{P}$ 23. $b = \frac{2A - ah}{h}$, or $\frac{2A}{h} - a$

24. $a = \frac{2A - bh}{h}$, or $\frac{2A}{h} - b$

25. $D^2 = \frac{2.5H}{N}$ 27. $S = \frac{360A}{\pi r^2}$ 28. $r^2 = \frac{360A}{\pi S}$

29. $t = \frac{R - 3.85}{-0.0075}$, or $\frac{3.85 - R}{0.0075}$ 30. $C = \frac{5F - 160}{9}$, or $\frac{5}{9}(F - 32)$

31. Not necessarily 33. A increases by 2h units

34. An increase in a 35. $b = c - ax$ 36. $a = \frac{c - b}{x}$

37. $x = -\frac{b}{a}$ 39. $s = \frac{t^2}{v}$ 40. $\frac{a}{c} = \frac{b}{d}$ 41. $a = \frac{g}{20} - 2n$, or $\frac{g - 40n}{20}$ 42. $f = 8h - 4r$ 43. $a = \frac{y}{1 - b}$

45. $f = \frac{1 - de}{d}$, or $\frac{1}{d} - e$

46. $y = \frac{xz^2}{t}$ 47. $b = \frac{m - ax^2 - c}{x}$ 48. No, a may be −b

49. 720° 51. $n = \frac{A + 360}{180}$, or $\frac{A}{180} + 2$ 52. $M = DV$

53. $V = \frac{M}{D}$ 54. 121.07 cm³ 55. 2722.5 grams

57. 19.3 grams per cubic centimeter. 58. 61.06 cm

Margin Exercises, Section 2.5

1. Consult an almanac, a reference librarian, or a government organization. 2. Open-ended answer. One might consider guessing various products such as 3·3, 3.1 × 3.1, 3.2 × 3.2, and so on. 3. 1424 4. 31
5. $\frac{81}{2}$ 6. 209,367,000 7. $3909

Exercise Set 2.5

1. 52 3. 142 4. 864 5. 319 6. 2236 7. 56
9. 72 10. 448 11. 3 12. 24 yd 13. 64 15. $67
16. 164,394 17. $0.26 18. $3.75 19. 667.5
21. 6.64 hr 22. 3145.3 km² 23. 64.8°C 24. 92 m
25. 383,631.7 km 27. 59°F 28. 120 29. 25.75¢
30. 7 31. 39.37 33. Salary now 34. Marked price of record 35. No solution. The absolute value of a number cannot be negative.

Margin Exercises, Section 2.6

1. 32% 2. 25% 3. 11.04 4. 10 5. 225 6. 50
7. 111,416 sq mi 8. $8400 9. $225

Exercise Set 2.6

1. 25% 3. 24% 4. 19% 5. 150 6. 85 7. 2.5
9. 546 10. 10,000 11. 125% 12. 2050% 13. 0.8
15. 5% 16. 2000 17. 86.4% 18. 52% 19. $800
21. 36 cm³, 436 cm³ 22. $7800 23. $7200 24. $1.50, $1.62 25. $16 27. $4400 28. $6540 29. About 3.6 billion 30. About 151 million 31. About 4538
33. 26% 34. 68% 35. 90% 36. 71% 37. 12, or 1200%
39. 345% 40. 98% 41. 2.5% 42. 3.7% 43. $20
45. 40%, 70%, 95% 46. 62.5% 47. 20% 48. $90.72
49. 16, 4 51. $9.17 not $9.10

Margin Exercises, Section 2.7

1. 3 ft, 5 ft 2. 5 3. 18, 20 4. ≈ 241 mi
5. Length is 20 m; width is 10 m 6. 30°, 90°, 60°

Exercise Set 2.7

1. 19 3. -10 4. -68 5. 40 6. 36 7. 20 m, 40 m, 120 m 9. 37, 39 10. 41, 43 11. 56, 58
12. 52, 54 13. 35, 36, 37 15. 61, 63, 65
16. 83, 85, 87 17. Length is 90 m; width is 65 m
18. Length 96 m; width is 56 m 19. Length is 49 m; width is 27 m 21. 22.5° 22. 25.625° 23. 19°C
24. $560.51 25. 670 mph 27. 450.5 mi 28. 460.5 mi
29. 28°, 84°, 68° 30. 5°, 160°, 15° 31. 1984
33. 4 ft, 8 ft 34. 6 m, 4 m 35. 96°, 32°, 52°

36. 100°, 25°, 55° 37. 9, 11, 13
39. 46 cm, 55 cm 40. 86 41. 84 42. $100
43. 98 45. 1776 + (4s + 7) = 1863; a score is 20 yrs.
46. $1.50 47. 19¢ 48. 19 49. 12 cm, 9 cm
51. 7.5% 52. 76 53. 5 half dollars, 10 quarters, 20 dimes, 60 nickels 54. ≈ 0.65 in.
55. 25,599.9375 ft²

Summary and Review – Chapter 2

1. [2.1] -22 2. [2.1] 7 3. [2.1] -32 4. [2.1] 1
5. [2.1] $-\frac{7}{3}$ 6. [2.1] 25 7. [2.1] $\frac{1}{2}$ 8. [2.1] $-\frac{15}{64}$
9. [2.1] 9.99 10. [2.2] -8 11. [2.2] -5
12. [2.2] $-\frac{1}{3}$ 13. [2.2] 4 14. [2.2] 3 15. [2.2] 4
16. [2.2] 16 17. [2.3] 6 18. [2.3] -3
19. [2.3] 12 20. [2.3] 4 21. [2.1] -12, 12
22. [2.2] $-\frac{41}{7}$, 7 23. [2.4] $d = \frac{C}{\pi}$ 24. [2.4] $B = \frac{3V}{h}$
25. [2.4] $a = 2A - b$ 26. [2.5] $591 27. [2.5] 27
28. [2.5] 1. Familiarize; 2. Translate; 3. Carry out; 4. Check; 5. State 29. [2.5] 67 30. [2.5] 30
31. [2.5] 32 32. [2.6] 250 33. [2.6] 250
34. [2.6] $14,200 35. [2.7] 3 m, 5 m 36. [2.7] 9
37. [2.7] 57, 59 38. [2.7] w = 11 cm, ℓ = 17 cm
39. [2.7] $220 40. [2.7] $26,087
41. [2.7] 35°, 85°, 60° 42. [2.7] Amazon: 6437 km; Nile: 6671 km 43. [2.7] 140 44. [2.6] $14,150
45. [2.6] (a) 250; (b) 226 46. [2.6] 0.0000006%
47. [2.6] 40% 48. [2.6] No difference
49. [2.7] No solution 50. [2.4] $a = \frac{y - 3}{2 - b}$

CHAPTER 3

ASVAB Test – Arithmetic Reasoning   Chapter 3

1. D 2. D 3. A 4. C

Margin Exercises, Section 3.1

1. $x^2 + 2x - 8$; $x^3 + x^2 - 2x - 5$; $y - 7$; answers may vary 2. -19 3. -104 4. -11 5. 8 6. 210
7. 100π mi² 8. $-9x^3 + (-4x^5)$ 9. $-2y^3 + 3y^7 + (-7y)$
10. $3x^2$; $6x$; $\frac{1}{2}$ 11. $-4y^5$; $7y^2$; $-3y$; -2 12. $4x^3$ and $-x^3$ 13. $4t^4$ and $-7t^4$; $-9t^3$ and $10t^3$ 14. $8x^2$
15. $2t^3 + 7$ 16. $-\frac{1}{4}y^5 + 2y^2$ 17. $-4y^3$ 18. $7x^3$
19. $25 - 3m^5$ 20. $6y$ 21. $4x^3 + 4$
22. $-\frac{1}{4}p^3 + 4p^2 + 7$ 23. $3t^2 + t^3 + 9$

Exercise Set 3.1

1. -18 3. 19 4. 59 5. -12 6. 51 7. Approx 449 9. 2 10. 8 11. 4 12. -10 13. 11 15. 68 ft 16. 114 m 17. 1024 ft 18. 144 ft 19. 400 gal 21. 2, -3x, x² 22. 2x², 3x, -4 23. 6x² and -3x² 24. 3x² and -2x² 25. 2x⁴ and -3x⁴, 5x and -7x
27. -3x 28. 10x² 29. -8x 30. -4x 31. 11x³ + 4
33. $x^3 - x$ 34. $4a^4$ 35. $4b^5$ 36. $6x^2 - 3x$
37. $\frac{3}{4}x^5 - 2x - 42$ 39. x^4 40. $-x^3$
41. $\frac{15}{16}x^3 - \frac{7}{6}x^2$ 42. 0 43. $3x^2 + 2x + 1$
45. $3x^6$ 46. For s = 18: 99, -99; For s = 25: 50, -50; For s = 32: 99, -99 47. 50

Margin Exercises, Section 3.2

1. $6x^7 + 3x^5 - 2x^4 + 4x^3 + 5x^2 + x$ 2. $7x^5 - 5x^4 + 2x^3 + 4x^2 - 3$ 3. $14t^7 - 10t^5 + 7t^2 - 14$ 4. $-2x^2 - 3x + 2$ 5. $10x^4 - 8x - \frac{1}{2}$ 6. 4, 2, 1, 0; 4 7. 5, 6, 1, -1, 4 8. x 9. x^3, x^2, x, x^0 10. x^2, x 11. x^3
12. Monomial 13. None of these 14. Binomial
15. Trinomial

Exercise Set 3.2

1. $x^5 + 6x^3 + 2x^2 + x + 1$ 3. $15x^9 + 7x^8 + 5x^3 - x^2 + x$
4. $x^5 - 5x^4 + 6x^3 + 9x - 5$ 5. $-5y^8 + y^7 + 9y^6 + 8y^3 - 7y^2$ 6. $p^8 - 7p^4 + p^2 + p - 4$ 7. $x^6 + x^4$ 9. $13x^3 - 9x + 8$ 10. $x^2 - 4x + 1$ 11. $-5x^2 + 9x$ 12. $-4x^3 - 6x$

13. $12x^4 - 2x + \frac{1}{4}$ 15. 1, 0; 1 16. 0, 1; 1 17. 2, 1, 0; 2 18. 3, 2, 0; 3 19. 3, 2, 1, 0; 3 21. 2, 1, 6, 4; 6 22. 1, 2, 0, 3; 3 23. -3, 6 24. 2, -4 25. 5, 3, 3 27. -7, 6, 3, 7 28. 5, 1, -1, 2 29. -5, 6, -3, 8, -2 30. 7, -4, -4, 5 31. x^2, x 33. x^3, x^2, x^0 34. x^3, x^2 35. None missing 36. x^2, x, x^0 37. Trinomial 39. None of these 40. Binomial 41. Binomial 42. None of these 43. Monomial 45. \$0.125 46. 27 47. Any ax^5 plus 3 terms of degree < 5 48. Answers may vary. $\frac{1}{2}y^4 + 4.7y^3 - 2$ 49. 10 51. $ax^3 + (a - 3)x^2 + 3(a - 3)x + (a + 2)$, $a + a - 3 + 3(a - 3) + a + 2 = -4$, $a = 1$, $x^3 - 2x^2 - 6x + 3$ is the polynomial

Margin Exercises, Section 3.3
1. $x^2 + 7x + 3$ 2. $-4y^5 + 7y^4 + 3y^3 + 2y^2 + 4$
3. $24x^4 + 5x^5 + x^2 + 1$ 4. $2x^3 + \frac{10}{3}$ 5. $2x^2 - 3x - 1$
6. $8t^3 - 2t^2 - 8t + \frac{5}{2}$ 7. $-8x^4 + 4x^3 + 12x^2 + 5x - 8$
8. $-x^3 + x^2 + 3x + 3$ 9. $\frac{7}{2}x^2$ 10. 224

Exercise Set 3.3
1. $-x + 5$ 3. $x^2 - 5x - 1$ 4. $6x^4 + 3x^3 + 4x^2 - 3x + 2$ 5. $3x^5 + 13x^2 + 6x - 3$ 6. $7x^3 + 6x - 6$ 7. $-4x^4 + 6x^3 + 6x^2 + 2x + 4$ 9. $6 + 12x^2$ 10. $3x^4 - 4x^3 + x^2 + x + 4$ 11. $5x^4 - 2x^3 - 7x^2 - 5x$ 12. $8x^5 - 4x^4 - 3x^3 - x + 5$ 13. $9x^8 + 8x^7 - 3x^4 + 2x^2 - 2x + 5$
15. $-\frac{1}{2}x^4 + \frac{2}{3}x^3 + x^2$ 16. $\frac{2}{15}x^9 - \frac{2}{5}x^5 + \frac{1}{4}x^4 + \frac{1}{4}x^2 + \frac{15}{2}$
17. $0.01x^5 + x^4 - 0.2x^3 + 0.2x + 0.06$ 18. $0.10x^6 + 0.02x^3 + 0.22x + 0.55$ 19. $-3x^4 + 3x^2 + 4x$ 21. $3x^5 - 3x^4 - 3x^3 + x^2 + 3x$ 22. $4x^5 - 4x^4 - 3x^3 + 2x^2 + 2x$ 23. $5x^3 - 9x^2 + 4x - 7$ 24. $3x^3 - 9x^2 + 7x - 2$ 25. $\frac{1}{4}x^4 - \frac{1}{4}x^3 + \frac{3}{2}x^2 + 6\frac{3}{4}x + \frac{1}{4}$ 27. $-x^4 + 3x^3 + 2x + 1$ 28. $-2x^4 + 3x^3 + 5x^2 + 3x + 2$ 29. $x^4 + 4x^2 + 12x - 1$ 30. $2x^4 + 10x - 2$ 31. $x^5 - 6x^4 + 4x^3 - x^2 + 1$ 33. $7x^4 - 2x^3 + 7x^2 + 4x + 9$ 34. $5x^4 - 3x^3 + 4x^2 + 3x + 6$ 35. $-6 + 3x + x^2 + 3x^3 + x^4 + 3x^5$ 36. $3 + 2x + 3x^2 + 2x^3 - 2x^4 - 2x^5$ 37. $1.05x^4 + 0.36x^3 + 14.22x^2 + x + 0.97$ 39. (a) $5x^2 + 4x$; (b) 57, 352 40. (a) $r^2\pi + 13\pi$; (b) 38π, 140.69π 41. $14y + 17$ 42. $11\frac{1}{2}a + 9$ 43. $9r + 99 + r^2 + 11r = r^2 + 20r + 99$ 45. $(x + 3)^2 = x^2 + 3x + 3x + 3^2 = x^2 + 6x + 9$ 46. $48.544x^6 - 0.795x^5 + 890x$ 47. (a) Compare $(ax + b) + (cx + d)$ and $(cx + d) + (ax + b)$. $(ax + b) + (cx + d) = ax + b + cx + d = (a + c)x + (b + d)$ (collecting like terms); $(cx + d) + (ax + b) = cx + d + ax + b = (c + a)x + (d + b) = (a + c)x + (b + d)$ (commutative property of real numbers). Thus, addition of these kind of binomials is commutative. (b) Compare $(ax^2 + bx + c) + (dx^2 + ex + f)$ and $(dx^2 + ex + f) + (ax^2 + bx + c)$ as in part (a). 48. $n^2 + 2n - 3$ 49. 20, 25, 30

Margin Exercises, Section 3.4
1. $-(12x^4 - 3x^2 + 4x)$, $-12x^4 + 3x^2 - 4x$ 2. $-(-4x^4 + 3x^2 - 4x)$, $4x^4 - 3x^2 + 4x$ 3. $-(-13x^6 + 2x^4 - 3x^2 + x - \frac{5}{13})$, $13x^6 - 2x^4 + 3x^2 - x + \frac{5}{13}$ 4. $-(-7y^3 + 2y^2 - y + 3)$, $7y^3 - 2y^2 + y - 3$ 5. $-4x^3 + 6x - 3$ 6. $-5x^4 - 3x^2 - 7x + 5$ 7. $-14x^{10} + \frac{1}{2}x^5 - 5x^3 + x^2 - 3x$ 8. $2x^3 + 2x + 8$ 9. $y^2 - 6y - 2$ 10. $-8x^4 - 5x^3 + 8x^2 - 1$ 11. $x^3 - x^2 - \frac{4}{3}x - 0.9$ 12. $2t^3 + 5t^2 - 2t - 5$ 13. $-x^5 - 2x^3 + 3x^2 - 2x + 2$ 14. $4x^2 - \pi x^2$, or $(4 - \pi)x^2$

Exercise Set 3.4
1. $-(-5x)$, $5x$ 3. $-(-x^2 + 10x - 2)$, $x^2 - 10x + 2$ 4. $-(-4x^3 - x^2 - x)$, $4x^3 + x^2 + x$ 5. $-(12x^4 - 3x^3 + 3)$, $-12x^4 + 3x^3 - 3$ 6. $-(4x^3 - 6x^2 - 8x + 1)$, $-4x^3 +$

$6x^2 + 8x - 1$ 7. $-3x + 7$ 9. $-4x^2 + 3x - 2$ 10.
$6a^3 - 2a^2 + 9a - 1$ 11. $4x^4 + 6x^2 - \frac{3}{4}x + 8$ 12. $5x^4 - 4x^3 + x^2 - 0.9$ 13. $2x^2 + 14$ 15. $-2x^5 - 6x^4 + x + 2$ 16. $-x^2 - 2x + 4$ 17. $9x^2 + 9x - 8$ 18. $7x^3 + 3x^2 + 2x - 1$ 19. $\frac{3}{4}x^3 - \frac{1}{2}x$ 21. $0.06x^3 - 0.05x^2 + 0.01x + 1$ 22. $0.1x^4 - 0.9$ 23. $3x + 6$ 24. $-x^2 + 1$ 25. $4x^3 - 3x^2 + x + 1$ 27. $11x^4 + 12x^3 - 9x^2 - 8x - 9$ 28. $5x^4 - 6x^3 - x^2 + 5x + 15$ 29. $-4x^5 + 9x^4 + 6x^2 + 16x + 6$ 30. $-4x^5 - 6x^3 + 8x^2 - 5x - 2$ 31. $x^4 - x^3 + x^2 - x$ 33. $\pi r^2 - 9\pi$ 34. $m^2 - 28$ 35. $z^2 - 24(z - 3) - (24 \cdot 3) - 3(z - 24) = z^2 - 27z + 72$ 36. $(\pi - 2)x^2$ 37. $144 - 4x^2$ 39. $y - 9$ 40. $12y^2 - 23y + 21$ 41. $11a^2 - 18a - 4$ 42. $5x^2 - 9x - 1$ 43. $-10y^2 - 2y - 10$ 45. $-3y^4 - y^3 + 5y - 2$ 46. $2 + x + 2x^2 + 4x^3$ 47. $569.607x^3 - 15.168x$ 48. No, $-3(-x)^2 = -3x^2$

Margin Exercises, Section 3.5
1. $-15x$ 2. $-x^2$ 3. x^2 4. $-x^5$ 5. $12x^7$ 6. $-8y^{11}$ 7. $7y^5$ 8. 0 9. $8x^2 + 16x$ 10. $-15t^3 + 6t^2$ 11. $x^2 + 13x + 40$ 12. $x^2 + x - 20$ 13. $5x^2 - 17x - 12$ 14. $6x^2 - 19x + 15$ 15. $x^4 + 3x^3 + x^2 + 15x - 20$ 16. $6y^5 - 20y^3 + 15y^2 + 14y - 35$ 17. $3x^3 + 13x^2 - 6x + 20$ 18. $20x^4 - 16x^3 + 32x^2 - 32x - 16$ 19. $6x^4 - x^3 - 18x^2 - x + 10$

Exercise Set 3.5
1. $42x^2$ 3. x^4 4. $-x^6$ 5. $-x^8$ 6. x^8 7. $6x^6$ 9. $28t^8$ 10. $30a^4$ 11. $-0.02x^{10}$ 12. $-0.12x^9$ 13. $\frac{1}{15}x^4$ 15. 0 16. $-3x^2 + 15x$ 17. $8x^2 - 12x$ 18. $12x^3 + 24x^2$ 19. $-10x^3 + 5x^2$ 21. $-4x^4 + 4x^3$ 22. $18y^6 + 24y^5$ 23. $4y^7 - 24y^6$ 24. $42x^{54} + 60x^{15} + 18x^{61} + 180x^{19}$ 25. $20x^{38} - 50x^{25} + 25x^{14}$ 27. $-66y^{108} + 42y^{58} - 66y^{49} + 360y^{12} - 54y^8$ 28. $x^2 + 9x + 18$ 29. $x^2 + 7x + 10$ 30. $x^2 + 3x - 10$ 31. $x^2 + 4x - 12$ 33. $x^2 - 10x + 21$ 34. $x^2 - 9$ 35. $x^2 - 36$ 36. $25 - 15x + 2x^2$ 37. $18 + 12x + 2x^2$ 39. $9x^2 - 24x + 16$ 40. $9y^2 - 16$ 41. $4y^2 - 1$ 42. $x^2 - \frac{21}{10}x - 1$ 43. $x^2 + \frac{17}{6}x + 2$ 45. $x^3 + x^2 + 4$ 46. $4x^3 + 14x^2 + 8x + 1$ 47. $12x^3 - 10x^2 - x + 1$ 48. $3y^4 - 6y^3 - 7y^2 + 18y - 6$ 49. $3y^4 + 18y^3 - 18y - 3$ 51. $x^6 - 2x^5 + 2x^4 - x^3$ 52. $-10x^5 - 9x^4 + 7x^3 + 2x^2 - x$ 53. $-20x^5 + 25x^4 - 4x^3 - 5x^2 - 2$ 54. $1 - 2x - x^2 + x^4$ 55. $1 - 2x + 3x^2 - 2x^3 + x^4$ 57. $4x^4 - 12x^3 - 5x^2 + 17x + 6$ 58. $6t^4 + t^3 - 16t^2 - 7t + 4$ 59. $6a^4 - 19a^3 + 31a^2 - 26a + 8$ 60. $-4x^4 + 6x^3 - 2x^2 - 13x + 10$ 61. $-4x^4 + 4x^3 + 36x^2 - 20x + 2$ 63. $3x^3 + 3x^7 + 3x^{11}$ 64. $x^4 - 1$ 65. $x^4 - 3x^3 + 3x^2 - 4x + 4$ 66. $x^4 - 2x^3 - 4x^2 + 9$ 67. $x^4 + 3x^3 - 5x^2 + 16$ 69. $a^2 - 2ab + b^2$ 70. $4x^2 + 12x + 9$ 71. $25y^2 + 60y + 36$ 72. $84y^2 - 30y$ 73. $84t^2 + 32t - (6t^2 - 8t) = 78t^2 + 40t$ 75. $x^3 - 5x^2 + 8x - 4$ 76. h = 10 ft, b = 6 ft 77. 8 ft × 16 ft 78. (a) $2x^2 + 18x + 36$; (b) 0 79. (a) $2x^2 - 18x + 28$; (b) 0 81. (a) $2x^2 - 112$; (b) $-6x$

Margin Exercises, Section 3.6
1. $8x^3 - 12x^2 + 16x$ 2. $10y^6 + 8y^5 - 10y^4$ 3. $x^2 + 7x + 12$ 4. $x^2 - 2x - 15$ 5. $2t^2 + 9t + 4$ 6. $2x^3 - 4x^2 - 3x + 6$ 7. $12x^5 + 6x^2 + 10x^3 + 5$ 8. $y^6 - 49$ 9. $-2x^7 + x^5 + x^3$ 10. $x^2 - \frac{16}{25}$ 11. $x^5 + 0.5x^3 - 0.5x^2 - 0.25$ 12. $8 + 2x^2 - 15x^4$ 13. $30y^5 - 3y^4 - 6y^3$

Exercise Set 3.6
1. $4x^2 + 4x$ 3. $-3x^2 + 3x$ 4. $5x^2 + 5x$ 5. $x^5 + x^2$ 6. $-2x^5 + 2x^3$ 7. $6x^3 - 18x^2 + 3x$ 9. $x^3 + 3x + x^2 + 3$ 10. $x^3 - x^2 - 3x + 3$ 11. $x^4 + x^3 + 2x + 2$ 12. $x^5 + 12x^4 + 2x + 24$ 13. $y^2 - y - 6$ 15. $9x^2 + 15x + 6$ 16. $8x^2 + 10x + 2$ 17. $5x^2 + 4x - 12$ 18. $x^2 - 64$ 19. $9t^2 - 1$ 21. $4x^2 - 6x + 2$ 22. $6x^2 - x - 1$ 23. $p^2 - \frac{1}{16}$ 24. $q^2 + \frac{3}{2}q + \frac{9}{16}$ 25. $x^2 - 0.01$

27. $2x^3 +$
$2x^2 + 6x + 6$ 28. $4x^3 - 2x^2 + 6x - 3$ 29. $-2x^2 -$
$11x + 6$ 30. $6x^2 - 4x - 16$ $\underline{31}$. $a^2 + 14a + 49$ 33.
$1 - x - 6x^2$ 34. $-3x^2 - 5x - 2$ 35. $x^5 - x^2 + 3x^3 - 3$
36. $2x^5 + x^4 - 6x - 3$ $\underline{37}$. $x^3 - x^2 - 2x + 2$ 39. $3x^6 -$
$6x^2 - 2x^4 + 4$ 40. $x^{20} - 9$ 41. $6x^7 + 18x^5 + 4x^2 + 12$
42. $1 + 3x^2 - 2x - 6x^3$ $\underline{43}$. $8x^6 + 65x^3 + 8$ 45. $4x^3 -$
$12x^2 + 3x - 9$ 46. $14x^2 - 53x + 14$ 47. $4y^6 + 4y^5 +$
$y^4 + y^3$ 48. $10y^{12} + 16y^9 + 6y^6$ $\underline{49}$. $8y^3 + 72y^2 +$
$160y$ 51. $-2x^4 + x^3 + 5x^2 - x - 2$ $\underline{52}$. $16x^4 - 1$
53. -7 54. 0 $\underline{55}$. $-\frac{5}{2}$ 57. $w(w + 1)(w + 2) = w^3 +$
$3w^2 + 2w$ 58. $\ell(\ell - 1)(\ell + 1) = \ell^3 - \ell$ 59.
$h(h - 1)(h - 2) = h^3 - 3h^2 + 2h$
60. $Q^2 - 14Q - 5(Q - 14)$ or $(Q - 14)(Q - 5) = Q^2 -$
$19Q + 70$ $\underline{61}$. $F^2 - (F - 17)(F - 7)$ or $17F + 7(F - 17)$
$= 24F - 119$.   63. Each trip is $0.70 + (3\frac{6}{7})(7)(0.10) =$
3.40; $3.40(11x + 6) = 37.40x + 20.40$

Margin Exercises, Section 3.7
1. $x^2 - 25$ 2. $4y^2 - 9$ 3. $t^2 - 4$ 4. $x^2 - 49$ 5.
$16 - 25t^2$ 6. $4x^6 - 1$ 7. $x^2 + 16x + 64$ 8. $x^2 -$
$10x + 25$ 9. $x^2 + 4x + 4$ 10. $a^2 - 8a + 16$ 11.
$4x^2 + 20x + 25$ 12. $16x^4 - 24x^3 + 9x^2$ 13. $49 + 14y +$
y^2 14. $9x^4 - 30x^2 + 25$ 15. $x^2 + 11x + 30$ 16. $t^2 -$
16 17. $-8x^5 + 20x^4 + 40x^2$ 18. $81x^4 + 18x^2 + 1$
19. $4a^2 + 6a - 40$ 20. $25x^2 + 5x + \frac{1}{4}$ 21. $4x^2 - 2x + \frac{1}{4}$

Exercise Set 3.7
$\underline{1}$. $x^2 - 16$ 3. $4x^2 - 1$ 4. $x^4 - 1$ 5. $25m^2 - 4$
$\overline{6}$. $9x^8 - 4$ $\underline{7}$. $4x^4 - 9$ 9. $9x^8 - 16$ 10. $t^4 - 0.04$
11. $x^{12} - x^4$ 12. $4x^6 - 0.09$ $\underline{13}$. $x^8 - 9x^2$
15. $x^{24} - 9$ 16. $144 - 9x^4$ 17. $4y^{16} - 9$
18. $m^2 - \frac{4}{9}$ $\underline{19}$. $x^2 + 4x + 4$ 21. $9x^4 + 6x^2 + 1$
22. $9x^2 + \frac{9}{2}x + \frac{9}{16}$
23. $a^2 - a + \frac{1}{4}$ 24. $4a^2 - \frac{4}{5}a + \frac{1}{25}$
$\underline{25}$. $9 + 6x + x^2$ 27. $x^4 + 2x^2 + 1$ 28. $64x^2 - 16x^3 +$
x^4 29. $4 - 12x^4 + 9x^8$ 30. $36x^6 - 24x^3 + 4$ $\underline{31}$. $25 +$
$60t^2 + 36t^4$ 33. $9 - 12x^3 + 4x^6$ 34. $x^2 - 8x^4 + 16x^6$
35. $4x^3 + 24x^2 - 12x$ 36. $-8x^6 + 48x^3 + 72x$ $\underline{37}$.
$4x^4 - 2x^2 + \frac{1}{4}$ 39. $-1 + 9p^2$ 40. $-9q^2 + 4$ 41.
$15t^5 - 3t^4 + 3t^3$ 42. $-6x^5 - 48x^3 + 54x^2$ $\underline{43}$. $36x^8 +$
$48x^4 + 16$ 45. $12x^3 + 8x^2 + 15x + 10$ 46. $6x^4 - 3x^2 - 63$
47. $64 - 96x^4 + 36x^8$ 48. $\frac{3}{25}x^4 + 4x^2 - 63$
$\underline{49}$. $4567.0564x^2 + 435.891x + 10.400625$
51. $(100 - 7)(100 + 7) = 10,000 - 49 = 9951$
52. $16x^4 - 1$ 53. $a^2 + 10a + 24$ 54. $5a^2 + 12a - 9$
$\underline{55}$. $x^2 + 10x + 25$ 57. 6 58. -10
59. $\frac{1}{17}$ 60. (a) ac, ad, bc, bd; (b) $ac + ad + bc + bd$;
(c) $ac + ad + bc + bd$ $\underline{61}$. (a) $a^2 + ab$; (b) $ab + b^2$;
(c) $a^2 + ab - (b^2 + ab)$; (d) $a(a - b) =$
$b(a - b) = a^2 - b^2$ 63. $100(x^2 + x) + 25$. Add the
first digit to its square, multiply by 100, and add 25.

Summary and Review - Chapter 3
1. [3.1] -17 2. [3.1] 10 3. [3.1] -26 4. [3.1] 0
5. [3.1] 0 6. [3.1] -20 7. [3.1] 50 8. [3.1] 0
9. [3.1] 0 10. [3.1] 100 11. [3.1] 0 12. [3.1] 0
13. [3.1] $3x^2$, $6x$, $\frac{1}{2}$ 14. [3.1] $-4y^5$, $7y^2$, $-3y$, -2
15. [3.2] x^2, x^0 16. [3.2] 6, 17 17. [3.2] 4, 6,
-5, $\frac{5}{3}$ 18. [3.2] $3,1,0;3$ 19. [3.2] $0,4,9,6,3$; 9
20. [3.2] Binomial   21. [3.2] None of these   22.
[3.2] Monomial   23. [3.1] $-x^2 + 9x$ 24. [3.1] $-\frac{1}{4}x^3 +$
$4x^2 + 7$ 25. [3.1] $-3x^5 + 25$ 26. [3.2] $-2x^2 - 3x + 2$
27. [3.2] $10x^4 - 7x^2 - x - \frac{1}{2}$ 28. [3.3] $x^5 - 2x^4 +$
$6x^3 + 3x^2 - 9$ 29. [3.3] $2x^5 - 6x^4 + 2x^3 - 2x^2 + 2$

30. [3.3] $\frac{3}{4}x^4 + \frac{1}{4}x^3 - \frac{1}{3}x^2 - \frac{7}{4}x + \frac{3}{8}$ 31. [3.4] $2x^2 -$
$4x - 6$ 32. [3.4] $x^5 - 3x^3 - 2x^2 + 8$ 33. [3.4]
$-x^5 + x^4 - 5x^3 - 2x^2 + 2x$ 34. (a) [3.3] $4L + 8$; (b)
[3.4] $L^2 + 4L$ 35. [3.5] $-12x^3$ 36. [3.7] $49x^2 +$
$14x + 1$ 37. [3.6] $x^2 + \frac{7}{6}x + \frac{1}{3}$ 38. [3.6] $0.3x^2 +$
$0.65x - 8.45$ 39. [3.5] $12x^3 - 23x^2 + 13x - 2$ 40.
[3.7] $x^2 - 18x + 81$ 41. [3.5] $15x^7 - 40x^6 + 50x^5 +$
$10x^4$ 42. [3.6] $x^2 - 3x - 28$ 43. [3.6] $x^2 - 1.05x +$
0.225 44. [3.5] $x^7 + x^5 - 3x^4 + 3x^3 - 2x^2 + 5x - 3$
45. [3.7] $9y^4 - 12y^3 + 4y^2$ 46. [3.6] $2t^4 - 11t^2 - 21$
47. [3.5] $4x^5 - 5x^4 - 8x^3 + 22x^2 - 15x$ 48. [3.7]
$9x^4 - 16$ 49. [3.7] $4 - x^2$ 50. [3.6] $13x^2 - 172x +$
39 51. [3.5-3.7] (a) 3; (b) 2; (c) 1 52. [3.1]
$-28x^8$ 53. [3.2] $8x^4 + 4x^3 + 5x - 2$ 54. [3.6] $-4x^6 +$
$3x^4 - 20x^3 + x^2 - 16$ 55. [3.7] $\frac{94}{13}$

ASVAB Test - Numerical Operations   Chapter 3
1. B   2. A   3. C   4. B   5. C   6. D   7. B   8. B
9. C   10. C   11. B   12. C   13. A   14. C   15. D
16. B   17. C   18. A   19. D   20. A   21. B   22. C
23. C   24. B   25. D   26. A   27. B   28. D   29. A
30. D   31. C   32. D   33. A   34. C   35. D   36. B
37. C   38. D   39. A   40. D   41. B   42. C   43. A
44. C   45. C   46. D   47. A   48. B   49. D   50. C

CHAPTER 4

Margin Exercises, Section 4.1
1. (a) $12x^2$; (b) $3x\cdot6x$, $2x\cdot6x$, answers may vary   2.
(a) $16x^3$; (b) $(2x)(8x^2)$, $(4x)(4x^2)$, answers may vary
3. $8x\cdot x^3$; $4x^2\cdot2x^2$; $2x^3\cdot4x$, answers may vary   4. $7x\cdot3x$,
$(-7x)(-3x)$, $(21x^2)(x)$, answers may vary   5. $6x^4\cdot x$,
$(-2x^3)(-3x^2)$, $(3x^3)(2x^2)$, answers may vary   6. (a)
$3x + 6$; (b) $3(x + 2)$ 7. (a) $2x^3 + 10x^2 + 8x$; (b)
$2x(x^2 + 5x + 4)$ 8. $x(x + 3)$ 9. $x^2(3x^4 - 5x + 2)$
10. $3x^2(3x^2 - 5x + 1)$ 11. $\frac{1}{4}(3x^3 + 5x^2 + 7x + 1)$
12. $7x^3(5x^4 - 7x^3 + 2x^2 - 9)$ 13. $(x + 2)(x + 5)$
14. $(y + 3)(y - 4)$ 15. $(2x - 3)(2x^2 - 3)$
16. $(4t + 1)(2t^2 + 3)$ 17. $(3m^3 + 2)(m^2 - 5)$

Exercise Set 4.1
$\underline{1}$. Answers may vary. $(6x)(x^2)$, $(3x^2)(2x)$, $(2x^2)(3x)$
$\overline{3}$. Answers may vary. $(-3x^2)(3x^3)$, $(-x)(9x^4)$,
$(3x^2)(-3x^3)$ 4. Answers may vary. $(-4x)(3x^5)$,
$(-6x^2)(2x^4)$, $(12x^3)(-x^3)$ 5. Answers may vary.
$(6x)(4x^3)$, $(-3x^2)(-8x^2)$, $(2x^3)(12x)$ 6. Answers may
vary. $(3x^2)(5x^3)$, $(15x)(x^4)$, $(-5x^4)(-3x)$ $\underline{7}$. $x(x - 4)$
9. $2x(x + 3)$ 10. $3x(x - 1)$ 11. $x^2(x + 6)$
12. $x^2(4x^2 + 1)$ $\underline{13}$. $8x^2(x^2 - 3)$ 15. $2(x^2 + x - 4)$
16. $3(2x^2 + x - 5)$ 17. $17x(x^4 + 2x^2 + 3)$
18. $16x(x^5 - 2x^4 - 3)$ $\underline{19}$. $x^2(6x^2 - 10x + 3)$
21. $x^2(x^3 + x^2 + x - 1)$ $\underline{22}$. $x^3(x^6 - x^4 + x + 1)$
23. $2x^3(x^4 - x^3 - 32x^2 + 2)$ 24. $5(2x^3 + 5x^2 + 3x - 4)$
$\underline{25}$. $0.8x(2x^3 - 3x^2 + 4x + 8)$
27. $\frac{1}{3}x^3(5x^3 + 4x^2 + x + 1)$ 28. $\frac{1}{7}x(5x^6 + 3x^4 - 6x^2 - 1)$
29. $(y + 1)(y + 4)$ 30. $(x + 2)(x + 5)$
$\underline{31}$. $(x - 1)(x - 4)$ 33. $(2x + 3)(3x + 2)$
$\overline{34}$. $(x + 1)(3x - 2)$ 35. $(x - 4)(3x - 4)$
36. $(6 - 5y)(4 - 3y)$ $\underline{37}$. $(5x + 3)(7x - 8)$
39. $(2x - 3)(2x + 3)$ $\overline{40}$. $(2x^2 - 5)(x^2 - 3)$
41. $(2x^2 + 5)(x^2 + 3)$ 42. $(2x^2 - 3)(2x^2 - 3)$
$\underline{43}$. $(2x^2 + 1)(x + 3)$ 45. $(2x - 3)(4x^2 + 3)$
$\overline{46}$. $(2x - 5)(5x^2 + 2)$ 47. $(3x - 4)(4x^2 + 1)$
48. $(6x - 7)(3x^2 + 5)$ $\underline{49}$. $(x + 8)(x^2 - 3)$
51. No   52. Yes   53. No   54. Yes $\underline{55}$. No   57. Yes
58. Yes   59. No   60. No $\underline{61}$. $(2x^2 + 3)(2x^3 + 3)$
63. $(x^4 + 1)(x^2 + 1)$ 64. $(x^7 + 1)(x^6 + 1)$
65. $x^2(x + 1)^2 - (x^2 + 1)^2 = x^4 + 2x^3 + x^2 -$
$(x^4 + 2x^2 + 1) = 2x^3 - 2x^2 + x^2 - 1 = 2x^2(x - 1) +$
$(x + 1)(x - 1) = (2x^2 + x + 1)(x - 1)$

Margin Exercises, Section 4.2

1. (a), (e), (f), (g)   2. $(x - 3)(x + 3)$
3. $(t + 8)(t - 8)$ 4. $8y^2(2 + y^2)(2 - y^2)$
5. $x^4(8 + 5x)(8 - 5x)$ 6. $5(1 + 2x^3)(1 - 2x^3)$
7. $(9x^2 + 1)(3x + 1)(3x - 1)$ 8. $t^4(7 + 5t^3)(7 - 5t^3)$

Exercise Set 4.2

1. Yes   3. No   4. No   5. No   6. No   7. Yes   9. No
10. $(x - 2)(x + 2)$ 11. $(x - 6)(x + 6)$
12. $(x - 3)(x + 3)$ 13. $(x - 1)(x + 1)$
15. $(5x - 2)(5x + 2)$ 16. $(2x - 5)(2x + 5)$
17. $(3a - 4)(3a + 4)$ 18. $2(2x - 7)(2x + 7)$
19. $6(2x - 3)(2x + 3)$ 21. $x(4 - 9x)(4 + 9x)$
22. $x^2(4 - 5x)(4 + 5x)$ 23. $x^2(x^7 - 3)(x^7 + 3)$
24. $(7a^2 - 9)(7a^2 + 9)$ 25. $(5a^2 - 3)(5a^2 + 3)$
27. $(11a^4 - 10)(11a^4 + 10)$ 28. $(9y^3 + 5)(9y^3 - 5)$
29. $(10y^3 - 7)(10y^3 + 7)$ 30. $(x^2 + 1)(x + 1)(x - 1)$
31. $(x^2 + 4)(x + 2)(x - 2)$ 33. $5(x^2 + 4)(x + 2)(x - 2)$
34. $(1 + y^4)(1 + y^2)(1 + y)(1 - y)$
35. $(x^4 + 1)(x^2 + 1)(x + 1)(x - 1)$
36. $(x^6 + 4)(x^3 + 2)(x^3 - 2)$
37. $(x^4 + 9)(x^2 + 3)(x^2 - 3)$ 39. $(\frac{1}{5} + x)(\frac{1}{5} - x)$
40. $(5 + \frac{1}{7}x)(5 - \frac{1}{7}x)$ 41. $(2 + \frac{1}{3}y)(2 - \frac{1}{3}y)$
42. $(4 + t^2)(2 + t)(2 - t)$
43. $(1 + a^2)(1 + a)(1 - a)$ 45. $100°, 25°, 55°$
46. $4x^2(x - 1)(x + 1)$ 47. $3x^3(x + 2)(x - 2)$
48. $3(x - \frac{1}{3})(x + \frac{1}{3})$ 49. $2x(3x + \frac{2}{5})(3x - \frac{2}{5})$
51. $x(x + \frac{1}{1.3})(x - \frac{1}{1.3})$ 52. $(1.8x - 0.9)(1.8x + 0.9)$
53. $(0.8x + 1.1)(0.8x - 1.1)$ 54. $2(0.8x - 1)(0.8x + 1)$
55. $x(x + 6)$ 57. $3(a - 1)(3a + 11)$ 58. $4(y - 4)(y - 3)$
59. $(y^4 + 16)(y^2 + 4)(y + 2)(y - 2)$
60. $(x^8 + 1)(x^4 + 1)(x^2 + 1)(x + 1)(x - 1)$
61. $(x - \frac{1}{x})(x + \frac{1}{x})$ or $\frac{1}{x^2}(x^2 + 1)(x + 1)(x - 1)$
63. Irreducible   64. Irreducible   65. Prime,
irreducible   66. $16x^3 - 9x = x(16x^2 - 9) =$
$x(4x - 3)(4x + 3)$, Not irreducible, not prime
67. Irreducible

Margin Exercises, Section 4.3

1. (a), (b), (d), (f)   2. $(x + 1)^2$ 3. $(x - 1)^2$
4. $(t + 2)^2$ 5. $(5x - 7)^2$ 6. $(7 - 4y)^2$
7. $3(4m + 5)^2$

Exercise Set 4.3

1. Yes   3. No   4. No   5. No   6. No   7. No   9. No
10. $(x - 7)^2$ 11. $(x - 8)^2$ 12. $(x + 8)^2$
13. $(x + 7)^2$ 15. $(x + 1)^2$ 16. $(x + 2)^2$
17. $(x - 2)^2$ 18. $(y - 3)^2$ 19. $(y + 3)^2$
21. $2(x - 10)^2$ 22. $x(x - 9)^2$ 23. $x(x + 12)^2$
24. $5(2x + 5)^2$ 25. $3(2x + 3)^2$ 27. $(8 - 7x)^2$
28. $5(y^2 + 1)^2$ 29. $(a^2 + 7)^2$ 30. $(y^3 + 13)^2$
31. $(y^3 - 8)^2$ 33. $(3x^5 + 2)^2$ 34. $(2x^2 + 1)^2$
35. $(1 - a^3)^2$ 36. $(\frac{1}{9}x^3 + \frac{4}{3})^2$ 37. $(\frac{1}{3}a + \frac{1}{2})^2$
39. Length: 144.5 m, width: 125.5 m   40. Not possible
41. $x(27x^2 - 13)$ 42. $(x + 11)^2$ 43. Not possible
45. $2x(3x + 1)^2$ 46. $7(9x - 4)$ 47. $2(81x^2 - 41)$
48. $(x^2 + 3)(x^2 - 3)$ 49. $-\frac{1}{10}(9x - 8)(9x + 8)$
51. $(9 + x^2)(3 + x)(3 - x)$ 52. $(y + 4)^2$ 53. $(a + 3)^2$
54. $(2a + 15)^2$ 55. $(7x + 4)^2$ 57. $(a + 1)^2$ 58. Yes.
$(y + 2)^2(y - 2)^2 = [(y + 2)(y - 2)]^2 = (y^2 - 4)^2 =$
$y^4 - 8y^2 + 16$ 59. No. $(x + 3)^2(x - 3)^2 =$
$[(x + 3)(x - 3)]^2 = (x^2 - 9)^2 = x^4 - 18x^2 + 81$
60. $(3x^9 + 8)^2$ 61. $(x^n + 5)^2$ 63. $(y + 3)^2 - (x + 4)^2$
$= (y + x + 7)(y - x - 1)$ 64. 9   65. 16
66. $(x + 1)^2 - x^2 = x^2 + 2x + 1 - x^2 = 2x + 1$ 67.
$x^2 + a^2x + a^2 = x^2 + 2ax + a^2, a^2x = 2ax, a^2 = 2a$; then
$a = 0$ or $a = 2$.

Margin Exercises, Section 4.4

1. $(x + 3)(x + 4)$ 2. $(x + 9)(x + 4)$ 3. $(x - 5)(x - 3)$
4. $(t - 5)(t - 4)$ 5. $(x + 6)(x - 2)$ 6. $(y - 6)(y + 2)$
7. $(t + 7)(t - 2)$ 8. $(x - 6)(x + 5)$

Exercise Set 4.4

1. $(x + 5)(x + 3)$ 3. $(x + 4)(x + 3)$ 4. $(x + 1)(x + 8)$
5. $(x - 3)^2$ 6. $(y + 4)(y + 7)$ 7. $(x + 7)(x + 2)$
9. $(b + 4)(b + 1)$ 10. $(x - \frac{1}{5})^2$ 11. $(x + \frac{1}{3})^2$
12. $(z - 1)(z - 7)$ 13. $(d - 5)(d - 2)$
15. $(y - 10)(y - 1)$ 16. $(x + 3)(x - 5)$
17. $(x + 7)(x - 6)$ 18. $(x - 3)(x + 5)$
19. $(x + 2)(x - 9)$ 21. $(x + 2)(x - 8)$
22. $(x + 6)(x - 7)$ 23. $(y + 5)(y - 9)$
24. $(x + 5)(x - 12)$ 25. $(x + 9)(x - 11)$
27. $(c + 8)(c - 7)$ 28. $(b - 3)(b + 8)$
29. $(a + 7)(a - 5)$ 30. $(1 - x)(2 + x)$
31. $(x + 10)^2$ 33. $(x - 25)(x + 4)$ 34. $(x - 8)(x - 12)$
35. $(x - 24)(x + 3)$ 36. $4(x + 5)^2$ 37. $(x - 16)(x - 9)$
39. $(a + 12)(a - 11)$ 40. $(a - 6)(a + 15)$
41. $(x - 15)(x - 8)$ 42. $(d + 6)(d + 16)$
43. $(12 + x)(9 - x)$ 45. 15, -15, 27, -27, 51, -51
46. 49, -49, 23, -23, 5, -5   47. $(x + \frac{1}{4})(x - \frac{3}{4})$
48. $(x - \frac{1}{2})(x + \frac{1}{4})$ 49. $(x + 5)(x - \frac{5}{7})$
51. $2x^2(4 - \pi)$ 52. $x^2(\pi - 1)$

Margin Exercises, Section 4.5

1. $(2x + 5)(x - 3)$ 2. $(4x + 1)(3x - 5)$
3. $(3x - 4)(x - 5)$ 4. $2(5x - 4)(2x - 3)$
5. $(2x + 1)(3x + 2)$ 6. $3(2x + 3)(x + 1)$
7. $2(y + 3)(y - 1)$ 8. $2(2t + 3)(t - 1)$
9. $(2x - 1)(3x - 1)$

Exercise Set 4.5

1. $(2x + 1)(x - 4)$ 3. $(5x - 9)(x + 2)$
4. $(3x + 5)(x - 3)$ 5. $(2x + 7)(3x + 1)$
6. $(2x + 3)(3x + 2)$ 7. $(3x + 1)(x + 1)$
9. $(2x + 5)(2x - 3)$ 10. $(3x - 2)(3x + 4)$
11. $(2x + 1)(x - 1)$ 12. $(3x - 5)(5x + 2)$
13. $(3x + 8)(3x - 2)$ 15. $(3x + 1)(x - 2)$
16. $(6x - 5)(3x + 2)$ 17. $(3x + 4)(4x + 5)$
18. $(3x + 5)(5x - 2)$ 19. $(7x - 1)(2x + 3)$
21. $(3x + 4)(3x + 2)$ 22. $(2 - 3x)(3x - 1)$
23. $(7 - 3x)^2$ 24. $(5x + 4)^2$ 25. $(x + 2)(24x - 1)$
27. $(7x + 4)(5x - 11)$ 28. $(3a - 1)(3a + 5)$
29. $2(5 - x)(2 + x)$ 30. $(5 + 2x)(3 - x)$
31. $4(3x - 2)(x + 3)$ 33. $6(5x - 9)(x + 1)$
34. $5(4x - 1)(x - 1)$ 35. $2(3x + 5)(x - 1)$
36. $3(2x - 3)(3x + 1)$ 37. $(3x - 1)(x + 1)$
39. $4(3x + 2)(x - 3)$ 40. $3(2x - 1)(x - 5)$
41. $(2x + 1)(x - 1)$ 42. $(5x - 3)(3x - 2)$
43. $(3x - 8)(3x + 2)$ 45. $5(3x + 1)(x - 2)$
46. $(6x + 5)(3x - 2)$ 47. $x(3x + 4)(4x + 5)$
48. $x(5x - 2)(3x + 5)$ 49. $x^2(2x + 3)(7x - 1)$
51. $3x(8x - 1)(7x - 1)$ 52. 6366 km, 3947 miles
53. Lamps: 500 watts; air conditioner 2000 watts;
television: 50 watts   54. $(3x^2 + 2)(3x^2 + 4)$
55. $(3 - 2x)(2 - 3x)$ 57. $(5x^2 - 3)(3x^2 - 2)$
58. $2x(3x + 5)(x - 1)$ 59. $3x(2x - 3)(3x + 1)$
60. Not factorable   61. Not factorable
63. $(3x + 7)(3x - 7)(3x - 7)$ 64. $(10x^n + 3)(2x^n + 1)$
65. $(-3x^m + 4)(5x^m - 2)$ 66. $(x^{3a} - 1)(3x^{3a} + 1)$
67. $x(x^n - 1)^2$

Margin Exercises, Section 4.6

1. $3(m^2 + 1)(m - 1)(m + 1)$ 2. $(x^3 + 4)^2$
3. $2x^2(x + 1)(x + 3)$ 4. $(3x^2 - 2)(x + 4)$
5. $8x(x - 5)(x + 5)$

Exercise Set 4.6

1. $2(x + 8)(x - 8)$ 3. $(a - 5)^2$ 4. $(y + 7)^2$
5. $(2x - 3)(x - 4)$ 6. $(2y - 5)(4y + 1)$ 7. $x(x + 12)^2$
9. $(x - 2)(x + 2)(x + 3)$ 10. $(x - 3)(x^2 + x + 5)$
11. $6(2x + 3)(2x - 3)$ 12. $2(2x + 7)(2x - 7)$
13. $4x(x - 2)(5x + 9)$ 15. Not factorable   16. Not
factorable   17. $x(x - 3)(x^2 + 7)$ 18. $m(m^2 + 8)(m + 8)$
19. $x^3(x - 7)^2$ 21. $-2(x - 2)(x + 5)$
22. $-3(2x - 5)(x + 5)$ 23. Not factorable   24. Not
factorable   25. $4(x^2 + 4)(x + 2)(x - 2)$
27. $(y^4 + 1)(y^2 + 1)(y + 1)(1 - y)$
28. $(t^4 + 1)(t^2 + 1)(t + 1)(t - 1)$

29. $x^3(x - 3)(x - 1)$ 30. $x^4(x^2 - 2x + 7)$

<u>31</u>. $(6a - \frac{5}{4})^2$ 33. $(a + 1)^2(a - 1)^2$

34. Not factorable   35. $(3.5x - 1)^2$

36. $\frac{1}{5}(x - 4)(x - 1)$ <u>37</u>. $(x + 1.8)(5x + 4)$

39. $(y - 2)(y + 3)(y - 3)$ 40. $-(x^2 + 2)(x + 3)(x - 3)$

41. $(a + 4)(a^2 + 1)$ 42. $(x + 2)(x - 2)(x + 1)$

<u>43</u>. $(x + 3)(x - 3)(x^2 + 2)$ 45. $(x + 2)(x - 2)(x - 1)$

<u>46</u>. $(y - 7)(y + 3)(y + 1)$ 47. $(y - 1)^3$

48. $(x^4 + 16)(x^2 + 4)(x + 2)(x - 2)$

<u>49</u>. $(a^2 + 9)(a + 3)(a - 3)$

Margin Exercises, Section 4.7

1. $3, -4$ 2. $7, 3$ 3. $-\frac{1}{4}, \frac{2}{3}$ 4. $0, \frac{17}{3}$ 5. $-2, 3$

6. $7, -4$ 7. 3 8. $0, 4$ 9. $\frac{4}{5}, -\frac{4}{5}$

Exercise Set 4.7

<u>1</u>. $-8, -6$ 3. $3, -5$ 4. $-9, 3$ 5. $-12, 11$

<u>6</u>. $13, -53$ <u>7</u>. $0, -5$ 9. $0, 13$ 10. $0, 4$ 11. $0, -10$

12. $0, 21$ <u>13</u>. $-\frac{5}{2}, -4$ 15. $\frac{1}{3}, -2$ 16. $3, -3$

17. $-\frac{1}{5}, 3$ 18. $-\frac{9}{4}, \frac{1}{2}$ <u>19</u>. $4, \frac{1}{4}$ 21. $0, \frac{2}{3}$

22. $0, \frac{9}{8}$ 23. $0, 18$ 24. $0, 8$ <u>25</u>. $\frac{1}{9}, \frac{1}{10}$ 27. $2, 6$

28. $\frac{1}{21}, \frac{18}{11}$ 29. $\frac{1}{3}, 20$ 30. $3, 50$ <u>31</u>. $0, \frac{2}{3}, \frac{1}{2}$

33. $-5, -1$ 34. $-6, -1$ 35. $2, -9$ 36. $-7, 3$

<u>37</u>. $5, 3$ 39. $0, 8$ 40. $0, 3$ 41. $0, -19$ 42. $0, -12$

<u>43</u>. $4, -4$ 45. $\frac{2}{3}, -\frac{2}{3}$ 46. $\frac{3}{2}, -\frac{3}{2}$ 47. -3 48. -5

<u>49</u>. 4 51. $0, \frac{6}{5}$ 52. $0, \frac{8}{7}$ 53. $-1, \frac{5}{3}$ 54. $4, -\frac{5}{3}$

<u>55</u>. $\frac{2}{3}, -\frac{1}{4}$ 57. $7, -2$ 58. $\frac{2}{3}, -1$ 59. $\frac{9}{8}, -\frac{9}{8}$

60. $\frac{7}{10}, -\frac{7}{10}$ <u>61</u>. $-3, 1$ 63. $\frac{4}{5}, \frac{3}{2}$ 64. $\frac{5}{3}, -\frac{1}{4}$

65. $4, -5$ 66. 4 <u>67</u>. $9, -3$ 69. $\frac{1}{8}, -\frac{1}{8}$ 70. $\frac{5}{6}, -\frac{5}{6}$

71. $4, -4$ 72. $\frac{5}{9}, -\frac{5}{9}$ <u>73</u>. (a) $x^2 + 2x - 3 = 0$;

(b) $x^2 - 2x - 3 = 0$; (c) $x^2 - 4x + 4 = 0$; (d) $x^2 - 7x + 12 = 0$; (e) $x^2 + x - 12 = 0$; (f) $x^2 - x - 12 = 0$; (g) $x^2 + 7x + 12 = 0$; (h) $x^2 - x + \frac{1}{4} = 0$ or $4x^2 - 4x + 1 = 0$; (i) $x^2 - 25 = 0$; (j) $x^3 - \frac{7}{20}x^2 + \frac{1}{40}x = 0$ or $40x^3 - 14x^2 + x = 0$ 75. $-2000, -51.546392$

76. (a) i; (b) k; (c) g; (d) h; (e) j; (f) ℓ

Margin Exercises, Section 4.8

1. $5, -5$ 2. $7, 8$ 3. $0, 1$ 4. $-4, 5$ 5. Length: 5 cm; width: 3 cm   6. 9 7. 21 and 22; -22 and -21

Exercise Set 4.8

<u>1</u>. $-\frac{3}{4}, 1$ 3. $2, 4$ 4. $3, 5$ 5. 13 and 14, -13 and -14 6. 7 and 8, -8 and -7 <u>7</u>. 12 and 14, -12 and -14 9. 15 and 17, -15 and -17 10. 11 and 13, -13 and -11 11. Length: 12 m; width: 8 m   12. Length: 12 cm; width: 7 cm   <u>13</u>. 5 ft   15. Height: 4 cm; base: 14 cm   16. Height: 2 m; base: 10 m   17. 6 m   18. 4 km   <u>19</u>. 5 and 7 21. 506 22. 182 23. 12 24. 10 <u>25</u>. 780

27. 20 28. 25 29. 5 ft   30. 4 sec, $7\frac{1}{4}$ sec

<u>31</u>. (a) 2 sec; (b) 4.2 sec   33. 7 m   34. 30 cm by 15 cm   35. 5 in.   36. 100 cm^2, 225 cm^2

Margin Exercises, Section 4.9

1. -7940 2. -176 3. 32 4. $-3, 3, -2, 1, 2$ 5. $3, 7, 1, 1, 0; 7$ 6. $2x^2y + 3xy$ 7. $5pq + 4$ 8. $-4x^3 + 2x^2 - 4x + 2$ 9. $14x^3y + 7x^2y - 3xy - 2y$ 10. $-5p^2q^4 + 2p^2q^2 + 3p^2q + 6pq^2 + 3q + 5$ 11. $-8s^4t + 6s^3t^2 + 2s^2t^3 - s^2t^2$ 12. $-9p^4q + 10p^3q^2 - 4p^2q^3 - 9q^4$ 13. $x^5y^5 + 2x^4y^2 + 3x^3y^3 + 6x^2$ 14. $p^5q - 4p^3q^3 + 3pq^3 + 6q^4$ 15. $3x^3y + 6x^2y^3 + 2x^3 + 4x^2y^2$ 16. $2x^2 - 11xy + 15y^2$ 17. $16x^2 + 40xy + 25y^2$

18. $9x^4 - 12x^3y^2 + 4x^2y^4$ 19. $4x^2y^4 - 9x^2$

20. $16y^2 - 9x^2y^4$ 21. $9y^2 + 24y + 16 - 9x^2$

22. $4a^2 - 25b^2 - 10bc - c^2$ 23. $x^2y(x^2y + 2x + 3)$

24. $2p^4q^2(5p^2 - 2pq + q^2)$ 25. $(a - b)(2x + 5 + y^2)$

26. $(a + b)(x^2 + y)$ 27. $(x^2 + y^2)^2$ 28. $-(2x - 3y)^2$

29. $(xy + 4)(xy + 1)$ 30. $2(x^2y^3 + 5)(x^2y^3 - 2)$

31. $t^3(t - 2m)(t + 2m)$

Exercise Set 4.9

<u>1</u>. -1 3. -7 4. -1 5. \$12,597.12   6. \$12,250.43   <u>7</u>. 44.4624 in^2 9. Coefficients: $1, -2, 3, -5$; degrees: $4, 2, 2, 0; 4$ 10. Coefficients: $5, -1, 15, 1$; degrees: $3, 2, 1, 0; 3$ 11. Coefficients: $17, -3,$ -7; degrees: $5, 5, 0; 5$ 12. Coefficients: $6, -1, 8, -1$; degrees: $0, 2, 4, 5; 5$ <u>13</u>. $-a - 2b$ 15. $3x^2y - 2xy^2 + x^2$ 16. $m^3 + 2m^2n - 3m^2 + 3mn^2$ 17. $8u^2v - 5uv^2$ 18. $-2x^2 - 4xy - 2y^2$ <u>19</u>. $-8au + 10av$ 21. $x^2 - 4xy + 3y^2$ 22. $6 - z$ 23. $3r + 7$ 24. $-a^3b^2 - 3a^2b^3 + 5ab + 3$ <u>25</u>. $-x^2 - 8xy - y^2$ 27. $2ab$ 28. $y^4x^2 + y + 2x$ <u>29</u>. $-2a + 10b - 5c + 8d$ 30. $-5b$ <u>31</u>. $6z^2 + 7zu - 3u^2$ 33. $a^4b^2 - 7a^2b + 10$ 34. $x^2y^2 + 3xy - 28$ 35. $a^4 + a^3 - a^2y - ay + a + y - 1$ 36. $tvx^2 + stx + rvx + rs$ <u>37</u>. $a^6 - b^2c^2$ 39. $y^6x + y^4x + y^4 + 2y^2 + 1$ 40. $a^3 - b^3$ 41. $12x^2y^2 + 2xy - 2$ 42. $m^6n^2 + 2m^3n - 48$ <u>43</u>. $12 - c^2d^2 - c^4d^4$ 45. $m^3 + m^2n - mn^2 - n^3$ 46. $0.4p^2q^2 - 0.02pq - 0.02$ 47. $x^9y^9 - x^6y^6 + x^5y^5 - x^2y^2$ 48. $x^2 + xy^3 - 2y^6$ <u>49</u>. $x^2 + 2xh + h^2$ 51. $r^6t^4 - 8r^3t^2 + 16$ 52. $9a^4b^2 - 6a^2b^3 + b^4$ 53. $p^8 + 2m^2n^2p^4 + m^4n^4$ 54. $a^2b^2 + 2abcd + c^2d^2$ <u>55</u>. $4a^6 - 2a^3b^3 + \frac{1}{4}b^6$ 57. $3a^3 - 12a^2b + 12ab^2$ <u>58</u>. $a^4 + 2a^2b + 4a^2 + b^2 + 4b + 4$ 59. $4a^2 - b^2$ 60. $x^2 - y^2$ <u>61</u>. $c^4 - d^2$ 63. $a^2b^2 - c^2d^4$ 64. $x^2y^2 - p^2q^2$ <u>65</u>. $x^2 + 2xy + y^2 - 9$ 66. $p^2 + 2pq + q^2 - 16$ <u>67</u>. $x^2 - y^2 - 2yz - z^2$ 69. $a^2 - b^2 - 2bc - c^2$ 70. $9x^2 + 12x + 4 - 25y^2$ 71. $12n^2(1 + 2n)$ 72. $a(x^2 + y^2)$ <u>73</u>. $9xy(xy - 4)$ 75. $2\pi r(h + r)$ 76. $5p^2q^2(2p^2q^2 + 7pq + 2)$ 77. $(a + b)(2x + 1)$ 78. $(5c - 1)(a^3 + b)$ <u>79</u>. $(x - 1)(x + 1)$ 81. $(n + p)(n + 2)$ 82. $(a + y)(a - 3)$ 83. $(2x + z)(x - 2)$ 84. $(3y + p)(2y - 1)$ <u>85</u>. $(x - y)^2$ 87. $(3c + d)^2$ 88. $(4x + 3y)^2$ 89. $(7m^2 - 8n)^2$ 90. $(2xy + 3z)^2$ <u>91</u>. $(y^2 + 5z^2)^2$ 93. $(\frac{1}{2}a + \frac{1}{3}b)^2$ 94. $p(2p + q)^2$ 95. $(a + b)(a - 2b)$ 96. $(3b + a)(b - 6a)$ <u>97</u>. $(m + 20n)(m - 18n)$ 99. $(mn - 8)(mn + 4)$ 100. $(pq + 6)(pq + 1)$ 101. $a^3(ab + 5)(ab - 2)$ 102. $n^4(mn + 8)(mn - 4)$ <u>103</u>. $a^3(a - b)(a + 5b)$ 105. $(x^3 - y)(x^3 + 2y)$ <u>106</u>. $(a^2 - bc)(a^2 + 2bc)$ 107. $(x - y)(x + y)$ 108. $(pq - r)(pq + r)$ <u>109</u>. $7(p^2 + q^2)(p + q)(p - q)$ 111. $(9a^2 + b^2)(3a - b)(3a + b)$ 112. $(1 + 4x^6y^6)(1 + 2x^3y^3)(1 - 2x^3y^3)$ 113. $4xy - 4y^2$ 114. $2\pi ab - \pi b^2$ <u>115</u>. $2xy + \pi x^2$ 117. $(A + B)^3 = A^3 + 3A^2B + 3AB^2 + B^3$ <u>118</u>. $[3(x - 1) - y][2(x - 1) + 3y]$ 119. $[(y + 4) + x]^2$ 120. $(2a + b + 4)(a - b + 5)$

Margin Exercises, Section 4.10

1. $(x - 2)(x^2 + 2x + 4)$ 2. $(4 - y)(16 + 4y + y^2)$ 3. $(3x + y)(9x^2 - 3xy + y^2)$ 4. $(2y + z)(4y^2 - 2yz + z^2)$ 5. $2xy(2x^2 + 3y^2)(4x^4 - 6x^2y^2 + 9y^4)$ 6. $(3x + 2y)(9x^2 - 6xy + 4y^2)(3x - 2y)(9x^2 + 6xy + 4y^2)$ 7. $(x - 0.3)(x^2 + 0.3x + 0.09)$

Exercise Set 4.10

<u>1</u>. $(x + 2)(x^2 - 2x + 4)$ 3. $(y - 4)(y^2 + 4y + 16)$ <u>4</u>. $(z - 1)(z^2 + z + 1)$ 5. $(w + 1)(w^2 - w + 1)$ 6. $(x + 5)(x^2 - 5x + 25)$ <u>7</u>. $(2a + 1)(4a^2 - 2a + 1)$ 9. $(y - 2)(y^2 + 2y + 4)$ <u>10</u>. $(p - 3)(p^2 + 3p + 9)$ 11. $(2 - 3b)(4 + 6b + 9b^2)$ 12. $(4 - 5x)(16 + 20x + 25x^2)$ <u>13</u>. $(4y + 1)(16y^2 - 4y + 1)$ 15. $(2x + 3)(4x^2 - 6x + 9)$

16. $(3y + 4)(9y^2 - 12y + 16)$ 17. $(a - b)(a^2 + ab + b^2)$

18. $(x - y)(x^2 + xy + y^2)$ **19.** $(a + \frac{1}{2})(a^2 - \frac{1}{2}a + \frac{1}{4})$

21. $2(y - 4)(y^2 + 4y + 16)$ 22. $3(z - 1)(z^2 + z + 1)$
23. $3(2a + 1)(4a^2 - 2a + 1)$ 24. $2(3x + 1)(9x^2 - 3x + 1)$
25. $r(s + 4)(s^2 - 4s + 16)$
27. $5(x - 2z)(x^2 + 2xz + 4z^2)$
28. $2(y - 3z)(y^2 + 3yz + 9z^2)$
29. $(x + 0.1)(x^2 - 0.1x + 0.01)$
30. $(y + 0.5)(y^2 - 0.5y + 0.25)$
31. $8(2x^2 - t^2)(4x^4 + 2x^2t^2 + t^4)$ 33. 1969
34. \$140 35. $(x^{2a} + y^{2b})(x^{4a} - x^{2a}y^b + y^{2b})$
36. $(ax - by)(a^2x^2 + abxy + b^2y^2)$
37. $3(x^a + 2y^b)(x^{2a} - 2x^ay^b + 4y^{2b})$

39. $\frac{1}{3}(\frac{1}{2}xy + z)(\frac{1}{4}x^2y^2 - \frac{1}{2}xyz + z^2)$

40. $\frac{1}{2}(\frac{1}{2}x^a + y^{2a}z^{3b})(\frac{1}{4}x^{2a} - \frac{1}{2}x^ay^{2a}z^{3b} + y^{4a}z^{6b})$

41. $x^2 - 16$ 42. $3\pi r^2 - 9\pi$

Summary and Review – Chapter 4

1. [4.1] Answers may vary. $(-5x)(2x)$, $(-x)(10x)$, $(-2x)(5x)$ 2. [4.1] Answers may vary. $(4x^2)(9x^3)$, $(18x)(2x^4)$, $(-6x^3)(-6x^2)$ 3. [4.2] $5(1 + 2x^3)(1 - 2x^3)$
4. [4.1] $x(x - 3)$ 5. [4.2] $(3x + 2)(3x - 2)$ 6. [4.4] $(x + 6)(x - 2)$ 7. [4.3] $(x + 7)^2$
8. [4.1] $3x(2x^2 + 4x + 1)$ 9. [4.1] $(x^2 + 3)(x + 1)$
10. [4.5] $(3x - 1)(2x - 1)$
11. [4.2] $(x^2 + 9)(x + 3)(x - 3)$
12. [4.5] $3x(3x - 5)(x + 3)$ 13. [4.2] $2(x + 5)(x - 5)$
14. [4.1] $(x^3 - 2)(x + 4)$
15. [4.2] $(4x^2 + 1)(2x + 1)(2x - 1)$
16. [4.1] $4x^4(2x^2 - 8x + 1)$ 17. [4.3] $3(2x + 5)^2$
18. [4.2] Not factorable 19. [4.4] $x(x - 6)(x + 5)$
20. [4.2] $(2x + 5)(2x - 5)$ 21. [4.3] $(3x - 5)^2$
22. [4.4] $2(3x + 4)(x - 6)$ 23. [4.3] $(x - 3)^2$
24. [4.5] $(2x + 1)(x - 4)$ 25. [4.3] $2(3x - 1)^2$
26. [4.2] $3(x + 3)(x - 3)$ 27. [4.4] $(x - 5)(x - 3)$
28. [4.3] $(5x - 2)^2$ 29. [4.7] 1, -3 30. [4.7] -7, 5

31. [4.7] -4, 3 32. [4.7] $\frac{2}{3}$, 1 33. [4.7] $\frac{3}{2}$, -4

34. [4.7] 8, -2 35. [4.8] 3, -2 36. [4.8] -18 and -16, 16 and 18 37. [4.8] -19 and -17, 17 and 19

38. [4.8] $\frac{5}{2}$, -2 39. [4.9] 49 40. [4.9] Coefficients: 1, -7, 9, -8; degrees 6, 2, 2, 0; 6 41. [4.9] Coefficients: 1, -1, 1; degrees: 16, 40, 23; 40
42. [4.9] $9w - y - 5$ 43. [4.9] $m^6 - 2m^2n + 2m^2n^2 + 8n^2m - 6m^3$ 44. [4.9] $-9xy - 2y^2$
45. [4.9] $11x^3y^2 - 8x^2y - 6x - 6x^2 + 6$ 46. [4.9] $p^3 - q^3$
47. [4.9] $9a^8 - 2a^4b^3 + \frac{1}{9}b^6$ 48. [4.9] $(xy + 4)(xy - 3)$
49. [4.9] $3(2a + 7b)^2$ 50. [4.9] $(m + t)(m + 5)$
51. [4.9] $32(x^2 - 2y^2z^2)(x^2 + 2y^2z^2)$
52. [4.10] $(3x + 2)(9x^2 - 6x + 4)$
53. [4.10] $(t - 4m)(t^2 + 4mt + 16m^2)$
54. [4.10] $(0.4b - 0.5c)(0.16b^2 + 0.2bc + 0.25c^2)$

55. [4.8] $2\frac{1}{2}$ cm 56. [4.8] 0, 2 57. [4.8] $\ell = 12$, $w = 6$ 58. [4.2, 4.7] No real solution

59. [4.7] 2, -3, $\frac{5}{2}$

The ASVAB Test – Arithmetic Reasoning   Chapter 4

1. B 2. A 3. D 4. B 5. A 6. D 7. A 8. C
9. D 10. A 11. B 12. B 13. D 14. B

CHAPTER 5

Margin Exercises, Section 5.1

1-8.

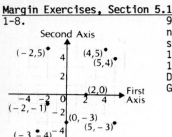

9. Both are negative numbers.   10. First pos.; second neg.   11. I
12. III   13. IV   14. II
15. B(-3,5); C(-4,-3); D(2,-4); E(1,5); F(-2,0); G(0,3)

Exercise Set 5.1

1.

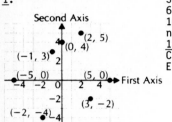

3. II   4. II   5. IV
6. IV   7. III   9. I
10. I   11. Negative; negative   12. Second, first
13. A: (3,3), B: (0,-4), C: (-5,0), D: (-1,-1), E: (2,0)

15.

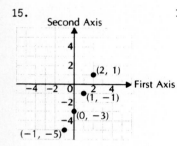

16.

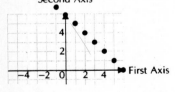

17. 44,330 mi   18. 20 in.
22. II or IV   23. (-1,-5)
25. Answers may vary

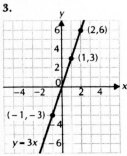

19. I or IV   21. I or III
24. (5,2), (-7,2), (3,-8)
27. 26   28. 32½
29. Latitude 32.5° north, longitude 64.5° west
30. Latitude 27° north, longitude 81° west

Margin Exercises, Section 5.2

1. No   **2.** Yes

3.

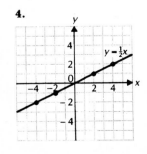

4.

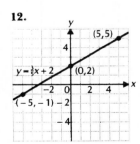

5.

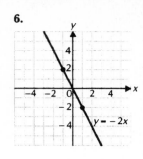

6.

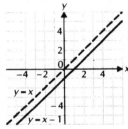

7.

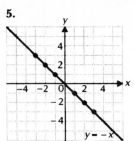

8.

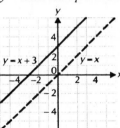

9. $y = x + 3$ looks like $y = x$ moved *up* 3 units.

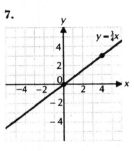

10. $y = x - 1$ looks like $y = x$ moved *down* 1 unit.

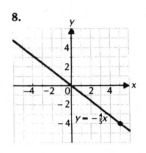

11. $y = 2x + 3$ looks like $y = 2x$ moved *up* 3 units.

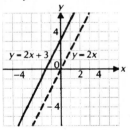

12.

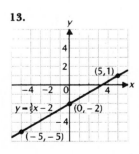

13.

14.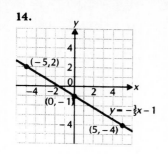

15.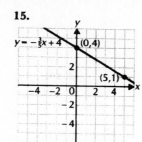

17.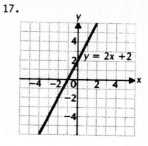

18.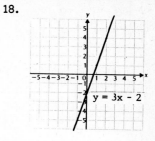

Exercise Set 5.2

<u>1</u>. Yes   3. No   4. No   5. No   6. No

<u>7</u>.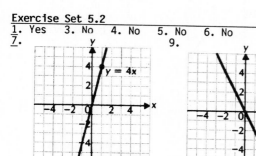

9.

19.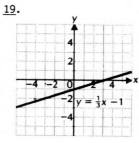

21.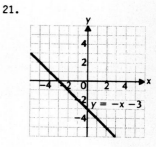

10.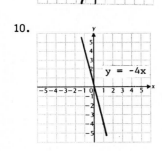

11.

22.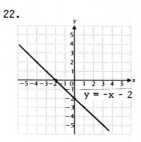

23.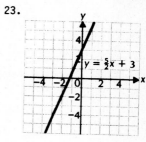

12.

13.

24.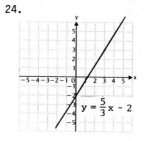

25.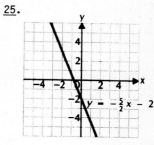

15.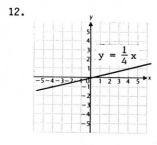

16.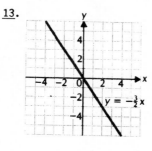

27.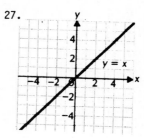

28.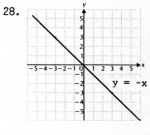

29.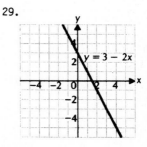

30.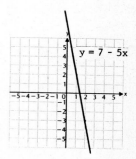

31.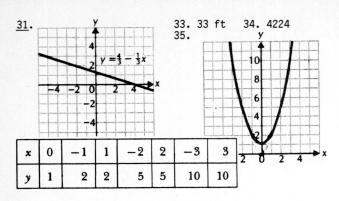

$y = \frac{4}{3} - \frac{1}{3}x$

x	0	−1	1	−2	2	−3	3
y	1	2	2	5	5	10	10

33. 33 ft   **34.** 4224

35.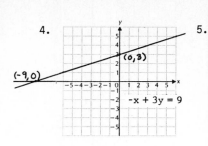

36. (0,6), (1,5), (2,4), (3,3), (4,2), (5,1), (6,0)

37. (15,0), (12,1), (9,2), (6,3), (3,4), (0,5)

39. 5n + 25q = 235; (n,q) form, (27,4), (7,8), (47,0), answers may vary   **40.** Answers may vary. (−3,3), (2,2), (0,0)   **41.** Answers may vary. (−3,4), (2,3), (0,1)

Margin Exercises, Section 5.3

1. (a) (4, 0); (b) (0, 3)

2. *x*-intercept is (3, 0); *y*-intercept is (0, 2)

3. *x*-intercept is (−3, 0); *y*-intercept is (0, 4)

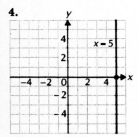

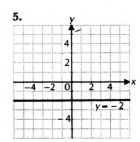

4.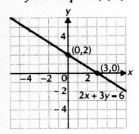

5.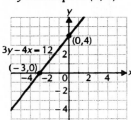

6.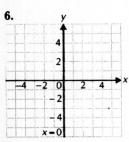

7.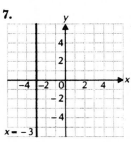

Exercise Set 5.3

1.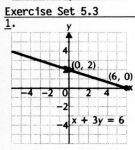

3.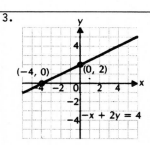

4.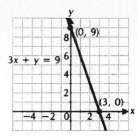

5.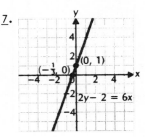

6.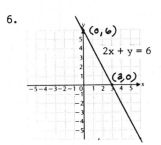

7.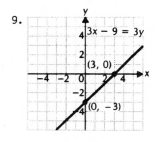

9.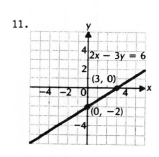

10.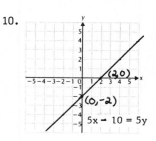

11.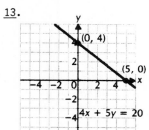

12.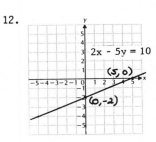

13.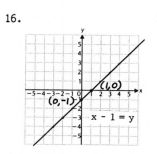

15.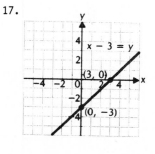

16.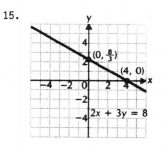

17.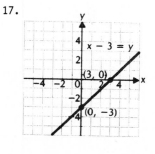

18.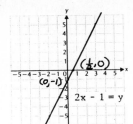

$2x - 1 = y$

19.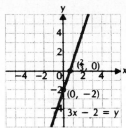

$3x - 2 = y$

21.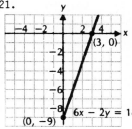

$6x - 2y = 18$

22.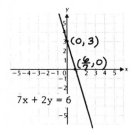

$7x + 2y = 6$

23.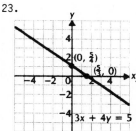

$3x + 4y = 5$

24.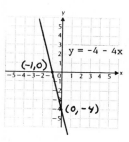

$y = -4 - 4x$

25.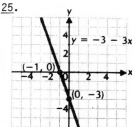

$y = -3 - 3x$

27.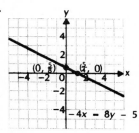

$-4x - 8y - 5$

28.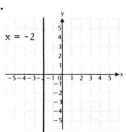

$x = -2$

29.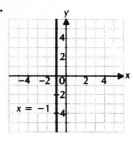

$x = -1$

30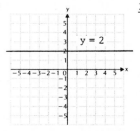

$y = 2$

31.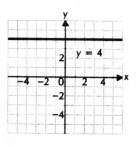

$y = 4$

33.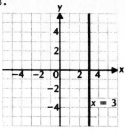

$x = 3$

34.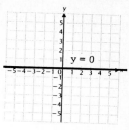

$y = 0$

35.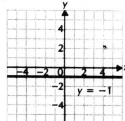

$y = -1$

36.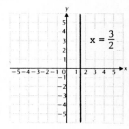

$x = \frac{3}{2}$

37.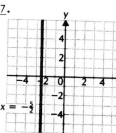

$x = -\frac{5}{2}$

39. $b = 7$ m, $h = 2$ m
40. $x = 0$ **41.** $y = 0$
42. $(-3,6)$ **43.** $y = -5$

45. $y = 2.8$ **46.** $-\frac{3}{2}$

47. -8 **48.** $3250,
$2762.50, $2275, $1787.50,
$1300

Margin Exercises, Section 5.4

1. $\frac{2}{5}$ 2. $-\frac{5}{3}$ 3. $\frac{2}{5}$ 4. $-\frac{5}{3}$ 5. 0 6. No slope
7. No slope 8. 0 9. -8 10. $-\frac{4}{5}$ 11. $\frac{1}{4}$ 12. $\frac{5}{4}$

13. 5, (0,0) 14. $-\frac{3}{2}$, (0,-6) 15. 2, $(0,-\frac{17}{2})$

16. $-\frac{3}{4}$, $(0,\frac{15}{4})$ 17. $-\frac{7}{5}$, $(0, -\frac{22}{5})$ 18. $y = 5x - 18$

19. $y = -3x - 5$ 20. $y = 6x - 13$ 21. $y = -\frac{2}{3}x + \frac{14}{3}$
22. $y = x + 2$ 23. $y = 2x + 4$

Exercise Set 5.4

1. 0 3. $-\frac{4}{5}$ 4. $-\frac{5}{6}$ 5. 7 6. $\frac{2}{3}$ 7. $-\frac{2}{3}$ 9. -2
10. $-\frac{11}{8}$ 11. 0 12. 0 13. No slope 15. No slope
16. No slope 17. 0 18. 0 19. No slope 21. 0
22. 0 23. $-\frac{3}{2}$ 24. 4 25. $-\frac{1}{4}$ 27. 2 28. 5
29. $\frac{4}{3}$ 30. $\frac{1}{2}$ 31. $\frac{1}{3}$ 33. $\frac{1}{2}$ 34. $\frac{5}{7}$ 35. -4, (0,-9)
36. -3, (0,-5) 37. 1.8, (0,0) 39. $-\frac{2}{3}$, (0,3)
40. $-\frac{5}{4}$, (0,3) 41. $-\frac{8}{7}$, (0,3) 42. $-\frac{2}{9}$, $(0, -\frac{13}{9})$
43. 3, $(0,-\frac{5}{3})$ 45. $-\frac{3}{2}$, $(0, -\frac{1}{2})$ 46. 0, (0,-17)
47. $y = 5x - 5$ 48. $y = -2x - 6$ 49. $y = \frac{3}{4}x + \frac{5}{2}$
51. $y = x - 8$ 52. $y = 6x - 26$ 53. $y = -3x - 9$
54. $y = -3x + 3$ 55. $y = \frac{2}{3}x + \frac{8}{3}$ 57. $y = \frac{1}{4}x + \frac{5}{2}$
58. $y = x + 4$
59. $y = -\frac{1}{2}x + 4$ 60. $y = \frac{1}{2}x$ 61. $y = -\frac{3}{2}x + \frac{13}{2}$
63. $y = \frac{2}{5}x - 2$ 64. $y = \frac{5}{3}x + \frac{4}{3}$ 65. $y = \frac{3}{4}x - \frac{5}{2}$
66. $y = -4x - 7$ 67. $y = 3x - 9$ 69. $y = \frac{3}{2}x - 2$

Margin Exercises, Section 5.5

1. Slope is $\frac{2}{3}$; y-intercept is (0,0)

2. Slope is -6; y-intercept is (0,0)   3. Slope is -2; y-intercept is (0,9)   4. Slope is 0.7; y-intercept is (0,-3)

5.
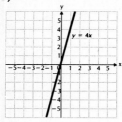

6.
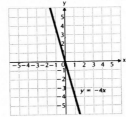

7.
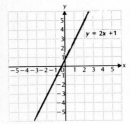

8.
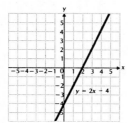

9.
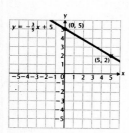

10.
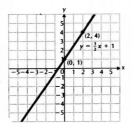

11.
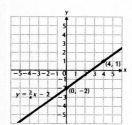

12.
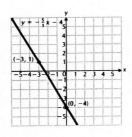

4.
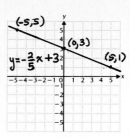

5.
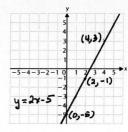

6
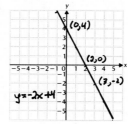

7.
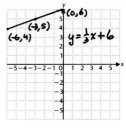

9.
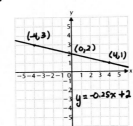

10.
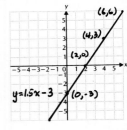

11.
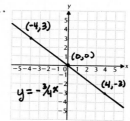

12.
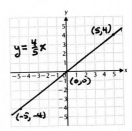

Summary and Review - Chapter 5

1-3. [5.1]
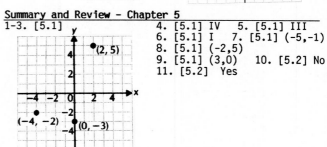

4. [5.1] IV   5. [5.1] III
6. [5.1] I   7. [5.1] (-5,-1)
8. [5.1] (-2,5)
9. [5.1] (3,0)   10. [5.2] No
11. [5.2] Yes

12. [5.2]
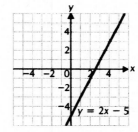

13. [5.2]
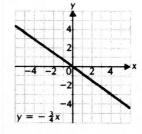

Exercise Set 5.5

1.
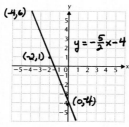

3.

14. [5.2]

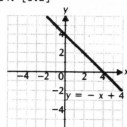

15. [5.2]

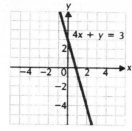

16. [5.3]

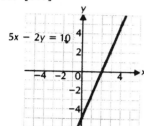

17. [5.3]

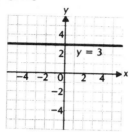

18. [5.3]

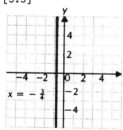

19. [5.3]

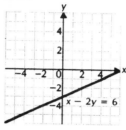

20. [5.4] $\frac{3}{2}$ 21. [5.4] 0   22. [5.4] No slope
23. [5.4] 2   24. [5.4] 0   25. [5.4] No slope
26. [5.4] $-\frac{4}{3}$ 27. [5.4] -9, (0,46)   28. [5.4] -1,
(0,9)   29. [5.4] $\frac{3}{5}$, $(0,-\frac{4}{5})$ 30. [5.4] $y = 3x - 1$

31. [5.4] $y = \frac{2}{3}x - \frac{11}{3}$ 32. [5.4] $y = -2x - 4$
33. [5.4] $y = x + 2$ 34. [5.4] $y = \frac{1}{2}x - 1$

35. [5.5]

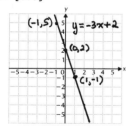

36. [5.5]

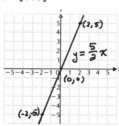

37. [5.5]

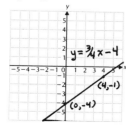

38. [5.2] -1   39. [5.2] 19
40. [5.1] Area = 45,
perimeter = 28   41. [5.2]
(0,4), (1,3), (-1,3), answers
may vary   42. [5.2] $y = -x$
43. [5.4] $m = -b/a$;
x-intercept is (a,0);
y-intercept is (0,b)

CHAPTER 6

Margin Exercises, Section 6.1

1. $x + y = 115$, $x - y = 21$, where x is one number and y is the other   2. $31.95 + 0.33m = c$, $34.95 + 0.29m = c$, where m = mileage and c = cost.   3. $2\ell + 2w = 86$, $\ell = w + 19$, where ℓ = length and w = width

Exercise Set 6.1

1. $x + y = 58$, $x - y = 16$, x is one number and y is the other.   3. $2\ell + 2w = 400$, $w = \ell - 40$, ℓ = length, w = width   4. $2\ell + 2w = 76$, $w = \ell - 17$, ℓ = length, w = width   5. $53.95 + 0.30m = c$, $54.95 + 0.20m = c$, m = mileage, c = cost   6. $45.95 + 0.40m = c$, $46.95 + 0.20m = c$, m = mileage, c = cost   7. $x - y = 16$, $3x = 7y$, x = larger number, y = smaller number   9. $x + y = 180$, $y = 3x + 8$, x and y are the angles   10. $x + y = 180$, $y = 2x + 30$, x and y are the angles   11. $x + y = 90$, $x - y = 34$, x and y are the angles   12. $x + y = 90$, $y = \frac{1}{2}x + 42$, x and y are the angles   13. $x + y = 820$, $x = y + 140$, x = hectares of Chardonnay, y = hectares of Riesling   15. $x - y = 18$, $2y + 3x = 74$, x = larger number, y = smaller   16. $2\ell + 2w = 400$, $\ell = w + 40$, ℓ = length, w = width   17. $2\ell + 2w = 76$, $\ell = w + 17$, ℓ = length, w = width   18. $2\ell + 2w = 306\frac{2}{3}$, $\ell = w + 46\frac{2}{3}$, ℓ = length, w = width   19. $x = 0.2y$, $x + 20 = 0.52(y + 20)$, x = Patrick's age, y = his father's age   21. $\frac{1}{3}(b + 2) = h - 1$, $\frac{1}{2}(b + 2)(h - 1) = 24$, b = base, h = height

Margin Exercises, Section 6.2

1. Yes   2. No   3. (2,-3)   4. No solution, lines are parallel

Exercise Set 6.2

1. Yes   3. No   4. Yes   5. Yes   6. Yes   7. Yes
9. Yes   10. Yes   11. (2,1)   12. (4,2)   13. (-12,11)
15. (4,3)   16. (3,2)   17. (-3,-3)   18. (-6,-2)
19. No solution   21. (2,2)   22. (1,-3)   23. (5,3)
24. $(\frac{1}{3},1)$ 25. Unlimited number of solutions   27. A = 2, B = 2   28. No   29. Yes   30. Three lines intersecting in one point. The solution is one point
31. The solution is $(\frac{2}{3},\frac{3}{7})$, but would be very difficult to discover by graphing. The check would tell you whether the possible pair you found by graphing is a solution.

Margin Exercises, Section 6.3

1. (3,2)   2. (3,-1)   3. $(\frac{24}{5}, -\frac{8}{5})$ 4. Length: $27\frac{1}{2}$ cm; width: $10\frac{1}{2}$ cm

Exercise Set 6.3

1. (1,3)   3. (1,2)   4. (2,-4)   5. (4,3)   6. (-1,-1)
7. (-2,1)   9. (-1,-3)   10. (2,-4)   11. $(\frac{17}{3},\frac{16}{3})$
12. $(\frac{17}{3},\frac{2}{3})$ 13. $(\frac{25}{8}, -\frac{11}{4})$ 15. (-3,0)
16. (4,-2)   17. (6,3)   18. (1,2)   19. No solution
21. 15, 12   22. 19, 17   23. 37, 21   24. 37, 29   25.
28, 12   27. 120 m, 80 m   28. $27\frac{1}{2}$ cm, $10\frac{1}{2}$ cm
29. 100 yd, $53\frac{1}{3}$ yd   30. (4.382,4.328)   31. (7,-1)
33. (2,-1)   34. No   35. $\ell = \frac{P + 10}{4}$ 36. $w = \frac{P - 16}{4}$

37. (2,-1,3)   39. The x-terms drop out and leave 12 = 14. The lines are parallel, so there is no solution
40. The x-terms drop out and leave 15 = 15. The lines coincide, so all solutions to one equation are solutions of the system.   41. Answers may vary, x = 10 - z, y = 10 - z, so (10 - z) + (10 - z) + z = 10 or z = 10. x = 10 - z, so x = 0, y = 10 - z, so y = 0; then x + y = 0; then x + y = 0. But we know x + y = 10, so there is no solution.

Margin Exercises, Section 6.4

1. (3,2)   2. (1,-1)   3. (1,4)   4. (1,1)   5. (1,-1)
6. $(\frac{17}{13}, -\frac{7}{13})$ 7. No solution   8. (1,-1)   9. 75 mi

Exercise Set 6.4

1. (9,1)   3. (3,5)   4. (5,1)   5. (3,0)   6. (2,7)
7. $(-\frac{1}{2},3)$ 9. $(-1,\frac{1}{5})$ 10. (-2,3)   11. No solution
12. Infinitely many solutions   13. (-3,-5)   14. (-1,-6)   15. (4,5)   16. (3,1)   17. (4,1)   18. (10,3)
19. (4,3)   21. (1,-1)   22. (4,-1)   23. (-3,-1)
24. (2,5)   25. (2,-2)   27. (5,½)   28. (50,18)   29. (2,-1)   30. (10,-2)   31. 10 mi   33. 43° and 137°
34. 50° and 130°   35. 62° and 28°   36. 58° and 32°
37. 480 hectares Chardonnnay, 340 hectares Riesling
39. 23 pheasants, 12 rabbits   40. 6 yr, 30 yr   41. 45 yr, 10 yr   43. (5,2)   45. (1,-1)   46. (0,-1)
47. (525,1000)   48. (4,0)   49. $\left[\frac{b-c}{1-a}, \frac{b-ac}{1-a}\right]$ 51. (4,3)   52. (4,3)   53. (a) $x = \frac{ce-bf}{ae-bd}$, $y = \frac{af-cd}{ae-bd}$, (b) (2700,0.5)

Margin Exercises, Section 6.5

1. 20, 16   2. 4   3. 135 adults, 31 children   4. 22.5 L of 50%; 7.5 L of 70%   5. 30 lb of A, 20 lb of B

Exercise Set 6.5

1. 350 cars, 160 trucks   3. Sammy; 44, daughter: 22
4. Ann: 28, Son: 10   5. Marge: 18, Consuelo: 9   6. Andy: 24, Wendy: 6   7. 70 dimes, 33 quarters   9. 300 nickels, 100 dimes   10. 32 nickels, 13 dimes   11. 203 adults, 226 children   12. 236 adults, 342 children
13. 130 adults, 70 students   15. 40 g of A, 60 g of B
16. $55\frac{5}{9}$ L of A, $44\frac{4}{9}$ L of B   17. 43.75 L   18. 2666.67 L
19. 80 L of 30%, 120 L of 50%   21. 6 kg of cashews, 4 kg of pecans   22. 100 kg of Brazilian, 200 kg of Turkish   23. $12,500 at 12%, $14,500 at 13%   24. 54
25. 75   27. 10 at $20 per day, 5 at $25 per day   28. 9%, 10.5%   29. 74   30. 126   31. Glove: $79.95, bat: $14.50, ball: $4.55

Margin Exercises, Section 6.6

1. 168 km   2. 275 km/h   3. 324 mi   4. 3 hr

Exercise Set 6.6

1. 2 hr   3. 4.5 hr   4. 4.6 hr   5. $7\frac{1}{2}$ hr after first train leaves   6. $2\frac{1}{2}$ hr after private plane leaves
7. 14 km/h   9. 384 km   10. 1911 km   11. 330 km/h
12. 9 km/h   13. 15 mi   15. ≈ 317.03 km/h   16. Wind: 50 mph, Plane: 200 mph   17. 180 mi, 96 mi   18. Wind: 40 mph, Plane: 620 mph   19. 144 mi   21. 40 min
22. ≈ 3680 mi

Summary and Review – Chapter 6

1. [6.2] No   2. [6.2] Yes   3. [6.2] Yes   4. [6.2] No
5. [6.2] (6,-2)   6. [6.2] (6,2)   7. [6.2] (0,5)   8. [6.2] No solution; lines are parallel   9. [6.3] (0,5)
10. [6.3] (-2,4)   11. [6.3] (1,-2)   12. [6.3] (-3,9)
13. [6.3] (1,4)   14. [6.3] (3,-1)   15. [6.4] (3,1)
16. [6.4] (1,4)   17. [6.4] (5,-3)   18. [6.4] (-4,1)
19. [6.4] (-2,4)   20. [6.4] (-2,-6)   21. [6.4] (3,2)
22. [6.4] (2,-4)   23. [6.3] 10, -2   24. [6.4] 12, 15
25. [6.3] ℓ = $37\frac{1}{2}$ cm, w = $10\frac{1}{2}$ cm   26. [6.6] 135 km/h
27. [6.5] 297 orchestra seats, 211 balcony seats
28. [6.5] 40 L of each   29. [6.5] Jeff: 27, Son: 9
30. [6.2] C = 1, D = 3   31. [6.3] (2,1,-2)   32. [6.4] (2,0)   33. [6.5] 24   34. [6.5] $96

CHAPTER 7

Margin Exercises, Section 7.1

1. (a) Yes; (b) Yes; (c) Yes; (d) No; (e) No 2. (a) Yes; (b) No; (c) No; (d) Yes; (e) No 3. $\{x|x > 2\}$
4. $\{x|x < 13\}$ 5. $\{x|x < -3\}$ 6. $\left\{x\middle|x \leqslant \frac{2}{15}\right\}$
7. $\{y|y \leqslant -3\}$

Exercise Set 7.1

1. (a) No; (b) No; (c) No; (d) Yes 3. (a) No; (b) No; (c) Yes; (d) Yes 4. (a) Yes; (b) Yes; (c) Yes; (d) No 5. (a) No; (b) No; (c) Yes; (d) No 6. (a) Yes; (b) No; (c) Yes; (d) Yes 7. (a) Yes; (b) Yes; (c) Yes; (d) No
9. $\{x|x > -5\}$ 10. $\{x|x > -3\}$ 11. $\{y|y > 3\}$
12. $\{y|y > 2\}$ 13. $\{x|x \leqslant -18\}$ 15. $\{a|a < -6\}$
16. $\{a|a < -12\}$ 17. $\{x|x \leqslant 16\}$ 18. $\{x|x \leqslant 17\}$
19. $\{x|x > 8\}$ 21. $\{y|y > -5\}$ 22. $\{y|y > -6\}$
23. $\{x|x > 2\}$ 24. $\{x|x > 3\}$ 25. $\{x|x \leqslant -3\}$
27. $\{x|x \geqslant 13\}$ 28. $\{x|x \geqslant 20\}$ 29. $\{x|x < 4\}$
30. $\{x|x < -1\}$ 31. $\{y|y \geqslant -11\}$ 33. $\{c|c > 0\}$
34. $\{c|c > 18\}$ 35. $\{y|y \leqslant \frac{1}{4}\}$ 36. $\{y|y \leqslant \frac{1}{2}\}$
37. $\{x|x > \frac{7}{12}\}$ 39. $\{x|x > 0\}$ 40. $\{x|x > 3\}$
41. 0.6 km 42. Hunting: 19,000; Fishing 24,500
43. $\{r|r < -2\}$ 45. $\{x|x \geqslant 1\}$ 46. $\{x|x \leqslant 1\}$
47. $\{x|x > -11.8\}$ 48. $\{x|x > -4\}$ 49. $\{x|x \leqslant -1.2\}$
51. $\{x|x \leqslant -18,058,999\}$ 52. Yes

Margin Exercises, Section 7.2

1. $\{x|x < 8\}$ 2. $\{y|y \geqslant 32\}$ 3. $\{x|x \geqslant -6\}$
4. $\{y|y < -\frac{13}{5}\}$ 5. $\{x|x > -\frac{1}{4}\}$ 6. $\{x|x \leqslant -1\}$
7. $\{y|y \leqslant \frac{19}{9}\}$

Exercise Set 7.2

1. $\{x|x < 7\}$ 3. $\{y|y \leqslant 9\}$ 4. $\{x|x > 24\}$
5. $\{x|x < \frac{13}{7}\}$ 6. $\{y|y < \frac{17}{8}\}$ 7. $\{x|x > -3\}$
9. $\{y|y > -\frac{2}{5}\}$ 10. $\{x|x > -\frac{4}{7}\}$ 11. $\{x|x \geqslant -6\}$
12. $\{y|y \geqslant -5\}$ 13. $\{y|y \leqslant 4\}$ 15. $\{x|x > \frac{17}{3}\}$
16. $\{y|y < \frac{23}{5}\}$ 17. $\{y|y < -\frac{1}{14}\}$ 18. $\{x|x \geqslant -\frac{1}{36}\}$
19. $\{x|\frac{3}{10} \geqslant x\}$ 21. $\{x|x < 8\}$ 22. $\{y|y < 8\}$
23. $\{y|y \geqslant 6\}$ 24. $\{x|x \geqslant 8\}$ 25. $\{x|x \leqslant 6\}$
27. $\{y|y > 4\}$ 28. $\{y|y > 4\}$ 29. $\{x|x < -3\}$
30. $\{y|y < -6\}$ 31. $\{x|x \geqslant -2\}$ 33. $\{y|y < -3\}$
34. $\{x|x < -2\}$ 35. $\{x|x > -3\}$ 36. $\{y|y > -5\}$
37. $\{y|y < -\frac{10}{3}\}$ 39. $\{x|x \leqslant 7\}$ 40. $\{x|x \leqslant 4\}$
41. $\{x|x > -10\}$ 42. $\{x|x > -13\}$ 43. $\{y|y < 2\}$
45. $\{y|y \geqslant 3\}$ 46. $\{y|y \geqslant 2\}$ 47. $\{y|y > -2\}$
48. $\{y|y > -3\}$ 49. $\{y|y > -2\}$ 51. $\{y|y \leqslant \frac{33}{7}\}$
52. $\{y|y \geqslant \frac{30}{13}\}$ 53. $\{x|x < \frac{9}{5}\}$ 54. $\{y|y < \frac{61}{28}\}$
55. 140°, 28°, 12° 57. $\{t|t \leqslant 0\}$ 58. $\{z|z < \frac{65}{16}\}$
59. $\{y|y < \frac{2 \cdot 2}{7}\}$ 60. $\{y|y > 1.8\}$ 61. $\{x|x \leqslant 9\}$
63. $\{y|y \leqslant -3\}$ 64. $\{x|x \geqslant -25\}$ 65. $\{x|x > \frac{8}{3}\}$
66. $\{x|x < \frac{30}{13}\}$ 67. $\{x|x \leqslant -4a\}$ 69. $\{x|x \geqslant \frac{3y}{2}\}$
70. $\{x|x > \frac{y-b}{a}\}$ 71. All nonzero real numbers
72. All real numbers 73. Yes 75. No 76. No
77. $x \geqslant 6$ 78. $a \geqslant b$ 79. $x \leqslant y$

Margin Exercises, Section 7.3

1. $\{w|w \leqslant 11\frac{2}{3}\}$ 2. $\{t|t > 64.3\}$; that is, after 1984

Exercise Set 7.3

1. $\{s|s \geqslant 97\}$ 3. $\{t|t > 4\frac{2}{3}\}$ 4. $\{n|n > 5\}$
5. $\{n|n \leqslant 0\}$ 6. $\{m|m \leqslant 341.4 \text{ mi}\}$ 7. $\{m|m \leqslant 525.8 \text{ mi}\}$
9. $\{\ell|\ell \geqslant 16.5 \text{ yd}\}$ 10. $\{w|w > 14 \text{ cm}\}$
11. $\{\ell|\ell \geqslant 92 \text{ ft}\}$; $\{\ell|\ell \leqslant 92 \text{ ft}\}$ 12. $\{b|b > 6 \text{ cm}\}$

13. $\{c|c \geqslant 20\}$ 15. $\{s|s \leqslant \$18.05\}$ The most the student can spend for the sweater is $18.05.
16. $\{\ell|\ell \geqslant 64 \text{ km}\}$ 17. $\{b|b \leqslant 4\}$
18. $\{s|s \leqslant 8 \text{ cm and } s \text{ is positive}\}$ 19. 47 and 49
21. $\{s|s < \$20,000\}$ 22. $\{n|n > 54.5 \text{ hr}\}$

Margin Exercises, Section 7.4

1.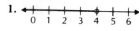
2.
3.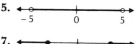
4.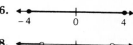
5.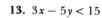
6.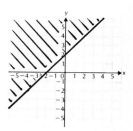
7.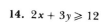
8.

9. No
10. $y < x$
11. $y \geqslant x + 2$

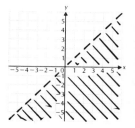

12. $2x + 4y < 8$
13. $3x - 5y < 15$

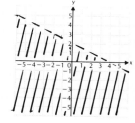

14. $2x + 3y \geqslant 12$

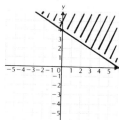

Exercise Set 7.4

1.  3.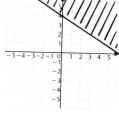
4.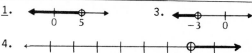
5.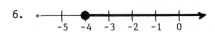
6.

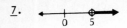

7.
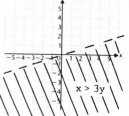

9.

10.

11.

12.

13.

15.

16.

17.

18.

19.

21.

22.

23.

24.

25.

27.

28.

29.

30. No

31. Yes   33. Yes

34. 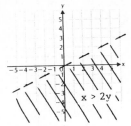 x > 2y

35. x > 3y

36. 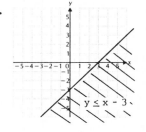 y ≤ x - 3

37. 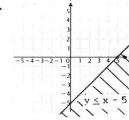 y ≤ x - 5

39. 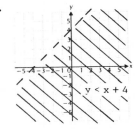 y < x + 4

40. 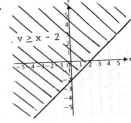 y ≥ x - 2

41. 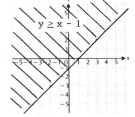 y ≥ x - 1

42. 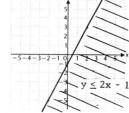 y ≤ 2x - 1

43. 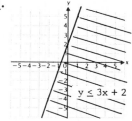 y ≤ 3x + 2

45. 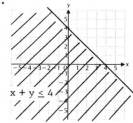 x + y ≤ 4

46. 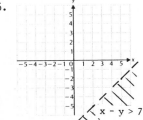 x - y > 7

47. 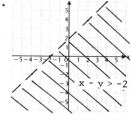 x - y > -2

48.

49.

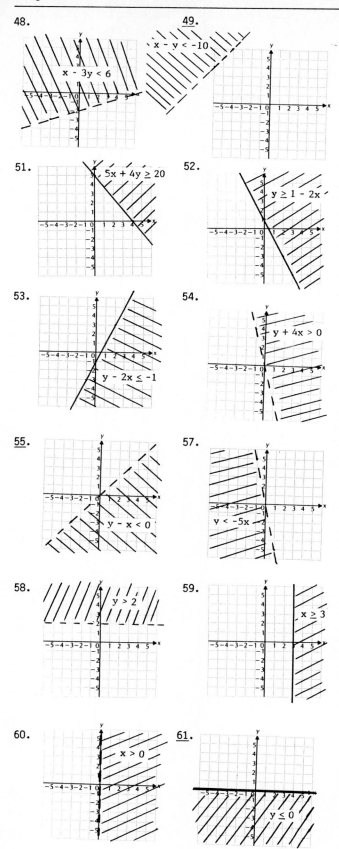

x - 3y < 6

x - y < -10

51.

5x + 4y ≥ 20

52.

y ≥ 1 - 2x

53.

y - 2x ≤ -1

54.

y + 4x > 0

55.

y - x < 0

57.

y < -5x - 2

58.

y > 2

59.

x ≥ 3

60.

x > 0

61.

y ≤ 0

63. No   64. No   65. Yes   66. Yes   67. No   69. Yes

Margin Exercises, Section 7.5

1. (3,5)

2.

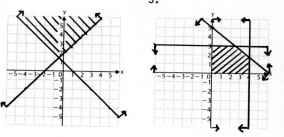

3.

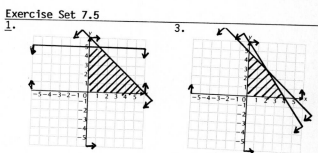

4.   Max = 15 at x = 4, y = 0; Min = -5 at x = 0, y = 3

Exercise Set 7.5

1.

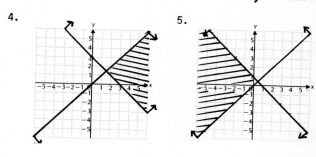

3.

4.

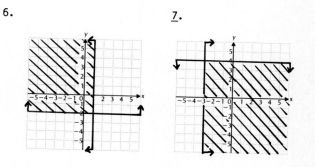

5.

6.

7.

9. Max = 11 at x = 1, y = 5; Min = 0 at x = 0, y = 0
10. Max = 4 at x = 4, y = 0; Min = -2 at x = 0, y = 4
11. Max = 22 at x = 2, y = 3; Min = 0 at x = 0, y = 0
12. Max = 8 at x = 4, y = 0; Min = 0 at x = 0, y = 0
13. Maximum income = $3600 by making 30 suits and 0 dresses   15.   Maximum profit = $10,040 by producing 60 motorcycles and 100 bicylces
16. Maximum profit = $23.70 by selling 50 hot dogs and 40 hamburgers   17. 120 of A1 and 400 of A2
18. Maximum score = 102 by doing 8 questions of Type A and 10 questions of Type B.
19.   Maximum score = 425 by doing 5 questions of Type A and 15 questions of Type B.

Summary and Review - Chapter 7

1. [7.1] Yes   2. [7.1] No   3. [7.1] Yes

4. [7.1] $\{y \mid y \geqslant -\frac{1}{2}\}$ 5. [7.2] $\{x \mid x \geqslant 7\}$ 6. [7.2]

$\{y \mid y > 2\}$ 7. [7.2] $\{y \mid y \leqslant -4\}$ 8. [7.2] $\{x \mid x < -11\}$

9. [7.2] $\{y \mid y > -7\}$ 10. [7.2] $\{x \mid x > -6\}$ 11. [7.2]

$\{x \mid x > -\frac{9}{11}\}$ 12. [7.2] $\{y \mid y \leqslant 7\}$ 13. [7.2]

$\{x \mid x \geqslant -\frac{1}{12}\}$ 14. [7.3] 86   15. [7.3] $\{w \mid w > 17 \text{ cm}\}$

16. [7.4] No   17. [7.4] No   18. [7.4] Yes

19. [7.4] 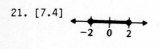   20. [7.4]

21. [7.4] 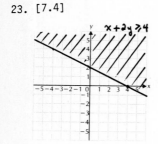   22. [7.4]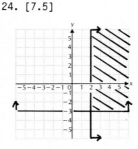

23. [7.4] 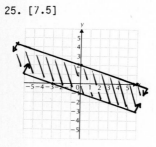   24. [7.5]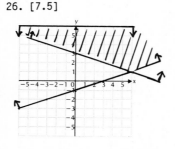

25. [7.5]   26. [7.5]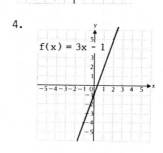

27. [7.5]   Max = 38 at $x = 7$, $y = 0$; Min = 3 at $x = 0$,

$y = 7$ 28. [7.2] $\{y \mid y > \frac{8}{3}\}$ 29. [7.2] $\{x \mid x \leqslant \frac{15}{2}\}$

30. [7.3] 63, 65, 67

CHAPTER 8

Margin Exercises, Section 8.1

1. Yes   2. Yes   3. No   4. 3, $1\frac{2}{5}$, $1\frac{3}{5}$, 2   5. 8, -5, 1

6. 0, -10, 2   7. 11, 35, 5   8. -1, 5, 20

9. (a) 209.36 cm; (b) 203.48 cm

Exercise Set 8.1

1. Yes   3. Yes   4. Yes   5. No   6. Yes   7. 8, 12,
-4   9. -6, 15, 72   10. -24, 2, -80   11. 6, -10, 16
12. 19, 19, 19   13. 2, 7, 4   15. 4, 5, 3   16. 6, 1,
10   17. 1, 91, 98   18. 4, -122, 3   19. -3, -2, 78

21. 1.606, 1.909, 4.03   22. $166\frac{2}{3}$, $666\frac{2}{3}$, $833\frac{1}{3}$

23. 70°; 220°; 10,020°   24. 1.792 cm, 2.8 cm, 11.2 cm

25. $45.15   27. $54.24   28. $39.82   29. 5, 8, 11, 14

30. -5, -4, -1, 4   31. 0, 2   33. 40   34. 100

35. -186   36. 880   37. $f(x) = \frac{15}{4}x - \frac{13}{4}$ 39. No. One

flip may correspond to 0 or 1 heads.

Margin Exercises, Section 8.2

1.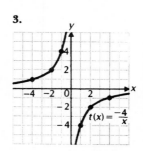

2.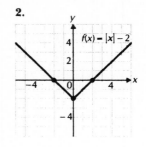

3.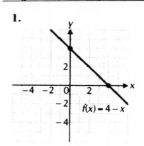

4. Yes
5. No
6. No

Exercise Set 8.2

1.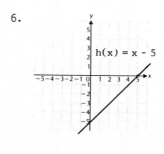

3.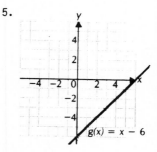

4.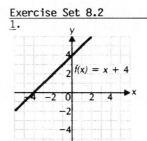

5.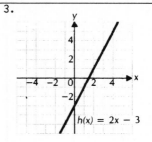

6.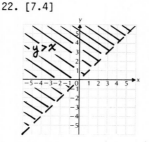

7.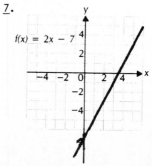

9.

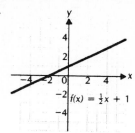

$f(x) = \frac{1}{2}x + 1$

10.

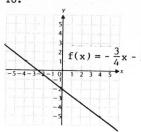

$f(x) = -\frac{3}{4}x - 2$

24. No

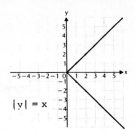

$|y| = x$

11.

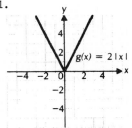

$g(x) = 2|x|$

12.

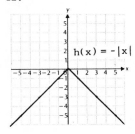

$h(x) = -|x|$

25. No

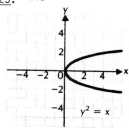

$y^2 = x$

13.

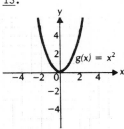

$g(x) = x^2$

15.

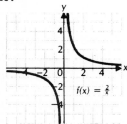

$f(x) = \frac{2}{x}$

16.

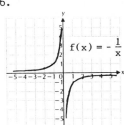

$f(x) = -\frac{1}{x}$

17. No   **18.** Yes
19. Yes   **21.** 390 km

22. 24 m
23. Answers may vary Here are two possible answers.

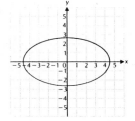

Margin Exercises, Section 8.3

1. $y = 7x$ **2.** $y = 0.625x$ **3.** $46\frac{2}{3}¢$, $1\frac{17}{18}¢$ **4.** 79.2 kg

Exercise Set 8.3

1. $y = 4x$ **3.** $y = 1.75x$ **4.** $y = 1.6x$ **5.** $y = 3.2x$
6. $y = 3.6x$ **7.** $y = \frac{2}{3}x$ **9.** $183.75   **10.** 16,445
11. $22\frac{6}{7}$ **12.** 93.3 servings   **13.** $16\frac{2}{3}$ kg   **15.** 68.4 kg
16. $4500   **17.** $P = kS$ (k = number of sides)
18. $C = kr$ (k = $2\pi \approx 6.28$)   **19.** $B = kN$
21. If $p = kq$, then $q = \frac{1}{k}p$. Since k is a constant, so
is $\frac{1}{k}$. Thus, q varies directly as p.   **22.** π
23. $S = kV^6$ **24.** $P^2 = kt$ **25.** $P = kRI^2$

Margin Exercises, Section 8.4

1. $y = \frac{63}{x}$ **2.** $y = \frac{900}{x}$ **3.** 8 hr   **4.** $7\frac{1}{2}$ hr

Exercise Set 8.4

1. $y = \frac{75}{x}$ **3.** $y = \frac{80}{x}$ **4.** $y = \frac{70}{x}$ **5.** $y = \frac{1}{x}$ **6.** $y = \frac{1}{x}$
7. $y = \frac{1050}{x}$ **9.** $y = \frac{0.06}{x}$ **10.** $y = \frac{0.24}{x}$ **11.** $5\frac{1}{3}$ hr
12. $4\frac{1}{2}$ hr   **13.** 320 cm³   **15.** 54 min   **16.** 250 cm,
1000 cm²   **17.** 2.4 ft   **18.** $7\frac{1}{2}$ hr   **19.** $C = \frac{k}{N}$
21. $I = \frac{k}{R}$ **22.** $D = \frac{k}{V}$ **23.** $I = \frac{k}{d^2}$ **24.** No   **25.** Yes
27. No   **28.**   **29.** $F = \frac{kS^2m}{r}$

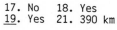

$y = \frac{6}{x}$

30. $L = \frac{kwt^2}{d}$

Summary and Review – Chapter 8

1. [8.1] No   2. [8.1] Yes   3. [8.1] 2, -4, -7
4. [8.1] 0, 2, -3   5. [8.1] -7, 1, 2   6. [8.1] 2700 calories
7. [8.2]

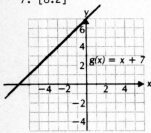

8. [8.2]

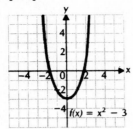

9. [8.2]

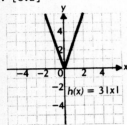

10. [8.2] No
11. [8.2] Yes
12. [8.3] $y = 3x$
13. [8.3] $y = \frac{1}{2}x$
14. [8.3] $y = \frac{4}{5}x$
15. [8.4] $y = \frac{30}{x}$
16. [8.4] $y = \frac{1}{x}$
17. [8.4] $y = \frac{0.65}{x}$

18. [8.3] $247.50   19. [8.4] 1 hr
20. [8.1] (a) 115 ft; (b) 179 ft   21. [8.1] -84
22. [8.2] No

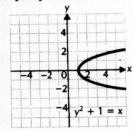

The ASVAB Test – Mathematics Knowledge   Chapter 8

1. D   2. D   3. B   4. C   5. B   6. C   7. A   8. C
9. C   10. D   11. A   12. C   13. C   14. A   15. B
16. D

CHAPTER 9

Margin Exercises, Section 9.1

1. (a) -1; (b) -4; (c) $\frac{2}{3}$; (d) does not exist, because the denominator is 0   2. $\frac{(x - 2)(x + 2)}{5(x + 4)}$
3. $\frac{(x + y)(x + y)}{(x + 3)(x - 3)}$
4. $\frac{x(3x + 2y)}{x(5x + 4y)}$ 5. $\frac{(2x^2 - y)(3x + 2)}{(3x + 4)(3x + 2)}$ 6. $\frac{-1(2a - 5)}{-1(a - b)}$
7. 7x   8. 2a + 3   9. $\frac{3x + 2}{2x + 4}$ 10. $\frac{2y + 4}{y - 1}$
11. $\frac{3x^2 - 6xy + 3y^2}{x + y}$ 12. a -b   13. $\frac{x - 5}{x + 3}$ 14. $\frac{1}{x + 7}$
15. $y^3 - 9$ 16. $\frac{x + 5}{2x - 10}$ 17. $\frac{2a^2b + 2ab^2}{a - b}$

Exercise Set 9.1

1. $v(0) = \frac{2}{3}$, $v(3) = \frac{23}{6}$, $v(7) = \frac{163}{10}$ 3. $r(0) = 0$, $r(4)$ = -184, $r(5)$ does not exist   4. $f(2) = 12.56$, $f(5) = \frac{78.5 + 31.4h}{4}$, $f(1)$ does not exist   5. $\frac{3x(x + 1)}{3x(x + 3)}$
6. $\frac{(4 - y^2)(-1)}{(6 - y)(-1)}$ 7. $\frac{(t - 3)(t + 3)}{(t + 2)(t + 3)}$ 9. $\frac{3y}{5}$ 10. $\frac{x}{3}$
11. a - 3   12. a - 2   13. $\frac{y - 3}{y + 3}$ 15. $\frac{t + 4}{t - 4}$
16. $\frac{p - 5}{p + 5}$ 17. $\frac{(x + 4)(x - 4)}{x(x + 3)}$ 18. $\frac{y + 5}{y - 3}$ 19. $\frac{y + 4}{2}$
21. $\frac{(x + 5)(2x + 3)}{7x}$ 22. $\frac{1}{y + 1}$ 23. c - 2
24. $\frac{x^2 - 6x + 9}{x + 3}$ 25. $\frac{1}{x + y}$ 27. 3   28. 6x²
29. $\frac{(y - 3)(y + 2)}{y}$ 30. $\frac{(x + 2)(x + 4)}{x}$ 31. $\frac{2a + 1}{a + 2}$
33. $\frac{(x + 4)(x + 2)}{3(x - 5)}$ 34. $\frac{(y + 6)(y + 3)}{3(y - 4)}$ 35. $\frac{y(y^2 + 3)}{(y + 3)(y - 2)}$
36. $\frac{a^3 + 4a}{a^2 + 7a + 12}$ 37. $\frac{x^2 + 4x + 16}{(x + 4)^2}$ 39. $\frac{2s}{r + 2s}$
40. $\frac{d^2 - 6d + 5}{5d^2 + 25d}$ 41. $\frac{21,934.2x^2}{y^2 - 182,499.84}$
42. $\frac{246,636}{x^2 - 8811.5769}$ 43. $\frac{132.3t^2}{t^2 - 0.000049}$
45. $\frac{x - 3}{(x + 1)(x + 3)}$ 46. $\frac{-4}{x - 2}$ 47. $\frac{m - t}{m + t + 1}$
48. $\frac{a^2 + 2}{a^2 - 3}$ 49. $\frac{x^2 + xy + y^2 + x + y}{x - y}$ 51. $\frac{-2x}{x - 1}$
52.   Answers may vary

Margin Exercises, Section 9.2

1. 90   2. 72   3. $\frac{47}{60}$ 4. $\frac{97}{72}$ 5. 5a³b²
6. $(y + 3)(y + 4)(y + 4)$ 7. $2x^2(x - 3)(x + 3)(x + 2)$
8. $2(a + b)(a - b)$, or $2(a + b)(b - a)$

Exercise Set 9.2

1. 2·2·3·3, or 36   3. 2·2·2·2·3·3, or 144
4. 2·3·3·3·5, or 270   5. 2·2·2·3·3, or 72
6. 2·3·5·5, or 150   7. 3·3·5, or 45   9. 2·2·2·3·3·7, or 504   10. 2·2·2·3·5·7, or 840   11. $\frac{11}{10}$ 12. $\frac{25}{72}$
13. $\frac{43}{36}$ 15. $\frac{79}{60}$ 16. $\frac{89}{60}$ 17. 24x³   18. 24y³
19. 12x²y   21. 30a³b²   22. 90x³y³   23. $(a + b)(a - b)$
24. $(x - 4)(x + 4)$ 25. $6(y - 2)$, or $6(2 - y)$
27. $3(y + 3)(y - 3)$ 28. $b(a + b)(a - b)$
29. $5(y - 3)(y - 3)$ 30. $4(x - 4)^2$
31. $(a + 1)(a - 1)(a - 1)$ 33. $(x + 2)(x - 2)$, or $(x + 2)(2 - x)$ 34. $(y + 3)(y - 3)$, or $(y + 3)(3 - y)$
35. $(x + 5)(x + 5)(x - 3)$ 36. $(y + 4)^2(y - 7)$
37. $(2r + 3)(r - 4)(3r - 1)$ 39. 15 ft long, 12 ft wide
40. True for all such even numbers
41. $(2x + 1)(x - 3)(x - 3)(x - 1)$
42. $(3x - 2)(x + 2)(2x + 3)(x - 2)^2$
43. $x^4(x^2 + 1)(x - 1)(x + 1)(x^2 + x + 1)(x^2 - x + 1)$

Margin Exercises, Section 9.3

1. $\frac{12 + y}{y}$ 2. 3x + 1   3. $\frac{a - b}{b + 2}$ 4. $\frac{y + 12}{x^2 + y^2}$
5. $\frac{1 - b^2}{3}$ 6. $\frac{2x^2 + 11}{x - 5}$ 7. $\frac{3 + 7x}{4y}$ 8. $\frac{11x^2}{2x - y}$
9. $\frac{9x^2 + 28y}{21x}$ 10. $\frac{3}{x + y}$ 11. $\frac{34y^2 + 21y - 1}{3y(y - 1)(y + 1)}$
12. $\frac{a + 12}{a^2 + 3a}$ 13. $\frac{3y^2 + 12y + 3}{(y - 4)(y - 3)(y + 5)}$ 14. $\frac{2}{x - 1}$

Exercise Set 9.3

1. 2   3. $\dfrac{3y + 5}{y - 2}$ 4. $\dfrac{2t + 6}{t - 4}$ 5. $a + b$ 6. $-(r + s)$

7. $\dfrac{11}{x}$ 9. $\dfrac{1}{x + 5}$ 10. $\dfrac{-2}{y^2 - 16}$ 11. $\dfrac{2y^2 + 22}{y^2 - y - 20}$

12. $\dfrac{2x^2 - x + 14}{x^2 - x - 12}$ 13. $\dfrac{x + y}{x - y}$ 15. $\dfrac{3x - 4}{x^2 - 3x + 2}$

16. $\dfrac{3y^2 + 7y + 14}{(2y - 5)(y + 2)(y - 1)}$ 17. $\dfrac{8x + 1}{x^2 - 1}$ 18. $\dfrac{4y + 17}{y^2 - 4}$

19. $\dfrac{2x - 14}{15x + 75}$ 21. $\dfrac{-a^2 + 7ab - b^2}{a^2 - b^2}$ 22. $\dfrac{-x^2 + 4xy - y^2}{x^2 - y^2}$

23. $\dfrac{y}{(y - 2)(y - 3)}$ 24. $\dfrac{2x^2 + 21x}{(x - 2)(x - 4)(x + 3)}$

25. $\dfrac{3y - 10}{y^2 - y - 20}$ 27. $\dfrac{3y^2 - 3y - 29}{(y - 3)(y + 8)(y - 4)}$

28. $\dfrac{5y^2 - 11y - 6}{(y - 2)(y - 5)(y - 3)}$ 29. $\dfrac{2x^2 - 13x + 7}{(x + 3)(x - 1)(x - 3)}$

30. $\dfrac{2p^2 + 7p + 10}{(p - 4)(p + 6)(p + 4)}$ 31. 0   33. $\dfrac{-3x^2 - 3x - 4}{x^2 - 1}$

34. $\dfrac{-14y^2 - 3y + 3}{4y^2 - 1}$ 35. About 56.8 yrs.   36. 4 thirty

minute tapes, 8 sixty minute tapes   37. $\dfrac{2y^2 + 3 - 7x^3y}{x^2y^2}$

39. $\dfrac{5y + 23}{5 - 2y}$ 40. $\dfrac{x^2 - y^2}{xy}$ 41. $\dfrac{x - y + x^2 + xy}{x + y - x^2 + xy}$

Margin Exercises, Section 9.4

1. $\dfrac{7(2y + 1)}{2(7y - 1)}$ 2. $\dfrac{x}{x + 1}$ 3. $\dfrac{b + a}{b - a}$ 4. $\dfrac{a^2b^2}{b^2 + ba + a^2}$

Exercise Set 9.4

1. $\dfrac{1 + 4x}{1 - 3x}$ 3. $\dfrac{x^2 - 1}{x^2 + 1}$ 4. $\dfrac{y^2 + 1}{y^2 - 1}$ 5. $\dfrac{3y + 4x}{4y - 3x}$

6. $\dfrac{2z + 5y}{z - 4y}$ 7. $\dfrac{x + y}{x}$ 9. $\dfrac{a^2(b - 3)}{b^2(a - 1)}$ 10. $\dfrac{3}{3x + 2}$

11. $\dfrac{1}{a - b}$ 12. $x - y$ 13. $\dfrac{-1}{x(x + h)}$ 15. $\dfrac{y - 3}{y + 5}$

16. $\dfrac{(x - 4)(x - 7)}{(x - 5)(x + 6)}$ 17. Perimeter of the first
rectangle is 15; perimeter of the second is 29
18. $4.25   19. $\dfrac{5}{6}$ 21. $\dfrac{11x + 8}{4x + 3}$ 22. $\dfrac{x^4}{81}$ 23. $\dfrac{b - a}{ab}$

24. $\dfrac{x}{x^3 - 1}$ 25. a 27. $\dfrac{3x + 2}{5x + 3}$ 28. $\dfrac{x - 1}{x}$; x

29. $\dfrac{-6x - 3h}{x^2(x + h)^2}$ 30. $\dfrac{-5}{x(x + h)}$ 31. $\dfrac{1}{(1 - x - h)(1 - x)}$

Margin Exercises, Section 9.5

1. $\dfrac{2}{3}$ 2. -31 3. No solution   4. 3   5. -3 6. 4, -3
7. 7   8. -13

Exercise Set 9.5

1. $\dfrac{51}{2}$ 3. -2 4. $\dfrac{40}{9}$ 5. 144   6. $-\dfrac{225}{2}$ 7. -5, -1

9. 2   10. $-\dfrac{1}{2}$ 11. $\dfrac{17}{4}$ 12. 14   13. 11   15. No
solution   16. No solution   17. 2   18. 3   19. $\dfrac{3}{5}$

21. 5   22. -1 23. -145 24. -23 25. $-\dfrac{10}{3}$ 27. -3

28. 4   29. -6, 5   30. 3   31. No solution   33. 8.75
liters of A, 26.25 liters of B   34. 16 and 18, -18 and
-16 35. All real numbers except 1 and -1 36. $-\dfrac{7}{2}$

37. 0.947   39. All real numbers except -2 40. All
real numbers except 3   41. Yes   42. Yes   43. No
45. Yes   46. Yes

Margin Exercises, Section 9.6

1. 1, -7 2. $2\dfrac{2}{5}$ hr   3. A: 32 hr, B: 96 hr

4. 280 mi   5. 35.5 mph

Exercise Set 9.6

1. $\dfrac{35}{12}$ 3. -3, -2 4. -3, -7 5. $3\dfrac{3}{14}$ hr   6. $1\dfrac{5}{7}$ hr

7. $8\dfrac{4}{7}$ hr   9. $3\dfrac{9}{52}$ hr   10. 2.475 hr   11. A: 10 days;

B: 40 days   12. A: $1\dfrac{1}{3}$ hr; B: 4 hr   13. 7 mph

15. 375 km   16. 450 km   17. Train A: 46 mph; Train B:
58 mph   18. Passenger: 80 mph; Freight: 66 mph

19. 9 km/h   21. $21\dfrac{9}{11}$ min after 4:00   22. $8\dfrac{2}{11}$ min after

10:30, or $21\dfrac{9}{11}$ min until 11:00   23. Boat: 14 km/h;
Stream: 10 km/h   24. 700 mi   25. 3.75 km/h
27. 48 km/h   28. 50 mph   29. $t = \dfrac{2}{3}$ hr   30. 12 mi

Margin Exercises, Section 9.7

1. $\dfrac{x^2}{2} + 8x + 3$ 2. $5y^3 - 2y^2 + 6y$ 3. $\dfrac{x^2}{2} + 5x + 8$

4. $x + 5$ 5. $3y^3 - 2y^2 + 6y - 4$ 6. $y^2 - 8y - 24$,

R -66; or $y^2 - 8y - 24 + \dfrac{-66}{y - 3}$ 7. $x^2 + 10x + 10$, R 5;

8. $y - 11$, R $3y - 27$ or $y - 11 + \dfrac{3y - 27}{y^2 - 3}$

Exercise Set 9.7

1. $6x^4 - 3x^2 + 8$ 3. $-2a^2 + 4a - 3$ 4. $-8x^3 - 6x^2 - 3x$
5. $y^3 - 2y^2 + 3y$ 6. $12a^2 + 14a - 10$
7. $-6x^5 + 3x^3 + 2x$ 9. $1 - ab^2 - a^3b^4$
10. $x - xy - x^2$ 11. $-2pq + 3p - 4q$
12. $4z - 2y^2z^3 + 3y^4z^2$ 13. $x + 7$ 15. $a - 12$, R 32
16. $y - 5$, R -50 17. $y - 5$ 18. $a + 9$
19. $y^2 - 2y - 1$, R -8 21. $a^2 + 4a + 15$, R 72
22. $x^2 - 2x + 3$ 23. $4x^2 - 6x + 9$ 24. $16y^2 + 8y + 4$
25. $x^2 + 6$ 27. $x^3 + x^2 - 1$, R 1
28. $y^3 - y^2 - 1$, R 4   29. $2y^2 + 2y - 1 + \dfrac{8}{5y - 2}$

30. $3x^2 - x + 4$, R 10   31. $2x^2 - x - 9 + \dfrac{3x + 12}{x^2 + 2}$

33. 12, 13, 14   34. $\dfrac{2}{3}$ 35. $x^2 + 2y$ 36. $a^2 + ab$

37. $x^3 + x^2y + xy^2 + y^3$ 39. $\dfrac{14}{3}$ 40. $-\dfrac{3}{2}$

Margin Exercises, Section 9.8

1. $T = \dfrac{PV}{k}$ 2. $T = \dfrac{A + Vnt}{Vn}$ 3. 6 cm

Exercise Set 9.8

1. $d_1 = \dfrac{d_2W_1}{W_2}$ 3. $t = \dfrac{2s}{v_1 + v_2}$ 4. $v_1 = \dfrac{2s - v_2t}{t}$

5. $r_1 = \dfrac{Rr_2}{r_2 - R}$ 6. $R = \dfrac{r_1r_2}{r_1 + r_2}$ 7. $s = \dfrac{Rg}{g - R}$

9. $r = \dfrac{2V - IR}{2I}$ 10. $R = \dfrac{2V - 2Ir}{I}$ 11. $f = \dfrac{pq}{q + p}$

12. $p = \dfrac{fq}{q - f}$ 13. $r = \dfrac{nE - IR}{In}$ 15. $H = m(t_1 - t_2)S$

16. $t_1 = \dfrac{H + Smt_2}{Sm}$ 17. $e = \dfrac{Er}{R + r}$ 18. $r = \dfrac{eR}{E - e}$

19. $a = \dfrac{S - Sr}{1 - r^n}$ 21. 12 runs   22. 5.2 ohms

23. $5\dfrac{5}{9}$ ohms   24. 6 cm   25. (a) $t = \dfrac{ab}{b + a}$;

(b) $a = \dfrac{tb}{b - t}$;   (c) $b = \dfrac{ta}{a - t}$

Summary and Review – Chapter 9

1. [9.1] $\dfrac{y - 8}{2}$ 2. [9.1] $\dfrac{(x - 2)(x + 5)}{x - 5}$

3. [9.1] $\dfrac{3a - 1}{a - 3}$ 4. [9.1] $\dfrac{x^3 - 64}{(x + 6)(x + 2)}$

5. [9.2] $15y^2$ 6. [9.2] $48x^3$ 7. [9.2] $3(x^2 - 49)$
8. [9.2] $(x - 2)(x + 5)(x - 4)$
9. [9.3] $\dfrac{x^2 + 11xy + y^2}{x^2 - y^2}$

10. [9.3] $\dfrac{2(x^3 + x^2y + xy^2 - y^3)}{x^2 - y^2}$

11. [9.3] $\dfrac{-y}{(y + 4)(y - 1)}$ 12. [9.4] $\dfrac{1 - x^2}{1 + x^2}$

13. [9.4] $\dfrac{b^2 - ab + a^2}{ab(b - a)}$ 14. [9.4] $\dfrac{y - 6}{y - 9}$ 15. [9.5] 2

16. [9.5] $\dfrac{28}{11}$ 17. [9.5] 6   18. [9.6] $5\frac{1}{7}$ hr

19. [9.6] 742.5 mi   20. [9.7] $4s^2 - 3s - 2rs^2$

21. [9.7] $y - 3$ 22. [9.7] $y^2 + 3y + 9$

23. [9.7] $4x + 3 + \dfrac{-9x - 5}{x^2 + 1}$ 24. [9.7] $x^2 + 6x + 20$,

R 54   25. [9.7] $4x^2 - 6x + 18$, R -59

26. [9.8] $g = \dfrac{Rs}{s - R}$ 27. [9.8] $m = \dfrac{H}{S(t_1 - t_2)}$

28. [9.4] $\dfrac{2}{a}$ 29. [9.3] $\dfrac{-3x^2 + 9x - 7}{(x - 1)^2(x - 2)}$

Something Extra: Metric-American Conversion   Chapter 9
Some variance will occur depending on whether one converts dircetly from ft to m, or from in to m.

1. 708.66   2. 1504.8   3. 762   4. 310.5   5. 220
6. 643.6   7. 342.47   8. 0.91   9. 17.78   10. 454.55
11. 33.82   12. 40.91   13. 30.3   14. 90.91
15. 338.24   16. 354.78   17. 0.47   18. 118.18
19. 264   20. 110   21. 90.91   22. 100   23. 804.5
24. 42.16   25. 381,472.98   26. 88.5   27. 149.04
28. 700.35   29. 88.67   30. 238.05   31. 289.86
32. 61.18   33. 1618.12   34. 0.0016
35. 4,625,000   36. 0.61

CHAPTER 10

Margin Exercises, Section 10.1
1. 13, -13   2. 16   3. -10   4. Irrational
5. Rational   6. Irrational   7. Irrational
8. Rational   9. Rational   10. Rational
11. Irrational   12. 2.645   13. 2.646   14. 8.485
15. 20, 16

Exercise Set 10.1
1. 1, -1   3. 4, -4   4. 3, -3   5. 10, -10   6. 11,
-11   7. 13, -13   9. 2   10. 1   11. -3   12. -5
13. -8   15. -15   16. 20   17. 19   18. 21
19. Irrational   21. Irrational   22. Irrational
23. Rational   24. Rational   25. Irrational
26. Irrational   27. Rational   28. Rational
29. Irrational   30. Irrational   31. Rational
33. Rational   34. Rational   35. Rational
36. Rational   37. Rational   39. Rational
40. Irrational   41. Irrational   42. Irrational
43. Irrational   45. 2.449   46. 4.123   47. 4.359
48. 9.644   49. 6.557   51. 13, 24   52. 20, 25
53. 2   54. 5   55. -5, -6   57. 14.071   58. 30.496
59. 70.228   60. 32.309   61. $\dfrac{x + y}{2}$ 63. $3\frac{3}{8}$ 64. $-1\frac{1}{15}$
65. $\dfrac{3z + w}{4}$ 66. $-\frac{7}{6}$ 67. $-\frac{19}{3}$ 69. 21   70. $-\frac{4}{9}$
71. $-\frac{3}{4}$ 72. 0.3   73. 0.6   75. 0.12

Margin Exercises, Section 10.2
1. $45 + x$ 2. $\dfrac{x}{x + 2}$ 3. No   4. Yes   5. Yes   6. No
7. Yes   8. No   9. $a \geqslant 0$ 10. $x \geqslant 3$ 11. $x \geqslant \frac{5}{2}$
12. All real-number replacements   13. $|xy|$ 14. $|xy|$
15. $|x - 1|$ 16. $|x + 4|$ 17. $5|y|$ 18. $\frac{1}{2}|t|$ 19. 4
20. -5   21. $2(x + 2)$ 22. $-3(y - 4)$ 23. $-2x$
24. $3x + 2$ 25. (a) 3; (b) -3; (c) No real-number fourth roots   26. $2|x - 2|$ 27. $|x|$ 28. $|x + 3|$

Exercise Set 10.2
1. $a - 4$ 3. $t^2 + 1$ 4. $x^2 + 5$ 5. $\dfrac{3}{x + 2}$ 6. $\dfrac{a}{a - b}$
7. Yes   9. No   10. No   11. $x \geqslant 0$ 12. $y \geqslant 0$ 13.
$t \geqslant 5$ 15. $y \geqslant -8$ 16. $x \geqslant -6$ 17. $x \geqslant -20$
18. $m \geqslant 18$ 19. $y \geqslant \frac{7}{2}$ 21. Any real number   22. Any
real number   23. $|t|$ 24. $|x|$ 25. $3|x|$ 27. 7
28. 5   29. $4|d|$ 30. $3|b|$ 31. $|x + 3|$ 33. $|a - 5|$
34. $|x + 1|$ 35. 3   36. -4   37. $-4x$ 39. -6
40. 10   41. $0.7(x + 1)$ 42. $0.02(y - 2)$ 43. 5
45. -1   46. 2   47. $-\frac{2}{3}$ 48. $-\frac{1}{2}$ 49. $|x|$ 51. $5|a|$
52. $7|b|$ 53. 6   54. 10   55. $|a + b|$ 57. $|y|$
58. -6   59. $x - 2$ 60. $2xy$ 61. $5330   63. 6, -6
64. No value   65. 3, -3   66. 7, -7   67. $3|a|$
69. $2\dfrac{x^4}{|y^3|}$ 70. $\dfrac{y^6}{90}$ 71. $\dfrac{13}{m^8}$ 72. $\dfrac{|p|}{60}$
73. $m \geqslant 0$ or $m \leqslant -3$ 75. $x \geqslant 2$ or $x \leqslant -3$
76. $x \leqslant -4$ or $x \geqslant -3$ 77. $x \geqslant 2$ or $x \leqslant -2$
78. $x \leqslant -\frac{1}{2}$ or $x \geqslant \frac{1}{2}$ 79. $\sqrt{20}$, 0, does not exist,
does not exist   81. -3, does not exist, -1, -5
82. Does not exist, $\sqrt{30}$, $\sqrt{30}$, $\sqrt{180}$, $\sqrt{180}$ 83. $\sqrt{26}$, $\sqrt{26}$,
1, $\sqrt{101}$, $\sqrt{101}$ 84. -2, 0, -4, -2, -6, -4   85. 2, 3,
-2, -4
87.   88.

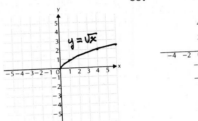

ASVAB Test - Mathematical Knowledge   Chapter 10
1. D   2. A   3. A   4. B   5. D   6. C   7. A   8. B
9. D   10. C   11. C   12. A   13. C   14. D   15. A
16. D   17. B   18. B

Margin Exercises, Section 10.3
1. $\sqrt{133}$ 2. $\sqrt{x^2 - 4y^2}$ 3. $\sqrt[4]{2821}$ 4. $\sqrt[3]{8x^5 + 40x}$
5. $4\sqrt{2}$ 6. $2\sqrt[3]{10}$ 7. $10\sqrt{3}$ 8. $6y$ 9. $(x + 2)\sqrt{3}$
10. $2bc\sqrt{3ab}$ 11. $2\sqrt[3]{2}$ 12. $3xy^2\sqrt[3]{3xy^2}$
13. $(a + b)\sqrt[3]{a + b}$ 14. $3\sqrt{2}$ 15. $6y\sqrt{7}$ 16. $3x\sqrt[3]{4y}$
17. $7\sqrt{3ab}$ 18. $2(y + 5)\sqrt[3]{(y + 5)^2}$ 19. Rational
20. Rational   21. Rational   22. Rational
23. Irrational   24. Irrational   25. Rational
26. 15.9   27. 20.1   28. 2.175×10^7
29. 5.605×10^{-8}

Exercise Set 10.3
1. $\sqrt{6}$ 3. $\sqrt[3]{10}$ 4. $\sqrt[3]{14}$ 5. $\sqrt[4]{72}$ 6. $\sqrt[4]{18}$
7. $\sqrt{30ab}$ 9. $\sqrt[5]{18t^3}$ 10. $\sqrt[5]{80y^4}$ 11. $\sqrt{x^2 - a^2}$
12. $\sqrt{y^2 - b^2}$ 13. $\sqrt[3]{0.06x^2}$ 15. $\sqrt[4]{x^3 - 1}$
16. $\sqrt[5]{(x - 2)^3}$ 17. $\sqrt{\dfrac{6y}{5x}}$ 18. $\sqrt{\dfrac{7s}{11t}}$ 19. $2\sqrt{2}$
21. $2\sqrt{6}$ 22. $2\sqrt{5}$ 23. $6x^2\sqrt{5}$ 24. $5y^3\sqrt{7}$
25. $3x^2\sqrt[3]{2x^2}$ 27. $2x^2\sqrt[3]{10x^2}$ 28. $3m\sqrt[3]{4m^2}$
29. $2\sqrt[4]{2}$ 31. $3cd\sqrt[4]{2d^2}$ 33. $(x + y)\sqrt[3]{x + y}$
34. $3\sqrt{2}$ 35. $5\sqrt{2}$ 36. $6\sqrt{5}$ 37. 8   39. $6\sqrt{7}$
40. $3\sqrt[3]{2}$ 41. $30\sqrt{3}$ 42. $5bc^2\sqrt{2b}$ 43. $2x^2y\sqrt{6}$
45. $25t^3\sqrt[3]{t}$ 46. $(b + 3)^2$ 47. $(x + y)^2\sqrt[3]{(x + y)^2}$
48. $4a^3b\sqrt{6ab}$ 49. Rational   51. Rational
52. Irrational   53. Rational   54. Irrational

55. Irrational   57. 15.9   58. 11.8   59. 13.7
60. 22.1   61. 18.7   63. 29.8   64. 19.4   65. 15.5
66. 19.2   67. 156,525   69. 684,251   70. 315,278
71. 418,832   72. 0.002152   73. 0.000199   75. 0.000707
76. 0.00318   77. 8   78. Length is 5, width is 20
79. (a) 20 mph; (b) 37.4 mph; (c) 42.4 mph

Margin Exercises, Section 10.4

1. $\frac{5}{6}$ 2. $\frac{x}{10}$ 3. $\frac{3x\,\sqrt[3]{2x^2}}{5}$ 4. 5   5. $56\sqrt{xy}$ 6. $5a$

7. $\frac{20}{7}$ 8. $\frac{2ab}{3}$ 9. 25   10. $6y\sqrt{6y}$

Exercise Set 10.4

1. $\frac{4}{5}$ 3. $\frac{4}{3}$ 4. $\frac{7}{8}$ 5. $\frac{7}{y}$ 6. $\frac{11}{x}$ 7. $\frac{5y\,\sqrt{y}}{x^2}$

9. $\frac{2x\,\sqrt[3]{x^2}}{3y}$ 10. $\frac{2x^2\,\sqrt[3]{x}}{3y^2}$ 11. $\sqrt{7}$ 12. $\sqrt{7}$ 13. 3

15. $y\sqrt{5y}$ 16. $2b\sqrt{2b}$ 17. $2\sqrt[3]{a^2b}$ 18. $3xy\sqrt[3]{y^2}$

19. $3\sqrt{xy}$ 21. $\sqrt{x^2 + xy + y^2}$ 22. $\sqrt{r^2 - rs + s^2}$

23. $6a\sqrt{6a}$ 24. $7y\sqrt{7y}$ 25. $4b\sqrt[3]{4b}$ 27. $54a^3b\sqrt{2b}$

28. $24x^3y\sqrt{3y}$ 29. $2c\sqrt[3]{18cd^2}$ 30. $3x\sqrt[3]{3xy^2}$

31. $5 per hr   33. (a) 1.62 sec; (b) 1.99 sec;

(c) 2.20 sec   34. a^3bxy^2 35. $9\sqrt[3]{9n^2}$ 36. $2yz\sqrt{2z}$

Margin Exercises, Section 10.5

1. $13\sqrt{2}$ 2. $10\sqrt[4]{5x} - \sqrt{7}$ 3. $19\sqrt{5}$

4. $(3y + 4)\sqrt[3]{y^2} + 2y^2$ 5. $2\sqrt{x - 1}$

Exercise Set 10.5

1. $8\sqrt{3}$ 3. $3\sqrt[3]{5}$ 4. $8\sqrt[5]{2}$ 5. $13\sqrt[3]{y}$ 6. $3\sqrt[4]{t}$

7. $7\sqrt{2}$ 9. $6\sqrt[3]{3}$ 10. $6\sqrt{7} + 4\sqrt[4]{11}$ 11. $21\sqrt{3}$

12. $41\sqrt{2}$ 13. $38\sqrt{5}$ 15. $122\sqrt{2}$ 16. $4\sqrt{5}$ 17. $9\sqrt[3]{2}$

18. -7 19. $29\sqrt{2}$ 21. $(1 + 6a)\sqrt{5a}$ 22. $(4x - 2)\sqrt{3x}$

23. $(2 - x)\sqrt[3]{3x}$ 24. $(3 - x)\sqrt[3]{2x}$ 25. $3\sqrt{2y - 2}$

27. $(x + 3)\sqrt{x - 1}$ 28. $(2 - x)\sqrt{x - 1}$ 29. $5\sqrt{2}$

30. $6\sqrt{3}$ 31. $15\sqrt[3]{4}$ 33. $-9\sqrt{5}$ 34. $8\sqrt{7}$ 35. 0

36. $-8\sqrt{5x^2 + 4}$ 37. 36   39. $a = 0$ and $b \geqslant 0$, or $a \geqslant 0$ and $b = 0$; any pair of numbers that does not satisfy the preceding conditions   40. $-28\sqrt{3} - 22\sqrt{5}$

41. $(25x - 30y - 9xy)\sqrt{2xy}$

42. $ac(3ac\sqrt{ab} - 2c\sqrt[3]{ab} + 2\sqrt{ab})$ 43. $(7x^2 - 2y^2)\sqrt{x + y}$

Margin Exercises, Section 10.6

1. $5\sqrt{6} + 3\sqrt{14}$ 2. $a\sqrt[3]{3} - \sqrt[3]{2a^2}$ 3. $-4 - 9\sqrt{6}$
4. $3\sqrt{ab} - 4\sqrt{3a} + 6\sqrt{3b} - 24$ 5. -3 6. $20 - 4y\sqrt{5} + y^2$
7. $58 + 12\sqrt{6}$

Exercise Set 10.6

1. $2\sqrt{6} - 18$ 3. $\sqrt{6} - \sqrt{10}$ 4. $5 - \sqrt{10}$ 5. $2\sqrt{15} - 6\sqrt{3}$
6. $6\sqrt{5} - 4$ 7. -6 9. $3a\sqrt[3]{2}$ 10. $-2x\sqrt[3]{3}$ 11. 1
13. -12 15. $a - b$ 16. $x - y$ 17. $1 + \sqrt{5}$
18. $2 + 2\sqrt{6}$ 19. $7 + 3\sqrt{3}$ 21. -6 22. $24 - 7\sqrt{15}$
23. $a + \sqrt{3a} + \sqrt{2a} + \sqrt{6}$ 24. $2 - 3\sqrt{x} + x$
25. $2\sqrt[3]{9} - 3\sqrt[3]{6} - 2\sqrt[3]{4}$ 27. $7 + 4\sqrt{3}$ 28. $6 + 2\sqrt{5}$
29. 28, 82   30. 4 hr   31. (a) 19.485281;
(b) 19.485281   33. $2x - 2\sqrt{x^2 - 4}$
34. $14 + 2\sqrt{15} − 6\sqrt{2} - 2\sqrt{30}$ 35. $\sqrt[3]{y} - y$ 36. $\sqrt[3]{91}$
37. $12 + 6\sqrt{3}$ 39. $16(a + 3) + 38a\sqrt[3]{3(a + 3)} - 15a^2$
40. $2(24\sqrt{21} - 8\sqrt{15} + 15\sqrt{14} - 5\sqrt{10})$ 41. $2\sqrt{34}$ 42. 12

Margin Exercises, Section 10.7

1. $\frac{\sqrt{6}}{3}$ 2. $\frac{\sqrt{140}}{7}$ 3. $\frac{\sqrt[3]{4}}{2}$ 4. $\frac{2\sqrt{3ab}}{3b}$ 5. $\frac{2x^2\,\sqrt{3xy}}{3y^2}$

6. $\frac{\sqrt[3]{28}}{2}$ 7. $\frac{x\,\sqrt[3]{12x^2y^2}}{2y}$ 8. $\frac{11}{\sqrt{66}}$ 9. $\frac{7}{\sqrt[3]{196}}$

10. $\frac{\sqrt{3} + y}{\sqrt{3} + y}$ 11. $\frac{\sqrt{2} - \sqrt{3}}{\sqrt{2} - \sqrt{3}}$ 12. $-5(1 + \sqrt{2})$

13. $\frac{5 + \sqrt{7}}{5 + \sqrt{7}}$ 14. $\frac{4}{3\sqrt{2} - \sqrt{10} - 3\sqrt{6} + \sqrt{30}}$

Exercise Set 10.7

1. $\frac{\sqrt{30}}{5}$ 3. $\frac{\sqrt{70}}{7}$ 4. $\frac{\sqrt{66}}{3}$ 5. $\frac{2\sqrt{15}}{5}$ 6. $\frac{\sqrt{6}}{5}$ 7. $\frac{2\sqrt[3]{6}}{3}$

9. $\frac{\sqrt[3]{75ac^2}}{5c}$ 10. $\frac{\sqrt[3]{63xy^2}}{3y}$ 11. $\frac{y\,\sqrt[3]{9yx^2}}{3x^2}$ 12. $\frac{a\,\sqrt[3]{147ab}}{7b}$

13. $\frac{\sqrt[3]{x^2y^2}}{xy}$ 14. $\frac{\sqrt[3]{a^2b^2}}{ab}$ 15. $\frac{7}{\sqrt{21x}}$ 16. $\frac{6}{\sqrt{30x}}$ 17. $\frac{2}{\sqrt{6}}$

18. $\frac{2}{\sqrt{5}}$ 19. $\frac{52}{\sqrt[3]{91}}$ 21. $\frac{7}{\sqrt[3]{98}}$ 22. $\frac{5}{\sqrt[3]{100}}$ 23. $\frac{7x}{\sqrt{21xy}}$

24. $\frac{3a}{\sqrt{3ab}}$ 25. $\frac{5y^2}{x\,\sqrt[3]{150x^2y^2}}$ 27. $\frac{ab}{\sqrt[3]{ab}}$ 28. $\frac{xy}{5\sqrt{xy}}$

29. $\frac{5(8 + \sqrt{6})}{58}$ 30. $\frac{7(9 - \sqrt{10})}{71}$ 31. $-2\sqrt{7}(\sqrt{5} + \sqrt{3})$

33. $\frac{\sqrt{15} + 20 - 6\sqrt{2} - 8\sqrt{30}}{-77}$

34. $\frac{-3\sqrt{2} - 2\sqrt{42} + 3\sqrt{15} + 6\sqrt{35}}{25}$ 35. $\frac{x - 2\sqrt{xy} + y}{x - y}$

36. $\frac{a + 2\sqrt{ab} + b}{a - b}$ 37. $\frac{3\sqrt{6} + 4}{2}$ 39. $\frac{-11}{4(\sqrt{3} - 5)}$

40. $\frac{7}{5(3 + \sqrt{2})}$ 41. $\frac{-22}{\sqrt{6} + 5\sqrt{2} + 5\sqrt{3} + 25}$

42. $\frac{-3}{3\sqrt{2} + 3\sqrt{3} + 7\sqrt{6} + 21}$ 43. $\frac{x - y}{x + 2\sqrt{xy} + y}$

45. $\frac{7}{43\sqrt{2} + 66}$ 46. $\frac{53}{75\sqrt{6} - 187}$ 47. $\frac{\sqrt{10} + 2\sqrt{5} - 2\sqrt{3}}{10}$

48. $\frac{15y + 6\sqrt{yz} + 20y\sqrt{z} + 8z\sqrt{y}}{25y - 4z}$ 49. $\frac{b^2 + \sqrt{b}}{b^2 + b + 1}$

51. $6a\sqrt[3]{36ab^2}$ 52. $\frac{1}{\sqrt{y + 18} + \sqrt{y}}$

53. $\frac{x - 19}{x + 31 + 10\sqrt{x + 6}}$ 54. $\frac{-3\sqrt{a^2 - 3}}{a^2 - 3}$

55. $\left[\frac{5x + 4y - 3}{xy}\right]\sqrt{xy}$ 57. $\frac{7\sqrt{3}}{39}$

Margin Exercises, Section 10.8

1. $\sqrt[4]{y}$ 2. $\sqrt{3a}$ 3. $\sqrt[4]{16}$, or 2   4. $\sqrt[3]{125}$, or 5
5. $\sqrt[5]{a^3b^2c}$ 6. $19^{1/3}$ 7. $(abc)^{1/2}$ 8. $\left[\frac{x^2y}{16}\right]^{1/5}$
9. $\sqrt{x^3}$ or $(\sqrt{x})^3$, which simplifies to $x\sqrt{x}$ 10. $\sqrt[3]{8^2}$ or $(\sqrt[3]{8})^2$, which simplifies to 4   11. $\sqrt{4^5}$ or $(\sqrt{4})^5$, which simplifies to 32   12. $(7abc)^{4/3}$ 13. $6^{7/5}$ 14. $\frac{1}{2}$
15. $\frac{1}{(3xy)^{7/8}}$ 16. $7^{14/15}$ 17. $5^{2/6}$ or $5^{1/3}$ 18. $9^{2/5}$
19. $\sqrt{a}$ 20. x 21. $\sqrt{2}$ 22. ab^2 23. xy^3
24. $\sqrt[4]{xy^2}$ 25. $\sqrt[4]{63}$ 26. $\sqrt[4]{a - b}$ 27. $\sqrt[6]{x^4y^3z^5}$
28. $\sqrt[4]{ab}$

Exercise Set 10.8

1. $\sqrt[4]{x}$ 3. 2   4. 4   5. $\sqrt[5]{a^2b^2}$ 6. $\sqrt[4]{x^3y^3}$ 7. $\sqrt[3]{a^2}$
9. 8   10. 128   11. $20^{1/3}$ 12. $19^{1/3}$ 13. $17^{1/2}$
15. $(cd)^{1/4}$ 16. $(xy)^{1/5}$ 17. $(xy^2z)^{1/5}$
18. $(x^3y^2z^2)^{1/7}$ 19. $(3mn)^{3/2}$ 21. $(8x^2y)^{5/7}$

22. $(2a^5b)^{7/6}$ 23. $\dfrac{1}{x^{1/3}}$ 24. $\dfrac{1}{y^{1/4}}$ 25. $\dfrac{1}{(2rs)^{3/4}}$

27. $10^{2/3}$ 28. $8^{3/4}$ 29. $x^{2/3}$ 30. $x^{5/6}$ 31. $5^{7/8}$

33. $7^{1/4}$ 34. $9^{2/11}$ 35. $8.3^{7/20}$ 36. $3.9^{7/20}$

37. $10^{6/25}$ 39. $\sqrt[3]{a^2}$ 40. $\sqrt[3]{y}$ 41. $2y^2$ 42. x^2y^3

43. $2\sqrt[4]{2}$ 45. $\sqrt[3]{2x}$ 46. $2x\sqrt{y}$ 47. $2c^2d^3$ 48. $2x^3y^4$

49. $\dfrac{m^2n^4}{2}$ 51. $\sqrt[4]{r^2s}$ 52. $\sqrt{2ts}$ 53. $3ab^3$ 54. $3x^2y^2$

55. $\sqrt[6]{392}$ 57. $\sqrt[6]{4x^5}$ 58. $\sqrt[15]{27y^8}$

59. $\sqrt[6]{x^5 - 4x^4 + 4x^3}$ 60. $\sqrt[4]{3xy^2 + 24xy + 48x}$

61. $\sqrt[6]{a + b}$ 63. $\sqrt[12]{a^8b^9}$ 64. $\sqrt[12]{x^4y^3z^2}$ 65. $t\sqrt[4]{s}$
66. $y\sqrt[5]{x}$ 67. $\sqrt[10]{x^9y^7}$ 69. $(a + b)\sqrt[12]{(a + b)^{11}}$

70. $\sqrt[4]{2xy^2}$ 71. $343w^4$ 72. $\sqrt[6]{p + q}$

73. $\sqrt[3]{9} + \sqrt[3]{6} + \sqrt[3]{4}$ 75. $x^5y^7z(\sqrt[p]{yz^3})$
76. (a) 27.4 m; (b) 45.9 m; (c) 21.9 m; (d) 79.4 m

Margin Exercises, Section 10.9

1. 100   2. No solution   3. 4   4. 17   5. 9   6. 5
7. 13.038   8. 93.81   9. 22.2 mi

Exercise Set 10.9

1. 2   3. 168   4. 11   5. 3   6. 29   7. 19   9. $\dfrac{80}{3}$

10. 84   11. 4   12. $-\dfrac{1}{16}$ 13. -27   15. 397   16. 117

17. No solution   18. No solution   19. $\dfrac{1}{64}$ 21. -6

22. $-\dfrac{5}{3}$ 23. 5   24. 2   25. $-\dfrac{1}{4}$ 27. 3   28. -1

29. 9   30. No solution   31. 7   33. $\dfrac{80}{9}$ 34. $\dfrac{15}{4}$

35. -1   36. $\dfrac{1}{3}$, -1   37. 6, 2   39. No solution

40. 2   41. 5   42. 6   43. 12   45. 10   46. 15

47. $\sqrt{58} \approx 7.62$ 48. $\sqrt{23} \approx 4.80$ 49. $\sqrt{2} \approx 1.41$

51. $\sqrt{12} \approx 3.46$ 52. $\sqrt{63} \approx 7.94$ 53. Both triangles
have the same area.   54. 102.8 ft   55. $\ell = 2\sqrt{s^2 - h^2}$

57. 208 mi   58. 10 mi

59. $-\dfrac{8}{9}$ 60. 6912   61. -8, 8   63. -4, 3

64. -1, 6   65. 1, 8   66. 2   67. $\dfrac{1}{36}$, 36   69. 25

70. 2   71. 3   72. 3   73. 3   75. No

Extra Practice on Solving Radical Equations

1. 25   2. 49   3. 38.44   4. 18.49   5. 397   6. 117
7. $\dfrac{621}{2}$ 8. 84   9. 5   10. 52   11. 3   12. $\dfrac{4}{5}$ 13. $\dfrac{17}{4}$

14. $\dfrac{14}{3}$ 15. No solution   16. No solution   17. No
solution   18. No solution   19. 2, -2   20. 0   21. 4

22. $\dfrac{1}{2}$ 23. 9   24. 5   25. $\dfrac{15}{4}$ 26. 0, 4

Summary and Review – Chapter 10

1. [10.2] 48   2. [10.2] $5|x|$ 3. [10.2] $\dfrac{-3}{4x}$

4. [10.2] -2   5. [10.2] $5|x|$ 6. [10.3] $15x\sqrt{2}$

7. [10.3] m^2n^3 8. [10.3] $3ab^2\sqrt[3]{2a}$ 9. [10.3] 15.7
10. [10.3] 25.3   11. [10.3] 250,998
12. [10.3] 0.00174   13. [10.4] 3   14. [10.4] $\dfrac{2\sqrt{5}x^2}{y^3}$

15. [10.4] $\dfrac{3a\sqrt{5a}}{5}$ 16. [10.5] $36\sqrt{2}$ 17. [10.5] $\sqrt[3]{2}$

18. [10.5] $(2x - 2)\sqrt[3]{4x}$ 19. [10.6] -41
20. [10.6] $105 - 24\sqrt{6}$ 21. [10.6] $6 + 5\sqrt{x} + x$

22. [10.7] $\dfrac{\sqrt{21}}{6}$ 23. [10.7] $\dfrac{48 - 8\sqrt{5}}{31}$

24. [10.7] $\sqrt{10} - 5\sqrt{5} + 2\sqrt{2} - 10$
25. [10.8] $x^{5/8}y^{1/2}z^{3/8}$ 26. [10.8] $5\sqrt[3]{5}$

27. [10.8] $\sqrt[5]{6}$ 28. [10.8] $3x^4y^5$ 29. [10.8] $\sqrt[4]{xy^2}$

30. [10.8] $\sqrt[6]{x^5}$ 31. [10.8] $\sqrt[20]{(a - b)^7}$ 32. [10.9] $\dfrac{7}{2}$

33. [10.9] 16   34. [10.1, 10.3] Irrational
35. [10.1, 10.3] Rational   36. [10.1, 10.3] Rational
37. [10.1, 10.3] Rational   38. [10.1, 10.3] Irrational
39. [10.1, 10.3] Irrational   40. [10.1, 10.3] Irrational

41. [10.6] $2x - 3\sqrt{xy} - 2y$ 42. [10.7] $\dfrac{x\sqrt{6y} + 3y\sqrt{x}}{2x - 3y}$

43. [10.8] $12xy^{-1/2}$

CHAPTER 11

Margin Exercises Section 11.1

1. $x^2 - 7x = 0$; $a = 1$, $b = -7$, $c = 0$
2. $x^2 + 9x - 3 = 0$; $a = 1$, $b = 9$, $c = -3$
3. $4x^2 + 2x + 4 = 0$; $a = 4$, $b = 2$, $c = 4$ 4. $\sqrt{5}$, $-\sqrt{5}$
5. 0   6. $\frac{\sqrt{6}}{2}$, $-\frac{\sqrt{6}}{2}$ 7. $\frac{3}{2}$, $-\frac{3}{2}$ 8. 9.3 sec

Exercise Set 11.1

1. $a = 1$, $b = -3$, $c = 2$ 3. $a = 2$, $b = 0$, $c = -3$
4. $a = 5$, $b = 0$, $c = -9$ 5. $a = 7$, $b = -4$, $c = 3$
6. $a = 9$, $b = -1$, $c = -5$ 7. $a = 2$, $b = -3$, $c = 5$
9. $a = 3$, $b = -2$, $c = 8$ 10. 11, -11   11. $\sqrt{10}$, $-\sqrt{10}$
12. $\sqrt{7}$, $-\sqrt{7}$ 13. $\sqrt{10}$, $-\sqrt{10}$ 15. $\frac{\sqrt{10}}{2}$, $-\frac{\sqrt{10}}{2}$
16. $\frac{5}{2}$, $-\frac{5}{2}$ 17. $\frac{2}{3}$, $-\frac{2}{3}$ 18. $\frac{7\sqrt{3}}{3}$, $-\frac{7\sqrt{3}}{3}$
19. $\frac{4\sqrt{5}}{5}$, $-\frac{4\sqrt{5}}{5}$ 21. $\frac{4}{7}$, $-\frac{4}{7}$ 22. $\frac{6}{5}$, $-\frac{6}{5}$
23. $2\sqrt{5}$, $-2\sqrt{5}$ 24. $\frac{\sqrt{5}}{10}$, $-\frac{\sqrt{5}}{10}$ 25. 7.9 sec   27. 3.3 sec
28. 4.4 sec   29. 49.9, -49.9   30. 6, -6   31. $\sqrt{3}$, $-\sqrt{3}$
33. $\sqrt{11}$, $-\sqrt{11}$ 34. $\frac{\sqrt{10}}{5}$, $-\frac{\sqrt{10}}{5}$ 35. 3, -3
36. $\frac{\sqrt{a(b^2 + 3b)}}{a}$, $-\frac{\sqrt{a(b^2 + 3b)}}{a}$ 37. $3\sqrt{1 + a}$, $-3\sqrt{1 + a}$

Margin Exercises, Section 11.2

1. 0, $-\frac{5}{3}$ 2. 0, $\frac{3}{5}$ 3. $\frac{2}{3}$, -1   4. 4, 1   5. (a) 14;
(b) 11

Exercise Set 11.2

1. 0, -7   3. 0, -2   4. 0, -2   5. 0, $\frac{2}{5}$ 6. 0, $\frac{7}{3}$
7. 0, -1   9. 0, 3   10. 0, 5   11. 0, $\frac{1}{5}$ 12. 0, $-\frac{1}{3}$
13. 0, $\frac{3}{14}$ 14. 0, $\frac{8}{17}$ 15. 0, $\frac{81}{2}$ 16. 12, 4
17. -12, 4   18. -6, -1   19. -5, -1   21. -9, 2
22. 7, 2   23. 5, 3   24. -5   25. -3   27. 4
28. $\frac{3}{2}$, 5   29. $-\frac{2}{3}$, $\frac{1}{2}$ 30. $-\frac{2}{3}$, 4   31. $\frac{1}{3}$, $\frac{4}{3}$
33. $\frac{5}{3}$, -1   34. -5, -1   35. $-\frac{1}{4}$, $\frac{2}{3}$ 36. $\frac{1}{3}$, $-\frac{1}{2}$
37. 7, -2   39. -5, 4   40. $\frac{3}{2}$, 3   41. 4   42. -2, 1
43. 3, -2   45. 9   46. 35   47. 7   48. 6   49. $-\frac{1}{3}$, 1
51. 0, $\frac{\sqrt{5}}{5}$ 52. 0, $-\frac{\sqrt{21}}{7}$ 53. $-\frac{5}{2}$, 1   54. $-\frac{1}{4}$, -3
55. 0, $-\frac{b}{a}$ 57. $\sqrt{2}$, $-\sqrt{2}$ 58. 1, 81

Margin Exercises, Section 11.3

1. 7, -1   2. $-3 \pm \sqrt{10}$ 3. $1 \pm \sqrt{5}$ 4. $x^2 - 8x + 16 =$ $(x - 4)^2$ 5. $x^2 + 20x + 100 = (x + 10)^2$
6. $x^2 + 7x + \frac{49}{4} = (x + \frac{7}{2})^2$ 7. $x^2 - 3x + \frac{9}{4} = (x - \frac{3}{2})^2$
8. $1299.60   9. 12.5%

Exercise Set 11.3

1. 9, -5   3. $-3 \pm \sqrt{21}$ 4. $3 \pm \sqrt{6}$ 5. $-13 \pm 2\sqrt{2}$
6. 21, 5   7. $7 \pm 2\sqrt{3}$ 9. $-9 \pm \sqrt{34}$ 10. 3, -7
11. $\frac{-3 \pm \sqrt{14}}{2}$ 12. $\frac{3 \pm \sqrt{17}}{4}$ 13. $x^2 - 2x + 1$
15. $x^2 + 18x + 81$ 16. $x^2 + 22x + 121$ 17. $x^2 - x + \frac{1}{4}$
18. $x^2 + x + \frac{1}{4}$ 19. $t^2 + 5t + \frac{25}{4}$ 21. $x^2 - \frac{3}{2}x + \frac{9}{16}$
22. $x^2 + \frac{4}{3}x + \frac{4}{9}$ 23. $m^2 + \frac{9}{2}m + \frac{81}{16}$ 24. $r^2 - \frac{2}{5}r + \frac{1}{25}$
25. 10%   27. 18.75%   28. 5%   29. 8%   30. 4%

31. 20%   33. -10, 8   34. -3, 5   35. $-2 \pm \sqrt{29}$
36. $-8 \pm \sqrt{15}$ 37. 2, -5   39. $\frac{4 \pm \sqrt{2}}{3}$ 40. $\frac{22}{9}$, $-\frac{26}{9}$
41. $\frac{7}{8}$, $-\frac{13}{8}$ 42. 41.42%   43. 12.6%   45. $2239.13
46. 4, -2   47. 3b, -b   48. $\frac{1}{3}$, -1   49. 0, $\frac{4}{5}$
51. $x^2 + (2b - 4)x + (b - 2)^2$ 52. $ax^2 + bx + \frac{b^2}{4a}$

Margin Exercises, Section 11.4

1. -2, -6   2. $5 \pm \sqrt{3}$ 3. $-3 \pm \sqrt{10}$ 4. 5, -2
5. -7, 2   6. $\frac{-3 \pm \sqrt{33}}{4}$ 7. $\frac{1 \pm \sqrt{10}}{3}$

Exercise Set 11.4

1. -2, 8   3. -21, -1   4. -15, 1   5. $1 \pm \sqrt{6}$
6. $2 \pm \sqrt{15}$ 7. $11 \pm \sqrt{19}$ 9. $-5 \pm \sqrt{29}$ 10. $5 \pm \sqrt{29}$
11. $\frac{7 \pm \sqrt{57}}{2}$ 12. $\frac{-7 \pm \sqrt{57}}{2}$ 13. -7, 4   15. $\frac{-3 \pm \sqrt{17}}{4}$
16. $\frac{3 \pm \sqrt{41}}{4}$ 17. $\frac{-3 \pm \sqrt{145}}{4}$ 18. $\frac{3 \pm \sqrt{17}}{4}$ 19. $\frac{-2 \pm \sqrt{7}}{3}$
21. $-\frac{1}{2}$, 5   22. $-\frac{3}{2}$, 4   23. $-\frac{7}{2}$, $\frac{1}{2}$ 24. $-\frac{5}{2}$, $\frac{2}{3}$
25. 5670   27. 12, -12   28. $\pm 2\sqrt{55}$ 29. $\pm 16\sqrt{2}$
30. 16, -16   31. $\pm 2\sqrt{c}$ 33. 3a, -2a
34. $-2b \pm \sqrt{4b^2 - 2b}$ 35. c + 1, -c   36. $\frac{b \pm \sqrt{b^2 - 12}}{6}$
37. $\frac{-2 \pm \sqrt{4 - 3a}}{a}$ 39. $\frac{-m \pm \sqrt{m^2 - 4nk}}{2k}$
40. $\frac{1 \pm \sqrt{1 - c^2}}{b}$

Margin Exercises, Section 11.5

1. $\frac{1}{2}$, -4   2. $\frac{4 \pm \sqrt{31}}{5}$ 3. -0.3, 1.9

Exercise Set 11.5

1. -3, 7   3. 3   4. 4   5. $-\frac{4}{3}$, 2   6. 1, $\frac{4}{3}$
7. $-\frac{7}{2}$, $\frac{1}{2}$ 9. -3, 3   10. -2, 2   11. $1 \pm \sqrt{3}$
12. $2 \pm \sqrt{11}$ 13. $5 \pm \sqrt{3}$ 15. $-2 \pm \sqrt{7}$ 16. $1 \pm \sqrt{5}$
17. $\frac{-4 \pm \sqrt{10}}{3}$ 18. $\frac{2 \pm \sqrt{10}}{3}$ 19. $\frac{5 \pm \sqrt{33}}{4}$ 21. $\frac{1 \pm \sqrt{2}}{2}$
22. $\frac{-1 \pm \sqrt{2}}{2}$ 23. $-\frac{5}{3}$, 0   24. 0, $\frac{2}{5}$
25. No real-number solutions   27. -5, 5   28. -4, 4
29. $\frac{5 \pm \sqrt{73}}{6}$ 30. $\frac{-3 \pm \sqrt{17}}{4}$ 31. $\frac{3 \pm \sqrt{29}}{2}$ 33. $\frac{-1 \pm \sqrt{10}}{3}$
34. -1.3, 5.3   35. -2.7, 0.7   36. -0.2, 6.2
37. -6.7, -3.3   39. -0.2, 1.2   40. -1.7, 0.4
41. 0.3, 2.4   42. -1.8, 0.8   43. 0, 2   45. $\frac{3 \pm \sqrt{5}}{2}$
46. $\frac{7 \pm \sqrt{69}}{10}$ 47. $\frac{-7 \pm \sqrt{61}}{2}$ 48. -8, 4   49. $\frac{-2 \pm \sqrt{10}}{2}$
51. No real-number solutions   52. No real-number
solutions   53. $\frac{-1 \pm \sqrt{3a + 1}}{a}$ 54. $\frac{5 \pm \sqrt{25 - 24b^2}}{4b}$
55. $\frac{c \pm d\sqrt{3}}{2}$ 57. $\frac{a}{b}$, $-\frac{c}{d}$ 58. -1, -2b + 1
59. (a) Yes, no; (b) yes, no; (c) when $b^2 \geqslant 4ac$
60. $-\frac{b}{a} = -\frac{5}{2}$, $\frac{c}{a} = -\frac{3}{2}$ 61. 0.3

The ASVAB Test - Arithmetical Reasoning   Chapter 11
1. D   2. D   3. B   4. D   5. B   6. A   7. D   8. C
9. A   10. B   11. C   12. D   13. C

Margin Exercises, Section 11.6

1. $w_2 = \frac{w_1}{A^2}$ 2. $r = \sqrt{\frac{V}{\pi h}}$ 3. $s = \frac{-R \pm \sqrt{R^2 - \frac{4L}{C}}}{2L}$

4. $b = \sqrt{\dfrac{p^2a^2}{1 + p^2}}$ 5. 5 cm

Exercise Set 11.6

1. $s = \sqrt{\dfrac{A}{6}}$ 3. $r = \sqrt{\dfrac{Gm_1m_2}{F}}$ 4. $s = \sqrt{\dfrac{kQ_1Q_2}{N}}$

5. $c = \sqrt{\dfrac{E}{m}}$ 6. $r = \sqrt{\dfrac{A}{\pi}}$ 7. $b = \sqrt{c^2 - a^2}$

9. $k = \dfrac{3 + \sqrt{9 + 8N}}{2}$ 10. $t = \dfrac{-v_0 + \sqrt{v_0^2 + 2gs}}{g}$

11. $r = \dfrac{-\pi h + \sqrt{\pi^2 h^2 + 2\pi A}}{2\pi}$ 12. $r = \dfrac{-\pi S + \sqrt{\pi^2 s^2 + 4\pi A}}{2\pi}$

13. $g = \dfrac{4\pi^2 \ell}{T^2}$ 15. $D = \sqrt{\dfrac{32LV}{g(P_1 - P_2)}}$, or $4\sqrt{\dfrac{2LV}{g(P_1 - P_2)}}$

16. $R = \sqrt{\dfrac{6.2A^2}{p(N + p)}}$, or $R = A\sqrt{\dfrac{6.2}{p(N + p)}}$

17. $v = \dfrac{c\sqrt{m^2 - m_0^2}}{m}$ 18. $c = \dfrac{vm}{\sqrt{m^2 - m_0^2}}$ 19. (a) 3.91 sec;

(b) 1.91 sec; (c) 79.6 m   20. (a) 10.1 sec;
(b) 7.49 sec; (c) 272.5 m   21. (a) 18.75%; (b) 10%;
(c) 11%; (d) 12.75%   22. 2 ft   23. 7 ft   24. 4.685 cm

25. $90\sqrt{2} \approx 127.28$ ft

Margin Exercises, 11.7

1. ± 3, ± 1 2. 4   3. 4, 2, -1, -3   4. $\dfrac{1}{2}$, $-\dfrac{1}{3}$

Exercise Set 11.7

1. 81, 1   3. $\pm\sqrt{5}$ 4. $\pm\sqrt{2}$, ± 1 5. -27, 8
6. 64, -8   7. 16   9. 7, -1, 5, 1   10. 1   11. 4, 1,
6, -1   12. $-\dfrac{3}{2}$, 1, $\dfrac{1}{2}$, -1   13. $\pm\sqrt{2 + \sqrt{6}}$; note that
$\pm\sqrt{2 - \sqrt{6}}$ is not a real-number solution because $\sqrt{6}$ is
larger than $\sqrt{2}$, so the radicand is negative.
15. $\dfrac{1}{3}$, $-\dfrac{1}{2}$ 16. $\dfrac{4}{5}$, -1   17. 2, -1   18. $-\dfrac{1}{10}$, 1

19. \$18   21. 28.5   22. ± 2.0 23. $3 \pm \sqrt{10}$, $-1 \pm \sqrt{2}$

24. 3.02   25. $1 \pm \sqrt{2}$, $\dfrac{-1 \pm \sqrt{5}}{2}$ 27. $\dfrac{5 \pm \sqrt{21}}{2}$, $\dfrac{3 \pm \sqrt{5}}{2}$

28. $\dfrac{100}{99}$ 29. $\dfrac{1}{3}$ 30. $-\dfrac{6}{7}$ 31. 1, 4   33. 19   34. 9

35. $\dfrac{2}{51 \pm 7\sqrt{61}}$, or $\dfrac{-51 \pm 7\sqrt{61}}{194}$

Margin Exercises, Section 11.8

1. $y = 7x^2$ 2. $y = \dfrac{9}{x^2}$ 3. $y = \dfrac{1}{2}xz$ 4. $y = \dfrac{5xz^2}{w}$
5. 490 m   6. 1 ohm

Exercise Set 11.8

1. $y = 15x^2$ 3. $y = \dfrac{0.0015}{x^2}$ 4. $y = \dfrac{54}{x^2}$ 5. $y = xz$

6. $y = \dfrac{5x}{z}$ 7. $y = \dfrac{3}{10}xz^2$ 9. $y = \dfrac{xz}{5wp}$ 10. $y = \dfrac{5xz}{4w^2}$

11. $355\dfrac{5}{9}$ ft   12. 624.24 ft^2 13. 94.03 kg   15. 97
16. 220 cm^3 17. y is tripled   18. y is multiplied
by $\dfrac{1}{3}$ 19. y is multiplied by $\dfrac{1}{n^2}$ 21. $\dfrac{\pi}{4}$ 22. \$7.20

23. y is multiplied by $\dfrac{1}{0.125}$, or 8   24. (a) $k = 0.001$,

$N = \dfrac{0.001P_1P_2}{d^2}$; (b) 1173.4 km

The ASVAB Test – Mathematical Knowledge   Chapter 11

1. D   2. C   3. A   4. B   5. A   6. B   7. A   8. C
9. C   10. A   11. C   12. A   13. A   14. B   15. B
16. D   17. C   18. C   19. A   20. D

Margin Exercises, Section 11.9

1. (a) Upward;
(b)
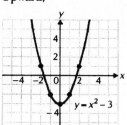
$y = x^2 - 3$

2. (a) Upward;
(b)
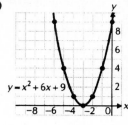
$y = x^2 + 6x + 9$

3. (a) Downward;
(b)
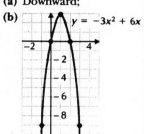
$y = -3x^2 + 6x$

4. 2   5. 0.2, -2.2

Exercise Set 11.9

1.
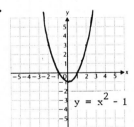
$y = x^2 + 1$

3.
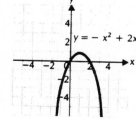
$y = -x^2$

4.
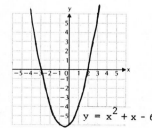
$y = x^2 - 1$

5.
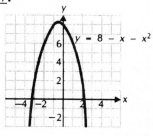
$y = -x^2 + 2x$

6.

$y = x^2 + x - 6$

7.
$y = 8 - x - x^2$

9.

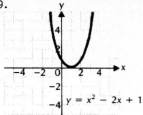

$y = x^2 - 2x + 1$

10.

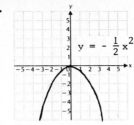

$y = -\frac{1}{2}x^2$

11.

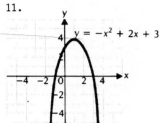

$y = -x^2 + 2x + 3$

12.

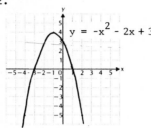

$y = -x^2 - 2x + 3$

13.

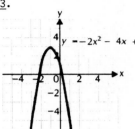

$y = -2x^2 - 4x + 1$

15.

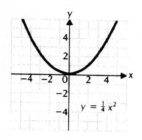

$y = \frac{1}{4}x^2$

16.

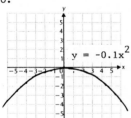

$y = -0.1x^2$

17.

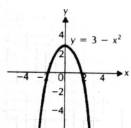

$y = 3 - x^2$

18.

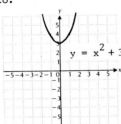

$y = x^2 + 3$

19.

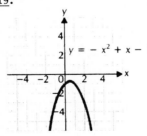

$y = -x^2 + x - 1$

21.

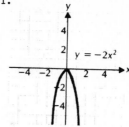

$y = -2x^2$

22.

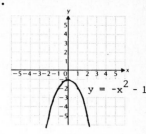

$y = -x^2 - 1$

23.

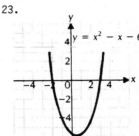

$y = x^2 - x - 6$

24.

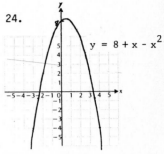

$y = 8 + x - x^2$

25. ± 2.2 27. $0, -2$ 28. $0, 2$ 29. $2.4, -3.4$
30. $-2.4, 3.4$ 31. -5 33. $-2.2, 0.2$ 34. $-2.2, 0.2$
35. Graph does not cross the x-axis, so there is no real-number solution   36. Graph does not cross the x-axis, so there is not real-number solution
37. (a) $3.4, -2.4$; (b) $2.3, -1.3$ 39. $(0, c)$
40. Graphs that open upward have a minimum point and that happens when the coefficient a in $y = ax^2 + bx + c$ is positive. Graphs that open downward have a maximum point and that happens when a is negative.
41. (1) $x = 0$; (2) $x = 0$; (3) $x = 0$; (4) $x = 0$;
(5) $x = 1$; (6) $x = -\frac{1}{2}$; (7) $x = -\frac{1}{2}$; (8) $x = -1$;
(9) $x = 1$; (10) $x = 0$ 43. 4 ft by 4 ft; 16 ft²

Margin Exercises, Section 11.10

1.

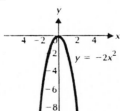

$y = \frac{1}{2}x^2$

2.

$y = 3x^2$

3.

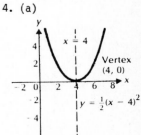

$y = -2x^2$

4. (a)

$x = 4$

Vertex $(4, 0)$

$y = \frac{1}{2}(x - 4)^2$

4. (b)

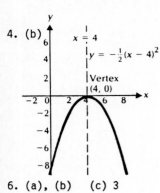

$x = 4$

$y = -\frac{1}{2}(x - 4)^2$

Vertex (4, 0)

5. (a), (b)　(c) −4

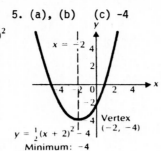

$x = -2$

$y = \frac{1}{2}(x + 2)^2 - 4$

Vertex (−2, −4)

Minimum: −4

3.

$f(x) = -4x^2$

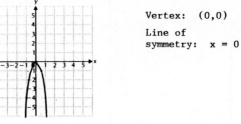

Vertex: (0,0)

Line of symmetry: x = 0

6. (a), (b)　(c) 3

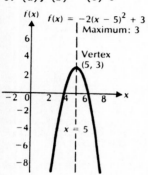

$f(x) = -2(x - 5)^2 + 3$

Maximum: 3

Vertex (5, 3)

$x = 5$

7. (a)

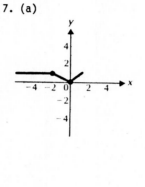

4. $f(x) = 2x^2$

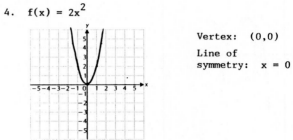

Vertex: (0,0)

Line of symmetry: x = 0

7. (b)

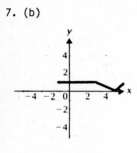

8. (a)

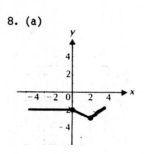

5. $f(x) = (x - 3)^2$

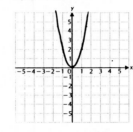
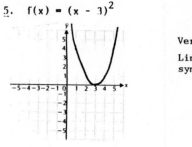

Vertex: (3,0)

Line of symmetry: x = 3

6. $f(x) = (x - 7)^2$

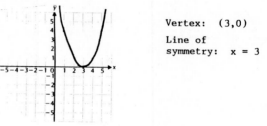

Vertex: (7,0)

Line of symmetry: x = 7

8. (b)

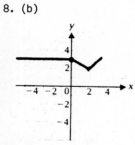
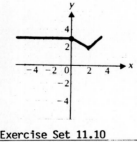

7. $f(x) = -(x + 4)^2$, or $f(x) = -[x - (-4)]^2$

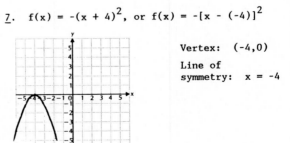

Vertex: (−4,0)

Line of symmetry: x = −4

1. $f(x) = x^2$

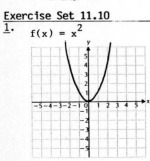
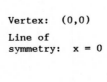

Vertex: (0,0)

Line of symmetry: x = 0

9. $f(x) = 2(x - 3)^2$

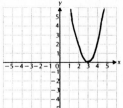

Vertex: (3,0)

Line of
symmetry: x = 3

16. $f(x) = -2(x + \frac{1}{2})^2$

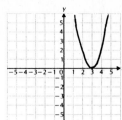

Vertex: $(-\frac{1}{2}, 0)$

Line of
symmetry: $x = -\frac{1}{2}$

10. $f(x) = -4(x - 7)^2$

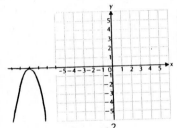

Vertex: (7,0)

Line of
symmetry: x = 7

17. $f(x) = \frac{1}{2}(x + 1)^2$, or $f(x) = \frac{1}{2}[x - (-1)]^2$

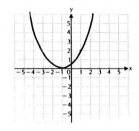

Vertex: (-1,0)

Line of
symmetry: x = -1

11. $f(x) = -2(x + 9)^2$, or $f(x) = -2[x - (-9)]^2$

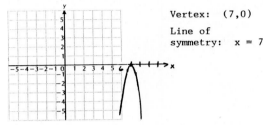

Vertex: (-9,0)

Line of
symmetry: x = -9

18. $f(x) = \frac{1}{3}(x - 2)^2$

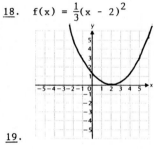

Vertex: (2,0)

Line of
symmetry: x = 2

12. $f(x) = 2(x + 7)^2$

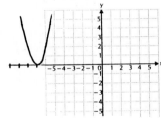

Vertex: (-7,0)

Line of
symmetry: x = -7

19.

$f(x) = (x - 3)^2 + 1$

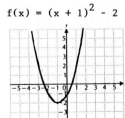

Vertex: (3,1)

Line of
symmetry: x = 3

Minimum: 1

13. $f(x) = 3(x - 1)^2$

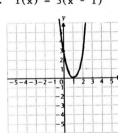

Vertex: (1,0)

Line of
symmetry: x = 1

21. $f(x) = (x + 1)^2 - 2$, or $f(x) = [x - (-1)]^2 + (-2)$

$f(x) = (x + 1)^2 - 2$

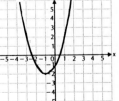

Vertex: (-1,-2)

Line of
symmetry: x = -1

Minimum: -2

15. $f(x) = -3(x - \frac{1}{2})^2$

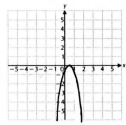

Vertex: $(\frac{1}{2},0)$

Line of
symmetry: $x = \frac{1}{2}$

22. $f(x) = (x - 1)^2 + 2$

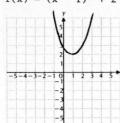

Vertex: (1,2)

Line of symmetry: x = 1

Minimum: 2

23. $f(x) = 2(x - 1)^2 - 3$, or $f(x) = 2(x - 1)^2 + (-3)$

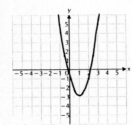

Vertex: (1,-3)

Line of symmetry: x = 1

Minimum: -3

24. $f(x) = 2(x + 1)^2 + 4$

Vertex: (-1,4)

Line of symmetry: x = -1

Minimum: 4

25. $f(x) = -3(x + 4)^2 + 1$, or $f(x) = -3[x - (-4)]^2 + 1$

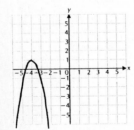

Vertex: (-4,1)

Line of symmetry: x = -4

Maximum: 1

27. Vertex: (9,5); Line of symmetry: x = 9; Minimum: 5
28. Vertex: (-5,-8); Line of symmetry: x = -5;

Minimum: -8   29. Vertex: $(-\frac{1}{4},-13)$; Line of symmetry:

$x = -\frac{1}{4}$; Minimum: -13   30. Vertex: $(\frac{1}{4},19)$; Line of

symmetry: $x = \frac{1}{4}$; Minimum: 19   31. Vertex: (10,-20);

Line of symmetry: x = 10; Maximum: -20 33. Vertex:
(-4.58,65π); Line of symmetry: x = -4.58; Minimum: 65π

34. Vertex: (38.2,$-\sqrt{34}$); Line of symmetry: x = 38.2;

Minimum: $-\sqrt{34}$

35.

y = h(x + 2)

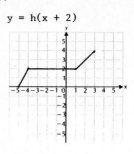

36.

y = h(x - 3)

37. y = h(x) + 3

39.

y = h(x - 4) - 2

40.

y = h(x - 2) + 3

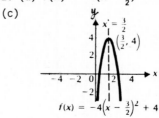

41. 16 12-oz sodas,
20 16-oz sodas
42. 11, 13, 15, 17
43. $f(x) = -2x^2 + 4$
45. $f(x) = 2(x - 6)^2$
46. $f(x) = -2x^2 + 3$
47. $f(x) = -2(x - 3)^2 + 8$
48. $f(x) = 2(x + 2)^2 + 3$
49. $f(x) = 2(x + 3)^2 + 6$
51. $f(x) = 2(x - 2)^2 - 3$
52. $f(x) = 6(x - 4)^2$
53. $f(x) = -\frac{1}{2}(x - 1)^2 - 6$

Margin Exercises, Section 11.11

1. (a) $f(x) = (x - 2)^2 + 3$; (b) Vertex; (2,3); Line of
symmetry: x = 2; (c)

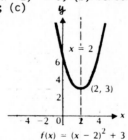

$f(x) = (x - 2)^2 + 3$

2. (a) $f(x) = -4(x - \frac{3}{2})^2 + 4$; (b) $x = \frac{3}{2}$;

(c)

$f(x) = -4\left(x - \frac{3}{2}\right)^2 + 4$

3. $(1 - \sqrt{6},0)$, $(1 + \sqrt{6},0)$ 4. (-1,0), (3,0) 5. (-4,0)
6. No x-intercepts

Exercise Set 11.11

1. $f(x) = (x - 1)^2 - 4$; Vertex: (1,-4); Line of
symmetry: x = 1; Minimum: -4   3. $f(x) = -(x - 2)^2 + 10$;
Vertex: (2,10); Line of symmetry: x = 2; Maximum: 10

Vertex: $(-\frac{5}{2}, -\frac{9}{4})$; Line of symmetry: $x = -\frac{5}{2}$; Minimum: $-\frac{9}{4}$ **7.** $f(x) = (x - \frac{9}{2})^2 - \frac{81}{4}$; Vertex: $(\frac{9}{2}, -\frac{81}{4})$; Line of symmetry: $x = \frac{9}{2}$; Minimum: $-\frac{81}{4}$

9. $f(x) = 3(x - 4)^2 + 2$; Vertex: $(4,2)$; Line of symmetry: $x = 4$; Minimum: 2

10. $f(x) = 4(x + 1)^2 - 7$; Vertex: $(-1,-7)$; Line of symmetry: $x = -1$; Minimum: -7

11. $f(x) = \frac{3}{4}(x + 6)^2 - 27$; Vertex: $(-6,-27)$; Line of symmetry: $x = -6$; Minimum: -27

12. $f(x) = -2(x - \frac{1}{2})^2 + \frac{3}{2}$; Vertex: $(\frac{1}{2},\frac{3}{2})$; Line of symmetry: $x = \frac{1}{2}$; Maximum: $\frac{3}{2}$ **13.** $(2 + \sqrt{3},0)$, $(2 - \sqrt{3},0)$ **15.** $(3,0)$, $(-1,0)$ **16.** $(1 - \sqrt{6},0)$, $(1 + \sqrt{6},0)$ **17.** $(4,0)$, $(-1,0)$ **18.** $(4 + \sqrt{11},0)$, $(4 - \sqrt{11},0)$ **19.** $(4,0)$, $(-1,0)$

21. $\left[\frac{-2 + \sqrt{6}}{2},0\right]$, $\left[\frac{-2 - \sqrt{6}}{2},0\right]$ **22.** No x-interecpts

23. No x-intercepts   **24.** $(-\frac{3}{2},0)$ **25.** No x-intercepts

27. Length is 12; width is 5   **28.** 4 hr, 48 min

29. Minimum: -6.95 **30.** Maximum: 7.014

31. $(62.75852,0)$, $(31.941480,0)$ **33.** No x-intercepts

34. $(1.272,0)$, $(-0.726,0)$ **35.** (a) 3.4, -2.4; (b) 2.3, -1.3 **36.** (a) -3, 1; (b) -4.5, 2.5; (c) -5.5, 3.5

37. $f(x) = a\left[x - \left[-\frac{b}{2a}\right]\right]^2 + \frac{4ac - b^2}{4a}$

39. $f(x) = |x^2 - 1|$

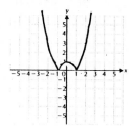

40. $f(x) = |3 - 2x - x^2|$

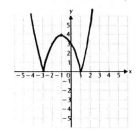

41. $y < x^2 - 4x - 1$ **42.** $y \geq x^2 + 3x - 4$

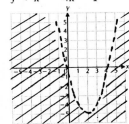
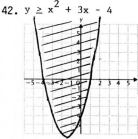

43. $y \leq x^2 + 5x + 6$

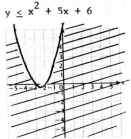

45. $y > 3x^2 + 6x + 2$

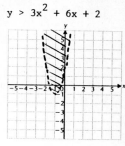

46. $y > 2x^2 + 4x - 2$

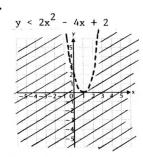

47. $y \geq 4x^2 + 8x + 3$

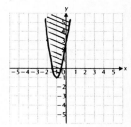

48.

$y < 2x^2 - 4x + 2$

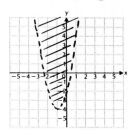

Margin Exercises, Section 11.12

1. 225   **2.** 25 m by 25 m   **3.** $f(x) = x^2 - 2x + 1$

4. (a) $f(x) = 0.625x^2 - 50x + 1150$; (b) 510

Exercise Set 11.12

1. 19 ft by 19 ft; 361 ft²   **3.** 121; 11 and 11

4. 506.25; 22.5 and 22.5   **5.** -4; 2 and -2

6. -9; 3 and -3 **7.** $-\frac{25}{4}$; $\frac{5}{2}$ and $-\frac{5}{2}$

9. $f(x) = 2x^2 + 3x - 1$ **10.** $f(x) = 3x^2 - x + 2$

11. $f(x) = -3x^2 + 13x - 5$ **12.** $f(x) = x^2 - 5x$

13. (a) $f(x) = -4x^2 + 40x + 2$; (b) \$98

15. (a) $f(x) = 0.05x^2 - 5.5x + 250$; (b) 100

16. (a) $f(x) = \frac{3}{16}x^2 - \frac{105}{4}x + 1150$;

(b) 306.25   **17.** Minimum: -7.0234008

18. Maximum: 6.6388161

19. $f(x) = 0.0499218x^2 - 2.7573651x + 29.571379$

21. 19 cm by 19 cm; 180.5 cm²   **22.** $11\sqrt{2}$ **23.** 4050 ft²

24. 30   **25.** \$6

Summary and Review – Chapter 11

1. [11.1] $-\sqrt{3}$, $\sqrt{3}$ **2.** [11.2] $\frac{3}{5}$, 1   **3.** [11.5] $1 \pm \sqrt{11}$

4. [11.2] $\frac{1}{3}$, -2 **5.** [11.3] $-8 \pm \sqrt{13}$ **6.** [11.1] 0

7. [11.2] 0, $\frac{7}{5}$ **8.** [11.5] $\frac{1 \pm \sqrt{10}}{3}$ **9.** [11.5] $-3 \pm 3\sqrt{2}$

10. [11.5] $\frac{2 \pm \sqrt{3}}{2}$ **11.** [11.5] $\frac{3 \pm \sqrt{33}}{2}$ **12.** [11.5] No real-number solution   **13.** [11.2] 0, $\frac{4}{3}$

14. [11.1] $-2\sqrt{2}$, $2\sqrt{2}$ **15.** [11.3] $\frac{5}{3}$, -1

16. [11.3] $\frac{5 \pm \sqrt{17}}{2}$ **17.** [11.5] 4.6, 4

18. [11.5] -1.9, -0.1
19. [11.9] Downward   20. [11.9] Upward

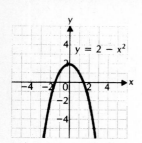

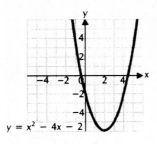

21. [11.9] -0.5, 4.5   22. [11.6] 1.7 m, 4.7 m
23. [11.3] 30%   24. [11.6] Width: 7 m, length: 10 m
25. [11.6] $t = \frac{b + \sqrt{b^2 + 4aN}}{2a}$ 26. [11.6] $p = \frac{9\pi^2}{N^2}$

27. [11.6] $T = \sqrt{\frac{3B}{2A}}$

28. [11.6] (a) $r = \frac{\sqrt{3A\pi h}}{\pi h}$; (b) 2.32 29. [11.7] $\pm 2, \pm 3$

30. [11.7] 3, -5   31. [11.7] $\pm \sqrt{7}, \pm \sqrt{2}$

32. [11.7] 25   33. [11.8] 500 watts   34. [11.8] $\frac{15}{2}$ m

35. [11.10] (a)   (b) (-2,4)
     x = -2   (c) x = -2
                                   (d) Maximum: 4

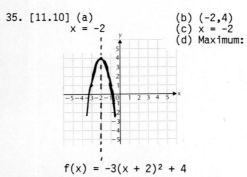

     $f(x) = -3(x + 2)^2 + 4$

36. [11.11] $f(x) = 2(x - 3)^2 + 5$ 37. [11.11] (7,0),
(2,0)   38. [11.12] -121; 11 and -11
39. [11.12] $f(x) = -x^2 + 6x - 2$ 40. [11.6] 31 and 32;
-32 and -31   41. [11.3] b = 14 or b = -14
42. [11.6] $s = 5\sqrt{\pi}$

APPENDIX A

Margin Exercises, Appendix A

1. (a) No; (b) Yes   2. (2,1,-1)   3. $(2,-2,\frac{1}{2})$

4. (20,30,50)   5. 2, -3, 4   6. 64°, 32°, 84°
7. A: 112; B: 90; C: 85

Exercise Set, Appendix A
1. Yes   2. No   3. (1,2,3)   4. (4,0,2)   5. (-1,5,-2)
6. (2,-2,2)   7. (3,1,2)   8. (3,-2,1)   9. (-3,-4,2)
10. (7,-3,-4)   11. (2,4,1)   12. (2,1,3) 13. (-3,0,4)
14. (2,-5,6)   15. (2,2,4)   16. (-2,-1,4)

17. $(\frac{1}{2},4,-6)$ 18. (3,-5,8)   19. $(\frac{1}{2},\frac{1}{3},\frac{1}{6})$ 20. $(\frac{3}{5},\frac{2}{3},-3)$

21. $(\frac{1}{2},\frac{2}{3},-\frac{5}{6})$ 22. $(4,\frac{1}{2},-\frac{1}{2})$ 23. 17, 9, 79

24. 16, 19, 22   25. 4, 2, -1   26. 8, 21, -3
27. 34°, 104°, 42°   28. 30°, 90°, 60°   29. 25°, 50°,
105°   30. 32°, 96°, 52°   31. $21 on Thurs., $18 on
Fri., $27 on Sat.   32. 20 qt on Mon., 35 qt on Tues.,
32 qt on Wed.   33. 74.5, 68.5, 82   34. Fred: 195;
Jane: 180; Mary: 200   35. (1,-1,2)   36. (1,1,1)

37. (1,-2,4,-1)   38. (-3,-1,0,4)   39. $(-1,\frac{1}{5},-\frac{1}{2})$

40. $(-\frac{1}{2},-1,-\frac{1}{3})$ 41. a = 2, b = -1, c = 3

42. A = 3, B = 4, C = 2; 3x + 4y + 2z = 12
43. b = 8, m = 2, n = 4; z = 8 - 2x - 4y
44. A: 1500; B: 1900; C: 2300   45. A: 2200; B: 2500;
C: 2700   46. A: 900; B: 1300; C: 1500   47. A: 10 ft;
B: 12 ft; C: 15 ft   48. 20   49. 464   50. 180°
51. Men: 5; Women: 1; Children: 94   52. 1869   53. 35
54. Adult: 2050; Senior citizens: 845; Children: 210
55. a = 1, b = -5, c = 6

APPENDIX B

Margin Exercises, Appendix B
1. (a) Inconsistent; (b) Consistent
2. (a) Inconsistent; (b) Consistent   3. (a) Dependent;
(b) Independent   4. (a) Dependent; (b) Independent

Exercise Set, Appendix B
1. Inconsistent   2. Inconsistent   3. Consistent
4. Consistent   5. Consistent   6. Consistent
7. Inconsistent   8. Inconsistent   9. Consistent
10. Consistent   11. Dependent   12. Dependent
13. Independent   14. Independent   15. Dependent
16. Dependent   17. Independent   18. Independent
19. Dependent   20. Dependent   21. No solution
22. a = 0, b = 3   23. Roland is 32, Jennifer is 8
24. No solution   25. No solution   26. Infinitely many
solutions; one solution is 12, 59, 0.   27. Infinitely
many solutions; (2,1), (3,4), and (1,-2)
28. No solution   29. No solution   30. (0,0,0)
31. No solution   32. Infinitely many solutions;
(1,1,1), (2,0,1), (0,2,1)
33. (27) Consistent and dependent; (29) Inconsistent
and independent; (32) Consistent and dependent

35. k = 2   36. k = 25   37. $(\frac{1 - 3y}{2},y)$

38. $(\frac{2y - 5}{3},y)$

Exercise Set 1.1

<u>1</u>. The integer 5 corresponds to winning 5 points.
The integer -12 corresponds to losing 12 points.

<u>7</u>. The integer 750 corresponds to the $750 deposit.
The integer -125 corresponds to the $125 withdrawal.

<u>13</u>. -9 is to the left of 5.   -9 < 5

<u>19</u>. -5 is to the right of -11.   -5 > -11

<u>25</u>. 120 is greater than -20.   120 > -20

<u>31</u>. The distance of -7 from 0 is 7, so $|-7|$ is 7.

<u>37</u>. The distance of 325 from 0 is 325, so $|325|$ is 325.

<u>43</u>. The number $\frac{10}{3}$ can be named $3\frac{1}{3}$ or $3.\overline{3}$.
Its graph is $\frac{1}{3}$ of the way from 3 to 4.

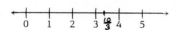

<u>49</u>. Using a calculator or Table 2 we find that
$-\sqrt{2} \approx -1.414$.   The graph is about 0.4 of
the way from -1 to -2.

<u>55</u>. $\frac{5}{3}$ means $5 \div 3$, so we divide:

```
      1.6 6 ...
  3 ⌐ 5.0 0
      3
      ___
      2 0
      1 8
      ___
        2 0
        1 8
        ___
          2
```

Decimal notation for $\frac{5}{3}$ is 1.66 ···, or $1.\overline{6}$.

<u>61</u>. $\frac{1}{2}$ means $1 \div 2$, so we divide:

```
      0.5
  2 ⌐ 1.0
      1 0
      ___
        0
```

Decimal notation for $\frac{1}{2}$ is 0.5.

<u>67</u>. $0.14\underline{5} = \frac{145}{1000}$, or $\frac{29}{200}$

The last decimal place is <u>thousandths</u> so the
denominator is 1000.

<u>73</u>. 2.14 is to the right of 1.24.   2.14 > 1.24

<u>79</u>. $\frac{5}{12} = 0.41\overline{6}$ $\frac{11}{25} = 0.44$

$\frac{5}{12}$, or $0.41\overline{6}$, is to the left of $\frac{11}{25}$, or 0.44.

$\frac{5}{12} < \frac{11}{25}$

<u>85</u>. The distance of 3 from 0 is 3, so $|3| = 3$.

<u>91</u>. The negative number indicates a score less than
par or a weight less than a certain weight, which
is better than scoring over par or being overweight.

<u>97</u>. $|-8| = 8$, $|8| = 8$
Thus, $|-8| = |8|$.

Exercise Set 1.2

<u>1</u>. 1×21, 3×7 <u>7</u>. 1×93, 3×31

<u>13</u>. We begin by factoring 18 any way we can and
continue factoring until each factor is prime.
$2 \times 9 = 2 \times 3 \times 3$ or $3 \times 6 = 3 \times 2 \times 3$

<u>19</u>. $45 = \underline{3 \cdot 3} \cdot 5$
$72 = \underline{2 \cdot 2 \cdot 2} \cdot 3 \cdot 3$
LCM = $2 \cdot 2 \cdot 2 \cdot 3 \cdot 3 \cdot 5$, or 360

<u>25</u>. $18 = 2 \cdot \underline{3 \cdot 3}$
$24 = \underline{2 \cdot 2 \cdot 2} \cdot 3$
LCM = $2 \cdot 2 \cdot 2 \cdot 3 \cdot 3$, or 72

<u>31</u>. $24 = \underline{2 \cdot 2 \cdot 2} \cdot 3$
$36 = 2 \cdot 2 \cdot \underline{3 \cdot 3}$
$12 = 2 \cdot 2 \cdot 3$
LCM = $2 \cdot 2 \cdot 2 \cdot 3 \cdot 3$, or 72

<u>37</u>. $\frac{18}{45} = \frac{2 \cdot 9}{5 \cdot 9} = \frac{2}{5} \cdot \frac{9}{9} = \frac{2}{5}$

<u>43</u>. $\frac{56}{7} = \frac{8 \cdot 7}{1 \cdot 7} = \frac{8}{1} \cdot \frac{7}{7} = \frac{8}{1} = 8$

<u>49</u>. $\frac{9}{5} \div \frac{4}{5} = \frac{9}{5} \cdot \frac{5}{4} = \frac{9 \cdot 5}{5 \cdot 4} = \frac{9}{4} \cdot \frac{5}{5} = \frac{9}{4}$

<u>55</u>. The LCM of the denominators is 60.
$\frac{5}{12} + \frac{4}{15} = \frac{5}{12} \cdot \frac{5}{5} + \frac{4}{15} \cdot \frac{4}{4} = \frac{25}{60} + \frac{16}{60} = \frac{41}{60}$

<u>61</u>. $8\frac{3}{4} + 5\frac{5}{6} = 8\frac{9}{12} + 5\frac{10}{12} = 13\frac{19}{12} = 14\frac{7}{12}$

<u>67</u>. $21 - 8\frac{3}{4} = 20\frac{4}{4} - 8\frac{3}{4} = 12\frac{1}{4}$

73. $4\frac{7}{10} \cdot 5\frac{3}{10} = \frac{47}{10} \cdot \frac{53}{10} = \frac{2491}{100} = 24\frac{91}{100}$

79. $1\frac{7}{8} \div 1\frac{2}{3} = \frac{15}{8} \div \frac{5}{3} = \frac{15}{8} \cdot \frac{3}{5} = \frac{45}{40} = \frac{9}{8} = 1\frac{1}{8}$

85. $\frac{1}{2} + \frac{1}{2} = \frac{1+1}{2} = \frac{2}{2} = 1$

91. $\frac{5}{4} - \frac{3}{4} = \frac{5-3}{4} = \frac{2}{4} = \frac{1}{2}$

97. $\frac{7}{6} \div \frac{3}{5} = \frac{7}{6} \cdot \frac{5}{3} = \frac{35}{18}$

103. $\dfrac{\frac{13}{12}}{\frac{39}{5}} = \frac{13}{12} \cdot \frac{5}{39} = \frac{13 \cdot 5}{12 \cdot 13 \cdot 3} = \frac{5}{12 \cdot 3} = \frac{5}{36}$

109. $\dfrac{1 - \frac{3}{4}}{1 + \frac{9}{16}} = \dfrac{\frac{4}{4} - \frac{3}{4}}{\frac{16}{16} + \frac{9}{16}} = \dfrac{\frac{1}{4}}{\frac{25}{16}} = \frac{1}{4} \cdot \frac{16}{25} = \frac{16}{100} = \frac{4}{25}$

115. $\dfrac{5 - \frac{1}{3}}{3 + \frac{1}{4}} = \dfrac{\frac{15}{3} - \frac{1}{3}}{\frac{12}{4} + \frac{1}{4}} = \dfrac{\frac{14}{3}}{\frac{13}{4}} = \frac{14}{3} \cdot \frac{4}{13} = \frac{56}{39}$

121. $100\% = 100 \times 0.01 = 1$

127. $78.6\% = 78.6 \times \frac{1}{100}$

$= \frac{78.6}{100} = \frac{78.6}{100} \times \frac{10}{10} = \frac{786}{1000}$, or $\frac{393}{500}$

133. $4.54 = 4.54 \times 1$

$= 4.54 \times (100 \times 0.01)$

$= (4.54 \times 100) \times 0.01$

$= 454 \times 0.01$

$= 454\%$

139. $0.072 = 0.072 \times 1$

$= 0.072 \times (100 \times 0.01)$

$= (0.072 \times 100) \times 0.01$

$= 7.2 \times 0.01$

$= 7.2\%$

145. $\frac{17}{100} = 17 \times \frac{1}{100} = 17\%$

151. $\frac{2}{3} = \frac{2}{3} \times 1 = \frac{2}{3} \times (100 \times \frac{1}{100})$

$= (\frac{2}{3} \times 100) \times \frac{1}{100}$

$= \frac{200}{3} \times \frac{1}{100}$

$= \frac{200}{3}\%$, or $66\frac{2}{3}\%$

157. Answers may vary.

$0.39 \div 0.02 = \frac{0.39}{0.02} = \frac{0.39}{0.02} \times \frac{100}{100} = \frac{39}{2} \approx \frac{40}{2} = 20$

163. Answers may vary.

$2359 - 1402 \approx 2400 - 1400 = 1000$

169. $8 = \underline{2 \cdot 2 \cdot 2}$

$12 = 2 \cdot 2 \cdot \underline{3}$

The greatest number of times the 2 occurs in the factorizations of 8 and 12 is three times. The greatest number of times the 3 occurs is once. Thus, the LCM must contain only three factors of 2 and one factor of 3.   LCM $= 2 \cdot 2 \cdot 2 \cdot 3$

a) No,   b) No,   c) No,   d) Yes

175.   Jupiter: $12 = \underline{2 \cdot 2 \cdot 3}$

Saturn: $30 = 2 \cdot 3 \cdot \underline{5}$

Uranus: $84 = 2 \cdot 2 \cdot 3 \cdot \underline{7}$

The LCM of 12, 30, and 84 is $2 \cdot 2 \cdot 3 \cdot 5 \cdot 7$, or 420. Jupiter, Saturn, and Uranus will appear in the same direction every 420 years.

Exercise Set 1.3

1. Start at -9.   Move 2 units to the right.

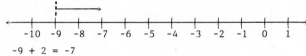

$-9 + 2 = -7$

7. Start at -3.   Move 5 units to the left.

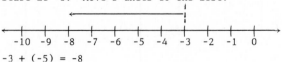

$-3 + (-5) = -8$

13. Numbers have the same absolute value. Make the answer 0. $17 + (-17) = 0$

19. Numbers have the same absolute value. Make the answer 0. $-18 + 18 = 0$

25. $13 + (-6)$ The absolute values are 13 and 6.   The difference is $13 - 6$, or 7.   The positive addend, 13, has the larger absolute value, so the answer is positive. $13 + (-6) = 7$

31. $-15 + (-7)$ Two negatives
Add the absolute values: $15 + 7 = 22$
Make the answer negative: $-15 + (-7) = -22$

37. 63 + (-18)   The absolute values are 63 and 18. The difference is 63 - 18, or 45. The positive addend, 63, has the larger absolute value, so the answer is positive.   63 + (-18) = 45

43. $-\frac{3}{5} + \frac{2}{5}$ The absolute values are $\frac{3}{5}$ and $\frac{2}{5}$. The difference is $\frac{3}{5} - \frac{2}{5}$, or $\frac{1}{5}$. The negative addend, $-\frac{3}{5}$, has the larger absolute value, so the answer is negative. $-\frac{3}{5} + \frac{2}{5} = -\frac{1}{5}$

49. $-\frac{3}{7} + (-\frac{2}{5})$ Two negatives
Add the absolute values: $\frac{3}{7} + \frac{2}{5} = \frac{15}{35} + \frac{14}{35} = \frac{29}{35}$
Make the answer negative: $-\frac{3}{7} + (-\frac{2}{5}) = -\frac{29}{35}$

55. $-8.5 + 7.9 + (-3.7)$
$= -8.5 + (-3.7) + 7.9$
$= -12.2 + 7.9$
$= -4.3$

61. $-44 + (-\frac{3}{8}) + 95 + (-\frac{5}{8})$
$= -44 + 95 + (-\frac{3}{8}) + (-\frac{5}{8})$
$= 51 + (-1)$
$= 50$

67. The inverse of -9 is 9 because -9 + 9 = 0.

73. If $x = -\frac{14}{3}$, then $-x = -(-\frac{14}{3}) = \frac{14}{3}$.

79. If $x = \frac{5}{3}$, then $-(-x) = -(-\frac{5}{3}) = \frac{5}{3}$.

85. $-(-14) = 14$

91. $1012 + (-6) + 3 + (-14) + 4$
$= 1012 + 3 + 4 + (-6) + (-14)$
$= 1019 + (-20)$
$= 999$
The pressure was 999 mb.

97. -8 - (-2) = -8 + 2 = -6

103. 7 - (-7) = 7 + 7 = 14

109. -4 - (-9) = -4 + 9 = 5

115. 8 - (-10) = 8 + 10 = 18

121. -7 - 14 = -7 + (-14) = -21

127. 7 - (-5) = 7 + 5 = 12

133. -71 - 2 = -71 + (-2) = -73

139. $\frac{3}{8} - \frac{5}{8} = \frac{3}{8} + (-\frac{5}{8}) = -\frac{2}{8} = -\frac{1}{4}$

145. $-\frac{5}{8} - (-\frac{3}{4}) = -\frac{5}{8} + \frac{3}{4} = -\frac{5}{8} + \frac{6}{8} = \frac{1}{8}$

151. 0.99 - 1 = 0.99 + (-1) = -0.01

157. -2.8 - 0 = -2.8 + 0 = -2.8

163. $-\frac{4}{7} - (-\frac{10}{7}) = -\frac{4}{7} + \frac{10}{7} = \frac{6}{7}$

169. $18 - (-15) - 3 - (-5) + 2$
$= 18 + 15 + (-3) + 5 + 2$
$= 40 + (-3)$
$= 37$

175. $-93 - (-84) - 41 - (-56)$
$= -93 + 84 + (-41) + 56$
$= -134 + 140$
$= 6$

181. -40 - (-156) = -40 + 156 = 116
Lake Assal is 116 m lower than Valdes Peninsula.

Exercise Set 1.4

1. $-8 \cdot 2 = -16$ 7. $-9 \cdot 8 = -72$

13. $15 \cdot (-8) = -120$ 19. $-3.5 \cdot (-28) = 98$

25. $-6 \cdot (-4) = 24$

31. $-\frac{3}{8} \cdot (-\frac{2}{9}) = \frac{6}{72} = \frac{1 \cdot 6}{12 \cdot 6} = \frac{1}{12} \cdot \frac{6}{6} = \frac{1}{12}$

37. $7 \cdot (-4) \cdot (-3) \cdot 5 = (-28) \cdot (-15) = 420$

43. $-2 \cdot (-5) \cdot (-3) \cdot (-5) = 10 \cdot 15 = 150$

49. $-4 \cdot (-1.8) \cdot 7 = (7.2) \cdot 7 = 50.4$

55. $0.07 \cdot (-7) \cdot 6 \cdot (-6) = 0.07 \cdot 6 \cdot (-7) \cdot (-6)$
$$= 0.42(42)$$
$$= 17.64$$

61. $(-8)(-9)(-10) = 72(-10) = -720$

67. $\frac{26}{-2} = = -13$ 73. $\frac{-72}{9} = -8$ 79. $\frac{200}{-25} = -8$

85. The reciprocal of $\frac{15}{7}$ is $\frac{7}{15}$ because $\frac{15}{7} \cdot \frac{7}{15} = 1$.

91. The reciprocal of 4.3 is $\frac{1}{4.3}$ (or $\frac{10}{43}$) because
$4.3 \cdot \frac{1}{4.3} = 1$.

97. $\frac{6}{-13} = 6(-\frac{1}{13})$

103. $-\frac{5}{4} \div (-\frac{3}{4}) = -\frac{5}{4} \cdot (-\frac{4}{3}) = \frac{20}{12} = \frac{4 \cdot 5}{4 \cdot 3} = \frac{4}{4} \cdot \frac{5}{3} = \frac{5}{3}$

109. $-6.6 \div 3.3 = -\frac{6.6}{3.3} = -2$

115. $7(-9) + 3 = -63 + 3 = -60$

121. $\frac{2}{5}(\frac{8}{9} + 4\frac{2}{3}) = \frac{2}{5}(\frac{8}{9} + 4\frac{6}{9})$
$$= \frac{2}{5}(4\frac{14}{9}) = \frac{2}{5}(5\frac{5}{9})$$
$$= \frac{2}{5}(\frac{50}{9}) = \frac{100}{45}$$
$$= \frac{5 \cdot 20}{5 \cdot 9} = \frac{20}{9}$$

127. $\frac{-3(-9) + 7}{-4} = \frac{27 + 7}{-4} = \frac{34}{-4} = -\frac{34}{4} = -\frac{17}{2}$

133. $-3[(-8) + (-6)](-\frac{1}{7})$
$$= -3(-14)(-\frac{1}{7})$$
$$= -3(2)$$
$$= -6$$

139. $3\pi + \pi = 3\pi + 1\pi = (3 + 1)\pi = 4\pi$

Exercise Set 1.5

1. Substitute 14 for x in x - 4.   14 - 4 = 10
Substitute 29 for x in x - 4.   29 - 4 = 25
Substitute 52 for x in x - 4.   52 - 4 = 48

7. Substitute 9 for x and 3 for y.
$\frac{x}{y} = \frac{9}{3} = 3$

13. Phrase: 6 more than m
Algebraic expression: 6 + m, m + 6

19. Phrase: b more than a
Algebraic expression: b + a, or a + b

25. Phrase: The sum of r and s
Algebraic expression: r + s, or s + r

31. Phrase: The difference of 3 and b
Algebraic expression: 3 - b or b - 3

37. Phrase: A number x plus three times y
Algebraic expression: x + 3y, or 3y + x

43. n^1 means n, for any number n

49. $(10pq)^0$ means 1, for any numbers p and q, except 0

55. $3y \cdot 3y \cdot 3y \cdot 3y = (3y)^4$

61. Substitute 4 for x.
$x^4 = 4^4 = 4 \cdot 4 \cdot 4 \cdot 4 = 256$

67. Substitute 24m for s.
$A = s^2 = (24m)^2 = 24m \cdot 24m$
$$= 24 \cdot 24 \cdot m \cdot m$$
$$= 576 \ m^2$$

73. Phrase: Age two years ago, if b represents age now.
(Two less than b)
Algebraic expression: b - 2

79. x = 9, y = 3x = 3·9 = 27
Substitute 9 for x and 27 for y.
$\frac{y - x}{3} = \frac{27 - 9}{3} = \frac{18}{3} = 6$

85. $(v + 2) + (v - 2)$
$$= v + 2 + v + (-2)$$
$$= v + v + 2 + (-2)$$
$$= 2v + 0$$
$$= 2v$$

91. $\frac{2^6}{8^2} = \frac{2^6}{8 \cdot 8} = \frac{2 \cdot 2 \cdot 2 \cdot 2 \cdot 2 \cdot 2}{2 \cdot 2 \cdot 2 \cdot 2 \cdot 2 \cdot 2} = 1$

97. Substitute -4 for n.
$(3n)^3 = [3 \cdot (-4)]^3 = [-12]^3 = -12 \cdot (-12) \cdot (-12) = -1728$
$3n^3 = 3(-4)^3 = 3(-4)(-4)(-4) = 3(-64) = -192$

103. $x \cdot x \cdot x \cdot y \cdot y \cdot y = x^3 y^3$

109. Answers may vary. For a 10-digit readout, 6^{13} and larger will be too large.

115. No, $x \cdot x \cdot x \cdot x = x^4$

Exercise Set 1.6

1. $y + 8 = 8 + y$, Commutative Law of Addition

7. $ab + c = c + ab$, Commutative Law of Addition
$ab + c = ba + c$, Commutative Law of Multiplication
$ab + c = c + ba$, Both Commutative Laws

13. $19 - 5 \times 3 + 3 = 19 - 15 + 3 = 4 + 3 = 7$

19. $(6 \cdot 3)^2 = 18^2 = 324$

25. $6 + 4^2 = 6 + 16 = 22$

31. $2 \times 10^3 - 500 = 2 \times 1000 - 500 = 2000 - 500 = 1500$

37. Substitute 7 for w.
$(12 - w)^3 = (12 - 7)^3 = 5^3 = 125$

43. Substitute 5 for x.
$\dfrac{(4x) + 2}{2x} = \dfrac{(4 \cdot 5) + 2}{2 \cdot 5} = \dfrac{20 + 2}{10} = \dfrac{22}{10} = \dfrac{11}{5}$

49. $(5 + x) + y = 5 + (x + y)$, Assoc. Law of Addition

55. $\quad 6 + (x + y)$
$= (x + y) + 6$, Commutative Law of Addition
$= x + (y + 6)$, Associative Law of Addition
$= (y + 6) + x$, Commutative Law of Addition

61. $4x + 3z$ The terms are $4x$ and $3z$.

67. $7(4 - 3) = 7 \cdot 4 - 7 \cdot 3 = 28 - 21 = 7$

73. $7(x - 2) = 7 \cdot x - 7 \cdot 2 = 7x - 14$

79. $-4(x - 3y - 2z) = -4 \cdot x - (-4) \cdot 3y - (-4) \cdot 2z$
$= -4x - (-12y) - (-8z)$
$= -4x + 12y + 8z$

85. $32 - 4y = 4 \cdot 8 - 4 \cdot y = 4(8 - y)$

91. $ax - ay - az = a \cdot x - a \cdot y - a \cdot z = a(x - y - z)$

97. $9x + 2y - 5x = (9 - 5)x + 2y = 4x + 2y$

103. $\dfrac{1}{5}x + \dfrac{4}{5}y + \dfrac{2}{5}x - \dfrac{1}{5}y$
$= (\dfrac{1}{5} + \dfrac{2}{5})x + (\dfrac{4}{5} - \dfrac{1}{5})y$
$= \dfrac{3}{5}x + \dfrac{3}{5}y$

109. $14y + y = 14y + 1y = (14 + 1)y = 15y$

115. $5x - 3x + 8x = (5 - 3 + 8)x = 10x$

121. $4x - 7 + 18x + 25 = (4 + 18)x + (-7 + 25)$
$= 22x + 18$

127. $5m + 6 = 6 + 5m$, Commutative Law of Addition

133. For any number except 0, the two expressions are not equivalent. Let $x = 4$.
$3x^2 = 3 \cdot 4^2 = 3 \cdot 16 = 48$
$(3x)^2 = (3 \cdot 4)^2 = 12^2 = 144$

139. Let x represent the number. Then $x + 7$ represents the sum and $(x + 7)^2$ represents the square of the sum.

145. $2\pi r + \pi rs = \pi r \cdot 2 + \pi r \cdot s = \pi r(2 + s)$

151. Phrase: Three times the sum of a and b, decreased by 7a
Algebraic expression: $3(a + b) - 7a$
$= 3a + 3b - 7a$
$= 3b - 4a$ (Simplified)

157. $\dfrac{33sba}{2(11a)} = \dfrac{3sb \cdot 11a}{2 \cdot 11a} = \dfrac{3sb}{2}$

163. $\dfrac{s}{20} = \dfrac{s}{20} \cdot 1 = \dfrac{s}{20} \cdot \dfrac{t}{t} = \dfrac{st}{20t}$

Exercise Set 1.7

1. $-(2x + 7) = -2x - 7$

7. $-(6x - 8y + 5) = -6x + 8y - 5$

13. $9x - (4x + 3) = 9x - 4x - 3 = 5x - 3$

19. $2x - 4y - 3(7x - 2y) = 2x - 4y - 21x + 6y$
$= -19x + 2y$

25. $8[7 - 6(4 - 2)]$

$= 8[7 - 6(2)]$

$= 8[7 - 12]$

$= 8[-5]$

$= -40$

31. $[7(x + 5) - 19] - [4(x - 6) + 10]$

$= [7x + 35 - 19] - [4x - 24 + 10]$

$= [7x + 16] - [4x - 14]$

$= 7x + 16 - 4x + 14$

$= 3x + 30$

37. $16(-24) + 50 = -384 + 50 = -334$

43. $3000 \cdot (1 + 0.16)^3$

$= 3000 \cdot (1.16)^3$

$= 3000(1.560896)$

$= 4682.688$

49. $(2 * (5 - 3))\hat{\ }2 = (2 * 2)\hat{\ }2 = 4\hat{\ }2 = 16$

55. $\dfrac{a}{b} - \dfrac{c}{d} = A/B - C/D$

61. $(A + B)\hat{\ }2 = (a + b)^2$

67. $z - \{2z - [3z - (4z - 5z) - 6z] - 7z\} - 8z$

$= z - \{2z - [3z - 4z + 5z - 6z] - 7z\} - 8z$

$= z - \{2z - [-2z] - 7z\} - 8z$

$= z - \{2z + 2z - 7z\} - 8z$

$= z - \{-3z\} - 8z$

$= z + 3z - 8z$

$= -4z$

73.   False, $-n - m = -(n + m)$

79.   Yes. $(-a)(-b) = (-1)(a)(-1)(b) = (-1)^2(ab) =$
$1ab = ab$

1. 3^{-2} means $\dfrac{1}{3^2}$, or $\dfrac{1}{3 \cdot 3}$, or $\dfrac{1}{9}$

7. $\dfrac{1}{x^3} = x^{-3}$ **13.** $7^{-3} = \dfrac{1}{7^3}$ **19.** $z^{-n} = \dfrac{1}{z^n}$

25. $x^{-2} \cdot x = x^{-2} \cdot x^1 = x^{-2+1} = x^{-1}$

31. $t^8 \cdot t^{-8} = t^{8+(-8)} = t^0 = 1$

37. $\dfrac{x^7}{x^{-2}} = x^{7-(-2)} = x^{7+2} = x^9$

43. $\dfrac{m^{-9}}{m^{-9}} = m^{-9-(-9)} = m^{-9+9} = m^0 = 1$

49. $(x^{-3})^{-4} = x^{-3(-4)} = x^{12}$

55. $(3x^3y^{-8}z^{-3})^2 = 3^2x^{3 \cdot 2}y^{-8 \cdot 2}z^{-3 \cdot 2} = 9x^6y^{-16}z^{-6}$

61. $A = P(1 + r)^t$

$A = 10,400(1 + 8.75\%)^5$ $(8\tfrac{3}{4}\% = 8.75\%)$

$= 10,400(1 + 0.0875)^5$

$= 10,400(1.0875)^5$

$= 10,400(1.52105994)$

$\approx 15,819.02$

At the end of 5 years, $15,819.02 is in the
account.

67. $\dfrac{(5^{12})^2}{5^{25}} = \dfrac{5^{24}}{5^{25}} = \dfrac{1}{5^{25-24}} = \dfrac{1}{5^1} = \dfrac{1}{5}$

73. $\dfrac{(\tfrac{1}{2})^4}{(\tfrac{1}{2})^5} = \dfrac{1}{(\tfrac{1}{2})^{5-4}} = \dfrac{1}{(\tfrac{1}{2})^1} = \dfrac{1}{\tfrac{1}{2}} = 1 \cdot \dfrac{2}{1} = 2$

79. $\dfrac{(9x)^{12}}{(9x)^{14}} = \dfrac{1}{ax^2}$

$\dfrac{1}{(9x)^{14-12}} = \dfrac{1}{ax^2}$

$\dfrac{1}{(9x)^2} = \dfrac{1}{ax^2}$

$\dfrac{1}{9^2x^2} = \dfrac{1}{ax^2}$

$\dfrac{1}{81x^2} = \dfrac{1}{ax^2}$

$\dfrac{ax^2}{81x^2} = 1$

$ax^2 = 81x^2$

$a = \dfrac{81x^2}{x^2}$

$a = 81$

85. $(xy)^m = x^m y^m$; True

91. $\dfrac{(0.2)(0.3)^3}{(0.03)(0.5)^2} = \dfrac{(0.2)(0.027)}{(0.03)(0.25)} = \dfrac{0.0054}{0.0075}$

$= \dfrac{54}{75} = \dfrac{18}{25}$, or 0.72

97. $\dfrac{-12x^{a+1}}{4x^{2-a}} = -3x^{(a+1)-(2-a)} = -3x^{a+1-2+a} = -3x^{2a-1}$

103. $(12^{3-a})^{2b} = 12^{(3-a)2b} = 12^{6b-2ab}$

109. $(-2)^0 - (-2)^3 - (-2)^{-1} + (-2)^4 - (-2)^{-2}$

$= 1 - (-8) - (-\tfrac{1}{2}) + 16 - \tfrac{1}{4}$

$= 1 + 8 + \tfrac{1}{2} + 16 - \tfrac{1}{4}$

$= 25\tfrac{1}{4}$

115. $\left[\dfrac{(-3x^{-2a}y^{5b})^{-2}}{(2x^{4a}y^{-8b})^{-3}}\right]^2 = \dfrac{(-3x^{-2a}y^{5b})^{-4}}{(2x^{4a}y^{-8b})^{-6}}$

$= \dfrac{(2x^{4a}y^{-8b})^6}{(-3x^{-2a}y^{5b})^4}$

$= \dfrac{2^6(x^{4a})^6(y^{-8b})^6}{(-3)^4(x^{-2a})^4(y^{5b})^4}$

$= \dfrac{64x^{24a}y^{-48b}}{81x^{-8a}y^{20b}}$

$= \dfrac{64}{81}x^{32a}y^{-68b}$, or $\dfrac{64x^{32a}}{81y^{68b}}$

1. $47{,}000{,}000{,}000$

$= 47{,}000{,}000{,}000 \times (10^{-10} \times 10^{10})$

$= (47{,}000{,}000{,}000 \times 10^{-10}) \times 10^{10}$

$= 4.7 \times 10^{10}$

7. 0.00000000007

$= 0.00000000007 \times (10^{11} \times 10^{-11})$

$= (0.00000000007 \times 10^{11}) \times 10^{-11}$

$= 7 \times 10^{-11}$

13. 6.73×10^8

$= 673{,}000{,}000$ (Moving decimal point 8 places to the right)

19. $(2.3 \times 10^6)(4.2 \times 10^{-11})$

$= (2.3 \times 4.2)(10^6 \times 10^{-11})$

$= 9.66 \times 10^{-5}$

25. $(3.01 \times 10^{-5})(6.5 \times 10^7)$

$= (3.01 \times 6.5)(10^{-5} \times 10^7)$

$= 19.565 \times 10^2$

$= (1.9565 \times 10^1) \times 10^2$

$= 1.9565 \times (10^1 \times 10^2)$

$= 1.9565 \times 10^3$

31. $\dfrac{12.6 \times 10^8}{4.2 \times 10^{-3}}$

$= \dfrac{12.6}{4.2} \times \dfrac{10^8}{10^{-3}}$

$= 3 \times 10^{11}$

37. $\dfrac{780{,}000{,}000 \times 0.00071}{0.000005}$

$= \dfrac{(7.8 \times 10^8) \times (7.1 \times 10^{-4})}{5 \times 10^{-6}}$

$= \dfrac{7.8 \times 7.1}{5} \times \dfrac{10^8 \times 10^{-4}}{10^{-6}}$

$\approx \dfrac{8 \times 7}{5} \times \dfrac{10^4}{10^{-6}}$

$\approx \dfrac{56}{5} \times 10^{10}$

$\approx 11 \times 10^{10}$

$\approx (1.1 \times 10^1) \times 10^{10}$

$\approx 1.1 \times (10^1 \times 10^{10})$

$\approx 1.1 \times 10^{11}$

43.   3,014,000 × 0.0000072

≈ $(3 \times 10^6) \times (7 \times 10^{-6})$

= $21 \times 10^0 = 21 = 2.1 \times 10^1$

The order of magnitude is 1.

49. Familiarize:

When the roll of plastic is unrolled, the dimensions are 30 m by 1 m by 0.8 mm (or 0.0008 m).
The volume is found by multiplying the length
times the width times the height.

Translate:

Volume = Length × Width × Height

Volume =   30   ·   1   · 0.0008

Carry out:

Volume = 30(1)(0.0008)

= 0.024

Check: We repeat the above calculations.

State:

The volume of plastic in a roll is 0.024 m^3,
or 2.4×10^{-2} m^3.

55. 1 hour = 60 minutes = 60 × 60, or 3600 seconds

$4{,}200{,}000 \ \dfrac{ft^3}{sec}$ · 3600 sec = 15,120,000,000 ft^3

or 1.512×10^{10} ft^3

61. $\dfrac{2.4 \times 10^{13}}{5.88 \times 10^{12}} \approx 0.408 \times 10^1 = 4.08 \times 10^{-1} \times 10^1$

$= 4.08 \times 10^0$

$= 4.08$

Alpha Centauri is approximately 4 light years
from the earth.

67. Solve $a = 4 \cdot 3^b$ for 3^b.

$a = 4 \cdot 3^b$

$\dfrac{a}{4} = 3^b$

Substitute a/4 for 3^b in $c = 2 - 3^{-b}$.

$c = 2 - 3^{-b}$

$c = 2 - \dfrac{1}{3^b}$

$c = 2 - \dfrac{1}{a/4} = 2 - \dfrac{4}{a}$, or $\dfrac{2a - 4}{a}$

Exercise Set 2.1

1. $x + 2 = 6$ Check: $\underline{x + 2 = 6}$

$x + 2 + (-2) = 6 + (-2)$ $4 + 2 \;|\; 6$

$x + 0 = 6 + (-2)$ $6 \;|$

$x = 4$

7. $x + 16 = -2$ Check: $\underline{x + 16 = -2}$

$x + 16 + (-16) = -2 + (-16)$ $-18 + 16 \;|\; -2$

$x + 0 = -2 + (-16)$ $-2 \;|$

$x = -18$

13. $5 + t = 7$ Check: $\underline{5 + t = 7}$

$-5 + 5 + t = -5 + 7$ $5 + 2 \;|\; 7$

$t = 2$ $7 \;|$

19. $r + \frac{1}{3} = \frac{8}{3}$ Check: $\underline{r + \frac{1}{3} = \frac{8}{3}}$

$r + \frac{1}{3} + (-\frac{1}{3}) = \frac{8}{3} + (-\frac{1}{3})$ $\frac{7}{3} + \frac{1}{3} \;\big|\; \frac{8}{3}$

$r = \frac{7}{3}$ $\frac{8}{3} \;\big|$

25. $-\frac{1}{5} + z = -\frac{1}{4}$

$\frac{1}{5} + (-\frac{1}{5}) + z = \frac{1}{5} + (-\frac{1}{4})$

$0 + z = \frac{4}{20} + (-\frac{5}{20})$

$z = -\frac{1}{20}$

Check: $\underline{-\frac{1}{5} + z = -\frac{1}{4}}$

$-\frac{1}{5} + (-\frac{1}{20}) \;\big|\; -\frac{1}{4}$

$-\frac{4}{20} + (-\frac{1}{20}) \;\big|\; -\frac{5}{20}$

$-\frac{5}{20} \;\big|$

31. $-9.7 = -4.7 + y$

$4.7 + (-9.7) = 4.7 + (-4.7) + y$

$-5 = y$

Check: $\underline{-9.7 = -4.7 + y}$

$-9.7 \;\big|\; -4.7 + (-5)$

$\big|\; -9.7$

37. $6x = 36$ Check: $\underline{6x = 36}$

$\frac{1}{6} \cdot 6x = \frac{1}{6} \cdot 36$ $6 \cdot 6 \;|\; 36$

$1 \cdot x = 6$ $36 \;|$

$x = 6$

43. $-x = 40$ Check: $\underline{-x = 40}$

$-1 \cdot x = 40$ $-(-40) \;|\; 40$

$-1 \cdot (-1 \cdot x) = -1 \cdot 40$ $40 \;|$

$1 \cdot x = -40$

$x = -40$

49. $-12x = 72$ Check: $\underline{-12x = 72}$

$-\frac{1}{12} \cdot (-12x) = -\frac{1}{12} \cdot 72$ $-12(-6) \;|\; 72$

$1 \cdot x = -\frac{72}{12}$ $72 \;|$

$x = -6$

55. $\frac{3}{4}x = 27$ Check: $\underline{\frac{3}{4}x = 27}$

$\frac{4}{3} \cdot \frac{3}{4}x = \frac{4}{3} \cdot 27$ $\frac{3}{4} \cdot 36 \;|\; 27$

$1 \cdot x = 36$ $27 \;|$

$x = 36$

61. $-\frac{3}{5}r = -\frac{9}{10}$ Check: $\underline{-\frac{3}{5}r = -\frac{9}{10}}$

$-\frac{5}{3} \cdot (-\frac{3}{5}r) = -\frac{5}{3} \cdot (-\frac{9}{10})$ $-\frac{3}{5} \cdot \frac{3}{2} \;\big|\; -\frac{9}{10}$

$1 \cdot r = \frac{45}{30}$ $-\frac{9}{10} \;\big|$

$r = \frac{3}{2}$

67. $-3.1y = 21.7$

$-\frac{1}{3.1} \cdot (-3.1y) = -\frac{1}{3.1} \cdot (21.7)$

$1 \cdot y = -\frac{21.7}{3.1}$

$y = -7$

Check: $\underline{-3.1y = 21.7}$

$-3.1(-7) \;|\; 21.7$

$21.7 \;|$

73. $|x| = 5$

$x = -5$ or $x = 5$ (Absolute Value Principle)

The solutions are -5 and 5.

79. $|5x| = 40$

$5x = -40$ or $5x = 40$

$x = -8$ or $x = 8$

The solutions are -8 and 8.

85. $16 + x - 22 = -16$ **91.** $x - 4 = a$

$x + 16 - 22 = -16$ $x - 4 + 4 = a + 4$

$x - 6 = -16$ $x = a + 4$

$x - 6 + 6 = -16 + 6$

$x = -10$

97.
$$-0.2344m = 2028.732$$
$$-\frac{1}{0.2344} \cdot (-0.2344m) = -\frac{1}{0.2344} \cdot (2028.732)$$
$$1 \cdot m = -\frac{2028.732}{0.2344}$$
$$m = -8655$$

103.
$$\frac{a}{b}x = 4$$
$$\frac{b}{a} \cdot \frac{a}{b}x = \frac{b}{a} \cdot 4$$
$$x = \frac{4b}{a}$$

109.
$$-\frac{2}{3} - \frac{4}{15} = x + \frac{4}{5}$$
$$-\frac{10}{15} - \frac{4}{15} = x + \frac{12}{15}$$
$$-\frac{14}{15} = x + \frac{12}{15}$$
$$-\frac{26}{15} = x$$

Exercise Set 2.2

1.
$$5x + 6 = 31$$
$$5x + 6 + (-6) = 31 + (-6)$$
$$5x = 25$$
$$\frac{1}{5} \cdot 5x = \frac{1}{5} \cdot 25$$
$$x = 5$$

Check:
$5x + 6 = 31$	
$5 \cdot 5 + 6$	31
$25 + 6$	
31	

7.
$$3x - 9 = 33$$
$$3x - 9 + 9 = 33 + 9$$
$$3x = 42$$
$$\frac{1}{3} \cdot 3x = \frac{1}{3} \cdot 42$$
$$x = 14$$

Check:
$3x - 9 = 33$	
$3 \cdot 14 - 9$	33
$42 - 9$	
33	

13.
$$-4x + 7 = 35$$
$$-4x + 7 + (-7) = 35 + (-7)$$
$$-4x = 28$$
$$-\frac{1}{4} \cdot (-4x) = -\frac{1}{4} \cdot 28$$
$$x = -7$$

Check:
$-4x + 7 = 35$	
$-4(-7) + 7$	35
$28 + 7$	
35	

19.
$$5x + 7x = 72$$
$$12x = 72$$
$$\frac{1}{12} \cdot 12x = \frac{1}{12} \cdot 72$$
$$x = 6$$

Check:
$5x + 7x = 72$	
$5 \cdot 6 + 7 \cdot 6$	72
$30 + 42$	
72	

25.
$$4y - 2y = 10$$
$$2y = 10$$
$$\frac{1}{2} \cdot 2y = \frac{1}{2} \cdot 10$$
$$y = 5$$

Check:
$4y - 2y = 10$	
$4 \cdot 5 - 2 \cdot 5$	10
$20 - 10$	
10	

31.
$$10.2y - 7.3y = -58$$
$$2.9y = -58$$
$$\frac{1}{2.9} \cdot 2.9y = \frac{1}{2.9} \cdot (-58)$$
$$y = -\frac{58}{2.9}$$
$$y = -20$$

Check:
$10.2y - 7.3y = -58$	
$10.2(-20) - 7.3(-20)$	-58
$-204 + 146$	
-58	

37.
$$4x - 7 = 3x$$
$$4x = 3x + 7$$
$$4x - 3x = 7$$
$$x = 7$$

Check:
$4x - 7 = 3x$	
$4 \cdot 7 - 7$	$3 \cdot 7$
$28 - 7$	21
	21

43.
$$6x + 3 = 2x + 11$$
$$6x - 2x = 11 - 3$$
$$4x = 8$$
$$x = \frac{8}{4}$$
$$x = 2$$

Check:
$6x + 3 = 2x + 11$	
$6 \cdot 2 + 3$	$2 \cdot 2 + 11$
$12 + 3$	$4 + 11$
15	15

49.
$$4y - 4 + y = 6y + 20 - 4y$$
$$5y - 4 = 2y + 20$$
$$5y - 2y = 20 + 4$$
$$3y = 24$$
$$y = \frac{24}{3}$$
$$y = 8$$

Check:
$4y - 4 + y = 6y + 20 - 4y$	
$4 \cdot 8 - 4 + 8$	$6 \cdot 8 + 20 - 4 \cdot 8$
$32 - 4 + 8$	$48 + 20 - 32$
$28 + 8$	$68 - 32$
36	36

55.
$$\frac{2}{3} + 3y = 5y - \frac{2}{15}, \quad \text{LCM} = 15$$
$$15\left(\frac{2}{3} + 3y\right) = 15\left(5y - \frac{2}{15}\right)$$
$$15 \cdot \frac{2}{3} + 15 \cdot 3y = 15 \cdot 5y - 15 \cdot \frac{2}{15}$$
$$10 + 45y = 75y - 2$$
$$10 + 2 = 75y - 45y$$
$$12 = 30y$$
$$\frac{12}{30} = y$$
$$\frac{2}{5} = y$$

55. (continued)

Check: $\dfrac{2}{3} + 3y = 5y - \dfrac{2}{15}$

$$\begin{array}{c|c} \dfrac{2}{3} + 3 \cdot \dfrac{2}{5} & 5 \cdot \dfrac{2}{5} - \dfrac{2}{15} \\[2mm] \dfrac{2}{3} + \dfrac{6}{5} & 2 - \dfrac{2}{15} \\[2mm] \dfrac{10}{15} + \dfrac{18}{15} & \dfrac{30}{15} - \dfrac{2}{15} \\[2mm] \dfrac{28}{15} & \dfrac{28}{15} \end{array}$$

61. $1.03 - 0.62x = 0.71 - 0.22x$

$100(1.03 - 0.62x) = 100(0.71 - 0.22x)$

$100(1.03) - 100(0.62x) = 100(0.71) - 100(0.22x)$

$103 - 62x = 71 - 22x$

$32 = 40x$

$\dfrac{32}{40} = x$

$\dfrac{4}{5} = x$

67. $\dfrac{4}{5}x - \dfrac{3}{4}x = \dfrac{3}{10}x - 1$, LCM = 20

$20\left(\dfrac{4}{5}x - \dfrac{3}{4}x\right) = 20\left(\dfrac{3}{10}x - 1\right)$

$20 \cdot \dfrac{4}{5}x - 20 \cdot \dfrac{3}{4}x = 20 \cdot \dfrac{3}{10}x - 20 \cdot 1$

$16x - 15x = 6x - 20$

$x = 6x - 20$

$20 = 6x - x$

$20 = 5x$

$\dfrac{20}{5} = x$

$4 = x$

Check: $\dfrac{4}{5}x - \dfrac{3}{4}x = \dfrac{3}{10}x - 1$

$$\begin{array}{c|c} \dfrac{4}{5} \cdot 4 - \dfrac{3}{4} \cdot 4 & \dfrac{3}{10} \cdot 4 - 1 \\[2mm] \dfrac{16}{5} - 3 & \dfrac{6}{5} - 1 \\[2mm] \dfrac{16}{5} - \dfrac{15}{5} & \dfrac{6}{5} - \dfrac{5}{5} \\[2mm] \dfrac{1}{5} & \dfrac{1}{5} \end{array}$$

73. $0.008 + 9.62x - 42.8 = 0.944x + 0.0083 - x$

$9.62x - 42.792 = -0.056x + 0.0083$

$9.62x + 0.056x = 0.0083 + 42.792$

$9.676x = 42.8003$

$x = \dfrac{42.8003}{9.676}$

$x \approx 4.42$

79. $0 = y - (-14) - (-3y)$

$0 = y + 14 + 3y$

$0 = 4y + 14$

$-14 = 4y$

$\dfrac{-14}{4} = y$

$-\dfrac{7}{2} = y$

85. $\dfrac{4 - 3x}{7} = \dfrac{2 + 5x}{49} - \dfrac{x}{14}$

$98 \cdot \left(\dfrac{4 - 3x}{7}\right) = 98\left(\dfrac{2 + 5x}{49} - \dfrac{x}{14}\right)$

$14(4 - 3x) = 2(2 + 5x) - 7x$

$56 - 42x = 4 + 10x - 7x$

$56 - 42x = 4 + 3x$

$52 = 45x$

$\dfrac{52}{45} = x$

Exercise Set 2.3

1. $3(2y - 3) = 27$ Check: $\dfrac{3(2y - 3) = 27}{}$

$6y - 9 = 27$ $\begin{array}{c|c} 3(2 \cdot 6 - 3) & 27 \\ 3(12 - 3) & \\ 3 \cdot 9 & \\ 27 & \end{array}$

$6y = 27 + 9$

$6y = 36$

$y = \dfrac{36}{6}$

$y = 6$

7. $5r - (2r + 8) = 16$ Check: $\dfrac{5r - (2r + 8) = 16}{}$

$5r - 2r - 8 = 16$ $\begin{array}{c|c} 5 \cdot 8 - (2 \cdot 8 + 8) & 16 \\ 40 - (16 + 8) & \\ 40 - 24 & \\ 16 & \end{array}$

$3r - 8 = 16$

$3r = 16 + 8$

$3r = 24$

$r = \dfrac{24}{3}$

$r = 8$

13. $5(d + 4) = 7(d - 2)$

$5d + 20 = 7d - 14$

$20 + 14 = 7d - 5d$

$34 = 2d$

$\dfrac{34}{2} = d$

$17 = d$

19. $3(r - 6) + 2 = 4(r + 2) - 21$

$3r - 18 + 2 = 4r + 8 - 21$

$3r - 16 = 4r - 13$

$13 - 16 = 4r - 3r$

$-3 = r$

25. $2[4 - 2(3 - x)] - 1 = 4[2(4x - 3) + 7] - 25$

$2[4 - 6 + 2x] - 1 = 4[8x - 6 + 7] - 25$

$2[-2 + 2x] - 1 = 4[8x + 1] - 25$

$-4 + 4x - 1 = 32x + 4 - 25$

$4x - 5 = 32x - 21$

$16 = 28x$

$\frac{16}{28} = x$

$\frac{4}{7} = x$

31. $\frac{3}{4}(3x - \frac{1}{2}) - \frac{2}{3} = \frac{1}{3}$

$\frac{3}{4}(3x - \frac{1}{2}) = 1$ (Adding $\frac{2}{3}$)

$3(3x - \frac{1}{2}) = 4$ (Multiplying by 4)

$9x - \frac{3}{2} = 4$ (Removing parentheses)

$9x = \frac{11}{2}$ (Adding $\frac{3}{2}$)

$x = \frac{11}{18}$ (Multiplying by $\frac{1}{9}$)

37. $475(54x + 7856) + 9762 = 402(83x + 975)$

$25,650x + 3,731,600 + 9762 = 33,366x + 391,950$

$25,650x + 3,741,362 = 33,366x + 391,950$

$3,741,362 - 391,950 = 33,366x - 25,650x$

$3,349,412 = 7716x$

$\frac{3,349,412}{7716} = x$

$\frac{837,353}{1929} = x$

43. $2(x - 3) + 5 = 3(x - 2) + 5$

$2x - 6 + 5 = 3x - 6 + 5$

$2x - 1 = 3x - 1$

$2x = 3x$

The equation $2x = 3x$ is only true when $x = 0$. It is not true for all sensible replacements. Thus, it is not an identity.

Exercise Set 2.4

1. a) $\ell = 17$ ft, $w = 4$ ft; $A = \ell w = 17 \cdot 4 = 68$ ft^2

   b) $A = \ell w$

$\frac{1}{\ell} \cdot A = \frac{1}{\ell} \cdot \ell w$

$\frac{A}{\ell} = w$

   c) $A = 48$ cm^2, $\ell = 6$ cm; $w = \frac{A}{\ell} = \frac{48}{6} = 8$ cm

7. $d = rt$

$d \cdot \frac{1}{t} = rt \cdot \frac{1}{t}$

$\frac{d}{t} = r$

13. $P = 2\ell + 2w$

$P - 2\ell = 2w$

$\frac{1}{2}(P - 2\ell) = \frac{1}{2} \cdot 2w$

$\frac{P - 2\ell}{2} = w$

or $\frac{1}{2}P - \ell = w$

19. $A = \frac{a + b + c}{3}$

$3 \cdot A = 3 \cdot \frac{a + b + c}{3}$

$3A = a + b + c$

$3A - a - c = b$

25. $H = \frac{D^2 N}{2.5}$

$2.5 \cdot H = 2.5 \cdot \frac{D^2 N}{2.5}$

$2.5H = D^2 N$

$2.5H \frac{1}{N} = D^2 N \cdot \frac{1}{N}$

$\frac{2.5H}{N} = D^2$

31. $P = 2a + 2b$

$P = 2(a + b)$

$2 \cdot P = 2 \cdot [2(a + b)]$

If P doubles, the sum of a and b doubles. It is not necessary for both a and b to double. Let $a_1 = 6$ and $b_1 = 8$. Then $P_1 = 2 \cdot 6 + 2 \cdot 8 = 28$. Let $a_2 = 2$ and $b_2 = 26$. Then $P_2 = 2 \cdot 2 + 2 \cdot 26 = 56$. $P_2 = 56$ is double $P_1 = 28$, but a_2 and b_2 are not double a_1 and b_1 respectively. Note that the sum of a_2 and b_2, $2 + 26 = 28$, is double the sum of a_1 and b_1, $6 + 8 = 14$.

37. $ax + b = 0$

$ax = 0 - b$

$ax = -b$

$x = -\frac{b}{a}$

43. $y = a - ab$

$y = a(1 - b)$

$\frac{y}{1 - b} = a$

49. $A = 180°(n - 2) = 180°(6 - 2) = 180° \cdot 4 = 720°$

55. $D = \frac{M}{V}$

$7.5 = \frac{M}{363}$ (Substituting 7.5 for D and 363 for V)

$7.5(363) = M$

$2722.5 = M$

The mass of a piece of iron which has a volume of 363 cm^3 is 2722.5 grams.

Exercise Set 2.5

<u>1.</u> What number plus 60 is 112?

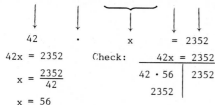

$$x + 60 = 112$$

$x + 60 = 112$ 　　Check: $\underline{x + 60 = 112}$

$\quad x = 112 - 60$ 　　　　　　$52 + 60 \mid 112$

$\quad x = 52$ 　　　　　　　　　$112 \mid$

The answer is 52.

<u>7.</u> Forty-two times what number is 2352?

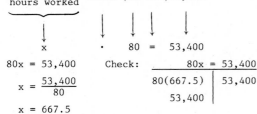

$$42 \cdot x = 2352$$

$42x = 2352$ 　Check: $\underline{42x = 2352}$

$x = \dfrac{2352}{42}$ 　　　　$42 \cdot 56 \mid 2352$

$\quad\quad\quad\quad\quad\quad 2352 \mid$

$x = 56$

The answer is 56.

<u>13.</u> The number of games the Twins won plus 37 is the number of games the Yankees won.

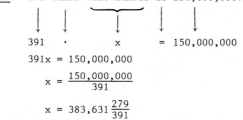

$$x + 37 = 101$$

$x + 37 = 101$ 　Check: $\underline{x + 37 = 101}$

$\quad x = 101 - 37$ 　　　　$64 + 37 \mid 101$

$\quad x = 64$ 　　　　　　　$101 \mid$

The Twins won 64 games.

<u>19.</u> The number of hours worked times $80 is $53,400?

$$x \cdot 80 = 53{,}400$$

$80x = 53{,}400$ 　Check: $\underline{80x = 53{,}400}$

$x = \dfrac{53{,}400}{80}$ 　　　$80(667.5) \mid 53{,}400$

$\quad\quad\quad\quad\quad\quad\quad 53{,}400 \mid$

$x = 667.5$

The consultant worked 667.5 hours.

<u>25.</u> 391 times what number is 150,000,000?

$$391 \cdot x = 150{,}000{,}000$$

$391x = 150{,}000{,}000$

$x = \dfrac{150{,}000{,}000}{391}$

$x = 383{,}631\dfrac{279}{391}$

The distance from the earth to the moon is about 383,632 km.

<u>31.</u> 2.54 times the number of inches in a meter is the total number of centimeters in a meter.

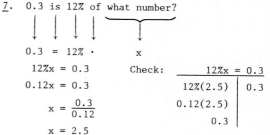

$$2.54 \cdot x = 100$$

$2.54x = 100$

$x = \dfrac{100}{2.54}$

$x \approx 39.37$

The number of inches in a meter is approximately 39.37.

Exercise Set 2.6

<u>1.</u> What percent of 68 is 17?

$$x \quad \% \quad \cdot \quad 68 = 17$$

$x\% \cdot 68 = 17$ 　Check: $\underline{x\% \cdot 68 = 17}$

$x \cdot 0.01 \cdot 68 = 17$ 　　　$25\% \cdot 68 \mid 17$

$x(0.68) = 17$ 　　　　$0.25(68)$

$x = \dfrac{17}{0.68}$ 　　　　　　17

$x = 25$

The answer is 25%

<u>7.</u> 0.3 is 12% of what number?

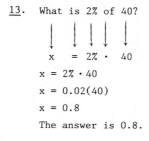

$$0.3 = 12\% \cdot x$$

$12\%x = 0.3$ 　Check: $\underline{12\%x = 0.3}$

$0.12x = 0.3$ 　　　　$12\%(2.5) \mid 0.3$

$x = \dfrac{0.3}{0.12}$ 　　　　$0.12(2.5)$

$x = 2.5$ 　　　　　　　0.3

The answer is 2.5.

<u>13.</u> What is 2% of 40?

$$x = 2\% \cdot 40$$

$x = 2\% \cdot 40$

$x = 0.02(40)$

$x = 0.8$

The answer is 0.8.

19. $208 is 26% of what amount?

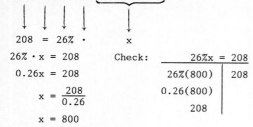

$208 = 26\% \cdot x$

$26\% \cdot x = 208$

$0.26x = 208$

$x = \dfrac{208}{0.26}$

$x = 800$

Check:

$$\begin{array}{c|c} 26\%x = 208 \\ \hline 26\%(800) & 208 \\ 0.26(800) & \\ 208 & \end{array}$$

The monthly income was $800.

25. Marked price - Reduction = Sale price

$$x \quad - \quad 40\%x \quad = \quad \$9.60$$

$x - 0.4x = 9.60$

$(1 - 0.4)x = 9.60$

$0.6x = 9.60$

$x = \dfrac{9.60}{0.6}$

$x = 16$

Check:

$$\begin{array}{c|c} x - 40\%x = 9.60 \\ \hline 16 - 0.4(16) & 9.60 \\ 16 - 6.40 & \\ 9.60 & \end{array}$$

The marked price was $16.

31. 59 is 1.3% of what number?

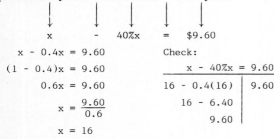

$59 = 1.3\% \cdot x$

$59 = 0.013x$

$\dfrac{59}{0.013} = x$

$4538 \approx x$

About 4538 bottles were produced.

37. $12 \times 100\%$

$= 12 \times (100 \times 0.01)$

$= 12 \times 1 = 12$

43. What is 6% of $16.41?

$x = 6\% \times 16.41 = 0.06 \times 16.41 = 0.9846$

The sales tax is $0.98.
The sum of the meal and the tax is $16.41 + $0.98, or $17.39.

What is 15% of $17.39?

$y = 15\% \times 17.39 = 0.15 \times 17.39 = 2.6085$

The tip is $2.61.

The total paid is $17.39 + $2.61, or $20.

49. Let x represent the larger number.

Then 25%x, or 0.25x, represents the smaller number.

The larger number is 12 more than the smaller.

$x = 12 + 0.25x$

$0.75x = 12$

$x = \dfrac{12}{0.75}$

$x = 16$

If x = 16, then 0.25x = 0.25(16) = 4. The values check, 16 = 12 + 4. The numbers are 16 and 4.

1. Familiarize:

Let x be a certain number. Then 6x is 6 times that number.

Translate:

18 subtracted from six times a certain number is 96.

$$6x - 18 \qquad = 96$$

Carry out: We solve the equation.

$6x - 18 = 96$

$6x = 96 + 18$

$6x = 114$

$x = \dfrac{114}{6}$

$x = 19$

Check:

Six times 19 is 114. Subtracting 18 from 114 is 96. This checks.

State:

The number is 19.

7. Familiarize: First draw a picture.

We use x for the first length, 2x for the second length, and $3 \cdot 2x$, or 6x, for the third length.

Translate:

Length of 1st piece plus Length of 2nd piece plus Length of 3rd piece is 180

$$x \quad + \quad 2x \quad + \quad 6x \quad = 180$$

Carry out: We solve the equation.

$x + 2x + 6x = 180$

$9x = 180$

$x = \dfrac{180}{9}$

$x = 20$

Check:

If the first piece is 20 m long, then the second is $2 \cdot 20$ m, or 40 m and the third is $6 \cdot 20$ m, or 120 m. The lengths of these pieces add up to 180 m $(20 + 40 + 120 = 180)$. This checks.

State:

The first piece measures 20 m. The second measures 40 m, and the third measures 120 m.

13. Familiarize:

Let x be the smallest integer. Then x + 1 and x + 1 + 1, or x + 2, are the next two consecutive integers.

Translate:

First integer	plus	Second integer	plus	Third integer	is 108.

$$x \quad + \quad (x + 1) \quad + \quad (x + 2) \quad = 108$$

Carry out: We solve the equation.

$$x + (x + 1) + (x + 2) = 108$$
$$3x + 3 = 108$$
$$3x = 108 - 3$$
$$3x = 105$$
$$x = \frac{105}{3}$$
$$x = 35$$

Check:

If the smallest integer is 35, then the second is 35 + 1, or 36, and the third is 35 + 1 + 1, or 37. They are consecutive integers. Their sum, 35 + 36 + 37, is 108. This checks.

State:

The integers are 35, 36, and 37.

19. Familiarize: First draw a picture.

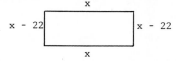

Let x represent the length. Then x - 22 represents the width.

Translate:

Width	+	Width	+	Length	+	Length	is 152.

$$(x - 22) + (x - 22) + \quad x \quad + \quad x \quad = 152$$

Carry out: We solve the equation.

$$(x - 22) + (x - 22) + x + x = 152$$
$$4x - 44 = 152$$
$$4x = 152 + 44$$
$$4x = 196$$
$$x = \frac{196}{4}$$
$$x = 49$$

The width which is 22 less than the length is 49 - 22, or 27.

Check:

The perimeter is 27 + 27 + 49 + 49, which is 152. This checks.

State:

The width is 27 m, and the length is 49 m.

25. Let x = the speed of the first supersonic flight. The speed of Apollo 10 is thirty-seven times the speed of the 1947 flight.

$$24{,}790 = 37x \qquad \text{(Translating)}$$
$$\frac{1}{37} \cdot 24{,}790 = \frac{1}{37} \cdot 37x$$
$$\frac{24{,}790}{37} = 1 \cdot x$$
$$670 = x$$

The speed of the first supersonic flight was 670 miles per hour.

31. R = -0.028t + 20.8

R stands for the record in seconds, and t stands for the number of years since 1920.

Substitute 19.0 for R and solve for t.

$$R = -0.028t + 20.8$$
$$19.0 = -0.028t + 20.8$$
$$0.028t = 20.8 - 19.0$$
$$0.028t = 1.8$$
$$t = \frac{1.8}{0.028}$$
$$t \approx 64.3$$

In approximately 64.3 years since 1920, the record will be 19.0 seconds.

37. Familiarize:

Consecutive odd integers are next to each other such as -11, -9 and 21, 23. If x represents the first odd integer, then x + 2 and x + 4 represent the next two consecutive odd integers.

Translate:

$$x = \text{1st odd integer}$$
$$x + 2 = \text{2nd odd integer}$$
$$x + 4 = \text{3rd odd integer}$$

First integer	+	2 times the second integer	+	3 times the third integer	= 70

$$x \quad + \quad 2(x + 2) \quad + \quad 3(x + 4) \quad = 70$$

Carry out: We solve the equation.

$$x + 2(x + 2) + 3(x + 4) = 70$$
$$x + 2x + 4 + 3x + 12 = 70$$
$$6x + 16 = 70$$
$$6x = 54$$
$$x = 9$$

If x = 9, then x + 2 = 9 + 2 = 11 and x + 4 = 9 + 4 = 13.

Check:

The numbers 9, 11, and 13 are consecutive odd integers. The sum of the first, 9, plus twice the second, 2·11 (or 22), and three times the third, 3·13 (or 39) is 9 + 22 + 39, or 70. The numbers check.

State:

The consecutive odd integers are 9, 11, and 13.

43. Familiarize:

After the next test, there will be six test scores. The average of the six scores is their sum divided by six.

Translate:

We let x represent the score on the next test. The sum of the six scores is represented by

93 + 89 + 72 + 80 + 96 + x.

The average of the six scores is

$$\frac{93 + 89 + 72 + 80 + 96 + x}{6}.$$

The average of the six scores is 88.

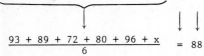

$$\frac{93 + 89 + 72 + 80 + 96 + x}{6} = 88$$

Carry out:

93 + 89 + 72 + 80 + 96 + x = 6·88

                            (Multiplying by 6)

           430 + x = 528

                  x = 98

Check:

The sum of the six scores is

93 + 89 + 72 + 80 + 96 + 98 = 528.

The average of the six scores is 528/6, or 88. The value checks.

State:

The score on the sixth test must be 98.

49. First draw pictures.

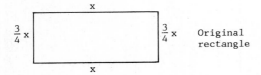

We let x represent the length.   Then $\frac{3}{4}$x represents the width.

The length of the new rectangle is 2 cm greater than the length of the original rectangle. The width of the new rectangle is 2 cm greater than the width of the original rectangle. We let x + 2 represent the length and $\frac{3}{4}$x + 2 represent the width.

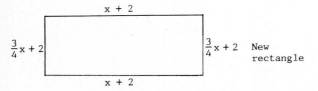

49. (continued)

The perimeter of the new rectangle is 50 cm.

Length + Length +   Width   +   Width   = 50

$(x + 2) + (x + 2) + (\frac{3}{4}x + 2) + (\frac{3}{4}x + 2) = 50$

$(x + 2) + (x + 2) + (\frac{3}{4}x + 2) + (\frac{3}{4}x + 2) = 50$

$$\frac{7}{2}x + 8 = 50$$

$$2(\frac{7}{2}x + 8) = 2(50)$$

$$7x + 16 = 100$$

$$7x = 100 - 16$$

$$7x = 84$$

$$x = \frac{84}{7}$$

$$x = 12$$

Possible dimensions of the original rectangle are x = 12 and $\frac{3}{4}$x = 9. When these dimensions are each increased by 2, the new dimensions are 14 and 11. The perimeter of the new rectangle is

14 + 14 + 11 + 11, or 50 cm. This checks.

The dimensions of the original rectangle are 12 cm by 9 cm.

55. First draw a picture.

               x + 6

x $\boxed{}$ x

               x + 6

We let x represent the width and x + 6 the length in inches.

The perimeter is 640 ft, or 7680 in.

x + x + x + 6 + x + 6 = 7680

          4x + 12 = 7680

              4x = 7668

               x = 1917

If the width, x, is 1917 in, the length, x + 6, is 1917 + 6, or 1923 in. The perimeter is 2·1917 + 2·1923, or 7680 in.

A = length × width

A = 1917 in × 1923 in = 3,686,391 in^2

                 or 25,599.9375 ft^2

Exercise Set 3.1

<u>1</u>. $x = 4$; $-5x + 2 = -5 \cdot 4 + 2$
$$= -20 + 2$$
$$= -18$$

<u>7</u>. $a = 18$; $0.4a^2 - 40a + 1039 = 0.4(18^2) - 40 \cdot 18 + 1039$
$$= 0.4(324) - 40 \cdot 18 + 1039$$
$$= 129.6 - 720 + 1039$$
$$= 1168.6 - 720$$
$$= 448.6$$

The average number of daily accidents involving 18 year-old drivers is approximately 449.

<u>13</u>. $x = -1$; $-3x^3 + 7x^2 - 3x - 2$
$$= -3(-1)^3 + 7(-1)^2 - 3(-1) - 2$$
$$= -3(-1) + 7(1) - 3(-1) - 2$$
$$= 3 + 7 + 3 - 2$$
$$= 13 - 2$$
$$= 11$$

<u>19</u>. $t = 0$ before the water begins draining.
$$400 - 200t + 25t^2$$
$$= 400 - 200 \cdot 0 + 25 \cdot 0^2$$
$$= 400 - 0 + 0$$
$$= 400$$

The amount of water in the tub before the water begins draining is 400 gallons.

<u>25</u>. $2x^4 + 5x - 7x - 3x^4$

Like terms: $2x^4$ and $-3x^4$

Like terms: $5x$ and $-7x$

<u>31</u>. $5x^3 + 6x^3 + 4$
$$= (5 + 6)x^3 + 4$$
$$= 11x^3 + 4$$

<u>37</u>. $\frac{1}{4}x^5 - 5 + \frac{1}{2}x^5 - 2x - 37$
$$= \left(\frac{1}{4} + \frac{1}{2}\right)x^5 - 2x + (-5 - 37)$$
$$= \frac{3}{4}x^5 - 2x - 42$$

<u>43</u>. $3x^2 + 2x - 2 + 3x^0$
$$= 3x^2 + 2x - 2 + 3 \cdot 1$$
$$= 3x^2 + 2x - 2 + 3$$
$$= 3x^2 + 2x + 1$$

Exercise Set 3.2

<u>1</u>. $x^5 + x + 6x^3 + 1 + 2x^2$
$$= x^5 + x^1 + 6x^3 + 1x^0 + 2x^2$$
$$= x^5 + 6x^3 + 2x^2 + x^1 + 1x^0$$
$$= x^5 + 6x^3 + 2x^2 + x + 1$$

<u>7</u>. $3x^4 - 5x^6 - 2x^4 + 6x^6$
$$= x^4 + x^6$$
$$= x^6 + x^4$$

<u>13</u>. $-x + \frac{3}{4} + 15x^4 - x - \frac{1}{2} - 3x^4$
$$= -2x + \frac{1}{4} + 12x^4$$
$$= 12x^4 - 2x + \frac{1}{4}$$

<u>19</u>. $-7x^3 + 6x^2 + 3x + 7 = -7x^3 + 6x^2 + 3x^1 + 7x^0$

The degree of $-7x^3$ is 3.
The degree of $6x^2$ is 2.
The degree of $3x$ is 1.
The degree of 7 is 0.

The degree of the polynomial is 3, the largest exponent.

<u>25</u>. $5x^2 + 3x + 3$

The coefficient of $5x^2$, the first term, is 5.
The coefficient of $3x$, the second term, is 3.
The coefficient of 3, the third term, is 3.

<u>31</u>. In the polynomial $x^3 - 27$, there are no x^2 or x terms. The x^2 term (or second degree term) and the x term (or first degree term) are missing.

<u>37</u>. The polynomial $x^2 - 10x + 25$ is a <u>trinomial</u> because it has only three terms.

43. The polynomial 40x is a <u>monomial</u> because it has only one term.

49. $(5m^5)^2 = 5^2m^{10} = 25m^{10}$

The degree of $25m^{10}$ is 10.

Exercise Set 3.3

1. $(3x + 2) + (-4x + 3)$

$= (3 - 4)x + (2 + 3)$

$= -x + 5$

7. $(-4x^4 + 6x^2 - 3x - 5) + (6x^3 + 5x + 9)$

$= -4x^4 + 6x^3 + 6x^2 + (-3 + 5)x + (-5 + 9)$

$= -4x^4 + 6x^3 + 6x^2 + 2x + 4$

13. $(9x^8 - 7x^4 + 2x^2 + 5) + (8x^7 + 4x^4 - 2x)$

$= 9x^8 + 8x^7 + (-7 + 4)x^4 + 2x^2 - 2x + 5$

$= 9x^8 + 8x^7 - 3x^4 + 2x^2 - 2x + 5$

19. $-3x^4 + 6x^2 + 2x - 1$

$\underline{\qquad - 3x^2 + 2x + 1}$

$-3x^4 + 3x^2 + 4x + 0$

$-3x^4 + 3x^2 + 4x$

25. $-\frac{1}{2}x^4 - \frac{3}{4}x^3 \qquad + 6x \qquad (x^2 = 1x^2)$

$\qquad \frac{1}{2}x^3 + x^2 + \frac{1}{4}x \qquad (1\frac{1}{2} = \frac{3}{2})$

$\underline{\frac{3}{4}x^4 \qquad + \frac{1}{2}x^2 + \frac{1}{2}x + \frac{1}{4}} \qquad (6\frac{3}{4} = \frac{27}{4})$

$\frac{1}{4}x^4 - \frac{1}{4}x^3 + \frac{3}{2}x^2 + \frac{27}{4}x + \frac{1}{4}$

31. $3x^5 - 6x^4 + 3x^3 \qquad - 1$

$\qquad 6x^4 - 4x^3 + 6x^2$

$3x^5 \qquad + 2x^3$

$\qquad -6x^4 \qquad - 7x^2$

$\underline{-5x^5 \qquad + 3x^3 \qquad + 2}$

$1x^5 - 6x^4 + 4x^3 - 1x^2 + 1$

$x^5 - 6x^4 + 4x^3 - x^2 + 1$

37. $0.15x^4 + 0.10x^3 - 0.90x^2 \qquad\qquad (-0.9 = -0.90)$

$\qquad -0.01x^3 + 0.01x^2 + x \qquad (15 = 15.00)$

$1.25x^4 \qquad + 0.11x^2 \qquad + 0.01$

$\qquad 0.27x^3 \qquad\qquad + 0.99$

$\underline{-0.35x^4 \qquad + 15.00x^2 \qquad - 0.03}$

$1.05x^4 + 0.36x^3 + 14.22x^2 + x + 0.97$

43.

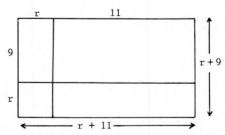

The area of the figure can be found by finding the sum of the areas of the four rectangles A, B, C, and D. The area of a rectangle is the product of the length and the width.

Area + Area + Area + Area
of A   of B   of C   of D

$= 9 \cdot r + 11 \cdot 9 + r \cdot r + 11 \cdot r$

$= 9r + 99 + r^2 + 11r$

$= r^2 + 20r + 99$

An algebraic expression for the area of the figure is $r^2 + 20r + 99$.

The length and width of the figure can be expressed as $r + 11$ and $r + 9$. The area of this figure (a rectangle) is the product of the length and width. An algebraic expression for the area is $(r + 11) \cdot (r + 9)$.

Both algebraic expressions $r^2 + 20r + 99$ and $(r + 11) \cdot (r + 9)$ represent the same area.

49. Let x represent the youngest brother's age <u>now</u>.

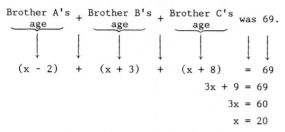

Brothers' Ages Now	Brothers' Ages Two Years Ago
A x	x - 2
B x + 5	x + 5 - 2 or x + 3
C x + 5 + 5, or x + 10	x + 10 - 2 or x + 8

The sum of their ages two years ago was 69.

$$\underbrace{\text{Brother A's}}_{\text{age}} + \underbrace{\text{Brother B's}}_{\text{age}} + \underbrace{\text{Brother C's}}_{\text{age}} \text{ was } 69.$$

$$(x - 2) \quad + \quad (x + 3) \quad + \quad (x + 8) \quad = \quad 69$$
$$3x + 9 = 69$$
$$3x = 60$$
$$x = 20$$

If the youngest brother is 20, the other two are 25 and 30. The ages are consecutive multiples of 5. Their ages two years ago were 18, 23, and 28. The sum of the ages two years ago, 18 + 23 + 28, was 69. These numbers check.

The brothers' ages now are 20, 25, and 30.

Exercise Set 3.4

1. Two equivalent expressions for the additive inverse of -5x are -(-5x) and 5x.

   -(-5x) = 5x

7. We change the sign of every term inside parentheses.

   -(3x - 7) = -3x + 7

13. $(5x^2 + 6) - (3x^2 - 8)$

    $= (5x^2 + 6) + (-3x^2 + 8)$

    $= 2x^2 + 14$

19. $(\frac{5}{8}x^3 - \frac{1}{4}x - \frac{1}{3}) - (-\frac{1}{8}x^3 + \frac{1}{4}x - \frac{1}{3})$

    $= (\frac{5}{8}x^3 - \frac{1}{4}x - \frac{1}{3}) + (\frac{1}{8}x^3 - \frac{1}{4}x + \frac{1}{3})$

    $= \frac{6}{8}x^3 - \frac{2}{4}x + 0$

    $= \frac{3}{4}x^3 - \frac{1}{2}x$

25. Subtract:

$$\begin{array}{l} x^4 \qquad\quad - 3x^2 + x + 1 \\ \underline{x^4 - 4x^3} \end{array}$$

Think: Add

$$\begin{array}{l} x^4 \qquad\quad - 3x^2 + x + 1 \\ \underline{-x^4 + 4x^3} \\ \qquad 4x^3 - 3x^2 + x + 1 \end{array}$$

Subtracting the polynomial $x^4 - 4x^3$ is the same as adding the polynomial $-x^4 + 4x^3$.

31. Subtract:

$$\begin{array}{l} x^5 \qquad\qquad\qquad - 1 \\ \underline{x^5 - x^4 + x^3 - x^2 + x - 1} \end{array}$$

Think: Add

$$\begin{array}{l} x^5 \qquad\qquad\qquad - 1 \\ \underline{-x^5 + x^4 - x^3 + x^2 - x + 1} \\ \qquad x^4 - x^3 + x^2 - x \end{array}$$

Subtracting the polynomial $x^5 - x^4 + x^3 - x^2 + x - 1$ is the same as adding the polynomial $-x^5 + x^4 - x^3 + x^2 - x + 1$.

37.

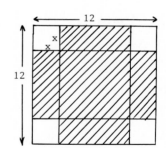

The area of the square is 12 · 12, or 144.

The area of each corner that is not colored is x · x, or x^2.

The sum of the areas of all four corners is $4x^2$.

$$\begin{array}{ccc} \text{Area of} & \text{Area} & \text{Area of} \\ \text{colored} = & \text{of} & - \text{four} \\ \text{section} & \text{square} & \text{corners} \\ \downarrow & \downarrow & \downarrow \\ \text{Area of} & & \\ \text{colored} = & 144 & - \quad 4x^2 \\ \text{section} & & \end{array}$$

A polynomial for the colored area is $144 - 4x^2$.

43. $(-8y^2 - 4) - (3y + 6) - (2y^2 - y)$

   $= -8y^2 - 4 - 3y - 6 - 2y^2 + y$

   $= -10y^2 - 2y - 10$

Exercise Set 3.5

1. $(6x^2)(7)$

   $= (6 \cdot 7)x^2$

   $= 42x^2$

7. $(3x^4)(2x^2)$

   $= (3 \cdot 2)(x^4 \cdot x^2)$

   $= 6x^{4+2}$

   $= 6x^6$

13. $(-\frac{1}{5}x^3)(-\frac{1}{3}x)$

   $= (-\frac{1}{5})(-\frac{1}{3})(x^3 \cdot x)$

   $= \frac{1}{15}x^4$

19. $(5x^2)(-2x + 1)$

   $= (5x^2)(-2x) + (5x^2)(1)$

   $= -10x^3 + 5x^2$

25. $(5x^6)(4x^{32} - 10x^{19} + 5x^8)$

   $= (5x^6)(4x^{32}) - (5x^6)(10x^{19}) + (5x^6)(5x^8)$

   $= 20x^{38} - 50x^{25} + 25x^{14}$

31. $(x + 6)(x - 2)$

   $= (x + 6)(x) - (x + 6)(2)$

   $= (x \cdot x + 6 \cdot x) - (x \cdot 2 + 6 \cdot 2)$

   $= x^2 + 6x - 2x - 12$

   $= x^2 + 4x - 12$

37. $(3 + x)(6 + 2x)$

   $= (3 + x)(6) + (3 + x)(2x)$

   $= 3 \cdot 6 + x \cdot 6 + 3 \cdot 2x + x \cdot 2x$

   $= 18 + 6x + 6x + 2x^2$

   $= 18 + 12x + 2x^2$

43. $(x + \frac{4}{3})(x + \frac{3}{2})$

   $= (x + \frac{4}{3})(x) + (x + \frac{4}{3})(\frac{3}{2})$

   $= x \cdot x + \frac{4}{3} \cdot x + x \cdot \frac{3}{2} + \frac{4}{3} \cdot \frac{3}{2}$

   $= x^2 + \frac{4}{3}x + \frac{3}{2}x + 2$

   $= x^2 + (\frac{4}{3} + \frac{3}{2})x + 2$

   $= x^2 + (\frac{8}{6} + \frac{9}{6})x + 2$

   $= x^2 + \frac{17}{6}x + 2$

49.
$$
\begin{array}{l}
y^2 + 6y + 1 \\
\underline{3y^2 - 3} \\
3y^4 + 18y^3 + 3y^2 \qquad \text{(Multiplying by } 3y^2) \\
\underline{\qquad - 3y^2 - 18y - 3} \quad \text{(Multiplying by -3)} \\
3y^4 + 18y^3 \qquad - 18y - 3 \quad \text{(Adding)}
\end{array}
$$

55.
$$
\begin{array}{l}
1 - x + x^2 \\
\underline{1 - x + x^2} \\
1 - x + x^2 \qquad\qquad \text{(Multiplying by 1)} \\
\quad - x + x^2 - x^3 \qquad \text{(Multiplying by -x)} \\
\underline{\qquad\quad x^2 - x^3 + x^4} \quad \text{(Multiplying by } x^2) \\
1 - 2x + 3x^2 - 2x^3 + x^4 \quad \text{(Adding)}
\end{array}
$$

61.
$$
\begin{array}{l}
3x^2 - 8x + 1 \\
\underline{-2x^2 - 4x + 2} \\
-6x^4 + 16x^3 - 2x^2 \\
\qquad\quad - 12x^3 + 32x^2 - 4x \\
\underline{\qquad\qquad\qquad 6x^2 - 16x + 2} \\
-6x^4 + 4x^3 + 36x^2 - 20x + 2
\end{array}
$$

67.
$$
\begin{array}{l}
x^3 - x^2 - x + 4 \\
\underline{x + 4} \\
x^4 - x^3 - x^2 + 4x \\
\underline{\quad 4x^3 - 4x^2 - 4x + 16} \\
x^4 + 3x^3 - 5x^2 \qquad + 16
\end{array}
$$

73.

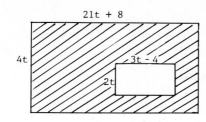

The area of a rectangle is the product of the length and width.

The area of the large rectangle is $(4t)(21t + 8)$.

The area of the small rectangle is $(2t)(3t - 4)$.

$$\underset{\text{Area}}{\text{Shaded}} = \underset{\text{rectangle}}{\text{Area of large}} - \underset{\text{rectangle}}{\text{Area of small}}$$

$$\underset{\text{Area}}{\text{Shaded}} = (4t)(21t + 8) - (2t)(3t - 4)$$

$$= (84t^2 + 32t) - (6t^2 - 8t)$$

$$= 84t^2 + 32t - 6t^2 + 8t$$

$$= 78t^2 + 40t$$

79. a) $(x - 2)(x - 7) + (x - 2)(x - 7)$

$$= (x^2 - 9x + 14) + (x^2 - 9x + 14)$$

$$= 2x^2 - 18x + 28$$

  b) $(x - 6)(x - 2) - (x - 6)(x - 2)$

$$= (x^2 - 8x + 12) - (x^2 - 8x + 12)$$

$$= x^2 - 8x + 12 - x^2 + 8x - 12$$

$$= 0$$

Exercise Set 3.6

1. $4x(x + 1) = 4x^2 + 4x$

7. $3x(2x^2 - 6x + 1) = 6x^3 - 18x^2 + 3x$

13. $(y + 2)(y - 3)$
$$\quad \text{F}\qquad \text{O}\qquad \text{I}\qquad \text{L}$$
$$= y \cdot y + y \cdot (-3) + 2 \cdot y + 2 \cdot (-3)$$
$$= y^2 - 3y + 2y - 6$$
$$= y^2 - y - 6$$

19. $(3t - 1)(3t + 1)$
$$\quad \text{F}\qquad \text{O}\qquad \text{I}\qquad \text{L}$$
$$= 3t \cdot 3t + 3t \cdot 1 + (-1) \cdot 3t + (-1) \cdot 1$$
$$= 9t^2 + 3t - 3t - 1$$
$$= 9t^2 - 1$$

25. $(x - 0.1)(x + 0.1)$
$$\quad \text{F}\qquad \text{O}\qquad \text{I}\qquad \text{L}$$
$$= x \cdot x + x \cdot (0.1) + (-0.1) \cdot x + (-0.1)(0.1)$$
$$= x^2 + 0.1x - 0.1x - 0.01$$
$$= x^2 - 0.01$$

31. $(a + 7)(a + 7)$
$$\quad \text{F}\quad \text{O}\quad \text{I}\quad \text{L}$$
$$= a^2 + 7a + 7a + 49$$
$$= a^2 + 14a + 49$$

37. $(x^2 - 2)(x - 1)$
$$\quad \text{F}\quad \text{O}\quad \text{I}\quad \text{L}$$
$$= x^3 - x^2 - 2x + 2$$

43. $(8x^3 + 1)(x^3 + 8)$
$$= 8x^6 + 64x^3 + x^3 + 8$$
$$= 8x^6 + 65x^3 + 8$$

49. $4y(y + 5)(2y + 8)$
$$= 4y(2y^2 + 8y + 10y + 40)$$
$$= 4y(2y^2 + 18y + 40)$$
$$= 8y^3 + 72y^2 + 160y$$

55. $(x + 1)(x + 2) = (x + 3)(x + 4)$
$$x^2 + 2x + x + 2 = x^2 + 4x + 3x + 12$$
$$x^2 + 3x + 2 = x^2 + 7x + 12$$
$$3x + 2 = 7x + 12$$
$$2 - 12 = 7x - 3x$$
$$-10 = 4x$$
$$-\frac{10}{4} = x$$
$$-\frac{5}{2} = x$$

61.

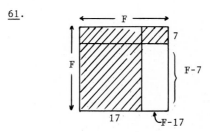

The area of the entire figure is $F \cdot F$, or F^2.

The area of the section not shaded is $(F - 7)(F - 17)$

61. (continued)

$$\underbrace{\text{Area of shaded}}_{\downarrow} = \underbrace{\text{Area of entire}}_{\downarrow} - \underbrace{\text{Area of section}}_{\downarrow}$$
region   figure   not shaded.

Area of shaded
region $= F^2 - (F - 7)(F - 17)$

$= F^2 - (F^2 - 24F + 119)$

$= F^2 - F^2 + 24F - 119$

$= 24F - 119$

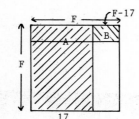

$$\underbrace{\text{Area of shaded}}_{\downarrow} = \underbrace{\text{Area of}}_{\downarrow} + \underbrace{\text{Area of}}_{\downarrow}$$
region   A   B

Area of shaded
region $= 17F + 7(F - 17)$

$= 17F + 7F - 119$

$= 24F - 119$

Exercise Set 3.7

1. $(x + 4)(x - 4)$

$= x^2 - 4^2$

$= x^2 - 16$

7. $(2x^2 + 3)(2x^2 - 3)$

$= (2x^2)^2 - 3^2$

$= 4x^4 - 9$

13. $(x^4 + 3x)(x^4 - 3x)$

$= (x^4)^2 - (3x)^2$

$= x^8 - 9x^2$

19. $(x + 2)^2$

$= x^2 + 2 \cdot x \cdot 2 + 2^2$

$= x^2 + 4x + 4$

25. $(3 + x)^2$

$= 3^2 + 2 \cdot 3 \cdot x + x^2$

$= 9 + 6x + x^2$

31. $(5 + 6t^2)^2$

$= 5^2 + 2 \cdot 5 \cdot 6t^2 + (6t^2)^2$

$= 25 + 60t^2 + 36t^4$

37. $(2x^2 - \frac{1}{2})(2x^2 - \frac{1}{2})$

$= (2x^2)^2 - 2 \cdot 2x^2 \cdot \frac{1}{2} + (\frac{1}{2})^2$

(Using Method 2)

$= 4x^4 - 2x^2 + \frac{1}{4}$

43. $(6x^4 + 4)^2$

$= (6x^4)^2 + 2 \cdot 6x^4 \cdot 4 + 4^2$

(Using Method 1)

$= 36x^8 + 48x^4 + 16$

49. $(67.58x + 3.225)^2$

$= (67.58x)^2 + 2(67.58x)(3.225) + (3.225)^2$

$= 4567.0564x^2 + 435.891x + 10.400625$

55. $[(x + 3) + 2]^2$

$= (x + 3)^2 + 2 \cdot (x + 3) \cdot 2 + 2^2$

$= (x^2 + 6x + 9) + (4x + 12) + 4$

$= x^2 + 10x + 25$

61. a)

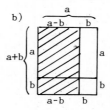

The area of the entire rectangle is $a(a + b)$, or $a^2 + ab$.

b)

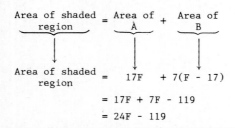

The sum of the area of the two uncolored rectangles is $ab + b \cdot b$, or $ab + b^2$.

c)   Part a) -   Part b)

$(a^2 + ab) - (ab + b^2)$

$= a^2 + ab - ab - b^2$

$= a^2 - b^2$

d) The area of the colored region is

$a(a - b) + b(a - b)$

$= a^2 - ab + ba - b^2$

$= a^2 - b^2$

The area of the entire figure minus the area of the uncolored region is equal to the area of the colored region.

Exercise Set 4.1

1. $6x^3 = (6x)(x^2) = (3x^2)(2x) = (2x^2)(3x)$

   Answers may vary.

7. $x^2 - 4x$
   $= x \cdot x - x \cdot 4$
   $= x(x - 4)$

13. $8x^4 - 24x^2$
    $= 8x^2 \cdot x^2 - 8x^2 \cdot 3$
    $= 8x^2(x^2 - 3)$

19. $6x^4 - 10x^3 + 3x^2$
    $= x^2 \cdot 6x^2 - x^2 \cdot 10x + x^2 \cdot 3$
    $= x^2(6x^2 - 10x + 3)$

25. $1.6x^4 - 2.4x^3 + 3.2x^2 + 6.4x$
    $= 0.8x(2x^3) - 0.8x(3x^2) + 0.8x(4x) + 0.8x(8)$
    $= 0.8x(2x^3 - 3x^2 + 4x + 8)$

31. $x^2 - 4x - x + 4$
    $= (x^2 - 4x) + (-x + 4)$
    $= x(x - 4) - 1(x - 4)$
    $= (x - 1)(x - 4)$

37. $35x^2 - 40x + 21x - 24$
    $= (35x^2 - 40x) + (21x - 24)$
    $= 5x(7x - 8) + 3(7x - 8)$
    $= (5x + 3)(7x - 8)$

43. $2x^3 + 6x^2 + x + 3$
    $= (2x^3 + 6x^2) + (x + 3)$
    $= 2x^2(x + 3) + 1(x + 3)$
    $= (2x^2 + 1)(x + 3)$

49. $x^3 + 8x^2 - 3x - 24$
    $= (x^3 + 8x^2) + (-3x - 24)$
    $= x^2(x + 8) - 3(x + 8)$
    $= (x^2 - 3)(x + 8)$

55. $7a = 7 \cdot \underline{a}$, $a = 1 \cdot \underline{a}$
    a is a common factor of 7a and a.
    7a and a __are not__ relatively prime.

61. $4x^5 + 6x^3 + 6x^2 + 9$

    $= (4x^5 + 6x^2) + (6x^3 + 9)$
    $= 2x^2(2x^3 + 3) + 3(2x^3 + 3)$
    $= (2x^2 + 3)(2x^3 + 3)$

Exercise Set 4.2

1. $x^2 - 4$

   The first expression is a square: x^2
   The second expression is a square: $4 = 2^2$
   There is a minus sign between x^2 and 4.
   $x^2 - 4$ is a difference of squares.

7. $16x^2 - 25$

   The first expression is a square: $16x^2 = (4x)^2$
   The second expression is a square: $25 = 5^2$
   There is a minus sign between $16x^2$ and 25.
   $16x^2 - 25$ is a difference of squares.

13. $x^2 - 1$
    $= x^2 - 1^2$
    $= (x - 1)(x + 1)$

19. $24x^2 - 54$
    $= 6(4x^2 - 9)$
    $= 6[(2x)^2 - 3^2]$
    $= 6(2x - 3)(2x + 3)$

25. $25a^4 - 9$
    $= (5a^2)^2 - 3^2$
    $= (5a^2 - 3)(5a^2 + 3)$

31. $x^4 - 16$
    $= (x^2 + 4)(x^2 - 4)$
    $= (x^2 + 4)(x + 2)(x - 2)$

37. $x^8 - 81$
    $= (x^4 + 9)(x^4 - 9)$
    $= (x^4 + 9)(x^2 + 3)(x^2 - 3)$

43. $1 - a^4$
    $= (1 + a^2)(1 - a^2)$
    $= (1 + a^2)(1 + a)(1 - a)$

49. $18x^3 - \frac{8}{25}x$

$= 2x(9x^2 - \frac{4}{25})$

$= 2x[(3x)^2 - (\frac{2}{5})^2]$

$= 2x(3x - \frac{2}{5})(3x + \frac{2}{5})$

55. $(x + 3)^2 - 9$

$= (x + 3)^2 - 3^2$

$= [(x + 3) - 3][(x + 3) + 3]$

$= x(x + 6)$

61. $x^2 - (\frac{1}{x})^2$ 　　or　 $x^2 - (\frac{1}{x})^2$

$= (x - \frac{1}{x})(x + \frac{1}{x})$ 　　$= x^2 - \frac{1}{x^2}$

$= x^2 \cdot \frac{x^2}{x^2} - \frac{1}{x^2}$

$= \frac{x^4}{x^2} - \frac{1}{x^2}$

$= \frac{x^4 - 1}{x^2}$

$= \frac{1}{x^2}(x^4 - 1)$

$= \frac{1}{x^2}(x^2 + 1)(x^2 - 1)$

$= \frac{1}{x^2}(x^2 + 1)(x + 1)(x - 1)$

67. $-25y^2 - 49$ cannot be factored except for removing the common constant factor, -1.

$-25y^2 - 49 = -1(25y^2 + 49)$

The polynomial $-25y^2 - 49$ is <u>irreducible</u>.

The coefficient of the leading term is not 1.
The irreducible polynomial is not prime.

Exercise Set 4.3

1. $x^2 - 14x + 49$

x^2 and 49 are squares. There is no minus sign before x^2 or 49. If we multiply the square roots, x and 7, and double the product, we get 14x, the additive inverse of the remaining term, $-14x$. Thus $x^2 - 14x + 49$ <u>is</u> a trinomial square.

7. $8x^2 + 40x + 25$

Two of the terms must be squares. This polynomial <u>is not</u> a trinomial square because only one term, 25, is a square.

13. $x^2 + 14x + 49 = x^2 + 2 \cdot x \cdot 7 + 7^2 = (x + 7)^2$

$A^2 + 2 \quad A \quad B + B^2 = (A + B)^2$

19. $y^2 + 6y + 9 = y^2 + 2 \cdot y \cdot 3 + 3^2 = (y + 3)^2$

$A^2 + 2 \quad A \quad B + B^2 = (A + B)^2$

25. $12x^2 + 36x + 27$

$= 3(4x^2 + 12x + 9)$

$= 3[(2x)^2 + 2 \cdot 2x \cdot 3 + 3^2]$

$= 3(2x + 3)^2$

31. $y^6 - 16y^3 + 64$

$= (y^3)^2 - 2 \cdot y^3 \cdot 8 + 8^2$

$= (y^3 - 8)^2$

37. $\frac{1}{9}a^2 + \frac{1}{3}a + \frac{1}{4}$

$= (\frac{1}{3}a)^2 + 2 \cdot \frac{1}{3}a \cdot \frac{1}{2} + (\frac{1}{2})^2$

$= (\frac{1}{3}a + \frac{1}{2})^2$

43. $4x^2 + 9$ cannot be factored. This polynomial is a sum of two squares.

49. $8.1x^2 - 6.4$

$= \frac{1}{10}(81x^2 - 64)$

$= \frac{1}{10}(9x - 8)(9x + 8)$

55. $49(x + 1)^2 - 42(x + 1) + 9$

Think: Let $y = x + 1$. $49y^2 - 42y + 9 = (7y - 3)^2$

$49(x + 1)^2 - 42(x + 1) + 9 = [7(x + 1) - 3]^2$

$= (7x + 7 - 3)^2$

$= (7x + 4)^2$

61. $x^{2n} + 10x^n + 25$

$= (x^n)^2 + 2 \cdot x^n \cdot 5 + 5^2$

$= (x^n + 5)^2$

67. $x^2 + a^2x + a^2 = (x + a)^2$

$x^2 + a^2x + a^2 = x^2 + 2ax + a^2$

$a^2x = 2ax$

$a^2 = 2a$

Either $a = 0$ or $a = 2$.

Exercise Set 4.4

1. $x^2 + 8x + 15$

The leading term is x^2. The first term of each binomial factor is x: $(x + \_)(x + \_)$. Since the constant term is positive, and the coefficient of the middle term is positive, we look for a factorization of 15 in which both factors are positive. Their sum must be 8.

Pairs of factors	Sums of factors
1, 15	16
* 3, 5	8

The numbers we want are 3 and 5.

$x^2 + 8x + 15 = (x + 3)(x + 5)$

7. $x^2 + 9x + 14$

Since the constant term is positive and the coefficient of the middle term is positive, we look for a factorization of 14 in which both factors are positive. Their sum must be 9.

Pairs of factors	Sums of factors
1, 14	15
* 2, 7	9

The numbers we want are 2 and 7.

$x^2 + 9x + 14 = (x + 2)(x + 7)$.

13. $d^2 - 7d + 10$

Since the constant term is positive and the coefficient of the middle term is negative, we look for a factorization of 10 in which both factors are negative. Their sum must be -7.

Pairs of factors	Sums of factors
-1, -10	-11
* -2, -5	-7

The numbers we want are -2 and -5.

$d^2 - 7d + 10 = (d - 2)(d - 5)$.

19. $x^2 - 7x - 18$

Since the constant term is negative, we look for a factorization of -18 in which one factor is positive and one factor is negative. Their sum must be -7, the coefficient of the middle term.

Pairs of factors	Sums of factors
-1, 18	17
1, -18	-17
-2, 9	7
* 2, -9	-7
-3, 6	3
3, -6	-3

The numbers we want are 2 and -9.

$x^2 - 7x - 18 = (x + 2)(x - 9)$.

25. $x^2 - 2x - 99$

Since the constant term is negative, we look for a factorization of -99 in which one factor is positive and one factor is negative. Their sum must be -2, the coefficient of the middle term.

Pairs of factors	Sums of factors
-1, 99	98
1, -99	-98
-3, 33	30
3, -33	-30
-9, 11	2
* 9, -11	-2

The numbers we want are 9 and -11.

$x^2 - 2x - 99 = (x + 9)(x - 11)$.

31. $x^2 + 20x + 100$

We look for two factors, both positive, whose product is 100 and whose sum is 20.

They are 10 and 10. $10 \cdot 10 = 100$ and $10 + 10 = 20$.

$x^2 + 20x + 100 = (x + 10)(x + 10)$, or $(x + 10)^2$.

37. $x^2 - 25x + 144$

We look for two factors, both negative, whose product is 144 and whose sum is -25.

They are -9 and -16. $-9 \cdot (-16) = 144$ and $-9 + (-16) = -25$.

$x^2 - 25x + 144 = (x - 9)(x - 16)$

43. First write the polynomial in descending order and factor out -1.

$$108 - 3x - x^2$$

$$= -x^2 - 3x + 108$$

$$= -1(x^2 + 3x - 108)$$

Now we factor the polynomial $x^2 + 3x - 108$. We look for two factors, one positive and one negative, whose product is -108 and whose sum is 3.

They are -9 and 12. $-9 \cdot 12 = -108$ and $-9 + 12 = 3$.

$$x^2 + 3x - 108 = (x - 9)(x + 12)$$

The final answer must include the -1 which was factored out above.

$-x^2 - 3x + 108 = -1(x - 9)(x + 12)$.
$-1(x - 9)(x + 12)$ can also be expressed as $(9 - x)(12 + x)$.

49. $x^2 + \dfrac{30}{7}x - \dfrac{25}{7}$

We look for two factors, one positive and one negative, whose product is $-\dfrac{25}{7}$ and whose sum is $\dfrac{30}{7}$.

They are 5 and $-\dfrac{5}{7}$.

$5 \cdot (-\dfrac{5}{7}) = -\dfrac{25}{7}$ and $5 + (-\dfrac{5}{7}) = \dfrac{35}{7} + (-\dfrac{5}{7}) = \dfrac{30}{7}$.

$x^2 + \dfrac{30}{7}x - \dfrac{25}{7} = (x + 5)(x - \dfrac{5}{7})$

Exercise Set 4.5

1. $2x^2 - 7x - 4$

   a) First look for a common factor. There is none (other than 1).

   b) Multiply the leading coefficient and the constant, 2 and -4: $2(-4) = -8$.

   c) Try to factor -8 so that the sum of the factors is -7.

Pairs of factors	Sums of factors
-1, 8	7
* 1, -8	-7
-2, 4	2
2, -4	-2

   d) Split the middle term: $-7x = 1x - 8x$

   e) Factor by grouping:

   $$2x^2 - 7x - 4 = 2x^2 + x - 8x - 4$$
   $$= (2x^2 + x) - (8x + 4)$$
   $$= x(2x + 1) - 4(2x + 1)$$
   $$= (x - 4)(2x + 1)$$

7. $3x^2 + 4x + 1$

   a) First look for a common factor. There is none (other than 1).

   b) Multiply the leading coefficient and the constant, 3 and 1: $3 \cdot 1 = 3$.

   c) Try to factor 3 so that the sum of the factors is 4. The numbers we want are 1 and 3: $1 \cdot 3 = 3$ and $1 + 3 = 4$.

   d) Split the middle term: $4x = 1x + 3x$

   e) Factor by grouping:

   $$3x^2 + 4x + 1 = 3x^2 + x + 3x + 1$$
   $$= x(3x + 1) + 1(3x + 1)$$
   $$= (x + 1)(3x + 1)$$

13. $9x^2 + 18x - 16$

   a) First look for a common factor. There is none (other than 1).

   b) Multiply the leading coefficient and the constant, 9 and -16: $9(-16) = -144$.

   c) Try to factor -144 so that the sum of the factors is 18.

Pairs of factors	Sums of factors
-1, 144	143
1, -144	-143
-2, 72	70
2, -72	-70
-3, 48	45
3, -48	-45
-4, 36	32
4, -36	-32
* -6, 24	18
6, -24	-18
-8, 18	10
8, -18	-10
-9, 16	7
9, -16	-7
-12, 12	0

   d) Split the middle term: $18x = -6x + 24x$

   e) Factor by grouping:

   $$9x^2 + 18x - 16 = 9x^2 - 6x + 24x - 16$$
   $$= 3x(3x - 2) + 8(3x - 2)$$
   $$= (3x + 8)(3x - 2)$$

19. $14x^2 + 19x - 3$

   a) First look for a common factor. There is none (other than 1).

   b) Multiply the leading coefficient and the constant, 14 and -3: $14(-3) = -42$.

19. (continued)

c) Try to factor -42 so that the sum of the factors is 19.

Pairs of factors	Sums of factors
-1, 42	41
1, -42	-41
* -2, 21	19
2, -21	-19
-3, 14	11
3, -14	-11
-6, 7	1
6, -7	-1

d) Split the middle term: $19x = -2x + 21x$

e) Factor by grouping:

$$14x^2 + 19x - 3 = 14x^2 - 2x + 21x - 3$$
$$= 2x(7x - 1) + 3(7x - 1)$$
$$= (2x + 3)(7x - 1)$$

25. $24x^2 + 47x - 2$

a) First look for a common factor. There is none (other than 1).

b) Multiply the leading coefficient and the constant, 24 and -2: $24(-2) = -48$.

c) Try to factor -48 so that the sum of the factors is 47. The numbers we want are 48 and -1: $48(-1) = -48$ and $48 + (-1) = 47$.

d) Split the middle term: $47x = 48x - 1x$

e) Factor by grouping:

$$24x^2 + 47x - 2 = 24x^2 + 48x - x - 2$$
$$= 24x(x + 2) - 1(x + 2)$$
$$= (24x - 1)(x + 2)$$

31. $12x^2 + 28x - 24$

Method 1:

a) We first factor out the common factor, 4.

$$12x^2 + 28x - 24 = 4(3x^2 + 7x - 6)$$

b) Now we factor the trinomial $3x^2 + 7x - 6$. Multiply the leading coefficient and the constant, 3 and -6: $3(-6) = -18$.

c) Try to factor -18 so that the sum of the factors is 7.

Pairs of factors	Sums of factors
-1, 18	17
1, -18	-17
* -2, 9	7
2, -9	-7
-3, 6	3
3, -6	-3

31. (continued)

d) Split the middle term: $7x = -2x + 9x$

e) Factor by grouping:

$$3x^2 + 7x - 6 = 3x^2 - 2x + 9x - 6$$
$$= x(3x - 2) + 3(3x - 2)$$
$$= (x + 3)(3x - 2)$$

We must include the common factor to get a factorization of the original trinomial.

$$12x^2 + 28x - 24 = 4(x + 3)(3x - 2)$$

Method 2:

$$12x^2 + 28x - 24$$

First we look for a common factor. The number 4 is a common factor, so we factor it out.

$$4(3x^2 + 7x - 6)$$

Next we factor the trinomial $3x^2 + 7x - 6$. We look for pairs of numbers whose product is 3. It is common practice that both of these factors should be positive. Thus, the only pair to consider is 1 and 3.

We have this possibility:

$$(x \quad)(3x \quad)$$

Next we look for pairs of numbers whose product is -6. These are:

-1, 6	2, -3
1, -6	-2, 3

Then we list possibilities for factorization using these pairs of numbers.

(x - 1)(3x + 6)	(x + 2)(3x - 3)
(x + 6)(3x - 1)	(x - 3)(3x + 2)
(x + 1)(3x - 6)	(x - 2)(3x + 3)
(x - 6)(3x + 1)	(x + 3)(3x - 2)

We multiply and find that the desired factorizarion is $(x + 3)(3x - 2)$. The complete factorization is $4(x + 3)(3x - 2)$.

37. $3x^2 - 4x + 1$

a) First look for a common factor. There is none (other than 1).

b) Multiply the leading coefficient and the constant, 3 and 1: $3 \cdot 1 = 3$.

c) Try to factor 3 so that the sum of the factors is -4. The numbers we want are -1 and -3: $-1 \cdot (-3) = 3$ and $-1 + (-3) = -4$.

d) Split the middle term: $-4x = -1x - 3x$

e) Factor by grouping:

$$3x^2 - 4x + 1 = 3x^2 - x - 3x + 1$$
$$= x(3x - 1) - 1(3x - 1)$$
$$= (x - 1)(3x - 1)$$

43. $9x^2 - 18x - 16$

    a) First look for a common factor. There is none (other than 1).

    b) Multiply the leading coefficient and the constant, 9 and -16: $9(-16) = -144$.

    c) Try to factor -144 so that the sum of the factors is -18. From the table in Exercise 13 we find that the numbers we want are 6 and -24: $6(-24) = -144$ and $6 + (-24) = -18$.

    d) Split the middle term: $-18x = 6x - 24x$

    e) Factor by grouping:

$$9x^2 - 18x - 16 = 9x^2 + 6x - 24x - 16$$
$$= 3x(3x + 2) - 8(3x + 2)$$
$$= (3x - 8)(3x + 2)$$

49. $14x^4 + 19x^3 - 3x^2$

We first factor out the common factor, x^2.

$$14x^4 + 19x^3 - 3x^2 = x^2(14x^2 + 19x - 3)$$

In Exercise 19 we factored $14x^2 + 19x - 3$.

$$14x^2 + 19x - 3 = (2x + 3)(7x - 1)$$

We must include the common factor to get a factorization of the original trinomial.

$$14x^4 + 19x^3 - 3x^2 = x^2(2x + 3)(7x - 1)$$

55. $6 - 13x + 6x^2$

    a) First look for a common factor. There is none (other than 1).

    b) Multiply the leading coefficient and the constant, 6 and 6: $6 \cdot 6 = 36$.

    c) Try to factor 36 so that the sum of the factors is -13.

Pairs of factors	Sums of factors
1, 36	37
- 1, -36	-37
2, 18	20
-2, -18	-20
3, 12	15
- 3, -12	-15
4, 9	13
- 4, -9	-13*
6, 6	12
- 6, -6	-12

    d) Split the middle term: $-4x - 9x$

    e) Factor by grouping:

$$6 - 4x - 9x + 6x^2 = 2(3 - 2x) - 3x(3 - 2x)$$
$$= (2 - 3x)(3 - 2x)$$

61. $x^2 + 13x - 12$

We first look for a common factor. There is none (other than 1). We try to factor -12 so that the sum of the factors is 13.

Pairs of factors	Sums of factors
-1, 12	11
1, -12	-11
-2, 6	4
2, -6	-4
-3, 4	1
3, -4	-1

The sum of the factors is never 13.

$x^2 + 13x - 12$ cannot be factored.

67. $x^{2n+1} - 2x^{n+1} + x$

$$= x^{2n} \cdot x^1 - 2 \cdot x^n \cdot x^1 + x^1$$
$$= x(x^{2n} - 2x^n + 1)$$
$$= x(x^n - 1)(x^n - 1)$$
$$\text{or } x(x^n - 1)^2$$

Exercise Set 4.6

1. $2x^2 - 128$

$$= 2(x^2 - 64)$$
$$= 2(x - 8)(x + 8)$$

7. $x^3 + 24x^2 + 144x$

$$= x(x^2 + 24x + 144)$$
$$= x(x + 12)^2$$

13. $20x^3 - 4x^2 - 72x$

$$= 4x(5x^2 - x - 18)$$
$$= 4x(5x + 9)(x - 2)$$

19. $x^5 - 14x^4 + 49x^3$

$$= x^3(x^2 - 14x + 49)$$
$$= x^3(x - 7)^2$$

25. $4x^4 - 64$

$$= 4(x^4 - 16)$$
$$= 4(x^2 + 4)(x^2 - 4)$$
$$= 4(x^2 + 4)(x + 2)(x - 2)$$

31. $36a^2 - 15a + \frac{25}{16}$

$$= (6a)^2 - 2 \cdot 6a \cdot \frac{5}{4} + \left(\frac{5}{4}\right)^2$$
$$= \left(6a - \frac{5}{4}\right)^2$$

37. $5x^2 + 13x + 7.2$

There is no common factor (other than 1).
Multiply the leading coefficient and the constant,
5 and 7.2: $5(7.2) = 36$. Try to factor 36 so that
the sum of the factors is 13. The numbers we want
are 4 and 9: $4 \cdot 9 = 36$ and $4 + 9 = 13$. Split the
middle term and factor by grouping.

$$5x^2 + 13x + 7.2 = 5x^2 + 4x + 9x + \frac{36}{5}$$

$$= x(5x + 4) + \frac{9}{5}(5x + 4)$$

$$= (x + \frac{9}{5})(5x + 4)$$

$$= (x + 1.8)(5x + 4)$$

43. $x^4 - 7x^2 - 18$

Try to factor -18 so that the sum of the factors
is -7. the numbers we want are -9 and 2:
$-9 \cdot 2 = -18$ and $-9 + 2 = -7$. Split the middle
term and factor by grouping.

$$x^4 - 7x^2 - 18 = x^4 - 9x^2 + 2x^2 - 18$$

$$= x^2(x^2 - 9) + 2(x^2 - 9)$$

$$= (x^2 + 2)(x^2 - 9)$$

$$= (x^2 + 2)(x + 3)(x - 3)$$

49. $a^4 - 81$

$$= (a^2 + 9)(a^2 - 9)$$

$$= (a^2 + 9)(a + 3)(a - 3)$$

Exercise Set 4.7

1. $(x + 8)(x + 6) = 0$

$x + 8 = 0$ or $x + 6 = 0$

$x = -8$ or $x = -6$

Check:

For -8   For -6

$(x + 8)(x + 6) = 0$		$(x + 8)(x + 6) = 0$	
$(-8 + 8)(-8 + 6)$	0	$(-6 + 8)(-6 + 6)$	0
$0 \cdot (-2)$		$2 \cdot 0$	
0		0	

The solutions are -8 and -6.

7. $x(x + 5) = 0$

$x = 0$ or $x + 5 = 0$

$x = 0$ or $x = -5$

The solutions are 0 and -5.

13. $(2x + 5)(x + 4) = 0$

$2x + 5 = 0$ or $x + 4 = 0$

$2x = -5$ or $x = -4$

$x = -\frac{5}{2}$ or $x = -4$

The solutions are $-\frac{5}{2}$ and -4.

19. $(7x - 28)(28x - 7) = 0$

$7x - 28 = 0$ or $28x - 7 = 0$

$7x = 28$ or $28x = 7$

$x = 4$ or $x = \frac{7}{28} = \frac{1}{4}$

The solutions are 4 and $\frac{1}{4}$.

25. $(\frac{1}{3} - 3x)(\frac{1}{5} - 2x) = 0$

$\frac{1}{3} - 3x = 0$ or $\frac{1}{5} - 2x = 0$

$\frac{1}{3} = 3x$ or $\frac{1}{5} = 2x$

$\frac{1}{9} = x$ or $\frac{1}{10} = x$

The solutions are $\frac{1}{9}$ and $\frac{1}{10}$.

31. $9x(3x - 2)(2x - 1) = 0$

$9x = 0$ or $3x - 2 = 0$ or $2x - 1 = 0$

$x = 0$ or $3x = 2$ or $2x = 1$

$x = 0$ or $x = \frac{2}{3}$ or $x = \frac{1}{2}$

The solutions are 0, $\frac{2}{3}$, and $\frac{1}{2}$.

37. $x^2 - 8x + 15 = 0$

$(x - 5)(x - 3) = 0$

$x - 5 = 0$ or $x - 3 = 0$

$x = 5$ or $x = 3$

The solutions are 5 and 3.

43. $x^2 = 16$

$x^2 - 16 = 0$

$(x - 4)(x + 4) = 0$

$x - 4 = 0$ or $x + 4 = 0$

$x = 4$ or $x = -4$

The solutions are 4 and -4.

49.
$$x^2 + 16 = 8x$$
$$x^2 - 8x + 16 = 0$$
$$(x - 4)(x - 4) = 0$$
$$x - 4 = 0 \text{ or } x - 4 = 0$$
$$x = 4 \text{ or } \quad x = 4$$

There is only one solution, 4.

55.
$$12y^2 - 5y = 2$$
$$12y^2 - 5y - 2 = 0$$
$$(4y + 1)(3y - 2) = 0$$
$$4y + 1 = 0 \quad \text{or } 3y - 2 = 0$$
$$4y = -1 \quad \text{or} \quad 3y = 2$$
$$y = -\frac{1}{4} \text{ or} \qquad y = \frac{2}{3}$$

The solutions are $-\frac{1}{4}$ and $\frac{2}{3}$.

61.
$$3x^2 + 8x = 9 + 2x$$
$$3x^2 + 8x - 2x - 9 = 0$$
$$3x^2 + 6x - 9 = 0$$
$$3(x^2 + 2x - 3) = 0$$
$$3(x + 3)(x - 1) = 0$$
$$x + 3 = 0 \text{ or } x - 1 = 0$$
$$x = -3 \text{ or} \qquad x = 1$$

The solutions are -3 and 1.

67.
$$(t - 3)^2 = 36$$
$$t^2 - 6t + 9 = 36$$
$$t^2 - 6t - 27 = 0$$
$$(t - 9)(t + 3) = 0$$
$$t - 9 = 0 \text{ or } t + 3 = 0$$
$$t = 9 \text{ or} \qquad t = -3$$

The solutions are 9 and -3.

73. a) $(x - 1)(x + 3) = 0$ b) $(x - 3)(x + 1) = 0$
$$x^2 + 2x - 3 = 0 \qquad\quad x^2 - 2x - 3 = 0$$

c) $(x - 2)(x - 2) = 0$ d) $(x - 3)(x - 4) = 0$
$$x^2 - 4x + 4 = 0 \qquad\quad x^2 - 7x + 12 = 0$$

e) $(x - 3)(x + 4) = 0$ f) $(x + 3)(x - 4) = 0$
$$x^2 + x - 12 = 0 \qquad\quad x^2 - x - 12 = 0$$

g) $(x + 3)(x + 4) = 0$ h) $(x - \frac{1}{2})(x - \frac{1}{2}) = 0$
$$x^2 + 7x + 12 = 0 \qquad\quad x^2 - x + \frac{1}{4} = 0$$
$$\text{or}$$
$$4x^2 - 4x + 1 = 0$$

73. (continued)

i) $(x - 5)(x + 5) = 0$ j) $(x - 0)(x - 0.1)(x - \frac{1}{4}) = 0$
$$x^2 - 25 = 0 \qquad\qquad x(x - 0.1)(x - 0.25) = 0$$
$$x(x^2 - 0.35x + 0.025) = 0$$
$$x^3 - 0.35x^2 + 0.025x = 0$$
$$\text{or } x^3 - \frac{7}{20}x^2 + \frac{1}{40}x = 0$$

Exercise Set 4.8

1. Familiarize:

The problem is stated explicitly enough that we can go right to the translation.

Translate:

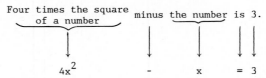

$$4x^2 \qquad - \qquad x \qquad = 3$$

Carry out: We solve the equation.
$$4x^2 - x = 3$$
$$4x^2 - x - 3 = 0$$
$$(4x + 3)(x - 1) = 0$$
$$4x + 3 = 0 \quad \text{or } x - 1 = 0$$
$$4x = -3 \quad \text{or} \qquad x = 1$$
$$x = -\frac{3}{4} \text{ or} \qquad x = 1$$

Check:

For $-\frac{3}{4}$ For 1

$4x^2 - x = 3$		$4x^2 - x = 3$	
$4(-\frac{3}{4})^2 - (-\frac{3}{4})$	3	$4 \cdot 1^2 - 1$	3
$4 \cdot \frac{9}{16} + \frac{3}{4}$		$4 - 1$	
$\frac{9}{4} + \frac{3}{4}$		3	
$\frac{9}{4} + \frac{3}{4}$			
$\frac{12}{4}$			
3			

Both numbers check.

State:

There are two such numbers, $-\frac{3}{4}$ and 1.

7. Familiarize:

Consecutive even integers are next to each other, such as 10 and 12, or -22 and -20. Let x represent the smaller even integer, then x + 2 represents the next even integer.

Translate:

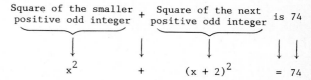

Smaller even integer times next even integer is 168.

$$x \cdot (x + 2) = 168$$

Carry out:

$$x(x + 2) = 168$$
$$x^2 + 2x = 168$$
$$x^2 + 2x - 168 = 0$$
$$(x + 14)(x - 12) = 0$$
$$x + 14 = 0 \quad \text{or} \quad x - 12 = 0$$
$$x = -14 \text{ or} \qquad x = 12$$

Check:

The solutions of the equation are -14 and 12. When x is -14, then x + 2 is -12 and -14(-12) = 168. The numbers -14 and -12 are consecutive even integers which are solutions to the problem. When x is 12, then x + 2 is 14 and 12 · 14 = 168. The numbers 12 and 14 are also consecutive even integers which are solutions to the problem.

State:

We have two solutions each of which consists of a pair of numbers: -14 and -12, and 12 and 14.

13. Familiarize: First draw a picture.

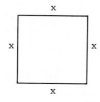

The area of the square is x · x, or x^2.
The perimeter of the square is x + x + x + x, or 4x.

Translate:

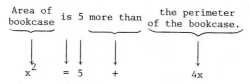

Area of bookcase is 5 more than the perimeter of the bookcase.

$$x^2 = 5 + 4x$$

Carry out:

$$x^2 = 5 + 4x$$
$$x^2 - 4x - 5 = 0$$
$$(x - 5)(x + 1) = 0$$
$$x - 5 = 0 \text{ or } x + 1 = 0$$
$$x = 5 \text{ or} \qquad x = -1$$

13. (continued)

Check:

The solutions of the equation are 5 and -1. The length of a side cannot be negative, so we only check 5. The area is 5 · 5, or 25. The perimeter is 5 + 5 + 5 + 5, or 20. The area, 25, is 5 more than the perimeter, 20. This checks.

State:

The length of a side is 5 ft.

19. Familiarize:

Consecutive odd positive integers are next to each other such as 21 and 23. Let x represent the smaller positive odd integer, then x + 2 represents the next positive odd integer.

Translate:

Square of the smaller positive odd integer + Square of the next positive odd integer is 74

$$x^2 + (x + 2)^2 = 74$$

Carry out:

$$x^2 + (x + 2)^2 = 74$$
$$x^2 + x^2 + 4x + 4 = 74$$
$$2x^2 + 4x - 70 = 0$$
$$2(x^2 + 2x - 35) = 0$$
$$2(x + 7)(x - 5) = 0$$
$$x + 7 = 0 \text{ or } x - 5 = 0$$
$$x = -7 \text{ or} \qquad x = 5$$

Check:

The solutions of the equation are -7 and 5. The problem asks for positive odd integers, so -7 cannot be a solution. When x is 5, x + 2 is 7. The numbers 5 and 7 are consecutive positive odd integers. The sum of their squares, 25 + 49, is 74. The numbers check.

State:

The integers are 5 and 7.

25. We substitute 40 for n and solve for N.

$$N = \frac{1}{2}(n^2 - n)$$
$$N = \frac{1}{2}(40^2 - 40)$$
$$= \frac{1}{2}(1600 - 40)$$
$$= \frac{1}{2}(1560)$$
$$= 780$$

The total number of possible handshakes is 780.

31. a) We substitute 20.6 for r and 21.6 for h and
         solve for t.

$$h = rt - 4.9t^2$$

$$21.6 = 20.6t - 4.9t^2$$

$$216 = 206t - 49t^2$$

$$49t^2 - 206t + 216 = 0$$

$$(49t - 108)(1t - 2) = 0$$

$$49t - 108 = 0 \quad \text{or} \quad t - 2 = 0$$

$$49t = 108 \quad \text{or} \qquad t = 2$$

$$t = \frac{108}{49} \quad \text{or} \qquad t = 2$$

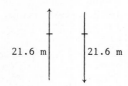

21.6 m   21.6 m

After 2 seconds, the ball (on its way up) will
reach a height of 21.6 m. After $2\frac{10}{49}$ seconds,
the ball (on its way down) will again reach a
height of 21.6 m.

  b) When the ball hits the ground, the height is
      0. We substitute 0 for h and 20.6 for r and
      solve for t.

$$h = rt - 4.9t^2$$

$$0 = 20.6t - 4.9t^2$$

$$0 = 206t - 49t^2$$

$$49t^2 - 206t = 0$$

$$t(49t - 206) = 0$$

$$t = 0 \quad \text{or} \quad 49t - 206 = 0$$

$$t = 0 \quad \text{or} \qquad 49t = 206$$

$$t = 0 \quad \text{or} \qquad t = \frac{206}{49}, \text{ or } \approx 4.2$$

Before the ball is thrown, t = 0 sec. After
4.2 seconds, the ball has returned to the
ground (h = 0).

Exercise Set 4.9

1. We replace each x by 3 and each y by -2.

$$x^2 - y^2 + xy$$

$$= 3^2 - (-2)^2 + 3(-2)$$

$$= 9 - 4 - 6$$

$$= -1$$

7. We replace each h by 4.7, each r by 1.2, and π
    by 3.14.

$$2\pi rh + 2\pi r^2$$

$$= 2(3.14)(1.2)(4.7) + 2(3.14)(1.2)^2$$

$$= 2(3.14)(1.2)(4.7) + 2(3.14)(1.44)$$

$$= 35.4192 + 9.0432$$

$$= 44.4624$$

The area is 44.4624 in^2.

13. $a + b - 2a - 3b$

$$= (1 - 2)a + (1 - 3)b$$

$$= -a - 2b$$

19. $6au + 3av - 14au + 7av$

$$= (6 - 14)au + (3 + 7)av$$

$$= -8au + 10av$$

25. $(2x^2 - 3xy + y^2) + (-4x^2 - 6xy - y^2) +$
                              $(x^2 + xy - y^2)$

$$= (2 - 4 + 1)x^2 + (-3 - 6 + 1)xy + (1 - 1 - 1)y^2$$

$$= -x^2 - 8xy - y^2$$

31. $(3z - u)(2z + 3u)$

          F   O   I   L

$$= 6z^2 + 9zu - 2uz - 3u^2$$

$$= 6z^2 + 7zu - 3u^2$$

37. $(a^3 + bc)(a^3 - bc)$

$$= (a^3)^2 - (bc)^2 \qquad [(A + B)(A - B) = A^2 - B^2]$$

$$= a^6 - b^2c^2$$

43. $(3 - c^2d^2)(4 + c^2d^2)$

          F   O   I   L

$$= 12 + 3c^2d^2 - 4c^2d^2 - c^4d^4$$

$$= 12 - c^2d^2 - c^4d^4$$

49. $(x + h)^2$

$$= x^2 + 2xh + h^2 \quad [(A + B)^2 = A^2 + 2AB + B^2]$$

55. $(2a^3 - \frac{1}{2}b^3)^2$

$$= (2a^3)^2 - 2 \cdot 2a^3 \cdot \frac{1}{2}b^3 + (\frac{1}{2}b^3)^2$$

$$[(A - B)^2 = A^2 - 2AB + B^2]$$

$$= 4a^6 - 2a^3b^3 + \frac{1}{4}b^6$$

61. $(c^2 - d)(c^2 + d)$

$= (c^2)^2 - d^2$

$= c^4 - d^2$

67. $[x + y + z][x - (y + z)]$

$= [x + (y + z)][x - (y + z)]$

$= x^2 - (y + z)^2$

$= x^2 - (y^2 + 2yz + z^2)$

$= x^2 - y^2 - 2yz - z^2$

73. The largest common factor is $9xy$.

$9x^2y^2 - 36xy$

$= 9xy \cdot xy - 9xy \cdot 4$

$= 9xy(xy - 4)$

79. The two terms have a common factor, $x + 1$.

$(x - 1)(x + 1) - y(x + 1)$

$= [(x - 1) - y](x + 1)$

$= (x - 1 - y)(x + 1)$

85. $x^2 + y^2 - 2xy$

$= x^2 - 2xy + y^2$

$= (x - y)^2 \qquad [A^2 - 2AB + B^2 = (A - B)^2]$

91. $y^4 + 10y^2z^2 + 25z^4$

$= (y^2)^2 + 2 \cdot y^2 \cdot 5z^2 + (5z^2)^2$

$= (y^2 + 5z^2)^2 \quad [A^2 + 2AB + B^2 = (A + B)^2]$

97. $2mn - 360n^2 + m^2$

$= m^2 + 2mn - 360n^2$

We look for two numbers whose product is -360 and whose sum is 2. They are -18 and 20:
$-18 \cdot 20 = -360$ and $-18 + 20 = 2$.

$m^2 + 2mn - 360n^2 = (m - 18n)(m + 20n)$

103. $a^5 + 4a^4b - 5a^3b^2$

$= a^3(a^2 + 4ab - 5b^2)$

$= a^3(a + 5b)(a - b)$

109. $7p^4 - 7q^4$

$= 7(p^4 - q^4)$

$= 7(p^2 + q^2)(p^2 - q^2)$

$= 7(p^2 + q^2)(p + q)(p - q)$

We factored a difference of squares twice.

115. It is helpful to add additional labels to the figure.

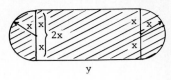

The two semicircles make a circle with radius x. The area of that circle is πx^2. The area of the rectangle is $2x \cdot y$. The sum of the two regions, $\pi x^2 + 2xy$, is the area of the shaded region.

Exercise Set 4.10

1. $x^3 + 8$

$= x^3 + 2^3$

$= (x + 2)(x^2 - x \cdot 2 + 2^2)$

$\qquad [A^3 + B^3 = (A + B)(A^2 - AB + B^2)]$

$= (x + 2)(x^2 - 2x + 4)$

7. $8a^3 + 1$

$= (2a)^3 + 1^3$

$= (2a + 1)[(2a)^2 - 2a \cdot 1 + 1^2]$

$\qquad [A^3 + B^3 = (A + B)(A^2 - AB + B^2)]$

$= (2a + 1)(4a^2 - 2a + 1)$

13. $64y^3 + 1$

$= (4y)^3 + 1^3$

$= (4y + 1)[(4y)^2 - 4y \cdot 1 + 1^2]$

$= (4y + 1)(16y^2 - 4y + 1)$

19. $a^3 + \frac{1}{8}$

$= a^3 + (\frac{1}{2})^3$

$= (a + \frac{1}{2})[a^2 - a \cdot \frac{1}{2} + (\frac{1}{2})^2]$

$= (a + \frac{1}{2})(a^2 - \frac{1}{2}a + \frac{1}{4})$

25. $rs^3 + 64r$

$= r(s^3 + 64)$

$= r(s^3 + 4^3)$

$= r(s + 4)(s^2 - 4s + 16)$

31. $64x^6 - 8t^6$

$= 8(8x^6 - t^6)$

$= 8[(2x^2)^3 - (t^2)^3]$

$= 8(2x^2 - t^2)(4x^4 + 2x^2t^2 + t^4)$

37. $3x^{3a} + 24y^{3b}$

$= 3(x^{3a} + 8y^{3b})$

$= 3[(x^a)^3 + (2y^b)^3]$

$= 3(x^a + 2y^b)(x^{2a} - 2x^ay^b + 4y^{2b})$

Exercise Set 5.1

1.

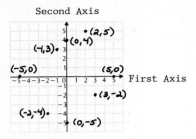

(2,5) is 2 units right and 5 units up.
(-1,3) is 1 unit left and 3 units up.
(3,-2) is 3 units right and 2 units down.
(-2,-4) is 2 units left and 4 units down.
(0,4) is 0 units left or right and 4 units up.
(0,-5) is 0 units left or right and 5 units down.
(5,0) is 5 units right and 0 units up or down.
(-5,0) is 5 units left and 0 units up or down.

7. Since both coordinates are negative, the point (-6,-29) is in the <u>third</u> quadrant.

13.

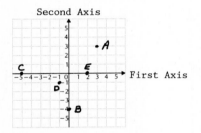

Point A is 3 units right and 3 units up.
The coordinates of A are (3,3).

Point B is 0 units left or right and 4 units down.
The coordinates of B are (0,-4).

Point C is 5 units left and 0 units up or down.
The coordinates of C are (-5,0).

Point D is 1 unit left and 1 unit down.
The coordinates of D are (-1,-1).

Point E is 2 units right and 0 units up or down.
The coordinates of E are (2,0).

19.

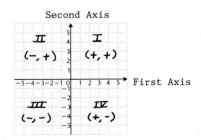

If the first coordinate is positive, then the point must be in either I or IV.

25. Answers may vary.

We select eight points such that the sum of the coordinates for each point is 6.

(-1,7)	-1 + 7 = 6
(0,6)	0 + 6 = 6
(1,5)	1 + 5 = 6
(2,4)	2 + 4 = 6
(3,3)	3 + 3 = 6
(4,2)	4 + 2 = 6
(5,1)	5 + 1 = 6
(6,0)	6 + 0 = 6

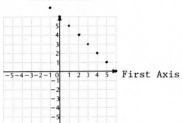

Exercise Set 5.2

1. (2,5); y = 3x - 1

We take the variables in alphabetical order. We replace x by the first coordinate and y by the second coordinate.

$$
\begin{array}{c|l}
\multicolumn{2}{l}{y = 3x - 1} \\
\hline
5 & 3 \cdot 2 - 1 \qquad \text{(Substituting 2 for x and 5 for y)} \\
  & 6 - 1 \\
  & 5
\end{array}
$$

The equation becomes true, (2,5) is a solution of y = 3x - 1.

7. y = 4x

We first make a table of values. We choose <u>any</u> number for x and then determine y by substitution.

When x = 0,   y = 4 · 0 = 0.
When x = -1, y = 4(-1) = -4.
When x = 1,   y = 4 · 1 = 4.

x	y
0	0
-1	-4
1	4

Since two points determine a line, that is all we really need to graph a line, but you should always plot a third point as a check.

Plot these points, draw the line they determine, and label the graph y = 4x.

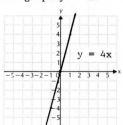

7. (continued)

You may have chosen different x values and thus have different ordered pairs in your table, but your graph will be the same.

In the equation $y = 4x$ the number 4, called the slope, tells us that the line slants up from left to right.

13. $y = -\frac{3}{2}x$

We first make a table of values. We choose <u>any</u> number for x and then determine y by substitution. Using multiples of 2 avoids fractions.

When x = 0, $y = -\frac{3}{2} \cdot 0 = 0.$

When x = 2, $y = -\frac{3}{2} \cdot 2 = -3.$

When x = -2, $y = -\frac{3}{2}(-2) = 3.$

x	y
0	0
2	-3
-2	3

Plot these points, draw the line they determine, and label the graph $y = -\frac{3}{2}x$.

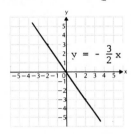

The number $-\frac{3}{2}$, called the slope, tells us how the line slants. For a negative slope, a line slants down from left to right.

19. $y = \frac{1}{3}x - 1$

We first make a table of values. We choose <u>any</u> number for x and then determine y by substitution. Using multiples of 3 avoids fractions.

When x = 0, $y = \frac{1}{3} \cdot 0 - 1 = 0 - 1 = -1.$

When x = -6, $y = \frac{1}{3}(-6) - 1 = -2 - 1 = -3.$

When x = 3, $y = \frac{1}{3} \cdot 3 - 1 = 1 - 1 = 0.$

x	y
0	-1
-6	-3
3	0

19. (continued)

Plot these points, draw the line they determine, and label the graph $y = \frac{1}{3}x - 1$.

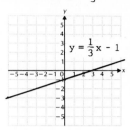

25. $y = -\frac{5}{2}x - 2$

We first make a table of values. Using multiples of 2 avoids fractions.

When x = 0, $y = -\frac{5}{2} \cdot 0 - 2 = 0 - 2 = -2.$

When x = -2, $y = -\frac{5}{2}(-2) - 2 = 5 - 2 = 3.$

When x = -4, $y = -\frac{5}{2}(-4) - 2 = 10 - 2 = 8.$

x	y
0	-2
-2	3
-4	8

Plot these points, draw the line they determine, and label the graph.

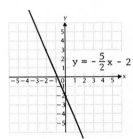

The graph of the equation $y = -\frac{5}{2}x - 2$ is a straight line that goes through the point (0,-2), the y-intercept, and has slope $-\frac{5}{2}$.

31. $y = \frac{4}{3} - \frac{1}{3}x$

We first make a table of values.

When x = 1, $y = \frac{4}{3} - \frac{1}{3} \cdot 1 = \frac{4}{3} - \frac{1}{3} = \frac{3}{3} = 1.$

When x = -2, $y = \frac{4}{3} - \frac{1}{3}(-2) = \frac{4}{3} + \frac{2}{3} = \frac{6}{3} = 2.$

When x = -5, $y = \frac{4}{3} - \frac{1}{3}(-5) = \frac{4}{3} + \frac{5}{3} = \frac{9}{3} = 3.$

When x = 0, $y = \frac{4}{3} - \frac{1}{3} \cdot 0 = \frac{4}{3} - 0 = \frac{4}{3}.$

31. (continued)

x	y
1	1
-2	2
-5	3
0	$\frac{4}{3}$

Plot these points, draw the line they determine, and label the graph.

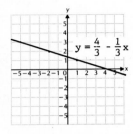

The graph of $y = \frac{4}{3} - \frac{1}{3}x$ ($y = -\frac{1}{3}x + \frac{4}{3}$) is a straight line that goes through the point $(0,\frac{4}{3})$, the y-intercept, and has slope $-\frac{1}{3}$.

37. $x + 3y = 15$

When x = 0, 0 + 3y = 15   When x = 15, 15 + 3y = 15

$3y = 15$ $3y = 0$

$y = 5$ $y = 0$

When x = 3, 3 + 3y = 15   When x = 6, 6 + 3y = 15

$3y = 12$ $3y = 9$

$y = 4$ $y = 3$

When x = 9, 9 + 3y = 15   When x = 12, 12 + 3y = 15

$3y = 6$ $3y = 3$

$y = 2$ $y = 1$

The whole number solutions are (0,5), (3,4), (6,3), (9,2), (12,1), (15,0).

Exercise Set 5.3

1. $x + 3y = 6$

To find the x-intercept, let y = 0.

$x + 3y = 6$

$x + 3 \cdot 0 = 6$

$x = 6$

Thus, (6,0) is the x-intercept.

To find the y-intercept, let x = 0.

$x + 3y = 6$

$0 + 3y = 6$

$3y = 6$

$y = 2$

Thus, (0,2) is the y-intercept.

1. (continued)

Plot these points and draw the line.

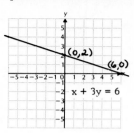

A third point should be used as a check. We substitute any value for x and solve for y.

We let x = 3. Then

$x + 3y = 6$

$3 + 3y = 6$

$3y = 3$

$y = 1$

The point (3,1) is on the graph, so the graph is probably correct.

7. $2y - 2 = 6x$

To find the x-intercept, let y = 0.

$2y - 2 = 6x$

$2 \cdot 0 - 2 = 6x$

$-2 = 6x$

$-\frac{1}{3} = x$

Thus, $(-\frac{1}{3},0)$ is the x-intercept.

To find the y-intercept, let x = 0.

$2y - 2 = 6x$

$2y - 2 = 6 \cdot 0$

$2y - 2 = 0$

$2y = 2$

$y = 1$

Thus, (0,1) is the y-intercept.

It is helpful to plot another point since the intercepts are so close together. This point can also serve as a check.

We let x = 1. Then

$2y - 2 = 6x$

$2y - 2 = 6 \cdot 1$

$2y - 2 = 6$

$2y = 8$

$y = 4$

7. (continued)

Plot the point (1,4) and the intercepts and draw the line.

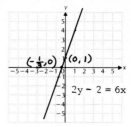

13. 4x + 5y = 20

To find the x-intercept, let y = 0.

$$4x + 5y = 20$$
$$4x + 5 \cdot 0 = 20$$
$$4x = 20$$
$$x = 5$$

Thus (5,0) is the x-intercept.

To find the y-intercept, let x = 0.

$$4x + 5y = 20$$
$$4 \cdot 0 + 5y = 20$$
$$5y = 20$$
$$y = 4$$

Thus (0,4) is the y-intercept.

Plot these points and draw the graph.

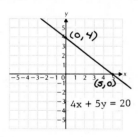

A third point should be used as a check. We substitute any value for x and solve for y.

We let x = 4.   Then

$$4x + 5y = 20$$
$$4 \cdot 4 + 5y = 20$$
$$16 + 5y = 20$$
$$5y = 4$$
$$y = \frac{4}{5}$$

The point $(4, \frac{4}{5})$ is on the graph, so the graph is probably correct.

19. 3x - 2 = y

To find the x-intercept, let y = 0.

$$3x - 2 = y$$
$$3x - 2 = 0$$
$$3x = 2$$
$$x = \frac{2}{3}$$

Thus, $(\frac{2}{3},0)$ is the x-intercept.

To find the y-intercept, let x = 0.

$$3x - 2 = y$$
$$3 \cdot 0 - 2 = y$$
$$-2 = y$$

Thus, (0,-2) is the y-intercept.

Plot these points and draw the line.

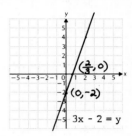

A third point should be used as a check.

We let x = 2.   Then

$$3x - 2 = y$$
$$3 \cdot 2 - 2 = y$$
$$6 - 2 = y$$
$$4 = y$$

The point (2,4) is on the graph, so the graph is probably correct.

25. y = -3 - 3x

To find the x-intercept, let y = 0.

$$y = -3 - 3x$$
$$0 = -3 - 3x$$
$$3x = -3$$
$$x = -1$$

Thus, (-1,0) is the x-intercept.

To find the y-intercept, let x = 0.

$$y = -3 - 3x$$
$$y = -3 - 3 \cdot 0$$
$$y = -3$$

Thus, (0,-3) is the y-intercept.

<u>25</u>. (continued)

Plot these points and draw the graph.

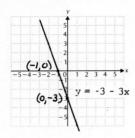

We use a third point as a check.

We let x = -2. Then

   y = -3 - 3x

   y = -3 - 3 · (-2)

   y = -3 + 6

   y = 3

The point (-2,3) is on the graph, so the graph is probably correct.

<u>31</u>. y = 4

Any ordered pair (x,4) is a solution. The variable y must be 4, but the x variable can be any number we choose. A few solutions are listed below. Plot these points and draw the line.

x	y
-3	4
0	4
2	4

<u>37</u>. x = -$\frac{5}{2}$

Any ordered pair (-$\frac{5}{2}$,y) is a solution. The variable x must be -$\frac{5}{2}$, but the y variable can be any number we choose. A few solutions are listed below. Plot these points and draw the line.

x	y
-$\frac{5}{2}$	-2
-$\frac{5}{2}$	0
-$\frac{5}{2}$	4

<u>43</u>.

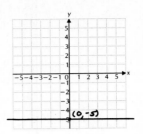

The line is parallel to the x-axis (horizontal) with y-intercept (0,-5). The equation is y = -5.

Exercise Set 5.4

<u>1</u>. (3,2) and (-1,2)

  m = $\frac{2 - 2}{3 - (-1)}$ = $\frac{2 - 2}{3 + 1}$ = $\frac{0}{4}$ = 0

<u>7</u>. (0,8) and (-3,10)

  m = $\frac{8 - 10}{0 - (-3)}$ = $\frac{8 - 10}{0 + 3}$ = $\frac{-2}{3}$ = -$\frac{2}{3}$

<u>13</u>. (9,-4) and (9,-7)

  m = $\frac{-4 - (-7)}{9 - 9}$ = $\frac{-4 + 7}{9 - 9}$ = $\frac{3}{0}$

Since division by 0 is not defined, this line has no slope.

<u>19</u>. The line x = 9 is a vertical line. A vertical line has <u>no</u> slope.

<u>25</u>. We solve for y.

  x + 4y = 8

    4y = -x + 8

     y = $\frac{1}{4}$(-x + 8)

     y = -$\frac{1}{4}$x + 2

The slope is -$\frac{1}{4}$.

<u>31</u>. We solve for y.

  x - 3y = -2

  -3y = -x - 2

    y = -$\frac{1}{3}$(-x - 2)

    y = $\frac{1}{3}$x + $\frac{2}{3}$

The slope is $\frac{1}{3}$.

<u>37</u>. y = 1.8x   (Think: y = 1.8x + 0)

The slope is 1.8 and the y-intercept is (0,0).

<u>43</u>. We solve for y.

     9x = 3y + 5

   9x - 5 = 3y

  $\frac{1}{3}$(9x - 5) = y

    3x - $\frac{5}{3}$ = y

The slope is 3 and the y-intercept is (0,-$\frac{5}{3}$).

<u>49</u>. $y - y_1 = m(x - x_1)$

We substitute $\frac{3}{4}$ for m, 2 for x_1, and 4 for y_1.

$y - 4 = \frac{3}{4}(x - 2)$

$y - 4 = \frac{3}{4}x - \frac{3}{2}$

$y = \frac{3}{4}x - \frac{3}{2} + 4$

$y = \frac{3}{4}x + \frac{5}{2}$

<u>55</u>. $y - y_1 = m(x - x_1)$

We substitute $\frac{2}{3}$ for m, 5 for x_1, and 6 for y_1.

$y - 6 = \frac{2}{3}(x - 5)$

$y - 6 = \frac{2}{3}x - \frac{10}{3}$

$y = \frac{2}{3}x - \frac{10}{3} + 6$

$y = \frac{2}{3}x + \frac{8}{3}$

<u>61</u>. (3,2) and (1,5)

First we find the slope.

$m = \frac{2 - 5}{3 - 1} = \frac{-3}{2} = -\frac{3}{2}$

Then we use the point-slope equation.

$y - y_1 = m(x - x_1)$

We substitute $-\frac{3}{2}$ for m, 3 for x_1, and 2 for y_1.

$y - 2 = -\frac{3}{2}(x - 3)$

$y - 2 = -\frac{3}{2}x + \frac{9}{2}$

$y = -\frac{3}{2}x + \frac{9}{2} + 2$

$y = -\frac{3}{2}x + \frac{13}{2}$

<u>67</u>. First find the slope of $3x - y + 4 = 0$

$3x - y + 4 = 0$

$3x + 4 = y$

The slope is 3.

Then find an equation of the line containing (2,-3) and having slope 3.

$y - y_1 = m(x - x_1)$

We substitute 3 for m, 2 for x_1, and -3 for y_1.

$y - (-3) = 3(x - 2)$

$y + 3 = 3x - 6$

$y = 3x - 9$

<u>1</u>. $y = \frac{5}{2}x + 1$

Slope is $\frac{5}{2}$; y-intercept is (0,1).

From the y-intercept, we go <u>up</u> 5 units and to the <u>right</u> 2 units. This gives us the point (2,6). We can now draw the graph.

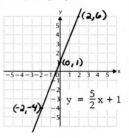

As a check, we can rename the slope and find another point.

$\frac{5}{2} = \frac{5}{2} \cdot \frac{-1}{-1} = \frac{-5}{-2}$

From the y-intercept, we go <u>down</u> 5 units and to the <u>left</u> 2 units. This gives us the point (-2,-4). Plot this point to see if it is on the line.

<u>7</u>. $y = \frac{1}{3}x + 6$

Slope is $\frac{1}{3}$; y-intercept is (0,6).

From the y-intercept, we go <u>up</u> 1 unit and to the <u>right</u> 3 units. This gives us the point (3,7). We can now draw the graph.

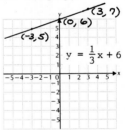

As a check, we can rename the slope and find another point.

$\frac{1}{3} = \frac{1}{3} \cdot \frac{-1}{-1} = \frac{-1}{-3}$

From the y-intercept, we go <u>down</u> 1 unit and to the <u>left</u> 3 units. This gives us the point (-3,5). Plot this point to see if it is on the line.

Exercise Set 6.1

1. Translate:

We translate the first statement:

The sum of two numbers is 58.

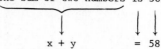

$$x + y = 58$$

We have used x and y for the numbers.

Now we translate the second statement:

The difference between two numbers is 16.

$$x - y = 16$$

The system of equations is

$$x + y = 58$$
$$x - y = 16$$

7. Translate:

We translate the first statement:

The difference between two numbers is 16.

$$x - y = 16$$

We have used x for the larger number and y for the smaller number.

Now we translate the second statement:

Three times the is seven times the
larger number   smaller number.

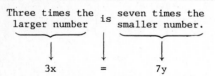

$$3x = 7y$$

The system of equations is

$$x - y = 16$$
$$3x = 7y$$

13. Translate:

We translate the first statement:

Hectares of   hectares of   820
Chardonnay plus   Riesling   totals   hectares.
grapes   grapes

$$x + y = 820$$

We used x for the number of hectares of Chardonnay grapes and y the number of hectares of Riesling grapes.

13. (continued)

Now we translate the second statement:

Hectares of   140   more than   hectares of
Chardonnay   is   hectares   Riesling
grapes   grapes.

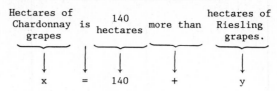

$$x = 140 + y$$

The system of equations is

$$x + y = 820$$
$$x = 140 + y$$

19. Familiarize: We first list the information in a chart.

	Ages now	Ages 20 years from now
Patrick	x	x + 20
Father	y	y + 20

We have used x for Patrick's age now and y for the father's age now. In 20 years they will be x + 20 and y + 20.

Translate:

We translate the first statement:

Patrick's age now is 20% of his father's age now.

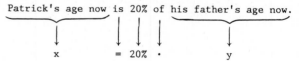

Now we translate the second statement:

Patrick's age   his father's age
twenty years   will be 52% of   twenty years
from now   from now.

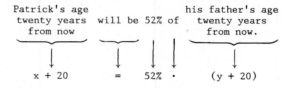

The system of equations is

$$x = 0.2y$$
$$x + 20 = 0.52(y + 20)$$

Exercise Set 6.2

1. Use alphabetical order of the variables. We substitute 3 for x and 2 for y.

$2x + 3y = 12$		$x - 4y = -5$	
$2 \cdot 3 + 3 \cdot 2$	12	$3 - 4 \cdot 2$	-5
$6 + 6$		$3 - 8$	
12		-5	

The ordered pair (3,2) is a solution of each equation. Therefore it _is_ a solution of the system of equations.

7. We substitute -1 for x and 1 for y.

x = -1	
-1	-1

x - y = -2	
-1 - 1	-2
	-2

The ordered pair (-1,1) is a solution of each equation. Therefore it <u>is</u> a solution of the system of equations.

13. First make a table of values for each equation.

x + 2y = 10   3x + 4y = 8

x	y
0	5
-2	6
4	3

x	y
0	2
-4	5
4	-1

Plot these points and draw the line each set of points determines.

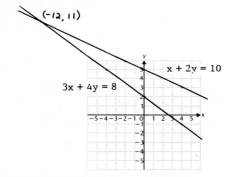

(-12, 11)

The point of intersection looks as if it has coordinates (-12,11).

Check:

x + 2y = 10	
-12 + 2 · 11	10
-12 + 22	
10	

3x + 4y = 8	
3(-12) + 4(11)	8
-36 + 44	
8	

The solution is (-12,11).

19. First make a table of values for each equation.

x = -y   x + y = 4

x	y
2	-2
0	0
-3	3

x	y
1	3
0	4
-2	6

19. (continued)

Plot these points and draw the line each set of points determines.

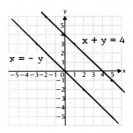

The lines are parallel. There is no solution.

25. First make a table of values for each equation.

x + y = 9   3x + 3y = 27

x	y
4	5
3	6
5	4

x	y
3	6
5	4
4	5

Plot these points and draw the line each set of points determines.

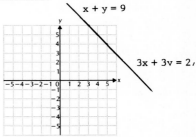

The lines coincide. There is an unlimited number of solutions.

31. The solution is $(\frac{2}{3},\frac{3}{7})$, but would be very difficult to discover by graphing. The check would tell you whether the possible pair you found by graphing is a solution.

Exercise Set 6.3

1. x + y = 4   (1)
   y = 2x + 1   (2)

   We substitute 2x + 1 for y in the first equation and solve for x.

   x + y = 4   (1)
   x + (2x + 1) = 4   (Substituting)
   3x + 1 = 4
   3x = 3
   x = 1

1. (continued)

Next we substitute 1 for x in either equation in the original system and solve for y.

x + y = 4   (1)

1 + y = 4   (Substituting)

y = 3

We check the ordered pair (1,3).

x + y = 4		y = 2x + 1	
1 + 3	4	?	2 · 1 + 1
4			2 + 1
			3

Since (1,3) checks in both equations, it is the solution.

7. x = -2y   (1)

x + 4y = 2   (2)

We substitute -2y for x in the second equation and solve for y.

x + 4y = 2   (2)

-2y + 4y = 2

2y = 2

y = 1

Next we substitute 1 for y in either equation of the original system and solve for x.

x = -2y   (1)

x = -2 · 1

x = -2

We check the ordered pair (-2,1).

x = -2y		3y - x = 5	
-2	-2 · 1	3 · 1 - (-2)	5
	-2	3 + 2	
		5	

Since (-2,1) checks in both equations, it is the solution.

13. 2x + 3y = -2   (1)

2x - y = 9   (2)

We solve the second equation for y.

2x - y = 9   (2)

2x - 9 = y

We substitute 2x - 9 for y in the first equation and solve for x.

2x + 3y = -2   (1)

2x + 3(2x - 9) = -2

2x + 6x - 27 = -2

8x - 27 = -2

8x = 25

$x = \frac{25}{8}$

13. (continued)

Now we substitute $\frac{25}{8}$ for x in either equation of the original system and solve for y.

2x - y = 9   (2)

$2 \cdot \frac{25}{8} - y = 9$ (Substituting)

$\frac{25}{4} - y = \frac{36}{4}$

$-y = \frac{11}{4}$

$y = -\frac{11}{4}$

We check the ordered pair $(\frac{25}{8}, -\frac{11}{4})$.

2x + 3y = -2		2x - y = 9	
$2 \cdot \frac{25}{8} + 3(-\frac{11}{4})$	-2	$2 \cdot \frac{25}{8} - (-\frac{11}{4})$	9
$\frac{25}{4} - \frac{33}{4}$		$\frac{25}{4} + \frac{11}{4}$	
$-\frac{8}{4}$		$\frac{36}{4}$	
-2		9	

Since $(\frac{25}{8}, -\frac{11}{4})$ checks in both equations, it is the solution.

19. x - 3y = 7   (1)

-4x + 12y = 28   (2)

We solve the first equation for x.

x - 3y = 7   (1)

x = 3y + 7

We substitute 3y + 7 for x in the second equation and solve for y.

-4x + 12y = 28   (2)

-4(3y + 7) + 12y = 28   (Substituting)

-12y - 28 + 12y = 28

-28 = 28

We obtain a false equation, -28 = 28, so there is no solution. The graphs of the equations are parallel lines. They do not intersect.

25. This problem has already been translated in Exercise Set 6.1, Problem 7.

Carry out: We solve the system of equations.

x - y = 16   (1)

3x = 7y   (2)

We solve the first equation for x.

x - y = 16   (1)

x = y + 16

25. (continued)

We substitute $y + 16$ for x in the second equation and solve for y.

$$3x = 7y \qquad (2)$$
$$3(y + 16) = 7y \qquad \text{(Substituting)}$$
$$3y + 48 = 7y$$
$$48 = 4y$$
$$12 = y$$

Next we substitute 12 for y in either of the original equations and solve for x.

$$x - y = 16 \qquad (1)$$
$$x - 12 = 16 \qquad \text{(Substituting)}$$
$$x = 28$$

Check:

The difference between 28 and 12, $28 - 12$, is 16. Three times the larger, $3 \cdot 28$ or 84, is seven times the smaller, $7 \cdot 12$ or 84. The numbers check.

State:

The numbers are 28 and 12.

31. $\frac{1}{4}(a - b) = 2 \qquad (1)$

$\frac{1}{6}(a + b) = 1 \qquad (2)$

We first clear of fractions.

$a - b = 8 \qquad (1a) \quad$ [Multiplying Equation (1) by 4]
$a + b = 6 \qquad (2a) \quad$ [Multiplying Equation (2) by 6]

We solve the first equation for a.

$$a - b = 8 \qquad (1a)$$
$$a = b + 8$$

We substitute $b + 8$ for a in Equation (2a) and solve for b.

$$a + b = 6 \qquad (2a)$$
$$(b + 8) + b = 6 \qquad \text{(Substituting)}$$
$$2b + 8 = 6$$
$$2b = -2$$
$$b = -1$$

Next we substitute -1 for b in Equation (1a) and solve for a.

$$a - b = 8$$
$$a - (-1) = 8$$
$$a + 1 = 8$$
$$a = 7$$

We check the ordered pair $(7,-1)$.

$\frac{1}{4}(a - b) = 2$		$\frac{1}{6}(a + b) = 1$	
$\frac{1}{4}[7 - (-1)]$	2	$\frac{1}{6}[7 + (-1)]$	1
$\frac{1}{4} \cdot 8$		$\frac{1}{6} \cdot 6$	
2		1	

Since $(7,-1)$ checks in both equations, it is the solution.

37. $x + y + z = 4 \qquad (1)$
$x - 2y - z = 1 \qquad (2)$
$y = -1 \qquad (3)$

Substitute -1 for y in the first and second equations.

$x + y + z = 4 \quad (1)$	$x - 2y - z = 1 \quad (2)$
$x + (-1) + z = 4$	$x - 2(-1) - z = 1$
$x + z = 5$	$x + 2 - z = 1$
	$x - z = -1$

We now have a system of two equations.

$x + z = 5 \qquad (4)$
$x - z = -1 \qquad (5)$

We solve Equation (5) for x.

$$x - z = -1 \qquad (5)$$
$$x = z - 1$$

We substitute $z - 1$ for x in Equation (4) and solve for z.

$$x + z = 5 \qquad (4)$$
$$(z - 1) + z = 5 \qquad \text{(Substituting)}$$
$$2z - 1 = 5$$
$$2z = 6$$
$$z = 3$$

Next we substitute -1 for y and 3 for z in the first equation and solve for x.

$$x + y + z = 4 \qquad (1)$$
$$x + (-1) + 3 = 4 \qquad \text{(Substituting)}$$
$$x + 2 = 4$$
$$x = 2$$

We check the ordered triple $(2,-1,3)$.

$x + y + z = 4$		$x - 2y - z = 1$		$y = -1$	
$2 + (-1) + 3$	4	$2 - 2(-1) - 3$	1	-1	-1
	4	$2 + 2 - 3$			
			1		

Since $(2,-1,3)$ checks in all three equations, it is the solution.

Exercise Set 6.4

1. $x + y = 10$

$\underline{x - y = 8}$
$2x + 0 = 18 \qquad \text{(Adding)}$
$2x = 18$
$x = 9$

Substitute 9 for x in one of the original equations and solve for y.

$$x + y = 10$$
$$9 + y = 10 \qquad \text{(Substituting)}$$
$$y = 1$$

1. (continued)

Check: For (9,1)

$x + y = 10$		$x - y = 8$	
$9 + 1$	10	$9 - 1$	8
10		8	

Since (9,1) checks, it is the solution.

7. $4a + 3b = 7$
$\underline{-4a + b = 5}$
$4b = 12$ (Adding)
$b = 3$

Substitute 3 for b in one of the original equations and solve for a.

$4a + 3b = 7$
$4a + 3 \cdot 3 = 7$ (Substituting)
$4a + 9 = 7$
$4a = -2$
$a = -\frac{1}{2}$

Check: For $(-\frac{1}{2}, 3)$

$4a + 3b = 7$		$-4a + b = 5$	
$4(-\frac{1}{2}) + 3 \cdot 3$	7	$-4(-\frac{1}{2}) + 3$	5
$-2 + 9$		$2 + 3$	
7		5	

Since $(-\frac{1}{2}, 3)$ checks, it is the solution.

13. $-x - y = 8$
$2x - y = -1$

We multiply by -1 on both sides of the first equation and then add.

$x + y = -8$ (Multiplying by -1)
$\underline{2x - y = -1}$
$3x = -9$ (Adding)
$x = -3$

Substitute -3 for x in one of the original equations and solve for y.

$2x - y = -1$
$2(-3) - y = -1$ (Substituting)
$-6 - y = -1$
$-y = 5$
$y = -5$

Check: For (-3,-5)

$-x - y = 8$		$2x - y = -1$	
$-(-3) - (-5)$	8	$2(-3) - (-5)$	-1
$3 + 5$		$-6 + 5$	
8		-1	

Since (-3,-5) checks, it is the solution.

19. $2w - 3z = -1$
$3w + 4z = 24$

We use the multiplication principle with both equations and then add.

$8w - 12z = -4$ (Multiplying by 4)
$\underline{9w + 12z = 72}$ (Multiplying by 3)
$17w = 68$ (Adding)
$w = 4$

Substitute 4 for w in one of the original equations and solve for z.

$3w + 4z = 24$
$3 \cdot 4 + 4z = 24$ (Substituting)
$12 + 4z = 24$
$4z = 12$
$z = 3$

Check: (4,3)

$2w - 3z = -1$		$3w + 4z = 24$	
$2 \cdot 4 - 3 \cdot 3$	-1	$3 \cdot 4 + 4 \cdot 3$	24
$8 - 9$		$12 + 12$	
-1		24	

Since (4,3) checks, it is the solution.

25. $3x - 2y = 10$
$5x + 3y = 4$

We use the multiplication principle with both equations and add.

$9x - 6y = 30$ (Multiplying by 3)
$\underline{10x + 6y = 8}$ (Multiplying by 2)
$19x = 38$ (Adding)
$x = 2$

Substitute 2 for x in one of the original equations and solve for v.

$5x + 3y = 4$
$5 \cdot 2 + 3y = 4$ (Substituting)
$10 + 3y = 4$
$3y = -6$
$y = -2$

Check: For (2,-2)

$3x - 2y = 10$		$5x + 3y = 4$	
$3 \cdot 2 - 2(-2)$	10	$5 \cdot 2 + 3(-2)$	4
$6 + 4$		$10 - 6$	
10		4	

Since (2,-2) checks, it is the solution.

31. The Translate step has been done in Exercise Set 6.4, Problem 5. The resulting system of equations is

$53.95 + 0.30m = c$

$54.95 + 0.20m = c$

where m represents the mileage and c the cost.

Carry out: We solve the system of equations.

We clear the system of decimals by multiplying on both sides by 100.

$5395 + 30m = 100c$

$5495 + 20m = 100c$

We multiply the first equation by -1 and then add.

$-5395 - 30m = -100c$

$\underline{5495 + 20m = 100c}$

$100 - 10m = 0$ (Adding)

$100 = 10m$

$10 = m$

Check:

For 10 mi, the cost of the Acme car is
53.95 + 0.30(10), or 53.95 + 3, or $56.95.

For 10 mi, the cost of the other car is
54.95 + 0.20(10), or 54.95 + 2, or $56.95,

so the costs are the same when the mileage is 10.

State:

When the cars are driven 10 miles, the cost will be the same.

37. The Translate step has been done in Exercise Set 6.4, Problem 13.

Carry out: We solve the system of equations.

$x + y = 820$ or $x + y = 820$

$x = 140 + y$ $\underline{x - y = 140}$ (Adding -y)

$2x = 960$ (Adding)

$x = 480$

Substitute 480 for x in one of the original equations and solve for y.

$x + y = 820$

$480 + y = 820$

$y = 340$

Check:

The sum of 480 hectares and 340 hectares is 820 hectares. The difference between 480 hectares and 340 hectares is 140 hectares. The numbers check.

State:

The vintner should plan 480 hectares of Chardonnay grapes and 340 hectares of Riesling grapes.

43. $3(x - y) = 9$ or $3x - 3y = 9$

$x + y = 7 x + y = 7$

Multiply the second equation by 3 and then add.

$3x - 3y = 9$

$\underline{3x + 3y = 21}$ (Multiplying by 3)

$6x = 30$

$x = 5$

Substitute 5 for x in one of the original equations and solve for y.

$x + y = 7$

$5 + y = 7$ (Substituting)

$y = 2$

The ordered pair (5,2) checks and is the solution.

49. $y = ax + b$

$y = x + c$

Substitute x + c for y in the first equation and solve for x.

$y = ax + b$

$x + c = ax + b$ (Substituting)

$x - ax = b - c$

$(1 - a)x = b - c$

$x = \dfrac{b - c}{1 - a}, \text{ or } \dfrac{c - b}{a - 1}$

Solve the second equation for x.

$y = x + c$

$y - c = x$

Substitute y - c for x in the first equation and solve for y.

$y = ax + b$

$y = a(y - c) + b$ (Substituting)

$y = ay - ac + b$

$y - ay = -ac + b$

$(1 - a)y = b - ac$

$y = \dfrac{b - ac}{1 - a}, \text{ or } \dfrac{ac - b}{a - 1}$

The ordered pair $\left(\dfrac{b - c}{1 - a}, \dfrac{b - ac}{1 - a}\right)$ checks and is the solution.

Exercise Set 6.5

1. Familiarize:

The total number of cars and trucks is 510. To assure maximum profits there must be 190 more cars than trucks.

1. (continued)

Translate:

We let x represent the number of cars and y represent the number of trucks.

The number of cars plus the number of trucks is 510.

$$x \quad + \quad y \quad = \quad 510$$

The number of cars is the number of trucks plus 190.

$$x \quad = \quad y \quad + \quad 190$$

Carry out: We solve the system of equations.

$x + y = 510$

$x = y + 190$

We substitute y + 190 for x in the first equation and solve for y.

$$x + y = 510$$
$$(y + 190) + y = 510 \qquad \text{(Substituting)}$$
$$2y + 190 = 510$$
$$2y = 320$$
$$y = 160$$

Next we substitute 160 for y in one of the original equations and solve for x.

$$x = y + 190$$
$$x = 160 + 190 \qquad \text{(Substituting)}$$
$$x = 350$$

Check:

If there are 350 cars and 160 trucks, then the total number of vehicles is 350 + 160, or 510. Since the number of trucks plus 190 is 160 + 190, or 350, we know that it is true that the number of cars is 190 more than the number of trucks.

State:

The firm should have 350 cars and 160 trucks.

7. Familiarize:

Let d represent the number of dimes and q the number of quarters. Then, 10d represents the value of the dimes in cents, and 25q represents the value of the quarters in cents. The total value is $15.25, or 1525¢. The total number of coins is 103.

Translate:

Number of dimes plus number of quarters is 103.

$$d \quad + \quad q \quad = \quad 103$$

Value of dimes plus value of quarters is $15.25

$$10d \quad + \quad 25q \quad = \quad 1525$$

7. (continued)

The resulting system is

$d + q = 103$

$10d + 25q = 1525$

Carry out: We use the addition method.

We multiply the first equation by -10 and then add.

$$-10d - 10q = -1030 \qquad \text{(Multiplying by -10)}$$
$$\underline{10d + 25q = 1525}$$
$$15q = 495 \qquad \text{(Adding)}$$
$$q = 33$$

Next we substitute 33 for q in one of the original equations and solve for d.

$$d + q = 103$$
$$d + 33 = 103 \qquad \text{(Substituting)}$$
$$d = 70$$

Check:

The number of dimes plus the number of quarters is 70 + 33, or 103. The total value in cents is $10 \cdot 70 + 25 \cdot 33$, or 700 + 825, or 1525. This is equal to $15.25. This checks.

State:

There are 70 dimes and 33 quarters.

13. Familiarize: List the information in a table.

We let x = number of student tickets sold and y = the number of adult tickets sold.

Ticket	Paid	Number sold	Money taken in
Student	$0.50	x	0.50x
Adult	$0.75	y	0.75y
	Totals	200	$132.50

Translate:

The total number of tickets sold was 200, so

$$x + y = 200$$

The total amount collected was $132.50, so

$$0.50x + 0.75y = 132.50$$
$$\text{or} \quad 50x + 75y = 13,250 \qquad \text{(Multiplying by 100)}$$

Carry out: We use the addition method.

$x + y = 200$

$50x + 75y = 13,250$

We multiply on both sides of the first equation by -50 and then add.

$$-50x - 50y = -10,000 \qquad \text{(Multiplying by -50)}$$
$$\underline{50x + 75y = 13,250}$$
$$25y = 3250$$
$$y = 130$$

13. (continued)

Next we substitute 130 for y in one of the original equations and solve for x.

$x + y = 200$

$x + 130 = 200$ (Substituting)

$x = 70$

Check:

The total number of tickets sold was 70 students plus 130 adults, or 200. The total receipts were $0.50(70) + 0.75(130)$. This amount is $35 + $97.50, or $132.50. The numbers check.

State:

Thus, 70 student tickets and 130 adult tickets were sold.

19. Familiarize:

We can arrange the information in a table. We let x represent the amount of 30% solution and y represent the amount of 50% solution.

Type of insecticide	Amount of solution	Percent of insecticide	Amount of insecticide in the solution
30% solution	x	30%	30%x
50% solution	y	50%	50%x
Mixture	200 liters	42%	42% × 200 or 84 liters

Translate:

Since the total is 200 liters, we have

$x + y = 200.$

The amount of insecticide in the mixture is to be 42% of 200, or 84 liters. The amounts of insecticide from the two solutions are 30%x and 50%y. Thus

$30\%x + 50\%y = 84$

or $0.3x + 0.5y = 84$

or $3x + 5y = 840$

Carry out: We use the addition method.

$x + y = 200$

$3x + 5y = 840$

We multiply the first equation by -3 and then add.

$-3x - 3y = -600$ (Multiplying by -3)

$\underline{3x + 5y = 840}$

$2y = 240$

$y = 120$

Next we substitute 120 for y in one of the original equations and solve for x.

$x + y = 200$

$x + 120 = 200$ (Substituting)

$x = 80$

19. (continued)

Check:

We consider $x = 80$ and $y = 120$. The sum is 200. Now 30% of 80 is 24 and 50% of 120 is 60. These add up to 84. The numbers check.

State:

Thus, 80 L of the 30% solution and 120 L of the 50% solution should be used.

25. Familiarize:

Let x represent the ten's digit and y the unit's digit. The sum of the digits, $x + y$, is 12. The number is $10x + y$. When the digits are reversed, the number is $10y + x$.

Translate:

The sum of the digits is 12.

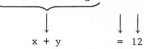

$x + y \qquad = 12$

The old number is the new number plus 18.

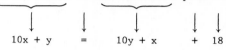

$10x + y \qquad = \qquad 10y + x \qquad + \qquad 18$

We can simplify the second equation.

$10x + y = 10y + x + 18$

$9x - 9y = 18$

$x - y = 2$

We add the resulting system.

$x + y = 12$

$\underline{x - y = 2}$

$2x = 14$

$x = 7$

Next we substitute 7 for x in one of the original equations and solve for y.

$x + y = 12$

$7 + y = 12$

$y = 5$

Check:

We consider the number 75. The sum of the digits is $7 + 5$, or 12. When the digits are reversed, the number is 57 which is 18 less than 75. The values check.

State:

The original number is 75.

<u>31.</u> Let x = cost of bat, y = cost of ball, and z = cost of glove.

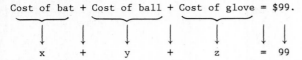

Cost of bat + Cost of ball + Cost of glove = $99.

x + y + z = 99

Cost of bat = $9.95 + Cost of ball

x = 9.95 + y

Cost of glove = $65.45 + Cost of bat

z = 65.45 + x

Solve the second equation for y.

$$x = 9.95 + y$$
$$x - 9.95 = y$$

Substitute x - 9.95 for y and x + 65.45 for z in the first equation and solve for x.

$$x + y + z = 99$$
$$x + (x - 9.95) + (x + 65.45) = 99 \quad \text{(Substituting)}$$
$$3x + 55.5 = 99$$
$$3x = 43.5$$
$$x = 14.5$$

If x = $14.50, then y = 14.50 - 9.95, or $4.55 and z = 65.45 + 14.50, or $79.95. The total cost is $99. The cost of the bat, $14.50, is $9.95 more than the cost of the ball, $4.55. The cost of the glove, $79.95, is $65.45 more than the cost of the bat, $14.50. The values check.

The bat costs $14.50, the ball costs $4.55, and the glove costs $79.95.

Exercise Set 6.6

<u>1.</u> Familiarize: First make a drawing.

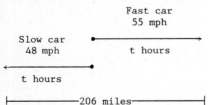

The sum of the distances is 206 miles. The times the cars travel are the same. We organize the information in a chart.

Car	Distance	Speed	Time
Slow car	Slow car distance	48	t
Fast car	Fast car distance	55	t
Total	206		

<u>1.</u> (continued)

Translate:

From the drawing we see that

Slow car distance + Fast car distance = 206

Then using d = rt in each row of the chart, we get

48t + 55t = 206

Carry out:

We solve this equation for t.

$$48t + 55t = 206$$
$$103t = 206$$
$$t = 2$$

Check:

If the time is 2 hr, then the distance the slow car travels is 48 · 2, or 96 mi. The fast car travels 55 · 2, or 110 mi. Since the sum of the distances, 96 + 110, is 206 mi, the problem checks.

State:

In 2 hours, the cars will be 206 miles apart.

<u>7.</u> Familiarize: We first make a drawing.

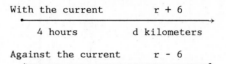

From the drawing we see that the distances are the same. Let d represent the distance. Let r represent the speed of the canoe in still water. Then, when the canoe is traveling with the current, its speed is r + 6. When it is traveling against the current, its speed is r - 6. We organize the information in a chart.

	Distance	Speed	Time
With current	d	r + 6	4
Against current	d	r - 6	10

Translate:

Using d = rt in each row of the chart, we get the following system of equations:

$$d = (r + 6)4$$
$$d = (r - 6)10$$

Carry out:

Substitute (r + 6)4 for d in the second equation and solve for r.

$$d = (r - 6)10$$
$$(r + 6)4 = (r - 6)10 \quad \text{(Substituting)}$$
$$4r + 24 = 10r - 60$$
$$84 = 6r$$
$$14 = r$$

7. (continued)

Check:

When r = 14, r + 6 = 20 and 20 · 4 = 80, the distance. When r = 14, r - 6 = 8 and 8 · 10 = 80. In both cases, we get the same distance.

State:

The speed of the canoe in still water is 14 km/h.

13. Familiarize: First make a drawing.

Home   t hr   45 mph |(2-t) hr 6 mph   Work

Motorcycle distance | Walking distance

←————25 miles————→

Let t represent the time the motorcycle was driven, then 2 - t represents the time the rider walked. We organize the information in a chart.

	Distance	Speed	Time
Motorcycling	Motorcycle distance	45	t
Walking	Walking distance	6	2 - t

Translate:

From the drawing we see that

Motorcycle distance + Walking distance = 25

Then using d = rt in each row of the chart we get

45t + 6(2 - t) = 25

Carry out:

We solve this equation for t.

$45t + 12 - 6t = 25$

$39t + 12 = 25$

$39t = 13$

$t = \frac{13}{39}$

$t = \frac{1}{3}$

Check:

The problem asks us to find how far the motorcycle went before it broke down. If $t = \frac{1}{3}$, then 45t (the distance the motorcycle traveled) = $45 \cdot \frac{1}{3}$, or 15 and 6(2 - t) (the distance walked) = $6(2 - \frac{1}{3})$ = $6 \cdot \frac{5}{3}$, or 10. The total of these distances is 25, so $\frac{1}{3}$ checks.

State:

The motorcycle went 15 miles before it broke down.

19. Familiarize: First we draw a picture.

Slow trip   32 mph

→

t hr   d miles

Fast trip   32 + 4, or 36 mph

→

$t - \frac{1}{2}$ hr   d miles

Let d represent the distance between the towns. Let t represent the time for the slower trip. Then $t - \frac{1}{2}$ represents the time for the faster trip. We organize the information in a chart.

	Distance	Speed	Time
Slow trip	d	32	t
Fast trip	d	32 + 4, or 36	$t - \frac{1}{2}$

Translate:

Using d = rt in each row of the chart, we get the following system of equations:

d = 32t

$d = 36(t - \frac{1}{2})$

Carry out:

We substitute 32t for d in the second equation and solve for t.

$d = 36(t - \frac{1}{2})$

$32t = 36(t - \frac{1}{2})$

$32t = 36t - 18$

$18 = 4t$

$\frac{18}{4} = t$

$4.5 = t$

Substitute 4.5 for t and solve for d.

d = 32t

d = 32(4.5)   (Substituting)

d = 144

Check:

If t = 4.5, then the distance of the slow trip is 32(4.5), or 144. If t = 4.5, then the distance of the fast trip is $36(4.5 - \frac{1}{2})$, or 36(4), or 144. The distances are the same. The values check.

State:

The towns are 144 miles apart.

Exercise Set 7.1

1. x > 4

    a) Since 4 > 4 is false, 4 <u>is not</u> a solution.

    b) Since 0 > 4 is false, 0 <u>is not</u> a solution.

    c) Since -4 > 4 is false, -4 <u>is not</u> a solution.

    d) Since 6 > 4 is true, 6 <u>is</u> a solution.

7. y ≥ -5

    a) Since 0 ≥ -5 is true, 0 <u>is</u> a solution.

    b) Since -4 ≥ -5 is true, -4 <u>is</u> a solution.

    c) Since -5 ≥ -5 is true, -5 <u>is</u> a solution.

    d) Since -6 ≥ -5 is false, -6 <u>is not</u> a solution.

13.   x + 8 ≤ -10

     x + 8 - 8 ≤ -10 - 8

            x ≤ -18

    The solution set is $\{x | x \le -18\}$.

19.   x - 6 > 2

     x - 6 + 6 > 2 + 6

            x > 8

    The solution set is $\{x | x > 8\}$.

25.   3x + 9 ≤ 2x + 6

     3x + 9 - 9 ≤ 2x + 6 - 9

            3x ≤ 2x - 3

     3x - 2x ≤ 2x - 3 - 2x

            x ≤ -3

    The solution set is $\{x | x \le -3\}$.

31.   3y + 4 ≥ 2y - 7

     3y + 4 - 4 ≥ 2y - 7 - 4

            3y ≥ 2y - 11

     3y - 2y ≥ 2y - 11 - 2y

            y ≥ -11

    The solution set is $\{y | y \ge -11\}$.

37. $x - \frac{1}{3} > \frac{1}{4}$

     $x - \frac{1}{3} + \frac{1}{3} > \frac{1}{4} + \frac{1}{3}$

            $x > \frac{3}{12} + \frac{4}{12}$

            $x > \frac{7}{12}$

    The solution set is $\{x | x > \frac{7}{12}\}$.

43.   3(r + 2) < 2r + 4

      3r + 6 < 2r + 4

      3r - 2r < 4 - 6

          r < -2

    The solution set is $\{r | r < -2\}$.

49.   12x + 1.2 ≤ 11x

      12x - 11x ≤ -1.2

            x ≤ -1.2

    The solution set is $\{x | x \le -1.2\}$.

Exercise Set 7.2

1.   5x < 35

    $\frac{1}{5} \cdot 5x < \frac{1}{5} \cdot 35$

        x < 7

    The solution set is $\{x | x < 7\}$.

7.   12x > -36

    $\frac{1}{12} \cdot 12x > \frac{1}{12} \cdot (-36)$

        x > -3

    The solution set is $\{x | x > -3\}$.

13.   -4y ≥ -16

    $-\frac{1}{4} \cdot (-4y) \le -\frac{1}{4} \cdot (-16)$

          y ≤ 4

    The solution set is $\{y | y \le 4\}$.

19. $-\frac{6}{5} \le -4x$

    $-\frac{1}{4} \cdot (-\frac{6}{5}) \ge -\frac{1}{4} \cdot (-4x)$

          $\frac{6}{20} \ge x$

          $\frac{3}{10} \ge x$, or $x \le \frac{3}{10}$

    The solution set is $\{x | x \le \frac{3}{10}\}$.

25.   3x - 5 ≤ 13

       3x - 5 + 5 ≤ 13 + 5

            3x ≤ 18

       $\frac{1}{3} \cdot 3x \le \frac{1}{3} \cdot 18$

            x ≤ 6

    The solution set is $\{x | x \le 6\}$.

31.
$$5x + 3 \geq -7$$
$$5x + 3 - 3 \geq -7 - 3$$
$$5x \geq -10$$
$$\frac{1}{5} \cdot 5x \geq \frac{1}{5} \cdot (-10)$$
$$x \geq -2$$

The solution set is $\{x \mid x \geq -2\}$.

37.
$$3 - 6y > 23$$
$$-3 + 3 - 6y > -3 + 23$$
$$-6y > 20$$
$$-\frac{1}{6} \cdot (-6y) < -\frac{1}{6} \cdot 20$$
$$y < -\frac{20}{6}$$
$$y < -\frac{10}{3}$$

The solution set is $\{y \mid y < -\frac{10}{3}\}$.

43.
$$6 - 4y > 4 - 3y$$
$$6 - 4y + 4y > 4 - 3y + 4y$$
$$6 > 4 + y$$
$$-4 + 6 > -4 + 4 + y$$
$$2 > y, \text{ or } y < 2$$

The solution set is $\{y \mid y < 2\}$.

49.
$$21 - 8y < 6y + 49$$
$$21 - 8y + 8y < 6y + 49 + 8y$$
$$21 < 14y + 49$$
$$21 - 49 < 14y + 49 - 49$$
$$-28 < 14y$$
$$\frac{1}{14} \cdot -28 < \frac{1}{14} \cdot 14y$$
$$-2 < y, \text{ or } y > -2$$

The solution set is $\{y \mid y > -2\}$.

55. Familiarize: First make a drawing.

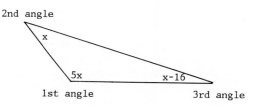

1st angle   3rd angle

Let x represent the measure of the second angle. Then 5x represents the first angle and x - 16 represents the third angle. The sum of the three angles is 180°.

Translate:

First angle + Second angle + Third angle = 180°

$$5x \quad + \quad x \quad + \quad (x - 16) = 180$$

55. (continued)

Carry out:
$$5x + x + x - 16 = 180$$
$$7x - 16 = 180$$
$$7x = 196$$
$$x = 28$$

Possible angle measures:

First angle: $5x = 5 \cdot 28 = 140°$
Second angle: $x = 28°$
Third angle: $x - 16 = 28 - 16 = 12°$

Check:

The sum of the measures, 140° + 28° + 12°, is 180°. The measure of the first angle is five times the second angle, $140 = 5 \cdot 28$. The measure of the third angle is sixteen less than the second, $28 - 16 = 12$. The values check.

State:

The measures of the angles are 140°, 28°, and 12°.

61.
$$\frac{x}{3} - 2 \leq 1$$
$$\frac{x}{3} \leq 1 + 2$$
$$\frac{x}{3} \leq 3$$
$$3 \cdot \frac{x}{3} \leq 3 \cdot 3$$
$$x \leq 9$$

The solution set is $\{x \mid x \leq 9\}$.

67.
$$-(x + 5) \geq 4a - 5$$
$$-x - 5 \geq 4a - 5$$
$$-x \geq 4a - 5 + 5$$
$$-x \geq 4a$$
$$-1(-x) \leq -1 \cdot 4a$$
$$x \leq -4a$$

The solution set is $\{x \mid x \leq -4a\}$.

73. Yes, $a + c < b + c$
$$a + c + (-c) < b + c + (-c) \quad \text{(-c does exist for every rational)}$$
$$a + 0 < b + 0$$
$$a < b$$

79. The largest x can be is y. Thus x is less than or equal to y, $x \leq y$.

Exercise Set 7.3

1. Familiarize:

The average of the five scores is their sum divided by the number of quizzes, 5.

Translate:

We let s represent the student's score on the last quiz. The average of the five scores is given by

$$\frac{73 + 75 + 89 + 91 + s}{5}.$$

Since this average must be <u>at least</u> 85, this means that it must be greater than or equal to 85. Thus, we can translate the problem to the inequality

$$\frac{73 + 75 + 89 + 91 + s}{5} \geq 85.$$

Carry out:

We first multiply by 5 to clear of fractions.

$$5(\frac{73 + 75 + 89 + 91 + s}{5}) \geq 5 \cdot 85$$

$$73 + 75 + 89 + 91 + s \geq 425$$

$$328 + s \geq 425$$

$$s \geq 425 - 328$$

$$s \geq 97$$

Check:

Suppose s is a score greater than or equal to 97. Then by successively adding 73, 75, 89, and 91 on both sides of the inequality we get

$$73 + 75 + 89 + 91 + s \geq 425$$

so

$$\frac{73 + 75 + 89 + 91 + s}{5} \geq \frac{425}{5}, \text{ or } 85$$

State:

Any score which is at least 97 will give an average quiz grade of 85.

{s|s ≥ 97}

7. Familiarize:

Let m represent the number of miles per day. Then the cost per day for those miles is $0.39m. The total cost is the daily rate plus the daily mileage cost. The total cost cannot exceed $250. In other words the total cost must be less than or equal to $250, the daily budget.

Translate:

Daily rate + Mileage cost ≤ Budget

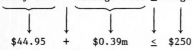

$$\$44.95 \quad + \quad \$0.39m \quad \leq \quad \$250$$

Carry out:

$$44.95 + 0.39m \leq 250$$

$$4495 + 39m \leq 25,000 \quad \text{(Clearing of decimals)}$$

$$39m \leq 20,505$$

$$m \leq \frac{20,505}{39}$$

$$m \leq 525.8 \quad \text{(Nearest tenth)}$$

7. (continued)

Check:

With inequalities it is impossible to check each solution. But we can check to see if the solution set we obtained seems reasonable.

When m = 526, the total cost is 44.95 + 0.39(526), or $250.09.

When m = 525.8, the total cost is 44.95 + 0.39(525.8), or $250.01.

When m = 525.7, the total cost is 44.95 + 0.39(525.7), or $249.97

From these calculations it would appear that m ≤ 525.8 (to the nearest tenth) is the correct solution.

State:

To stay within the daily budget, the number of miles the businessperson drives per day must not exceed 525.8.

{m|m ≤ 525.8 mi}

13. Familiarize:

The average number of calls per week is the sum of the calls for the three weeks divided by the number of weeks, 3.

Translate:

We let c represent the number of calls made during the third week. The average of the three weeks is given by

$$\frac{18 + 22 + c}{3}.$$

Since the average must be <u>at least</u> 20, this means that it must be greater than or equal to 20. Thus, we can translate the problem to the inequality

$$\frac{18 + 22 + c}{3} \geq 20$$

Carry out:

We first multiply by 3 to clear of fractions.

$$3(\frac{18 + 22 + c}{3}) \geq 3 \cdot 20$$

$$18 + 22 + c \geq 60$$

$$40 + c \geq 60$$

$$c \geq 20$$

Check:

Suppose c is a number greater than or equal to 20. Then by adding 18 and 22 on both sides of the inequality we get

$$18 + 22 + c \geq 18 + 22 + 20$$

$$18 + 22 + c \geq 60$$

so

$$\frac{18 + 22 + c}{3} \geq \frac{60}{3}, \text{ or } 20$$

State:

Any number of calls which is at least 20 will maintain an average of at least 20 for the three-week period.

{c|c ≥ 20}

19. Familiarize:

Consecutive odd integers are next to each other such as -9 and -7 or 15 and 17. Let x represent the first odd integer. Then x + 2 represents the next odd integer.

Translate:

First odd integer plus second odd integer is less than 100.

$$x + (x + 2) < 100$$

Carry out:

$$x + (x + 2) < 100$$
$$2x + 2 < 100$$
$$2x < 98$$
$$x < 49$$

Check:

If x is an odd integer less than 49, the largest such integer is 47. Then x + 2 is 49. The sum of 47 and 49 is 96 and 96 is less than 100. Let us check to see if x = 49 would be too large. If x = 49, then, x + 2 is 51. The sum of 49 and 51 is 100, but 100 is not less than 100.

State:

The largest pair of odd integers is 47 and 49.

Exercise Set 7.4

1. x < 5

We shade all points to the left of 5. The open circle at 5 indicates that 5 is not part of the graph.

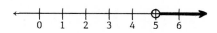

7. x + 2 > 7
$$x > 7 - 2$$
$$x > 5$$

We shade all points to the right of 5. The open circle at 5 indicates that 5 is not part of the graph.

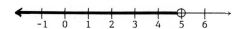

13. x - 8 ≥ 0
$$x \geq 8$$

We shade the point for 8 and all points to the right of 8. The closed circle at 8 indicates that 8 is part of the graph.

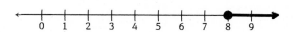

19. 4y + 9 > 11y - 12
$$9 + 12 > 11y - 4y$$
$$21 > 7y$$
$$\frac{1}{7} \cdot 21 > \frac{1}{7} \cdot 7y$$
$$3 > y, \text{ or } y < 3$$

We shade all points to the left of 3. The open circle at 3 indicates that 3 is not part of the graph.

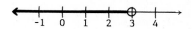

25. |x| ≤ 7

The absolute value of a number is its distance from 0 on a number line. For the absolute value of a number to be less than or equal to 7, its distance from 0 must be 7 or less. Thus the number must be -7, 7, or any number between -7 and 7. We shade the points for -7 and 7 and all the points between -7 and 7. The closed circles at -7 and 7 indicate that they are part of the graph.

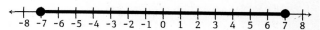

31. -2x + 4y ≤ -2; (5,-3)

We use alphabetical order of variables. We replace x by 5 and y by -3.

$-2x + 4y \leq -2$	
$-2 \cdot 5 + 4(-3)$	-2
$-10 - 12$	
-22	

Since -22 ≤ -2 is true, (5,-3) is a solution.

37. Graph y ≤ x - 5.

First graph the line y = x - 5. The intercepts are (5,0) and (0,-5). We use a solid line for the graph since we have ≤.

Then pick a point which does not belong to the line. The origin, (0,0), is an easy one to use. (If the line goes through the origin, then we test some other point.) Substitute (0,0) into y ≤ x - 5 to see if it is a solution.

$$y \leq x - 5$$
$$0 \leq 0 - 5 \quad \text{(Substituting)}$$
$$0 \leq -5$$

0 ≤ -5 is false, so the origin is not a solution. This means we shade the lower half-plane.

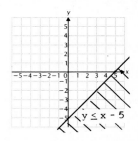

43. Graph $y \leq 3x + 2$.

First graph the line $y = 3x + 2$. The intercepts are $(-\frac{2}{3}, 0)$ and $(0, 2)$. We use a solid line for the graph since we have $\leq$.

Pick a point which does not belong to the line. The origin is easy to use. Substitute $(0,0)$ into $y \leq 3x + 2$ to see if it is a solution.

$y \leq 3x + 2$

$0 \leq 3 \cdot 0 + 2$ (Substituting)

$0 \leq 0 + 2$

$0 \leq 2$

$0 \leq 2$ is true, so the origin is a solution. This means we shade the lower half-plane.

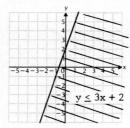

49. Graph $x - y < -10$.

First graph the line $x - y = -10$. The intercepts are $(-10, 0)$ and $(0, 10)$. We use a dashed line for the graph since we have $<$.

Pick a point which does not belong to the line. Let us choose $(0,0)$. Substitute $(0,0)$ into $x - y < -10$ to see if it is a solution.

$x - y < -10$

$0 - 0 < -10$ (Substituting)

$0 < -10$

$0 < -10$ is false, so the origin is not a solution. This means we shade the upper half-plane.

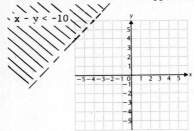

55. Graph $y - x < 0$.

First graph the line $y - x = 0$ ($y - x$). We make a table of values.

x	y
2	2
-4	-4
0	0

We plot these points and draw the line they determine. We use a dashed line since we have $<$.

55. (continued)

Since this line contains the origin, we must pick another point which does not belong to the line. Here we use $(2,5)$. Substitute $(2,5)$ into $y - x < 0$ to see if it is a solution.

$y - x < 0$

$5 - 2 < 0$ (Substituting)

$3 < 0$

$3 < 0$ is false, so $(2,5)$ is not a solution. This means we shade the lower half-plane.

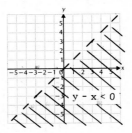

61. Graph $y \leq 0$. Think of this as $0 \cdot x + y \leq 0$.

First graph the line $y = 0$. This is a horizontal line with $(0,0)$ as the y-intercept (the x-axis). We use a solid line since we have $\leq$.

Pick a point which does not belong to the line. Substitute $(4,-3)$ to see if it is a solution.

$0 \cdot x + y \leq 0$

$0 \cdot 4 + (-3) \leq 0$ (Substituting)

$-3 \leq 0$

$-3 \leq 0$ is true, so $(4,-3)$ is a solution. This means we shade the lower half-plane.

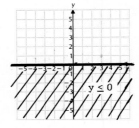

67. No, $(1,-3)$ is not a solution of $x - y < 1$.

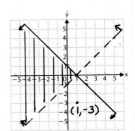

1. Graph this system of inequalities: x + y ≤ 6
                                                   y ≤ 5
                                                   x ≥ 0
                                                   y ≥ 0

We first graph the individual inequalities.

x + y ≤ 6   y ≤ 5

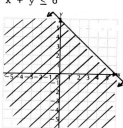

x ≥ 0   y ≥ 0

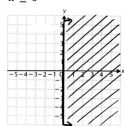

We now find the intersection of the four graphs.

x + y ≤ 6
   y ≤ 5
   x ≥ 0
   y ≥ 0

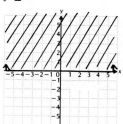

7. Graph this system of inequalities: x ≥ -3
                                                   y ≤ 4

We first graph the individual inequalities.

x ≥ -3   y ≤ 4

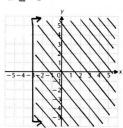
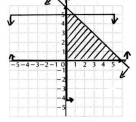

7. (continued)

We now find the intersection of the two graphs.

x ≥ -3
y ≤ 4

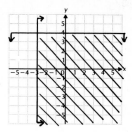

13. We organize the information in a table.

	Number made	Polyester required per garment	Wool required per garment	Price per garment
Suits	x	1 yd	4 yd	$120
Dresses	y	2 yd	3 yd	$75

	Total polyester required	Total wool required	Income
Suits	1x	4x	120x
Dresses	2y	3y	75y
Total	a)	b)	c)

a) Only 60 yd of polyester in stock: x + 2y ≤ 60

b) Only 120 yd of wool in stock: 4x + 3y ≤ 120

c) The total income is 120x + 75y.

We have used x to represent the number of suits and y to represent the number of dresses. We let I represent the total income. Expressed in terms of x and y:

I = 120x + 75y

We list the constraints.

x + 2y ≤ 60

4x + 3y ≤ 120

x ≥ 0 ⎫
         ⎬   (The number of garments made
y ≥ 0 ⎭   cannot be negative.)

The graph of the system of inequalities is as follows. The arrows indicate the half-planes which are the solution sets of each inequality.

13. (continued)

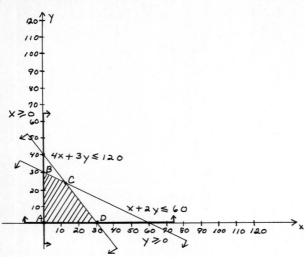

We find the coordinates of each vertex. Vertices A, B, and D are obviously

A (0,0)   B (0,30)   D (30,0)

To find C we solve the system:

$x + 2y = 60$

$4x + 3y = 120$

$-3x - 6y = -180$	(Multiplying by -3)
$8x + 6y = 240$	(Multiplying by 2)
$5x \qquad = 60$	(Adding)
$\qquad x = 12$	

Substitute 12 for x in either of the original equations and solve for y.

$x + 2y = 60$

$12 + 2y = 60$ (Substituting)

$\qquad 2y = 48$

$\qquad y = 24$

Thus, the coordinates of vertex C are (12,24).

We compute the total income for each vertex.

Vertex	Total Income I = 120x + 75y
A(0,0)	$I = 120 \cdot 0 + 75 \cdot 0 = 0$
B(0,30)	$I = 120 \cdot 0 + 75 \cdot 30 = 2250$
C(12,24)	$I = 120 \cdot 12 + 75 \cdot 24 = 1440 + 1880 = 3320$
D(30,0)	$I = 120 \cdot 30 + 75 \cdot 0 = 3600$

The maximum income of $3600 occurs when 30 suits and 0 dresses are made.

19. We organize the information in a table.

Type	Number of points for each	Number answered	Total points for type
A	10	x	10x
B	25	y	25y

We have used x to represent the number of Type A questions answered and y to represent the number of Type B questions answered. We let S represent the total score. Expressed in terms of x and y:

19. (continued)

$S = 10x + 25y$

We list the constraints:

$3 \leq x \leq 12$	The number of Type A questions answered must be at least 3 and no more than 12.
$4 \leq y \leq 15$	The number of Type B questions answered must be at least 4 and no more than 15.
$x + y \leq 20$	The total number of answered questions must not exceed 20.
$x \geq 0$ $\Big\}$ $y \geq 0$	The number of questions answered cannot be negative.

The graph of the system of inequalities is as follows. The arrows indicate the half-planes which are the solution sets of each inequality.

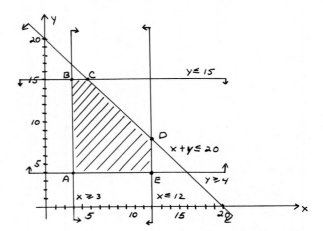

The coordinates of the vertices are obviously

A (3,4)

B (3,15)

C (5,15)

D (12,8)

E (12,4)

We compute the total score for each vertex.

Vertex	Total score S = 10x + 25y
A(3,4)	$S = 10 \cdot 3 + 25 \cdot 4 = 30 + 100 = 130$
B(3,15)	$S = 10 \cdot 3 + 25 \cdot 15 = 30 + 375 = 405$
C(5,15)	$S = 10 \cdot 5 + 25 \cdot 15 = 50 + 375 = 425$
D(12,8)	$S = 10 \cdot 12 + 25 \cdot 8 = 120 + 200 = 320$
E(12,4)	$S = 10 \cdot 12 + 25 \cdot 4 = 120 + 100 = 220$

The maximum score of 425 is obtained when 5 of Type A and 15 of Type B are answered.

Exercise Set 8.1

1. Yes, since each member of the domain is matched to only one member of the range.

7. $f(x) = x + 5$
   $f(3) = 3 + 5 = 8$
   $f(7) = 7 + 5 = 12$
   $f(-9) = -9 + 5 = -4$

13. $F(x) = 2x^2 - 3x + 2$
    $F(0) = 2 \cdot 0^2 - 3 \cdot 0 + 2 = 0 - 0 + 2 = 2$
    $F(-1) = 2(-1)^2 - 3(-1) + 2 = 2 + 3 + 2 = 7$
    $F(2) = 2 \cdot 2^2 - 3 \cdot 2 + 2 = 8 - 6 + 2 = 4$

19. $h(x) = x^4 - 3$
    $h(0) = 0^4 - 3 = 0 - 3 = -3$
    $h(-1) = (-1)^4 - 3 = 1 - 3 = -2$
    $h(3) = 3^4 - 3 = 81 - 3 = 78$

25. When the tread depth is 4 mm, the cost of replacing a defective tire whose regular price is $64.50 is 70% × 64.50, or $45.15.

31. $h(x) = |x| - x$
    The domain is the set {-1,0,1,2,3,4,5,6,7,8,9,10, 11,12,13,14,15,16,17,18,19,20}.
    $h(-1) = |-1| - (-1) = 1 + 1 = 2$
    $h(0) = |0| - 0 = 0 - 0 = 0$
    $h(1) = |1| - 1 = 1 - 1 = 0$
    $h(2) = |2| - 2 = 2 - 2 = 0$
    $h(3) = |3| - 3 = 3 - 3 = 0$
    .
    .
    .
    $h(20) = |20| - 20 = 20 - 20 = 0$
    The range is the set {0, 2}.

37. $f(-1) = -7$ gives us the ordered pair $(-1,-7)$.
    $f(3) = 8$ gives us the ordered pair $(3,8)$.

    The slope of the line determined by these points is
    $$m = \frac{-7 - 8}{-1 - 3} = \frac{-15}{-4} = \frac{15}{4}.$$

37. (continued)

    We substitute $\frac{15}{4}$ for m, 3 for x, and 8 for $f(x)$ and solve for b.

    $f(x) = mx + b$
    $8 = \frac{15}{4} \cdot 3 + b$
    $8 = \frac{45}{4} + b$
    $\frac{32}{4} - \frac{45}{4} = b$
    $-\frac{13}{4} = b$

    A linear equation for $f(x)$ is
    $$f(x) = \frac{15}{4}x - \frac{13}{4}.$$

Exercise Set 8.2

1. Graph $f(x) = x + 4$.
   Make a list of function values in a table.

   When $x = -2$, $f(-2) = -2 + 4 = 2$.
   When $x = 0$, $f(0) = 0 + 4 = 4$.
   When $x = 1$, $f(1) = 1 + 4 = 5$.

x	f(x)
-2	2
0	4
1	5

   We plot these points and connect them.

   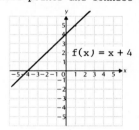

7. Graph $f(x) = 2x - 7$.
   Make a list of function values in a table.

x	f(x)
1	-5
2	-3
5	3

   When $x = 1$, $f(1) = 2 \cdot 1 - 7 = 2 - 7 = -5$.
   When $x = 2$, $f(2) = 2 \cdot 2 - 7 = 4 - 7 = -3$.
   When $x = 5$, $f(5) = 2 \cdot 5 - 7 = 10 - 7 = 3$.

   Plot these points and connect them.

   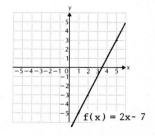

13. Graph g(x) = x².

Make a list of function values in a table.

x	g(x)
-2	4
-1	1
0	0
1	1
2	4

When x = -2, g(-2) = (-2)² = 4.

When x = -1, g(-1) = (-1)² = 1.

When x = 0, g(0) = 0² = 0.

When x = 1, g(1) = 1² = 1.

When x = 2, g(2) = 2² = 4.

Plot these points and connect them.

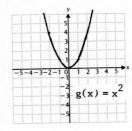

19. This graph <u>is</u> a graph of a function. No vertical line could cross the graph at more than one point.

25. Graph y² = x.

Make a table of values.

x	y
4	-2
1	-1
0	0
1	1
4	2

When y = -2, x = (-2)² = 4.

When y = -1, x = (-1)² = 1.

When y = 0, x = 0² = 0.

When y = 1, x = 1² = 1.

When y = 2, x = 2² = 4.

Plot these points and connect them.

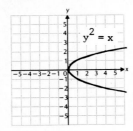

This graph <u>is not</u> a graph of a function. A vertical line does cross the graph at more than one point.

1. We substitute to find k.

y = kx

28 = k · 7   (Substituting 28 for y and 7 for x)

$\frac{28}{7}$ = k

4 = k   (k is the constant of variation.)

The <u>equation</u> of variation is y = 4x.

7. We substitute to find k.

y = kx

200 = k · 300   (Substituting 200 for y and 300 for x)

$\frac{200}{300}$ = k

$\frac{2}{3}$ = k

The <u>equation</u> of variation is y = $\frac{2}{3}$x.

13. Familiarize:

The problem states that we have direct variation between the variables M and E. Thus, an equation M = kE, k > 0, applies. As the weight on earth increases, the weight on moon increases.

Translate: We write an equation of variation.

Weight on moon varies directly as weight on earth.

This translates to M = kE.

Carry out:

a) First find an equation of variation.

M = kE

13 = k · 78   (Substituting 13 for M and 78 for E)

$\frac{13}{78}$ = k

$\frac{1}{6}$ = k

The equation of variation is M = $\frac{1}{6}$E.

b) Use the equation to find how much a 100 kg-person would weigh on the moon.

M = $\frac{1}{6}$E

M = $\frac{1}{6}$ · 100   (Substituting 100 for E)

M = $\frac{100}{6}$, or $16\frac{2}{3}$

Check:

In addition to repeating the computations, we can do some reasoning about the answer. The weight on the moon increased from 13 kg to $16\frac{2}{3}$ kg. Similarly, the weight on earth increased from 78 kg to 100 kg.

State:

A 100-kg person would weigh $16\frac{2}{3}$ kg on the moon.

19.   B = kN   25.   P = kRI^2

Exercise Set 8.4

1. We substitute to find k.

$$y = \frac{k}{x}$$

$$25 = \frac{k}{3} \quad \text{(Substituting 25 for y and 3 for x)}$$

$$75 = k \quad \text{(k is the constant of variation.)}$$

The _equation_ of variation is $y = \frac{75}{x}$.

7. We substitute to find k.

$$y = \frac{k}{x}$$

$$42 = \frac{k}{25} \quad \text{(Substituting 42 for y and 25 for x)}$$

$$1050 = k$$

The _equation_ of variation is $y = \frac{1050}{x}$.

13. Familiarize:

The problem states that we have inverse variation between the variables V and P. Thus, an equation $V = \frac{k}{P}$, k > 0, applies. As the pressure increases, the volume decreases.

Translate: We write an equation of variation. Volume varies inversely as pressure.

This translates to $V = \frac{k}{P}$.

Carry out:

a) First find an equation of variation.

$$V = \frac{k}{P}$$

$$200 = \frac{k}{32} \quad \text{(Substituting 200 for V and 32 for P)}$$

$$6400 = k$$

The equation of variation is $V = \frac{6400}{P}$.

b) Use the equation to find the volume of a gas under a pressure of 20 kg/cm^2.

$$V = \frac{6400}{P}$$

$$V = \frac{6400}{20} \quad \text{(Substituting 20 for P)}$$

$$V = 320$$

13. (continued)

Check:

Checking can be done by repeating the computations. We can also analyze the results. The pressure decreased from 32 km/cm^2 to 20 km/cm^2. The volume increased from 200 cm^3 to 320 cm^3. This is what we would expect with inverse variation.

State:

The volume is 320 cm^3 under a pressure of 20 km/cm^2.

19. $C = \frac{k}{N}$

25. Yes. As the runner's speed increases, the time decreases.

Exercise Set 9.1

1. $v(t) = \dfrac{4t^2 - 5t + 2}{t + 3}$

$v(0) = \dfrac{4 \cdot 0^2 - 5 \cdot 0 + 2}{0 + 3} = \dfrac{0 - 0 + 2}{0 + 3} = \dfrac{2}{3}$

$v(3) = \dfrac{4 \cdot 3^2 - 5 \cdot 3 + 2}{3 + 3} = \dfrac{36 - 15 + 2}{3 + 3} = \dfrac{23}{6}$

$v(7) = \dfrac{4 \cdot 7^2 - 5 \cdot 7 + 2}{7 + 3} = \dfrac{196 - 35 + 2}{7 + 3} = \dfrac{163}{10}$

7. $\dfrac{t - 3}{t + 2} \cdot \dfrac{t + 3}{t + 3} = \dfrac{(t - 3)(t + 3)}{(t + 2)(t + 3)}$

13. $\dfrac{4y - 12}{4y + 12} = \dfrac{4(y - 3)}{4(y + 3)} = \dfrac{4}{4} \cdot \dfrac{y - 3}{y + 3} = \dfrac{y - 3}{y + 3}$

19. $\dfrac{y^2 - 16}{2y + 6} \cdot \dfrac{y + 3}{y - 4}$

$= \dfrac{(y + 4)(y - 4)(y + 3)}{2(y + 3)(y - 4)}$

$= \dfrac{(y - 4)(y + 3)}{(y - 4)(y + 3)} \cdot \dfrac{y + 4}{2}$

$= \dfrac{y + 4}{2}$

25. $\dfrac{x^2 - y^2}{x^3 - y^3} \cdot \dfrac{x^2 + xy + y^2}{x^2 + 2xy + y^2}$

$= \dfrac{(x + y)(x - y)(x^2 + xy + y^2)}{(x - y)(x^2 + xy + y^2)(x + y)(x + y)}$

$= \dfrac{(x + y)(x - y)(x^2 + xy + y^2)}{(x + y)(x - y)(x^2 + xy + y^2)} \cdot \dfrac{1}{x + y}$

$= \dfrac{1}{x + y}$

31. $\dfrac{4a^2 - 1}{a^2 - 4} \div \dfrac{2a - 1}{a - 2}$

$= \dfrac{4a^2 - 1}{a^2 - 4} \cdot \dfrac{a - 2}{2a - 1}$

$= \dfrac{(2a + 1)(2a - 1)(a - 2)}{(a + 2)(a - 2)(2a - 1)}$

$= \dfrac{(2a - 1)(a - 2)}{(2a - 1)(a - 2)} \cdot \dfrac{2a + 1}{a + 2}$

$= \dfrac{2a + 1}{a + 2}$

37. $\dfrac{x^3 - 64}{x^3 + 64} \div \dfrac{x^2 - 16}{x^2 - 4x + 16}$

$= \dfrac{x^3 - 64}{x^3 + 64} \cdot \dfrac{x^2 - 4x + 16}{x^2 - 16}$

$= \dfrac{(x - 4)(x^2 + 4x + 16)(x^2 - 4x + 16)}{(x + 4)(x^2 - 4x + 16)(x + 4)(x - 4)}$

$= \dfrac{(x - 4)(x^2 - 4x + 16)}{(x - 4)(x^2 - 4x + 16)} \cdot \dfrac{x^2 + 4x + 16}{(x + 4)(x + 4)}$

$= \dfrac{x^2 + 4x + 16}{x^2 + 8x + 16}$

43. $\dfrac{0.0049t}{t + 0.007} \cdot \dfrac{27{,}000t}{t - 0.007}$

$= \dfrac{(0.0049t)(27{,}000t)}{(t + 0.007)(t - 0.007)}$

$= \dfrac{132.3t^2}{t^2 - 0.000049}$

49. $\dfrac{x^3 + x^2 - y^3 - y^2}{x^2 - 2xy + y^2}$

$= \dfrac{(x^3 - y^3) + (x^2 - y^2)}{x^2 - 2xy + y^2}$

$= \dfrac{(x - y)(x^2 + xy + y^2) + (x + y)(x - y)}{(x - y)^2}$

$= \dfrac{(x - y)[(x^2 + xy + y^2) + (x + y)]}{(x - y)(x - y)}$

$= \dfrac{x - y}{x - y} \cdot \dfrac{x^2 + xy + y^2 + x + y}{x - y}$

$= \dfrac{x^2 + xy + y^2 + x + y}{x - y}$

Exercise Set 9.2

1. $12 = 2 \cdot 2 \cdot 3$

$18 = 2 \cdot 3 \cdot 3$

$LCM = 2 \cdot 2 \cdot 3 \cdot 3$, or 36

7. $9 = 3 \cdot 3$

$15 = 3 \cdot 5$

$5 = 5$

$LCM = 3 \cdot 3 \cdot 5$, or 45

13. $\dfrac{7}{12} + \dfrac{11}{18}$, $LCM = 36$

$= \dfrac{7}{12} \cdot \dfrac{3}{3} + \dfrac{11}{18} \cdot \dfrac{2}{2}$

$= \dfrac{21}{36} + \dfrac{22}{36}$

$= \dfrac{43}{36}$

19. $12x^2y = 2 \cdot 2 \cdot 3 \cdot x \cdot x \cdot y$

$4xy = 2 \cdot 2 \cdot x \cdot y$

LCM $= 2 \cdot 2 \cdot 3 \cdot x \cdot x \cdot y$, or $12x^2y$

25. $3(y - 2) = 3(y - 2)$

$6(2 - y) = 2 \cdot 3(2 - y)$

LCM $= 2 \cdot 3 \cdot (y - 2) = 6(y - 2)$

or $= 2 \cdot 3(2 - y) = 6(2 - y)$

31. $(a + 1) = (a + 1)$

$(a - 1)^2 = (a - 1)(a - 1)$

$a^2 - 1 = (a + 1)(a - 1)$

LCM $= (a + 1)(a - 1)(a - 1)$

37. $2r^2 - 5r - 12 = (2r + 3)(r - 4)$

$3r^2 - 13r + 4 = (3r - 1)(r - 4)$

LCM $= (2r + 3)(r - 4)(3r - 1)$

43. $x^8 - x^4 = x^4(x^4 - 1)$

$\qquad = x^4(x^2 + 1)(x^2 - 1)$

$\qquad = x \cdot x \cdot x \cdot x(x^2 + 1)(x + 1)(x - 1)$

$x^5 - x^2 = x^2(x^3 - 1)$

$\qquad = x \cdot x(x - 1)(x^2 + x + 1)$

$x^5 - x^3 = x^3(x^2 - 1)$

$\qquad = x \cdot x \cdot x(x + 1)(x - 1)$

$x^5 + x^2 = x^2(x^3 + 1)$

$\qquad = x \cdot x(x + 1)(x^2 - x + 1)$

LCM $= x^4(x^2 + 1)(x + 1)(x - 1)(x^2 + x + 1)(x^2 - x + 1)$

Exercise Set 9.3

1. $\dfrac{a - 3b}{a + 5} + \dfrac{a + 5b}{a + b}$

$= \dfrac{a - 3b + a + 5b}{a + b}$

$= \dfrac{2a + 2b}{a + b} = \dfrac{2(a + b)}{a + b}$

$= \dfrac{2}{1} \cdot \dfrac{a + b}{a + b} = 2$

7. $\dfrac{3}{x} - \dfrac{8}{-x}$

$= \dfrac{3}{x} - \dfrac{-1}{-1} \cdot \dfrac{8}{-x}$

$= \dfrac{3}{x} - \dfrac{-8}{x}$

$= \dfrac{3 - (-8)}{x}$

$= \dfrac{11}{x}$

13. $\dfrac{4xy}{x^2 - y^2} + \dfrac{x - y}{x + y}$

$= \dfrac{4xy}{(x + y)(x - y)} + \dfrac{x - y}{x + y}$

LCM $= (x + y)(x - y)$

$= \dfrac{4xy}{(x + y)(x - y)} + \dfrac{x - y}{x + y} \cdot \dfrac{x - y}{x - y}$

$= \dfrac{4xy + x^2 - 2xy + y^2}{(x + y)(x - y)}$

$= \dfrac{x^2 + 2xy + y^2}{(x + y)(x - y)} = \dfrac{(x + y)(x + y)}{(x + y)(x - y)}$

$= \dfrac{x + y}{x + y} \cdot \dfrac{x + y}{x - y} = \dfrac{x + y}{x - y}$

19. $\dfrac{x - 1}{3x + 15} - \dfrac{x + 3}{5x + 25}$

$= \dfrac{x - 1}{3(x + 5)} - \dfrac{x + 3}{5(x + 5)}$

LCM $= 3 \cdot 5(x + 5)$, or $15(x + 5)$

$= \dfrac{x - 1}{3(x + 5)} \cdot \dfrac{5}{5} - \dfrac{x + 3}{5(x + 5)} \cdot \dfrac{3}{3}$

$= \dfrac{5x - 5 - (3x + 9)}{15(x + 5)}$

$= \dfrac{5x - 5 - 3x - 9}{15(x + 5)}$

$= \dfrac{2x - 14}{15x + 75}$

25. $\dfrac{y}{y^2 - y - 20} + \dfrac{2}{y + 4}$

$= \dfrac{y}{(y - 5)(y + 4)} + \dfrac{2}{y + 4}$

LCM $= (y - 5)(y + 4)$

$= \dfrac{y}{(y - 5)(y + 4)} + \dfrac{2}{y + 4} \cdot \dfrac{y - 5}{y - 5}$

$= \dfrac{y + 2y - 10}{(y - 5)(y + 4)}$

$= \dfrac{3y - 10}{y^2 - y - 20}$

31. $\dfrac{1}{x + 1} - \dfrac{x}{x - 2} + \dfrac{x^2 + 2}{x^2 - x - 2}$

$= \dfrac{1}{x + 1} - \dfrac{x}{x - 2} + \dfrac{x^2 + 2}{(x - 2)(x + 1)}$

LCM $= (x + 1)(x - 2)$

$= \dfrac{1}{x + 1} \cdot \dfrac{x - 2}{x - 2} - \dfrac{x}{x - 2} \cdot \dfrac{x + 1}{x + 1} + \dfrac{x^2 + 2}{(x - 2)(x + 1)}$

$= \dfrac{x - 2 - (x^2 + x) + x^2 + 2}{(x + 1)(x - 2)}$

$= \dfrac{x - 2 - x^2 - x + x^2 + 2}{(x + 1)(x - 2)}$

$= \dfrac{0}{(x + 1)(x - 2)} = 0$

37. $2x^{-2} + 3x^{-2}y^{-2} - 7xy^{-1}$

$= \dfrac{2}{x^2} + \dfrac{3}{x^2y^2} - \dfrac{7x}{y}$

LCM $= x^2y^2$

$= \dfrac{2}{x^2} \cdot \dfrac{y^2}{y^2} + \dfrac{3}{x^2y^2} - \dfrac{7x}{y} \cdot \dfrac{x^2y}{x^2y}$

$= \dfrac{2y^2 + 3 - 7x^3y}{x^2y^2}$

Exercise Set 9.4

1. $\dfrac{\dfrac{1}{x} + 4}{\dfrac{1}{x} - 3} = \dfrac{\dfrac{1}{x} + \dfrac{4}{1} \cdot \dfrac{x}{x}}{\dfrac{1}{x} - \dfrac{3}{1} \cdot \dfrac{x}{x}}$

$= \dfrac{\dfrac{1 + 4x}{x}}{\dfrac{1 - 3x}{x}}$

$= \dfrac{1 + 4x}{x} \cdot \dfrac{x}{1 - 3x}$

$= \dfrac{x}{x} \cdot \dfrac{1 + 4x}{1 - 3x}$

$= \dfrac{1 + 4x}{1 - 3x}$

7. $\dfrac{\dfrac{x^2 - y^2}{xy}}{\dfrac{x - y}{y}}$

$= \dfrac{x^2 - y^2}{xy} \cdot \dfrac{y}{x - y}$

$= \dfrac{(x + y)(x - y)y}{xy(x - y)}$

$= \dfrac{y(x - y)}{y(x - y)} \cdot \dfrac{x + y}{x}$

$= \dfrac{x + y}{x}$

13. $\dfrac{\dfrac{1}{x + h} - \dfrac{1}{x}}{h} = \dfrac{\dfrac{1}{x + h} \cdot \dfrac{x}{x} - \dfrac{1}{x} \cdot \dfrac{x + h}{x + h}}{h}$

$= \dfrac{\dfrac{x - (x + h)}{x(x + h)}}{\dfrac{h}{1}}$

$= \dfrac{-h}{x(x + h)} \cdot \dfrac{1}{h}$

$= \dfrac{h}{h} \cdot \dfrac{-1}{x(x + h)}$

$= \dfrac{-1}{x(x + h)}$

19. $\dfrac{5x^{-1} - 5y^{-1} + 10x^{-1}y^{-1}}{6x^{-1} - 6y^{-1} + 12x^{-1}y^{-1}}$

$= \dfrac{\dfrac{5}{x} \cdot \dfrac{y}{y} - \dfrac{5}{y} \cdot \dfrac{x}{x} + \dfrac{10}{xy}}{\dfrac{6}{x} \cdot \dfrac{y}{y} - \dfrac{6}{y} \cdot \dfrac{x}{x} + \dfrac{12}{xy}}$

$= \dfrac{\dfrac{5y - 5x + 10}{xy}}{\dfrac{6y - 6x + 12}{xy}}$

$= \dfrac{5y - 5x + 10}{xy} \cdot \dfrac{xy}{6y - 6x + 12}$

$= \dfrac{5y - 5x + 10}{6y - 6x + 12} = \dfrac{5(y - x + 2)}{6(y - x + 2)} = \dfrac{5}{6}$

25. $\dfrac{1 - \dfrac{1}{a}}{a - 1} = \dfrac{\dfrac{a - 1}{a}}{\dfrac{a - 1}{1}} = \dfrac{a - 1}{a} \cdot \dfrac{1}{a - 1} = \dfrac{1}{a}$ The reciprocal of $1/a$ is a.

31. $f(x) = \dfrac{1}{1 - x}, \quad f(x + h) = \dfrac{1}{1 - x - h}$

$\dfrac{f(x + h) - f(x)}{h} = \dfrac{\dfrac{1}{1 - x - h} - \dfrac{1}{1 - x}}{h}$

$= \dfrac{1 - x - (1 - x - h)}{(1 - x - h)(1 - x)} \cdot \dfrac{1}{h}$

$= \dfrac{1 - x - 1 + x + h}{(1 - x - h)(1 - x)} \cdot \dfrac{1}{h}$

$= \dfrac{1}{(1 - x - h)(1 - x)}$

Exercise Set 9.5

1. $\dfrac{2}{5} + \dfrac{7}{8} = \dfrac{y}{20}, \quad$ LCM $= 40 \qquad$ Check:

$40\left(\dfrac{2}{5} + \dfrac{7}{8}\right) = 40 \cdot \dfrac{y}{20}$

$40 \cdot \dfrac{2}{5} + 40 \cdot \dfrac{7}{8} = 40 \cdot \dfrac{y}{20}$

$16 + 35 = 2y$

$51 = 2y$

$\dfrac{51}{2} = y$

$$\begin{array}{c|c} \dfrac{2}{5} + \dfrac{7}{8} = \dfrac{y}{20} & \\ \hline \dfrac{16}{40} + \dfrac{35}{40} & \dfrac{\frac{51}{2}}{20} \\ \dfrac{51}{40} & \dfrac{51}{2} \cdot \dfrac{1}{20} \\ & \dfrac{51}{40} \end{array}$$

The solution is $\dfrac{51}{2}$.

7. $y + \dfrac{5}{y} = -6, \quad$ LCM $= y \qquad$ Check:

$y\left(y + \dfrac{5}{y}\right) = y(-6)$

$y \cdot y + y \cdot \dfrac{5}{y} = -6y$

$y^2 + 5 = -6y$

$y^2 + 6y + 5 = 0$

$(y + 5)(y + 1) = 0$

$y + 5 = 0 \quad$ or $\quad y + 1 = 0$

$y = -5 \quad$ or $\qquad y = -1$

For: -5

$$\begin{array}{c|c} y + \dfrac{5}{y} = -6 & \\ \hline -5 + \dfrac{5}{-5} & -6 \\ -5 - 1 & \\ -6 & \end{array}$$

For: -1

$$\begin{array}{c|c} y + \dfrac{5}{y} = -6 & \\ \hline -1 + \dfrac{5}{-1} & -6 \\ -1 - 5 & \\ -6 & \end{array}$$

The solutions are -5 and -1.

13.
$$\frac{3}{y + 1} = \frac{2}{y - 3}, \quad \text{LCM} = (y + 1)(y - 3)$$

$$(y + 1)(y - 3) \cdot \frac{3}{y + 1} = (y + 1)(y - 3) \cdot \frac{2}{y - 3}$$

$$3(y - 3) = 2(y + 1)$$

$$3y - 9 = 2y + 2$$

$$y = 11$$

Check:

$$\frac{\frac{3}{y + 1} = \frac{2}{y - 3}}{}$$

$$\frac{3}{11 + 1} \bigg| \frac{2}{11 - 3}$$

$$\frac{3}{12} \bigg| \frac{2}{8}$$

$$\frac{1}{4} \bigg| \frac{1}{4}$$

The solution is 11.

19. $\frac{2}{x} - \frac{3}{x} + \frac{4}{x} = 5, \quad \text{LCM} = x$ Check:

$$x\left(\frac{2}{x} - \frac{3}{x} + \frac{4}{x}\right) = x \cdot 5$$

$$2 - 3 + 4 = 5x$$

$$3 = 5x$$

$$\frac{3}{5} = x$$

$$\frac{\frac{2}{x} - \frac{3}{x} + \frac{4}{x} = 5}{}$$

$$\frac{2}{\frac{3}{5}} - \frac{3}{\frac{3}{5}} + \frac{4}{\frac{3}{5}} \bigg| \frac{15}{3}$$

$$\frac{10}{3} - \frac{15}{3} + \frac{20}{3}$$

$$\frac{15}{3}$$

The solution is $\frac{3}{5}$.

25.
$$\frac{7}{5x - 2} = \frac{5}{4x}, \quad \text{LCM} = 4x(5x - 2)$$

$$4x(5x - 2) \cdot \frac{7}{5x - 2} = 4x(5x - 2) \cdot \frac{5}{4x}$$

$$4x \cdot 7 = 5(5x - 2)$$

$$28x = 25x - 10$$

$$3x = -10$$

$$x = -\frac{10}{3}$$

Since $-\frac{10}{3}$ checks, it is the solution.

31.
$$\frac{2x + 3}{x - 1} = \frac{10}{x^2 - 1} + \frac{2x - 3}{x + 1}$$

$$\frac{2x + 3}{x - 1} = \frac{10}{(x - 1)(x + 1)} + \frac{2x - 3}{x + 1}$$

$$\text{LCM} = (x - 1)(x + 1)$$

$$(x-1)(x+1) \cdot \frac{2x+3}{x-1} = (x-1)(x+1)\left(\frac{10}{(x-1)(x+1)} + \frac{2x-3}{x+1}\right)$$

$$(x + 1)(2x + 3) = 10 + (x - 1)(2x - 3)$$

$$2x^2 + 5x + 3 = 10 + 2x^2 - 5x + 3$$

$$5x + 3 = 13 - 5x$$

$$10x = 10$$

$$x = 1$$

We know that 1 is not a solution of the original equation because it results in division by 0. The equation has no solution.

37.
$$\frac{2.315}{y} - \frac{12.6}{17.4} = \frac{6.71}{7} + 0.763$$

$$\text{LCM} = 7(17.4)y$$

$$7(17.4)y\left(\frac{2.315}{y} - \frac{12.6}{17.4}\right) = 7(17.4)y\left(\frac{6.71}{7} + 0.763\right)$$

$$7(17.4)(2.315) - 7(12.6)y = (17.4)(6.71)y + 7(17.4)(0.763)y$$

$$281.967 - 88.2y = 116.754y + 92.9334y$$

$$281.967 = 88.2y + 116.754y + 92.9334y$$

$$281.967 = 297.8874y$$

$$0.947 \approx y$$

Since 0.947 checks, it is the solution.

43. $(x + 3)^2 = x^2 + 9$

$$x^2 + 6x + 9 = x^2 + 9$$

$$6x = 0$$

$$x = \frac{0}{6}$$

$$x = 0$$

This equation is not an identity. It is not true for all sensible replacements. It is only true for x = 0.

Exercise Set 9.6

1. Translate:

Let x represent the number. Then the reciprocal of the number is $\frac{1}{x}$.

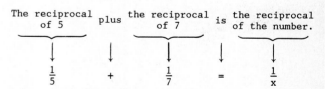

$$\frac{1}{5} \qquad + \qquad \frac{1}{7} \qquad = \qquad \frac{1}{x}$$

Carry out: We solve the equation.

$$\frac{1}{5} + \frac{1}{7} = \frac{1}{x}, \quad \text{LCM} = 35x$$

$$35x\left(\frac{1}{5} + \frac{1}{7}\right) = 35x \cdot \frac{1}{x}$$

$$7x + 5x = 35$$

$$12x = 35$$

$$x = \frac{35}{12}$$

Check:

The number to be checked is $\frac{35}{12}$. Its reciprocal is $\frac{12}{35}$. The sum of $\frac{1}{5} + \frac{1}{7}$ is $\frac{7}{35} + \frac{5}{35}$, or $\frac{12}{35}$, so the value checks.

State:

The number is $\frac{35}{12}$.

7. Familiarize and Translate:

The pool can be filled in 12 hours with only the pipe and in 30 hours with only the hose. Then in 1 hour, the pipe fills $\frac{1}{12}$ of the pool, and the hose fills $\frac{1}{30}$ of the pool. Using both the pipe and the hose, $\frac{1}{12} + \frac{1}{30}$ of the pool can be filled in 1 hour. We suppose that it takes t hours to fill the pool using both the pipe and hose so $\frac{1}{t}$ of the pool should be filled in 1 hour. This gives us an equation.

$\frac{1}{12} + \frac{1}{30} = \frac{1}{t}$,   LCM = 60t

Carry out: We solve the equation.

$60t(\frac{1}{12} + \frac{1}{30}) = 60t \cdot \frac{1}{t}$

$5t + 2t = 60$

$7t = 60$

$t = \frac{60}{7}$

Check:

The possible solution is $\frac{60}{7}$ hours. If the pipe is used $\frac{60}{7}$ hours, it fills $\frac{1}{12} \cdot \frac{60}{7}$, or $\frac{5}{7}$ of the pool. If the hose is used $\frac{60}{7}$ hours, it fills $\frac{1}{30} \cdot \frac{60}{7}$, or $\frac{2}{7}$ of the pool. Using both, $\frac{5}{7} + \frac{2}{7}$ of the pool, or all of it, will be filled.

State:

Using both, it will take $\frac{60}{7}$, or $8\frac{4}{7}$ hours, to fill the pool.

13. Familiarize: We first make a drawing.

Upstream   4 miles   r - 3 mph

10 miles   r + 3 mph   Downstream

We let r represent the speed of the boat in still water. Then r - 3 is the speed upstream and r + 3 is the speed downstream. The time is the same both upstream and downstream so we just use t for each time. We organize the information in a table.

	Distance	Speed	Time
Upstream	4	r - 3	t .
Downstream	10	r + 3	t

Translate:

Using $t = \frac{d}{r}$ we get two different equations from the rows of the table.

$t = \frac{4}{r - 3}$ and $t = \frac{10}{r + 3}$

Since the times are the same, we get

$\frac{4}{r - 3} = \frac{10}{r + 3}$

13. (continued)

Carry out:

We solve the equation. The LCM is (r - 3)(r + 3).

$(r - 3)(r + 3) \cdot \frac{4}{r - 3} = (r - 3)(r + 3) \cdot \frac{10}{r + 3}$

$4(r + 3) = 10(r - 3)$

$4r + 12 = 10r - 30$

$42 = 6r$

$7 = r$

Check:

If r = 7 mph, then r - 3 is 4 mph and r + 3 is 10 mph. The time upstream is $\frac{4}{4}$, or 1 hour. The time downstream is $\frac{10}{10}$, or 1 hour. The times are the same. The values check.

State:

The speed of the boat in still water is 7 mph.

19. Familiarize: We first make a drawing.

140 km   15 + r   t hours   Downstream

35 km   15 - r   t hours   Upstream

We let r represent the speed of the river. Then 15 + r is her speed downstream and 15 - r is her speed upstream. The times are the same. We organize the information in a table.

	Distance	Speed	Time
Downstream	140	15 + r	t
Upstream	35	15 - r	t

Translate:

Using $t = \frac{d}{r}$, we get two equations from the table.

$t = \frac{140}{15 + r}$ and $t = \frac{35}{15 - r}$

Since the times are the same, we have

$\frac{140}{15 + r} = \frac{35}{15 - r}$

Carry out:

We solve the equation. The LCM is (15 + r)(15 - r).

$(15 + r)(15 - r) \cdot \frac{140}{15 + r} = (15 + r)(15 - r) \cdot \frac{35}{15 - r}$

$140(15 - r) = 35(15 + r)$

$2100 - 140r = 525 + 35r$

$1575 = 175r$

$9 = r$

19. (continued)

Check:

If r = 9, then the speed downstream is 15 + 9. or 24 km/h and the speed upstream is 15 - 9, or 6 km/h. The time for the trip downstream is $\frac{140}{24}$, or $5\frac{5}{6}$ hours. The time for the trip upstream is $\frac{35}{6}$, or $5\frac{5}{6}$ hours. The times are the same. The values check.

State:

The speed of the river is 9 km/h.

25. Familiarize:

We let x represent the speed of the current and 3x represent the speed of the boat. Then the speed up the river is 3x - x, or 2x, and the speed down the river is 3x + x, or 4x. The total distance is 100 km; thus the distance each way is 50 km. Using $t = \frac{d}{r}$, we can use $\frac{50}{2x}$ for the time up the river and $\frac{50}{4x}$ for the time down the river.

Translate:

Since the total of the times is 10 hours, we have the following equation.

$$\frac{50}{2x} + \frac{50}{4x} = 10$$

Carry out:

We solve the equation. The LCM is 4x.

$$4x\left(\frac{50}{2x} + \frac{50}{4x}\right) = 4x \cdot 10$$

$$100 + 50 = 40x$$

$$150 = 40x$$

$$\frac{15}{5} = x$$

$$\text{or } x = 3\frac{3}{4}$$

Check:

If the speed of the current is $\frac{15}{4}$ km/h, then the speed of the boat is $3 \cdot \frac{15}{4}$, or $\frac{45}{4}$. The speed up the river is $\frac{45}{4} - \frac{15}{4}$, or $\frac{15}{2}$ km/h, and the time traveling up the river is $50 \div \frac{15}{2}$, or $6\frac{2}{3}$ hr. The speed down the river is $\frac{45}{4} + \frac{15}{4}$, or 15 km/h, and the time traveling down the river is $50 \div 15$, or $3\frac{1}{3}$ hr. The total time for the trip is $6\frac{2}{3} + 3\frac{1}{3}$, or 10 hr. The value checks.

State:

The speed of the current is $3\frac{3}{4}$ km/h.

1. $\dfrac{30x^8 - 15x^6 + 40x^4}{5x^4}$

$= \dfrac{30x^8}{5x^4} - \dfrac{15x^6}{5x^4} + \dfrac{40x^4}{5x^4}$

$= 6x^4 - 3x^2 + 8$

7. $(36x^6 - 18x^4 - 12x^2) \div -6x$

$= \dfrac{36x^6}{-6x} - \dfrac{18x^4}{-6x} - \dfrac{12x^2}{-6x}$

$= -6x^5 - (-3x^3) - (-2x)$

$= -6x^5 + 3x^3 + 2x$

13.

$$\begin{array}{r} x + 7 \\ x + 3 \enclose{longdiv}{x^2 + 10x + 21} \\ \underline{x^2 + 3x} \\ 7x + 21 \\ \underline{7x + 21} \\ 0 \end{array}$$

$(10x - 3x = 7x)$

The answer is x + 7.

19.

$$\begin{array}{r} y^2 - 2y - 1 \\ y - 2 \enclose{longdiv}{y^3 - 4y^2 + 3y - 6} \\ \underline{y^3 - 2y^2} \\ -2y^2 + 3y \\ \underline{-2y^2 + 4y} \\ -y - 6 \\ \underline{-y + 2} \\ -8 \end{array}$$

The answer is $y^2 - 2y - 1$ with R = -8, or $y^2 - 2y - 1 + \dfrac{-8}{y - 2}$.

25.

$$\begin{array}{r} x^2 + 6 \\ x^2 - 7 \enclose{longdiv}{x^4 - x^2 - 42} \\ \underline{x^4 - 7x^2} \\ 6x^2 - 42 \\ \underline{6x^2 - 42} \\ 0 \end{array}$$

The answer is $x^2 + 6$.

31.

$$
\begin{array}{r}
2x^2 - x - 9 \\
x^2 + 2 \overline{\smash{\big)}\ 2x^4 - x^3 - 5x^2 + x - 6} \\
\underline{2x^4\qquad\ + 4x^2} \\
-x^3 - 9x^2 + x \\
\underline{-x^3\qquad\ - 2x} \\
-9x^2 + 3x - 6 \\
\underline{-9x^2\qquad - 18} \\
3x + 12
\end{array}
$$

The answer is $2x^2 - x - 9$ with $R = 3x + 12$,

or $2x^2 - x - 9 + \dfrac{3x + 12}{x^2 + 2}$.

37.

$$
\begin{array}{r}
x^3 + x^2y + xy^2 + y^3 \\
x - y \overline{\smash{\big)}\ x^4 \qquad\qquad\qquad - y^4} \\
\underline{x^4 - x^3y} \\
x^3y \\
\underline{x^3y - x^2y^2} \\
x^2y^2 \\
\underline{x^2y^2 - xy^3} \\
xy^3 - y^4 \\
\underline{xy^3 - y^4} \\
0
\end{array}
$$

The answer is $x^3 + x^2y + xy^2 + y^3$.

Exercise Set 9.8

1.

$$\frac{W_1}{W_2} = \frac{d_1}{d_2}$$

$$\frac{d_2 W_1}{W_2} = d_1 \qquad \text{(Multiplying by } d_2)$$

7.

$$R = \frac{gs}{g + s}$$

$$R(g + s) = gs \qquad \text{(Multiplying by } g + s)$$

$$Rg + Rs = gs \qquad \text{(Removing parentheses)}$$

$$Rg = gs - Rs \qquad \text{(Adding } -Rs)$$

$$Rg = s(g - R) \qquad \text{(Factoring)}$$

$$\frac{Rg}{g - R} = s \qquad \text{(Multiplying by } \frac{1}{g - R})$$

13.

$$I = \frac{nE}{R + nr}$$

$$I(R + nr) = nE \qquad \text{(Multiplying by } R + nr)$$

$$IR + Inr = nE \qquad \text{(Removing parentheses)}$$

$$Inr = nE - IR \qquad \text{(Adding } -IR)$$

$$r = \frac{nE - IR}{In} \qquad \text{(Multiplying by } \frac{1}{In})$$

19.

$$S = \frac{a - ar^n}{1 - r}$$

$$S(1 - r) = a - ar^n \qquad \text{(Multiplying by } 1 - r)$$

$$S(1 - r) = a(1 - r^n) \qquad \text{(Factoring)}$$

$$\frac{S - Sr}{1 - r^n} = a \qquad \text{(Multiplying by } \frac{1}{1 - r^n})$$

25. Working alone Pam can do the job in a hours. Thus, Pam can do $\frac{1}{a}$ of the job in 1 hour. Elaine, working alone, can do the job in b hours. Thus, Elaine can do $\frac{1}{b}$ of the job in 1 hour. Working together it takes them t hours. Thus, they do $\frac{1}{t}$ of the job in 1 hour. We have the following equation.

$$\frac{1}{a} + \frac{1}{b} = \frac{1}{t}$$

or

$$bt + at = ab \qquad \text{(Multiplying by } abt)$$

a) $bt + at = ab$

$\qquad t(b + a) = ab$

$$t = \frac{ab}{b + a}$$

b) $bt + at = ab$

$\qquad bt = ab - at$

$\qquad bt = a(b - t)$

$$\frac{bt}{b - t} = a$$

c) $bt + at = ab$

$\qquad at = ab - bt$

$\qquad at = b(a - t)$

$$\frac{at}{a - t} = b$$

Exercise Set 10.1

1. The square roots of 1 are 1 and -1   **2.** 2, -2
because $1^2 = 1$ and $(-1)^2 = 1$.

7. The square roots of 169 are 13 and -13   **8.** 12, -12
because $13^2 = 169$ and $(-13)^2 = 169$.

13. $\sqrt{64} = 8$, so $-\sqrt{64} = -8$.

19. $\sqrt{2}$ is irrational, since 2 is not a perfect square.

25. $\sqrt{98}$ is irrational, since 98 is not a perfect square.

31. 4.23 is rational, since the decimal notation ends.

37. $\frac{2.3}{0.01}$ is rational since it can be expressed as the ratio of two integers ($\frac{2.3}{0.01} \times \frac{100}{100} = \frac{230}{1}$).

43. -63.030030003... is irrational, since the decimal notation neither ends nor repeats.

49. $\sqrt{43} \approx 6.557$

55. $-\sqrt{36} < -\sqrt{33} < -\sqrt{25}$
$-6 < -\sqrt{33} < -5$
$-\sqrt{33}$ is between -6 and -5.

61. Consider 10 and 20. The number halfway between 10 and 20 is 15, or $\frac{10 + 20}{2}$.

The number halfway between x and y is $\frac{x + y}{2}$.

67. $\sqrt{\frac{361}{9}} = \sqrt{\left(\frac{19}{3}\right)^2} = \frac{19}{3}$, so $-\sqrt{\frac{361}{9}} = -\frac{19}{3}$

73. $\sqrt{0.36} = \sqrt{(0.6)^2} = 0.6$

Exercise Set 10.2

1. a - 4   **7.** Yes, $\sqrt{-16}$ is meaningless.

13. The radicand must be greater than or equal to 0.
We solve: $t - 5 \geq 0$
$t \geq 5$
Any number greater than or equal to 5 is sensible.

19. The radicand must be greater than or equal to 0.
We solve: $2y - 7 \geq 0$
$2y \geq 7$
$y \geq \frac{7}{2}$

Any number greater than or equal to $\frac{7}{2}$ is sensible.

25. $\sqrt{9x^2} = \sqrt{(3x)^2} = |3x| = |3| \cdot |x| = 3|x|$

31. $\sqrt{(x + 3)^2} = |x + 3|$

37. $\sqrt[3]{-64x^3} = -4x$ $[(-4x)^3 = -64x^3]$

43. $\sqrt[4]{625} = 5$ $[5^4 = 625]$

49. $\sqrt[6]{x^6} = |x|$
Since k is even, we use absolute value.

55. $\sqrt[414]{(a + b)^{414}} = |a + b|$
Since k is even, we use absolute value.

61. Familiarize:

This problem states that we have direct variation between F and I. Thus, an equation F = kI, k > 0, applies. As the income increases, the amount spent on food increases.

Translate: We write an equation of variation.
Amount spent on food varies directly as the income.
This translates to F = kI.

Carry out:

a) First find an equation of variation.
$$F = kI$$
$$5096 = k \cdot 19{,}600$$ (Substituting 5096 for F and 19,600 for I)
$$\frac{5096}{19{,}600} = k$$
$$0.26 = k$$
The equation of variation is F = 0.26I.

b) We use the equation to find how much a family spends on food when their income is $20,500.
$$F = 0.26I$$
$$F = 0.26(\$20{,}500)$$ (Substituting $20,500 for I)
$$F = 5330$$

Check:

Let us do some reasoning about the answer. The income increased from $19,600 to $20,500. Similarly, the amount spent on food increased from $5096 to $5330. This is what we would expect with direct variation.

State:

The amount spent on food is $5330.

67. $\sqrt{(3a)^2} = |3a| = |3| \cdot |a| = 3|a|$

73. We solve: $m(m + 3) \geq 0$

If $m(m + 3) = 0$, then $m = 0$ or -3.

If $m(m + 3) > 0$, then either

both m and m + 3 are positive

or

both m and m + 3 are negative.

If $m > 0$ and $m + 3 > 0$ (or $m > -3$), then $m > 0$.

If $m < 0$ and $m + 3 < 0$ (or $m < -3$), then $m < -3$.

Thus, the sensible replacements are $m \geq 0$ or $m \leq -3$.

79. $f(y) = \sqrt{5y - 10}$

$f(6) = \sqrt{5 \cdot 6 - 10} = \sqrt{20}$

$f(2) = \sqrt{5 \cdot 2 - 10} = \sqrt{0} = 0$

$f(0)$ and $f(-3)$ do not exist.

$\quad f(-3) = \sqrt{5(-3) - 10} = \sqrt{-25}$. Negative numbers do not have real-number square roots.

85. $f(x) = \sqrt[3]{x + 1}$

$f(7) = \sqrt[3]{7 + 1} = \sqrt[3]{8} = 2$

$f(26) = \sqrt[3]{26 + 1} = \sqrt[3]{27} = 3$

$f(-9) = \sqrt[3]{-9 + 1} = \sqrt[3]{-8} = -2$

$f(-65) = \sqrt[3]{-65 + 1} = \sqrt[3]{-64} = -4$

Exercise Set 10.3

1. $\sqrt{3} \sqrt{2} = \sqrt{3 \cdot 2} = \sqrt{6}$

7. $\sqrt{3a} \sqrt{10b} = \sqrt{3a \cdot 10b} = \sqrt{30ab}$

13. $\sqrt[3]{0.3x} \sqrt[3]{0.2x} = \sqrt[3]{0.3x \cdot 0.2x} = \sqrt[3]{0.06x^2}$

19. $\sqrt{8} = \sqrt{4 \cdot 2} = \sqrt{4} \sqrt{2} = 2\sqrt{2}$

25. $\sqrt[3]{54x^8} = \sqrt[3]{27 \cdot 2 \cdot x^6 \cdot x^2} = \sqrt[3]{27x^6} \sqrt[3]{2x^2} = 3x^2 \sqrt[3]{2x^2}$

31. $\sqrt[4]{162c^4d^6} = \sqrt[4]{81 \cdot 2 \cdot c^4 \cdot d^4 \cdot d^2} = \sqrt[4]{81c^4d^4} \sqrt[4]{2d^2}$

$\qquad\qquad\qquad = 3cd \sqrt[4]{2d^2}$

37. $\sqrt{2} \sqrt{32} = \sqrt{64} = 8$

43. $\sqrt{2x^3y} \sqrt{12xy} = \sqrt{24x^4y^2}$

$\qquad\qquad\qquad = \sqrt{4 \cdot 6 \cdot x^4 \cdot y^2}$

$\qquad\qquad\qquad = \sqrt{4x^4y^2} \sqrt{6}$

$\qquad\qquad\qquad = 2x^2y\sqrt{6}$

49. $-\frac{5}{6}$ is rational.   55. 2.101001... is irrational.

61. $\sqrt{350} = \sqrt{25 \cdot 14} = 5\sqrt{14} \approx 5(3.742) \approx 18.7$

67. $\sqrt{24,500,000,000} = \sqrt{245 \times 10^8}$

$\qquad\qquad\qquad = \sqrt{245} \times \sqrt{10^8}$

$\qquad\qquad\qquad \approx 15.6525 \times 10^4$

$\qquad\qquad\qquad \approx 156,525$

73. $\sqrt{0.0000000395} = \sqrt{3.95 \times 10^{-8}}$

$\qquad\qquad\qquad = \sqrt{3.95} \times \sqrt{10^{-8}}$

$\qquad\qquad\qquad \approx 1.98746 \times 10^{-4}$

$\qquad\qquad\qquad \approx 0.000199$

79. a) $r = 2\sqrt{5L}$
    $\quad r = 2\sqrt{5 \cdot 20}$
    $\quad = 2\sqrt{100}$
    $\quad = 2 \cdot 10$
    $\quad = 20$ mph

   b) $r = 2\sqrt{5L}$
    $\quad r = 2\sqrt{5 \cdot 70}$
    $\quad = 2\sqrt{350}$
    $\quad \approx 2(18.708)$
    $\quad \approx 37.4$ mph

   c) $r = 2\sqrt{5L}$
    $\quad r = 2\sqrt{5 \cdot 90}$
    $\quad = 2\sqrt{450}$
    $\quad \approx 2(21.213)$
    $\quad \approx 42.4$ mph

Exercise Set 10.4

1. $\sqrt{\frac{16}{25}} = \frac{\sqrt{16}}{\sqrt{25}} = \frac{4}{5}$

7. $\sqrt{\frac{25y^3}{x^4}} = \frac{\sqrt{25y^3}}{\sqrt{x^4}} = \frac{\sqrt{25y^2 \cdot y}}{\sqrt{x^4}} = \frac{5y\sqrt{y}}{x^2}$

13. $\frac{\sqrt[3]{54}}{\sqrt[3]{2}} = \sqrt[3]{\frac{54}{2}} = \sqrt[3]{27} = 3$

19. $\frac{\sqrt{72xy}}{2\sqrt{2}} = \frac{1}{2}\sqrt{\frac{72xy}{2}} = \frac{1}{2}\sqrt{36xy} = \frac{1}{2} \cdot 6\sqrt{xy} = 3\sqrt{xy}$

25. $(\sqrt[3]{16b^2})^2$

$= \sqrt[3]{16b^2}\,\sqrt[3]{16b^2}$

$= \sqrt[3]{256b^4}$

$= \sqrt[3]{64b^3 \cdot 4b}$

$= 4b\sqrt[3]{4b}$

31. Familiarize and Translate:

We let x represent his salary per hour the first year. Then x + 10%x represents his salary per hour the second year, and (x + 10%x) + 10%(x + 10%x) represents his salary per hour the third year. Since his salary per hour the third year was $6.05, we have the following equation.

(x + 10%x) + 10%(x + 10%x) = 6.05

Carry out:

x + 0.1x + 0.1x + 0.01x = 6.05

1.21x = 6.05

$x = \dfrac{6.05}{1.21}$

x = 5

Check:

First year: $5 per hour
Second year: 5 + 10%·5 = 5 + 0.5 = $5.50 per hour
Third year: 5.50 + 10%(5.50) = 5.50 + 0.55
                                = $6.05

State:

Bob earned $5 per hour the first year.

Exercise Set 10.5

1. $6\sqrt{3} + 2\sqrt{3} = (6 + 2)\sqrt{3} = 8\sqrt{3}$

7. $8\sqrt{2} - 6\sqrt{2} + 5\sqrt{2} = (8 - 6 + 5)\sqrt{2} = 7\sqrt{2}$

13. $8\sqrt{45} + 7\sqrt{20}$

$= 8\sqrt{9 \cdot 5} + 7\sqrt{4 \cdot 5}$

$= 8 \cdot 3\sqrt{5} + 7 \cdot 2\sqrt{5}$

$= 24\sqrt{5} + 14\sqrt{5}$

$= 38\sqrt{5}$

19. $2\sqrt{128} - \sqrt{18} + 4\sqrt{32}$

$= 2\sqrt{64 \cdot 2} - \sqrt{9 \cdot 2} + 4\sqrt{16 \cdot 2}$

$= 2 \cdot 8\sqrt{2} - 3\sqrt{2} + 4 \cdot 4\sqrt{2}$

$= 16\sqrt{2} - 3\sqrt{2} + 16\sqrt{2}$

$= 29\sqrt{2}$

25. $\sqrt{8y - 8} + \sqrt{2y - 2}$

$= \sqrt{4(2y - 2)} + \sqrt{2y - 2}$

$= 2\sqrt{2y - 2} + 1\sqrt{2y - 2}$

$= 3\sqrt{2y - 2}$

31. $5\sqrt[3]{32} - \sqrt[3]{108} + 2\sqrt[3]{256}$

$= 5\sqrt[3]{8 \cdot 4} - \sqrt[3]{27 \cdot 4} + 2\sqrt[3]{64 \cdot 4}$

$= 5 \cdot 2\sqrt[3]{4} - 3\sqrt[3]{4} + 2 \cdot 4\sqrt[3]{4}$

$= 10\sqrt[3]{4} - 3\sqrt[3]{4} + 8\sqrt[3]{4}$

$= 15\sqrt[3]{4}$

37. Familiarize: We first make a drawing.

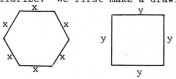

We let x represent the length of a side of the regular hexagon and y represent the length of a side of the square.

Translate:

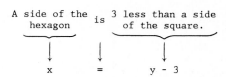

The perimeter of the hexagon | is equal to | the perimeter of the square.

6x = 4y

A side of the hexagon | is | 3 less than a side of the square.

x = y - 3

Carry out:

We solve the resulting system.

6x = 4y

x = y - 3

We substitute y - 3 for x in the first equation and solve for y.

6(y - 3) = 4y

6y - 18 = 4y

2y = 18

y = 9

Check:

If y = 9, then x = 9 - 3, or 6, and 6 is three less than 9. The perimeter of the hexagon is 6·6, or 36. The perimeter of the square is 4·9, or 36. The perimeters are the same.

State:

The perimeter of each polygon is 36.

43. $7x\sqrt{(x + y)^3} - 5xy\sqrt{x + y} - 2y\sqrt{(x + y)^3}$

$= 7x(x + y)\sqrt{x + y} - 5xy\sqrt{x + y} - 2y(x + y)\sqrt{x + y}$

$= [7x(x + y) - 5xy - 2y(x + y)]\sqrt{x + y}$

$= (7x^2 + 7xy - 5xy - 2xy - 2y^2)\sqrt{x + y}$

$= (7x^2 - 2y^2)\sqrt{x + y}$

Exercise Set 10.6

1. $\sqrt{6}(2 - 3\sqrt{6})$
$= \sqrt{6} \cdot 2 - \sqrt{6} \cdot 3\sqrt{6}$
$= 2\sqrt{6} - 3 \cdot 6$
$= 2\sqrt{6} - 18$

7. $\sqrt[3]{2}(\sqrt[3]{4} - 2\sqrt[3]{32})$
$= \sqrt[3]{2}\,\sqrt[3]{4} - \sqrt[3]{2} \cdot 2\sqrt[3]{32}$
$= \sqrt[3]{8} - 2\sqrt[3]{64}$
$= 2 - 2 \cdot 4$
$= 2 - 8$
$= -6$

13. $(\sqrt{8} + 2\sqrt{5})(\sqrt{8} - 2\sqrt{5})$
$= (\sqrt{8})^2 - (2\sqrt{5})^2$
$= 8 - 4 \cdot 5$
$= 8 - 20$
$= -12$

19. $(\sqrt{3} + 1)(2\sqrt{3} + 1)$
$= \sqrt{3} \cdot 2\sqrt{3} + \sqrt{3} \cdot 1 + 1 \cdot 2\sqrt{3} + 1^2$
$= 2 \cdot 3 + \sqrt{3} + 2\sqrt{3} + 1$
$= 7 + 3\sqrt{3}$

25. $(2\sqrt[3]{3} + \sqrt[3]{2})(\sqrt[3]{3} - 2\sqrt[3]{2})$
$= 2\sqrt[3]{3} \cdot \sqrt[3]{3} - 2\sqrt[3]{3} \cdot 2\sqrt[3]{2} + \sqrt[3]{2}\,\sqrt[3]{3} - \sqrt[3]{2} \cdot 2\sqrt[3]{2}$
$= 2\sqrt[3]{9} - 4\sqrt[3]{6} + \sqrt[3]{6} - 2\sqrt[3]{4} = 2\sqrt[3]{9} - 3\sqrt[3]{6} - 2\sqrt[3]{4}$

31. a) $(3 + \sqrt{2})^2 \approx (3 + 1.414214)^2$
$\approx (4.414214)^2$
≈ 19.485281

   b) $(3 + \sqrt{2})^2 = 3^2 + 2 \cdot 3\sqrt{2} + (\sqrt{2})^2$
$= 9 + 6\sqrt{2} + 2$
$= 11 + 6\sqrt{2}$
$\approx 11 + 6(1.414214)$
≈ 19.485281

37. $[\sqrt{3 + \sqrt{2 + \sqrt{1}}}]^4$
$= [\sqrt{3 + \sqrt{2 + 1}}]^4$
$= [\sqrt{3 + \sqrt{3}}]^4$
$= \sqrt{(3 + \sqrt{3})^4}$
$= (3 + \sqrt{3})^2$
$= 3^2 + 6\sqrt{3} + (\sqrt{3})^2$
$= 9 + 6\sqrt{3} + 3$
$= 12 + 6\sqrt{3}$

Exercise Set 10.7

1. $\sqrt{\dfrac{6}{5}} = \sqrt{\dfrac{6 \cdot 5}{5 \cdot 5}} = \sqrt{\dfrac{30}{25}} = \dfrac{\sqrt{30}}{\sqrt{25}} = \dfrac{\sqrt{30}}{5}$

7. $\sqrt[3]{\dfrac{16}{9}} = \sqrt[3]{\dfrac{16}{9} \cdot \dfrac{3}{3}} = \sqrt[3]{\dfrac{48}{27}} = \dfrac{\sqrt[3]{8 \cdot 6}}{\sqrt[3]{27}} = \dfrac{2\sqrt[3]{6}}{3}$

13. $\dfrac{1}{\sqrt[3]{xy}} = \dfrac{1}{\sqrt[3]{xy}} \cdot \dfrac{\sqrt[3]{x^2 y^2}}{\sqrt[3]{x^2 y^2}} = \dfrac{\sqrt[3]{x^2 y^2}}{\sqrt[3]{x^3 y^3}} = \dfrac{\sqrt[3]{x^2 y^2}}{xy}$

19. $\dfrac{4\sqrt{13}}{3\sqrt{7}} = \dfrac{4\sqrt{13}}{3\sqrt{7}} \cdot \dfrac{\sqrt{13}}{\sqrt{13}} = \dfrac{4 \cdot 13}{3\sqrt{91}} = \dfrac{52}{3\sqrt{91}}$

25. $\dfrac{\sqrt[3]{5y^4}}{\sqrt[3]{6x^5}} = \dfrac{\sqrt[3]{5y^4}}{\sqrt[3]{6x^5}} \cdot \dfrac{\sqrt[3]{5^2 y^2}}{\sqrt[3]{5^2 y^2}} = \dfrac{\sqrt[3]{5^3 y^6}}{\sqrt[3]{150x^5 y^2}} = \dfrac{5y^2}{x\sqrt[3]{150x^2 y^2}}$

31. $\dfrac{-4\sqrt{7}}{\sqrt{5} - \sqrt{3}} = \dfrac{-4\sqrt{7}}{\sqrt{5} - \sqrt{3}} \cdot \dfrac{\sqrt{5} + \sqrt{3}}{\sqrt{5} + \sqrt{3}}$
$= \dfrac{-4\sqrt{7}(\sqrt{5} + \sqrt{3})}{5 - 3}$
$= \dfrac{-4\sqrt{7}(\sqrt{5} + \sqrt{3})}{2}$
$= -2\sqrt{7}(\sqrt{5} + \sqrt{3})$

37. $\dfrac{5\sqrt{3} - 3\sqrt{2}}{3\sqrt{2} - 2\sqrt{3}} = \dfrac{5\sqrt{3} - 3\sqrt{2}}{3\sqrt{2} - 2\sqrt{3}} \cdot \dfrac{3\sqrt{2} + 2\sqrt{3}}{3\sqrt{2} + 2\sqrt{3}}$
$= \dfrac{15\sqrt{6} + 10 \cdot 3 - 9 \cdot 2 - 6\sqrt{6}}{9 \cdot 2 - 4 \cdot 3}$
$= \dfrac{12 + 9\sqrt{6}}{6}$
$= \dfrac{3(4 + 3\sqrt{6})}{3 \cdot 2}$
$= \dfrac{4 + 3\sqrt{6}}{2}$

43. $\dfrac{\sqrt{x} - \sqrt{y}}{\sqrt{x} + \sqrt{y}} = \dfrac{\sqrt{x} - \sqrt{y}}{\sqrt{x} + \sqrt{y}} \cdot \dfrac{\sqrt{x} + \sqrt{y}}{\sqrt{x} + \sqrt{y}}$
$= \dfrac{(\sqrt{x})^2 - (\sqrt{y})^2}{(\sqrt{x})^2 + 2\sqrt{x}\,\sqrt{y} + (\sqrt{y})^2}$
$= \dfrac{x - y}{x + 2\sqrt{xy} + y}$

49. $\dfrac{b + \sqrt{b}}{1 + b + \sqrt{b}} = \dfrac{b + \sqrt{b}}{(1 + b) + \sqrt{b}} \cdot \dfrac{(1 + b) - \sqrt{b}}{(1 + b) - \sqrt{b}}$
$= \dfrac{(b + \sqrt{b})(1 + b - \sqrt{b})}{(1 + b)^2 - (\sqrt{b})^2}$
$= \dfrac{b + b^2 - b\sqrt{b} + \sqrt{b} + b\sqrt{b} - b}{1 + 2b + b^2 - b}$
$= \dfrac{b^2 + \sqrt{b}}{b^2 + b + 1}$

55. $5\sqrt{\dfrac{x}{y}} + 4\sqrt{\dfrac{y}{x}} - \dfrac{3}{\sqrt{xy}}$

$= 5\sqrt{\dfrac{x}{y} \cdot \dfrac{y}{y}} + 4\sqrt{\dfrac{y}{x} \cdot \dfrac{x}{x}} - \dfrac{3}{\sqrt{xy}} \cdot \dfrac{\sqrt{xy}}{\sqrt{xy}}$

$= \dfrac{5}{y}\sqrt{xy} + \dfrac{4}{x}\sqrt{xy} - \dfrac{3}{xy}\sqrt{xy}$

$= (\dfrac{5}{y} + \dfrac{4}{x} - \dfrac{3}{xy})\sqrt{xy}$

$= (\dfrac{5x + 4y - 3}{xy})\sqrt{xy}$

Exercise Set 10.8

1. $x^{1/4} = \sqrt[4]{x}$ **7.** $a^{2/3} = \sqrt[3]{a^2}$

13. $\sqrt{17} = 17^{1/2}$ **19.** $(\sqrt{3mn})^3 = (3mn)^{3/2}$

25. $(2rs)^{-3/4} = \dfrac{1}{(2rs)^{3/4}}$

31. $5^{3/4} \cdot 5^{1/8} = 5^{3/4+1/8} = 5^{6/8+1/8} = 5^{7/8}$

37. $(10^{3/5})^{2/5} = 10^{(3/5)(2/5)} = 10^{6/25}$

43. $\sqrt[4]{32} = \sqrt[4]{2^5} = 2^{5/4} = 2^{4/4} \cdot 2^{1/4} = 2\sqrt[4]{2}$

49. $\sqrt[6]{\dfrac{m^{12}n^{24}}{64}} = (\dfrac{m^{12}n^{24}}{2^6})^{1/6} = \dfrac{m^{12/6}n^{24/6}}{2^{6/6}} = \dfrac{m^2n^4}{2}$

55. $\sqrt[3]{7} \cdot \sqrt{2} = 7^{1/3} \cdot 2^{1/2} = 7^{2/6} \cdot 2^{3/6}$

$= (7^2 \cdot 2^3)^{1/6}$

$= (49 \cdot 8)^{1/6}$

$= 392^{1/6}$

$= \sqrt[6]{392}$

61. $\dfrac{\sqrt[3]{(a + b)^2}}{\sqrt{(a + b)}} = \dfrac{(a + b)^{2/3}}{(a + b)^{1/2}} = \dfrac{(a + b)^{4/6}}{(a + b)^{3/6}}$

$= (a + b)^{4/6 - 3/6}$

$= (a + b)^{1/6}$

$= \sqrt[6]{a + b}$

67. $\sqrt[5]{yx^2}\sqrt{xy} = (yx^2)^{1/5}(xy)^{1/2} = (yx^2)^{2/10}(xy)^{5/10}$

$= [(yx^2)^2(xy)^5]^{1/10}$

$= (y^2x^4 \cdot x^5y^5)^{1/10}$

$= (x^9y^7)^{1/10}$

$= \sqrt[10]{x^9y^7}$

73. $\dfrac{1}{\sqrt[3]{3} - \sqrt[3]{2}} = \dfrac{1}{\sqrt[3]{3} - \sqrt[3]{2}} \cdot \dfrac{\sqrt[3]{9} + \sqrt[3]{6} + \sqrt[3]{4}}{\sqrt[3]{9} + \sqrt[3]{6} + \sqrt[3]{4}}$

$= \dfrac{\sqrt[3]{9} + \sqrt[3]{6} + \sqrt[3]{4}}{(\sqrt[3]{3})^3 - (\sqrt[3]{2})^3}$

$= \dfrac{\sqrt[3]{9} + \sqrt[3]{6} + \sqrt[3]{4}}{3 - 2}$

$= \sqrt[3]{9} + \sqrt[3]{6} + \sqrt[3]{4}$

Exercise Set 10.9

1. $\sqrt{2x - 3} = 1$ Check:

$(\sqrt{2x - 3})^2 = 1^2$ $\sqrt{2x - 3} = 1$

$2x - 3 = 1$ $\sqrt{2 \cdot 2 - 3} \mid 1$

$2x = 4$ $\sqrt{4 - 3}$

$x = 2$ $\sqrt{1}$

1

The solution is 2.

7. $\sqrt[4]{y - 3} = 2$ Check:

$(\sqrt[4]{y - 3})^4 = 2^4$ $\sqrt[4]{y - 3} = 2$

$y - 3 = 16$ $\sqrt[4]{19 - 3} \mid 2$

$y = 19$ $\sqrt[4]{16}$

2

The solution is 19.

13. $\sqrt[3]{x} = -3$ Check:

$(\sqrt[3]{x})^3 = (-3)^3$ $\sqrt[3]{x} = -3$

$x = -27$ $\sqrt[3]{-27} \mid -3$

-3

The solution is -27.

19. $8 = \dfrac{1}{\sqrt{x}}$ Check:

$8\sqrt{x} = 1$ $8 = \dfrac{1}{\sqrt{x}}$

$(8\sqrt{x})^2 = 1^2$ $8 \; \Big| \; \dfrac{1}{\sqrt{\dfrac{1}{64}}}$

$64x = 1$

$x = \dfrac{1}{64}$ $\dfrac{1}{\dfrac{1}{8}}$

8

The solution is $\dfrac{1}{64}$.

25. $2\sqrt{1 - x} = \sqrt{5}$ Check:

$(2\sqrt{1 - x})^2 = (\sqrt{5})^2$ $2\sqrt{1 - x} = \sqrt{5}$

$4(1 - x) = 5$ $2\sqrt{1 - (-\dfrac{1}{4})} \; \Big| \; \sqrt{5}$

$4 - 4x = 5$

$-4x = 1$ $2\sqrt{\dfrac{5}{4}}$

$x = -\dfrac{1}{4}$ $\dfrac{2\sqrt{5}}{2}$

$\sqrt{5}$

The solution is $-\dfrac{1}{4}$.

31. $3 + \sqrt{z - 6} = \sqrt{z + 9}$

$(3 + \sqrt{z - 6})^2 = (\sqrt{z + 9})^2$

$9 + 6\sqrt{z - 6} + z - 6 = z + 9$

$6\sqrt{z - 6} = 6$

$\sqrt{z - 6} = 1$

$(\sqrt{z - 6})^2 = 1^2$

$z - 6 = 1$

$z = 7$

The number 7 checks, so it is the solution.

37. $\sqrt{4y + 1} - \sqrt{y - 2} = 3$

$\sqrt{4y + 1} = 3 + \sqrt{y - 2}$

$(\sqrt{4y + 1})^2 = (3 + \sqrt{y - 2})^2$

$4y + 1 = 9 + 6\sqrt{y - 2} + y - 2$

$3y - 6 = 6\sqrt{y - 2}$

$y - 2 = 2\sqrt{y - 2}$

$(y - 2)^2 = (2\sqrt{y - 2})^2$

$y^2 - 4y + 4 = 4(y - 2)$

$y^2 - 4y + 4 = 4y - 8$

$y^2 - 8y + 12 = 0$

$(y - 6)(y - 2) = 0$

$y - 6 = 0$ or $y - 2 = 0$

$y = 6$ or $\qquad y = 2$

The numbers 6 and 2 check, so they are the solutions.

43. $a^2 + b^2 = c^2$ **49.** $a^2 + b^2 = c^2$

$16^2 + b^2 = 20^2$ $1^2 + 1^2 = c^2$

$256 + b^2 = 400$ $1 + 1 = c^2$

$b^2 = 144$ $2 = c^2$

$b = \sqrt{144}$ $\sqrt{2} = c$

$b = 12$ $1.41 \approx c$

55.

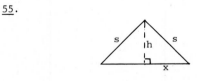

The triangle is isosceles. Thus, h is the perpendicular bisector of the third side. We let x represent half of the third side and then use the Pythagorean property.

$h^2 + x^2 = s^2$

$x^2 = s^2 - h^2$

$x = \sqrt{s^2 - h^2}$

Thus, $2x = 2\sqrt{s^2 - h^2}$.

The length of the base is $2\sqrt{s^2 - h^2}$.

61. $\sqrt[4]{z^2 + 17} = 3$

$(\sqrt[4]{z^2 + 17})^4 = 3^4$

$z^2 + 17 = 81$

$z^2 - 64 = 0$

$(z + 8)(z - 8) = 0$

$z + 8 = 0$ or $z - 8 = 0$

$z = -8$ or $z = 8$

The numbers -8 and 8 check, so
they are the solutions.

67. $6\sqrt{y} + 6y^{-1/2} = 37$

$6\sqrt{y} + \dfrac{6}{\sqrt{y}} = 37$

$\sqrt{y}(6\sqrt{y} + \dfrac{6}{\sqrt{y}}) = \sqrt{y} \cdot 37$

$6y + 6 = 37\sqrt{y}$

$(6y + 6)^2 = (37\sqrt{y})^2$

$36y^2 + 72y + 36 = 1369y$

$36y^2 - 1297y + 36 = 0$

$(36y - 1)(y - 36) = 0$

$36y - 1 = 0$ or $y - 36 = 0$

$36y = 1$ or $y = 36$

$y = \dfrac{1}{36}$ or $y = 36$

Both $\dfrac{1}{36}$ and 36 check, so they are
the solutions.

73. $\sqrt{2x - 2} + \sqrt{7x + 4} = \sqrt{13x + 10}$

$(2x - 2) + 2\sqrt{2x - 2}\sqrt{7x + 4} + (7x + 4) = 13x + 10$

$2\sqrt{2x - 2}\sqrt{7x + 4} = 4x + 8$

$(\sqrt{2x - 2}\sqrt{7x + 4})^2 = (2x + 4)^2$

$14x^2 - 6x - 8 = 4x^2 + 16x + 16$

$10x^2 - 22x - 24 = 0$

$5x^2 - 11x - 12 = 0$

$(5x + 4)(x - 3) = 0$

$5x + 4 = 0$ or $x - 3 = 0$

$5x = -4$ or $x = 3$

$x = -\dfrac{4}{5}$ or $x = 3$

The number 3 checks but $-\dfrac{4}{5}$ does not, so the
solution is 3.

Exercise Set 11.1

<u>1.</u> $x^2 - 3x + 2 = 0$

This equation is already in standard form.

$a = 1, b = -3, c = 2$

<u>7.</u> $5 = -2x^2 + 3x$

$2x^2 - 3x + 5 = 0$

$a = 2, b = -3, c = 5$

<u>13.</u> $3x^2 = 30$

$x^2 = 10$

$|x| = \sqrt{10}$

$x = \sqrt{10}$ or $x = -\sqrt{10}$

Check:

For $\sqrt{10}$ For $-\sqrt{10}$

$3x^2 = 30$		$3x^2 = 30$	
$3(\sqrt{10})^2$	30	$3(-\sqrt{10})^2$	30
$3 \cdot 10$		$3 \cdot 10$	
30		30	

The solutions are $\sqrt{10}$ and $-\sqrt{10}$.

<u>19.</u> $5x^2 - 16 = 0$

$5x^2 = 16$

$x^2 = \frac{16}{5}$

$|x| = \sqrt{\frac{16}{5}}$

$x = \sqrt{\frac{16}{5}}$ or $x = -\sqrt{\frac{16}{5}}$

$x = \sqrt{\frac{16}{5} \cdot \frac{5}{5}}$ or $x = -\sqrt{\frac{16}{5} \cdot \frac{5}{5}}$

$x = \frac{\sqrt{80}}{5}$ or $x = -\frac{\sqrt{80}}{5}$

$x = \frac{4\sqrt{5}}{5}$ or $x = -\frac{4\sqrt{5}}{5}$

Both check. The solutions are $\frac{4\sqrt{5}}{5}$ and $-\frac{4\sqrt{5}}{5}$.

<u>25.</u> $s = 16t^2$

$1000 = 16t^2$ (Substituting 1000 for s)

$\frac{1000}{16} = t^2$

$62.5 = t^2$

$\sqrt{62.5} = |t|$

$7.9 \approx |t|$

$7.9 \approx t$ or $-7.9 \approx t$

The number -7.9 cannot be a solution, since time cannot be negative in this situation. We substitute 7.9 in the original equation:

$s = 16(7.9)^2 = 16(62.41) = 998.56$. This is close. We have a check. It takes about 7.9 sec for the body to fall.

<u>31.</u> $1 = \frac{1}{3}x^2$

$3 = x^2$

$\sqrt{3} = |x|$

$x = \sqrt{3}$ or $x = -\sqrt{3}$

Both check. The solutions are $\sqrt{3}$ and $-\sqrt{3}$.

<u>37.</u> $x^2 + 9a^2 = 9 + ax^2$

$9a^2 - 9 = ax^2 - x^2$

$9(a^2 - 1) = x^2(a - 1)$

$\frac{9(a - 1)(a + 1)}{a - 1} = x^2$

$9(a + 1) = x^2$

$\sqrt{9(a + 1)} = |x|$

$3\sqrt{a + 1} = |x|$

$x = 3\sqrt{a + 1}$ or $x = -3\sqrt{a + 1}$

Both check. The solutions are $3\sqrt{a + 1}$ or $-3\sqrt{a + 1}$.

Exercise Set 11.2

<u>1.</u> $x^2 + 7x = 0$

$x(x + 7) = 0$

$x = 0$ or $x + 7 = 0$

$x = 0$ or $x = -7$

The solutions are 0 and -7.

7. $4x^2 + 4x = 0$

$4x(x + 1) = 0$

$4x = 0$ or $x + 1 = 0$

$x = 0$ or $\quad x = -1$

The solutions are 0 and -1.

13. $\qquad 14t^2 = 3t$

$14t^2 - 3t = 0$

$t(14t - 3) = 0$

$t = 0$ or $14t - 3 = 0$

$t = 0$ or $\quad 14t = 3$

$t = 0$ or $\quad t = \dfrac{3}{14}$.

The solutions are 0 and $\dfrac{3}{14}$.

19. $\quad 5 + 6x + x^2 = 0$

$(5 + x)(1 + x) = 0$

$5 + x = 0$ or $1 + x = 0$

$x = -5$ or $\quad x = -1$

The solutions are -5 and -1.

25. $\quad x^2 + 6x + 9 = 0$

$(x + 3)(x + 3) = 0$

$x + 3 = 0$ or $x + 3 = 0$

$x = -3$ or $\quad x = -3$.

The solution is -3.

31. $15b - 9b^2 = 4$

$0 = 9b^2 - 15b + 4$

$0 = (3b - 1)(3b - 4)$

$3b - 1 = 0$ or $3b - 4 = 0$

$3b = 1$ or $\quad 3b = 4$

$b = \dfrac{1}{3}$ or $\quad b = \dfrac{4}{3}$

The solutions are $\dfrac{1}{3}$ and $\dfrac{4}{3}$.

37. $\qquad t(t - 5) = 14$

$t^2 - 5t = 14$

$t^2 - 5t - 14 = 0$

$(t + 2)(t - 7) = 0$

$t + 2 = 0$ or $t - 7 = 0$

$t = -2$ or $\quad t = 7$

The solutions are -2 and 7.

43. $(x - 2)(x + 2) = x + 2$

$x^2 - 4 = x + 2$

$x^2 - x - 6 = 0$

$(x - 3)(x + 2) = 0$

$x - 3 = 0$ or $x + 2 = 0$

$x = 3$ or $\quad x = -2$

The solutions are 3 and -2.

49. $\qquad 4m^2 - (m + 1)^2 = 0$

$4m^2 - (m^2 + 2m + 1) = 0$

$4m^2 - m^2 - 2m - 1 = 0$

$3m^2 - 2m - 1 = 0$

$(3m + 1)(m - 1) = 0$

$3m + 1 = 0$ or $m - 1 = 0$

$3m = -1$ or $\quad m = 1$

$m = -\dfrac{1}{3}$ or $\quad m = 1$

The solutions are $-\dfrac{1}{3}$ and 1.

55. $\quad ax^2 + bx = 0$

$x(ax + b) = 0$

$x = 0$ or $ax + b = 0$

$x = 0$ or $\quad ax = -b$

$x = 0$ or $\quad x = -\dfrac{b}{a}$

The solutions are 0 and $-\dfrac{b}{a}$.

Exercise Set 11.3

1. $(x - 2)^2 = 49$

$|x - 2| = \sqrt{49}$

$|x - 2| = 7$

$x - 2 = 7$ or $x - 2 = -7$

$x = 9$ or $\quad x = -5$

The solutions are 9 and -5.

7. $(x - 7)^2 = 12$

$|x - 7| = \sqrt{12}$

$|x - 7| = 2\sqrt{3}$

$x - 7 = 2\sqrt{3}\quad$ or $x - 7 = -2\sqrt{3}$

$x = 7 + 2\sqrt{3}$ or $\quad x = 7 - 2\sqrt{3}$

The solutions are $7 + 2\sqrt{3}$ and $7 - 2\sqrt{3}$, or $7 \pm 2\sqrt{3}$.

13. $x^2 - 2x$ $\qquad (\frac{-2}{2})^2 = (-1)^2 = 1$

The trinomial $x^2 - 2x + 1$ is the square of $x - 1$.

19. $t^2 + 5t$ $\qquad (\frac{5}{2})^2 = \frac{25}{4}$

The trinomial $t^2 + 5t + \frac{25}{4}$ is the square of $t + \frac{5}{2}$.

25. $\qquad A = P(1 + r)^t$

$1210 = 1000(1 + r)^2$ $\qquad$ (Substituting 1210 for A,
$\qquad\qquad\qquad\qquad\qquad\qquad$ 1000 for P and 2 for t)

$\frac{1210}{1000} = (1 + r)^2$

$1.21 = (1 + r)^2$

$\sqrt{1.21} = |1 + r|$

$1.1 = |1 + r|$

$1 + r = 1.1$ or $1 + r = -1.1$

$\qquad r = 0.1$ or $\qquad r = -2.1$

Since the interest rate cannot be negative, we only check 0.1.

$1000(1 + 0.1)^2 = 1000(1.1)^2 = 1000(1.21) = 1210$

The interest rate must be 10% for \$1000 to grow to \$1210 in 2 years.

31. $\qquad A = P(1 + r)^t$

$3600 = 2500(1 + r)^2$ $\qquad$ (Substituting 3600 for A,
$\qquad\qquad\qquad\qquad\qquad\qquad$ 2500 for P, and 2 for t)

$\frac{3600}{2500} = (1 + r)^2$

$\frac{36}{25} = (1 + r)^2$

$\sqrt{\frac{36}{25}} = |1 + r|$

$\frac{6}{5} = |1 + r|$

$1 + r = \frac{6}{5}$ $\quad$ or $1 + r = -\frac{6}{5}$

$\qquad r = \frac{1}{5}$ $\quad$ or $\qquad r = -\frac{11}{5}$

$\qquad r = 0.2$ or $\qquad r = -2.2$

Since the interest rate cannot be negative, we only check 0.2.

$2500(1 + 0.2)^2 = 2500(1.2)^2 = 2500(1.44) = 3600$

The interest rate must be 20% for \$2500 to grow to \$3600 in 2 years.

37. $t^2 + 3t + \frac{9}{4} = \frac{49}{4}$

$\qquad (t + \frac{3}{2})^2 = \frac{49}{4}$

$\qquad |t + \frac{3}{2}| = \sqrt{\frac{49}{4}}$

$\qquad |t + \frac{3}{2}| = \frac{7}{2}$

37. (continued)

$t + \frac{3}{2} = \frac{7}{2}$ or $t + \frac{3}{2} = -\frac{7}{2}$

$\qquad t = \frac{4}{2}$ or $\qquad t = -\frac{10}{2}$

$\qquad t = 2$ or $\qquad t = -5$

The solutions are 2 and -5.

43. $\qquad A = P(1 + r)^t$

$1267.88 = 1000(1 + r)^2$

$1.26788 = (1 + r)^2$

$\sqrt{1.26788} = |1 + r|$

$1.126 \approx |1 + r|$

$1 + r \approx 1.126$ or $1 + r \approx -1.126$

$\qquad r \approx 0.126$ or $\qquad r \approx -2.126$

The interest rate cannot be negative. We only check 0.126.

$1000(1.126)^2 = 1000(1.267876) = 1267.88$

The interest rate must be 12.6%.

49. $5(5x - 2)^2 - 7 = 13$

$\qquad 5(5x - 2)^2 = 20$

$\qquad (5x - 2)^2 = 4$

$\qquad |5x - 2| = \sqrt{4}$

$\qquad |5x - 2| = 2$

$5x - 2 = 2$ or $5x - 2 = -2$

$\qquad 5x = 4$ or $\qquad 5x = 0$

$\qquad x = \frac{4}{5}$ or $\qquad x = 0$

The solutions are $\frac{4}{5}$ and 0.

Exercise Set 11.4

1. $x^2 - 6x - 16 = 0$

$\qquad x^2 - 6x \qquad = 16$

$\qquad x^2 - 6x + 9 = 16 + 9$

$\qquad (x - 3)^2 = 25$

$\qquad |x - 3| = \sqrt{25}$

$\qquad |x - 3| = 5$

$x - 3 = 5$ or $x - 3 = -5$

$\qquad x = 8$ or $\qquad x = -2$

The solutions are 8 and -2.

7. $x^2 - 22x + 102 = 0$

$x^2 - 22x = -102$

$x^2 - 22x + 121 = -102 + 121$

$(x - 11)^2 = 19$

$|x - 11| = \sqrt{19}$

$x - 11 = \sqrt{19}$ or $x - 11 = -\sqrt{19}$

$x = 11 + \sqrt{19}$ or $x = 11 - \sqrt{19}$

The solutions are $11 \pm \sqrt{19}$.

13. $x^2 + 3x - 28 = 0$

$x^2 + 3x = 28$

$x^2 + 3x + \frac{9}{4} = 28 + \frac{9}{4}$

$(x + \frac{3}{2})^2 = \frac{121}{4}$

$|x + \frac{3}{2}| = \sqrt{\frac{121}{4}}$

$|x + \frac{3}{2}| = \frac{11}{2}$

$x + \frac{3}{2} = \frac{11}{2}$ or $x + \frac{3}{2} = -\frac{11}{2}$

$x = \frac{8}{2}$ or $x = -\frac{14}{2}$

$x = 4$ or $x = -7$

The solutions are 4 and -7.

19. $3x^2 + 4x - 1 = 0$

$x^2 + \frac{4}{3}x - \frac{1}{3} = 0$

$x^2 + \frac{4}{3}x = \frac{1}{3}$

$x^2 + \frac{4}{3}x + \frac{4}{9} = \frac{1}{3} + \frac{4}{9}$

$(x + \frac{2}{3})^2 = \frac{7}{9}$

$|x + \frac{2}{3}| = \sqrt{\frac{7}{9}}$

$|x + \frac{2}{3}| = \frac{\sqrt{7}}{3}$

$x + \frac{2}{3} = \frac{\sqrt{7}}{3}$ or $x + \frac{2}{3} = -\frac{\sqrt{7}}{3}$

$x = \frac{-2 + \sqrt{7}}{3}$ or $x = \frac{-2 - \sqrt{7}}{3}$

The solutions are $\frac{-2 \pm \sqrt{7}}{3}$.

25. Familiarize:

Let x represent the number of people who bought their tickets in advance and y represent the number who bought their tickets at the door. Let us organize the information in a chart.

	Number of people	Price of ticket	Receipts
Advance	x	6.50	6.50x
At door	y	7	7y
Totals	12,000		81,165

Translate:

From the chart we can write two equations.

The total number of people was 12,000.

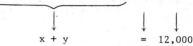

$x + y = 12,000$

The total receipts were $81,165.

$6.50x + 7y = 81,165$

Carry out: We solve the system.

$x + y = 12,000$

$6.5x + 7y = 81,165$

We solve the first equation for y and then use substitution.

$x + y = 12,000$

$y = 12,000 - x$

Next we substitute $12,000 - x$ for y in the second equation and solve for x.

$6.5x + 7y = 81,165$

$6.5x + 7(12,000 - x) = 81,165$ (Substituting)

$6.5x + 84,000 - 7x = 81,165$

$84,000 - 0.5x = 81,165$

$2835 = 0.5x$

$5670 = x$

Check:

If x = 5670, then y = 12,000 - 5670, or 6330. The sum of 5670 and 6330 is 12,000. The total receipts are 6.5(5670) + 7(6330), or 36,855 + 44,310, or $81,165. The values check.

State:

Thus, 5670 people bought tickets in advance.

31. $x^2 + bx + c$

$(\frac{b}{2})^2 = c$

$\frac{b^2}{4} = c$

$b^2 = 4c$

$|b| = \sqrt{4c}$

$|b| = 2\sqrt{c}$

$b = 2\sqrt{c}$ or $b = -2\sqrt{c}$

The trinomial $x^2 + 2\sqrt{c}\,x + c$ is the square of $x + \sqrt{c}$.

The trinomial $x^2 - 2\sqrt{c}\,x + c$ is the square of $x - \sqrt{c}$.

Thus, b can be $2\sqrt{c}$ or $-2\sqrt{c}$.

37. $ax^2 + 4x + 3 = 0$

$x^2 + \frac{4}{a}x + \frac{3}{a} = 0$

$x^2 + \frac{4}{a}x \qquad = -\frac{3}{a}$

$x^2 + \frac{4}{a}x + \frac{4}{a^2} = -\frac{3}{a} + \frac{4}{a^2}$

$(x + \frac{2}{a})^2 = \frac{-3a + 4}{a^2}$

$|x + \frac{2}{a}| = \sqrt{\frac{4 - 3a}{a^2}}$

$|x + \frac{2}{a}| = \frac{\sqrt{4 - 3a}}{a}$

$x + \frac{2}{a} = \frac{\sqrt{4 - 3a}}{a}$ or $x + \frac{2}{a} = -\frac{\sqrt{4 - 3a}}{a}$

$x = \frac{-2 + \sqrt{4 - 3a}}{a}$ or $x = \frac{-2 - \sqrt{4 - 3a}}{a}$

The solutions are $\frac{-2 \pm \sqrt{4 - 3a}}{a}$.

Exercise Set 11.5

1. $x^2 - 4x = 21$

$x^2 - 4x - 21 = 0$

$(x - 7)(x + 3) = 0$

$x - 7 = 0$ or $x + 3 = 0$

$x = 7$ or $\quad x = -3$

The solutions are 7 and -3.

7. $4x^2 + 12x = 7$

$4x^2 + 12x - 7 = 0$

$(2x - 1)(2x + 7) = 0$

$2x - 1 = 0$ or $2x + 7 = 0$

$2x = 1$ or $\quad 2x = -7$

$x = \frac{1}{2}$ or $\quad x = -\frac{7}{2}$

The solutions are $\frac{1}{2}$ and $-\frac{7}{2}$.

13. $y^2 - 10y + 22 = 0$

$a = 1 \qquad b = -10 \qquad c = 22$

$y = \frac{-(-10) \pm \sqrt{(-10)^2 - 4 \cdot 1 \cdot 22}}{2 \cdot 1}$

$y = \frac{10 \pm \sqrt{100 - 88}}{2}$

$y = \frac{10 \pm \sqrt{12}}{2}$

$y = \frac{10 \pm 2\sqrt{3}}{2}$

$y = \frac{2(5 \pm \sqrt{3})}{2}$

$y = 5 \pm \sqrt{3}$

19. $2x^2 - 5x = 1$

$2x^2 - 5x - 1 = 0$

$a = 2 \qquad b = -5 \qquad c = -1$

$x = \frac{-(-5) \pm \sqrt{(-5)^2 - 4 \cdot 2 \cdot (-1)}}{2 \cdot 2}$

$x = \frac{5 \pm \sqrt{25 + 8}}{4}$

$x = \frac{5 \pm \sqrt{33}}{4}$

25. $2t^2 + 6t + 5 = 0$

$a = 2 \qquad b = 6 \qquad c = 5$

The discriminant is $b^2 - 4ac$.

$b^2 - 4ac = 6^2 - 4 \cdot 2 \cdot 5 = 36 - 40 = -4$

Since $b^2 - 4ac < 0$, the equation has no real-number solutions.

31. $2y^2 - 6y = 10$

$2y^2 - 6y - 10 = 0$

$y^2 - 3y - 5 = 0$

$a = 1 \qquad b = -3 \qquad c = -5$

$y = \dfrac{-(-3) \pm \sqrt{(-3)^2 - 4 \cdot 1 \cdot (-5)}}{2 \cdot 1}$

$y = \dfrac{3 \pm \sqrt{9 + 20}}{2}$

$y = \dfrac{3 \pm \sqrt{29}}{2}$

37. $y^2 + 10y + 22 = 0$

$a = 1 \qquad b = 10 \qquad c = 22$

$y = \dfrac{-10 \pm \sqrt{10^2 - 4 \cdot 1 \cdot 22}}{2 \cdot 1}$

$y = \dfrac{-10 \pm \sqrt{100 - 88}}{2}$

$y = \dfrac{-10 \pm \sqrt{12}}{2}$

$y = \dfrac{-10 \pm 2\sqrt{3}}{2}$

$y = \dfrac{2(-5 \pm \sqrt{3})}{2}$

$y = -5 \pm \sqrt{3}$

Using a calculator or Table 2, we see that $\sqrt{3} \approx 1.732$.

$-5 + \sqrt{3} \approx -5 + 1.732 \approx -3.268 \approx -3.3$

$-5 + \sqrt{3} \approx -5 - 1.732 \approx -6.732 \approx -6.7$

The solutions are -3.3 and -6.7.

43. $5x + x(x - 7) = 0$

$5x + x^2 - 7x = 0$

$x^2 - 2x = 0$

$x(x - 2) = 0$

$x = 0$ or $x - 2 = 0$

$x = 0$ or $\qquad x = 2$

The solutions are 0 and 2.

49. $x^2 + (x + 2)^2 = 7$

$x^2 + x^2 + 4x + 4 = 7$

$2x^2 + 4x - 3 = 0$

$a = 2 \qquad b = 4 \qquad c = -3$

49. (continued)

$x = \dfrac{-4 \pm \sqrt{4^2 - 4 \cdot 2 \cdot (-3)}}{2 \cdot 2}$

$x = \dfrac{-4 \pm \sqrt{16 + 24}}{4}$

$x = \dfrac{-4 \pm \sqrt{40}}{4}$

$x = \dfrac{-4 \pm 2\sqrt{10}}{4}$

$x = \dfrac{2(-2 \pm \sqrt{10})}{2 \cdot 2} = \dfrac{-2 \pm \sqrt{10}}{2}$

55. $4x^2 - 4cx + c^2 - 3d^2 = 0$

$a = 4 \qquad b = -4c \qquad c = c^2 - 3d^2$

Note: These c's are not the same.

$x = \dfrac{-(-4c) \pm \sqrt{(-4c)^2 - 4 \cdot 4 \cdot (c^2 - 3d^2)}}{2 \cdot 4}$

$x = \dfrac{4c \pm \sqrt{16c^2 - 16c^2 + 48d^2}}{8}$

$x = \dfrac{4c \pm 4d\sqrt{3}}{4 \cdot 2}$

$x = \dfrac{c \pm d\sqrt{3}}{2}$

61. The product of the solutions is $\dfrac{c}{a}$, or $-\dfrac{3}{2}$.

Let x represent the unknown solution.

Then $-5 \cdot x = -\dfrac{3}{2}$

$-\dfrac{1}{5}(-5x) = -\dfrac{1}{5}\left(-\dfrac{3}{2}\right)$

$x = \dfrac{3}{10}$

The other solution is $\dfrac{3}{10}$.

Exercise Set 11.6

1. $A = 6s^2$

$\dfrac{A}{6} = s^2$

$\sqrt{\dfrac{A}{6}} = s$

7. $a^2 + b^2 = c^2$

$b^2 = c^2 - a^2$

$b = \sqrt{c^2 - a^2}$

13. $T = 2\pi\sqrt{\dfrac{\ell}{g}}$

$\dfrac{T}{2\pi} = \sqrt{\dfrac{\ell}{g}}$ (Multiplying by $\dfrac{1}{2\pi}$)

$\dfrac{T^2}{4\pi^2} = \dfrac{\ell}{g}$ (Squaring)

$gT^2 = 4\pi^2\ell$ (Multiplying by $4\pi^2 g$)

$g = \dfrac{4\pi^2\ell}{T^2}$ (Multiplying by $\dfrac{1}{T^2}$)

19. a) $s(t) = 4.9t^2 + v_0 t$

$75 = 4.9t^2 + 0\cdot t$

[Substituting 75 for $s(t)$
and 0 for v_0]

$75 = 4.9t^2$

$\dfrac{75}{4.9} = t^2$

$\sqrt{\dfrac{75}{4.9}} = t$

$3.9 \approx t$

It takes approximately 3.9 sec to reach the ground.

b) $s(t) = 4.9t^2 + v_0 t$

$75 = 4.9t^2 + 30t$

[Substituting 75 for $s(t)$
and 30 for v_0]

$0 = 4.9t^2 + 30t - 75$

$a = 4.9, \quad b = 30, \quad c = -75$

$t = \dfrac{-30 \pm \sqrt{30^2 - 4(4.9)(-75)}}{2(4.9)}$

$= \dfrac{-30 \pm \sqrt{900 + 1470}}{9.8}$

$\approx \dfrac{-30 + 48.68}{9.8}$

≈ 1.9

It takes approximately 1.9 sec to reach the ground.

c) $s(t) = 4.9t^2 + v_0 t$

$s(2) = 4.9(2)^2 + 30(2)$

(Substituting 2 for t and 30 for v_0)

$= 19.6 + 60$

$= 79.6$

An object will fall 79.6 m in 2 sec.

25.   We first make a drawing and label the unknown distance d.

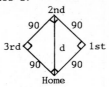

We then use the Pythagorean property.

$90^2 + 90^2 = d^2$

$8100 + 8100 = d^2$

$16{,}200 = d^2$

$127.3 \approx d$

The direct distance from second to home is approximately 127.3 ft.

Exercise Set 11.7

1. $x - 10\sqrt{x} + 9 = 0$

Let $u = \sqrt{x}$ and think of x as $(\sqrt{x})^2$.

$u^2 - 10u + 9 = 0$ (Substituting u for $\sqrt{x}$)

$(u - 9)(u - 1) = 0$

$u - 9 = 0$ or $u - 1 = 0$

$u = 9$ or $u = 1$

Now we substitute $\sqrt{x}$ for u and solve these equations:

$\sqrt{x} = 9$ or $\sqrt{x} = 1$

$x = 81$ or $x = 1$

The numbers 81 and 1 both check. They are the solutions.

7. $z^{1/2} - z^{1/4} - 2 = 0$

Let $u = z^{1/4}$ and think of $z^{1/2}$ as $(z^{1/4})^2$.

$u^2 - u - 2 = 0$ (Substituting u for $z^{1/4}$)

$(u - 2)(u + 1) = 0$

$u - 2 = 0$ or $u + 1 = 0$

$u = 2$ or $u = -1$

Now we substitute $z^{1/4}$ for u and solve these equations.

$z^{1/4} = 2$ or $z^{1/4} = -1$

$\sqrt[4]{z} = 2$ $\sqrt[4]{z} = -1$

$z = 16$ This equation has no real
solution since principal
fourth roots are never negative.

The number 16 checks, so it is the solution.

13. $w^4 - 4w^2 - 2 = 0$

Let $u = w^2$ and think of w^4 as $(w^2)^2$.

$u^2 - 4u - 2 = 0$ (Substituting u for w^2)

$$u = \frac{-(-4) \pm \sqrt{(-4)^2 - 4 \cdot 1 \cdot (-2)}}{2 \cdot 1}$$

$$= \frac{4 \pm \sqrt{16 + 8}}{2}$$

$$= \frac{4 \pm \sqrt{24}}{2}$$

$$= \frac{4 \pm 2\sqrt{6}}{2}$$

$$= 2 \pm \sqrt{6}$$

Now we substitute w^2 for u and solve these equations.

$w^2 = 2 + \sqrt{6}$ or $w^2 = 2 - \sqrt{6}$

$w = \pm\sqrt{2 + \sqrt{6}}$ or $w = \pm\sqrt{2 - \sqrt{6}}$

The numbers $\sqrt{2 + \sqrt{6}}$, $-\sqrt{2 + \sqrt{6}}$, $\sqrt{2 - \sqrt{6}}$, and $-\sqrt{2 - \sqrt{6}}$ check. Note that $\sqrt{2 - \sqrt{6}}$ is nonreal.

19. Familiarize and Translate:

Let x represent the original price of the toaster and x + 1.50 represent the original price of the drill.

The sale prices are 80%x for the toaster (20% reduction) and 75%(x + 1.50) for the drill (25% reduction).

The sale prices are the same. This gives us an equation.

80%x = 75%(x + 1.50)

Carry out:

0.80x = 0.75x + 0.75(1.50)

0.05x = 1.125

$$x = \frac{1.125}{0.05}$$

x = 22.50

Check:

When x = 22.50, x + 1.50 = 24.00. The sale prices, which are 80%(22.50), or $18, and 75%(24.00), or $18, are the same. The value checks.

State:

During the sale each appliance costs $18.

25. $\left(\frac{x^2 - 1}{x}\right)^2 - \left(\frac{x^2 - 1}{x}\right) - 2 = 0$

Let $u = \frac{x^2 - 1}{x}$.

$u^2 - u - 2 = 0$

$(u - 2)(u + 1) = 0$

$u - 2 = 0$ or $u + 1 = 0$

$u = 2$ or $u = -1$

We substitute $\frac{x^2 - 1}{x}$ for u and solve these equations.

$\frac{x^2 - 1}{x} = 2$ or $\frac{x^2 - 1}{x} = -1$

$x^2 - 1 = 2x$ $x^2 - 1 = -x$

$x^2 - 2x - 1 = 0$ $x^2 + x - 1 = 0$

$x = \frac{2 \pm \sqrt{8}}{2}$ $x = \frac{-1 \pm \sqrt{5}}{2}$

$x = 1 \pm \sqrt{2}$

The numbers $1 \pm \sqrt{2}$ and $\frac{-1 \pm \sqrt{5}}{2}$ check. They are the solutions.

31. $9x^{3/2} - 8 = x^3$

$0 = x^3 - 9x^{3/2} + 8$

Let $u = x^{3/2}$.

$0 = u^2 - 9u + 8$

$0 = (u - 8)(u - 1)$

$u - 8 = 0$ or $u - 1 = 0$

$u = 8$ or $u = 1$

We substitute $x^{3/2}$ for u and solve these equations.

$x^{3/2} = 8$ or $x^{3/2} = 1$

$x^3 = 64$ or $x^3 = 1$

$x = 4$ or $x = 1$

Both 4 and 1 check. They are the solutions.

Exercise Set 11.8

1. $y = kx^2$

   We first find k.

   $0.15 = k(0.1)^2$ (Substituting 0.15 for y
                          and 0.1 for x)

   $0.15 = 0.01k$

   $\dfrac{0.15}{0.01} = k$

   $15 = k$

   The equation of variation is $y = 15x^2$.

7. $y = kxz^2$

   We first find k.

   $105 = k \cdot 14 \cdot 5^2$ (Substituting 105 for y,
                                     14 for x, and 5 for z)

   $105 = 350k$

   $\dfrac{105}{350} = k$

   $0.3 = k$

   The equation of variation is $y = 0.3xz^2$.

13. $W = \dfrac{k}{d^2}$

    We first find k.

    $100 = \dfrac{k}{(6400)^2}$ (Substituting 100 for W
                                      and 6400 for d)

    $100 = \dfrac{k}{40,960,000}$

    $4,096,000,000 = k$

    The equation of variation is $W = \dfrac{4,096,000,000}{d^2}$

    Substitute 6600(6400 + 200) for d and solve for W.

    $W = \dfrac{4,096,000,000}{(6600)^2} = \dfrac{4,096,000,000}{43,560,000} = 94\dfrac{34}{1089}$

    When the astronaut is 200 km above the surface of
    the earth, his weight is $94\dfrac{34}{1089}$ kg.

19. $y = \dfrac{k}{x^2}$

    $y = \dfrac{k}{(nx)^2}$ (Substituting nx for x)

    $= \dfrac{k}{n^2x^2}$

    $= \dfrac{1}{n^2} \cdot \dfrac{k}{x^2}$

    Thus, y is multiplied by $\dfrac{1}{n^2}$ when x is multiplied
    by n.

Exercise Set 11.9

1. $y = x^2 + 1$

   We choose some numbers for x and then compute the
   corresponding values of y.

   When x = -2, $y = (-2)^2 + 1 = 4 + 1 = 5$.

   When x = -1, $y = (-1)^2 + 1 = 1 + 1 = 2$.

   When x = 0, $y = 0^2 + 1 = 0 + 1 = 1$.

   When x = 1, $y = 1^2 + 1 = 1 + 1 = 2$.

   When x = 2, $y = 2^2 + 1 = 4 + 1 = 5$.

   We plot the ordered pairs resulting from the
   computations and connect the points with a
   smooth curve.

x	y
-2	5
-1	2
0	1
1	2
2	5

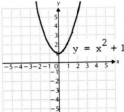

7. $y = 8 - x - x^2$

   We choose some numbers for x and then compute the
   corresponding values of y.

   When x = -4, $y = 8 - (-4) - (-4)^2 = 8 + 4 - 16 = -4$.

   When x = -3, $y = 8 - (-3) - (-3)^2 = 8 + 3 - 9 = 2$.

   When x = -2, $y = 8 - (-2) - (-2)^2 = 8 + 2 - 4 = 6$.

   When x = -1, $y = 8 - (-1) - (-1)^2 = 8 + 1 - 1 = 8$.

   When $x = -\dfrac{1}{2}$, $y = 8 - (-\dfrac{1}{2}) - (-\dfrac{1}{2})^2$

   $= 8 + \dfrac{1}{2} - \dfrac{1}{4}$

   $= 8\dfrac{1}{4}$

   When x = 0, $y = 8 - 0 - 0^2 = 8 - 0 - 0 = 8$.

   When x = 1, $y = 8 - 1 - 1^2 = 8 - 1 - 1 = 6$.

   When x = 2, $y = 8 - 2 - 2^2 = 8 - 2 - 4 = 2$.

   When x = 3, $y = 8 - 3 - 3^2 = 8 - 3 - 9 = -4$.

   We plot the ordered pairs resulting from the
   computations and connect the points with a
   smooth curve.

x	y
-4	-4
-3	2
-2	6
-1	8
$-\dfrac{1}{2}$	$8\dfrac{1}{4}$
0	8
1	6
2	2
3	-4

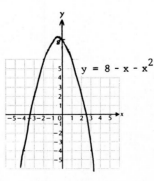

13. $y = -2x^2 - 4x + 1$

We choose some values for x and then compute the corresponding values of y.

When $x = -3$, $y = -2(-3)^2 - 4(-3) + 1$
$\qquad = -18 + 12 + 1 = -5.$

When $x = -2$, $y = -2(-2)^2 - 4(-2) + 1$
$\qquad = -8 + 8 + 1 = 1.$

When $x = -1$, $y = -2(-1)^2 - 4(-1) + 1$
$\qquad = -2 + 4 + 1 = 3.$

When $x = 0$, $y = -2 \cdot 0^2 - 4 \cdot 0 + 1 = 0 - 0 + 1 = 1.$

When $x = 1$, $y = -2 \cdot 1^2 - 4 \cdot 1 + 1 = -2 - 4 + 1 = -5.$

We plot the resulting ordered pairs and connect them with a smooth curve.

x	y
-3	-5
-2	1
-1	3
0	1
1	-5

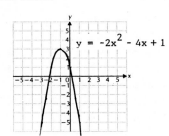

19. $y = -x^2 + x - 1$

We choose some values for x and then compute the corresponding values of y.

When $x = -1$, $y = -(-1)^2 + (-1) - 1$
$\qquad = -1 - 1 - 1 = -3.$

When $x = 0$, $y = -0^2 + 0 - 1 = -0 + 0 - 1 = -1.$

When $x = \frac{1}{2}$, $y = -(\frac{1}{2})^2 + \frac{1}{2} - 1$
$\qquad = -\frac{1}{4} + \frac{1}{2} - 1 = -\frac{3}{4}.$

When $x = 1$, $y = -1^2 + 1 - 1 = -1 + 1 - 1 = -1.$

When $x = 2$, $y = -2^2 + 2 - 1 = -4 + 2 - 1 = -3.$

We plot the resulting ordered pairs and connect them with a smooth curve.

x	y
-1	-3
0	-1
$\frac{1}{2}$	$-\frac{3}{4}$
1	-1
2	-3

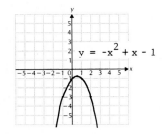

25. Graph $y = x^2 - 5$.

We choose some numbers for x and then compute the corresponding values of y.

When $x = -3$, $y = (-3)^2 - 5 = 9 - 5 = 4.$

When $x = -2$, $y = (-2)^2 - 5 = 4 - 5 = -1.$

When $x = -1$, $y = (-1)^2 - 5 = 1 - 5 = -4.$

When $x = 0$, $y = 0^2 - 5 = 0 - 5 = -5.$

When $x = 1$, $y = 1^2 - 5 = 1 - 5 = -4.$

When $x = 2$, $y = 2^2 - 5 = 4 - 5 = -1.$

When $x = 3$, $y = 3^2 - 5 = 9 - 5 = 4.$

We plot the ordered pairs resulting from the computations and connect the points with a smooth curve.

x	y
-3	4
-2	-1
-1	-4
0	-5
1	-4
2	-1
3	4

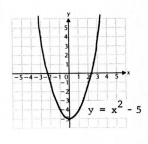

The graph crosses the x-axis at about $(-2.2, 0)$ and $(2.2, 0)$. So the solutions of the equation $x^2 - 5 = 0$ are about -2.2 and 2.2.

31. Graph $y = x^2 + 10x + 25$
$\qquad$ or $y = (x + 5)^2$

We choose some numbers for x and then compute the corresponding values of y.

When $x = -7$, $y = (-7 + 5)^2 = (-2)^2 = 4.$

When $x = -6$, $y = (-6 + 5)^2 = (-1)^2 = 1.$

When $x = -5$, $y = (-5 + 5)^2 = 0^2 = 0.$

When $x = -4$, $y = (-4 + 5)^2 = 1^2 = 1.$

When $x = -3$, $y = (-3 + 5)^2 = 2^2 = 4.$

We plot the ordered pairs resulting from the computations and connect the points with a smooth curve.

x	y
-7	4
-6	1
-5	0
-4	1
-3	4

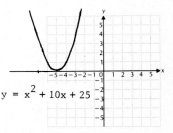

The graph intersects the x-axis at $(-5, 0)$. The solution of the equation $x^2 + 10x + 25 = 0$ is -5.

37. $y = x^2 - x - 6$

We choose some numbers for x and then compute the corresponding values of y.

When $x = -3$, $y = (-3)^2 - (-3) - 6 = 9 + 3 - 6 = 6$.

When $x = -2$, $y = (-2)^2 - (-2) - 6 = 4 + 2 - 6 = 0$.

When $x = -1$, $y = (-1)^2 - (-1) - 6 = 1 + 1 - 6 = -4$.

When $x = 0$, $y = 0^2 - 0 - 6 = 0 - 0 - 6 = -6$.

When $x = \frac{1}{2}$, $y = (\frac{1}{2})^2 - \frac{1}{2} - 6 = \frac{1}{4} - \frac{1}{2} - 6 = -6\frac{1}{4}$.

When $x = 1$, $y = 1^2 - 1 - 6 = 1 - 1 - 6 = -6$.

When $x = 2$, $y = 2^2 - 2 - 6 = 4 - 2 - 6 = -4$.

When $x = 3$, $y = 3^2 - 3 - 6 = 9 - 3 - 6 = 0$.

When $x = 4$, $y = 4^2 - 4 - 6 = 16 - 4 - 6 = 6$.

We plot the ordered pairs resulting from the computations and connect the points with a smooth curve.

x	y
-3	6
-2	0
-1	-4
0	-6
$\frac{1}{2}$	$-6\frac{1}{4}$
1	-6
2	-4
3	0
4	6

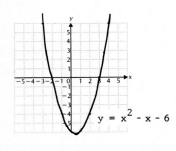

a) We graph $y = 2$ on the same set of axes as $y = x^2 - x - 6$.

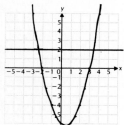

The graph of $y = 2$ intersects the graph of $y = x^2 - x - 6$ at about (-2.4, 2) and (3.4, 2). The solutions of the equation $x^2 - x - 6 = 2$ are about -2.4 and 3.4.

37. (continued)

b) We graph $y = -3$ on the same set of axes as $y = x^2 - x - 6$.

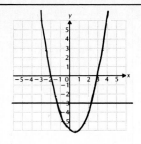

The graph of $y = -3$ intersects the graph of $y = x^2 - x - 6$ at about (-1.3, -3) and (2.3, -3). The solutions of the equation $x^2 - x - 6 = -3$ are about -1.3 and 2.3.

43. $A = \ell \cdot w$

$A = \ell(8 - \ell)$ (Substituting $8 - \ell$ for w)

$A = 8\ell - \ell^2$

We graph $y = 8x - x^2$ and find its maximum.

We choose some numbers for x and then compute the corresponding values of y.

When $x = -1$, $y = 8(-1) - (-1)^2 = -8 - 1 = -9$.

When $x = 0$, $y = 8 \cdot 0 - 0^2 = 0 - 0 = 0$.

When $x = 1$, $y = 8 \cdot 1 - 1^1 = 8 - 1 = 7$.

When $x = 2$, $y = 8 \cdot 2 - 2^2 = 16 - 4 = 12$.

When $x = 3$, $y = 8 \cdot 3 - 3^2 = 24 - 9 = 15$.

When $x = 4$. $y = 8 \cdot 4 - 4^2 = 32 - 16 = 16$.

When $x = 5$, $y = 8 \cdot 5 - 5^2 = 40 - 25 = 15$.

When $x = 6$, $y = 8 \cdot 6 - 6^2 = 48 - 36 = 12$.

When $x = 7$, $y = 8 \cdot 7 - 7^2 = 56 - 49 = 7$.

When $x = 8$, $y = 8 \cdot 8 - 8^2 = 64 - 64 = 0$.

When $x = 9$, $y = 8 \cdot 9 - 9^2 = 72 - 81 = -9$.

We plot the ordered pairs resulting from the computations and connect the points with a smooth curve.

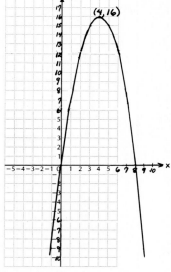

The maximum point on the graph is (4, 16). The largest rectangular area that can be enclosed with 16 feet of fence is 16 square feet. The dimensions of the rectangle must be 4 ft by 4 ft.

Exercise Set 11.10

1. $f(x) = x^2$

We choose some values of x and compute f(x).

x	0	1	-1	2	-2	3	-3
f(x)	0	1	1	4	4	9	9

We plot these ordered pairs and connect them with a smooth curve.

$f(x) = x^2$

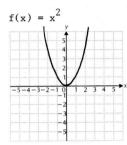

Vertex: (0,0)

Line of symmetry: x = 0

7. $f(x) = -(x + 4)^2$, or $f(x) = -[x - (-4)]^2$

We choose some values of x and compute f(x).

x	-4	-3	-5	-2	-6
f(x)	0	-1	-1	-4	-4

We plot these ordered pairs and connect them with a smooth curve.

$f(x) = -(x + 4)^2$

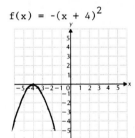

Vertex: (-4,0)

Line of symmetry: x = -4

The graph of $f(x) = -(x + 4)^2$ looks like the graph of $f(x) = x^2$, except that it is moved four units to the left and opens downward.

13. $f(x) = 3(x - 1)^2$

We choose some values of x and compute f(x).

x	1	2	0	3	-1
f(x)	0	3	3	12	12

13. (continued)

We plot these ordered pairs and connect them with a smooth curve.

$f(x) = 3(x - 1)^2$

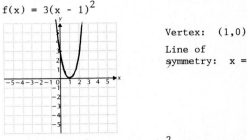

Vertex: (1,0)

Line of symmetry: x = 1

The graph of $f(x) = 3(x - 1)^2$ looks like the graph of $f(x) = 3x^2$ except that it is moved one unit to the right.

19. $f(x) = (x - 3)^2 + 1$

We know that the graph looks like the graph of $f(x) = x^2$ but moved three units right and one unit up.

We can also graph by choosing some values of x and computing f(x).

x	3	4	2	5	1
f(x)	1	2	2	5	5

We plot these ordered pairs and connect them with a smooth curve.

$f(x) = (x - 3)^2 + 1$

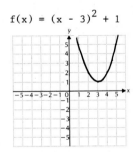

Vertex: (3,1)

Line of symmetry: x = 3

Minimum: 1

25. $f(x) = -3(x + 4)^2 + 1$, or $f(x) = -3[x - (-4)]^2 + 1$

We know that the graph of $f(x) = -3(x + 4)^2 + 1$ looks like the graph of $f(x) = 3x^2$ but moved four units left and one unit up and now opens downward.

We can also graph by choosing some values for x and computing f(x).

x	-4	-3	-5	-2	-6
f(x)	1	-2	-2	-11	-11

25. (continued)

We plot these ordered pairs and connect them with a smooth curve.

$f(x) = -3(x + 4)^2 + 1$

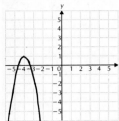

Vertex: (-4,1)

Line of symmetry: x = -4

Maximum: 1

31. $f(x) = -7(x - 10)^2 - 20$

$f(x) = -7(x - 10)^2 + (-20)$

Vertex: (10,-20)

Line of symmetry: x = 10

Since -7 < 0, -20 is the maximum function value.

37. y - 3 = h(x)

y = h(x) + 3

The graph of y = h(x) + 3 looks like the graph of y = h(x) but is moved three units in the positive y-direction.

y = h(x)   y = h(x) + 3

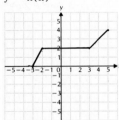

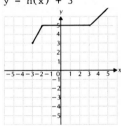

43. The vertex of the new function is (0,4), thus h = 0 and k = 4. THe coefficient must be negative since the function has a maximum value. The new function is

$f(x) = -2(x - 0)^2 + 4$

$f(x) = -2x^2 + 4$

49. The vertex of the new function if (-3,6), thus h = -3 and k = 6. The coefficient remains positive since the function has a minimum value. The new function is

$f(x) = 2[x - (-3)]^2 + 6$

$f(x) = 2(x + 3)^2 + 6$

1. $f(x) = x^2 - 2x - 3$

$= (x^2 - 2x) - 3$

We complete the square inside parentheses.

$f(x) = (x^2 - 2x + 1 - 1) - 3$ (Adding 1 - 1 inside parentheses)

$= (x^2 - 2x + 1) - 1 - 3$

$= (x - 1)^2 - 4$

$= (x - 1)^2 + (-4)$

Vertex: (1,-4)

Line of symmetry: x = 1

Minimum value: -4   (The graph opens upward.)

7. $f(x) = x^2 - 9x$

We first complete the square.

$f(x) = x^2 - 9x + \frac{81}{4} - \frac{81}{4}$ (Adding $\frac{81}{4} - \frac{81}{4}$)

$= (x^2 - 9x + \frac{81}{4}) - \frac{81}{4}$

$= (x - \frac{9}{2})^2 - \frac{81}{4}$

$= (x - \frac{9}{2})^2 + (-\frac{81}{4})$

Vertex: $(\frac{9}{2}, -\frac{81}{4})$

Line of symmetry: $x = \frac{9}{2}$

Minimum value: $-\frac{81}{4}$ (The graph opens upward.)

13. $f(x) = x^2 - 4x + 1$

We solve the equation $0 = x^2 - 4x + 1$ using the quadratic formula.

$x = \frac{-(-4) \pm \sqrt{(-4)^2 - 4 \cdot 1 \cdot 1}}{2 \cdot 1}$

$= \frac{4 \pm \sqrt{16 - 4}}{2}$

$= \frac{4 \pm \sqrt{12}}{2}$

$= \frac{4 \pm 2\sqrt{3}}{2}$

$= 2 \pm \sqrt{3}$

The x-intercepts are $(2 + \sqrt{3}, 0)$ and $(2 - \sqrt{3}, 0)$

19. $f(x) = -x^2 + 3x + 4$

We solve the following equation.

$0 = -x^2 + 3x + 4$

$0 = x^2 - 3x - 4$

$0 = (x - 4)(x + 1)$

$x - 4 = 0$ or $x + 1 = 0$

$x = 4$ or $x = -1$

The x-intercepts are $(4,0)$ and $(-1,0)$.

25. $f(x) = -x^2 - 3x - 3$

We solve the following equation using the quadratic formula.

$0 = -x^2 - 3x - 3$

$0 = x^2 + 3x + 3$

$x = \dfrac{-3 \pm \sqrt{3^2 - 4 \cdot 1 \cdot 3}}{2 \cdot 1}$

$= \dfrac{-3 \pm \sqrt{9 - 12}}{2}$

$= \dfrac{-3 \pm i\sqrt{3}}{2}$

The solutions are not real. Thus, there are no x-intercepts.

31. $f(x) = 0.05x^2 - 4.735x + 100.23$

We solve the equation $0 = 0.05x^2 - 4.735x + 100.23$ using the quadratic formula.

$x = \dfrac{-(-4.735) \pm \sqrt{(-4.735)^2 - 4(0.05)(100.23)}}{2(0.05)}$

$= \dfrac{4.735 \pm \sqrt{2.374225}}{0.1}$

$\approx \dfrac{4.735 \pm 1.541}{0.1}$

$x \approx \dfrac{6.276}{0.1}$ or $x \approx \dfrac{3.194}{0.1}$

$x \approx 62.8$ or $x \approx 31.9$

The x-intercepts are $(62.8,0)$ and $(31.9,0)$.

37. $f(x) = ax^2 + bx + c$

$f(x) = a(x^2 + \dfrac{b}{a}x) + c$

$f(x) = a(x^2 + \dfrac{b}{a}x + \dfrac{b^2}{4a^2} - \dfrac{b^2}{4a^2}) + c$

$f(x) = a(x^2 + \dfrac{b}{a}x + \dfrac{b^2}{4a^2}) - \dfrac{b^2}{4a} + c$

$f(x) = a(x + \dfrac{b}{2a})^2 - \dfrac{b^2}{4a} + \dfrac{4ac}{4a}$

$f(x) = a[x - (-\dfrac{b}{2a})]^2 + \dfrac{4ac - b^2}{4a}$

43. $y \le x^2 + 5x + 6$

We first graph the parabola $y = x^2 + 5x + 6$. Since the inequality symbol is $\le$, we draw it solid.

x	y
-5	6
-4	2
-3	0
-2	0
-1	2
0	6

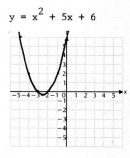

$y = x^2 + 5x + 6$

We determine whether to shade above or below the parabola by trying some point off the parabola. The point $(0,0)$ is easy to check.

$0 \le 0^2 + 5 \cdot 0 + 6$ (Substituting 0 for x and 0 for y)

$0 \le 6$

Since $0 \le 6$ is true, the point is in the graph. Thus, we shade below the parabola. The graph consists of the region below the parabola and the solid parabola as well.

$y \le x^2 + 5x + 6$

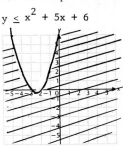

Exercise Set 11.12

1. Familiarize:

We make a drawing and label it.

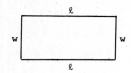

Perimeter = $2\ell + 2w$
Area = $\ell \cdot w$

Translate:

We have a system of equations.

$2\ell + 2w = 76$

$A = \ell w$

Carry out:

We first solve $2\ell + 2w = 76$ for ℓ.

$2\ell + 2w = 76$

$\ell + w = 38$

$\ell = 38 - w$

Next we substitute $38 - w$ for ℓ in the second equation.

$A = \ell w$

$A = (38 - w)w$

$\quad = -w^2 + 38w$

Completing the square, we get

$A = -(w^2 - 38w)$

$\quad = -(w^2 - 38w + 361 - 361)$

$\quad = -(w^2 - 38w + 361) + 361$

$\quad = -(w - 19)^2 + 361$

The maximum function value is 361 when w is 19. When w = 19, $\ell = 38 - 19$, or 19.

Check:

We check a function value for w less than 19 and for w greater than 19.

$A(18) = -(18)^2 + 38 \cdot 18 = 360$

$A(20) = -(20)^2 + 38 \cdot 20 = 360$

Since 361 is greater than these numbers, it looks as though we have a maximum.

State:

The maximum area is 361 ft^2 when the dimensions are 19 ft by 19 ft.

7. Familiarize and Translate:

We let x and y represent the numbers.

We have two equations.

$x - y = 5$

$P = xy$

Carry out:

We first solve $x - y = 5$ for y.

$x - y = 5$

$x - 5 = y$

Next we substitute $x - 5$ for y in the second equation.

$P = xy$

$P = x(x - 5)$

$\quad = x^2 - 5x$

Completing the square, we get

$P = x^2 - 5x + \frac{25}{4} - \frac{25}{4}$

$\quad = (x - \frac{5}{2})^2 - \frac{25}{4}$

The minimum function value is $-\frac{25}{4}$ when x is $\frac{5}{2}$.

When $x = \frac{5}{2}$, $y = \frac{5}{2} - 5$, or $-\frac{5}{2}$.

Check:

We check a function value for x less than $\frac{5}{2}$ and for x greater than $\frac{5}{2}$.

$P(2) = 2^2 - 5 \cdot 2 = -6$

$P(3) = 3^2 - 5 \cdot 3 = -6$

Since $-\frac{25}{4}$ is less than these numbers, it looks as though we have a minimum.

State:

The minimum product is $-\frac{25}{4}$ when the numbers are $\frac{5}{2}$ and $-\frac{5}{2}$.

13. a) $f(x) = ax^2 + bx + c$

We substitute some values for x and f(x).

$38 = a(1)^2 + b(1) + c$

$66 = a(2)^2 + b(2) + c$

$86 = a(3)^2 + b(3) + c$

or

$a + b + c = 38$

$4a + 2b + c = 66$

$9a + 3b + c = 86$

Solving this system, we get

$a = -4$ $b = 40$ and $c = 2$.

Therefore the function we are looking for is

$f(x) = -4x^2 + 40x + 2$

b) $f(x) = -4x^2 + 40x + 2$

$f(4) = -4(4)^2 + 40(4) + 2$

$= -64 + 160 + 2$

$= 98$

The earnings for the fourth week will probably be $98.

19. $f(x) = ax^2 + bx + c$

We substitute some values for x and f(x).

$-5.86 = a(20.34)^2 + b(20.34) + 20.34$

$-6.02 = a(34.67)^2 + b(34.67) + 34.67$

$-8.46 = a(28.55)^2 + b(28.55) + 28.55$

Solving this system we get

$a = 0.0499218$, $b = -2.7573651$, and $c = 29.571379$

Therefore, the function we are looking for is

$f(x) = 0.0499218x^2 - 2.7573651x + 29.571379$.

25. Familiarize:

Admission price	Average attendance
$1.90	101
2.00	100
2.10	99
2.20	98
2.00 + x	100 - x

Let x represent the number of 10¢ increases in the admission price. Then 2.00 + 0.1x represents the admission price and 100 - x represents the corresponding averge attendance.

Translate:

Since total revenue is admission price times the attendance, we have the following function for the revenue.

$R(x) = (2.00 + 0.1x)(100 - x)$

$R(x) = 200 + 8x - 0.1x^2$

Carry out:

To find the maximum function value we complete the square.

$R(x) = -0.1x^2 + 8x + 200$

$= -0.1(x^2 - 80x) + 200$

$= -0.1(x^2 - 80x + 1600 - 1600) + 200$

$= -0.1(x^2 - 80x + 1600) + 160 + 200$

$= -0.1(x - 40)^2 + 360$

The maximum function value is 360 when x is 40. If x = 40, then the price should increase 0.1(40), or $4.00. The new price would be 2.00 + 4.00, or $6.00.

Check:

We check a function value for x less than 40 and for x greater than 40.

$R(39) = -0.1(39)^2 + 8(39) + 200 = 359.9$

$R(41) = -0.1(41)^2 + 8(41) + 200 = 359.9$

Since 360 is greater than these numbers, it looks as though we have a maximum.

State:

In order to maximize revenue the theater owner should charge $6.00 for admission.

INDEX